# Microstructural Principles of Food Processing and Engineering, Second Edition

Food Engineering Series
Series Editor
Gustavo V. Barbosa-Cánovas, Washington State University

G. V. Barbosa-Cánovas and H. Vega-Mercado, *Dehydration of Foods*
P. Fito, E. Ortega-Rodríguez, and G. V. Barbosa-Cánovas, *Food Engineering 2000*
P. J. Fryer, D. L. Pyle, and C. D. Rielly, *Chemical Engineering for the Food Industry*
D. Holdsworth, *Thermal Processing of Packaged Foods*
M. J. Lewis and N. J. Heppell, *Continuous Thermal Processing of Foods*
R. Moreira, M. E. Castell-Perez, and M. A. Barrufet, *Deep-Fat Frying: Fundamentals and Applications*
M. A. Rao, *Rheology of Fluid and Semisolid Foods*

# Microstructural Principles of Food Processing and Engineering, Second Edition

**José Miguel Aguilera**
Eng., MSc, MBA, PhD
Professor of Chemical and Food Engineering
Department of Chemical and Bioprocess Engineering
Universidad Católica de Chile
Santiago, Chile

**David W. Stanley**
BSc, MSc, PhD
Professor (Retired), Adjunct Professor
University of Guelph
Guelph, Ontario, Canada

A Chapman & Hall Food Science Book

**AN ASPEN PUBLICATION®**
Aspen Publishers, Inc.
Gaithersburg, Maryland
1999

**Library of Congress Cataloging-in-Publication Data**

Aguilera, José Miguel.
Microstructural principles of food processing and engineering /
Jose Miguel Aguilera and David W. Stanley.
p. cm.
Includes bibliographical references.
ISBN 0-8342-1256-0 (alk. paper)
1. Food—Analysis. 2. Electron microscopy. I. Stanley, David W.
II. Title.
TX543.A382 1999 99-31202
664′.07—dc21 CIP



Editorial Services: Kathleen Ruby
Library of Congress Catalog Card Number: 99-31202
ISBN: 0-8342-1256-0

*Printed in the United States of America*

1 2 3 4 5

# DEDICATION

We gratefully dedicate this volume to our families in the hope that it will provide for them an understanding of where we go when we slip away into the micropores of our minds.

# Table of Contents

# Preface

It has been only a decade since the publication of the first edition of this book, but what a period it has been. Nothing less than a true paradigm shift has occurred in the way we obtain, process and interpret structural information from food materials.

A cornucopia of new instrumentation has become available to probe the microstructure of foods; among the most obvious are confocal laser scanning microscopy, atomic force microscopy, environmental scanning microscopy, magnetic resonance imaging, advanced differential scanning calorimetry and mechanical spectroscopy. Consequently, these and other techniques as well as images, data and applications are presented in Chapter One. Also, the remaining text has been updated and expanded in light of these advances.

As a result of giant strides in computing and software capabilities, image analysis has developed into a serious and indispensable tool. A new Chapter Two attempts to cover the introductory material needed to understand image analysis and its possible applications. It is important to realize that qualitative descriptions of images are no longer adequate; image analysis capability, the genesis of quantitative microscopy, is now a necessity in every research facility.

Food materials science has become an area of rapid growth in the general field of food science during this decade as important scientific concepts have been introduced and accepted. This discipline is concerned with the properties and processing of materials and has changed the way we look at food structures. In response to this important trend Chapter Three presents the fundamentals of polymer, colloid and materials science as they relate to foods.

The microstructural approach is gaining acceptance as it recognizes that most properties and phenomena occurring in foods are rooted at the microstructural level as explained in Chapter Ten. We conceive the microstructural approach as a problem-solving undertaking and an integrative effort to bring together information from various sources, not only microscopy, to understand structure-property relationships in food materials and derive appropriate functional models. Coupling microscopy with physicochemical and rheological techniques opens a whole new field to examine structure-property relationships in foods.

Thus, our aim has been to achieve a fusion of somewhat disparate disciplines into what is a most useful practical tool for food scientists. We call it *Food Material Science,* and it is aimed at solving problems faced by those seeking to understand how and why foods behave as they do. We would like to think that this approach will enable researchers to continue to provide the world with a safe, nutritious, affordable and enjoyable supply of food. How well it succeeds will be measured by the bringing of previously unavailable insights to the disciplines of food science and food engineering.

We have been aided in this effort by many colleagues who have allowed us to cite their research and provided us with images and other materials used in this edition. Enlightening discussions with Professors H. Watzke, P. J. Lillford, M. Karel and

V. Tolstoguzov are appreciated. Mr. Ken Baker again ably performed microscopic studies and prepared illustrations. Mistakes that remain are only attributable to ourselves.

As we did previously, we would like to thank the sponsors of our research programs. These include the International Development Research Centre of Canada, the Canadian Natural Sciences and Engineering Research Council and the Ontario Ministry of Agriculture, Food and Rural Affairs (D. W. Stanley) and FONDECYT, Nestle Research Centre (Lausanne), Unilever Research Laboratories, the Agency for International Development and the J. S. Guggenheim Foundation (J. M. Aguilera). Finally, as always, we acknowledge with gratitude the contribution to this work made by our past and present students: it has been a pleasure to have you in our laboratories.

*J. M Aguilera*
*D. W. Stanley*

## Chapter 1

# Examining Food Microstructure

### 1.1 INTRODUCTION

Food science became a legitimate profession following World War II. Prior to that, the vast majority of foods in the world were prepared locally, and so consumers had no option but to visit butchers, bakeries, dairies, greengrocers, and other purveyors close to their homes and purchase mainly unbranded goods of questionable quality. Technologies developed in the early and middle 20th century allowed the manufacture and distribution of canned, chilled, and frozen foods and furnished consumers with electric-driven refrigerators and freezers. At this point, it became imperative for food processors to provide consistently high-quality branded products, and thus they needed to understand the materials and ingredients used in the manufacture of these products. A study of food microstructure is an important requisite for understanding food materials. This chapter describes how microstructural studies may be carried out. The chapter has grown since the first edition of this book because of the many new methodologies that have become available to those wishing to examine food structure, including some techniques that provide images and some that do not. The knowledge of how to analyze the images obtained and how to better interface structure-obtaining instrumentation with computers has also grown enormously, as discussed in Chapter 2.

### 1.2 HISTORY OF FOOD MICROSTRUCTURE STUDIES

Although systematic investigation of food structure began recently, scientists have been looking through lenses or down microscopes at foods for much longer than that. The unaided eye can detect gross structural formations in animal and plant tissue, but more finely grained organization is wholly invisible without microscopic assistance. The development, around 1600, of the compound microscope—in which an objective lens is used to form an enlarged image that is then magnified by an eyepiece lens—led during the 17th and 18th centuries to a great deal of descriptive anatomy of biological matter and the concomitant development of cellular theory. Advances made in this area were carried forward in the 19th century. (The kinds of images seen during this era are indicated by exquisite hand-drawn diagrams [Figure 1–1]; photomicrography was not well established until the 1850s.)

Before the 20th century, adulteration of food in Europe and North America was common. According to Clayton (1909),

> Contemptible and gross adulteration of all conceivable kinds were everywhere the rule. The dough of bread was mingled with alum, carbonate of lime, bone ash, potatoes and beans. The pigments used by the sweetmeat-maker contained lead, chromium, mercury,

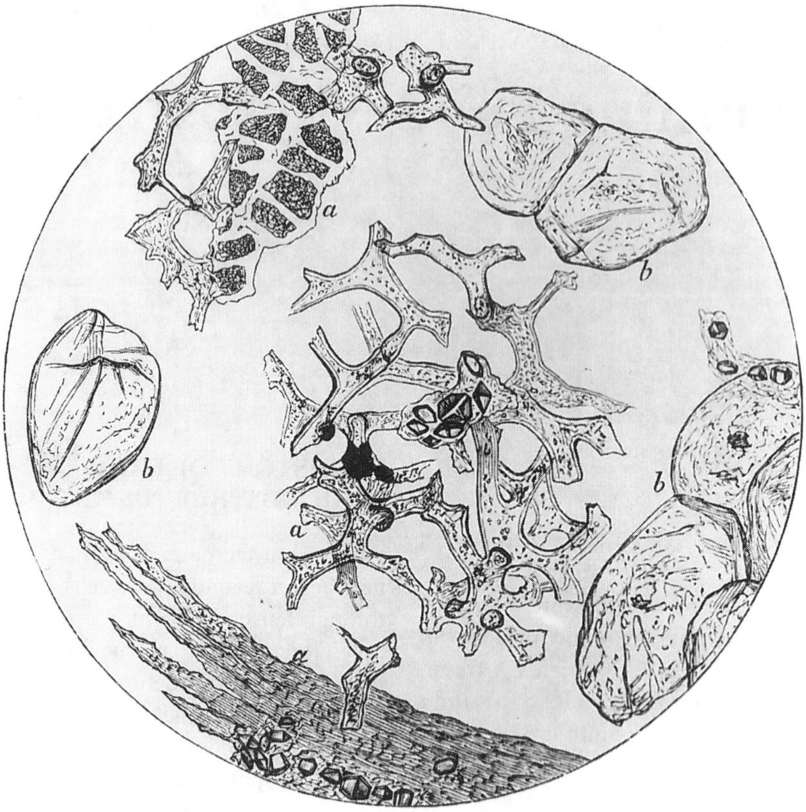


**Figure 1–1** Early hand-drawn line drawing of microscopic field. Probably from camera lucida. Original legend: "Orange marmalade, containing apple or turnip. ×100. a,a,a, Tissues of orange: above, angular cells of epicarp, or superficial layer, showing parenchyma of mesocarp. Underneath: center, loose parenchyma, with intercellular spaces, constituting the woody part of the rind, or mesocarp, and showing crystals and spherical masses of the glucoside hesperidin: below, fibro-vascular bundles with similar crystals. b,b,b, parenchymatous cells of added ingredient." *Source:* Clayton, 1909.

copper, sometimes even arsenic. The unwholesome hues of preserved green fruits and vegetables were due to boiling in copper vessels, or to the addition of cupreous salts. Cayenne pepper and curry powder were beautified by the scarlet oxide of lead. Vinegar was fortified with sulphuric acid . . . the water supplied to the Metropolis carried solid and liquid filth from the sewers, reeked of odorous abominations, and teemed with offensive forms of plant and animal life.

Since the light microscope (LM) is well suited to the identification of small amounts of foreign or extraneous animal, plant, or mineral matter in admixture, not surprisingly it was employed to detect food adulteration. Advances in chemical analysis made its use even more instructive, and many books were published on light microscopy (several are listed at the end of the chapter).

It is interesting to note that microscopy has been used to examine not just modern foods but ancient ones as well. A recent paper (Samuel, 1996) reports on the usefulness of light and scanning electron microscopy in unearthing the methods employed by early Egyptians to bake bread and brew beer. Past descriptions of these processes have had to rely on incomplete pictorial renderings and written sources, since direct examination of foods was difficult owing to scarcity and decay. However, the arid climate of parts of Egypt preserved ancient organic materials and made possible the microscopic examination of small samples from desiccated bread loaves and the residue from beer vessels dating back to between 2000 and 1200 B.C. The examination focused on the structure of starch granules, which is a function of how they were processed. For example, gelatinization results in swelling and eventual disappearance, and thus evidence of these indicates the amount of water present; erosion, channeling, and pitting, on the other hand, indicate enzymatic treatment.

The analysis of ancient foods leads to the conclusion that the methods of their production are quite similar to those used currently. Bread makers used malt (germinated grain) as well as flour, and the presence of yeast cells implies that some loaves were leavened. Beer was made from a blend of imbibed heated and unheated malt, and the resulting liquid was strained to remove cereal husks before yeast was added.

The advent of the 20th century brought great changes to both the procedures employed and the research goals of food microscopists. Strong pure food and drug legislation limited adulteration and allowed scientists to pursue other avenues. New histological staining procedures and improved embedding media were developed, as was better optical equipment. The combining of engineering and science led to the commercialization of fabricated foods (products manufactured from a combination of raw materials). It soon became apparent that many food processing and food quality factors are governed by structural organization.

Basic research on electron beams early in the 1900s led to a revolution in the way biological microstructure was examined. By encasing an electron gun in a high vacuum and using electromagnetic lenses to focus the resulting beam, researchers could create an image by projecting electrons through a thin specimen. The first commercial transmission electron microscope (TEM) was made available around 1940, and suddenly the useful magnification upper limit leapt from about 1,000× to about 10,000×—and soon rose higher still. Cell inclusions not previously well defined could now be studied. This resulted in radical changes in the biological sciences that revolved around a microstructure-function axis and to several decades of unprecedented advances. Figure 1–2 shows TEM and LM images of biological tissue taken at roughly the same magnification. Comparison of the two illustrates that higher magnification is not the only advantage of electron microscopy; enhanced contrast and sharpness of details are also apparent.

Needless to say, food scientists, who routinely receive training in the biological sciences, profited from this technology. It was soon found, however, that the TEM, while producing enhanced resolution, had many limitations. These included the physical problems associated with high vacuum, the electron beam, specimen chemical preparation, and ultrasectioning and also the psychological tendency to neglect other suitable techniques for examining food microstructure. In addition, more enlargement alone is no guarantee of enhanced image detail, and even when the magnification is useful, structural organization is frequently more meaningful than the fine microstructure of individual components.

Thus, the scanning electron microscope (SEM) was welcomed with open arms by food researchers in the mid- to late 1960s. This instrument brought the promise of an apparently three-dimensional picture and a great depth of field, 500 times that of a light microscope at the same magnification. Depth of field is the distance along the lens axis in the object plane at which an image can be focused without loss of clarity.

The SEM also did away with the need to cut thin sections and allowed the examination of topographical details of external or internal surfaces. The SEM fills and overlaps the magnification gap between the LM and the TEM, from 20× to 100,000×. Shortly after the debut of the first commercial SEM, scientists began examining food specimens, and they have continued to do so at an increasing rate over the last several decades.

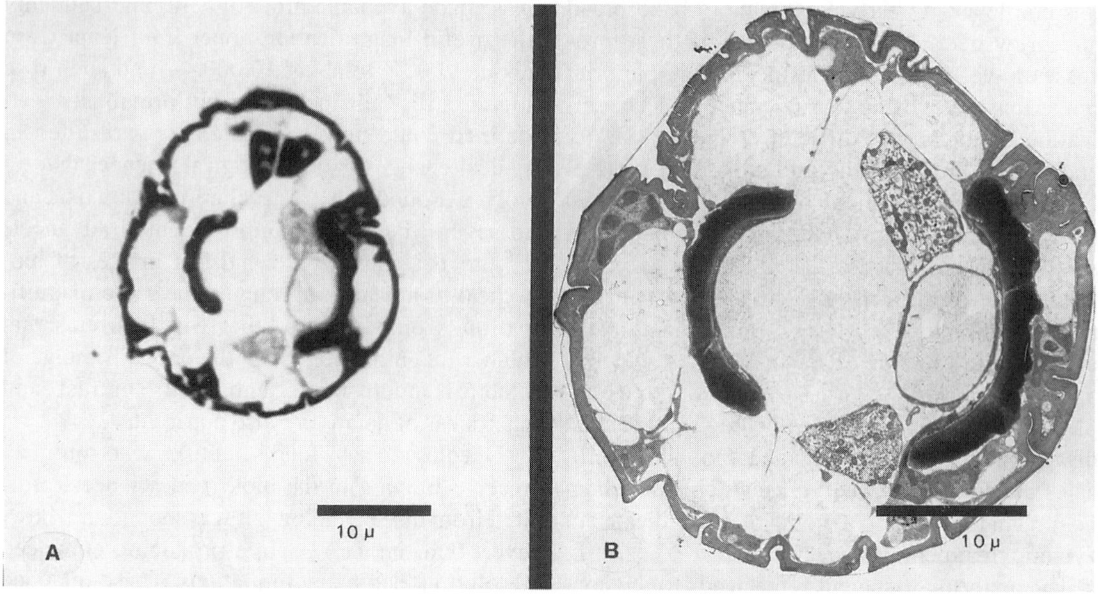


**Figure 1–2** Comparison of LM (A) and TEM (B) images of cross section of mosquito antenna. Note that photographic enlargement of LM image reveals only empty magnification as compared with TEM image. (A) 0.5 $\mu$m section embedded in Spurr's resin. (B) 0.1 $\mu$m section. *Source:* Courtesy of K.W. Baker.

Innovations continue to be made in microscope technology. In the case of light microscopy, recent advances include the application of lasers, video-enhanced imaging, confocal illumination, and tandem scanning microscopy. In the case of electron microscopy, advances include the medium-voltage TEM, low kV SEM operation, and several ancillary methodologies, such as backscattered electron analysis and X-ray microanalysis. As for new biological techniques, cryopreparative specimen handling merits particular mention. Adding to the effectiveness of these innovations are sophisticated computer-driven image analysis and storage devices for electronic images.

A critical realization in the examination of food materials was that many important organizational features may be studied—in reality, are probably better studied—through the application of nonmagnifying devices. The food scientist can now apply several techniques for revealing structural organization that complement microscopic studies. Some of these, such as light scattering and rheology, are discussed in this book; others, such as ultracentrifugation, and electron spin resonance, though not covered, can play a role in characterizing structure and should not be ignored.

In fact, the new approach to investigating food microstructure is to integrate structural data from many devices focused at different levels of organization. This approach has resulted in exciting new insights, especially insights into the relationship between microstructure and other important food characteristics, such as quality factors and food processing strategies.

As should be obvious, a wide variety of techniques exist for examining food microstructure. Their judicious application provides data capable of unmasking structural organization, if interpreted properly. The remainder of this chapter is devoted to a closer examination of these techniques and their application.

## 1.3 LIGHT MICROSCOPY

### 1.3.1 Introduction

Traditionally, the structure of foods was studied by means of images enhanced by glass lenses. The information thus obtained was used to augment that gained by direct macroscopic evaluation with the unaided eye and the sense of touch. Any time that an effort is made to gain knowledge from an image of a real object, there exists a serious risk of misunderstanding resulting from artifacts of magnification or sample preparation as well as psychological errors of interpretation. It is not within the scope of this work to detail the basic physical and optical principles that govern magnifying instruments, and the reader is referred to the numerous texts available on this subject. No instrument should be used before a thorough study is made of its theoretical basis.

Any examination of food structure should begin with the intelligent use of the human senses. The brain has an amazing facility to integrate sensory data to fashion information. Often the sensory data can be expanded usefully through such simple devices as a hand lens or a stereoscopic dissecting microscope. The latter, although capable of only limited magnification (its upper limit is about 100×) and operating with reflected light, can be of great use in structural investigation. It is stressed repeatedly in this work that higher magnification alone does not guarantee more information.

### 1.3.2 The Compound Microscope

The compound microscope has been an important tool in the study of food materials. It has a resolution about $10^3$ times smaller than the human eye and produces a magnified image of details unavailable to unaided vision (see Table 1–1). By resolution is meant the minimum linear distance between two points in the specimen at which they still appear as two points; beyond this limit the points will merge in the image and cannot be resolved. The following equation relates resolution to various parameters of a magnifying system:

$$r = \frac{0.61\lambda}{n \sin(u)} \qquad \textbf{Equation 1–1}$$

where $r$ is the minimum resolved separation, $\lambda$ is the wavelength of the radiation, $n$ is the refractive index of material between object and lens, and $u$ is the aperture angle of the lens.

**Table 1–1** Comparison of Microscopes

| | *LM* | *TEM* | *SEM* |
|---|---|---|---|
| General use | Surface structure and sections | Thin sections | Surface structure |
| Resolution (nm) | 200–500 | 0.2–1 | 3–6 |
| Magnification (×) | 10–1,500 | 200–500,000 | 20–100,000 |
| Depth of field at 500× (μm) | 2 | 800 | 1,000 |
| Illumination | Visible light | Hi-speed electrons | Hi-speed electrons |
| Lens | Glass or quartz | Electromagnetic | Electromagnetic |
| Specimen | | | |
| Preparation | Easy | Difficult | Easy |
| Thickness | Thick | Very thin | Reflectance |
| Environment | Versatile | Vacuum | Vacuum |
| Available space | Small | Small | Large |
| Image display | On eye, by lenses | On fluorescent screen | On CRT |

*Source:* Adapted from Stanley and Tung, 1976, and Flegler et al., 1993.

The quantity $n \sin(u)$ is called the numerical aperture of the system. For a conventional LM using blue light of wavelength 470 nm and having an oil immersion objective lens with a numerical aperture equal to 1.40, resolution would be limited to about 200 nm. Since the eye can comfortably resolve detail of about 0.2 mm, an LM is capable of magnification of about 1,000× using ideal specimens before experiencing empty magnification (increased magnification without improved resolution). As inspection of the above equation will show, improved resolution (decreased $r$) can be obtained by decreasing the wavelength of radiation employed. A TEM makes use of electrons that have a wavelength about 100,000 times shorter than visible light. Although the numerical aperture of a TEM is low compared with that of an LM, a resolution of about 0.5 nm is possible with a modern TEM (1 mm $= 1 \times 10^3$ $\mu$m $= 1 \times 10^6$ nm).

The advent of the electron lens caused the LM to be somewhat neglected in structural studies, but the introduction of innovations such as the confocal microscope has led to the realization that the versatility of light microscopy, combined with its ease of use (including ease of sample preparation), makes it an indispensable tool for the food scientist.

LMs now come in an almost infinite variety of configurations, but the essential sequence of components remains unchanged. Modern LMs have a built-in illumination source; the visible light produced travels through a diaphragm and condenser, which focus and control the intensity of the light beam before it is transmitted through the specimen. A glass slide and cover slip bracket the specimen on an adjustable $x/y$ direction stage; light then enters the objective lens set in a revolving nosepiece and travels up the tube and through an eyepiece to form an inverted, enlarged virtual image visualized by the eye of the operator or a real image that may be captured by photographic film or on video. Focusing is achieved by adjusting the focal plane by varying the tube length. Figure 1–3 shows a modern compound microscope.

### 1.3.3 Sample Preparation and Stains

Sample preparation for light microscopy is less complex than for electron microscopy but still incorporates a wide variety of operations. Food materials can be examined whole, but usually studies of cellular structure require the cutting of sections. Microtomes allow sections to be cut to a uniform thickness. In order to maintain structural integrity, tissues are often embedded with materials such as paraffin wax, plastics, or resins; alternatively, sectioning can be performed on frozen material.

Fixation is the primary preparatory step. Its purpose is to immobilize cellular components in such a way as to ensure that the resulting structure resembles the living state as closely as possible. Care must be taken to minimize osmotic damage and shrinkage while retaining cellular components *in situ*. Since proteins are the major reactive components requiring immobilization, most fixatives are directed toward these molecules. Aldehydes such as formaldehyde and glutaraldehyde have been the fixatives of choice, since they act to both denature and cross-link proteins.

In order to infiltrate biological samples with embedding media such as paraffin, it is necessary to transfer fixed tissue from a polar aqueous environment to the nonpolar wax. If the transfer is made too rapidly, microstructure can be damaged. The dehydration process must be gradual, and usually a series of increasing concentrations of ethanol are used in order to prevent distortion, although other organic solvents may be used as well.

In light microscopy it is common to apply a specific stain or dye in order to improve contrast or differentiate tissues (Figure 1–4). The chemistry of staining procedures is often obscure, and empirical knowledge is frequently relied upon. Myriad stains are available, and many are touted as specific for a given component. Moreover, interpretation is often difficult and proper controls are mandatory. Two or more stains are sometimes used to differentially color specific components and produce contrast between them, as in the well-known Gram's method used in microbiology. A common task in complex food product analysis is to discriminate protein, fat, and a complex carbo-

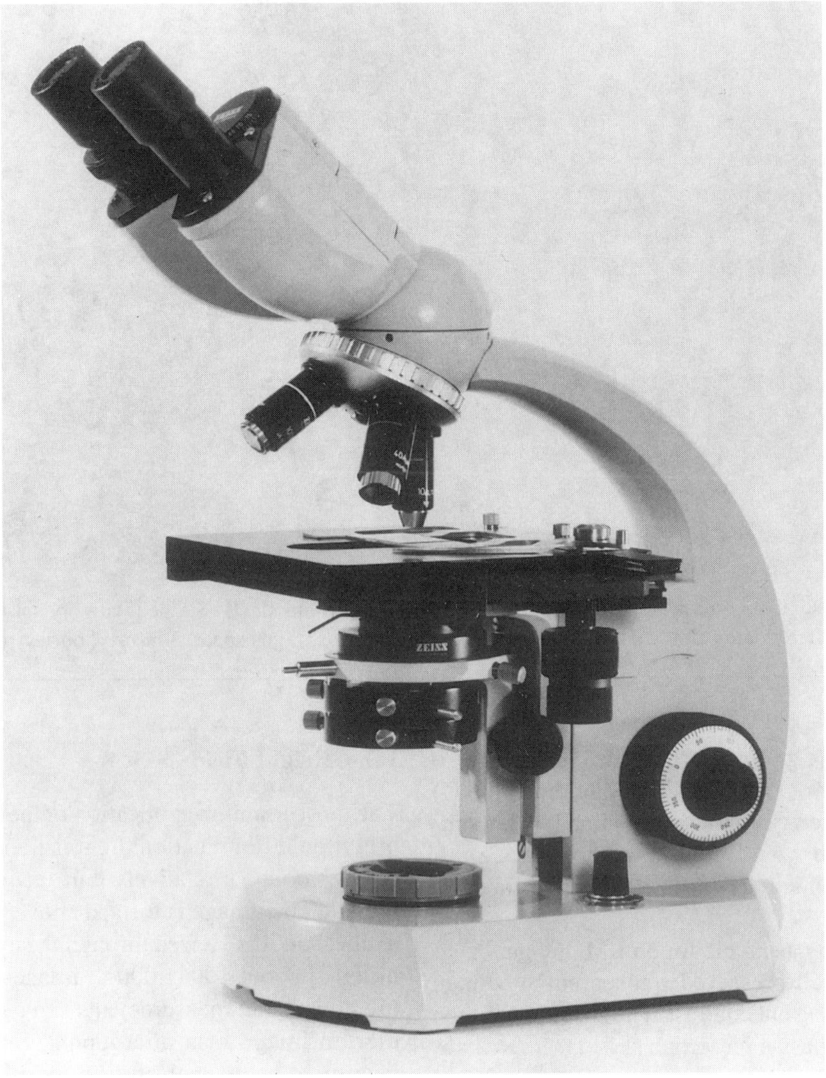

**Figure 1–3** A modern compound light microscope. Note the major components: eyepieces, binocular tube, rotary nosepiece with objectives, square mechanical stage, substage condenser, diaphragm insert, and base with built-in illuminator. Focusing is achieved with the co-rotating coarse and fine focusing knobs. *Source:* Courtesy of Carl Zeiss, Inc., Thornwood, NY.

hydrate (e.g., starch) in a sample. Obviously, the final result will depend on the sample's properties. Is the starch gelatinized? Is the fat in the form of an emulsion? Is it in the form of crystals? Has the product been heat treated? In any case, one could experiment with any of several combinations of various protein stains (e.g., FITC, eosin, saffranin, acridine orange, fast green, acid fuschsin), lipid stains (e.g., Nile red, Nile blue, osmium tetroxide, Sudan black B, oil red O), and

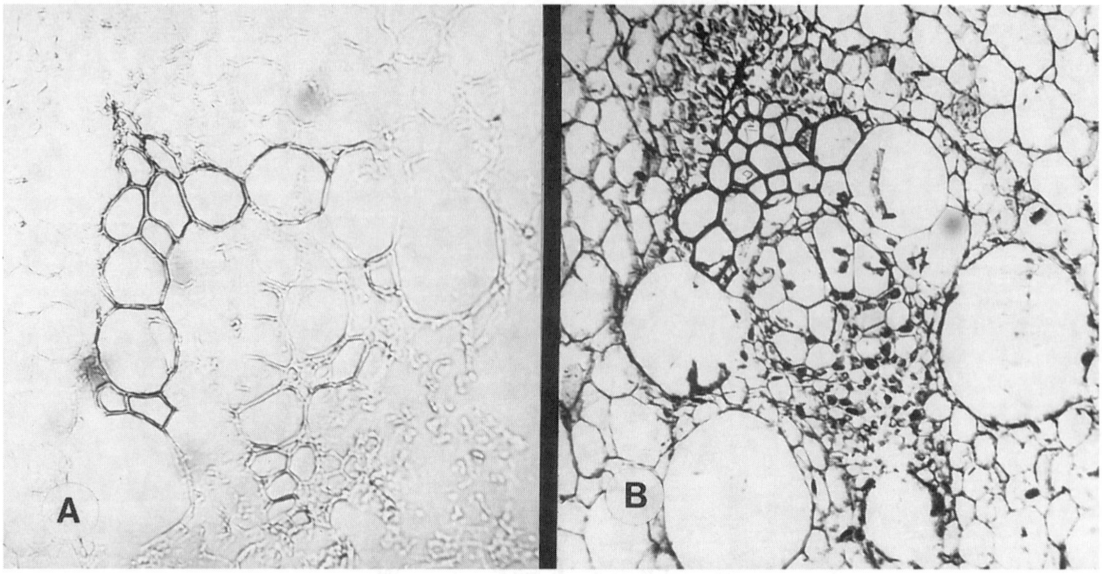


**Figure 1–4** Use of stain to improve LM image contrast. (A) Unstained. (B) Stained with crystal violet and erythrosin B to differentiate cell walls. Material is cross section of asparagus spear. *Source:* Courtesy of J.L. Smith.

starch stains (e.g., Congo red, Nile blue [may also stain fat], periodic acid, iodine, toluidine blue O) until a satisfactory result was obtained. A far simpler approach would be to use fluorescence microscopy, which is described later in this chapter.

In preparing a sample for an LM, the goal is to produce a well-preserved transparent specimen colored to bring out contrast. The specimen is then placed on a glass microscope slide (≈75 × 25 × 1 mm), a suitable mounting media is applied (Canada balsam or a synthetic), and a coverglass (≈22 mm dia. × 0.18 mm) is added.

In summary, sample preparation for the LM usually consists of (1) chemically fixing the tissue, (2) dehydration, (3) clearing to remove excess solvent, and (4) embedding. The result is a block containing the material from which sections can be cut. These are then adhered to a glass slide, stained, and mounted. Many variations exist, and unfortunately sample preparation is still as much of an art as a science. Trial and error remains the only way to arrive at optimal preparation procedures.

### 1.3.4 Bright Field

The most common application of the LM involves bright field illumination: light is transmitted from below through a relatively thin section or slice of material, the image is formed above the sample in a tube, and it is viewed through the eyepiece magnified at about 100–1,000×. Images can be photographed, and measurements can be made on either an image or a micrograph. Specimens are examined at normal atmospheric pressure and therefore do not have to be dehydrated, but care must be taken to prevent wet mounts from desiccating during prolonged observation. Thus, sample preparation is relatively easy. Alternatively, more permanent mounts can be achieved using fixed, dehydrated, and embedded tissue.

### 1.3.5 Phase/Differential Interference Contrast

A major advantage of light microscopy is its versatility. Staining is a useful procedure, because biological tissues are most often colorless and therefore lack contrast. Alternatively, contrast can be enhanced by phase contrast or differential interference contrast (Nomarski) optics, in which the phase of the light is altered and then recombined to yield improved differentiation (Figure 1–5). Although these two procedures produce similar images, they employ different mechanisms to modify the light path. To modify a bright field microscope for phase contrast, the condenser iris is exchanged for an annular diaphragm, and a phase plate is mounted above the objective lens. To modify a bright field microscope for interference microscopy, a polarizer and prism are added below the condenser and also above the objective lens.

The phase contrast image is characterized by enhanced contrast and visibility of unstained tissues, whereas the interference contrast image has a distinct relief appearance and a shallow depth of field (Figure 1–5). Contrast in the two methods depends upon the degree of difference between the refractive index of an object transparent in bright field illumination and that of the surrounding medium. Thus, in some cases phase contrast gives the best image, while in others differential interference contrast is the superior technique. The two methods, in other words, complement each other.

### 1.3.6 Polarizing Microscopy

Polarizing microscopy, another contrast-inducing technique, has many applications in the study of food structure. In this form of light microscopy, plane polarized light (light that vibrates in a single direction only) is allowed to impinge upon the specimen. If the material contains anisotropic or birefringent structures (i.e., structures capable of rotating the light plane), the emerging light beam will be altered by twisting and partially extinguished. On the other hand, isotropic substances have only one refractive index and will not rotate plane polarized light.

A bright field LM can be converted into a polarization microscope by inserting a polarizing prism or filter below the condenser (the polarizer) and one above the objective lens (the analyzer). If these two plates (called Nicol prisms) are arranged parallel to one another, plane polarized light is transmitted through to the eye. If, however, their axes are perpendicular, then no light is transmitted and a dark field results. When anisotropic material is placed between crossed prisms, it will rotate some of the plane polarized light and thus be visible. An isotropic sample will not disturb the extinction of the beam.

Common examples of the use of polarizing microscopy in the study of food microstructure include the following. Muscle fibers observed in a bright field LM exhibit transverse striations (Figure 4–12, part A). When viewed under polarizing light, these structures can be seen to result from alternating bands of isotropic and anisotropic muscle proteins, termed, appropriately, I and A bands (part B). Meat quality is often related to the degree of muscle contraction, and the distance between repeating bands (sarcomere length) can be measured by polarizing microscopy or through the use of phase contrast. It should be noted that, alternatively, a nonmagnifying method, based on the ability of striated muscle to act as a transmission diffraction grating, is now available. A laser is used as a source of coherent monochromatic light, and the spacing of the diffraction pattern is a function of sarcomere length.

Food starches have typical characteristics, sizes, and shapes that are observable using polarizing microscopy (Figure 1–6). Of particular importance in identifying the botanical origin of starch is the unique Maltese cross pattern produced by the crystalline nature of the starch granule. Also, the phenomenon of starch gelatinization can be followed by this technique. As gelatinization proceeds, the starch granules lose their crystal structure and hence their birefringence.

The microstructure of fats and emulsions has also been studied by polarized light microscopy. Because these materials contain crystalline triglycerides that occur in three major polymorphic forms, it is possible to differentiate them with the aid of polarized light microscopy. This is important, for the amount of each polymorphic form is related to the stability of physical properties. Such analyses are more useful when

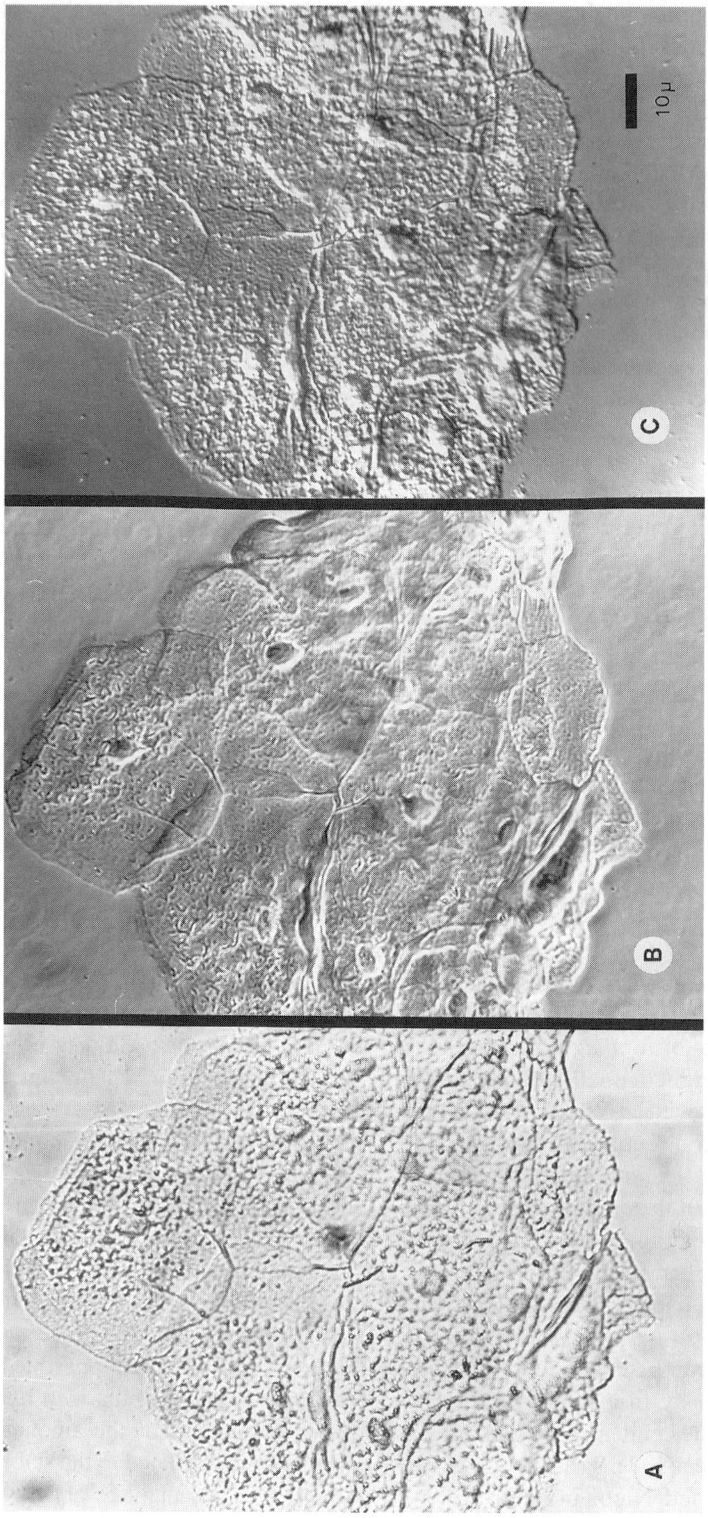


**Figure 1–5** Bright field (A), phase contrast (B), and Nomarski differential interference contrast (C) images of unstained human cheek epithelial cells. *Source:* Courtesy of K.W. Baker.

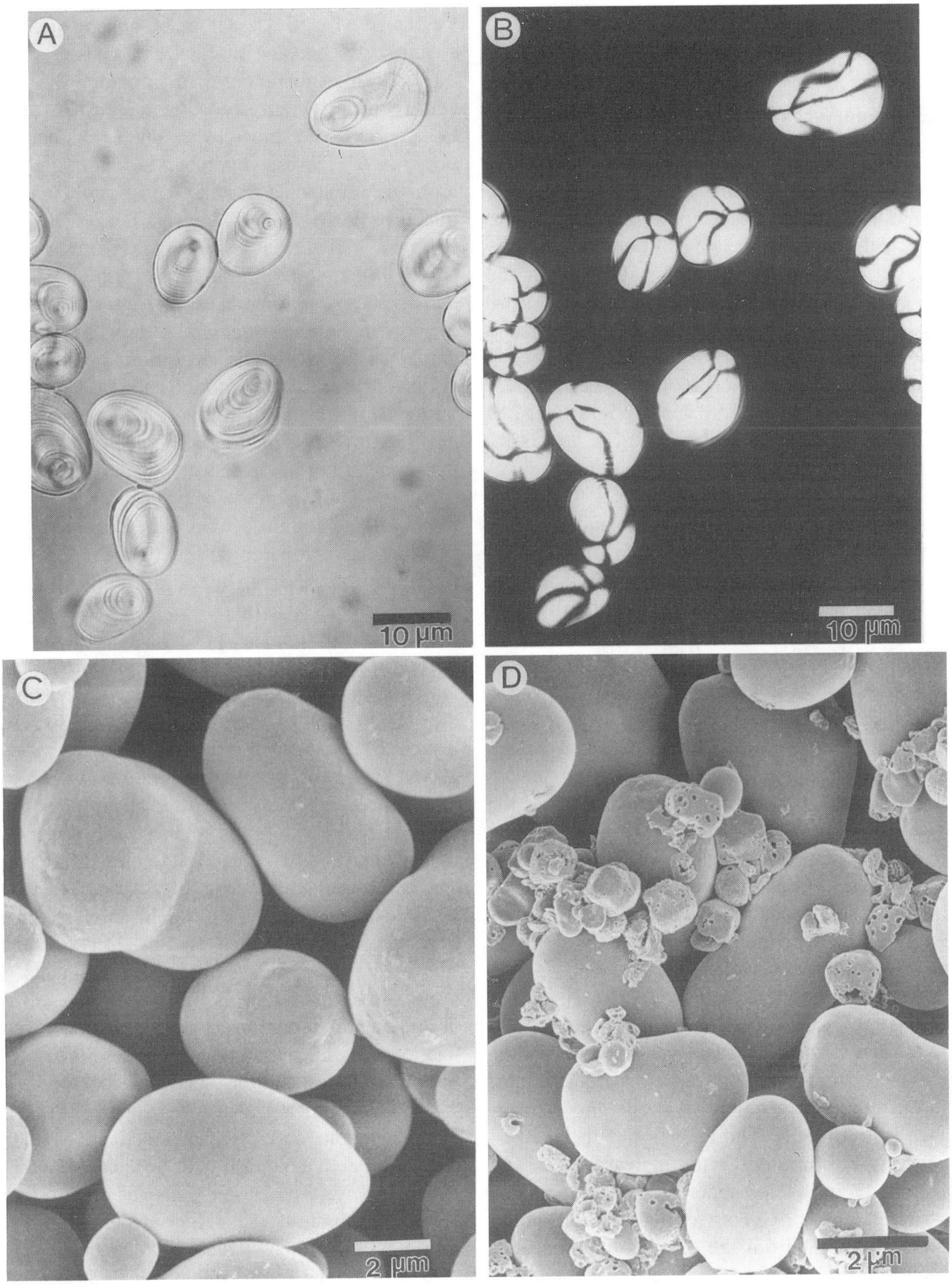


**Figure 1–6** Images of potato starch. (A) LM unpolarized. (B) Polarized, same field. (C) SEM untreated. (D) SEM amylase treated. *Source:* Courtesy of V. Barichello.

compared with the results of other techniques, such as X-ray diffraction.

### 1.3.7 Fluorescence Microscopy

Fluorescence is the luminescence of a substance excited by radiation. When radiation strikes a substance, some is absorbed, some is converted into heat, and some is reemitted as fluorescence in the form of light quanta at a longer wavelength and lower intensity. In fluorescence microscopy, samples that either fluoresce naturally or are caused to fluoresce through the use of fluorescent probes or dyes are examined microscopically. A fluorescence microscope differs from a bright field LM merely by the addition of two filters: one is inserted in the light beam prior to the sample to produce monochromatic illumination for the excitation of fluorescence (exciter filter), and the other is inserted following the sample to filter out the damaging short wavelengths while transmitting the longer wavelengths given off by the fluorescing specimen (barrier filter).

At this point it may be useful to discuss the role of filters in light microscopy. Widespread use is made of filters placed in the light path in order to control brightness and enhance contrast (see Section 2.3.1). In fluorescence microscopy, filters are used specifically to isolate certain regions of the visible spectrum (Color Plate 1), by transmitting light of particular wavelengths or by transmitting fluorescence emissions but preventing the transmission of the excitation wavelengths. High-quality fluorescence filter systems are available to cover a variety of wavelengths.

Currently, modern fluorescence microscopes utilize epifluorescence. That is, the specimen is illuminated by using a beam-splitter to direct the emission from a high-pressure mercury souce through the objective lens to excite only the surface layer of cells. The intensity of the fluorescence, often a limiting factor, increases as the objective magnification increases, since the same objective lens is used both for illumination and for collecting light from the sample. Fluorescence microscopy is useful to the food scientist because it can detect substances in low concentrations and thus allow visualization of materials not possible by other LM methods. Other advantages include specificity, speed and simplicity of analysis.

There are substances with inherent fluorescence capacity (autofluorescence). Most cells have an intrinsic fluorescence due to the natural presence of fluorescent molecules. These include aromatic amino acid residues in most proteins, reduced pyridine nucleotides (NADH, NADPH), flavins and flavin nucleotides (riboflavin, FMN, and FAD), and protoporphyrins. To the food scientist, materials of interest from a structural point of view include collagen and elastin fibers from animal tissue and lignins and various smaller phenolic compounds bound to the cell walls of plant tissue. Unfortunately, autofluorescing compounds have far lower extinction coefficients than most exogenous fluorophores used in fluorescence microscopy. Thus, high intensities of the excitation light will be required to produce a detectable emission, as compared with exogenous high-extinction fluorophores. In addition, autofluorescence is subject to fading, limiting detailed examination and photography. Thus, specimens are usually stained to ensure strong emissions. Fluorescence dyes also tend to be more specific, since they attach only to specific areas of the tissue and leave others unstained. They may bind to the target molecule via an inherent binding affinity by binding to another molecule or to an antibody that binds selectively to the target molecule. Illumination intensity must be carefully controlled, since overexcitation can cause bleaching, an irreversible chemical change of the fluorophores into nonfluorescent molecules. Table 1–2 gives useful fluorescent stains for several food components.

As an example of the usefulness of fluorescence in studying food microstructure, consider the investigation of lignification in asparagus described here. In order to follow this toughening reaction postharvest, cross-sectional samples were cut and fixed in picric acid solution. Following dehydration, tissue was embedded in paraffin wax. Sections were then cut and stained with crystal violet and erythrosin B to enhance fluorescence and differentiate lignified from nonlignified cell walls. Sections were examined with a microscope equipped with an epifluorescence condenser containing an exciter-barrier filter set with a maximum transmission of 365 nm,

**Table 1–2** Fluorescence Techniques Used for Food Components

| *Technique* | *Components* |
|---|---|
| Autofluorescence | Phenolic acids, lignin, seed coat, elastin, collagen |
| Acid fuchsin | Cereal proteins |
| Acridine orange | Casein, bacteria, cell nuclei |
| Acriflavine | Phytate, nucleic acids |
| Anilinonapthalene | Plant storage proteins |
| Calcofluor white | $\beta$-glucans, mucilage |
| Congo red | $\beta$-glucans, mucilage |
| Crystal violet/erythrosin B | Lignin |
| Fluorescein isothiocyanate | Proteins |
| Nile blue A, Nile red | Lipids, plant cuticle |
| Periodate, Schiff's | Starch, vicinal hydroxyl groups |
| Periodate, acriflavine | Starch, vicinal hydroxyl groups |
| Texas red | Proteins |
| Thiazine red R | Proteins |

*Source:* Data from R.G. Fulcher, Fluorescence Microscopy of Cereals, *Food Microstructure*, Vol. 1, pp. 167–176, © 1982, Scanning Microscopy International, Inc.; R.G. Fulcher and P.J. Wood, Identification of Cereal Carbohydrates by Fluorescence Microscopy, in *New Frontiers in Food Microstructure*, D.B. Bechtel, ed., pp. 111–128, © 1983, American Association of Cereal Chemists; H.M. Holz, *Worthwhile Facts about Fluorescence Microscopy*, © 1975, Carl Zeiss; J.L. Smith et al., Nonenzymic Lignification of Asparagus? *Journal of Texture Studies*, Vol. 18, pp. 339–358, © 1987, Food & Nutrition Press, Inc.; S.H. Yiu, Fluorescence Microscopy in Food Technology, *Zeiss Focus*, Vol. 4 No. 2, pp. 6–7, © 1987, Carl Zeiss Canada Ltd.; B.E. Brooker, Imaging Food Systems by Confocal Laser Scanning Microscopy, in *New Physico-Chemical Techniques for the Characterization of Complex Food Systems*, E. Dickinson, ed., pp. 53–68, © 1995, Blackie Academic and Professional; and M. Kalab, P. Allan-Wojtas, and S.S. Miller, Microscopy and Other Imaging Techniques in Food Structure Analysis. *Trends in Food Science Technology*, Vol. 6, pp. 177–186, © 1995, Elsevier Science Ltd.

combined with a high-pressure mercury-arc illuminator. Figure 1–7 shows typical results where increases in the width of the mechanical tissue layer and in the amount of fluorescence from cell walls indicate lignin deposition.

Another example involves the "hard-to-cook" textural defect associated with the storage of legumes, such as the common bean. This defect can lead to failure to germinate or imbibe water, extended cooking times, reduced nutritional value, and economic loss throughout the food chain. Although the defect is especially common in tropical climates, beans stored in temperate areas also will harden eventually, depending upon temperature and humidity. Hardened beans also often darken, causing further quality losses. Structurally, the hard-to-cook defect affects the cotyledons, rendering the cells unable to separate during cooking. A multiple mechanism based on oxidation and polymerization of phenolic compounds has been hypothesized (see Section 6.5.5). The idea is that a *reversible* hardening is produced by the enzymatic hydrolysis of phytate, rendering it unable to chelate divalent cations, which then migrate to the middle lamella and participate in cross-linking reactions with demethylated pectins. An *irreversible* hardening is caused by the strengthening of cell walls resulting from deposition of lignin-like material that impairs water imbibition during soaking and cooking.

In order to gather data on these reactions, a study was made using several common bean varieties previously stored under a range of temperatures and subjected to water activities for various lengths of time. To quantitate the degree of lignin deposition, which was present in amounts too small to assess chemically, fluorescence intensity of the cotyledons was determined microscopically by measuring the amount of autofluorescence. Figure 1–8 shows the relationship between the texture of the cooked beans and the microscopic examination of the uncooked material; coefficients of determination between irreversible hardness and autofluorescence for these samples

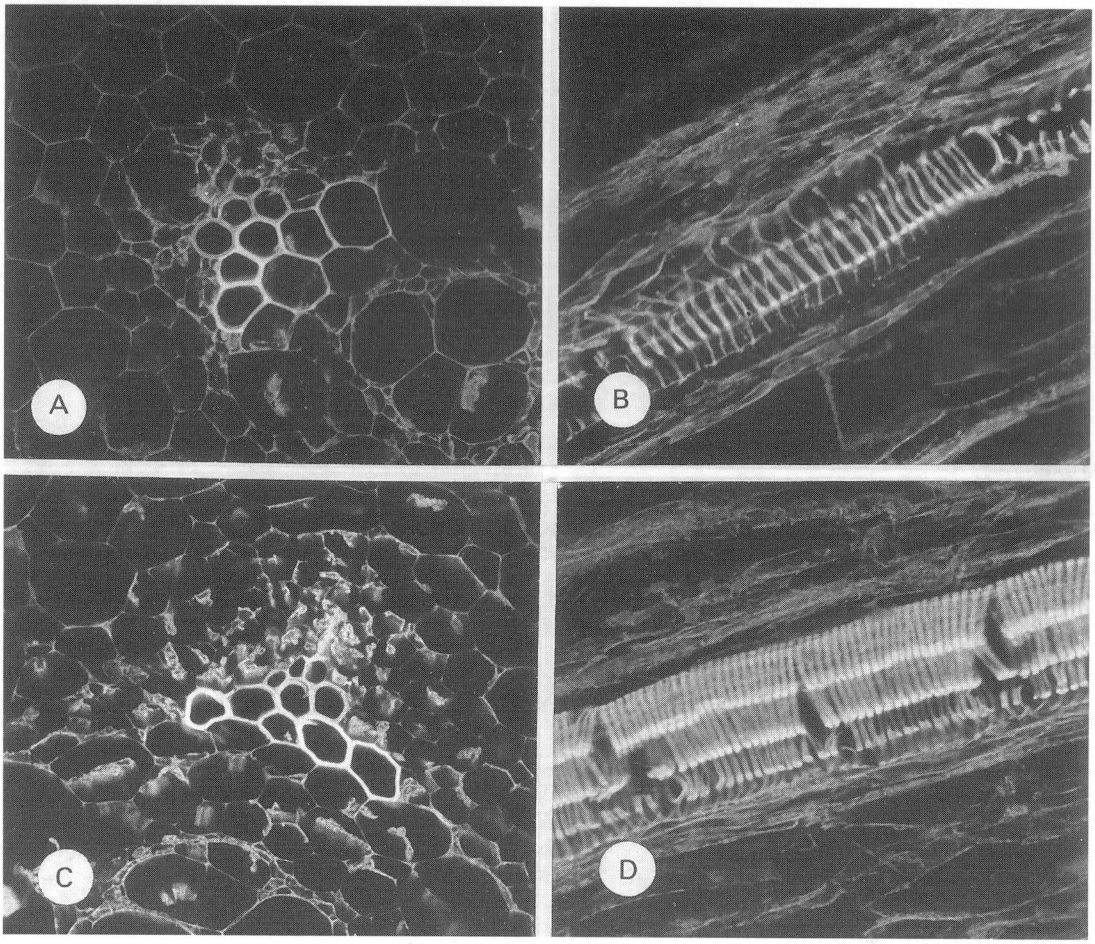


**Figure 1–7** Fluorescence light micrographs of asparagus lignification. Cross-sectional (A, C) and longitudinal (B, D) section of vascular bundles from asparagus at day 0 (A, B) and following 120 days storage at −10°C (C, D). Note increased fluorescence, indicating lignification of xylem vessels in stored tissue. *Source:* Smith et al., 1987.

ranged from 71.8% to 63.2% and were highly significant ($p \leq .01, n = 54$). Thus, not only was autofluorescence measured in this way useful for predicting texture but the data point to a mechanism involving lignin deposition.

### 1.3.8 Hot-Stage Microscopy

Many of the steps involved in food processing involve the controlled application of heat to food materials using unit operations such as blanching, pasteurization, canning, cooking, frying, and broiling. In addition to destroying microorganisms and enzymes, these processes make foods more tender and palatable. It would be useful, then, to have a technique for modeling these procedures at the microscopic level. A hot stage mounted directly on a microscope and controlled with a temperature programmer allows such modeling. Figure 1–9 shows the main elements of a hot stage. In general, data from image-creating techniques are synergestically augmented by data from nonimaging methods, as shown by the study of starch gelatinization presented below.

Irreversible swelling of starch granules by water above the gelatinization temperature is one of

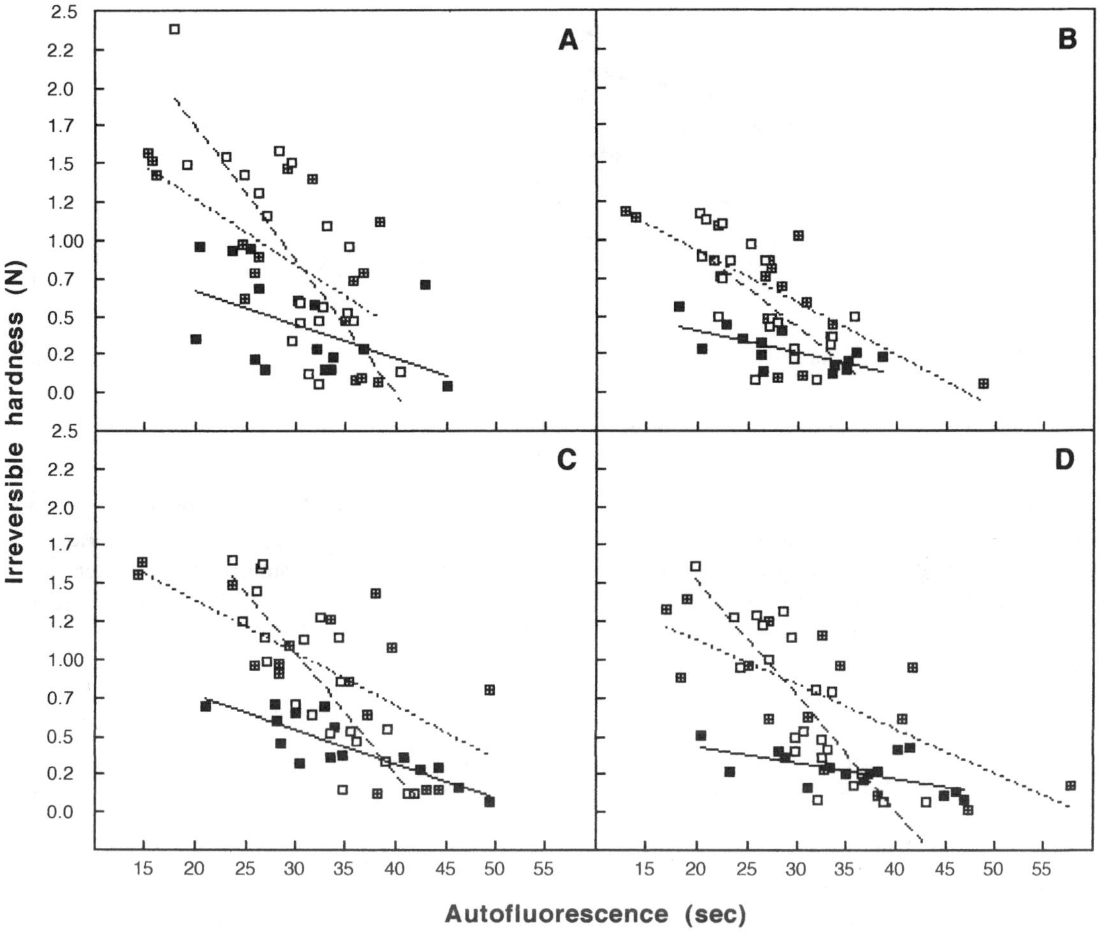


**Figure 1–8** Dependence of irreversible hardness of (A) whole and (B) dehulled black beans, and (C) whole and (D) dehulled white beans stored at (■—■) 15°C, (⊞—⊞) 30°C, or (□—□) 45°C on autofluorescence emanation. *Source:* Reprinted with permission from J.M. del Valle and D.W. Stanley, Reversible and Irreversible Components of Bean Hardening, *Food Research International*, Vol. 28, pp. 455–463, © 1995, Food & Nutrtion Press Inc.

the most important phenomena in structure formation of cooked, baked, or fried edible cereals or tubers and their products (see Section 4.5.3). When an aqueous starch suspension is heated at a temperature sufficient to cause gelatinization (greater than 65°C for most starches), the granules swell to many times their original size, and dramatic changes in the rheological properties are observed as a paste is being formed. Granule integrity during pasting is observed to remain at temperatures greater than 100°C under mild agitation. While gelatinization can be monitored by changes in the viscosity of the suspension or in the size of the granules, hot-stage microscopy and image analysis allow following the changes in the size distribution of starch granules kinetically.

In this study of starch gelatinization, the basic equipment consisted of a video camera mounted onto a light microscope or a stereo microscope, along with a monitor to display the image, an optical digitizer, and a PC computer with peripheral storage media and appropriate software for image enhancement and image analysis. Drops of a starch suspension were sandwiched between sealed cover slips and placed on a heating stage equipped with a temperature programmer. Filters

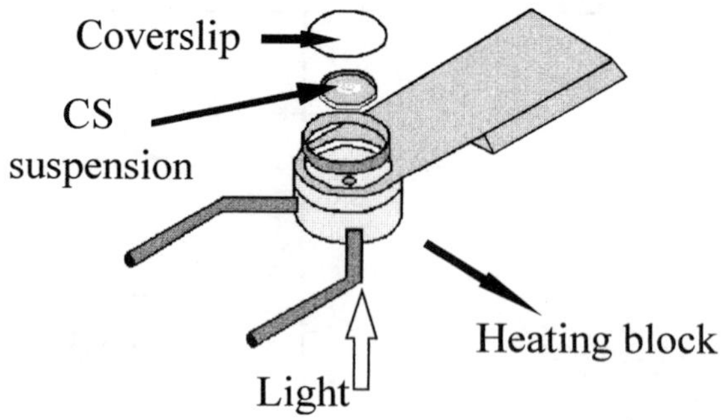


**Figure 1–9** Elements of a hot stage to be mounted under the objective lens of a light microscope. CS = cassava starch granules.

were used to obtain differential images or stains were added to the solution to enhance contrast. Heating rates and final heating temperatures were varied using the temperature controller. Images were videotaped in real time, and prints were obtained from selected images. Sizes, curvature, circularity, and shapes of starch granules were plotted as a function of time (see Section 2.6.3). Additional information on kinetics of granule swelling was obtained from differential scanning calorimetry.

Color Plate 2 shows a time sequence of swelling of cassava starch granules at 80°C and a heating rate of 40°C per minute. Swelling was not observed until a temperature of about 65°C was reached. There is a distribution of sizes among the granules, and larger granules started to swell first. The swelling phenomenon is completed in 2 minutes. There is a transfer of water into the granules, and a consequent decrease in remaining solvent water. The quantitative data of the physical changes revealed by hot-stage microscopy (mean diameter and volume fraction) correlated well with DSC (enthalpy) and oscillatory rheometry (G′) data as a function of heating time (Stanley, Aguilera, Baker, & Jackman, 1998). This shows that the microscopic, thermal, and rheological events associated with the gelatinization of starches coincide and underline the value of doing monitoring experiments of this type using different techniques. Figure 1–10 shows a sequence of binarized images of a potato cell during heating in oil in a hot stage. Again, the binary image permits identification of objects as well as determination of the major geometrical parameters (e.g., the area, perimeter, and circularity) of the cell as a function of "frying" time.

### 1.3.9 Microspectrophotometry

The integration of newer computer technology with advances in microscope instrumentation has led to a much wider role for microscopy in the food industry and food research. For example, a personal computer may be used to operate an LM for spectrophotometric scanning. The computer allows a sample to be scanned in a consistent manner and the data to be collected and analyzed automatically. Coupling a standard microscope to a computer results in a computer-assisted LM, essentially a programmable optical robot capable of many useful functions, one of the most important of which is mapping of the sites of specific chemical entities within cells.

The concept of microspectrophotometry arose in response to the need to better quantify levels of intracellular contents. Microspectrophotometry is similar to conventional spectrophotometry, in which the amount of monochromatic light passing through a colored solution is measured. In the microscopic procedure, monochromatic light is passed through a specimen and the amount transmitted is measured. Coupling the photomultiplier unit to a computer allows enhanced productivity

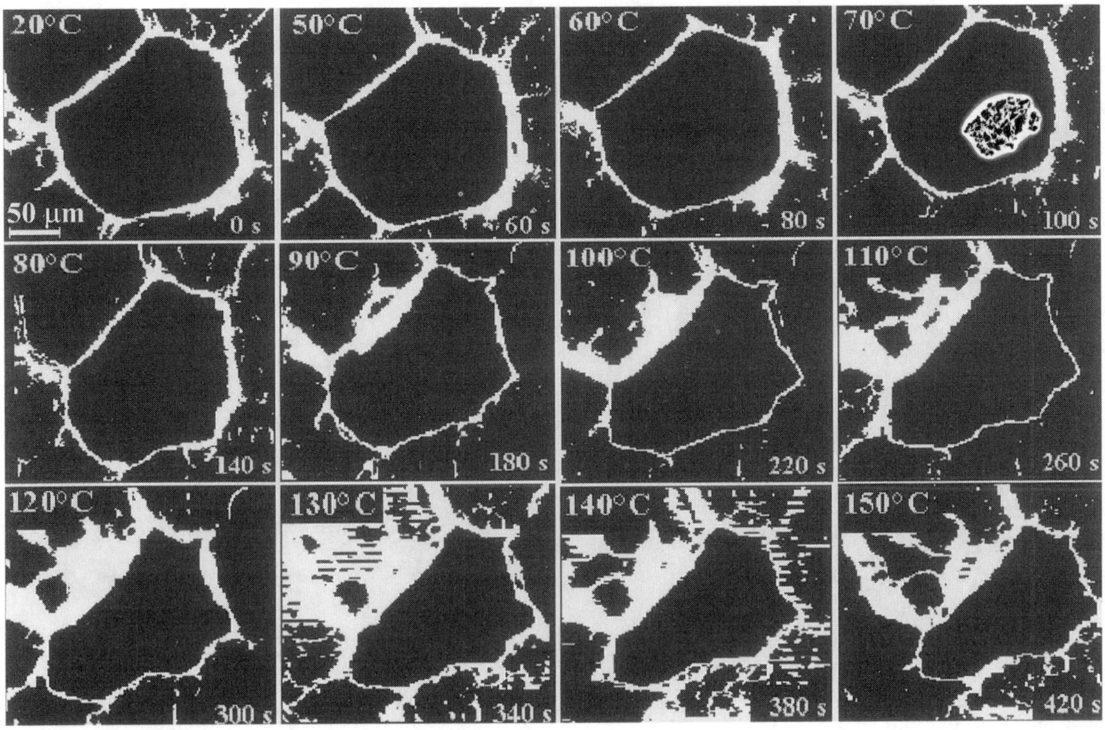


**Figure 1–10** Hot-stage video microscopy. Binarized image of a potato cell after heating in oil to 150°C (heating rate 40°C/min). *Source:* Courtesy of P. Bouchon.

and avoids operator errors during repetitive analyses. Cytochemistry applications have included determining the levels of nucleic acids, proteins, enzymes, pigments, and hormones in cells.

Microspectrophotometry has also been used to measure the rate of post-mortem glycolysis patterns in muscle tissue. The pH decline that normally accompanies post-mortem storage of meat has a major influence on the color and water-holding capacity of the final product. This drop in pH occurs as glycogen, the primary storage carbohydrate of muscle, is degraded biochemically to lactic acid, the final product of anaerobic glycolysis. Microscopically, glycogen appears in granules located in the sarcoplasm between myofibrils and under the plasma membrane, or sarcolemma (Figure 1–11). There is, however, a great deal of variation in the rate of post-mortem glycolysis patterns found among animals, and it is possible that this is the result of differences in glycogen storage patterns.

The work of Swatland (1990) illustrates how computer-assisted microscopy can be of use in investigating such questions. A light microscope was fitted with a scanning stage driven by computer software to enable the mapping of transverse sections of muscle stained for glycogen content and measured using an on-board photometer. Figure 1–12 shows a three-dimensional view of glycogen content in a single muscle fiber built up from absorbance values; it reveals a large core of glycogen-rich sarcoplasm. The initial intracellular pattern of glycogen distribution may thus be able to predict rates of post-mortem glycolysis.

Although many of the newer types of microscopy depend upon computerized directions—

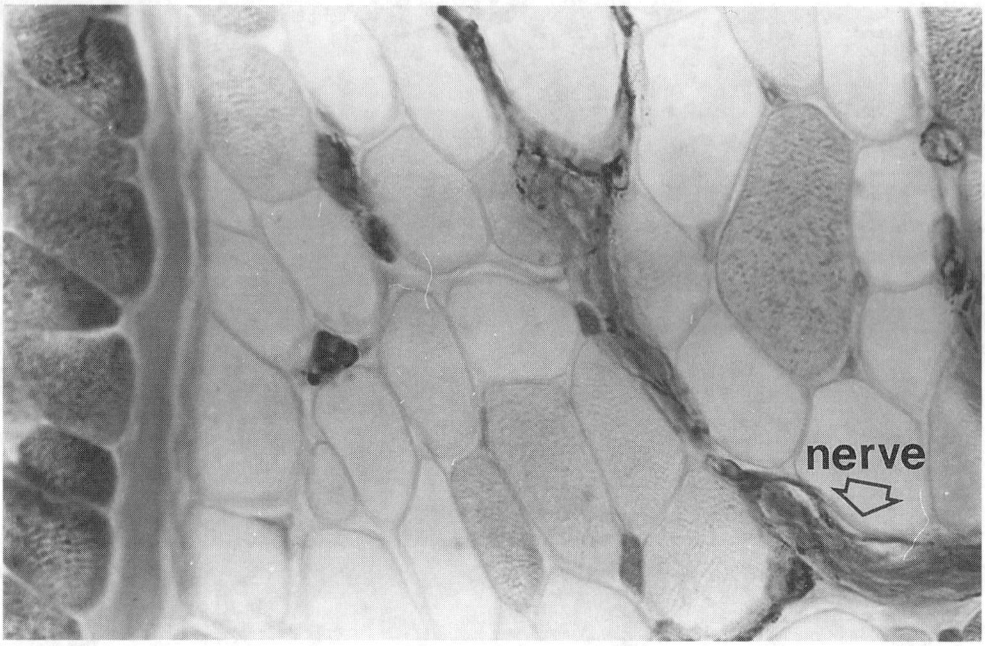


**Figure 1–11** Light micrograph of intracellular glycogen distribution in porcine longissimus dorsi muscle stained using the periodic acid–Schiff (PAS) reaction. Note dark-staining glycogen granules located in the sarcoplasm between myofibrils. *Source:* Courtesy of H.J. Swatland.

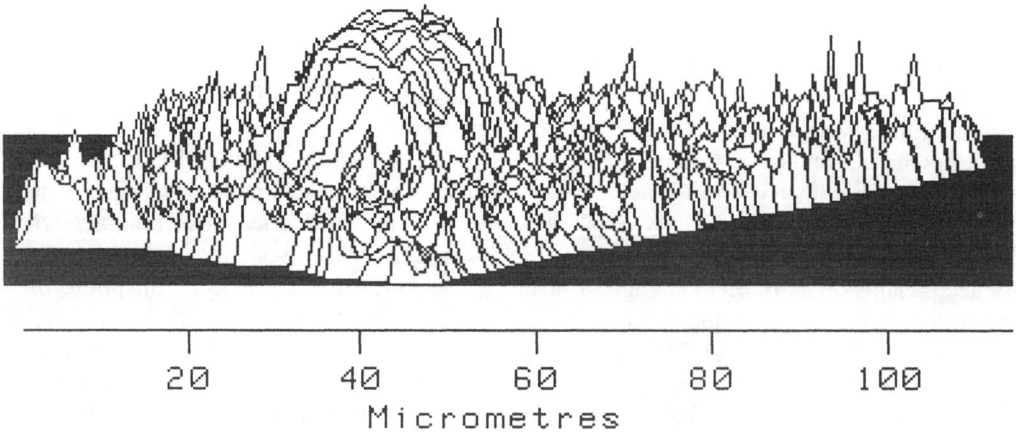


**Figure 1–12** Three-dimensional view of absorbance values (PAS reaction for glycogen) in a transverse section of muscle fiber with a large core of glycogen-rich sarcoplasm. Maximum absorbance at the core center was ≈0.6. *Source:* Reprinted with permission from H.J. Swatland, Intracellular Glycogen Distribution Examined Interactively with a Light Microscope Scanning Stage, *Journal of Computer Assisted Microscopy*, Vol. 2, pp. 233–237, © 1990, Plenum Publishing Corporation.

laser confocal microscopy for example—computers may also be used to operate a standard light microscope for spectrophotometry, fluorometry, polarimetry, and spatial scanning (Swatland, 1998).

### 1.3.10 Confocal Laser Scanning Microscopy

The confocal laser scanning microscope (CLSM) was conceived over 40 years ago but has recently become much more accessible to the average researcher and much easier to use. It is an excellent example of the marriage of old and new technologies to create an analytical instrument that far exceeds the sum of its parts. Advanced computer imaging technologies, fluorescent probe advances, and computer designed optics have been integrally linked with improved analytical light microscopes. The combination allows high-resolution volumetric imaging of light microscopic specimens, heretofore largely impossible. The accessibility and ease of use of CLSM has moreover resulted in its wide acceptance as an alternative or supplement for conventional wide-field light microscopy of thick, fluorescently labeled, or stained specimens.

While confocal laser scanning microscopy is an evolutionary form of light microscopy, the process by which an image is formed is very different. Figure 1–13 shows a generalized schematic diagram of how images are obtained. Most confocal microscopes allow the capture of nonconfocal, transmitted, and reflected laser light as well as confocal reflected laser light and epifluorescence confocal imaging. The laser light is focused by the objective lens to illuminate a single, precisely defined point in the specimen (the focal point). A scanning device deflects the beam in the $x/y$, $x/z$, or the $y/z$ dimension and so scans the focused spot on the specimen to create an image of the $x/y$, $x/z$, or $y/z$ focal plane. Reflected and fluorescent light returns via the illumination path and is then focused by the optics of the microscope at the confocal point at the center of a pinhole. Since the spot on the pinhole and the spot on the specimen are both located in the focal plane of the imaging lens, they are said to be confocal.

The pinhole permits passage of light only from the focal point and excludes light from other sources in the specimen. Light passing through the confocal pinhole is detected by a detector and, importantly, provides an image of only the in-focus plane: structures outside the focal plane are suppressed. Thus, when the $z$ dimension is varied, a confocal microscope acts as an "optical microtome," allowing a series of images to be constructed that mimic what would have been previously obtainable only by cutting thin physical sections. Rejection of out-of-focus light using the confocal concept enables the microscope to collect and combine a series of optical slices at different focus positions in order to generate a three-dimensional representation of the specimen. The depth to which the beam can penetrate is, however, limited; while it is difficult to generalize, because penetration is specimen dependent, depths beyond about 40 $\mu$m, achieved in 1 $\mu$m steps or less, become increasingly uninformative. There are many factors that influence depth of observation within the specimen and limit the total volume rendering that might be possible with any particular sample. The refractive index for mountant and immersion media, for example, affects the geometry of the final volume rendering. The emissive properties of fluorophores, objective lens quality, use of immersion fluids, the laser light source, cover slip thickness, and numerical aperture of the objective lens all affect the final image. Various manufacturers account for many of these factors and implement correction routines that facilitate the creation of very high-quality volume renderings from optical sections.

The CLSM is employed to best advantage when used to provide extraordinarily thin, in-focus, high-resolution optical sections through a thick specimen (see Figure 1–14 and Table 1–3). It does this by two-dimensionally scanning a fixed point of light through the specimen, a point of light that is well defined in all three dimensions (the $x$, $y$, and the $z$ axes). It is common to use the CLSM in the epifluorescence mode, since this mode allows a greater spatial resolution and signal-to-background ratio than are obtainable with

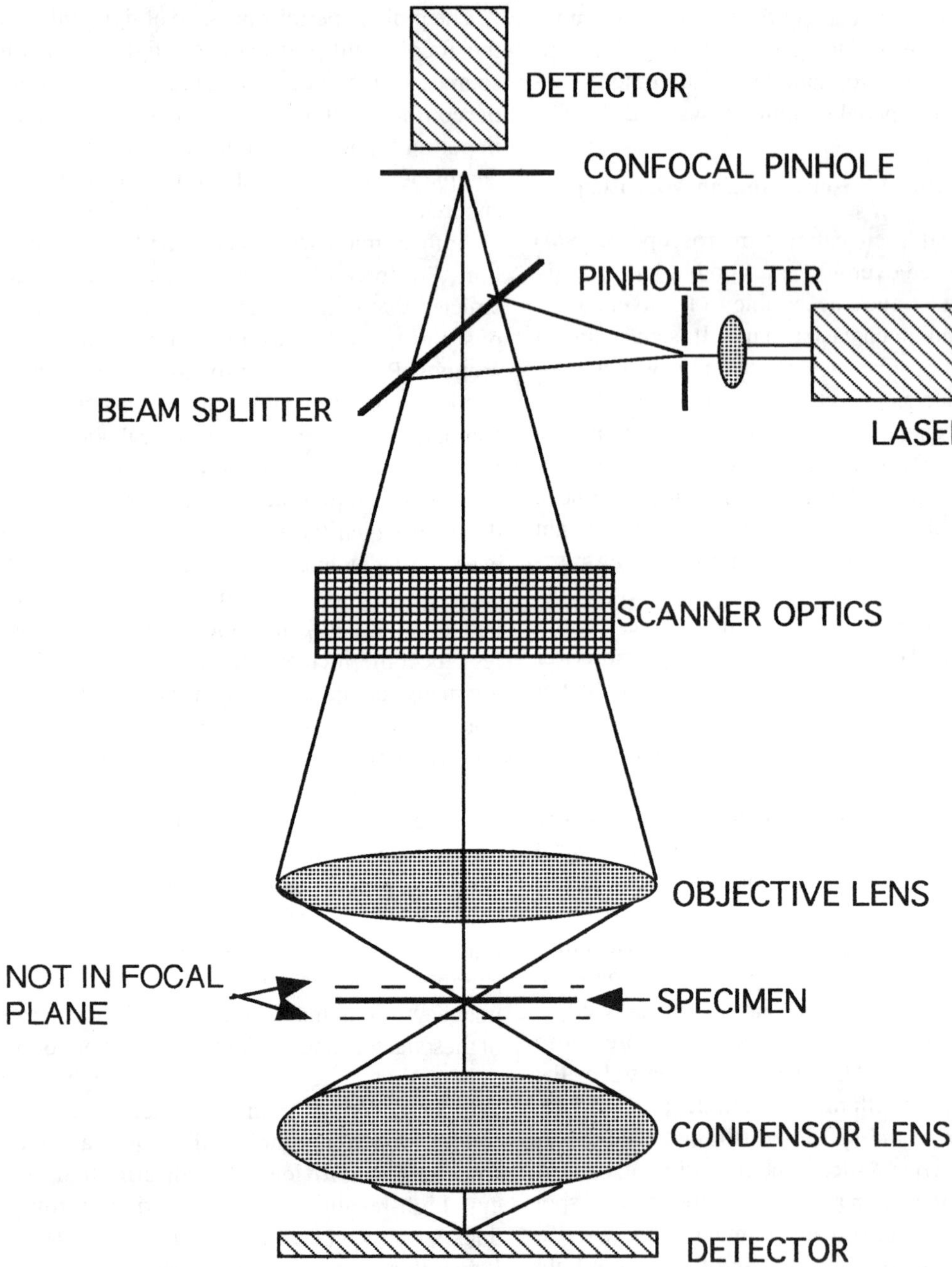


**Figure 1–13** Schematic optical paths of confocal laser scanning microscope (CLSM) with reflection, confocal, and nonconfocal transmission modes. *Source:* Reprinted with permission from H. Kitagawa, Theory and Principal Technologies of the Laser Scanning Confocal Microscope, in *Multidimensional Microscopy*, P.C. Cheng, T.H. Lin, W.L. Wu, and J.L. Wu, eds., pp. 53–71, © 1994, Springer-Verlag.

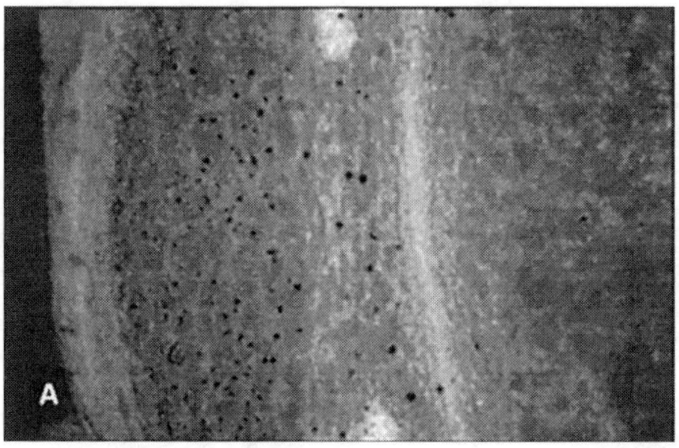

**Figure 1–14** Optical section of the fibrous sheath structure from a cross section of a canned green bean. Scanned 24-bit color transparency photograph of a full spectrum epifluorescence image obtained by CLSM. (This figure is upper part of Figure 2–6.) *Source:* Reprinted from D.W. Stanley, J.M. Aguilera, K.W. Baker, and R.L. Jackman, Structure/Property Relationships of Foods as Affected by Processing and Storage, in *Chemical, Structural, and Rheological Changes During Phase/State Transitions in Foods*, M.A. Rao and R. Hartel, eds., pp. 1–56, by courtesy of Marcel Dekker, Inc.

ordinary full-field fluorescence microscopes. Also, the light path of the CLSM is especially designed for confocal fluorescence microscopy. As in conventional fluorescence microscopy, fluorochromes are stimulated with laser light so that they emit fluorescence at a longer wavelength. The CLSM, however, has the enormous advantage that fluorescence emission from each focal point excludes out-of-focus light from above and below the focal plane. Fluorescence emission from each position is then gathered and intensified and converted to an electronic signal; finally, all points are reassembled to produce an image representing a single optical slice of the specimen precisely defined in the $z$ dimension.

Full-field epifluorescence, by comparison, acquires an image of the whole specimen simultaneously and must therefore include out-of-focus light emitted from the full depth of the specimen (i.e., from above and below the focal plane of the objective lens). These images lack resolution and clarity and are well known for the lack of focus and the glare associated with structures of interest. The CLSM, by acquiring and recombining multiple optical sections through the $z$ axis, can effectively reconstruct the three-dimensional volume of the specimen and in doing so improve on the optical resolution of standard full-field epifluorescent microscopes by as much as 30%. The volumetric data can, moreover, be very precisely quantitated. An example is found in the work of Travis, Murison, Perry, and Chesson (1997), who were able to successfully measure cell wall vol-

**Table 1–3** Advantages of Confocal Laser Scanning Microscopy

| *Advantage* | *Application* |
|---|---|
| Light from a point in the specimen outside the focal plane is blocked by the pinhole aperture in the objective lens | Optical sectioning to examine internal 3-D structure of thick specimens |
| Fluorescence capability | Multicomponent analysis<br>Immunochemical techniques |
| Higher resolution than LM | Improved imaging of structure |
| Minimal sample preparation | Examination of fragile structures |
| Physical sectioning not required | Examination of thicker or larger samples, with fewer artifacts |
| Image captured in digitized form | Image manipulation via computer software |

*Source:* Data from J.C.G. Blonk and H. van Aalst, Confocal Scanning Light Microscopy in Food Research, *Food Research International*, Vol. 26, pp. 297–311, © 1993, Elsevier Science Ltd; M. Kalab, P. Allan-Wojtas, and S.S. Miller, Microscopy and Other Imaging Techniques in Food Structure Analysis, *Trends in Food Science Technology*, Vol. 6, pp. 177–186, © 1995, Elsevier Science Ltd.; and Y. Vodovotz et al., Bridging the Gap: Use of Confocal Microscopy in Food Research, *Food Technology*, Vol. 50, No. 6, pp. 74–82, © 1996, Institute of Food Technologists.

umes using a CLSM. A three-dimensional reconstruction from a gallery of CLSM micrographs examining the presence of oil in the interior of the crust of fried potato is shown in Color Plate 8.

Among the other advantages of the CLSM is that the detector signal is digitized and recorded in an on-board computer. Computer storage allows contrast-improving enhancements, filtering out unwanted noise, reconstruction of three-dimensional views, and compilation of digital movies to show time sequences. Also, preparing samples for confocal microscopy is straightforward, much like sample preparation for conventional microscopy.

The CLSM has been particularly useful for examining lipid components because optical sectioning overcomes the tendency of fats to smear and migrate and because lipids are amenable to fluorescent staining (see Section 9.6.1, on the frying of foods). Figure 1–15, which is a confocal image of milk fat, shows the CLSM's advantages in this area. The CLSM has now been used in a number of new investigative areas, including research into food materials, fixed tissues, immunolabeled tissues, and live tissue physiologic responses. References are given at the end of this chapter to a number of food applications.

A difficulty does exist when the CLSM is operated in the fluorescence mode: the short wavelength light excites fluorescence in areas outside the target focal plane, leading to a cone-shaped pattern around the focal plane (Figure 1–16 and Color Plate 3) and potential photodamage or bleaching. This damage can be mitigated, though not eliminated, through the use of antioxidant mountants and more rapid scanning techniques. Recently, an innovative way around this difficulty has been found. If, instead of one short wavelength photon, two longer wavelength photons could be made to arrive simultaneously at the focal plane, the same degree of fluorescence would be excited but only negligible excitation would occur at any position except the focal point. Multiphoton fluorescence microscopy, based on the two-photon approach, seems to be near to providing a practical solution.

With confocal microscopy, it is now possible to obtain better image quality than conventional microscopy and to clearly visualize thin sections in the interior of dense light scattering or fluorescent samples. Because confocal microscopy is so well suited to the needs of food microscopists, the CLSM is expected to soon become a fixture in most microstructure laboratories.

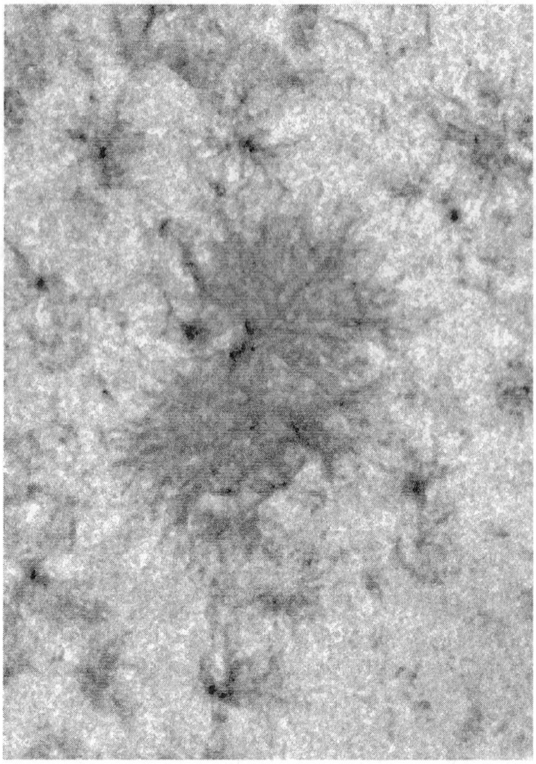

**Figure 1–15** Confocal laser scanning micrograph of spherulitic particles in the triglyceride crystal network of bovine milk fat at 18°C, 20% solids volume fraction. The fluorescent dye Nile blue was dissolved in the melted sample. Upon crystallization on the microscope slide, Nile blue partitioned into the liquid triglyceride phase, negatively staining the solid triglyceride crystal network. Magnification factor = 600×. *Source:* Copyright © Alejandro G. Marangoni, Ph.D.

## 1.4 TRANSMISSION ELECTRON MICROSCOPY

### 1.4.1 Introduction

In electron microscopy, a beam of electrons rather than light is used to form magnified images of specimens, the advantage being that electrons

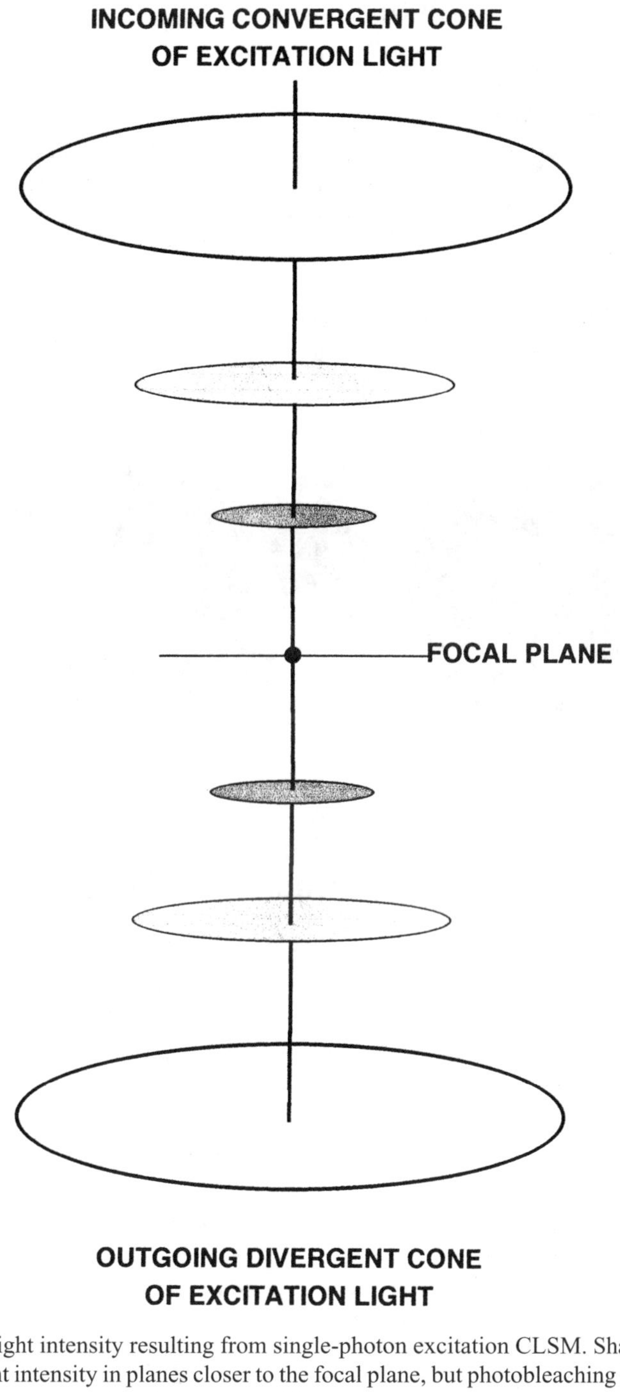


**Figure 1–16** Cones of light intensity resulting from single-photon excitation CLSM. Shading reflects light intensity; there is a higher light intensity in planes closer to the focal plane, but photobleaching is the same in every layer of the specimen. *Source:* Redrawn from Marelius, 1995.

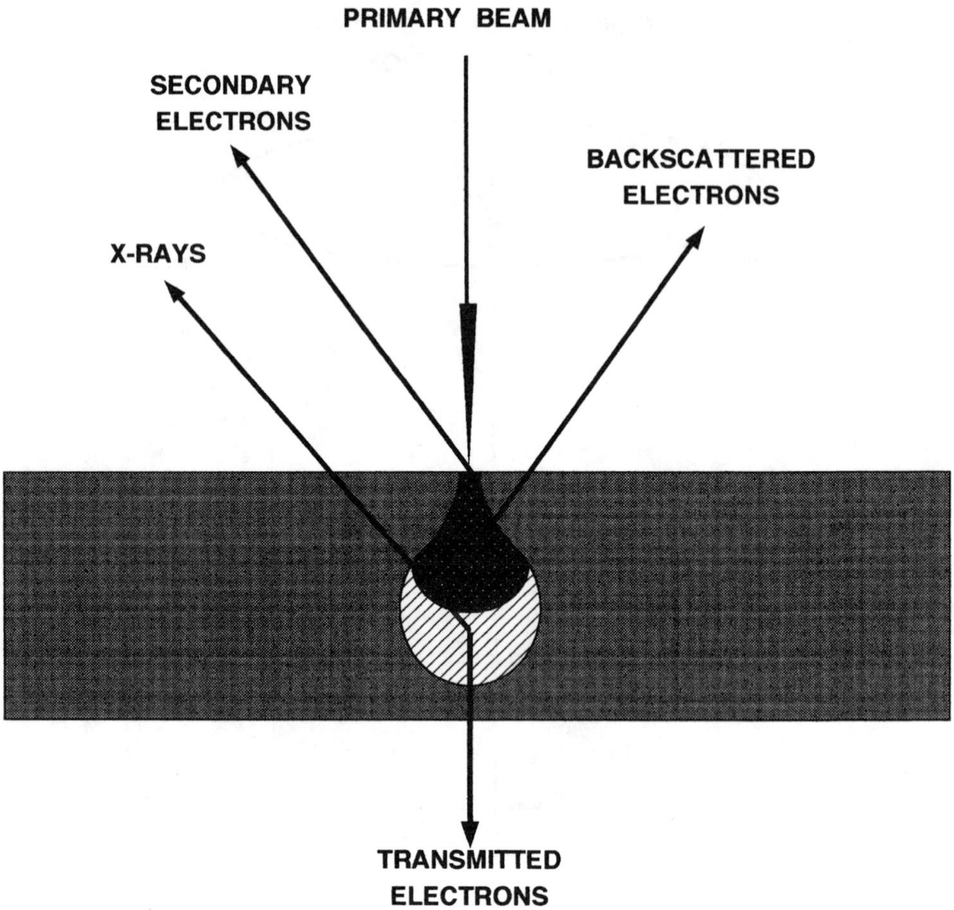


**Figure 1–17** Diagram of interactions of beam electrons and specimen. A variety of signals are generated; the most commonly used are transmitted (TEM) and secondary electrons (SEM), backscattered electrons, and X-rays. *Source:* Data from R. Johnson, *Environmental Scanning Electron Microscope*, © 1996, Robert Johnson Assocation.

provide as much as a thousandfold increase in resolving power. As an electron beam impinges on a sample, interactions occur that generate a variety of signals that can be captured to obtain images (Figure 1–17). The two basic types of electron microscopes are the TEM, in which transmitted electrons are captured, and the SEM, in which secondary electrons are captured. These instruments produce different but often complementary images. The TEM has been an extremely important tool in the study of biological structure. It is not an exaggeration to state that a major portion of our modern knowledge of subcellular structure and function is based on TEM data. Simple TEMs became available in the late 1930s, and it was soon apparent that they produced better resolution than the LM.

In its simplest form, the TEM resembles an inverted LM. An electron gun (tungsten filament) is heated and emits a narrow beam of electrons traveling at high speed. The voltage applied to achieve this acceleration is in the 40–100 kV range. The electron beam takes the place of the lamp in the LM and acts as the source of illumination. In focusing the electrons, use is made of the fact that an electron beam will be deflected by a magnetic field and magnetic lens can be employed like a converging glass lens. Since the human eye is not sensitive to electrons, the final image, formed by electrons that have passed through the specimen, is either focused on a fluorescent screen for viewing or onto a photographic plate. See Figure 1–18 for comparison of the LM and TEM.

A major difference between light and electron microscopes is that electrons need a high vacuum in order to travel the distances used in electron microscopy. A vacuum in the range of $10^{-4}$–$10^{-5}$ torr (1 torr = 1 mm Hg = 133 Pascal = 1.33 mBar) is required, and such a vacuum is produced by the action of a diffusion pump working in tandem with a rotary oil pump. The necessity of a high vacuum environment, coupled with the need for a powerful electron beam to pass through the viewed material, severely limits the types of specimens that can be examined; they must be bone dry, strong enough to resist beam damage, and extremely thin. Thus, great compromises must be made in order to take advantage of the high magnification possible with the TEM (≈300,000–500,000×).

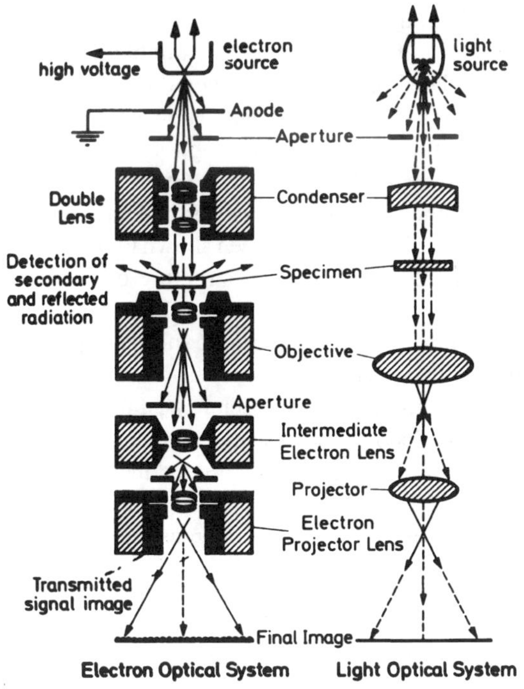


**Figure 1–18** Schematic representation of optical systems of transmission electron microscope (left) and light microscope (right). *Source:* Kessel and Shih, 1974.

### 1.4.2 Sample Preparation

Because of the requisites of electron microscopy, preparation of biological samples is much more complicated and difficult than for other types of microscopy. (Only general techniques are given here: the works cited at the end of this chapter provide detailed accounts.) Three approaches can be followed. If the sample is small or thin enough (rarely the case with foods), it can be mounted whole. More likely an ultrathin section (≤100 nm) must be cut with an ultramicrotome. In the third approach, a replica of the sample is made.

Since usually the material must be cut into thin sections, tissue fixation is the first step in sample preparation. This step is critical to all those that follow. The most commonly employed fixatives are the aldehydes and osmium tetroxide; the latter acts not only as a protein fixative but also as an electron stain (a major advantage for TEM samples), and it is uniquely suitable for fixation of lipids. The final choice of fixative is dictated by sample structure and composition, and in fact it is common to use various agents sequentially. Although chemical fixation methods are the most commonly used for TEM sample preparation, cryogenic fixation by ultrarapid freezing is an alternative (this process will be described subsequently). Following washing and dehydration, a suitable embedding medium is used in order to provide support for satisfactory sectioning. The materials most commonly used as embedments are epoxy resins, polyester resins, and methacrylates.

Biological materials must be cut extremely thin so that they may be penetrated by the electron beam. Section thicknesses of 50–100 nm are common, much thinner than the 500–1,000 nm paraffin thick sections cut for an LM. The cutting process is especially difficult in the case of food materials such as seeds, because they are packed with starch and proteins and little space is left for embedding media.

One factor limiting the thinness of sections is the accompanying decrease in contrast. Adequate contrast is mandatory if cellular components are to be differentiated. In electron microscopy, contrast is a consequence of differential electron opacity. Electron-dense areas of the sample are those that scatter electrons strongly. While traditional stains used in light microscopy fail (because an electron microscope cannot distinguish visible colors), substances that combine with specific components to increase their molecular density through enhanced electron scattering (called negative stains) are frequently used. Negative staining is usually performed on sections in order to achieve maximum differential electron opacity. Heavy metals, such as osmium, lead, tungsten, and uranyl salts, are considerably higher in atomic weight than the elements present in cellular organic molecules, and thus they are common electron microscope stains. Stains differ in specificity, and their differences can be used to improve contrast as well.

Procedures other than metal salt staining can be employed to localize desired components. Various organic dyes combined with heavy metals can be attached to enzymes while immunolabeling, and enzymatic digestion of thin sections has proven useful for cytochemical localization of components in the TEM.

The third general approach to TEM sample preparation is to make a replica of the material. In this approach, the surface details of a thick sample are replicated using a shadowcasting technique, and this step is followed by carbon coating. The interior details of a sample of various materials can be inspected if it is freeze-fractured or freeze-etched and split open (usually through the middle of biological membranes in the case of freeze-fracturing). The new surface can then be replicated and viewed in the microscope.

### 1.4.3 Transmission Electron Microscopy in Food Science

The benefits of the TEM cited above are balanced against certain drawbacks. One of the most important drawbacks is that sample preparation is difficult and invariably causes structural artifacts, often as a result of the drying or sectioning steps. Also, the high magnifications attainable often prove counterproductive to those studying generalized food microstructure. Researchers become enmeshed in the ultrastructure of materials and fail to see how this is related to overall structure formation and arrangement. The link between levels of organization visible with the LM and the TEM is frequently ignored, and the concept of correlative microscopy is not followed. The TEM seems to be most applicable in studies of certain food components such as proteins and how they interact with other components. Examples of such studies follow.

Structure in proteins was originally determined by chemical and physical methods, including amino acid analysis, circular dichroism, and X-ray crystallography. While these approaches allowed some insight into primary, secondary, and tertiary structures, knowledge of quaternary structure—how protein subunits interact—has remained elusive. Although X-ray crystallography has great resolution ($>0.3$ nm), it is unsuitable for many food-related proteins, since they do not form suitable crystals (Yada, Harauz, Marcone, Beniac, & Ottensmeyer, 1995). Electrophoresis, ultracentrifugation, and X-ray diffraction have all been used, although none of these provide a direct image. TEM, however, has been shown to be useful in examining the quaternary structure of food proteins. TEMs have always had sufficient resolution ($\approx 0.3$ nm) to differentiate large subunited proteins, and with the addition of modern sample preparation and image analysis techniques, they are now the tool of choice for this work. Previously, sample preparation, not resolution, limited their use.

Proteins are difficult to examine with electron microscopy for the following reasons:

- Proteins exist in aqueous form. Removing water as part of the sample preparation procedure may disrupt quaternary structure.
- Proteins are organic molecules prone to beam damage from the high kV excitation energies used.

- Because of their lack of heavier atoms, protein molecules do not generate sufficient contrast on their own. Adding heavy atom stains improves contrast but lowers resolution.
- Proteins have significant three-dimensional structure not captured with conventional electron microscopy.

These limitations may be overcome by these means:

- Newer methods of specimen preparation. Cryopreparation may be employed if a cryo-TEM is available. Instead of traditional negative staining with a heavy atom salt, the sample can be cryofrozen and sublimated. Or fully hydrated cryofrozen material can be visualized directly, providing it is kept at liquid nitrogen temperatures.
- Image analysis procedures. Computer software that uses reconstruction algorithms can generate three-dimensional structures even when the signal-to-noise ratio is low.
- Advanced microscopic techniques. Low kV, cryoelectron microscopy can be used to examine samples in a frozen hydrated state, which is close to their native state. Scanning transmission electron microscopes (STEMs) use very effective electron collection systems that allow the total exposure of samples to the potentially destructive beam to be minimized. Material is scanned in a method similar to that used by the SEM, and the image is resynthesized via computer.

The quaternary structure of food proteins was investigated by Marcone, Beniac, Harauz, & Yada (1994) and Yada et al. (1995). Amaranthin, a large (338 kD) subunited seed globulin from amaranth, was examined to determine if the quaternary structure was similar to the 12-subunit, two stacked hexagonal ring configuration previously reported for various legume globulins.

Sample preparation was done in accordance with the most common method of imaging a protein: the purified globulin was negatively stained with phosphotungstic acid and imaged in a TEM operating at 80 kV in the bright field mode. Selected fields were digitized and subjected to image analysis: a three-dimensional image was then created (Figure 1–19). The results establish that this protein has a dodecameric hexagonal ring structure about 9 nm in outer diameter stacked into a hollow cylinder, corresponding to that seen for other plant storage proteins. It is difficult to imagine how this structural information could have been gathered without the use of electron microscopy.

Food materials often present difficulties to the microscopist because of their heterogeneous composition and physical form (Heertje & Pâques, 1995). However, ingenious methods have been developed to handle troublesome samples. Particularly onerous are fats and fat-containing foods, since their microstructure is quite temperature dependent. Many of these products are emulsions that present their own peculiarities due to the membrane structures around emulsified droplets. While the polarizing LM and the SEM can both be used to good advantage for this type of work, fine details (less than $\approx$100 nm) in biological material require transmission electron microscopy.

One approach to sample preparation of food emulsions and suspensions for TEM is to use microencapsulation, which involves immobilizing the material by combining it with liquid agarose and allowing it to solidify in tubes. Small pieces are then fixed in glutaraldehyde, postfixed in osmium tetroxide, embedded in Spurr's resin (an epoxy resin), and thin-sectioned. Lipids are partially fixed during postfixation with osmium, which also prevents extraction of this component by organic solvents. Staining with uranyl acetate and lead citrate to enhance contrast is common.

Microencapsulation was used in a study of milk homogenization (see Section 7.3.2). Commercially homogenized milk presents an example of a stable emulsion formed from the breakdown of native fat globules of size 1–10 $\mu$m into new globules roughly of size 1 $\mu$m. Homogenization forces milk under pressure at high speed through a slit slightly larger than the globules, causing shearing, cavitation, and microturbulence and

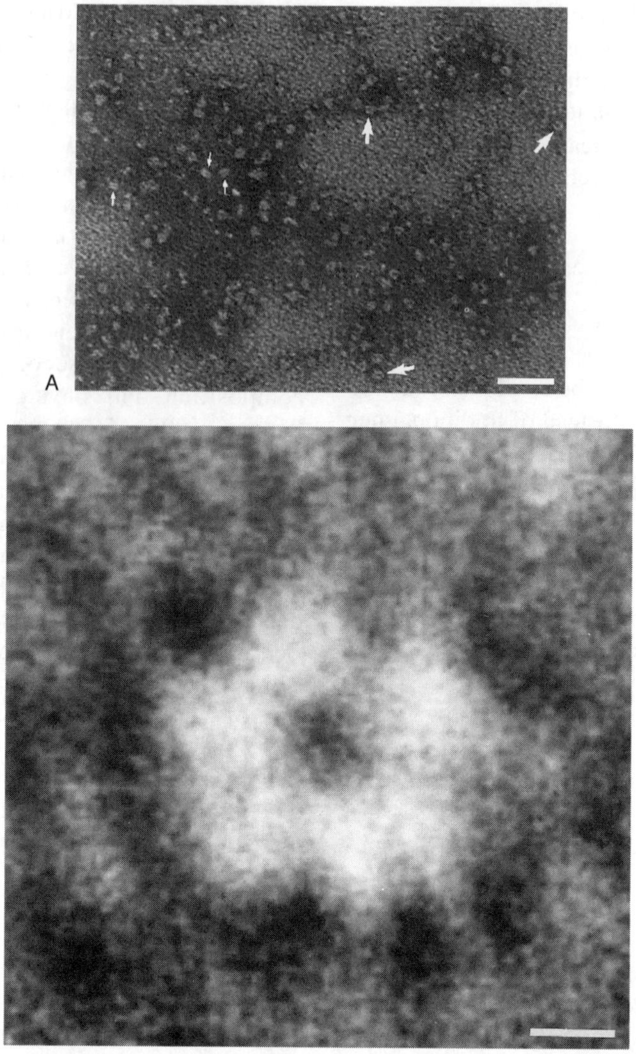


**Figure 1–19** Images of amaranthin, a major storage protein of amaranth. (A) TEM micrograph of a purified preparation of amaranth globulin, negatively contrasted with phosphotungstic acid. Some individual complexes indicated by arrows. Bar = 50 nm. *Source:* Reprinted with permission from M.F. Marcone, D.R. Beniac, G. Harauz, and R.Y. Yada, Quaternary Structure and Model for the Oligomeric Seed Globulin from Amaranthus Hypochondriacus K343, *Journal of Agricultural and Food Chemistry*, Vol. 42, pp. 2675–2678, © 1994, American Chemical Society. (B) Single globulin complex exhibiting six stain-excluding regions and a central stain-filled depression or hole (protein is white, stain is dark). Bar = 2.5 nm. *Source:* Reprinted from *Trends in Food Science Technology*, Vol. 6, R.Y. Yada, G. Harauz, M.F. Marcone, D.R. Beniac, and F.P. Ottensmeyer, Visions in the Mist: The Zeitgeist of Food Protein Imaging by Electron Microscopy, pp. 265–270, Copyright 1995, with permission from Elsevier Science.

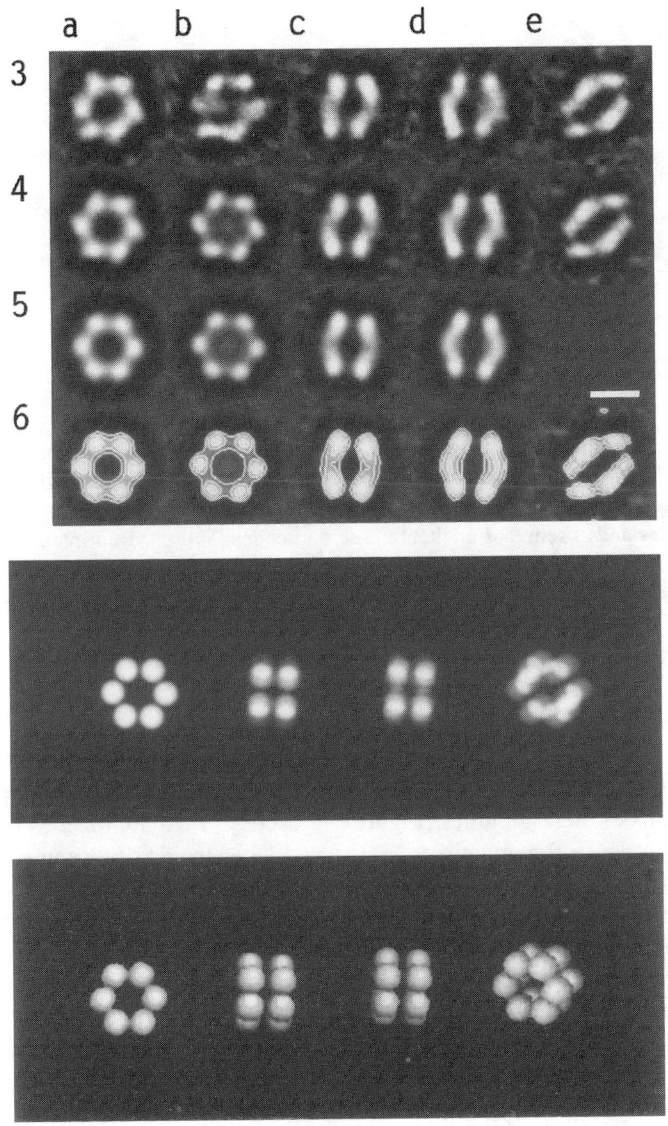


C

D

E

**Figure 1–19** continued. (C) Averaged images of single-particle electron image analysis and subsequent symmetrization. Images represent characteristic views or projections of the globulin complex and show end-on, oblique, and side-on orientations. Bar = 5 nm. *Source:* Reprinted from *Trends in Food Science Technology*, Vol. 6, R.Y. Yada, G. Harauz, M.F. Marcone, D.R. Beniac, and F.P. Ottensmeyer, Visions in the Mist: The Zeitgeist of Food Protein Imaging by Electron Microscopy, pp. 265–270, Copyright 1995, with permission from Elsevier Science. (D) Two-dimensional projections of a three-dimensional computational model composed of 12 equal spheres arranged in two parallel layers with a sixfold axis of symmetry. *Source:* Reprinted with permission from R.Y. Yada, G. Harauz, M.F. Marcone, D.R. Beniac, and F.P. Ottensmeyer, Visions in the Mist: The Zeitgeist of Food Protein Imaging by Electron Microscopy, *Trends in Food Science Technology*, Vol. 6, pp. 265–270, ©1995, Elsevier Science Ltd. (E) Shaded surface representation of a three-dimensional reconstruction derived from single-particle analysis. *Source:* Reprinted from *Trends in Food Science Technology*, Vol. 6, R.Y. Yada, G. Harauz, M.F. Marcone, D.R. Beniac, and F.P. Ottensmeyer, Visions in the Mist: The Zeitgeist of Food Protein Imaging by Electron Microscopy, pp. 265–270, Copyright 1995, with permission from Elsevier Science.

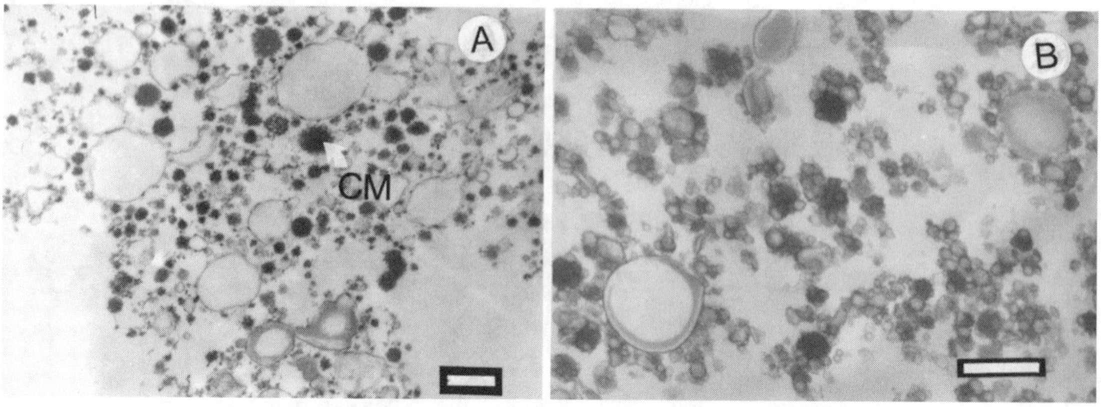


**Figure 1–20** TEM micrographs of commercially (A) and experimentally (B) homogenized milk. In the commercial sample (×16,000; bar = 0.5 μm), note the larger fat globules with attached, heavily stained whole casein micelles and also free casein micelles (CM). In the experimental sample (×26,000; bar = 0.5 μm), note the smaller but clustered fat globules and absence of unaltered casein micelles. *Source:* Reprinted from *Netherlands Milk & Dairy Journal*, Vol. 50, D.G. Dalgleish, S.M. Tosh, and S. West, Beyond Homogenization: The Formation of Very Small Emulsion Droplets During the Processing of Milk by a Microfluidizer, pp. 135–148, Copyright 1996, with permission from Elsevier Science.

also disruption of the fat globule membranes. To stabilize the surface of the new smaller globules, milk proteins, mainly casein, adsorb to their surface. TEM micrographs of these globules (Figure 1–20) show attached and also free micelles.

Comminuted meat products are also emulsion based. In this case, an O/W emulsion is entrapped in a gel formed by insoluble collagen proteins and muscle fibers. The fat globules, larger than those found in milk, are also coated with proteins (myofibrillar proteins extracted by salt). Figure 1–21 compares preparations made from muscle tissue at different pHs.

### 1.4.4 Scanning Transmission Electron Microscopy

It is theoretically possible to add scanning coils to a TEM, thus enabling scanning transmission electron microscopy. As in transmission electron microscopy, incident electrons are transmitted through the specimen, but the point source of electrons scans the specimen and the transmitted image is produced on a cathode ray tube, as in scanning electron microscopy. This type of microscopy dates back to around 1940. Recent advances in electron optics and computer interfacing have resulted in an instrument that provides a powerful alternative to other forms of electron microscopy. Two approaches have been taken; in a nondedicated STEM, samples are viewed in a conventional SEM with an attached detector below the specimen. If the sample is thin enough, sufficient numbers of electrons are transmitted to be collected and imaged. An SEM working in the transmission mode yields a relatively low resolution image compared with a dedicated STEM, in which a small-diameter beam of electrons is scanned across the section and the transmitted electrons are detected as a function of beam position.

Since the image produced is electronic in nature, it is possible to process it in order to emphasize certain features. Also, an image can be formed from the secondary electrons generated at the sample surface. A major advantage of the STEM when examining biological specimens is its ability to penetrate thicker samples (≈100 nm to 1 μm) than the ultrathin sections that must be used in the conventional TEM—without a loss in resolution. Thus, many of the artifacts produced during sample preparation can be avoided. Since the STEM provides information on internal struc-

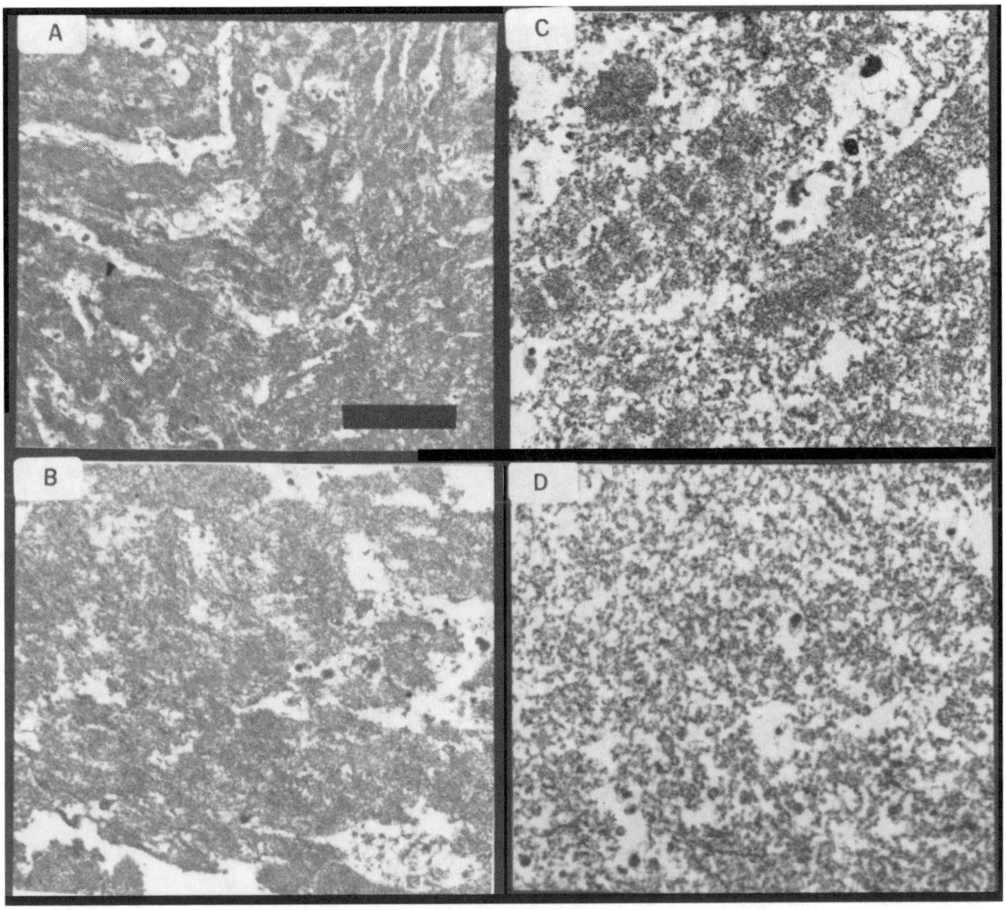


**Figure 1–21** TEM micrographs of comminuted turkey breast muscle cooked to 75°C. A = pH 4.5; B = pH 5.5; C = pH 6.5; D = pH 7.5. Bar = 2 $\mu$m. *Source:* Reprinted with permission from S. Barbut, Microstructure of White and Dark Turkey Meat Batters as Affected by pH, *British Poultry Science*, Vol. 38, pp. 175–182, © 1997, Carfax Publishing Co.

ture and the SEM provides similar information on surface structure, it is valuable to have both images. A sample STEM image is shown in Figure 1–37.

## 1.5 SCANNING ELECTRON MICROSCOPY

### 1.5.1 Introduction

The overriding need for food scientists to view a wide spectrum of structures made the appearance of the SEM welcome. In many ways, this apparatus combined the best features of the LM and TEM. Sample preparation is easier and introduces fewer artifacts, since no sectioning is required. Both surface and internal features can be studied (depending upon the preparative techniques used). A wide range of usable magnifications ($\approx$20–100,000$\times$) is possible, and the SEM can achieve a depth of field roughly 500 times that of the LM at equivalent magnifications (Figure 1–22). Also, the final data consist of electronic signals, not just a visible image, so that computer processing and storage are possible.

Drawbacks remain; the sample is still exposed to a high vacuum, meaning that total dehydration is necessary, and the material is bombarded by a

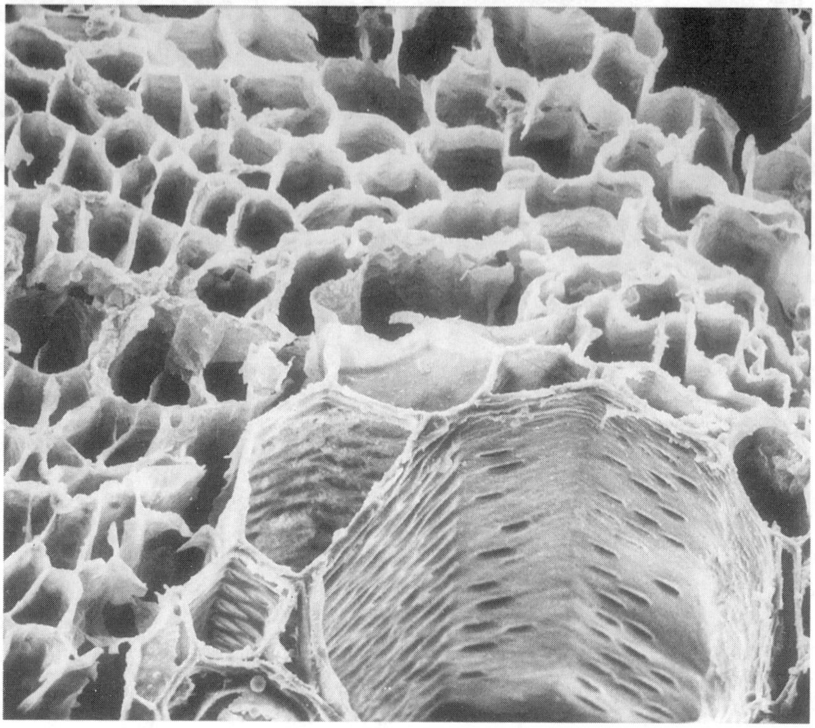

**Figure 1–22** Cross section of plant stem showing great depth of field possible with SEM. *Source:* Stanley and Tung, 1976.

potentially damaging electron beam. Nonetheless, the SEM is an important magnifying tool for examining food, and it has proven itself as the best sole instrument for microstructural studies. The first commercial SEM became available in the mid-1960s, and reports on its use in studies of food microstructure appeared in the scientific literature soon thereafter.

### 1.5.2 Principles

As mentioned previously, electrons can be made to serve the same function as light in the LM: they can be focused on a specimen to form an image. The primary advantage of using an electron beam is that electrons have vastly shorter wavelengths and therefore a potentially much greater resolving power than light. Otherwise, the two forms of microscopy are very similar: electrons are also generated by a heated tungsten filament, lenses (electromagnetic in this case) are also used to focus the beam onto the specimen, and an image is formed that can be recorded photographically.

The path traveled by the electron beam before it strikes the specimen, during which time it is accelerated through application of a high voltage, must be vacuum evacuated so that individual electrons will escape collision with molecules of gas. In an SEM, three types of electron sources are available, including the traditional tungsten hairpin filament, lanthanum hexaboride, and field emission cathodes. Although based on different physical principles, they all are designed to generate a stable electron beam. The tungsten filament delivers electrons as an electrical current is passed through it. Because the filament is heated, it is called a hot-cathode gun. Generated electrons are drawn to the anode by a voltage difference (called the accelerating voltage), where they pass through an aperture to produce the electron beam (Figure 1–23). The accelerating voltage determines the energy and wavelength of the electrons as they pass down the column. With TEMs, accelerating voltages range from around 60 kV to 400 kV, but most SEMs operate at between 5 kV and 30 kV. Lanthanum hexaboride and field emission cathodes are characterized by improved brightness and low energy spread, which leads to improved resolution, but the tungsten filament is still more commonly used because of its simplicity, low cost, and ease of operation.

As the electron beam passes down the column, it is focused first by the condenser lens and then by the objective lens, so it is cone shaped when it strikes the sample. The working distance (WD) is the distance between the objective lens and the sample. Most SEMs have a WD in the range of 10–40 mm.

When an electron beam strikes an ultrathin section ($\approx$100 nm), some of the incident electrons will be transmitted; these are used to form the image in the TEM. But the impinging beam also generates secondary electrons near the specimen surface that can escape. These electrons, sample atom electrons that have been ejected by interactions with the primary electrons of the beam, can be collected to form an image of the sample topography. In the SEM, a beam of electrons traverses an evacuated column and is focused obliquely on the specimen surface. The degree of obliqueness is termed the tilt. The beam then scans the surface repeatedly in a rectangular raster pattern and thereby liberates secondary electrons. The depth to which the primary beam penetrates the surface and promotes secondary emission is a function of the accelerating voltage and the density of the specimen.

The generated secondary electrons are gathered by a collector, conveyed to an amplifier, digitized, and then passed onto the screen of a cathode ray tube, where the raster of the electron beam is reproduced and an image is formed that is a magnified likeness of the exterior aspects of the sample.

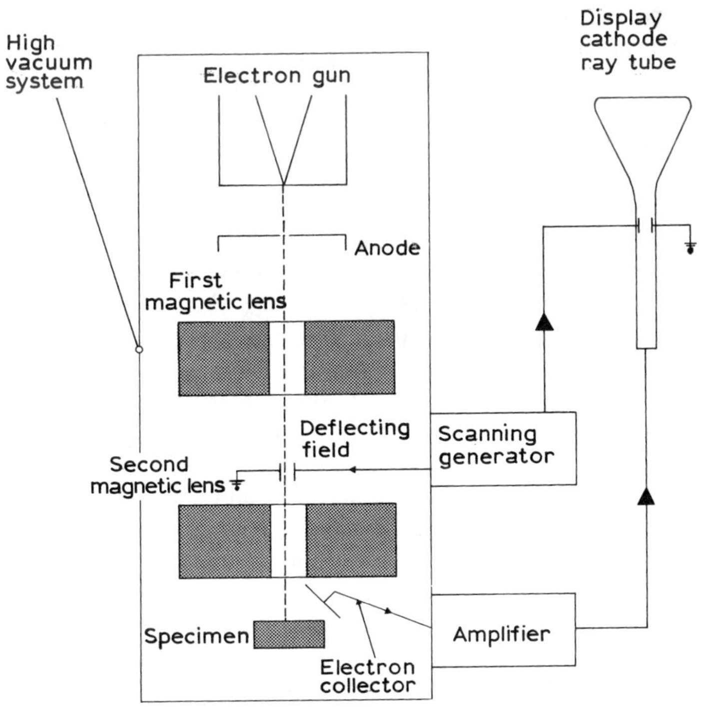


**Figure 1–23** Basic components of a scanning electron microscope. *Source:* Stanley and Tung, 1976.

Note that the electronic nature of the image allows it to be further processed.

As with the LM, an advantage of the SEM is that images can be obtained by more than one means. An example is the production of backscattered electrons. When the beam of primary electrons strikes the specimen, some of them (10–50%) are deflected through a large angle without significant energy loss. Electrons that reemerge from the surface are called backscattered electrons (they have been scattered back out of the sample by elastic collisions with the nuclei of sample atoms), and the image formed from their collection is characteristic of the atomic weight of the elements encountered in the sample. Thus, it is possible to gain knowledge of biological structure either by incorporating specific heavy metal stains into the tissue or by determining naturally occurring differences in atomic weight. A backscattered image is shown in Figure 1–24.

### 1.5.3 Sample Preparation

With the great advances that have been made in all types of microscopy instrumentation over the past few decades, this can no longer be considered the limiting factor in the examination of food materials. Rather, the success of microscopic examination now depends on the ability to prepare samples in such a way as to allow a faithful rendering of their structure. The principles of sample preparation for the SEM differ somewhat from those for the LM, although the major objective in each case is to make components visible while not inducing procedural artifacts. Besides mechanical stability and water removal, some arrangement must be made for conducting the absorbed electrons from the sample to the ground. If this is not done, the so-called charging phenomenon results, leading to a buildup of charges on the sample surface, with a consequent deflection of the electron beam. Since biological tissue is not an adequate conductor of electrons, a provision must be made for a low-resistance path for charges to escape to the earth. Surface charging is detrimental because it leads to serious image distortion and makes photography difficult (Figure 1–25). It may appear as lines on the screen or photograph, as abnormal contrast, or as breaks in the image.

The traditional way to overcome the problem of surface charging is to coat the specimen (which is mounted on metal holders called stubs) with a thin film of evaporated metal, such as gold or palladium, in a sputter coater so as to impart electrically conductivity to the sample. The coating makes an electrical connection between the tissue and the specimen stub, allowing the charge to be dissipated. A more attractive approach is to use low kV accelerating voltage during examination. This alternative not only minimizes electronic distortion but also saves the sample from beam damage. While reduced kV operation would previously have presented resolution problems, recent advances in instrumentation, including electron gun improvements (e.g., field emission) and better lens design, have mitigated these problems (Figures 1–26 and 1–27).

A major advantage of the use of SEMs is the relative simplicity of sample preparation—which is not to say that it should be done without care. Inevitably, the final results will reflect the quality of the preparative technique. An important fact to keep in mind in the face of continual improvements in electron microscopes is that, with biological material, the limiting factor is not the instrument but specimen preparation. Increased resolution provided by the builders of electron microscopes often leads only to empty magnification, not more structural information.

In general, the process of preparing biological material for the SEM is similar to the process for other types of microscopy. First, the decision must be made as to which surface of the sample to

**Figure 1–24** Backscattered electron image (A) versus secondary electron image (B) of fresh, unfixed leaf surface previously sprayed with copper-containing insecticide. Note the ease with which this element can be discriminated using the backscattered image. Note also the collapse of tissue that has occurred due to beam damage in the time between the micrographs. *Source:* Courtesy of A.K. Smith.

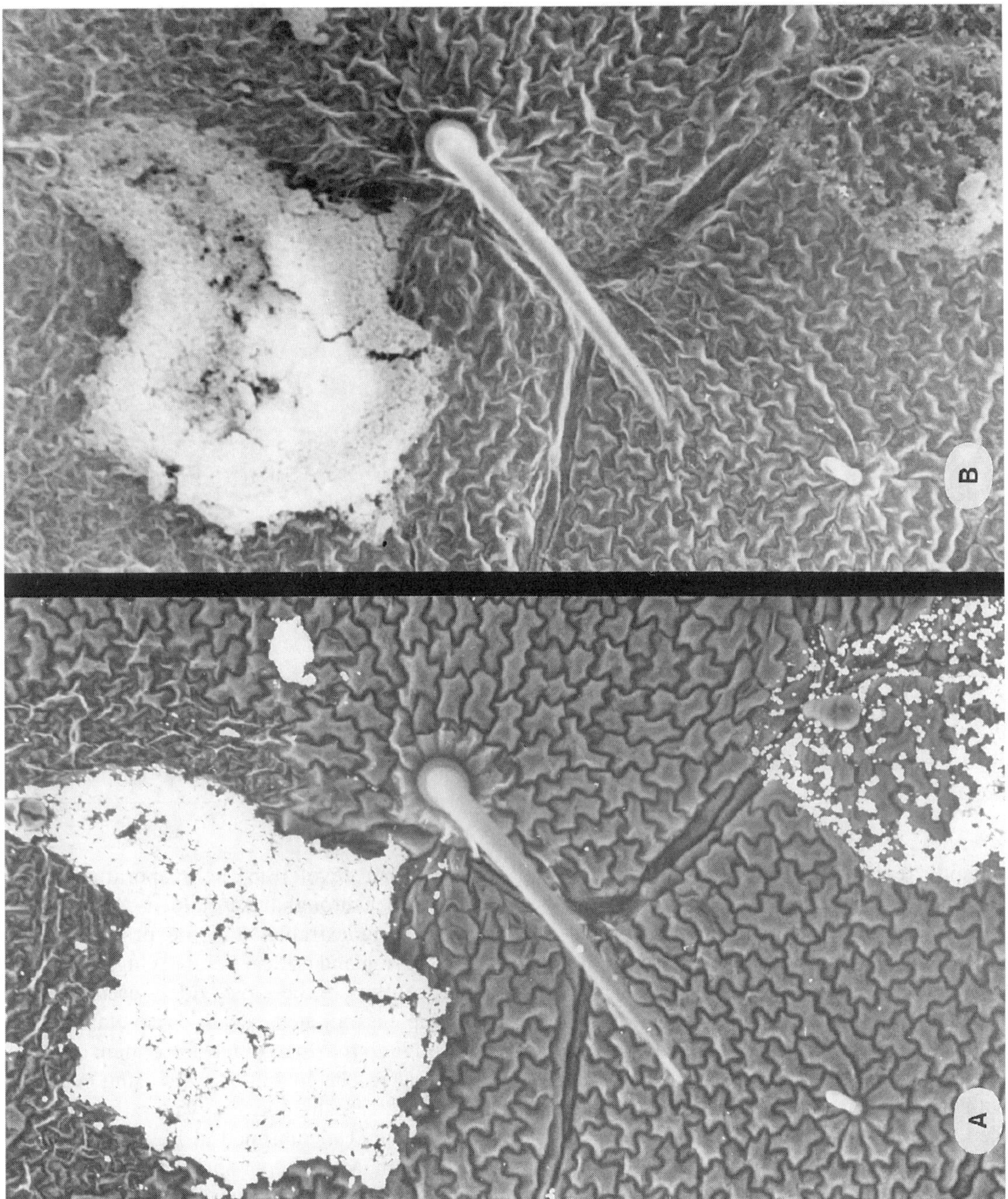
B
A

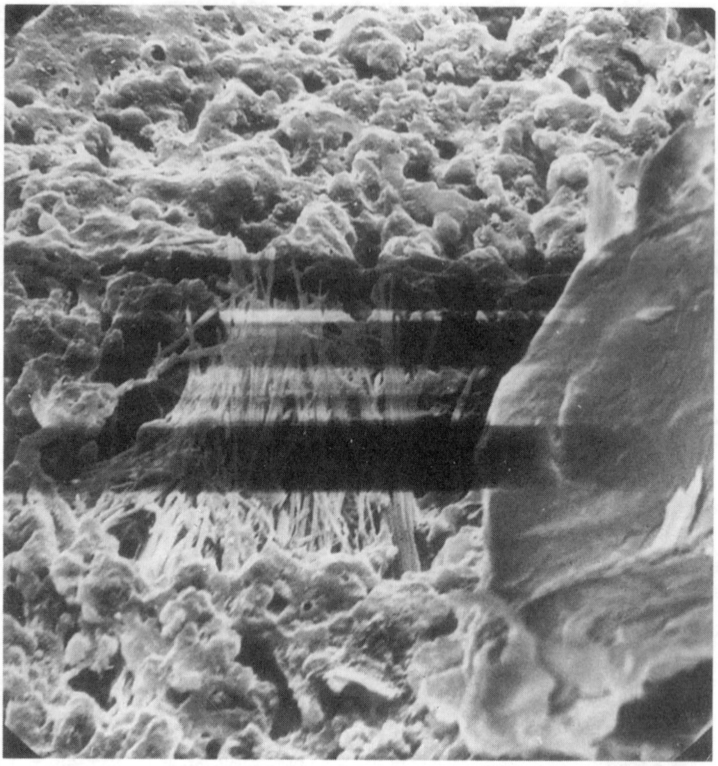

**Figure 1–25** Example of extreme charging in the SEM. In less pronounced cases, charging can be exhibited as bright flashes from areas in the specimen that have not been properly coated. *Source:* Courtesy of A.K. Smith.

examine. If an internal aspect is selected, then the material can be sectioned, cut, or fractured to expose the desired structure. In any case, the surface to be viewed should be cleaned—a simple step that is often overlooked.

Second, depending on the mechanical strength of the material, fixing may be necessary before cutting. Otherwise, fixing may occur afterwards, but it should be undertaken as rapidly as practicable to halt unwanted structurally degrading reactions, such as enzymatic proteolysis or wound response.

Third, unless the sample is already quite low in moisture content ($\approx$10%), the water must be removed, since the instrument is under high vacuum and endogenous moisture can contaminate the column. This step must be performed with great care in order to avoid distortion and shrinkage produced by forces occurring during phase changes that develop as the evaporating water front recedes through the specimen. Air drying is generally undesirable, and the procedures of freeze-drying and critical point drying, discussed subsequently, are the most widely used. Fourth, the sample is mounted, usually to a metal stub using conductive glue. If the specimen is to be metal coated, the coating is done using evaporative or sputter techniques. In evaporative coating, a heavy metal is sublimed at high temperature and vacuum. Metal vapor sprays the target but in a directional way, and often tilting or rotation is needed to achieve total and even coverage. Sputter coating requires less vacuum and no heat, since the metal ions are dislodged by a gas plasma, usually argon, and coat the specimen more uniformly.

A major problem that must be overcome by those seeking to employ electron microscopes for

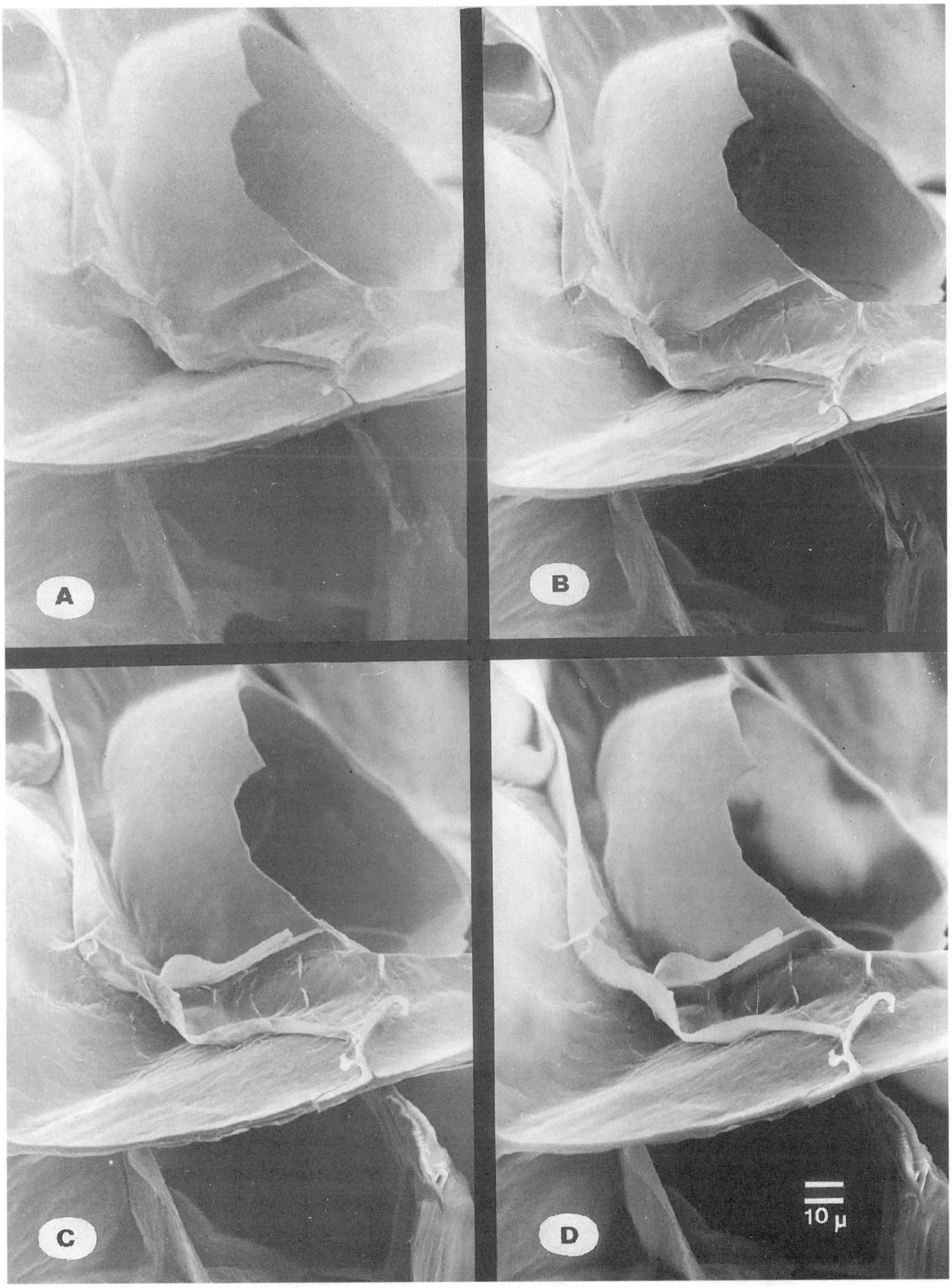


**Figure 1–26** Influence of accelerating voltage and metal coating on image. Extruded wheat snack, sputter coated with gold/palladium and photographed at different accelerating voltages (tilt = 25°, WD = 13 mm). (A) 1.0 kV. Note graininess and lack of fine detail due to insufficient electron emission. (B) 2.5 kV. Clarity of image improving with increased electron emission. (C) 10 kV. Good image clarity but some loss of subtle detail due to increased electron bombardment. (D) 20 kV. Best clarity, but notice loss of image in some areas previously showing detail. Bar = 10 $\mu$m. *Source:* Courtesy of K.W. Baker.

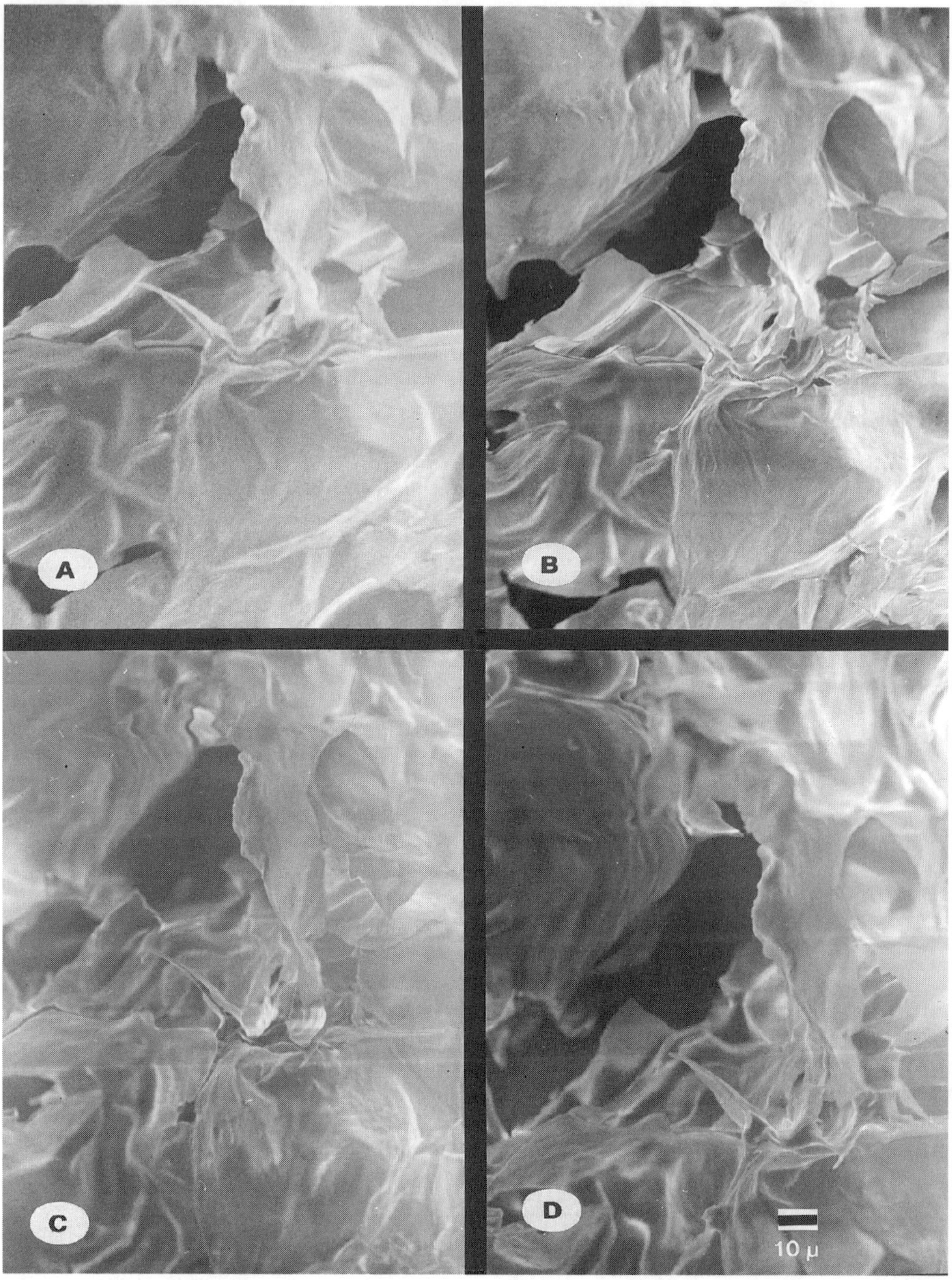


**Figure 1–27** Influence of accelerating voltage and metal coating on image. Extruded wheat snack, uncoated, photographed at different accelerating voltages (tilt = 25°, WD = 13 mm). (A). 1.0 kV. Similar to coated material (Figure 1–26) at same kV. (B) 2.5 kV. Good clarity. Similar to coated material at same kV but some charging evident. (C) 10 kV. Specimen disintegrating under beam, much charging. (D) 20 kV. More beam damage. Bar = 10 µm. *Source:* Courtesy of K.W. Baker.

food studies is that in almost all cases the material to be examined is mainly water. The water is present not only in bulk form as a dispersion medium for the various components, but it also interacts with and sometimes dictates the specific structures of macromolecules such as proteins and membranes. Since a hydrated sample placed in an electron microscope will immediately begin to desiccate, leading to structural distortion, and the liberated water will foul the column and cause a loss of vacuum, the material must be dried. Food samples also contain other volatile components, such as fats and oils, and these need to be removed as well.

One of the most common methods used for drying materials to be examined by an SEM is critical point drying (Figure 1–28). This technique is based on the existence of a critical point at which the density of a materials liquid phase equals the density of its vapor phase, thereby eliminating the phase boundary between the two that is responsible for generating the surface tension forces that cause sample distortion. In practice, a jacketed pressure vessel is employed in which are placed fixed samples that have been dehydrated, typically in a graded acetone series. The vessel is then charged with liquid $CO_2$ to a pressure of 73 atm, and the acetone is displaced. Hot water is then used to raise the temperature above the critical point (31°C), and the gaseous $CO_2$ is slowly bled off to avoid sudden decompression. The dried sample must be stored in a desiccator because of its deliquescent nature.

### 1.5.4 Cryomicroscopy

Even with critical point drying, artifacts will be produced in the sample. Dimensional changes and shrinkage of soft biological specimens can be severe. One approach gaining favor is to leave the water *in situ* but to lower the temperature to a point where the vapor pressure is reduced and the escape of water and other volatiles is negligible. During cooling at normal freezing rates, ice crystals form that can rupture cells and cause excessive structural damage. If the cooling rate exceeds a certain critical value, however, specimen water will solidify or vitrify without crystal growth. Even if this value is not reached, the size of the ice crystals formed will be so small as not to constitute a serious problem in most cases. However, microscopists always strive to achieve extremely high freezing rates at all locations inside a sample. The use of low temperatures to stabilize structures, termed cryogenic preparation, has several advantages:

- Delicate structures in high-moisture biological specimens are preserved.
- Metabolic activities that could disrupt structure are terminated.
- The procedure facilitates exposing internal surfaces by freeze-fracturing and preparing materials for X-ray microanalysis.
- Since the sample is viewed directly after freezing, not only is dehydration not required but chemical fixation is also unnecessary.
- Problems with low melting components, such as fat, are avoided.

In cryogenic preparation, the sample is frozen in subcooled ("slushy") nitrogen at −210°C or in liquid propane (which has better heat transfer properties). The goal is to freeze all the available water as fully as possible. The sample is then transferred to a vacuum chamber where it can, if desired, be fractured or etched and heated to remove ice that is coating the sample surface. Finally, the specimen, still in a frozen state, may be coated with a thin conducting layer of a metal, usually gold, to eliminate charging. The prepared sample can then be transferred directly to the cold stage of the SEM by an exchange air lock. Examination, which occurs with the sample still at a low temperature, approximately −180°C, can commence within a few minutes of freezing. (See Figure 1–29 for a summary of these steps.)

Cryogenic preparation results in a frozen hydrated specimen that has not undergone chemical fixation or drying. This is not to say that artifacts do not occur. These may include surface ice, cracking, and some deformation. Rapid freezing to a low temperature offers the highest probability of success. The conclusion to be drawn is clear: all methods of preparing biological material produce artifacts and influence its appearance in the microscope. Nonetheless, cryotechniques offer mi-

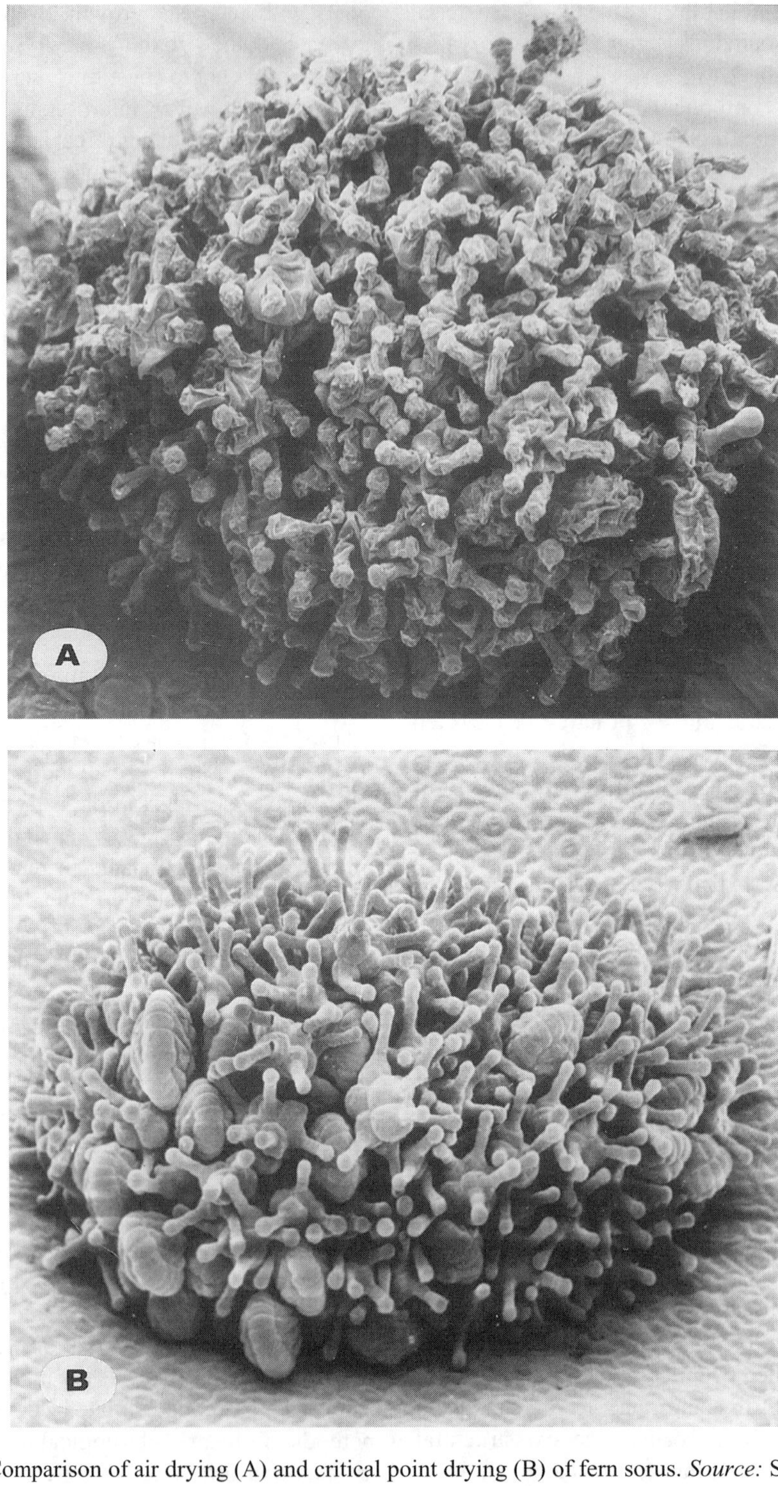


**Figure 1–28** Comparison of air drying (A) and critical point drying (B) of fern sorus. *Source:* Stanley and Tung, 1976.

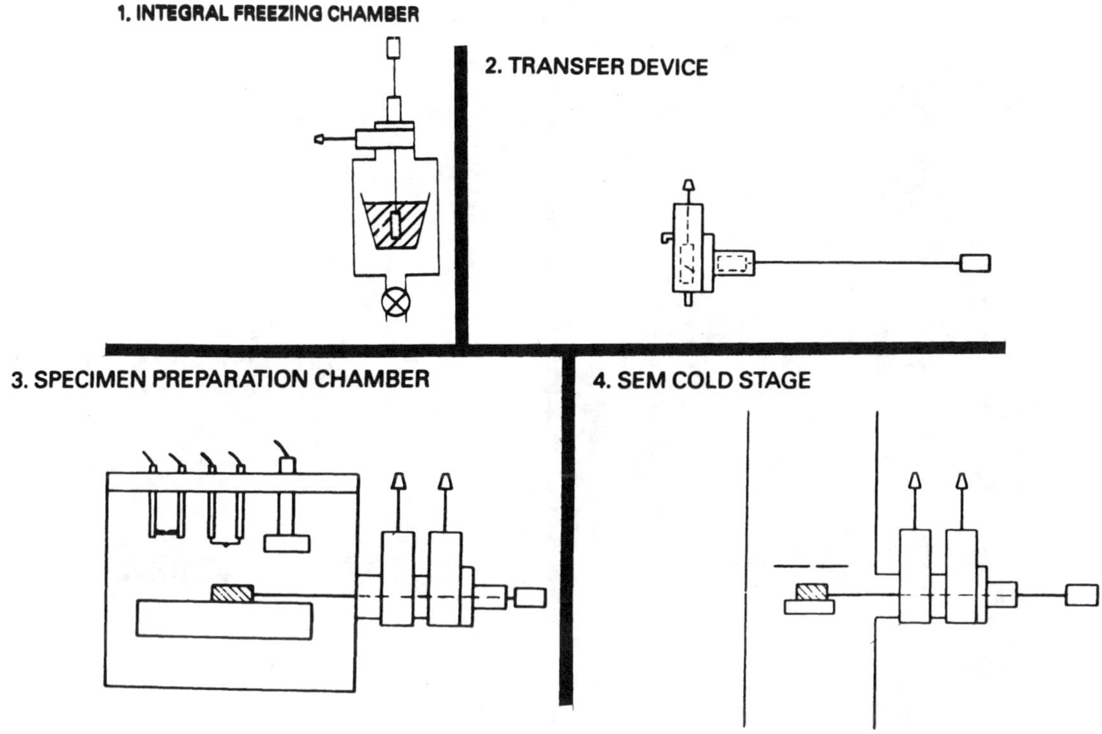


**Figure 1–29** Cryogenic sample preparation for the scanning electron microscope. Step 1: Samples are rapidly frozen by plunging into nitrogen slush at −210°C. Step 2: Specimen is transferred under vacuum to preparation chamber. Step 3: Sample preparation procedures can include fracturing, etching, and metal coating. Step 4: Specimen examination in microscope cold stage. *Source:* Courtesy of EMScope Laboratories Ltd., Kent, England.

croscopists the best chance of viewing microstructure in its natural state when using the SEM. (See Figure 1–30 for examples of cryoprepared samples.)

Worthy of mention here are the techniques of freeze-fracturing and freeze-etching. In freeze-fracturing, biological material is frozen and fractured at a low temperature. An important feature is that the division occurs along cleavage planes, usually membranes, to reveal internal facets normally difficult to examine. In the case of food samples, even when no biological membranes are present, freeze-fracturing is useful for examining emulsions such as margarines and dressings. One way to improve the image formed by cryopreparation is by freeze-etching. In this technique, the specimen temperature is raised to produce sublimation of surface ice and reveal underlying structure. Freeze-etching should not be confused with ion beam etching, in which ions of an inert gas are used to produce erosion of the specimen surface in the chamber of the SEM.

Note that the procedures discussed above are most often applicable to scanning and transmission electron microscopy. It is possible, however, to take advantage of cryotechniques in microscopy through the use of replicas. Several alternatives are possible: in one method, an exposed surface is coated with a solution of the replicating substance, and then the replica is removed and viewed. In another, contrast can be improved by shadow casting the replica with a film of electron dense material. For freeze-etched surfaces, metal replication is the technique of choice.

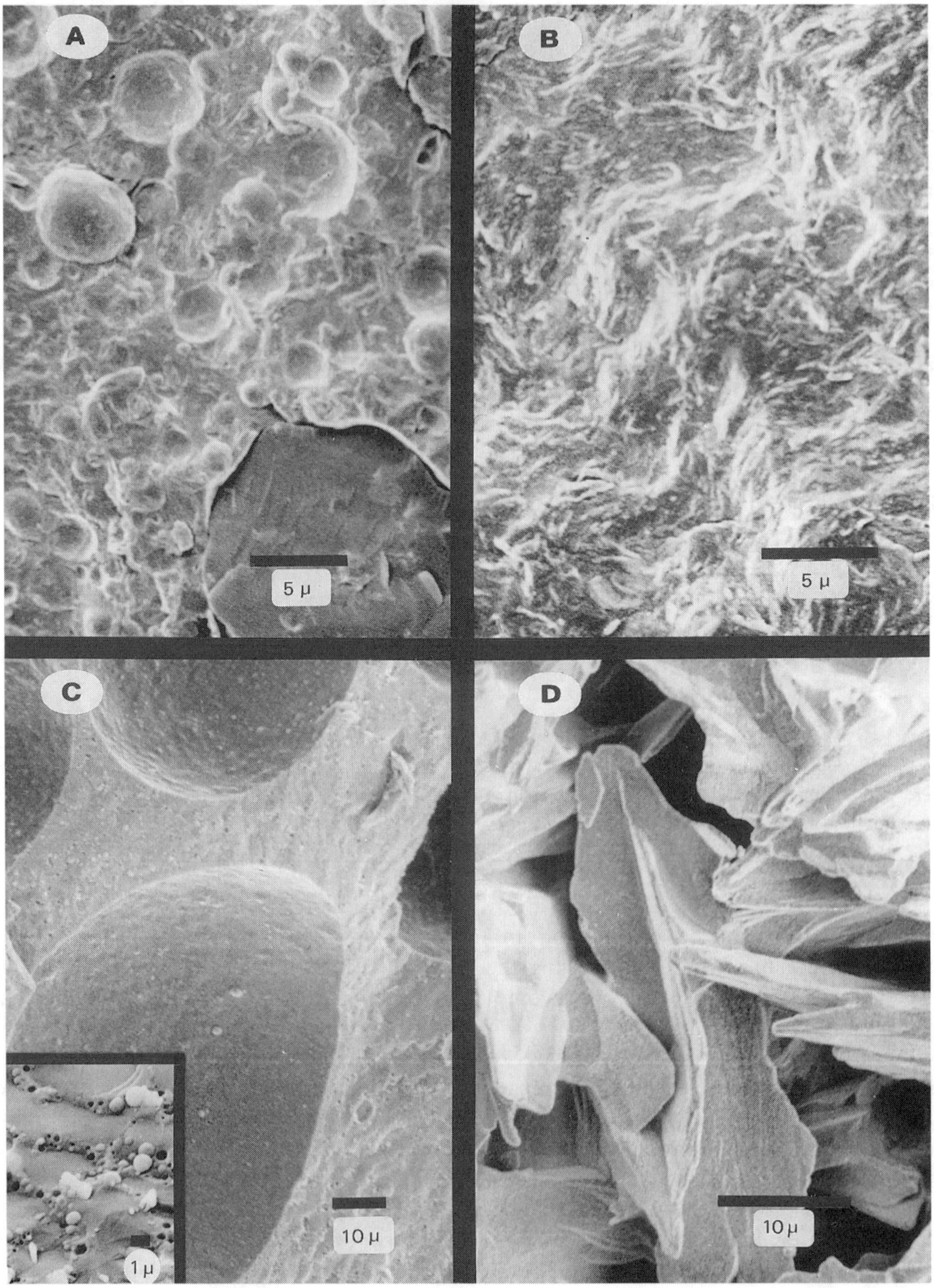


**Figure 1–30** Examples of cryoprepared material in a scanning electron microscope. Samples quick frozen in liquid nitrogen, sputter coated with gold/palladium, and examined in cold stage. (A) Bread. (B) Butter. (C) Ice cream. Inset: Higher magnification. (D) Chocolate showing "bloom" crystals. *Source:* Courtesy of K.W. Baker, A.K. Smith, and J.N.A. Lott.

Finally, cryosectioning, in which samples are sectioned at ultralow temperatures, provides thin sections that can be viewed microscopically.

### 1.5.5 Artifacts

As has been demonstrated, artifacts arise from both specimen processing and the generation of images in the various types of instruments. Instrument-induced effects result from exposing biological material to a high vacuum and an electron beam. The most frequently encountered are charging, beam damage, and vacuum damage. The first has already been discussed. The last, vacuum damage, is an inherent problem of exposing biological tissue to a high vacuum. It can lead to distortion in fragile specimens that have not been properly prepared.

Beam damage is common in electron microscopy. Surface structural alterations result from the interaction of the high-energy electron beam with the specimen, mainly because of local heating (see Figures 1–24, 1–26, 1–27, and 1–31). The degree of damage is a function of magnification (confinement of beam energy to a smaller area), exposure time, and beam current. As the electron beam interacts with the sample, a teardrop-shaped interaction volume is created whose dimensions vary directly with the accelerating voltage (Figure 1–32). The higher the beam voltage, the more damage results from the displacement or ionization of sample atoms. Recent improvements in lens quality and detection systems have reduced the chance of beam damage, since lower beam currents are able to produce high-resolution images. Effective beam voltages have been reduced by one order of magnitude or more; good results can be obtained at 1 kV, and voltages below unity are possible in modern SEMs. Low-voltage operation reduces beam damage and also helps to eliminate charging effects, raising the interesting possibility of viewing uncoated specimens.

The presence of artifacts can easily lead to errors in interpretation. Artifacts can be minimized by preparing specimens by various procedures and by using correlative microscopy. Although proper preparation takes time, it does improve the evaluation of microscopic data.

### 1.5.6 Environmental Scanning Electron Microscopy

A major limitation of conventional SEMs is the requirement for a high ($10^{-4}$–$10^{-5}$ torr) vacuum in the sample chamber. While a high vacuum is necessary to permit the use of electron detectors, it also imposes the requirements of vacuum tolerance and electrical conductivity. The specimens of interest to food scientists frequently do not have these properties. For example, many food materials cannot withstand a vacuum as high as that used in conventional SEM or the rigors of preparatory drying without undergoing structural collapse or losing volatile sample components. Also, delicate samples can lose structural definition during coating, and it is impossible to view liquids, weak gels, lipids, and other difficult materials.

The environmental SEM (ESEM), commercialized in the mid-1980s, overcomes these problems in several ways. A differential pumping system maintains the sample chamber at a vacuum of $10^{-1}$–20 torr, which is much lower than that in the column ($10^{-7}$ torr). A type of gaseous electron detector enables secondary electrons emitted from the surface of irradiated samples to be collected via an ionizing gas cascade, which amplifies the secondary electron signal (Danilatos, 1993). The set of possible imaging gases includes water vapor, nitrous oxide, carbon dioxide, nitrogen, and helium. Positive ions are a byproduct of ionizations during the gas cascade, and these fall toward the sample surface, helping to minimize charge buildup on insulating sample surfaces. Water vapor has so far been found to give the best secondary electron signal amplification, which is particularly useful since the amount of water vapor present in the sample chamber (and hence the relative humidity) can be varied by changing the sample temperature or chamber pressure. A Peltier-controlled stage is used for cooling and heating samples.

The ESEM allows the examination of many food samples, even liquid systems, in their natural state without the need for drying or coating (Stokes, Thiel, & Donald, 1998). Figure 3–13 shows an ESEM micrograph of a vegetable oil-in-water emulsion. Dehydration, hydration, freezing, freeze-drying, and melting processes may be viewed in

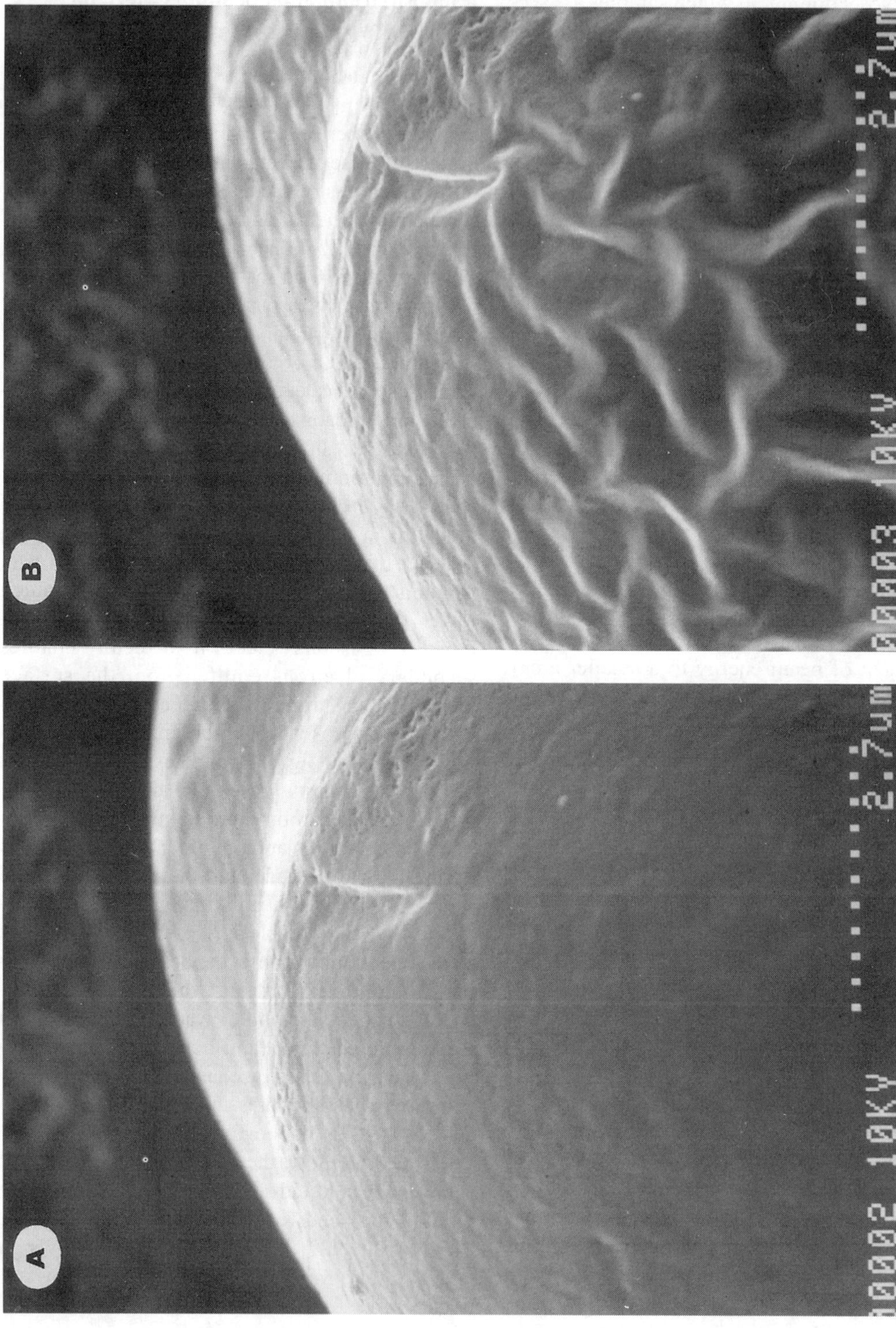


**Figure 1–31** Beam damage of starch granule due to sustained (1 min) exposure to high kV. (A) Undamaged. (B) Damaged. *Source:* Courtesy of K.W. Baker.

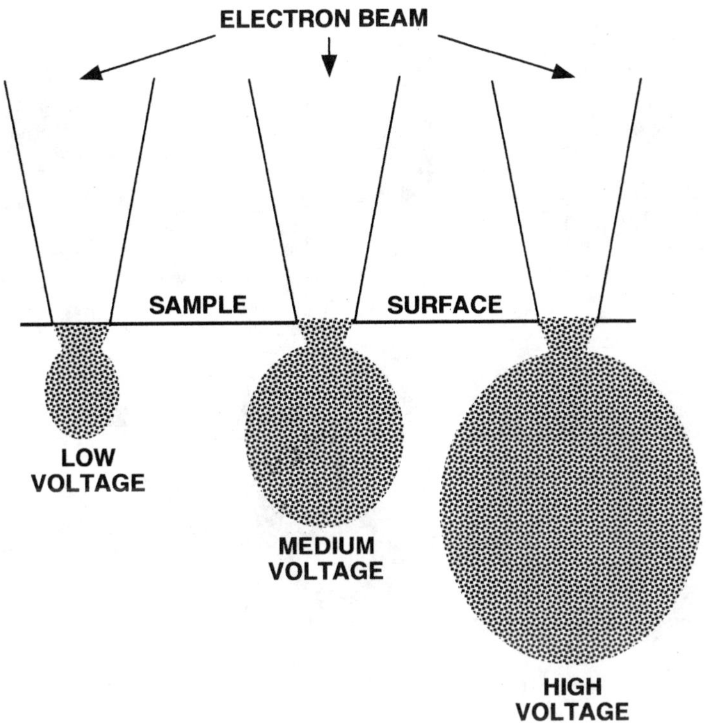


**Figure 1–32** Diagram of electron beam–sample interaction volume and its variation with accelerating voltage. Higher voltages lead to more displacement and/or ionization of specimen atoms. *Source:* Redrawn from Flegler et al., 1993.

real time by altering the chamber conditions. The dynamic mechanical behavior of dry or moist materials can be studied by using a tensile stage. X-ray analysis is also possible with the appropriate detector. The lack of charging artifacts and coating materials benefits these types of analyses and significantly broadens the range of materials that can be studied using electron microscopy. Figure 1–33 shows micrographs of food materials taken with an ESEM. Figure 10–3 demonstrates the capability of an ESEM to examine and dynamically test carrots *in situ* without a conducting coating.

## 1.6 OTHER INSTRUMENTATION AND TECHNIQUES

Many of the advances in microscopic instrumentation and techniques that have necessitated the second edition of this book are discussed in the following sections. Researchers studying food structure now have available a magnificent array of instrumentation with which to scrutinize their materials.

### 1.6.1 Scanning Probe Microscopy

The term *scanning probe microscope (SPM)* covers a wide range of instruments used to provide images of the surface topography of a specimen. All of these instruments operate by scanning a sharp probe closely over the sample surface and measuring some function of the distance between the material and the probe. As opposed to optical microscopes, these instruments provide estimates of distance assembled into an array that forms an image. The first of this family was the scanning

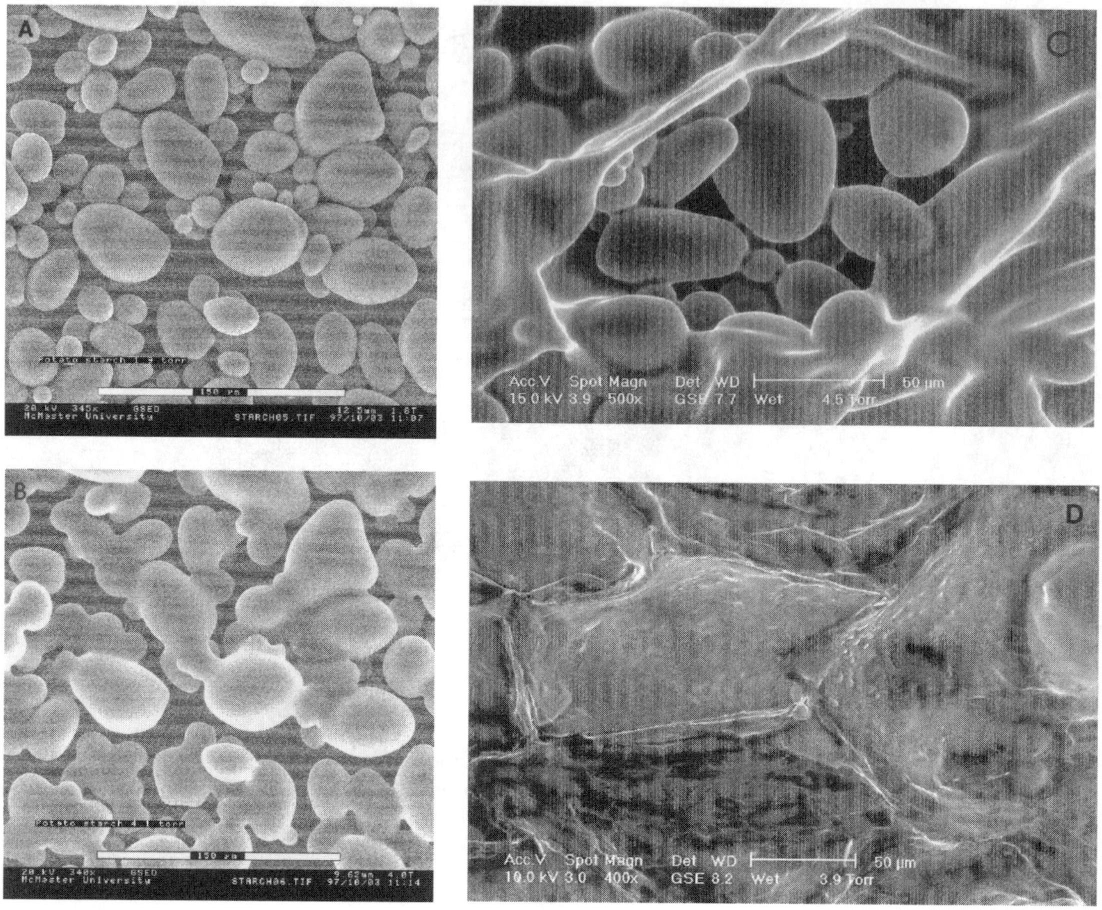


**Figure 1–33** Environmental SEM micrographs of potato starch under low vacuum. (A) Dry environment. (B) After the introduction of water vapor. Environmental SEM micrographs of raw (C) and cooked (D) potato using uncoated hydrated material. Samples were kept hydrated using the Peltier stage in combination with elevated pressure, indicated by the Torr readout on the databar. *Source:* Copyright © Ken Baker.

tunneling microscope (STM), initially described in 1981. This instrument measures the current of electrons that tunnel from atoms at the probe tip to the surface of the specimen ($\approx$1 nm) to form an image of the surface topography with atomic resolution as it scans the surface. Both the sample and the tip must be good conductors, eliminating most biological materials.

The first atomic force microscope (AFM) was introduced commercially in 1989. It represented a significant advance over the STM, because, although a sharp tip is used to scan the surface, there is no current drawn between the tip and the specimen. Rather, the tip is mounted on a cantilevered arm, and a constant but small spring force holds the probe against the sample. Vertical motion of the tip is detected by a system that senses the spacing between the probe and the sample and provides a correction signal. The strong dependence of the current on the tip-to-sample spacing makes it possible to use it in a feedback loop to control a precision motion device, called a piezoelectric scanner, in the *x, y,* and *z* dimensions. This system is used to keep the spacing constant: the most common type is called an optical lever (or beam deflection) system. It employs a laser shining onto and reflecting off the back of the cantilever and onto a segmented photodiode to measure the probe motion. Essentially, an AFM consists of (1) an optical beam with a scanning system that is protected against vibration and (2) an on-board computer. It creates an image by producing an amplified signal of the minute deflections of the cantilever. (See Table 1–4 for a comparison of characteristics of some common techniques for imaging and measuring surface morphology.)

The most recent advance in atomic force microscopy is the use of a tapping mode (Figure 1–34), which overcomes the tendency of the AFM probe to exceed the yield force or binding force of a feature on the surface, an especially common problem with wet biological samples. When the force is exceeded, surface features of interest can be deformed by the scanning probe. The tapping mode has several advantages over the traditional contact or noncontact modes used in atomic force microscopy, including the reduction of artifacts. In the tapping mode, the cantilever on which the tip is mounted is oscillated while separated from the sample surface. This oscillation is driven by a constant driving force, and the amplitude of its oscillation is monitored. The tip is brought toward the sample surface until it begins to touch the surface, which reduces the oscillation amplitude. The feedback loop of the system, controlled by the *z* component of the piezoelectric scanner, then maintains this new amplitude constant as the oscillating (tapping) tip traverses the surface. Thus, the tip height is adjusted for surface height variations as it scans across the sample surface. The tip moves across the surface at a slow rate ($\approx$1 sec/scan line) while tapping at a high rate ($\approx$50–500 kHz). This combination of rates reduces lateral, shear, or frictional forces that might damage the specimen surface, since the tip is prevented from being trapped by any adhesive meniscus forces. The use of intermittent stylus-sample contact has allowed scanning probe microscopy to be applied to soft, hydrated tissues and adhesive or fragile materials often found in food specimens, since it overcomes problems as-

**Table 1–4** Comparison of Common Techniques for Imaging and Measuring Surface Morphology

| | *LM* | *SEM* | *SPM* |
|---|---|---|---|
| Sample operating environment | Ambient; can be liquid or vacuum | Vacuum | Ambient; can be liquid or vacuum |
| Depth of field | Small | Large | Medium |
| Depth of focus | Medium | Small | Small |
| Resolution | | | |
| *x/y* | 1.0 $\mu$m | 5 nm | 0.1–1.0 nm |
| *z* | N/A | N/A | 0.01 nm |
| Magnification range | 1×–2 × 10$^3$× | 10×–10$^6$× | 5 × 10$^2$×–10$^8$× |
| Sample preparation | Little | Fixation, drying, coating | Little |
| Sample requirements | Sample must not be completely transparent to wavelength used | Surface must not build up charge and sample must be vacuum compatible | Sample must not have excessive variations in surface height |

*Source:* Reprinted from *American Laboratory*, Vol. 26, No. 5, p. 20, 1994. Copyright 1994 by International Scientific Communications, Inc.

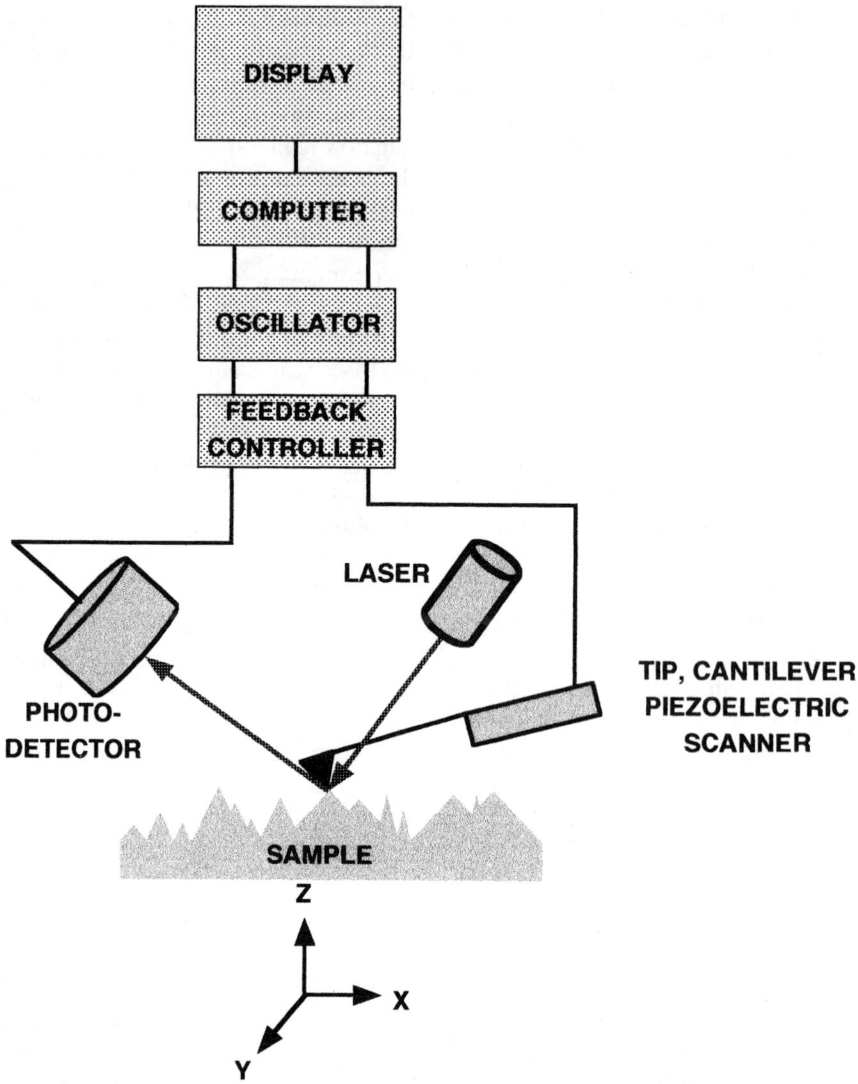


**Figure 1–34** Diagram of tapping mode atomic force microscope (AFM). The cantilever oscillation amplitude is maintained constant by a feedback loop. *Source:* Reprinted from *Biophotonics International*, 3(5), pp. 52–53. With permission from Laurin Publishing, Co. Inc.

sociated with friction, adhesion, and electrostatic forces.

Imaging the surface of food materials with AFM is growing in popularity as the instrumentation becomes more suitable for examining biological samples such as emulsions. Color Plate 4 presents images of a dairy emulsion produced by a tapping mode AFM.

### 1.6.2 X-Ray Microanalysis

X-ray spectra can provide a great deal of information about materials. For example, when high-energy, high-frequency X-rays strike a crystal, they are diffracted in a manner characteristic of the crystal structure, and this technique has been used to determine filament spacing in muscle proteins.

This section, however, considers only X-rays produced in electron microscopy.

When an electron beam strikes matter, interaction occurs in several ways. If the specimen is thin enough, some primary electrons are transmitted, and an image can be formed from them that is dependent upon the scattering power or diffraction of each point in the material (TEM). Secondary electrons are generated at the sample surface reflective of the surface topography as well (SEM). Backscattered electrons can also be measured. The basis for X-ray microanalysis, however, is related to the emission of X-rays produced by the electron beam that are characteristic of the elemental composition of the sample. These emitted X-rays, resulting from the interaction of incident electrons with inner shell electrons of the specimen atoms, can be analyzed to identify and quantify elements. X-ray microanalysis systems are available to be added on to all types of electron microscopes (SEM, TEM, and STEM). Spatial resolution of less than 10 nm is possible with thin-sectioning STEM equipment, and sensitivities of $10^{-18}$ g have been claimed. Biological specimens, however, rarely attain this standard.

Two types of X-ray microanalysis systems are available. Wavelength dispersive spectroscopy measures the wavelengths of X-rays produced when an electron beam hits a sample. It has the advantage of being able to detect lighter elements, including boron and upwards. The most commonly used technique is energy dispersive X-ray (EDX) analysis, in which the energy levels of X-rays entering the detector are measured. Its advantages include the simultaneous analysis of all detectable elements (most systems are capable of measuring elements with atomic numbers of 11 [sodium] or higher) and the ease of computer interfacing.

Sample preparation for X-ray microanalysis is difficult because of the need to ensure tissue integrity, retention of elements *in situ*, and exclusion of exogenous contaminants. Rapid cryopreparation is presently the method of choice, since soluble ions may be retained in position by quick freezing of unfixed tissue. Since sections are not cut, irregular surface topography can cause problems, as can beam damage and charging. Charging can be controlled through heavy metal coating, and chromium is a popular choice. Spatial resolution in frozen hydrated bulk tissue is in the range 5–10 $\mu$m, but this level of resolution is often satisfactory. Of course, the best resolution can be obtained with thin sections; unfortunately, thin sections introduce other problems.

X-ray microanalysis has proven useful for various types of microstructure-related research, including microstructure identification, qualification, and quantification of elemental distribution patterns and histochemical research. One study was of enzyme localization. Myrosinase is the generic name for a group of isoenzymes that hydrolyze glucosinolates to yield goitrogenic isothiocyanates. These enzymes are of importance to food scientists, since the use of rapeseed (canola) is limited by the presence of this group of toxic compounds, produced during the crushing step of oil extraction, which allows the enzyme to contact the substrate. Strategies to inactivate these enzymes during processing require locating myrosinases in the cell. An effort was made to locate them by taking advantage of the fact that bisulfite is a reaction product of the enzymatic hydrolysis of glucosinolates. This anion will complex with soluble lead to form electron-opaque precipitates of lead sulfate at the sites of myrosinase activity. Incubation of aldehyde-fixed seed in a reaction mixture containing substrate and lead nitrate resulted in the formation of deposits on the plasmalemma membrane in rapeseed embryo cells (Figure 1–35). The elemental composition of the precipitate was determined *in situ* by energy dispersive X-ray analysis, and it was found to contain only lead and sulfur (Figure 1–36).

In another study, attempts were made to characterize electron-dense granules found in thin sections of yogurt fixed with glutaraldehyde and osmium tetroxide (Figure 1–37). Samples of this material were mounted on carbon-coated aluminum grids and examined using an EDX detector and a STEM. Probe analysis was performed at 100 kV in the TEM mode with a probe diameter of 40 nm for a count time of 100 s. Examination of the granules by EDX showed osmium ac-

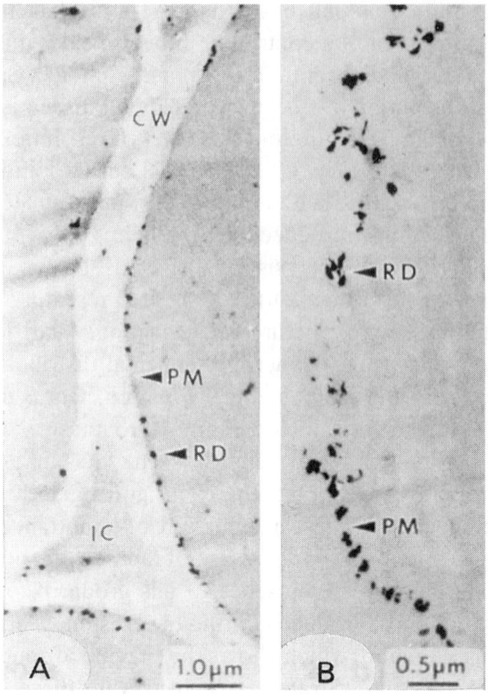


**Figure 1–35** TEM micrograph of rapeseed showing myrosinase catalyzed reaction deposits (lead sulfate [RD]) on the plasmalemma (PM) attached to (A) and dislodged from (B) the cell wall (CW). *Source:* Maheshwari et al., 1981.

counted for 83–90% of their composition, with chlorine making up the balance. X-ray dot maps demonstrated a high density of this element in corresponding electron-dense areas, but if the sample was treated with periodic acid, these granules were removed. It was concluded that the granules are fixation artifacts consisting of a complex of glutaraldehyde and osmium tetroxide. Studies such as this aid in the interpretation of microstructure by differentiating genuine components from artifactual material.

One of the more recent advances in electron microscopy X-ray analysis is position-tagged spectrometry (Friel, 1995). In this type of spectrometry, a complete spectrum of elemental analyses is collected for each point the electron beam scans. Thus, the final image is composed of voxels (volumetric picture elements) rather than pixels (planar picture elements). Once the scan is completed and the data properly digitized, the operator can elect to bring up images consisting of only those data of interest. The operator also has the ability to set Boolean operators that will allow limits to be placed on levels of individual elements so that, for example, some data will be repressed and others shown only if they surpass a given level. Or the operator can simply subtract a certain image from another. Position-tagged spectrometry integrates multielemental analysis and modern image analysis to create a fast and powerful tool. Color Plate 5 shows an example of position-tagged spectrometry.

### 1.6.3 Immunolocalization Techniques

Often the goal of food microscopy is to examine the structural composition of a material and how the various structural elements interact. More and more, however, the goal is to determine the spatial distribution of specific structures (e.g., particular macromolecules or elements). In the case of elemental analysis, the previously discussed X-ray microanalysis approach routinely provides results with a spatial resolution of a few cubic $\mu$m. For nonelemental analysis, localization probes are introduced during the sample preparation procedure. These probes range in specificity from the long-used stains and fluorescent dyes employed in light microscopy—stains and dyes that can identify classes of compounds, such as proteins, lipids, and carbohydrates—to molecule-specific labels. In the case of molecule-specific labels, immunolocalization techniques are by far the most commonly used. These may be employed in both light and electron microscopy modes, but if electron microscopy is to be used, the biological molecules must be tagged with a heavy metal, usually gold, to ensure sufficient electron density for visualization.

Immunolocalization techniques start with the preparation of antibodies having a strong binding affinity to the target molecule. Since immunological studies of food material are becoming more common, an extended line of commercially pro-

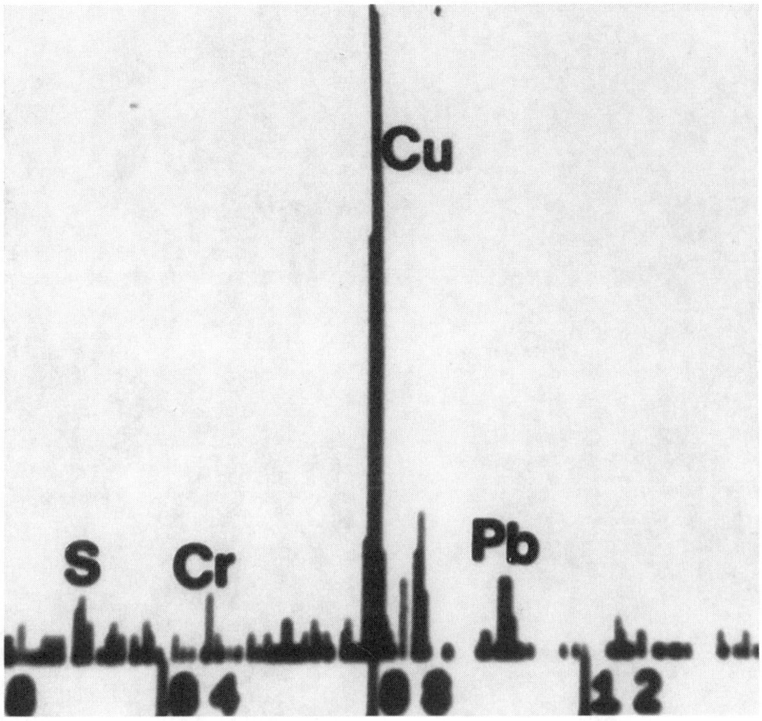


**Figure 1–36** EDX X-ray spectrum of electron opaque deposits along the rapeseed plasmalemma showing the presence of lead sulfate. Copper peak originates from copper grid, chromium from microscope pole piece. *Source:* Maheshwari et al., 1981.

duced antibodies is available. The antibodies then must be rendered visible by some means. For light microscopy, the usual approach involves fluorescent labeling using fluorescein or rhodamine. For electron microscopy, colloidal-gold probes attached to a secondary antibody can be used to label the target component (Figure 1–38). Experiments using immunolocalization techniques are specific in the extreme and allow researchers to localize with precision the presence of specific components.

### 1.6.4 Light Scattering

A basic necessity of those involved in food material science is the ability to measure accurately the size of microscopic particles and their distribution. Theory and practice often diverge, since the particles in foods are usually quite different from the uniform, nonporous spheres dealt with in theoretical treatises and equipment manufacturers' brochures.

Two approaches may be taken to solve the problem of size measurement in food materials having particles of various dimensions which may be highly concentrated and interact during the experimental period. The first is to use dynamic light scattering (DLS), a nondestructive analytical tool for the study of polymers, biopolymers, micelles, and colloids in solution. DLS is able to measure the average size and size distribution of particles in suspension as well as the z-average diffusion coefficient. It utilizes the scattering of light by diffusing particles. At any instant in time, suspended particles will have a specific set of positions within the scattering volume, each with different abilities to scatter

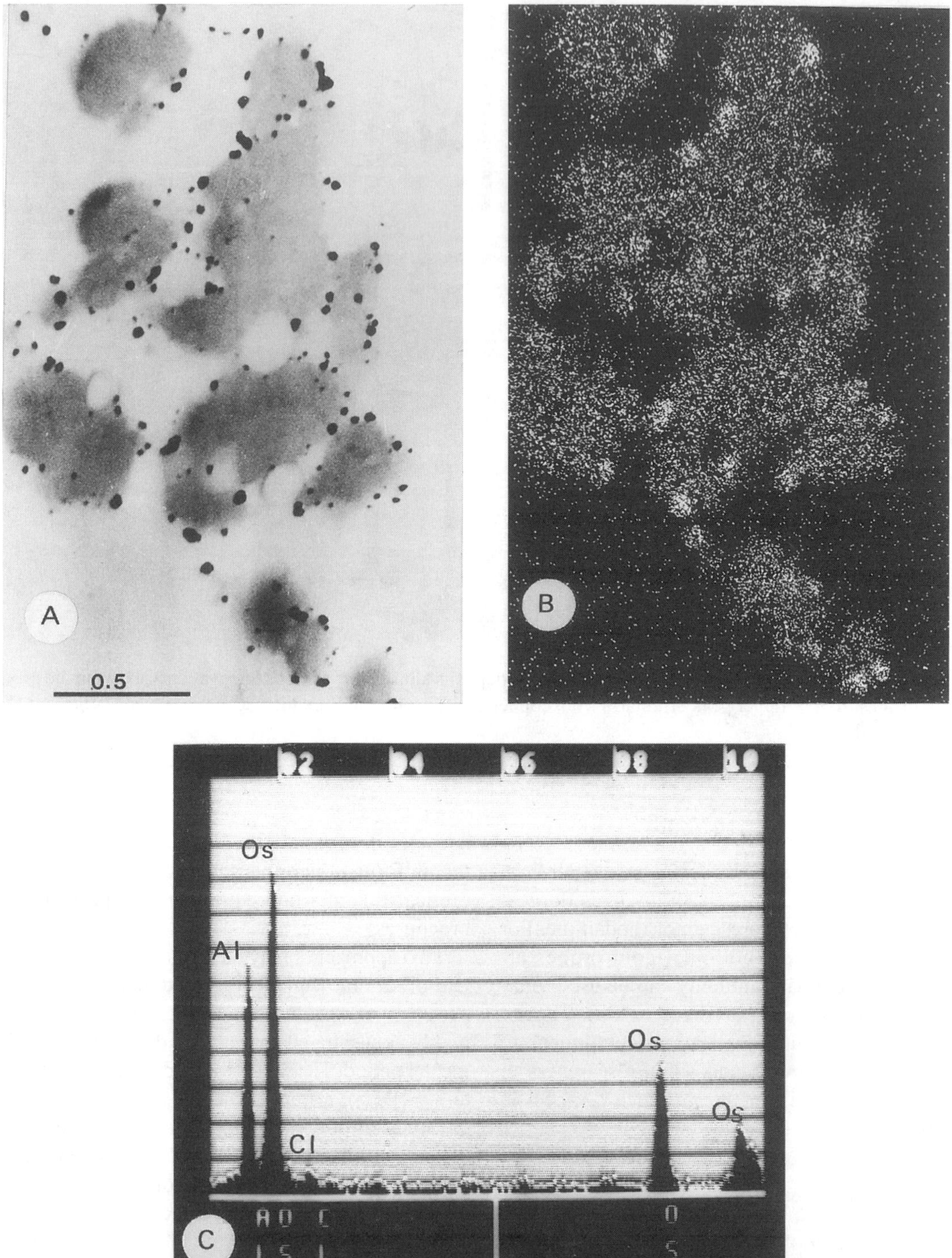


**Figure 1–37** (A) STEM micrograph of yogurt fixed in glutaraldehyde and osmium tetroxide. (B) X-ray dot map of osmium. (C) EDX spectrum of electron-dense granule in yogurt showing lines for Os and Cl. Al peak due to grid. *Source:* Parnell-Clunies et al., 1986.

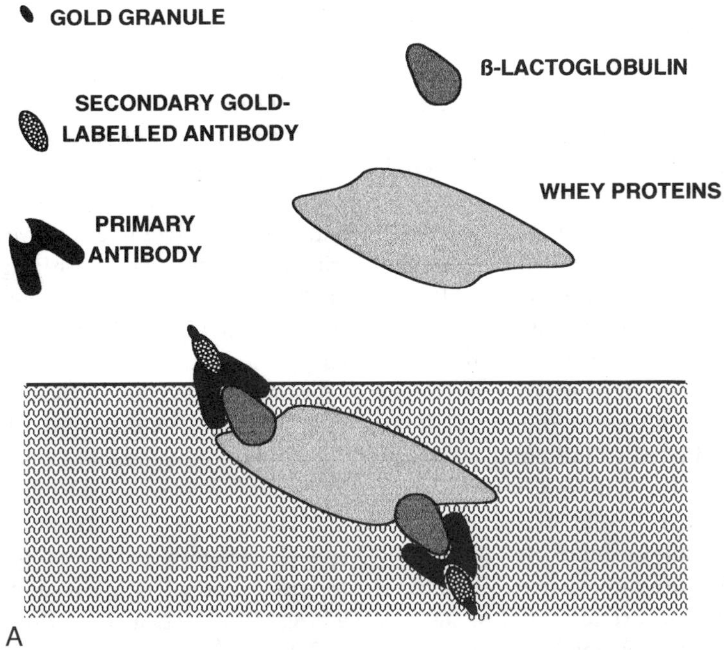


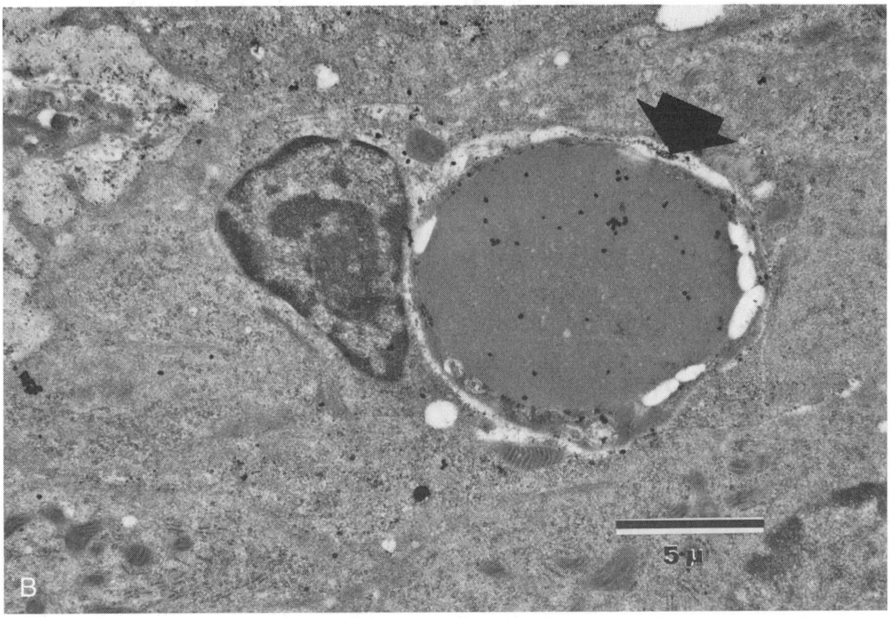


**Figure 1–38** (A) Example of immunolocalization used to visualize whey proteins in a meat product. Specific antibodies are obtained against the $\beta$-lactoglobulin protein component of whey. These are then applied to a thin section of the meat product, where they bind to any exposed $\beta$-lactoglobulin present. A secondary, nonspecific gold-labeled antibody is then applied to mark these sites with electron-dense gold. *Source:* Reprinted from *Trends in Food Science Technology*, Vol. 6, M. Kalab, P. Allan-Wojtas, and S.S. Miller, Microscopy and Other Imaging Techniques in Food Structure Analysis, pp. 177–186, Copyright 1995, with permission from Elsevier Science. (B) Example of immunogold labeling at the TEM level. A coccidial parasite within the intestinal epithelial cell of an infected chicken is indicated. 12 nm colloidal gold particles can be seen overlying the parasite and not the host tissues. *Source:* Copyright © Ken Baker.

light. The scattered light from all particles sets up an interference pattern at the detector at any instant. When the positions change, so does the interference pattern. Thus, the intensity of the light scattered is time dependent, and the fluctuations are analyzed. Experimentally, the fluctuations in incident laser light scattered are detected using a photon-counting photomultiplier and measured through autocorrelation analysis. In autocorrelation analysis, the fluctuations in light intensity are transformed into normalized electric field autocorrelation functions.

DLS has been shown to provide noninvasive, rapid particle size analysis of dilute solutions of rigid particles (diluted to the point where an incident photon can be scattered only once by a scattering particle en route to the detector, which usually means diluted to the point where the solutions are almost optically clear). Note that solutions of flexible particles, such as polymer chains, require specialized data treatment. Also, with biological samples characterized by wide variations in particle size, it is usually only possible to obtain average dimensions rather than the more informative size distribution function. Thus, as may be imagined, DLS has limited usefulness in analyzing real food systems, since dilution of normally concentrated food materials may significantly change their properties. Microscopic determination of particle size, with all its attendant problems (see Chapter 2), has been employed, often incorrectly, instead. However, a related form of dynamic light scattering, called diffusing wave spectroscopy (DWS), can provide measurements of mean particle size in concentrated suspensions or gels. The apparatus used for DWS is similar to that for DLS, except that the incident and the scattered light are conducted through a common fiber-optic bundle. The mathematical treatment of the resultant data, as could be expected, is more complex, but it is possible to obtain light scattering data for some food materials at their normal concentrations or at dilutions low enough not to disrupt the internal structure.

The second approach to size measurement is to use integrated light scattering (ILS). ILS is based on the fact that the amount of light scattered by a particle is a strong function of the angle at which scattering is measured. In this procedure, equipment capable of measuring light scattering over a wide range of scattering angles is used. With DLS studies, where light scattering at only a single angle is measured, inaccuracies are possible, particularly with very polydisperse samples. Data treatment also varies between DLS and ILS. DLS gives intensity-weighted volume distributions, with a bias toward larger particles, while analysis of the intensities of light scattered at different angles (ILS) gives a number size distribution and is biased toward smaller particles.

Sample preparation of food particles for ILS requires dilution, but the degree of dilution is usually not as great as required for DLS. As a rule of thumb, DLS and DWS are not suitable for particles bigger than about 2 $\mu$m. DLS is appropriate for measuring smaller changes in particle properties in samples of, for instance, small emulsion droplets having diameters less than or equal to 0.5 $\mu$m. More details about ILS theory and equipment may be obtained by consulting the works cited at the end of this chapter.

The application of light scattering systems to food materials is illustrated by the following two examples. The ILS approach was used by Agboola and Dalgleish (1995) to study simple O/W emulsions stabilized using milk proteins. The objective was to determine the effect of $Ca^{2+}$ on model system emulsions stabilized by casein, a $Ca^{2+}$ sensitive protein, or $\beta$-lactoglobulin, which is less responsive to $Ca^{2+}$.

On a macroscopic level, emulsion destabilization caused by aggregation of a protein emulsion can be easily followed by observation. However, to quantify this reaction, an ILS technique (Fraunhofer diffraction) was employed in conjunction with cryo-SEM in order to both measure and visualize the reaction. Figure 1–39 shows the effect of the presence of calcium on the particle sizes of emulsion droplets.

Indeed, under these conditions the addition of calcium to a casein-stabilized emulsion resulted in destabilization, reflected in a drastic increase in particle size. With $\beta$-lactoglobulin, a much smaller, yet still measurable, effect occurred. These results suggest a different mechanism of destabilization is involved with these two proteins. The particle size distributions of the emulsions with added calcium showed a significant contrast (Figure 1–40). The emulsions containing casein showed a reduction in

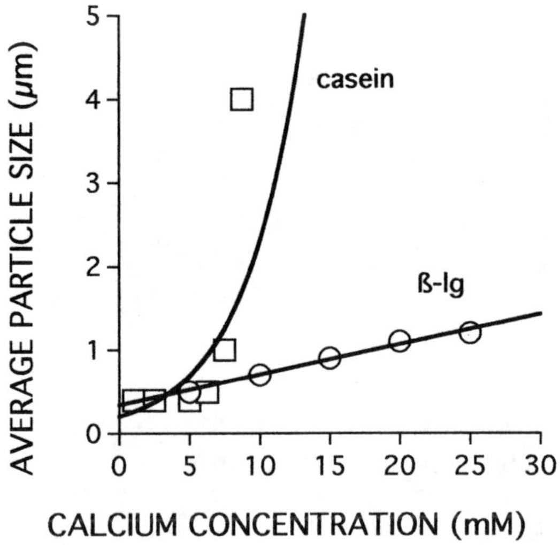


**Figure 1–39** The effect of $Ca^{2+}$ concentration on the particle size of emulsion droplets determined by ILS light scattering. Emulsions prepared from 2% protein (β-lactoglobulin [β-lg] or sodium caseinate [casein] homogenized with 20% soya oil). *Source:* Reprinted with permission from S.O. Agboola and D.G. Dalgleish, Calcium-Induced Destabilization of Oil-in-Water Emulsions Stabilized by Caseinate or by Beta-Lactoglobulin, *Journal of Food Science*, Vol. 60, pp. 399–404, © 1995, Institute of Food Technologists.

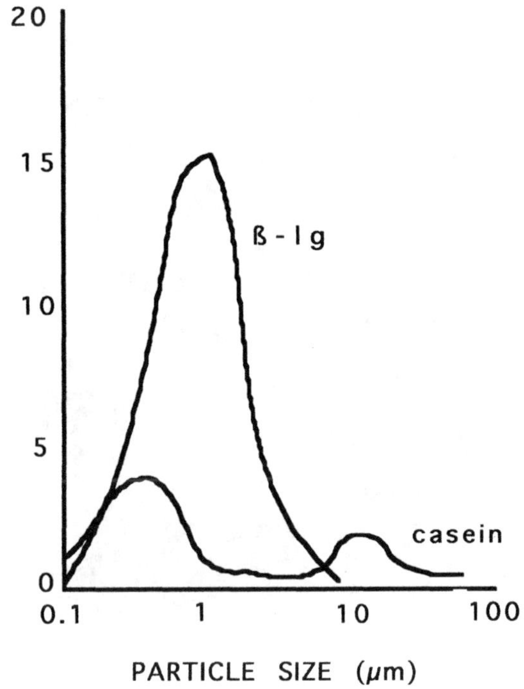


**Figure 1–40** Size distributions of the particles in emulsions containing 20% soya oil and 1% protein (β-lactoglobulin [β-lg] or sodium caseinate [casein]) following the addition of 10 mM $CaCl_2$. *Source:* Reprinted with permission from S.O. Agboola and D.G. Dalgleish, Calcium-Induced Destabilization of Oil-in-Water Emulsions Stabilized by Caseinate or by Beta-Lactoglobulin, *Journal of Food Science*, Vol. 60, pp. 399–404, © 1995, Institute of Food Technologists.

the proportion of emulsion droplets in the original size range but an increase in the proportion of larger particles. However, the emulsions with β-lactoglobulin exhibited monomodal size distributions that shifted to a slightly greater average size as the calcium concentration increased.

Cryo-SEM was used to provide a visual perspective of the mechanism of destabilization. Figure 1–41 shows a difference between the two proteins in the presence of calcium. Emulsions with caseinate formed amorphous aggregated particles, whereas those containing β-lactoglobulin produced a fine-stranded aggregate. The reason that only small aggregates were seen in the size distribution experiments (Figure 1–40) was that the fine aggregates of β-lactoglobulin were broken as they circulated through the sample cell while the more compact casein aggregates were not. Since casein binds calcium to a greater extent than β-lactoglobulin, these results are not unexpected. Yet the complementary light scattering and microscopic techniques revealed more about the nature and extent of the calcium-protein interaction mechanism than could either independently.

The second example, which comes from the same laboratory, is an extension of a study of milk homogenization described earlier (Section 1.4.3). The objective was to compare milk homogenization using traditional valve equipment and using an experimental unit. The experimental unit applied more pressure than the usual commercial two-stage valve homogenizer. Particle sizes in the emulsions resulting from the treatment of whole milk were analyzed using both DLS and ILS instruments.

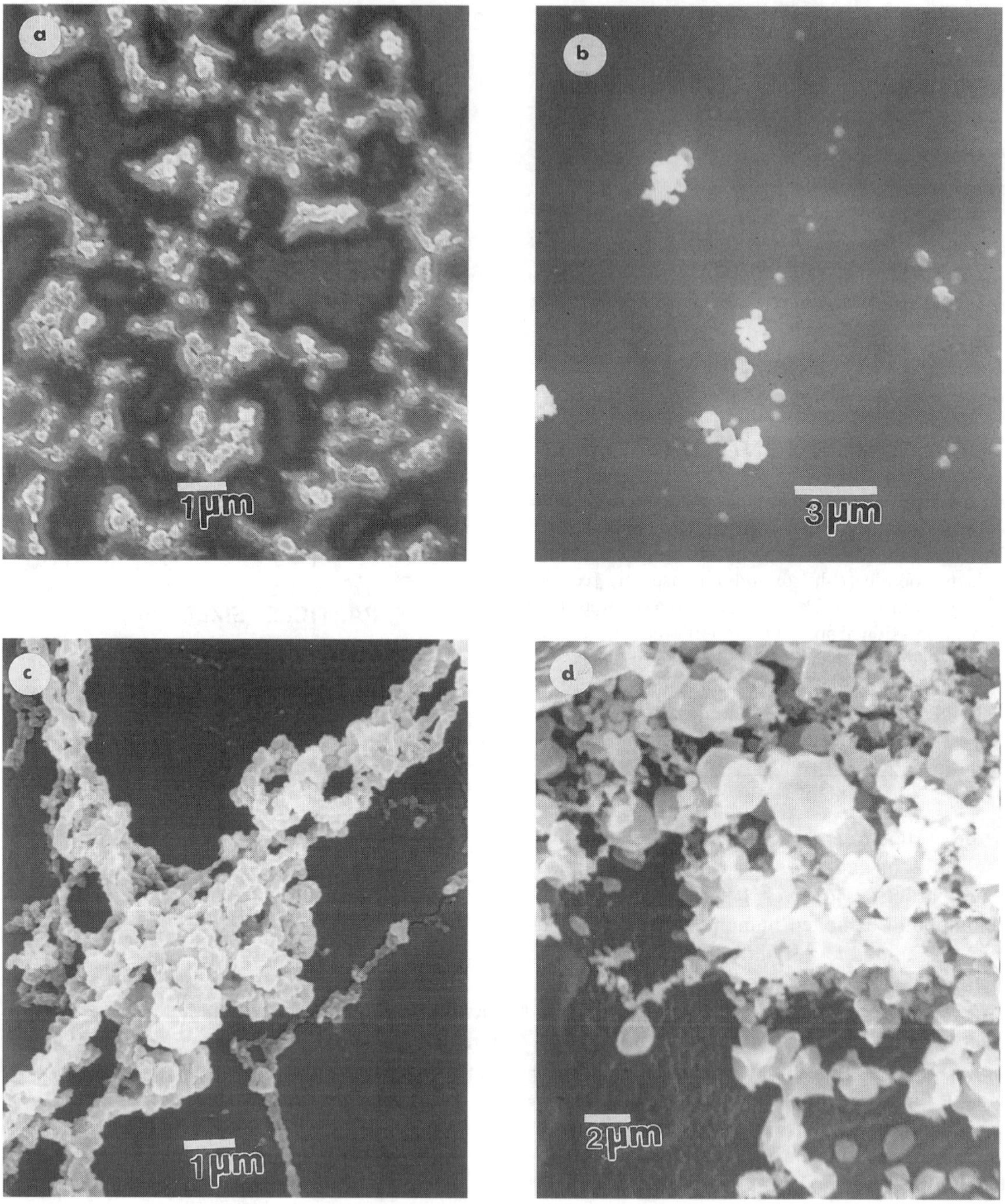


**Figure 1–41** Cryo-SEM micrographs of emulsions containing 20% soya oil and 1% protein stabilized by either $\beta$-lactoglobulin (a, c) or casein (b, d) in the presence (c, d) or absence (a, b) of 10 mM $CaCl_2$. *Source:* Reprinted with permission from S.O. Agboola and D.G. Dalgleish, Calcium-Induced Destabilization of Oil-in-Water Emulsions Stabilized by Caseinate or by Beta-Lactoglobulin, *Journal of Food Science*, Vol. 60, pp. 399–404, © 1995, Institute of Food Technologists.

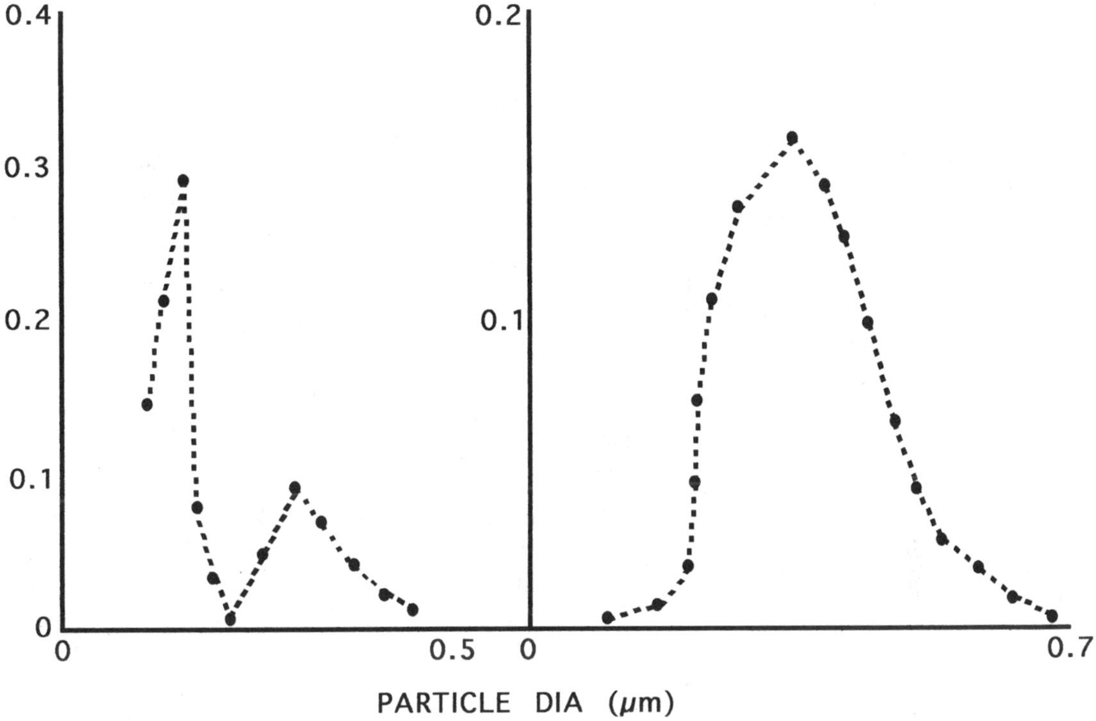


**Figure 1–42** ILS (left, showing number distribution) and DLS (right, showing intensity distribution) light-scattering data for the same fraction of milk following high-pressure homogenization. *Source:* Reprinted from *Netherlands Milk & Dairy Journal*, Vol. 50, D.G. Dalgleish, S.M. Tosh, and S. West, Beyond Homogenization: The Formation of Very Small Emulsion Droplets During the Processing of Milk by a Microfluidizer, pp. 135–148, © 1996, with permission from Elsevier Science.

Figure 1–42 shows the size distribution results of applying the two light scattering techniques to commercially and experimentally homogenized milk. The samples were diluted with buffer at a ratio of 1 to 2,000. Both light scattering techniques demonstrated that high pressure homogenization resulted in large quantities of emulsion droplets much smaller than those found with the traditional process ($\approx$1 $\mu$m). However, the (bimodal) distribution was better defined in the measurements obtained using ILS. The reason is that ILS provides better resolution of smaller particles because data are gathered over a wider range of scattering angles. The bimodal distribution suggests that a sizable number of the small globules may clump together, producing larger aggregates.

Micrographs of the experimental sample and the commercially homogenized sample were presented in Figure 1–20. The tendency for the small fat globules to cluster is apparent. These groups seem to be held together by protein, indicating that the high pressure homogenization treatment may have disrupted the casein micelle to the point where the increased number of casein particles would act not only as an efficient emulsifier but also to hold the small fat globules together as clusters. If only DLS data had been obtained and if transmission electron microscopy was not employed, it is quite likely that these clusters would have been overlooked, demonstrating the value of correlative studies.

### 1.6.5 Magnetic Resonance Imaging

Magnetic resonance imaging (MRI) provides researchers with yet another way to obtain representations of food materials. One type of magnetic resonance, electron spin resonance (ESR), measures an electron's reaction to an applied magnetic field. ESR has proven quite useful to food scientists seeking evidence for phase transitions in biological membranes, among other applications. In nuclear magnetic resonance (NMR), images are produced as a result of an interaction between atomic nuclei in the sample and an external magnetic field. When a material that contains nuclei exhibiting magnetic resonance (not all elements do) is placed in an external magnetic field and energy is added to the system in the form of a pulse of radio frequency radiation, the nuclei enter a nonequilibrium state. After a pulse of energy is administered, nuclei return to the equilibrium state: this relaxation is characterized by two relaxation time constants, $T_1$ and $T_2$. It is possible to weight these relaxation times so that the resulting magnetic resonance signal provides information on mobility, temperature, solute concentration, and moisture concentration. By varying the weighting given these two parameters and certain operating variables, measurement of the relaxation times provides information on such sample parameters as moisture content, fat content, and solids content. Applying a linear magnetic field gradient across the specimen allows spatial localization of points within the material from which the return signal originated. An image is created by Fourier transforming this signal. The signal for each voxel in the image is the weighted signal from this three-dimensional volume element and is a function of spin density, the two relaxation time constants, and experimental variables.

The resolving power or smallest voxel dimension of MRI instruments depends upon the magnetic field generated. Magnetic fields, measured in tesla ($1 \times 10^4$ gauss), vary from around 1–2T (units in this range are used for whole body imaging, where voxel dimensions are on the order of mm) to around 9–10T (units in this range can accept samples 2–5 cm in diameter and have a voxel dimension of around 100 $\mu$m). Instruments with very high magnetic fields have been constructed; their resolution approaches 5 $\mu$m, but the ultimate resolution for MRI is on the order of 2 $\mu$m and cannot be reduced. Thus, the best application of MRI technology in food science is not to produce high-resolution images of food materials but to produce nondestructional data on changes in internal structure as a function of various experimental variables, such as time, temperature, flow rate, and composition.

To see how MRI instruments can be used for studying food, consider this example. The hard-to-cook defect in beans curtails water imbibition into the cotyledon during the soaking process, as discussed in "Fluorescence Microscopy" early in this chapter, and researchers wanted to quantitate the kinetics of water uptake by determining diffusion constants of normal and defective beans. The imbibition process can be described using Fick's law of diffusion, according to which water uptake follows a gradient in moisture along the seed and is proportional to the diffusion coefficient or diffusivity of water. Water diffusivity, a complex function of the microstructure, chemical composition, moisture, and temperature of the seed, is thought to determine the rate of water uptake during soaking.

Values have been published for water absorption and swelling in dry bean seeds that had been stored under adverse conditions in order to induce the hard-to-cook defect (del Valle, Stanley, & Bourne, 1992). Although water absorption was significantly and negatively correlated with cooked bean hardness, diffusion coefficients between hard-to-cook and control samples were not significantly different. The method used to determine water absorption in this study entailed measuring weight gain of intact samples over an 8-hour period. Since beans are composed of cotyledons, free space between the seed coat and the cotyledon, and the cavity formed between adaxial cotyledon surfaces, it seems probable that the methodology employed could not differentiate these areas. These spaces collect water very quickly, much more rapidly than water permeates into the cotyledon tissue. In soft beans, the infusion of water into the spaces and cotyledons is more or less a continuous but slow process, while in hard-to-cook beans, the hardened cotyledon surface presents a barrier that slows further water uptake. The water initially rushes in (mediated by the seed coat) but is then stymied from additional migration. In order to determine water absorption in this complex system, a method is required that will allow measurement of diffusion coefficients in homogeneous domains.

Pulsed field MRI has the ability to measure the diffusion constants of water, since hydrogen atoms in the water have paramagnetic properties. In the study being described, diffusive water flow was measured in bean seeds and seed tissues using pulsed gradient spin-echo NMR spectroscopy and microimaging. Diffusion experiments were used to directly measure the random movement of water in beans through the effect of this motion on signal strength. Measurements were performed after 2.5 and 48 hours of imbibition of deionized water at room temperature.

Diffusion coefficients *(D)* were calculated from the relation:

$$-\ln [A(G)/A(0)] = \gamma^2 G_y^2 D\delta(\Delta - \delta/3)$$

**Equation 1–2**

where *ln[A(G)/A(0)]* is the natural log of data resulting from measuring the random movement of water in the sample at gradient strength (*G*) normalized as a fraction of the signal intensity at zero gradient strength, $\gamma$ is the gyromagnetic ratio for $^1H$ (4,257 Hz/gauss), $G_y$ is the diffusion gradient applied along the y orthogonal axis, $\delta$ is the duration time in ms, and $\Delta$ is the echo time in ms. Values for *D* were determined from the slope of the curve obtained when ln[A(G)/A(0)] was plotted against $G_y^2$. $^1H$ microimages were prepared using a PGSE pulse sequence superimposed on a spin-echo image. Each voxel had a spatial resolution of 60.2 $\mu$m × 60.2 $\mu$m × 500 $\mu$m (slice thickness). It was possible to expand and colorize the resulting images.

Cooked hardness of the beans used in this experiment is shown in Table 1–5. The magnitude of difference between soft and hard samples is indicative of the hard-to-cook defect. Diffusion coefficients calculated according to the above equation are given for the 2.5- and 48-hour imbibition times, along with data from the previous work (del Valle et al., 1992). The 8-hour data show no obvious effect of the hard-to-cook defect. By looking at both a shorter and a longer time period, the effect of hardening becomes more clear. At shorter times (2.5 h) the hard beans have a higher diffusion coefficient, while at longer times (48 h) the soft beans have a higher diffusion coefficient (close to 40% on average).

These data agree with what is known about the biological effect of bean hardening. Free spaces found within bean seeds would be expected to collect water very quickly, but the

**Table 1–5** Results of MRI Experiments with Dried Beans

| | | *Diffusion Coefficients* ($\times 10^{-10} m^2 s^{-1}$) | | |
|---|---|---|---|---|
| *Bean Sample* | *Cooked Force (N)* | *2.5 h** | *48 h** | *8 h†* |
| Control black | 0.33 | 0.76 | 1.73 | 2.31 |
| Hard-to-cook black | 6.53 | 1.53 | 1.03 | 2.89 |
| Control white | 0.24 | 1.18 | 1.73 | 4.84 |
| Hard-to-cook white | 5.03 | 1.35 | 1.11 | 4.48 |

* Obtained by MRI.

† Obtained by measuring weight gain of intact samples over an 8-hour period (del Valle et al., 1992).

*Source:* Data from E.J. Kendall, D.W. Stanley, and K.B. Chatson, A Comparison of Water Self-Diffusion in Fresh and Stored Beans, Abstract #510, in *Proceedings of the 77th Chemistry Society of Canada Conference*, © 1994, Chemistry Society of Canada.

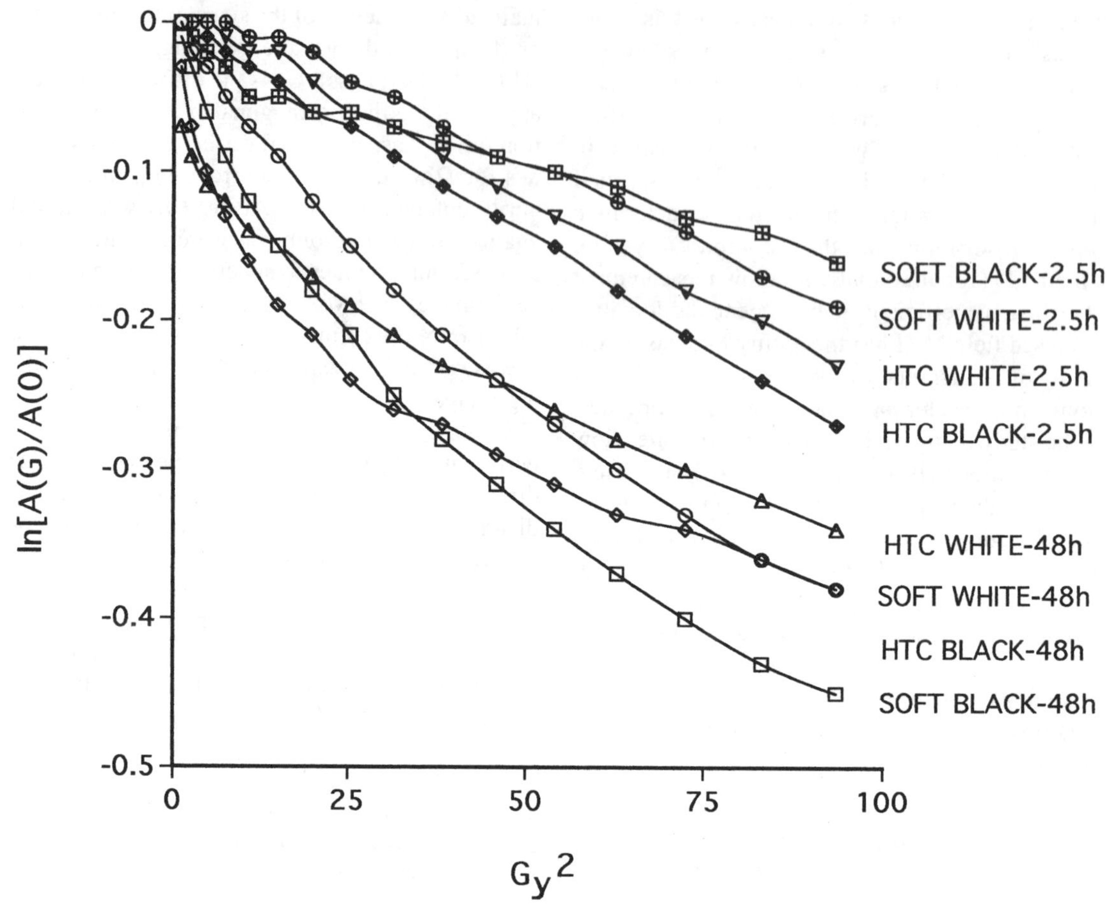


**Figure 1–43** Plot of natural log of normalized signal intensity versus diffusion gradient strength ($G_y^2$) for hard-to-cook (HTC) and control (soft) black and white beans soaked in distilled water at ambient temperature for 2.5 or 48 h. Diffusion coefficients were calculated from the slopes of these curves at the eight highest gradient settings. *Source:* Reprinted with permission from E.J. Kendall, D.W. Stanley, and K.B. Chatson, A Comparison of Water Self-Diffusion in Fresh and Stored Beans, in *Proceedings of the 77th Chemistry Society of Canada Conference,* Winnipeg, Man., May 29, 1994. Abs. #510.

progress of this water would soon encounter biological barriers more inhibitory in hard than soft beans. This notion can be investigated using MRI technology. Figure 1–43 shows the curves obtained when ln*[A(G)/A(0)]* was plotted against $G_y^2$. Note that, whereas all the 2.5-hour data are essentially straight lines, the 48-hour data are less linear. This is indicative of biological diffusion barriers and is particularly noticeable for the hard-to-cook samples. Further, diffusion coefficients obtained for different regions in the sample were obtained, and the cotyledon areas demonstrated lower values than the cavity areas. Sample diffusion attenuated images are shown in Color Plate 6.

### 1.6.6 Spectroscopy and Microscopy

Spectroscopy has been used as an analytical tool in food science for many decades. Coupling microscopy and spectroscopy allows simultaneous visualization and microanalysis of a particular area (as small as $10 \times 10\ \mu m$) in the field of view. The chemical information derived complements the information from X-ray microanalysis of chemical elements.

First, some principles of spectroscopy must be reviewed. An electromagnetic wave consists of oscillating electric and magnetic fields directed perpendicular to each other and to the direction of propagation. The basic relationship is

$$\bar{v} = \frac{1}{\lambda} = \frac{v}{c} \qquad \textbf{Equation 1–3}$$

where c is the speed of propagation of light in a vacuum ($2.9978 \times 10^{10}$ cm $s^{-1}$), $\lambda$ is the wavelength of radiation (units of distance, e.g., nm), $v$ is the frequency (cps or Hz) and $\bar{v}$ is the wave/number (waves/cm). The nature of all radiation is basically the same, but radiation differs in frequency and wavelength and in the effects it can produce in matter. Interaction between matter and radiation spans the entire spectrum of electromagnetic radiation, which can be conveniently divided according to the sources and detectors required.

Spectroscopy is the study of the interactions between electromagnetic radiation absorbed, scattered, or emitted by matter (atoms, molecules, etc.). These interactions are associated with changes in the energy states of chemical species, and since each species has characteristic energy states, spectroscopy can be used to provide qualitative and quantitative information. Table 1–6 lists the most commonly used spectroscopic techniques, the energy changes involved, and the type of information derived when samples are probed with radiation from different regions of the electromagnetic spectrum. Relevant to food science is the ability to combine the image-forming capabilities of microscopy and the illumination of samples using wavelengths of specific types to determine their structure and chemical makeup.

Infrared spectroscopy (IR) is used to study the vibrational movement of molecules. Mid-IR analysis has been used for the elucidation of bonds in organic structures and near-IR to obtain information from thicker samples and for single peaks. Raman spectroscopy (RS) is a branch of vibrational spectroscopy in which a sample is exposed to a laser beam and shifts in the wavelength or frequency resulting from inelastic scattering of photons are recorded. RS and IR techniques involve transitions between vibrational levels, but although polar groups such as C=O, N−H, and O−H are detected in IR, nonpolar groups such as C=C, C−C, and S−S have intense Raman bands. RS can be applied in food analysis to detect proteins, lipids, carbohydrates, and water. Raman microscopy is a unique tool for the selective analysis of spatially distinct entities, and applications for Raman imaging, confocal Raman microscopy, and polarization or orientation microscopy already exist. Further details of RS and its applications in food science can be found in Li-Chan (1996).

Infrared microscopy combines IS, microscopy, and computerized data processing (Richardson, 1997). It can be used to superimpose localized chemical information onto the microstructural information provided by light microscopy. Data acquisition is performed by driving a stage to cover a two-dimensional region in the sample (e.g., a 40 $\mu m \times 40\ \mu m$ window) and recording the interferogram with an IR microscope. Figure 1–44 shows a normal image obtained in the visible light mode of the microscope and the zone of analysis of a thin section of dehydrated potato tissue. Starch granules inside cells are clearly noticeable. The graph shows the collected infrared spectra of the sample, together with the reference curve for starch. *In situ* localization and simultaneous analysis of distinct microstructures, phases, and organelles can provide added dimensions to microstructural studies of foods (Wilson, 1995).

### 1.6.7 Electron Energy Loss Microscopy

When the electron beam in TEM interacts with a thin section of a sample, most electrons pass through the specimen to generate the image, and others experience elastic scattering. Some electrons interact *inelastically* with the specimen, suffering an energy loss but virtually no change in direction.

**Table 1–6** The Electromagnetic Spectrum and Spectroscopic Techniques

| | *Spectroscopic Technique* | | | | | | |
|---|---|---|---|---|---|---|---|
| | *γ-rays* | *X-rays* | *Circular Dichroism* | *UV/Visible* | *Infrared* | *Raman/ESR* | *NMR* |
| Information | | Elemental composition | Protein structure | Color/fluorescence | Organic bonds | Organic bonds/ radicals | Molecular structure and mobility |
| Energy changes involved | Electron ejection | Electron ejection | Light polarization | Transition of electrons | Molecular vibrations, stretching, bending | Molecular vibrations/ change in electron spin | Spin orientation (in magnetic fields) |
| Region in electromagnetic spectrum | | X-rays, Soft X-rays | Vacuum UV | Near UV Visible | Near IR, Mid-IR, Far -IR | Microwaves | Radio waves |
| Wavelengths (approx. range) | 0.01–1Å | 1 Å 10 Å | 100 Å | 200 mm 400–800 mm | 0.8–14 μm 1.4–25 μm 25–1,000 μm | 0.1–10 cm | 25 cm–100 cm |

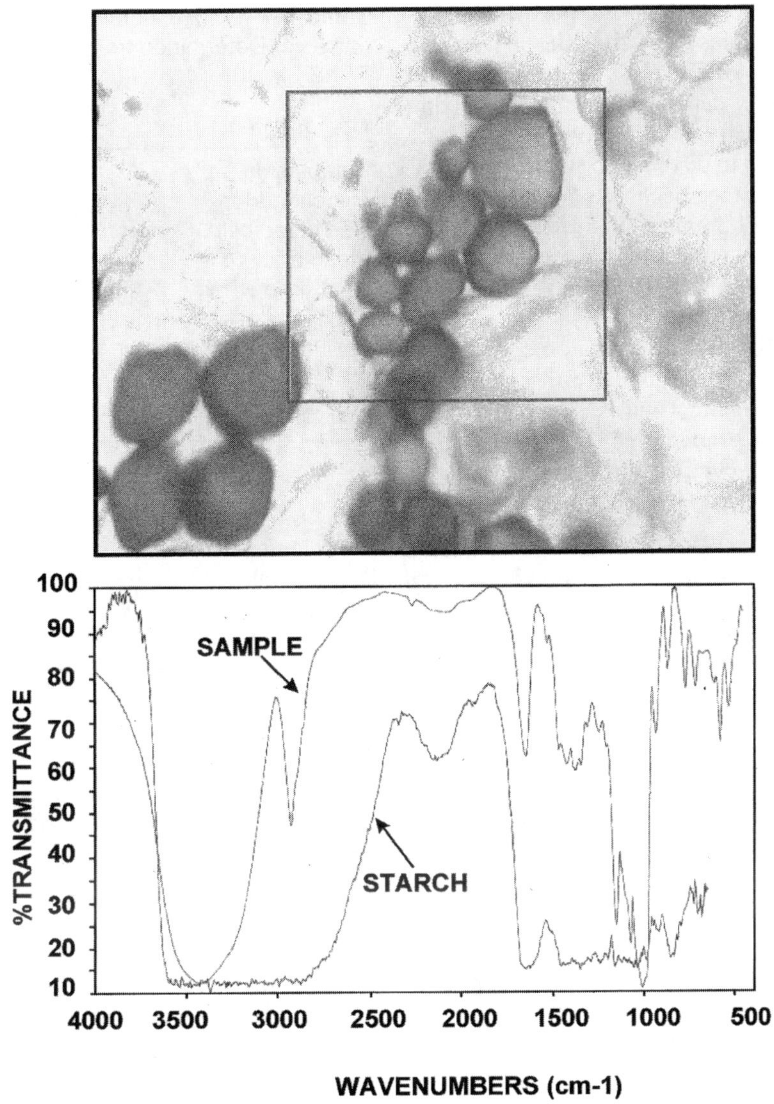


**Figure 1–44** Micrograph of a section of potato tissue showing starch inside cells (above). Infrared spectrum of the selected area (square) in sample and reference spectrum for starch (below).

These electrons can be deflected into an electron spectrometer, usually fitted below the electron microscope column. Images or maps can be produced from electrons that have suffered an energy loss or be interpreted in terms of the vibrational spectrum of the interacting species in the sample, a technique called electron energy loss spectroscopy (EELS).

Although still at the development stage, EELS is covered in this book because its ability, in combination with STEM, to map water distribution at the submicron scale may have important applications in food science. So far, food technologists have been restricted to using average moisture content to study chemical and microbial stability as well as structural phenomena in food materials. However, water in foods is likely to be compartmentalized in microregions or partitioned between different phases at the microstructural level. In the case of the glass-rubber transition, which is extremely dependent on the plasticizing effect of water, local rather than average moisture content is required to distinguish stable and unstable zones in foods (see Section 3.5.6).

A processed EELS image obtained at each pixel shows a strong contrast due to variations in water content. Figure 1–45 is a darkfield STEM micrograph superimposed on a water map obtained by EELS from a hydrated cryosection of rat liver. Black corresponds to 0% and white to 100% water (Sun, Shi, Hunt, & Leapman, 1995). The water content of identified structures, such as mitochondria (M), cytoplasm (C), red blood cells (R), and lipid droplets (L), is shown in Table 1–7.

**Table 1–7** Water Content in Compartments of Rat Liver Determined from 40 Pixel Regions in EELS Maps

| *Compartment* | *Percentage Water* |
|---|---|
| Cytoplasm | 75.4 ± 3.0 |
| Mitochondrion | 56.8 ± 2.0 |
| Red blood cell | 65.3 ± 1.8 |
| Plasma | 91.2 ± 2.3 |
| Lipid droplet | 0.0 |

*Source:* Reprinted with permission from S.Q. Sun, S.L. Shi, J.A. Hunt, and R.D. Leapman, Quantitative Water Mapping of Cryosectional Cells by Electron Energy-Loss Spectroscopy, *Journal of Microscopy*, Vol. 177, pp. 31–42, © 1995, Blackwell Scientific Publications Ltd.

### 1.6.8 Other Microscopy Techniques

Noninvasive microscopy techniques can be used to advantage when investigating hydrated, multicomponent, voluminous, and metastable materials such as foods, even though high resolution is sacrificed. Acoustic microscopy is worth reviewing for its potential to provide qualitative and quantitative information about mechanical properties of biomaterials and its ability to show very high contrast without the need of staining or other invasive preparation techniques (Hafsteinsson & Rizvi, 1984).

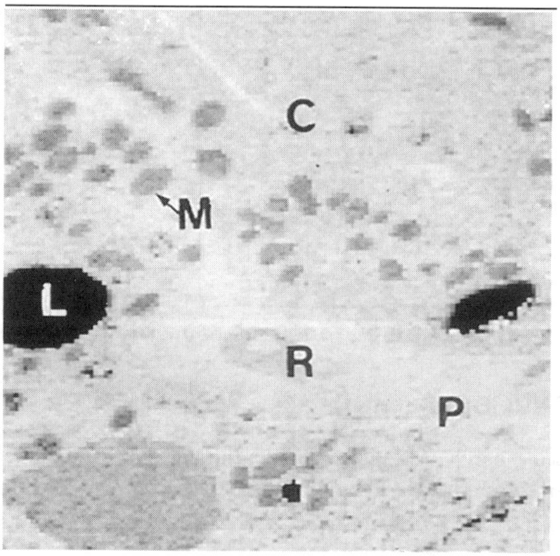


**Figure 1–45** Darkfield STEM micrograph superimposed on a water map obtained by EELS from a hydrated cryosection of rat liver. M = mitochondria, C = cytoplasm, R = red blood cells, P = plasma, L = lipid droplets. *Source:* Reprinted with permission from S.Q. Sun, S.L. Shi, J.A. Hunt, and R.D. Leapman, Quantitative Water Mapping of Cryosectioned Cells by Electron Energy-Loss Spectroscopy, *Journal of Microscopy*, Vol. 177, pp. 31–42, © 1995, Blackwell Scientific Publications Ltd.

Acoustic microscopy is based on totally different physical concepts than optical and electron microscopy. Still, it depends on the response of a sound wave as it is reflected, refracted, or scattered at an interface where changes in physical properties such as density, elasticity, or viscoelasticity exist. Two types of microscopes have been developed: the scanning laser acoustic microscope (SLAM) and the scanning acoustic microscope (SAM), which is operated in reflection mode. The SLAM operates at around 100 MHz and can examine thicker samples at low resolution, while the SAM operates at around 4.2 GHz and provides resolution at least five times better than the optical limit (Bereiter-Hahn, 1995). The main advantages of SAM over established microscopy techniques are that acoustic waves can penetrate the interior of opaque materials (allowing nondestructive examination) and that contrast in acoustic micrographs relates to variation in the mechanical properties of the specimen and thus gives information about the strength and texture of the material (Smith, Harvey, & Fathers, 1985).

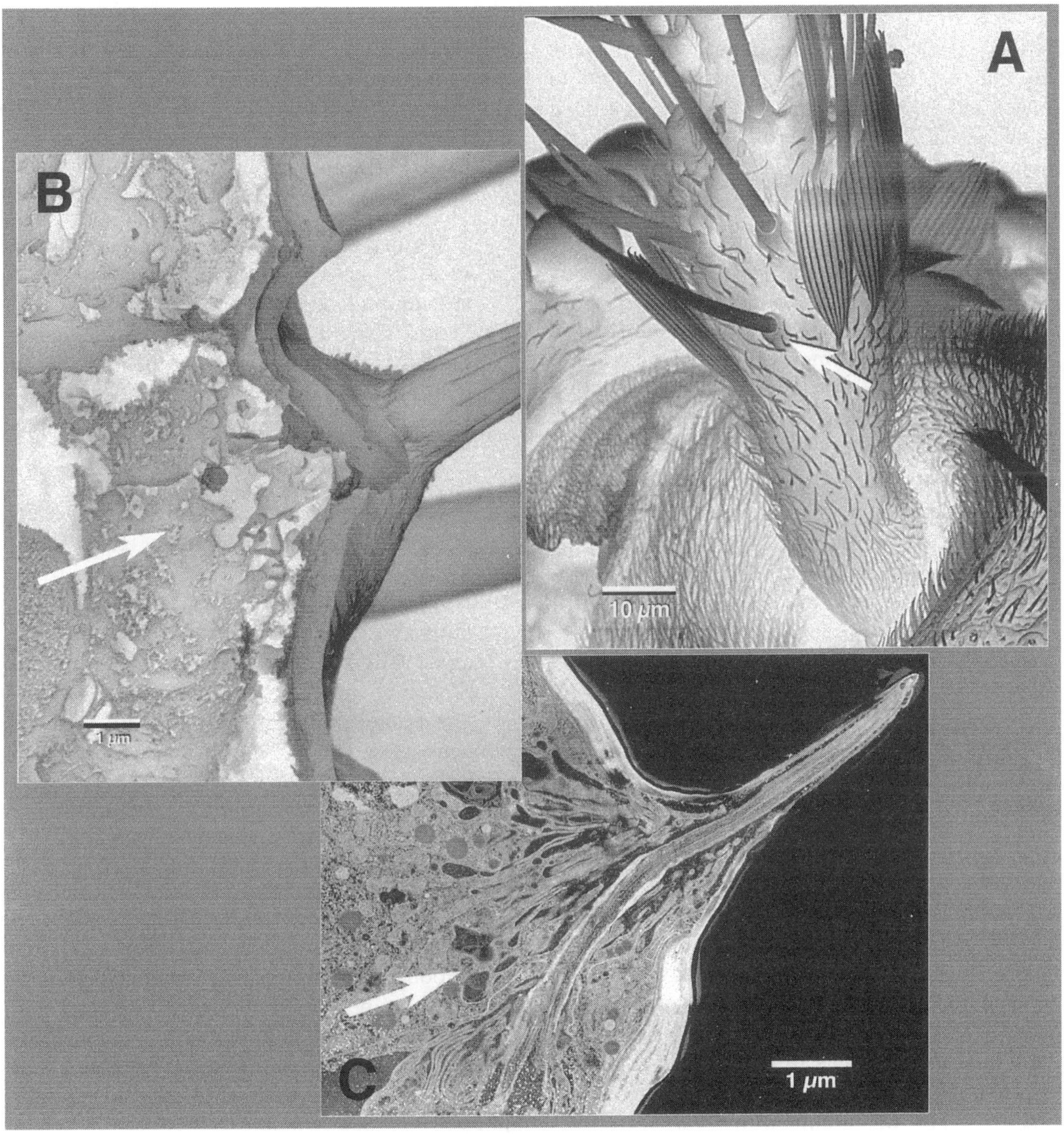


**Figure 1–46** Correlative scanning and transmission electron microscopy of a mosquito antenna. (A) SEM overview of basal segment of the antenna: the sensillum (arrow) is the same structure shown in (B) and (C). (B) SEM micrograph of freeze-fractured antenna preparation showing the internal orientation of dendrites as they pass through the sensillum (arrow). (C) The same structure cut in cross section and examined by TEM confirms the orientation and organization of neuronal dendrites and support cells of the sensillum (arrow). *Source:* Copyright © Ken Baker.

Microfocal (projection) X-ray microscopy also has the advantages of being nondestructive, being high in contrast, providing good resolution ($\mu$m scale), and requiring minimal sample preparation. It allows visualization of 3-D structures and offers the opportunity of subjecting samples to mechanical and thermal stimuli. It is at the development stage.

## 1.7 CONCLUSIONS

Examining food microstructure is always a difficult task because of the complexity of the material involved. One useful approach for differentiating structural features from artifacts is correlative microscopy—the practice of using a combination of microscopic techniques in order to unravel the confusion attendant upon image interpretation. Thus, it is desirable to confirm image data by analogy.

Another argument for correlative microscopy is that each instrument has its own range of magnification and attendant strengths and weaknesses. Researchers should regard these instruments as complementary and not competitive. Obviously, correlative procedures are time consuming, but they are necessary where doubts concerning structure persist. Figure 1–46 shows correlated SEM and TEM images. Seeing the same structure in different microscopes is plainly advantageous for discovering the proper explanations.

Throughout this chapter, emphasis has been placed on the great strides that have been made in instrumentation. It would be a mistake, however, to assume that perfecting microscopes automatically leads to an increase in knowledge. After all, structural information results from the ability of scientists to properly interpret microscopic images.

Lewis (1986), commenting on food microscopy, states that "the trouble with allowing fools to look down microscopes is that they are likely to come to foolish conclusions" (p. 379). This remark reflects the adage that the results are never better than the interpretation. To ensure the validity of structural data, the following safeguards are recommended:

- Use sample preparation procedures that minimize artifacts. Use several methods to check structural identification.
- Use techniques suitable to the information required. Technological overkill and empty magnification are costly and pointless.
- Use controls whenever possible so that interpretation can be based on differences between control and treatment samples.
- Use a sufficient number of preparations randomly obtained from representative treatment and control samples.
- Use correlative techniques, magnifying and nonmagnifying, to confirm conclusions and avoid falling into the trap of "If I can find it, it must be there."
- Use a range of magnifications in order to concentrate on structural organization rather than individual parts.
- Use objective interpretive procedures and statistical analysis whenever possible. Image analysis is covered in detail in the following chapter.

Although these safeguards will not totally eliminate "foolish conclusions," they will help prevent them. The study of food microstructure can be an exciting and rewarding undertaking, and as technology continues to improve, researchers will have excellent opportunities to achieve new insights in this important area.

## BIBLIOGRAPHY

Agboola, S.O., & Dalgleish, D.G. (1995). Calcium-induced destabilization of oil-in-water emulsions stabilized by caseinate or by $\beta$-lactoglobulin. *Journal of Food Science, 60,* 399–404.

Barbut, S. (1997). Microstructure of white and dark turkey meat batters as affected by pH. *British Poultry Science, 38,* 175–182.

Bereiter-Hahn, J. (1995). Probing biological cells and tissues with acoustic microscopy. In A. Briggs (Ed.), *Advances in acoustic microscopy* (Vol. 1) [pp. 79–115]. New York: Plenum Press.

Blonk, J.C.G., & van Aalst, H. (1993). Confocal scanning light microscopy in food research. *Food Research International, 26,* 297–311.

Brooker, B.E. (1995). Imaging food systems by confocal laser scanning microscopy. In E. Dickinson (Ed.), *New physico-chemical techniques for the characterization of complex food systems* (pp. 53–68). London: Blackie Academic and Professional.

Clayton, E.G. (1909). *A compendium of food-microscopy with a section on drugs, water and tobacco.* London: Bailliere, Tindell and Cox.

Dalgleish, D.G., Tosh, S.M., & West, S. (1996). Beyond homogenization: The formation of very small emulsion droplets during the processing of milk by a Microfluidizer. *Netherlands Milk and Dairy Journal, 50,* 135–148.

Danilatos, G.D. (1993). Introduction to the ESEM instrument. *Microscopy Research and Technology, 25,* 354–361.

del Valle, J.M., & Stanley, D.W. (1995). Reversible and irreversible components of bean hardening. *Food Research International, 28,* 455–463.

del Valle, J.M., Stanley D.W., & Bourne M.C. (1992). Water absorption and swelling in dry beans. *Journal of Food Processing and Preserving, 16,* 75–98.

Flegler, S.L., Heckman, J.W., Jr., & Klomparens, K.L. (1993). *Scanning and transmission electron microscopy: An introduction.* New York: W.H. Freeman.

Friel, J.J. (1995). *X-ray and image analysis in electron microscopy.* Princeton, NJ: Princeton Gamma-Tech.

Fulcher, R.G. (1982). Fluorescence microscopy of cereals. *Food Microstructure, 1,* 167–176.

Fulcher, R.G., & Wood, P.J. (1983). Identification of cereal carbohydrates by fluorescence microscopy. In D.B. Bechtel (Ed.), *New frontiers in food microstructure* (pp. 111–128). St. Paul, MN: American Association of Cereal Chemists.

Hafsteinsson, H., & Rizvi, S.S.H. (1984). Acoustic microscopy: Principles and applications in the study of biomaterial microstructure. *Scanning Electron Microscopy, III,* 1237–1247.

Heertje, I., & Pâques, M. (1995). Advances in electron microscopy. In E. Dickenson (Ed.), *New physico-chemical techniques for the characterization of complex food systems* (pp. 1–52). Glasgow: Blackie Academic & Professional.

Holz, H.M. (1975). *Worthwhile facts about fluorescence microscopy.* Oberkochen, West Germany: Carl Zeiss.

Johnson, R. (1996). *Environmental scanning electron microscopy.* El Dorado Hills, CA: Robert Johnson Associates.

Kalab, M., Allan-Wojtas, P., & Miller, S.S. (1995). Microscopy and other imaging techniques in food structure analysis. *Trends in Food Science and Technology, 6,* 177–186.

Kendall, E.J., Stanley, D.W., & Chatson, K.B. (1994). A comparison of water self-diffusion in fresh and stored beans. In *Proceedings of the 77th Chemistry Society of Canada Conference* (Abstract no. 510). Chemical Society of Canada.

Kessel, R.G., & Shih, C.Y. (1974). *Scanning electron microscopy in biology.* New York: Springer-Verlag.

Kitagawa, H. (1994). Theory and principal technologies of the laser scanning confocal microscope. In P.C. Cheng, T.H. Lin, W.L. Wu, & J.L. Wu (Eds.), *Multidimensional microscopy* (pp. 53–71). New York: Springer-Verlag.

Lewis, D.F. (1986). Features of food microscopy. *Food Microstructure, 5,* 1–18.

Lewis, D.F. (1988). An electron microscopist's view of foods. In J.M.V. Blanshard & J.R. Mitchell (Eds.), *Food structure: Its creation and evaluation* (pp. 367–384). London: Butterworths.

Li-Chan, E.C.Y. (1996). The application of Raman spectroscopy in food science. *Trends in Food Science Technology, 7,* 361–370.

Maheshwari, P.N., Stanley, D.W., Beveridge, T.J., & van de Voort, F.R. (1981). Localization of myrosinase (thioglucoside glucohydrolase, EC 3.2.3.1) in cotyledon cells of rapeseed. *Journal of Food Biochemistry, 5,* 39–61.

Marcone, M.F., Beniac, D.R., Harauz, G., & Yada, R.Y. (1994). Quaternary structure and model for the oligomeric seed globulin from Amaranthus hypochondriacus K343. *Journal of Agriculture Food Chemistry, 42,* 2675–2678.

Marelius, J. (1995). Autofluorescence imaging of living cells. Unpublished master's thesis, Uppsala University School of Engineering, Uppsala, Sweden.

Parnell-Clunies, E.M., Kakuda, Y., & Humphrey, R. (1986). Electron dense granules in yogurt: Characterization by x-ray microanalysis. *Food Microstructure, 5,* 295–302.

Richardson, T. (1997). Infrared light in the microscope: History, theory and practical aspects. *Proceedings of the Royal Microscopic Society, 32,* 229–235.

Samuel, D. (1996). Investigation of ancient Egyptian baking and brewing methods by correlative microscopy. *Science, 273,* 488–490.

Smith, I.R., Harvey, R.A., & Fathers, D.J. (1985). An acoustic microscope for industrial applications. *Institute of Electrical and Electronic Engineers Transactions Sonics and Ultrasonics, SU-32*(32), 274–288.

Smith, J.L., Stanley, D.W., & Baker, K.W. (1987). Nonenzymic lignification of asparagus. *Journal of Texture Studies 18,* 339–358.

Stanley, D.W., Aguilera, J.M., Baker, K.W., & Jackman, R.L. (1998). Structure/property relationships of foods as affected by processing and storage. In M.A. Rao & R. Hartel (Eds.), *Chemical, structural, and rheological changes during phase/state transitions in foods* (pp. 1–56). New York: Marcel Dekker.

Stanley, D.W., & Tung, M.A. (1976). Microstructure of food and its relation to texture. In J.M. deMan. P.W. Voisey,

V.F. Rasper, & D.W. Stanley (Eds.), *Rheology and texture in food quality* (pp. 28–78). Westport, CT: AVI Publishing Co.

Stokes, D.J., Thiel, B.L., & Donald, A.M. (1998). Direct observation of water-oil emulsion systems in the liquid state by environmental scanning electron microscopy. *Langmuir, 14,* 4402–4408.

Strausser, Y.E., & Heaton, M.G. (1994). Scanning probe microscopy. *American Laboratory, 26*(5), 15–21.

Sun, S.Q., Shi, S.-L., Hunt, J.A., & Leapman, R.D. (1995). Quantitative water mapping of cryosectioned cells by electron energy-loss spectroscopy. *Journal of Microscopy, 177,* 31–42.

Swatland, H.J. (1990). Intracellular glycogen distribution examined interactively with a light microscope scanning stage. *Journal of Computer-Assisted Microscopy, 2,* 233–237.

Swatland, H.J. (1998). *Computer operation for microscope photometry.* Boca Raton, FL: CRC Press.

Travis, A.J., Murison, S.D., Perry, P., & Chesson, A. (1997). Measurement of cell wall volume using confocal microscopy and its application to studies of forage degradation. *Annals of Botany, 80,* 1–11.

Troy, C.T., & Abrams, S.B. (1996). Scanning force microscopy. *Biophotonics International, 3*(5), 52–53.

Vodovotz, Y., Vittadini, E., Coupland, J., McClements, D.J., & Chinachoti, P. (1996). Bridging the gap: Use of confocal microscopy in food research. *Food Technology, 50*(6), 74–82.

Wilson, R.H. (1995). Recent developments in infrared spectroscopy and microscopy. In E. Dickinson (Ed.), *New physicochemical techniques for the characterization of complex food systems* (pp. 177–195). Glasgow: Blackie Academic & Professional.

Yada, R.Y., Harauz, G., Marcone, M.F., Beniac, D.R., & Ottensmeyer, F.P. (1995). Visions in the mist: The Zeitgeist of food protein imaging by electron microscopy. *Trends in Food Science Technology, 6,* 265–270.

Yiu, S.H. (1987). Fluorescence microscopy in food technology. *Zeiss Focus, 4*(2), 6–7.

## SUGGESTED READING

### History of Food Microstructure Studies

Davis, E.A., & Gordon, J. (1982). Food microstructure: An integrative approach. *Food Microstructure, 1,* 1–12.

Flint, O. (1994). *Food microscopy.* Oxford: Bios Scientific Publishers.

Holcomb, D.N., & Kalab, M. (Eds.). (1981). *Studies of food microstructure.* Chicago: Scanning Electron Microscopy, Inc.

Kalab, M. (1983). Electron microscopy of foods. In M. Peleg & E.B. Bagley (Eds.). *Physical properties of foods* (pp. 43–59). Westport, CT: AVI Publishing Co.

Möllring, F.K. (1981). *Microscopy from the very beginning.* Oberkochen, West Germany: Carl Zeiss.

O'Brien, T.P. (1983). Cereal structure: An historical perspective. In D.B. Bechtel (Ed.), *New frontiers in food microstructure* (pp. 3–26). St. Paul, MN: American Association of Cereal Chemists.

Stanley, D.W. (1994). Understanding the materials used in foods: Food materials science. *Food Research International, 27,* 135–144.

Swatland, H.J. (1985). Early research on the fibrous microstructure of meat. *Food Microstructure, 4,* 73–82.

Vaughn, J.G. (1979). *Food microscopy.* London: Academic Press.

Winton, A.L. (1917). *A course in food analysis.* New York: John Wiley & Sons.

Woodman, A.G. (1915). *Food analysis.* New York: McGraw-Hill.

### Light Microscopy

Cheng, P.C., Lin, T.H., Wu, W.L., & Wu, J.L. (Eds.). (1994). *Multidimensional microscopy.* New York: Springer-Verlag.

Cooke, P.M. (1996). Chemical microscopy. *Analytic Chemistry, 68,* 333R–378R.

Delly, J.G. (1988). *Photography through the microscope* (9th ed.). Rochester, NY: Eastman Kodak Co.

Flint, F.O. (1988). The evaluation of food structure by light microscopy. In J.M.V. Blanshard & J.R. Mitchell (Eds.), *Food structure: Its creation and evaluation* (pp. 351–380). London: Butterworths.

Flint, F.O. (1994). *Food microscopy.* Oxford: Bios Scientific Publishers.

McCrone, W.C. (1988). Future of light microscopy. *American Laboratory, 20*(4), 21–28.

McCrone, W.C., Drafty, R.G., & Delly, J.G. (1967). *The particle atlas.* Ann Arbor, MI: Ann Arbor Scientific Publishing.

McKenna, A.B. (1997). Examination of whole milk powder by confocal laser scanning microscopy. *Journal of Dairy Research, 64,* 423–432.

O'Brien, T.B., & McCully, M.E. (1981). *The study of plant structure. Principles and selected methods.* Melbourne: Termacarphi Proprietary Ltd.

Robinson, M.K. (1997). Multiphoton microscopy expands its reach. *Biophotonics International, 4*(5), 38–45.

Sanderson, J.B. (1994). *Biological microtechnique.* Oxford: Bios Scientific Publishers.

Stevenson, R. (1996). Bioapplications and instrumentation for light microscopy in the 1990s. *American Laboratory, 28*(6), 23–51.

Swatland, H.J. (1990). Questions in programming a fluorescence microscope. *Journal of Computer-Assisted Microscopy, 2,* 125–132.

### Transmission Electron Microscopy

Allen, R.M. (1985). Secondary electron imaging in the scanning transmission electron microscope. *Scanning Electron Microscopy, III,* 905–918.

Bechtel, D.B. (1983). From the farm to the table: Transmission electron microscope account of cereal structure and its relationship to end-use properties. In D.B. Bechtel (Ed.), *New Frontiers in Food Microstructure* (pp. 269–278). St. Paul, MN: American Association of Cereal Chemists.

Bullock, G.R. (1984). The current status of fixation for electron microscopy: A review. *Journal of Microscopy, 123,* 1–15.

Hayat, M.A. (1986). *Basic techniques for transmission electron microscopy.* Orlando, FL: Academic Press.

Heertje, I., & Pâques, M. (1995). Advances in electron microscopy. In E. Dickinson (Ed.), *New physico-chemical techniques for the characterization of complex food systems* (pp. 1–52). London: Blackie Academic & Professional.

Kalab, M. (1983). Electron microscopy of foods. In M. Peleg & E.B. Bagley (Eds.), *Physical properties of foods* (pp. 43–56). Westport, CT: AVI Publishing Co.

Revel, J.P., Barnard, T., & Haggis, G.H. (1984). *The science of biological specimen preparation for microscopy and microanalysis.* Chicago, IL: Scanning Electron Microscopy, Inc.

### Scanning Electron Microscopy

Beckett, A., & Read, N.D. (1986). Low-temperature scanning electron microscopy. In H.C. Aldrich & W.J. Todd (Eds.), *Ultrastructure techniques for microorganisms* (pp. 45–63). New York: Plenum Press.

Brooker, B.E. (1988). Food quality assessment using microscopy. In *Food technology international Europe 1988* (pp. 289–299). London: Stearling Publications.

Chabot, J.F. (1979). Preparation of food science samples for SEM. *Scanning Electron Microscopy, III,* 279–286, 298.

Griffith, E., & Newbury, D.E. (1996). Introduction to environmental scanning electron microscopy issue. *Scanning, 18,* 465.

Hayat, M.A. (1978). *Introduction to biological scanning electron microscopy.* Baltimore: University Park Press.

Hippe-Sanwald, S. (1995). Low temperature techniques as a tool in plant pathology. *Scanning Microscopy, 9*(3), 881–899.

Paul, R.N., & Egley, G.H. (1988). Preparation and staining of hard seed tissue for backscatter imaging. In G.W. Bailey (Ed.), *Proceedings of the 46th Annual Meeting of the Electron Microscopy Society of America* (pp. 412–415). San Francisco: San Francisco Press.

Read, N.D., Porter, R., & Beckett, A. (1983). A comparison of preparative techniques for the examination of the external morphology of fungal material with the scanning electron microscope. *Canadian Journal of Botany, 61,* 2059–2078.

Sargent, J.A. (1988). Low temperature scanning electron microscopy: Advantages and applications. *Scanning Microscopy, 2*(2), 835–849.

Steinbrecht, R.A., & Zierold, K. (1987). *Cryotechniques in biological electron microscopy.* Berlin: Springer-Verlag.

### Other Instrumentation and Techniques

Bottomley, L.A., Coury, J.E., & First, P.N. (1996). Scanning probe microscopy. *Analytic Chemistry, 68,* 185R–230R.

Callaghan, P.T. (1991). *Principles of nuclear magnetic resonance microscopy.* Oxford: Clarendon Press.

Chen, C.J. (1993). Scanning tunnelling microscopy: A chemical perspective. *Scanning Microscopy, 7*(3), 793–804.

Dahn, D.C., Cake, K., MacDonald, T.L., & Hale, L.R. (1995). Scanning tunneling microscopy studies of chloroplasts in solution. *Scanning Microscopy, 9*(2), 413–418.

Dalgleish, D.G., & Hallett, F.R. (1995). Dynamic light scattering: Applications to food systems. *Food Research International, 28,* 181–193.

Doyle, P., & Adams, C. (1996). Scanning probe microscopy and the study of lipids. *Lipid Technology, 8*(2), 39–42.

Goldstein, J.I., Newbury, D.E., Echlin, P., Joy, D.C., Jr., Romig, A.D., Lyman, C.E., Fiori, C., & Lifshin, E. (1992). *Scanning electron microscopy and x-ray microanalysis* (2nd ed.). New York: Plenum Press.

Heil, J.R., McCarthy, M.J., & Özilgen, M. (1992). Magnetic resonance imaging and modeling of water up-take in dry beans. *Lebensmittel-Wissenschaft und Technologie, 25,* 280–285.

Hills, B.P. (1995). Magnetic resonance imaging in food science. In E. Dickinson (Ed.), *New physico-chemical techniques for the characterization of complex food systems* (pp. 319–341). London: Blackie Academic & Professional.

Horne, D.S. (1995). Light scattering studies of colloid stability and gelation. In E. Dickinson (Ed.), *New physico-chemical techniques for the characterization of complex food systems* (pp. 240–267). London: Blackie Academic & Professional.

Kalab, M., Allan-Wojtas, P., & Shea Miller, S. (1995). Microscopy and other imaging techniques in food structure analysis. *Trends in Food Science and Technology, 6,* 177–186.

Kirby, A.R., Gunning, A.P., & Morris, V.J. (1995). Atomic force microscopy in food research: A new technique comes of age. *Trends in Food Science and Technology, 6,* 359–365.

Kirby, A.R., Gunning, A.P., & Morris, V.J. (1995). Imaging xanthan gum by atomic force microscopy. *Carbohydrate Research, 267,* 161–166.

Kirby, A.R., Gunning, A.P., Waldron, K.W., Morris, V.J., & Ng, A. (1996). Visualization of plant cell walls by atomic force microscopy. *Biophysics Journal, 70,* 1138–1143.

Marti, O., & Amrein, M. (Eds.). (1993). *STM and SFM in biology.* San Diego, CA: Academic Press.

McCarthy, M.J. (1994). *Magnetic resonance imaging in foods.* New York: Chapman & Hall.

McCarthy, M.J., & Kauten, R.J. (1990). Magnetic resonance imaging applications in food research. *Trends In Food Science and Technology, 1,* 134–139.

McCarthy, M.J., & Perez, E. (1990). Measurement of effective moisture diffusivities using magnetic resonance imaging. In *Engineering and food,* Vol. 1, *Physical properties and process control* (pp. 473–481). London: Elsevier Applied Science.

Miles, M.J., & McMaster, T.J. (1995). Scanning probe microscopy of food-related systems. In E. Dickinson (Ed.), *New physico-chemical techniques for the characterization of complex food systems* (pp. 69–85). London: Blackie Academic & Professional.

Morris, V.J., Kirby, A.R., & Gunning, A.P. (1995). Probe microscopies: Feeling their way. *Food Hydrocolloids, 9,* 273–280.

Naish, S.J. (Ed.). (1989). *Immunochemical staining methods.* Carpinteria, CA: DAKO Corp.

Neddermeyer, H. (Ed.). (1993). *Scanning tunnelling microscopy.* Dordrecht, Netherlands: Kluwer Academic Publishers.

Newbury, D.E., Joy, D.C., Echlin, P., Fiori, C.E., & Goldstein, J.I. (1986). *Advanced scanning electron microscopy and x-ray microanalysis.* New York: Plenum Press.

Prater, C.B., Maivald, P.G., Kjoller, K.J., & Heaton, M.G. (1995). Scanning probe microscopy. *American Laboratory, 27*(4), 50–54.

Shao, Z., & Zhang, Y. (1996). Biological cryo atomic force microscopy: A brief review. *Ultramicroscopy, 66,* 141–152.

Shortt, D.W., Roessner, D., & Wyatt, P.J. (1996). Absolute measurement of diameter distributions of particles using a multiangle light scattering photometer coupled with flow field-flow fractionation. *American Laboratory, 28*(17), 21–28.

Simoneau, C., McCarthy, M.J., & German, J.B. (1993). Magnetic resonance imaging and spectroscopy for food systems. *Food Research International 26,* 387–398.

### Web Sites

These sites are places to start. Some may be outdated or no longer exist. See also Mackenzie, R. (1997). Microscopy resources on the Internet. *Proceedings of the Royal Microscopy Society, 32*(1), 11.

*Analysis methods.* ⟨http://www.csu.edu.au/ci/vol3/complxb5/node5.html⟩.

*Biological microscopy.* ⟨www.biotech.ufl.edu⟩.

*Confocal microscopy.* ⟨www.science.uwaterloo.ca/research_groups/confocal⟩.

*Digital instruments.* Atomic force microscopy: ⟨http://www.di.com/⟩. Guide to microscopy and microanalysis on the Internet: ⟨http://www.mwrn.com/guide.htm⟩. Guide to microscopy on the Web: ⟨www.mwrn.com⟩.

*IFR Scanning Probe Microscopy Group.* ⟨http://www.ifrn.bbsrc.ac.uk/fb/spm/docs/spm_home.html⟩, ⟨listserver@ubvm.cc.buffalo.edu;listproc@soils.umn.edu⟩.

*Microscopical Society of Canada.* ⟨www.gause.biology.ualberta.ca/craig.hp⟩.

*Microscopy Society of America.* ⟨listserv@MSA.Microscopy.com⟩.

*National Institutes of Health.* Image analysis software: ⟨ftp://zippy.nih.gov/pub/nih-image⟩, ⟨http://rsb.info.nih.gov/nih-image⟩, ⟨www.pharm.Arizona.edu/centers/tox_center/swehsc/exp_path⟩.

*Royal Microscopy Society.* ⟨http://www.rms.org.uk⟩.

# CHAPTER 2

# Image Analysis

## 2.1 INTRODUCTION

Throughout the previous chapter emphasis was placed on the great strides that have been made in instrumentation. Significant improvements have occurred in resolving power and contrast enhancement for both light microscopes (LMs) and electron microscopes. While there is no doubt that these advances have resulted in concomitant structural advances, it would be a mistake to assume that perfecting the microscope automatically leads to an increase in knowledge. Rather, structural information results from the ability of the scientist to properly interpret microscopic images. In almost every case, microstructural research will benefit from quantifying images in some way. The current discussion focuses on means to accomplish this goal.

As mentioned previously, one of the most dangerous pitfalls of microscopy is our tendency to find what we are looking for. This is a very easy mistake for microscopists to make when examining images visually. Our human visual system is disposed to make biased subjective comparisons regarding the image features we see, but unfortunately our natural imaging system is not very well suited to making quantitative determinations. So-called "optical illusions" demonstrate that the eye is not to be trusted for objective assessments. In order to make valid judgments, we need to augment our imaging apparatus with unbiased image analysis tools so that we can obtain reliable quantitative information and numerical data from an image. Also, the computer can relieve us from the ennui and errors associated with repetitive measurements. As a sign in a university food research laboratory states, "Even a graduate student's time is worth something."

Image analysis relies heavily on computer technology to recognize, differentiate, and quantify images. Commercially available units combine optical and computer components to form powerful tools. Computing and software capabilities are constantly evolving, but the basic elements of an image analysis system are as shown in Figure 2–1. Nowhere else have such rapid advances been made as in the area of computerized image analysis. Enhanced computer capacity, new and better software, and the advent of dedicated on-board computers have all contributed to this progress. The study of image analysis has proliferated in recent years, and much of the knowledge in this field is based on advanced mathematics beyond the scope of this book. The following descriptions briefly introduce selected examples of analog and digital imaging techniques, leaving technically detailed, comprehensive descriptions to the instrumentation experts whose works are cited in the references at the end of the chapter. The reader is encouraged to consult the referenced texts and articles and to visit the various Internet addresses cited for discussion groups and locations dedicated to imaging technology, theory, communication, and teaching.

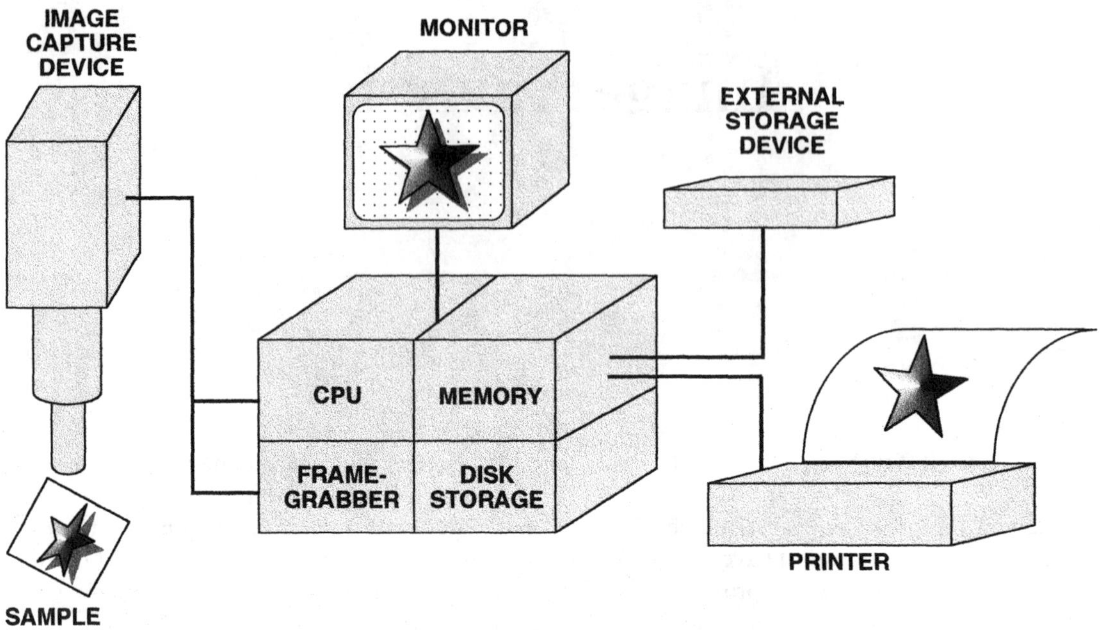


**Figure 2–1** Basic elements of an image analysis system. *Source:* Reprinted from C.A. Glasbey and G.W. Horgan, *Image Analysis for the Biological Sciences*. Copyright 1995, John Wiley & Sons Limited. Reproduced with permission.

## 2.2 IMAGE ACQUISITION

Image analysis starts with a picture obtained using one of the techniques described in the previous chapter. Depending upon the equipment, the information will be in digital or analog form. Modern image analysis systems begin by transforming an analog signal, such as a film-based "hard copy," into a digital "soft copy." The transformation may be achieved in a variety of ways, depending upon the resources available and the end use of the image, but in every case the result is a set of pixel values. The devices used for capturing images include cameras, scanners, and other equipment.

### 2.2.1 Video Cameras

The most common and straightforward devices for collecting digital images from microscopes are video cameras. There are two basic types, but both produce a composite analog signal that can be digitized for computer use by means of a wide variety of analog-to-digital converters (or capture boards). Monochrome cameras are the most common for scientific applications. They are much cheaper than color cameras and provide higher resolution, contrast, and sensitivity. They can even serve as color cameras by electronically merging sequentially captured red, green, and blue illuminated images using inexpensive filters, automated RGB filter wheels, or a liquid crystal interference filter electronically controlled by purpose-specific software. This approach to color imaging, though very effective for certain applications, can be cumbersome in practice and largely rules out motion and fluorescence imaging. The higher resolution and lower price may, nevertheless, justify the use of monochrome color cameras for image acquisition, particularly if motion or fluorescence fading are not important.

Dedicated color cameras, by comparison, are relatively expensive and often compromise image quality because the sensors must be separately sensitive to red, green, and blue signals. High-end color cameras, usually quite expensive, improve on image quality significantly through the use of a complete chip or tube for each color. Color cameras are obviously preferred for areas of microscopy in which color elements are the most important consideration. Color cameras thus have a role to play in microscopy but ought not be viewed as substitutes for high-quality monochrome cameras. The overall selection of a video camera for microscopy should be based on a variety of factors. It would make sense to consult some of the works listed at the end of this chapter before purchasing video equipment.

### 2.2.2 Scanners

Scanners resemble photocopiers: an object with a flat surface, such as a color or monochrome photograph, microscopic slide preparation, transparency, negative, or gel, is placed on a glass plate and scanned with a device employing a CCD array or single sensor to create a digital file. Scanning is a very useful and relatively inexpensive means of image acquisition, and scanners, either purchased for in-house use or accessed through a service bureau, should not be overlooked as a practical alternative to other image-capturing equipment. They usually come bundled with image-processing software, such as Adobe Photoshop™ or other proprietary software. Scanning equipment and array cameras communicate with the computer through SCSI devices. Hand-held scanners, although available at lower cost, usually produce lower quality images than flatbed scanners.

### 2.2.3 Other Image-Capturing Devices

Commercial scanning services are available in many locales. Using high-quality scanning devices and dedicated color management, a provider will, for a fee, permanently scan and store slides or negatives in the form of a CD ROM. Medical imaging equipment, such as magnetic resonance imagers and ultrasound scanners, also produce images in digital form as well.

## 2.3 IMAGE PROCESSING

The digital image in its original form, whether color or monochrome, is referred to as a gray scale image. With most modern equipment, 256 gray levels are available. Thus, in a typical image whose dimensions are 512 pixels × 512 pixels, each pixel has an integer value ranging from 0 to 255. The processing and analysis of acquired gray scale images basically follow the flowchart shown in Figure 2–2, although the precise nature of each step in image processing and analysis will be largely determined by study-specific analytical goals. Enhancement of the image proceeds by the application of one or more of the techniques described below. Note that the basis for image analysis is the conversion of pixel gray levels into numerical values to which various algorithms can be applied and that gray levels can be influenced by the physical features of the object and its orientation to the illumination source. Lighting-induced shadows can introduce artifacts into the image analysis process.

### 2.3.1 Filters

Once the gray scale image is obtained, one of several different avenues of approach must be chosen. Usually the first preprocessing step, no matter how high the original image quality, is the application of a filter or filters to remove unwanted noise or sharpen the edges of objects. These filters, mathematical algorithms implemented in software, serve to enhance images by applying transformations to individual or groups of pixels based on the level of surrounding pixels. Image processing of the gray level image encompasses a wide variety of pixel-based processing algorithms dedicated to such tasks as noise reduction, brightness and contrast enhancement, feature enhancement, sharpening, background correction/subtraction, image cropping, and so forth.

Median and Gaussian filters have the general effect of smoothing images. They are used to eliminate noise and background artifacts and to smooth sharp edges but also tend to remove some of the details in small objects. Sharpening filters

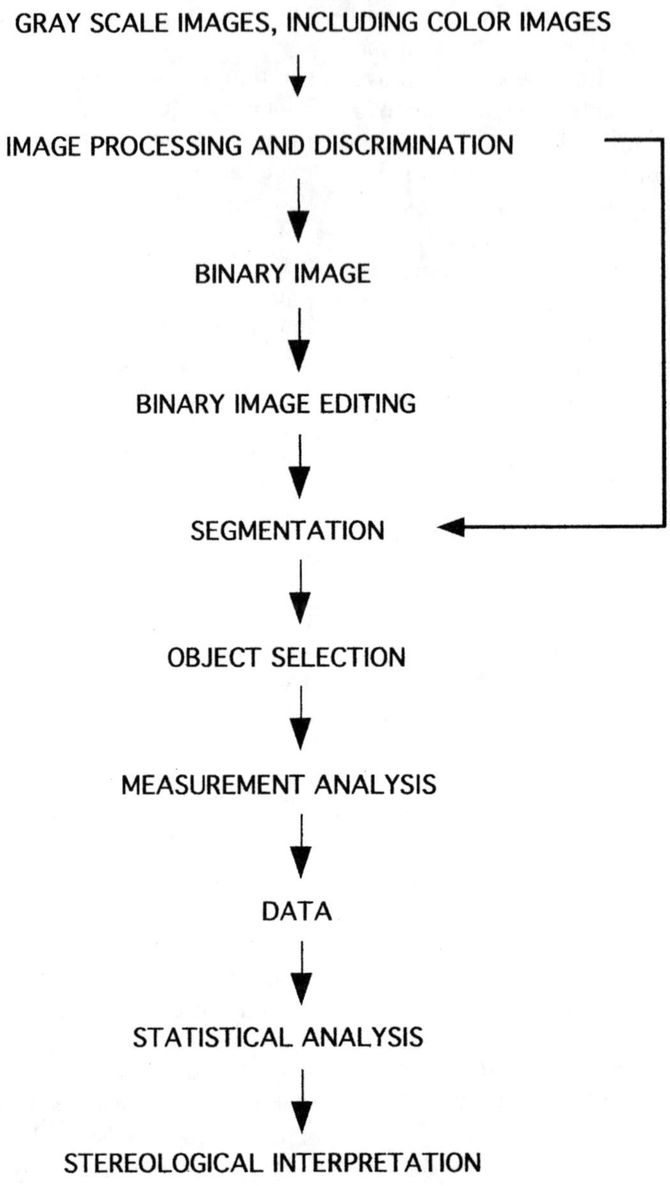


**Figure 2–2** Measurement of features in microscopic images. The newly acquired, digital "gray scale" image undergoes image processing for the discrimination of important features. The image can be additionally processed by thresholding to create a binary image that can be further processed by binary image editing. Segmentation divides the image into regions of structures intended for analysis. Object selection is followed by measurement and analysis and the collection of quantitative or qualitative data. The data are finally subjected to statistical analysis and, depending upon circumstances, used to support a stereologic interpretation and conclusions about the structures of interest. Not all steps are required for each image; for example, a binary image may not be employed. *Source:* Reprinted from D.W. Stanley, J.M. Aguilera, K.W. Baker, and R.L. Jackman, Structure/Property Relationships of Foods as Affected by Processing and Storage, in *Chemical, Structural, and Rheological Relationships During Phase/State Transitions in Foods,* M.A. Rao and R. Hartel, eds., pp. 1–56, by courtesy of Marcel Dekker, Inc.

can emphasize details but also highlight noise and other small artifacts. The application of sharpening filters is most useful when the image consists of fine structural components of a specimen or when edge enhancement is desired. The contrast and brightness of the image can be adjusted by changing the gray scale values. Histogram equalization can be used to improve contrast by a nonlinear mapping of the gray levels; this is most commonly applied when gray levels are concentrated in a small portion of the range of possible values. The application of filters can affect quantitative measurements of the resulting images. Thus, filters are often only used for displayed images, and quantitative measurements are made using the unprocessed data. Examples of filtered and unfiltered images are presented in Figure 2–3.

### 2.3.2 Binarization

Image processing might proceed by binarization instead of filtering. In binarization, the original gray level image is changed from a continuum of colors or gray levels into a black-and-white image by assigning to each pixel a value of black or white. The binary image, once created, can then be manually edited by selecting objects for removal or inclusion or by using a spectrum of well-documented and extraordinarily powerful binary image–editing techniques.

Binary image editing permits the selective removal of artifacts and noise, edge discrimination, skeletonizing, hole filling, application of Boolean operators using selected overlay or time-sequenced images, and other operations. The final edited binary image can be automatically measured using stereologic or morphometric methods. Feature sets can also be measured using similarly binarized stereologic overlays and Boolean "and" operators derived from computer overlays.

### 2.3.3 Segmentation

Images must then be segmented into measurable structures on the basis of color, brightness, edge discontinuities, elemental composition, temperature, or some other property that can be used to distinguish a feature from background. Segmentation refers to the process of extracting the desired object of interest from the image background. In image analysis, segmentation may be done by manual or automated methods and may be applied to an original image, to an image following filter transformation, or to a binary image. Examples of problematic subjects include chocolate chips in a cookie (dark brown structure on a lighter background), bubbles in a solution (same color as the surroundings but having a darker perimeter outline), textured phases in a composite (background with one pattern and the structure of interest with a distinctly different texture), ice cubes in a glass of liquid (colder temperature of subjects versus warmer temperature of background) (Russ, 1995). When segmentation is complete, every pixel in the image is included as an object (or part thereof) or as "background." Pixels contained in an object form a connected region in the image and have values similar to those of other pixels in that category but dissimilar to those of adjacent pixels in different categories.

There are several broad approaches to segmentation, each with its own family of algorithms. Many of these are automatic segmentation algorithms, but with complex images, such as occur with foods, manual intervention is often required. Segmentation can proceed by thresholding, edge-based methods, and region-based methods. In binarization, thresholding is the dynamic process of taking the original gray level image from a continuum of colors or gray levels and assigning to each pixel a value of white or black, but in this context it can mean a wider range of categories with more than one cutoff point. Thresholding, in other words, involves limiting the intensity values within an image to a certain bounded range.

Each pixel in an 8-bit gray scale image has a value between 0 (black) and 255 (white), and it may be decided that all pixels below a certain value do not contribute significantly to the object of interest and can be eliminated. This can be done by scanning the image one pixel at a time and keeping a pixel if it is at or above the selected intensity value or setting it to 0 (black) if it is below that value. This can be done either by manually

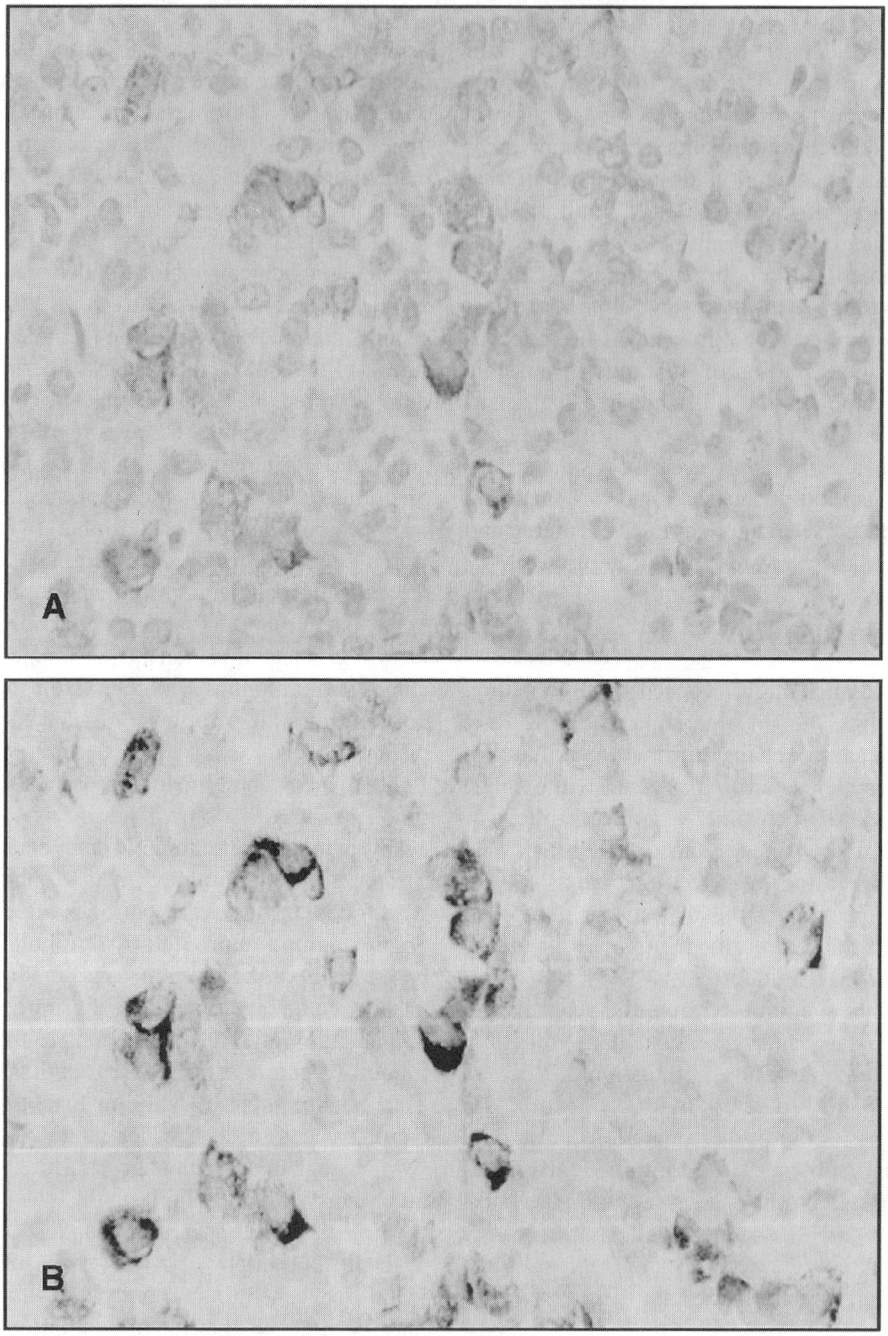


**Figure 2–3** Brightfield light microscope image of immunostained (orange) cells in a light blue background counterstain. Image acquired using a DAGE CCD-72 video camera and a Scion AG-5 capture board. (A) Field illuminated with unfiltered white light. (B) Field illuminated with blue filtered (#47B Wratten filter) white light. *Source:* Reprinted from D.W. Stanley, J.M. Aguilera, K.W. Baker, and R.L. Jackman, Structure/Property Relationships of Foods as Affected by Processing and Storage, in *Chemical, Structural, and Rheological Relationships During Phase/State Transitions in Foods*, M.A. Rao and R. Hartel, eds., pp. 1–56, by courtesy of Marcel Dekker, Inc.

tracing around the regions of interest with the mouse or by using an automated routine. Thresholding is the simplest and most commonly employed segmentation technique.

Edge-based segmentation separates pixels into those that are on an edge of a region and those that are not. Non-edge pixels that form connected regions are then allocated to the same category. Region-based methods use algorithms to group adjacent pixels having similar values and to divide groups of pixels that have dissimilar values.

Ideally, the processed gray level image in its final form will be better suited to the direct measurement of feature sets using either a manual or an automated feature analysis method.

## 2.4 MEASUREMENT ANALYSIS

Techniques by which numerical measurements are extracted from images vary considerably in technological complexity. At the simpler end of the scale are linear measurements performed in the microscope or taken from a photographic image. If the determination is to be made in an LM, the microscope must be provided with a measuring eyepiece and micrometer and a stage micrometer. Measurements made in an LM can provide accurate data, but they are limited by the resolution of the micrometer and the subjectivity of the operator. Also, remember that dimensional distortions and shrinkage are unavoidable consequences of chemical fixation. The same applies to measurements made in an EM, and note this additional caution: uneven surface topography makes specimens viewed in a scanning electron microscope extremely difficult to quantitate.

In lieu of any digital-imaging capability, conventional photographic prints can be scored or measured by combining a variety of digitizing tablets with any of the analytical software packages; measurements are then acquired using computer-based analytical software.

The simplest form of computer-driven measurement analysis requires enhancing the gray level image to facilitate simple counting or tracing of selected objects. Beyond this are a host of programs intended to gather data on the shape and size characteristics of the sample. The extraction of quantitative data from images is often the main goal of the researcher, and what has gone before are attempts to convert the image to a form in which measurements can be made easily and accurately. A partial list of possible measurements is contained in Table 2–1.

Many data extraction operations can be performed using commercial software packages, although usually with human involvement. That is, the researcher may be instructed as follows:

**Table 2–1** Partial List of Measurements Enabled by Image Analysis

| *Category* | *Measurement* | *Comment* |
|---|---|---|
| Size | Area | Calibration required; can measure straight or curved objects |
| | Length | |
| | Perimeter | |
| Shape | Elongation | Length, breadth |
| | Compactness | Object area, area of circle with same diameter |
| Boundary | Boundary descriptors | Useful for discriminating among similar objects |
| | Curvature | |
| Other | Count | |
| | Distance between objects | |

*Source:* Reprinted from C.A. Glasbey and G.W. Horgan, *Image Analysis for the Biological Sciences*. Copyright 1995, John Wiley & Sons Limited. Reproduced with permission.

> To make a manual area measurement, first outline a region of interest using the rectangular, oval, polygonal, or freehand selection tool. Then select the Measure command, which will compute the area, mean gray value, and the minimum and maximum gray value. Other measurements, such as perimeter, can be enabled using the Options command in the Analysis menu. Measure distances by making a straight, freehand, or segmented line selection, and then using the measure command. Use the angle tool to measure angles. The cross-hair tool counts objects, marks them, and records their *x* and *y* coordinates.

Public domain software may also be considered. Before purchasing software, potential buyers should make sure that it will perform all the operations required. Its capabilities are best ascertained by discussion with other users. Several Web sites that may be useful to researchers undertaking image analysis are listed at the end of this chapter.

## 2.5 EXAMPLES

The descriptive material given above is in no way comprehensive, nor can it directly aid researchers with image analysis problems. The following examples are given in order to provide the reader with some guidance as to the practical application of these principles and techniques.

The examples selected certainly do not exhaust the types of analysis possible in any given microscopic study. What they do is illustrate the range of existing possibilities. In all aspects of electronic imaging, there are usually many alternative methods for achieving the same result. It is worth emphasizing that capital investment is no guarantee that the end result will be more readily achieved or more reliable. In fact, often technological ignorance emboldened by expensive equipment creates the illusion of certainty.

### 2.5.1 Color versus Monochrome

Monochrome video images are usually digitized using 8-bit sampling that results in a range of 256 gray levels. Typically, such an image will be 640 × 480 pixels (8.89 in. × 6.69 in. at 72 pixels per inch, equaling about 300 kB of storage space). In comparison, a color image digitized using 8-bit sampling (8 bit red, 8 bit green, and 8 bit blue) ultimately results in the same resolution but requires 24 bits per pixel and 900 kB of storage space per image. Thus, 200 images would require approximately 200 Mb of storage media, a large amount even by today's standards. These 24-bit color images would also require a suitable video card for viewing the available color; image processing and printing needs would be equally demanding and expensive. The increase in size and complexity may be justified when analyzing certain color-related structures. However, if the structures of interest were, for example, orange stained on a light blue background, then image analysis requirements would be better achieved using smaller monochrome images acquired with the selective use of inexpensive color filters. For example, an orange stain may be easily discriminated using a blue filter to eliminate the blue counterstain (see Figure 2–3).

Using this simple and inexpensive alternative, the microscopist has eliminated the need for an expensive color camera and color digitizing device, has achieved a higher level of resolution, needs less computer storage and display space, and, because of smaller image sizes, has achieved faster image display, processing, and analysis speeds.

It should be noted that "false" color can be added to monochrome images to highlight regions of interest or to differentiate among parts of an object. The most common method of assigning color to intensity images is by thresholding. Various gray intensity ranges are assigned different colors, with the color in each range usually ramped from dark to light to reflect intensity.

### 2.5.2 Two-Dimensional Planimetry

Figure 2–4 shows a sequence of image-processing steps leading to a simple planimetric measurement taken from a light microscopic, gray level image of a cross section of a green bean. The digital image is a 640 × 480 monochrome image at 8-bit sampling and 256 gray levels. The image was processed by first averaging 16 video frames and subtracting an image of a blank field captured from a site immediately adjacent to the field of interest (A). A smoothing filter was applied to reduce background noise, and then the contrast and brightness values were adjusted for greater visibility of the structure of interest—in this case, the area and perimeter of cytoplasmic parenchymal cells. Once enhanced, the image was then thresholded and binarized (B and C). Unwanted cytoplasmic structures and cell walls were manually erased and traced, respectively, and the area filled and measured to provide an assessment of two-dimensional area and perimeter length (D and E). The calculated areas for individual cells are shown as a numbered gray profile (F). This basic approach is simple, direct, and often the easiest route to a quantitative conclusion based on digital images.

### 2.5.3 Stereology

In the next example (Figure 2–5), a portion of the previous gray level image (Figure 2–4D), thresholded and binarized (A), has been superimposed with a stereologic grid overlay (B). Here the structures of interest are the cell walls. All common intersections of overlaid points and the structure of interest are analyzed by considering their percentage of the total number of pixels (or line length or line area) constituting the original overlay (C). The percentage of counted points relative to the total number of overlaid points represents an accurate volumetric estimate of the area of the structure being evaluated, while the number of intersect points per total measured area provides an estimate of the surface or area of the cell walls in the sample. Classic stereologic techniques such as this can greatly facilitate quantitative assessment of structure.

### 2.5.4 Segmentation Analysis

Figure 2–6 shows an epifluorescence image of a green bean, which has already been presented in cross section (Figure 1–14). The original 24-bit color transparency (A) was scanned using a flatbed transparency scanner. This image was then processed to minimize noise, contrasted and brightened, and separated into its red, green, and blue image planes. The red plane (B) was selected, since it permitted the best discrimination of the fibrous sheath and vascular bundles. The gray level image was then thresholded so that all pixels at or above 210 on the gray scale were assigned a value of 0 (black) and those below 210 were assigned a value of 255 (white) (C). The image was then converted to a binary image (which by definition comprises white or black pixels). The binary image was edited using two erosions, the removal of two black pixel layers from any black structure to eliminate noise in the form of one- or two-pixel specks. This step was followed by two dilations, complementary to erosion, during which two pixel layers were added to any black structure to reconstitute the structures to be analyzed (D). After binary image editing, all remaining structures in excess of 10 pixels were analyzed. In this case, the fibrous sheath, effectively all that remains, was measured for area. This particular structure lends itself to automated analysis, with the possibility for many images to be evaluated and compared using the same thresholding and binary image–editing parameters.

## 2.6 ANALYZING PARTICLES IN FOODS

We live surrounded by particles—from the stellar dust to the dirt under our feet. A particle is a small portion of one phase surrounded by another phase. But what is meant by "small"? So-called "particles" in foods are really small objects and have several origins: fragments in powders or flours, granules, crystals, droplets, inclusions, bubbles, vesicles, and air cells, among others. Note that particles need not be solids; they can be liquid (droplets, aerosols), gas (air cells), or even composite (liposomes). For our purposes, particles

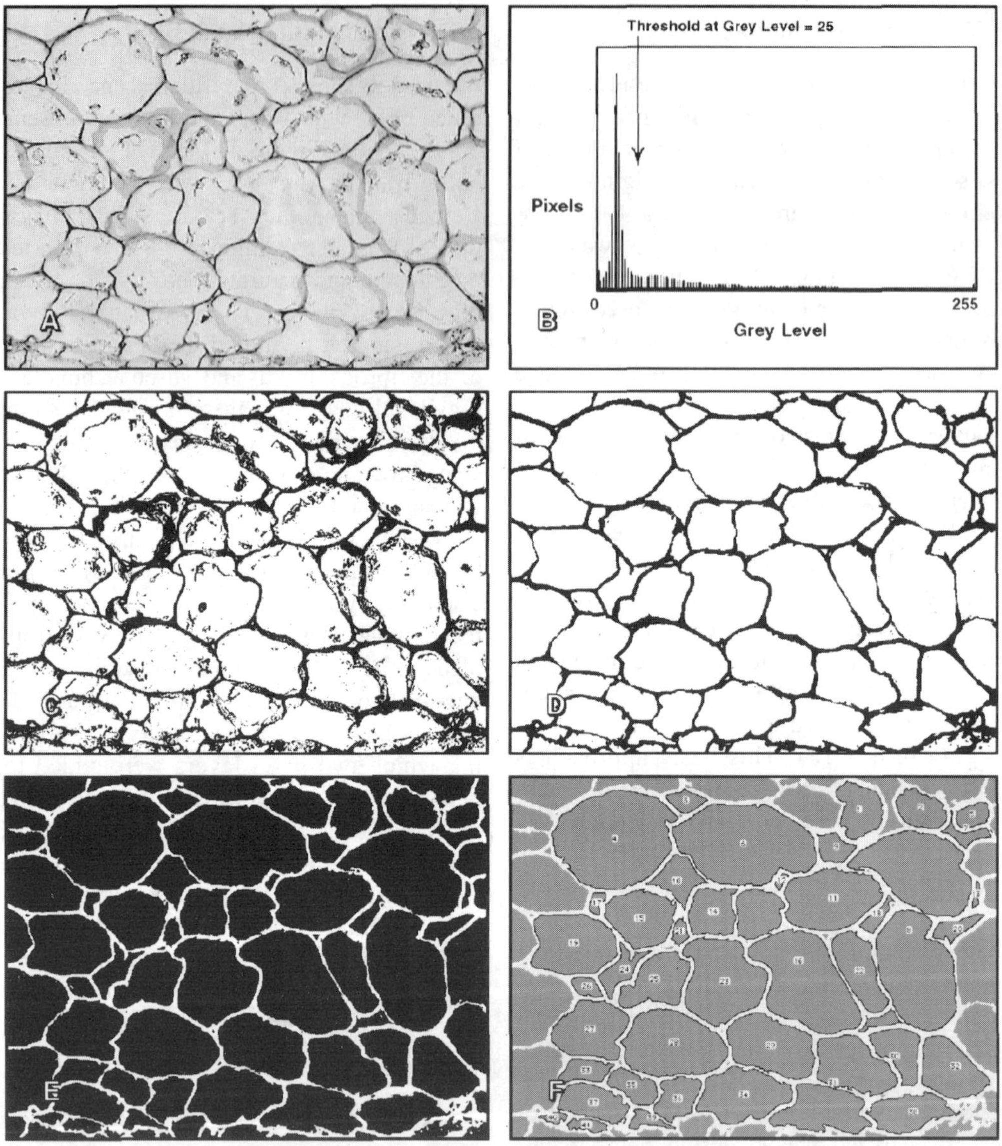


**Figure 2–4** Image analysis of a cross section of a heat-processed green bean using a 200 $\mu$m vibrating microtome section. (A) Enhanced gray level image after real-time averaging of 16 consecutive video frames, subtraction of a similarly captured background image, application of a single median filter, and increasing brightness and contrast. (B) Histogram of gray level distribution for all pixels composing the enhanced gray level image shown in (A). Vertical arrow indicates the thresholding gray level of 25 used to create the binary image. (C) Binary image produced by assigning all gray levels below 25 a value of 0 (white) and all gray levels above 25 a value of 255 (black). (D) Edited binary image. The transitional binary image, C, did not clearly reveal cell walls and cytoplasm as two distinct phases, thus requiring binary image editing. The image was manually edited (although automated routines could have been used) to produce a binary image suitable for analysis. (E) Inverted binary image. (F) Inverted binary image after analysis. All black objects not abutting the edge are measured and counted using automated analyses. *Source:* Reprinted from D.W. Stanley, J.M. Aguilera, K.W. Baker, and R.L. Jackman, Structure/Property Relationships of Foods as Affected by Processing and Storage, in *Chemical, Structural, and Rheological Relationships During Phase/State Transitions in Foods*, M.A. Rao and R. Hartel, eds., pp. 1–56, by courtesy of Marcel Dekker, Inc.

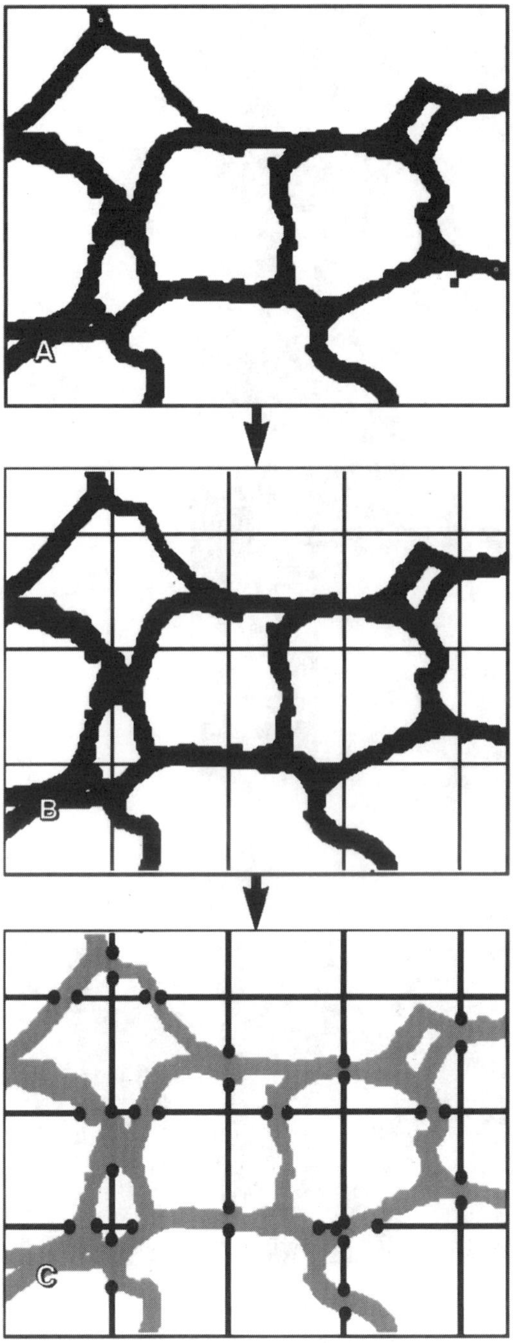


**Figure 2–5** Estimating area of the cell wall structure from a cross section of a heat-processed green bean. (A) A small portion of the binary image shown in Figure 2–4 selected for analysis. (B) Image with superimposition of typical square grid stereologic overlay. (C) Graphic representation showing line segments as black lines and points of intersection as solid black circles overlying the sample area. Line length, as a percentage of the total line length comprising the original overlay, is an estimate of volume; the number of intersected points per total measured area is an estimate of the surface area. *Source:* Reprinted from D.W. Stanley, J.M. Aguilera, K.W. Baker, and R.L. Jackman, Structure/Property Relationships of Foods as Affected by Processing and Storage, in *Chemical, Structural, and Rheological Relationships During Phase/State Transitions in Foods*, M.A. Rao and R. Hartel, eds., pp. 1–56, by courtesy of Marcel Dekker, Inc.

will be considered fragments of matter or specific objects better seen with the aid of a microscope.

### 2.6.1 Particle Size

The first problem is to characterize particle size. This may be unambiguous in the case of a spherical particle (i.e., it is the radius or the diameter) but vague if the particle is irregular. The problem is resolved if the particle size is related to the diameter of a sphere which is in some way equivalent to the particle. Such a sphere is called the *equivalent sphere* and the diameter is called an *equivalent diameter.* Table 2–2 shows some ways of defining particle diameters based on equivalent spheres, circles, or lengths.

It should be borne in mind that the listed equivalent diameters will generally be different for the same irregular particle. Consequently, the selected equivalent diameter has to be relevant to a property of interest. For example, if we wish to study the sedimentation of fat globules in an emulsion, we would select the Stokes' diameter as the descriptor. Alternatively, if we were interested in the covering power of a solid coating, it would be sensible to measure a size based on its projected area. Feret's and Martin's diameters are often determined by microscopy analysis, as they are based on projected area, while volume diameters are determined by instruments such as the Coulter Counter.

It is unusual to have to deal with only one particle. Virtually all problems need a description of the *distribution* of particle sizes. The usual way of summarizing data for a particulate system is to

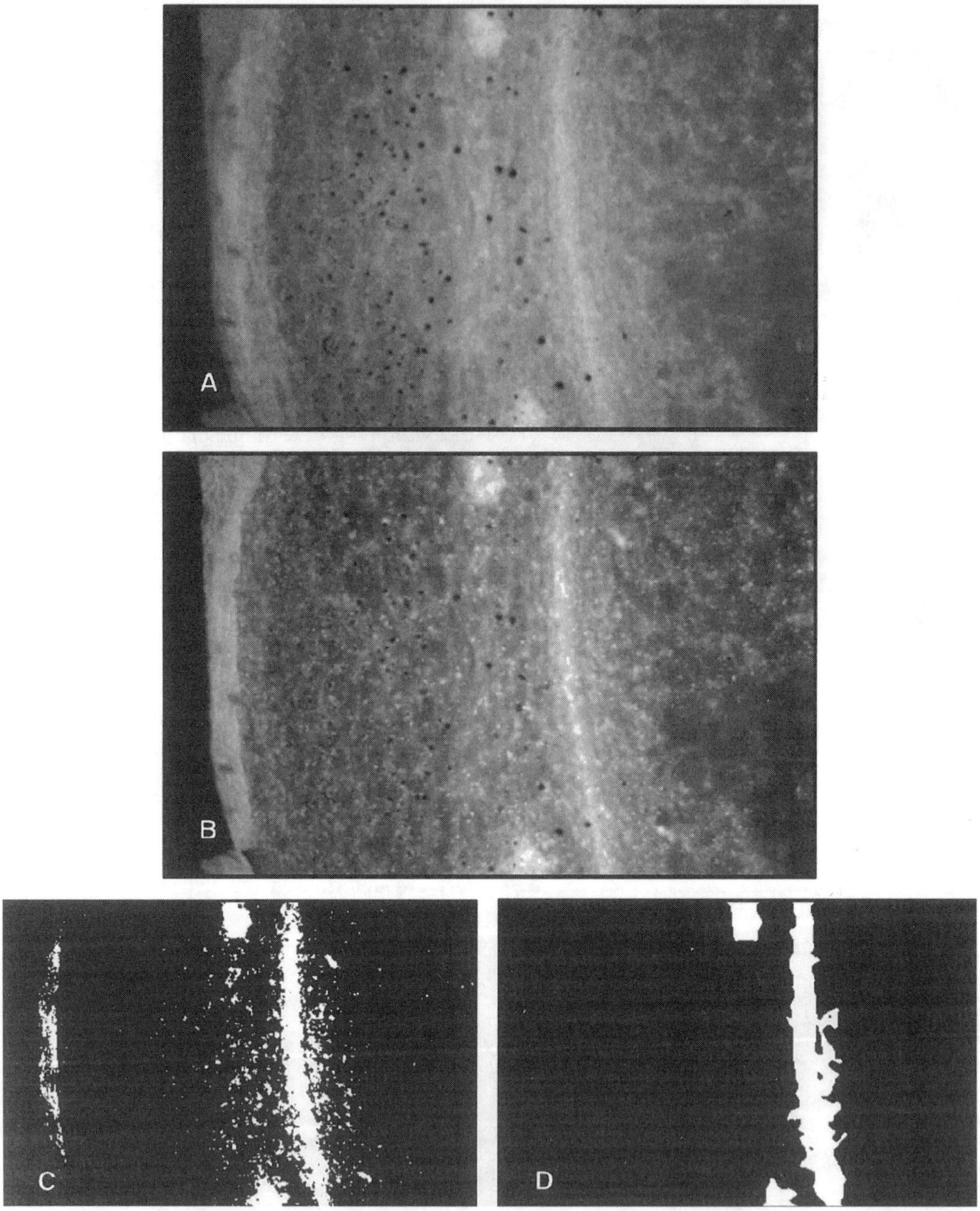


**Figure 2–6** Examining the fibrous sheath structure from a cross section of a heat-processed green bean. (A) Scanned 24-bit color transparency photograph of a full-spectrum epifluorescence image. The RGB color image was separated into its red, green, and blue color planes, and the red plane was selected as revealing the fibrous sheath in most detail. This plane was enhanced and then analyzed. (B) Red image plane separated from (A). (C) Thresholded binary image. (D) Thresholded binary image after editing by alternating erosion and dilation of pixels. *Source:* Reprinted from D.W. Stanley, J.M. Aguilera, K.W. Baker, and R.L. Jackman, Structure/Property Relationships of Foods as Affected by Processing and Storage, in *Chemical, Structural, and Rheological Relationships During Phase/State Transitions in Foods*, M.A. Rao and R. Hartel, eds., pp. 1–56, by courtesy of Marcel Dekker, Inc.

**Table 2–2** Particle Diameter Definitions

| *Symbol* | *Name* | *Definition* | *Formula* |
|---|---|---|---|
| $D_v$ | Volume diameter | Diameter of a sphere having the same volume as the particle | $V = \frac{\pi}{6} D_v^3$ |
| $D_s$ | Surface diameter | Diameter of a sphere having the same surface as the particle | $S = \pi D_S^2$ |
| $D_{sv}$ | Surface-volume diameter | Diameter of a sphere having the same external surface to volume ratio as the particle | $D_{sv} = (D_v^3 / D_S^2)$ |
| $D_d$ | Drag diameter | Diameter of a sphere having the same resistance to motion as the particle in a fluid of the same viscosity and at the same velocity ($D_d$ approaches $D_s$ when *Re* is small) | $F_d = 3\pi\eta D_d v$ |
| $D_f$, $D_{St}$ | Free-falling or Stokes' diameter | Diameter of a sphere having the same free-falling speed as a particle of the same density in a fluid of the same density and viscosity | $D_{St} = \sqrt{\frac{18\eta v}{\Delta\rho g}}$ |
| $D_a$ | Projected area diameter | Diameter of a circle having the same projected area as the particle in stable orientation | |
| $D_p$ | Projected area diameter | Diameter of a circle having the same projected area as the particle in random orientation (mean value of $D_p = D_s$ for convex particles) | |
| $D_c$ | Perimeter diameter | Diameter of a circle having the same perimeter as the projected outline of the particle | $P = \pi D_c$ |
| $D_A$ | Sieve diameter | Width of the minimum square aperture through which the particle will pass | Mesh size |
| $D_F$ | Feret's diameter | The distance between pairs of parallel tangents to the projected outline of the particle in some fixed direction | |
| $D_M$ | Martin's diameter | Length of chord, parallel to some fixed direction, that divides the particle projected outline into two equal areas | |
| $D_R$ | Unrolled diameter | Length of chord through the centroid of the particle outline | |

*Source:* Reprinted from T. Allen, *Particle Size Measurement,* Vol. 1, 5th ed., Table 2.2, © 1997, with kind permission from Kluwer Academic Publishers.

draw a frequency histogram of the *number* of particles in a certain size class. Another widely used method of depicting data is to calculate a cumulative distribution, which shows what percentage of material lies above or below a particular size. Other distributions can be made based on the *mass, volume,* or *surface.* For details about particle distributions and the statistics involved, see Allen (1997).

### 2.6.2 Using Microscopy Techniques

Microscopy techniques have the advantage that the particle contour, shape, state of the dispersion, and even color can be observed. Microscopy has been revolutionized by the introduction of computerized methods of image analysis. Light microscopy is most often used for the examination of particles from about 3 to 150 $\mu$m. Images seen in a microscope are projected areas whose dimensions depend on the particle orientation on the slide. Particles in their stable orientation tend to present the maximum area, and hence sizes measured by microscopy tend to be larger than those measured by other methods. Acceptable statistical diameters are Martin's, Feret's, and the projected area diameter. Automatic image analysis for particle characterization involves six steps: (1) sampling and image formation, (2) image scanning, (3) feature detection, (4) feature analysis, (5) data processing, and (6) data presentation (see Russ, 1995, for details).

### 2.6.3 Shapes and Shape Descriptors

Shapes are not something that human language can describe precisely. Exhibit 2–1 summarizes some of the most widely used shape descriptors. Selection of any one type is based on its relationship to changes observed in the shapes of features, since it is presumably being measured in order to quantify a comparison. For example, roundness can be used to follow shrinkage of cells during drying or to follow particle deformation during caking. Assigning numbers to sizes and shapes is the basis of quantitative microscopy and permits the generation of data for dynamic experiments. An excellent text for gaining a deeper understanding of characterizing and measuring objects, including examples from microscopy is Russ (1995).

An example of the application of microscopy and image analysis (using standard software) to a particulate system is presented in Figure 2–7. Particles of spray-dried coffee undergo flow and deformation when exposed to high relative humidity. Video stereomicroscopy can be used to follow changes in real time, since the process may take place in less than 1 hour at a relative humidity above 75% and ambient temperature. At any time we are interested in geometrical parameters (Feret's diameter, perimeter, and area) and shape descriptors (roundness and compactness). Results of the object analysis for a group of seven particles are shown at the bottom of Figure 2–7. Averages can be plotted as a function of time to derive the kinetics of the process. A similar procedure, including the appropriate microscopy technique, can be used to study and quantify the changes in ice crystals in ice cream, the coalescence of oil droplets in emulsions, the collapse of air cell in foams, among other processes.

**Exhibit 2–1** Representative Shape Descriptors

$$\text{Roundness} = \frac{4 \times \text{area}}{\pi \times \text{max diameter}^2}$$

$$\text{Aspect ratio} = \frac{\text{max diameter}}{\text{min diameter}}$$

$$\text{Curl} = \frac{\text{length}}{\text{fiber length}}$$

$$\text{Convexity} = \frac{\text{convex perimeter}}{\text{perimeter}}$$

$$\text{Compactness} = \frac{\sqrt{(4/\pi)\,\text{area}}}{\text{max diameter}}$$

$$\text{Extent} = \frac{\text{net area}}{\text{bounding rectangle}}$$

## 2.7 CONCLUSIONS

The proliferation of reliable low-cost data-processing equipment has reduced the tedium associated with iterative operations. Coupled with imaginative algorithms, it has led to modern image analysis capabilities suitable for both light and electron microscopy. Image analyzers use computer technology to obtain quantitative data by direct microscope interfacing or by measuring photomicrographs. Data can be collected on shapes, linear dimensions, and densities. Remember that the LM and transmission electron microscope provide only images. These images can now be converted to electronic signals by digitization and then further manipulated by several forms of image enhancement, such as magnifica-

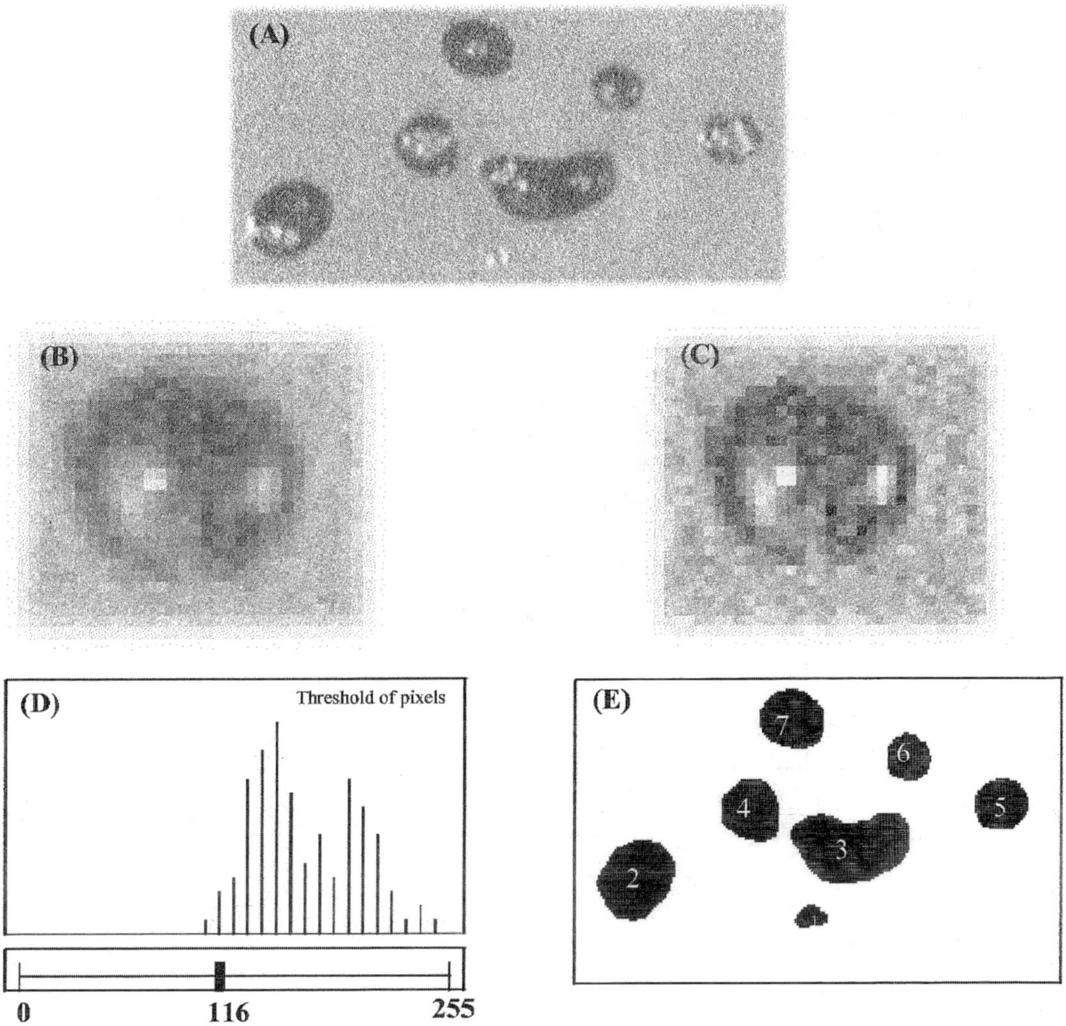


| Object | Feret Diam. | Perimeter | Area | Roundness | Compactness |
|---|---|---|---|---|---|
| 1 | 0.27 | 0.95 | 0.06 | 0.78 | 0.79 |
| 2 | 0.78 | 2.56 | 0.48 | 0.92 | 0.92 |
| 3 | 0.92 | 3.42 | 0.67 | 0.71 | 0.72 |
| 4 | 0.61 | 2.03 | 0.29 | 0.90 | 0.92 |
| 5 | 0.55 | 1.82 | 0.24 | 0.90 | 0.93 |
| 6 | 0.47 | 1.62 | 0.17 | 0.84 | 0.96 |
| 7 | 0.63 | 2.09 | 0.31 | 0.91 | 0.91 |
| Mean | 0.60 | 2.07 | 0.32 | 0.85 | 0.88 |
| Std. Dev. | 0.21 | 0.77 | 0.20 | 0.08 | 0.09 |

**Figure 2–7** Steps of image analysis for coffee particles. (A) Microscopy image. (B) Detail of one particle by zooming 6× to 300 × 300 pixels. (C) Detail of image sharpening using a filter. (D) Definition of threshold. (E) Same image as (A) after binarization. Bottom table: Results after object analysis. *Source:* Saragoni, 1998.

tion, subtraction, or addition of images; intensity differentiation; colorizing; and noise filtering. While these procedures can be extremely useful, especially for repetitive measurement, the data must be viewed critically, for they are no better than the limiting factors, such as specimen artifacts and instrumental restrictions. Certainly, image analysis has its limitations. For example, when we perform the operations needed to prepare an image for measurement, our eyes usually determine which is the "best" algorithm; unfortunately, it is often beyond the capability of our visual system to make this choice, and subjectivity still plays a role in our selection.

Image analysis and image measurement are driving the development of microscopy as a quantitative tool for processing and engineering applications. The combination of image acquisition in real time, image processing, and quantitation of relevant features is the basis of the modern food microscopy and fundamental for the microstructural approach.

Included in this chapter are two other topics: computer vision, an applied area of interest to food scientists that relies on image analysis, and fractal analysis, important to image analysis and other areas of food science.

## 2.8 COMPUTER VISION

For purposes of this discussion, we will treat computer vision as a series of computational steps performed on an industrial macro-image with the goal of extracting features to be used to control the processing operation via a feedback control loop. It is included here because many of the algorithms are similar to those used in image analysis of micro-images, which is an important topic for the food industry.

After the selected features have been captured, they must be recognized, inspected, verified, and measured. A system suitable for a food-processing plant must exhibit robustness of operation. The on-line vision system must work reliably and accurately under real-time constraints dictated by the manufacturing process, and it must save more money than it costs.

The proliferation of computer vision systems in food-processing plants results from the twin goals of maintaining quality and speed. Modern processing lines are concerned with issues of automated grading, quality control of visual appearance (including color and size), and portion control—issues amenable to the application of computer vision. A typical computer vision system is diagrammed in Figure 2–8.

Some recent applications of computer vision to food processing are listed in Exhibit 2–2. Although these systems have proven their worth in high-speed processing lines, numerous problems remain to be solved, including difficulties encountered when the lighting changes or the color of the object is altered. The fact is that, although many single algorithms have been generated to extract one particular type of information, human vision has the ability to integrate many of these algorithms in a split second and in all situations. Computer vision is just beginning to address its current limitations. Further information on computer vision may be obtained from the works listed at the end of this chapter.

**Exhibit 2–2** Food-Processing Applications of Computer Vision

- Beverage can codes (2,400 cans/min) (Connors, Giloi, & Wu, 1996)
- Bread crumb size and color (Locht, Mikkelsen, & Thomsen, 1996)
- Can manufacture (600 cans/min) (Schreiber & Loosely, 1997)
- Foreign object detection (Grimm, 1997)
- Meat cutting (Jacobs, 1996)
- Meat grading system (Jones, Lang, Tong, & Robertson, 1992; Swatland, 1995)
- Sorting pistachio nuts (Pearson, 1996)

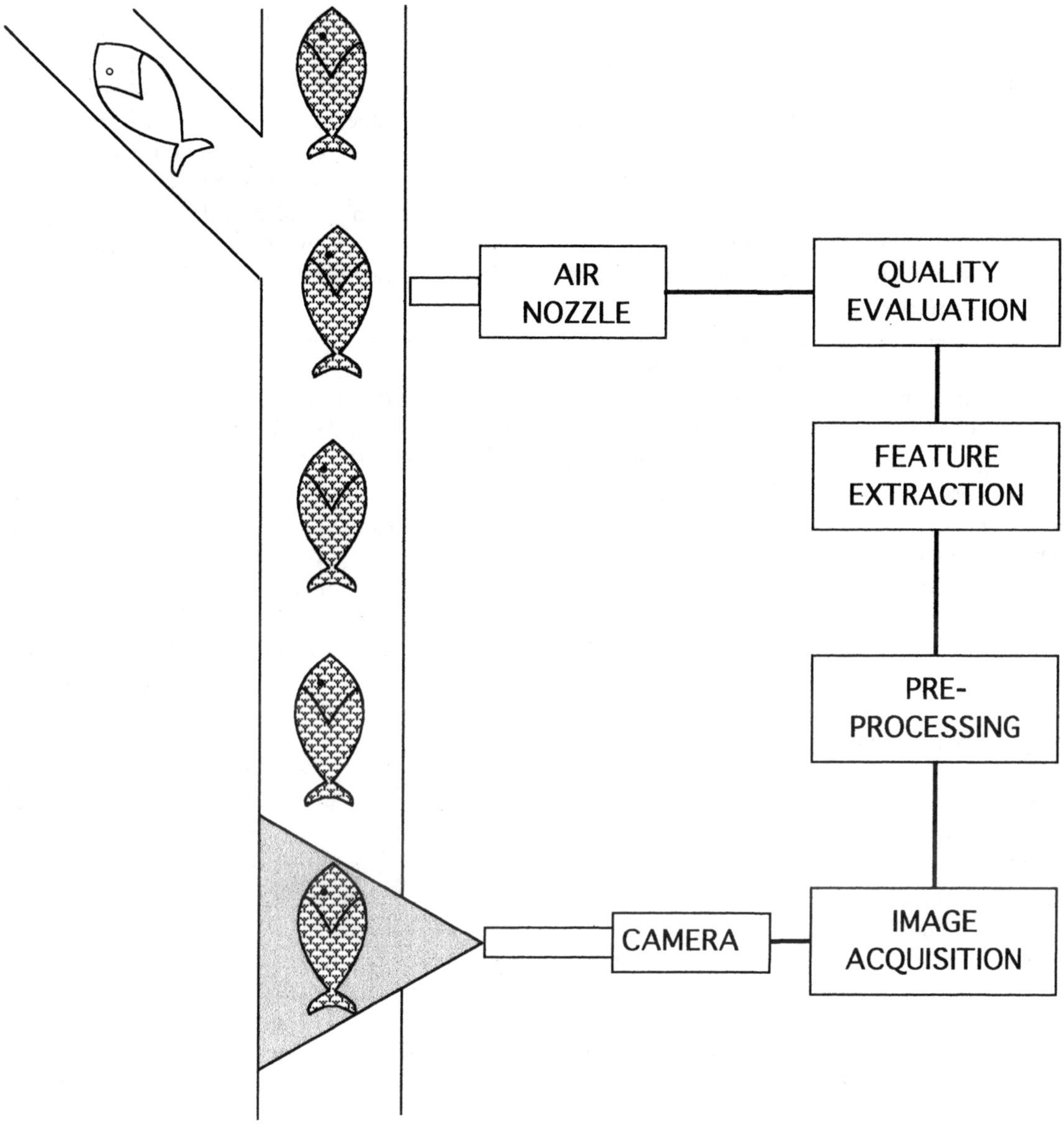


**Figure 2–8** Diagram of a computer vision system for use in food processing. *Source:* Reprinted from *Trends in Food Science Technology*, Vol. 7, S. Gunasekaran, Computer Vision Technology for Food Quality Assurance, pp. 245–256, Copyright 1996, with permission from Elsevier Science.

## 2.9 FRACTAL ANALYSIS

The general concept of fractals has percolated through the scientific community. The reason a brief discussion of fractals is included here is that determining the fractal dimensions of an object using image analysis provides a shape measure that has been shown to be predictive of several parameters important in food processing.

Readers will remember that naturally occurring complex and irregular shapes can be modeled using fractal geometry. A definition of an object's fractal dimension is a scale-free number that describes the roughness of a curve or surface. This number is scale free because the structures that have well-defined fractal dimension are generally "self-similar," since they have virtually identical roughness properties at any magnification. In real life, of course, such as in the examination of a microscopic sample, the range of magnification is limited. A dimension $D$ can be defined for Euclidean objects that is the ratio the log of its perimeter, its area, or its volume (the numbers 1, 2, and 3, respectively) to the log of the length, diameter, or side. Fractal objects, however, are those whose ratios (or $D$s) are not whole numbers but fractions (fractals) thereof. The real application of fractals in image analysis is based on the fact that the apparent fractal dimension (apparent because the object may not be rigorously or mathematically fractal but only apparently fractal) is a measure of the object's irregularity. Thus, as the jaggedness or convolution of a perimeter (where $D$ may vary from 1 to 2) or of a surface (where $D$ may vary from 2 to 3) increases, so does $D$.

It is possible to measure $D$ using image analysis tools. The most common approach to measuring the fractal dimension of an object is to measure the lengths or distances between points on the border of a binary image. The following description applies to a method employed by the NIH Image freeware referenced at the end of this chapter. This package uses the dilation method based on determining $D$ relative to the length of the border of an image. Convolution kernels of different sizes are convolved with the image border. Proper differentiation of edge pixels from background pixels is of great importance in this procedure. The resultant area is divided by the diameter of the kernel, and the log of that result is plotted against the log of the kernel diameter. A fractal object gives a straight line with slope $S$ and $D = 1 - S$. The software replaces each image border pixel with a $3 \times 3$ array of pixels. This operation is continued with successive passes over the cumulative image up to some final pass. After analysis of these data, a plot of log length versus log length is produced.

The characteristics most influencing the magnitude of $D$ are the profuseness of branching and the ruggedness or roughness of the border (increases in either lead to a larger $D$). Since many food materials are complex and irregular, their description through fractal geometry allows a quantification of these characteristics, although they are not differentiated.

Some applications of fractal analysis to foods are described by Peleg (1993) and by Barrett and Peleg (1995). These include relating fractal dimensions to physical properties, such as friability and dispersibility of irregular food particles; to stress-strain relationships and sensory qualities of foods such as extrudates; and to the sorption behavior of food products, including powdered milk. Moreover, fractal analysis may be used to characterize the surfaces of foods. It can also be applied to structures in semisolid systems such as gels and has been used to characterize colloidal aggregate structures and the aggregation mechanism.

An example of the latter application is given here. Fats can be thought of as a liquidlike system in which a continuous fat crystal network is embedded in a liquid matrix. The structural aspects of this system dictate important functional properties, including spreadability, graininess, smoothness, mouthfeel, water binding, and emulsion stability. Obviously, the microstructure of the fat crystal network will play an important role, but until recently it was impossible to obtain satisfactory images of this opaque, soft, temperature-sensitive material, since sample preparation destroyed the fine detail of the network structure. Recent application of confocal laser scanning microscopy has allowed an examination of the fat crystal network *in situ*. These studies (Marangoni & Hartel, 1998) revealed spherulitic particles interacting via a tenuous filamentous network. Thin tendrils extend from a more dense central core and interact with other spherulitic particles, as shown in the image of bovine milk fat presented in Figure 1–15. Image analysis was performed on images of these particles using an image analysis tool similar to that described above, and a fractal dimension of 1.97 was obtained. Much useful information can be derived from knowledge of the fractal dimensions of a colloidal aggregate. For example, fractal dimensions have been theoretically linked with certain mechanisms of aggregation and structure formation.

In studies of this type, it is always useful to employ an independent method of arriving at the fractal dimensions. Marangoni and Hartel used small deformation, controlled stress, and oscillatory measurements in the compression mode to measure a storage modulus for milk fat. The storage modulus in compression mode increased slightly as a function of increasing frequency, typical of a physical gel system with breakable or deformable cross-links. The fractal dimension of the butterfat was determined as indicated in Figure 2–9. This work has shown that triglyceride crystal networks are, in fact, fractal structures resembling colloidal aggregate networks. These fractal networks can profoundly influence the macroscopic properties of food materials. As the authors of this book argue on numerous occasions, food microstructure results primarily from the way in which various elements are joined, and it is their organization that is of most importance.

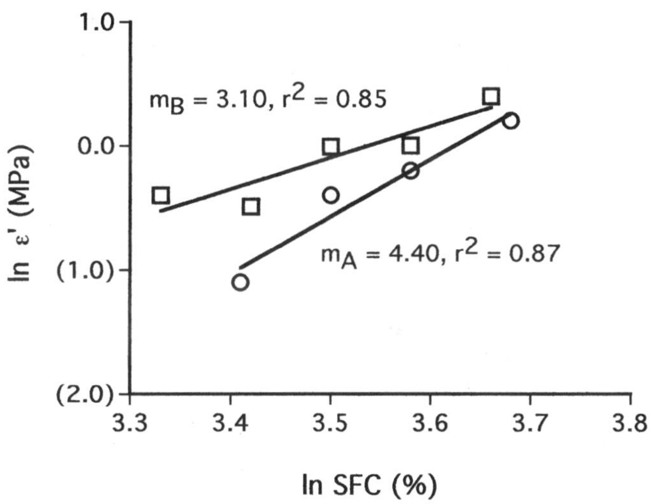


**Figure 2–9** Relationship between compression storage modulus ($\epsilon'$), determined by dynamic mechanical analysis, and solid fat content (SFC), determined by pulsed nuclear magnetic resonance spectroscopy, in two samples (A, B) of milk fat. Cylindrical samples of crystallized fat (10 mm high × 10 mm dia.) were compressed 5%, and a dynamic stress sweep was applied at 1 Hz from 100 to 1,000 Pa. The values for apparent $\epsilon'$ were taken at 1,000 Pa applied stress. $m$ is the slope of the log-log plot and $r^2$ is the coefficient of determination for two replicates. Fractal dimensions ($D$) were obtained from $m$ values using the relationship $m = 1/(3 - D)$, assuming a weak link regime associated with high volume fraction systems, where the primary particles are small and mechanically stronger than the links between them. *Source:* Reprinted with permission from A.G. Marangoni and R.W. Hartel, Visualization and Structural Analysis of Fat Crystal Networks, *Food Technology*, Vol. 52, No. 9, pp. 46–51, © 1998, Institute of Food Technologists.

## BIBLIOGRAPHY

Allen, T. (1997). *Particle size measurement* (5th ed., Vol. 1). London: Chapman & Hall.

Barrett, A.H., & Peleg, M. (1995). Applications of fractal analysis to food structure. *Lebensmittel-Wissenschaft und-Technologie, 28,* 553–563.

Connors, M., Giloi, W.K., & Wu, G. (1996). 2,400 cans-a-minute vision for the beverage industry. *Advanced Imaging, 11*(11), 18.

Glasbey, C.A., & Horgan, G.W. (1995). *Image analysis for the biological sciences.* Chichester, England: Wiley.

Grimm, L. (1997). X-ray imaging screens food for pebbles, glass and fishhooks. *Advanced Imaging, 12*(5), 57.

Gunasekaran, S. (1996). Computer vision technology for food quality assurance. *Trends in Food Science and Technology, 7,* 245–256.

Jacobs, J. (1996). Machine vision for meat processing. *Advanced Imaging, 11*(11), 21–24.

Jones, S.D.M., Lang, D., Tong, A.K.W., & Robertson, W. (1992). A commercial evaluation of video image analysis in the grading of beef carcasses. In *Proceedings of the 38th International Congress on Meat Science and Technology* (pp. 915–918).

Locht, P., Mikkelsen, P., & Thomsen, K. (1996). Advanced color image analysis for the food industry: It's here-now. *Advanced Imaging, 11*(11), 12–16.

Marangoni, A.G., & Hartel, R.W. (1998). Visualization and structural analysis of fat crystal networks. *Food Technology, 52*(9), 46–51.

Pearson, T. (1996). Machine vision system for automated detection of stained pistachio nuts. *Lebensmittel-Wissenschaft und-Technologie, 29,* 203–209.

Peleg, M. (1993). Fractals and foods. *Critical Reviews in Food Science and Nutrition*, *33*(2), 149–165.

Russ, J.C. (1995). *The image processing handbook* (2nd ed.). Boca Raton, FL: CRC Press.

Saragoni, P. (1998). Unpublished data.

Schreiber, M., & Loosely, G. (1997). Implementing distributed vision at H.J. Heinz. *Advanced Imaging, 12*(8), 12–14.

Stanley, D.W., Aguilera, J.M., Baker, K.W., & Jackman, R.L. (1998). Structure/property relationships of foods as affected by processing and storage. In M.A. Rao & R. Hartel (Eds.), *Chemical, structural, and rheological changes during phase/state transitions in foods* (pp. 1–56). New York: Marcel Dekker.

Swatland, H.J. (1995). *On-line evaluation of meat.* Lancaster, PA: Technomic Publishing.

## SUGGESTED READING

### Image Analysis

Aguilera, J.M., & Lillford, P. (1996). Microstructural and imaging analyses as related to food engineering. In P. Fito, E. Ortega, & G. Barbosa-Canovas (Eds.), *Food engineering 2000* (pp. 23–38). London: Chapman & Hall.

Anonymous (1995). A guide to selecting electronic cameras for light microscope-based imaging. *American Laboratory 27*(6), 25–40.

Ayers, J., & Fletcher, G. (1990). Color, motion analysis and biological imaging. *Advanced Imaging, 5*(11), 49–42.

Gonzalez, R.C., & Woods, R.E. (1992). *Digital image processing.* Reading, MA: Addison-Wesley.

Hansma, H.G., Kim, K.J., Laney, D.E., Garcia, R.A., Argaman, M., Allen, M.J., & Parsons, S.M. (1997). Properties of biomolecules measured from atomic force microscope images: A review. *Journal of Structural Biology, 119,* 99–108.

Inoué, S. (1986). *Video microscopy.* New York: Plenum Press.

Ladic, L.A., & Buchan, A.M.J. (1997). Internet graphics software for the processing analysis and display of digital microscopy data. *Proceedings of the Royal Microscopy Society, 32,* 171–174.

Lennard, P. (1990). Image analysis for all. *Nature, 347*(6), 103–104.

Miller, W. (1996). Video microscopy now: The eight greatest misconceptions. *Advanced Imaging, 11*(7), 47–49.

Rasbund, W.S., & Bright, D.S. (1995). NIH image: A public domain image processing program for the Macintosh. *Microbeam Analysis Society Journal, 4,* 137–149.

Russ, J.C. (1990). *Computer-assisted microscopy: The measurement and analysis of images.* New York: Plenum Press.

Russ, J.C. (1992). *The image processing handbook.* Boca Raton, FL: Chemical Rubber Co. Press.

Russ, J.C. (1995). Computer-assisted manual stereology. *Journal of Computer-Assisted Microscopy, 7*(1), 35–46.

Williams, M.A. (1977). Stereological techniques. In A.M. Glauert (Ed.), *Quantitative methods in biology* (pp. 5–80). New York: North-Holland.

### Computer Vision

Gunasekaran, S. (1996). Computerized automation/controls in dairy processing. In G.S. Mittal (Ed.), *Computerized control systems in the food industry* (pp. 407–449). New York: Marcel Dekker.

Gunasekaran, S., & Ding, K. (1994). Using computer vision for food quality evaluation. *Food Technology, 48*(6), 151–154.

Litchfield, J.B., Reid, J.F., & Schmidt, S.J. (1994). Machine vision microscopy and magnetic resonance microscopy. *Food Technology, 48*(6), 163–166.

Swatland, H.J. (1998). *Computer operation for microscope photometry.* Boca Raton, FL: CRC Press.

**Fractal Analysis**

Barrett, A.H., Normand, M.D., Peleg, M., & Ross, E.W. (1992). Characterization of the jagged stress-strain relationships of puffed extrudates using the fast Fourier transform. *Journal of Food Science, 57,* 227–232.

Barrett, A.H., Rosenberg, S., & Ross, W. (1994). Fracture intensity distributions during compression of puffed corn meal extrudates: Method for quantifying fracturability. *Journal of Food Science, 59,* 617–620.

Hagiwara, T., Kumagai, H., Matsunaga, T., & Nakamura, K. (1997). Analysis of aggregate structure in food protein gels with the concept of fractal. *Bioscience, Biotechnology, and Biochemistry, 61,* 1663–1667.

Mandelbrot, B.B. (1983). *The fractal geometry of nature.* New York: W.H. Freeman.

Marangoni, A.G., & Rousseau, D. (1996). Is plastic fat rheology is governed by the fractal nature of the fat crystal network: Interesterification decreases the fractal dimension of butterfat canola oil blends. *Journal of the American Oil Chemists Society, 73,* 991–994.

Peleg, M.A. (1994). A mathematical model of crunchiness and crispness loss in breakfast cereals. *Journal of Texture Studies, 25,* 403–410.

Suzuki, T., & Yano, T. (1990). Fractal structure analysis of some food materials. *Agricultural and Biological Chemistry, 54,* 3131–3135.

Suzuki, T., & Yano, T. (1991). Fractal surface structure of food materials recognized by different molecules. *Agricultural and Biological Chemistry, 55,* 967–971.

Vreeker, R., Hoekstra, L.L., den Boer, D.C., & Agteroff, W.G.M. (1992). Fractal aggregation of whey proteins. *Food Hydrocolloids, 6,* 423–435.

Vreeker, R., Hoekstra, L.L., den Boer, D.C., & Agteroff, W.G.M. (1992). The fractal nature of fat crystal networks. *Colloids and Surfactants, 65,* 185–189.

**Web Sites**

These sites are places to start. Some may be outdated or no longer exist.

*Image analysis.* Image analysis software: Amerinex ⟨www.aai.com⟩, Optimas ⟨www.optimas.com⟩, NIH ⟨ftp://zippy.nih.gov/pub/nih-image⟩, NIH Home Page ⟨rsb.info.nih.gov/nih-image/⟩, bulletin board ⟨listproc@soils.umn.edu⟩.

Image processing software: Research Systems ⟨www.rsinc.com⟩.

Imaging resources on the Web: ⟨www.precisionimages.com⟩, ⟨www.athenet.net/~jlindsay/imaging.html⟩.

Analysis methods: ⟨www.csu.edu.au/ci/vol3/complxb5/node5.html⟩, ⟨www.chemie.uni-marburg.de/~becker/image.html⟩, ⟨vims.ncsu.edu/Home/home.html⟩.

*Computer vision.* ⟨www.cogs.susx.ac.uk/users/davidy/teachvision⟩, ⟨www.arachnid.cm.cf.ac.uk/dave/vision_index.html⟩, ⟨www.cs.cmu.edu/afs/cs/project/cil/ftp/html/v-hardware.html⟩, ⟨www.cs.cmu.edu/afs/cs/project/cil/ftp/html/vision.html⟩, ⟨www.gel.ulaval.ca/~vision/⟩.

*Fractals.* ⟨www.edv.agrar.tu-muenchen.de/dvs/idolon/idolonhtml/idolon.html⟩, ⟨www.ncsa.uiuc.edu/Edu/Fractal/Fractal_Home.html⟩, ⟨www.vis.colostate.edu/~user1209/fractals/⟩, ⟨www.rialto.k12.ca.us/school/frisbie/fractals.html⟩, ⟨library.advanced.org/3288/links.html⟩.

# CHAPTER 3

# Fundamentals of Structuring: Polymer, Colloid, and Materials Science

## 3.1 INTRODUCTION

In discussing microstructural aspects of foods and their relationship to processing and fabrication, it is essential to have some basic understanding of the nature and properties of food materials. For the first time in history, the technological expertise exists to start with a human need and develop a material to satisfy it, literally, molecule by molecule. *Designer* or *engineered foods, nutraceuticals,* and *functional foods* are synonyms that reflect the food industry's new goal of fulfilling the rapidly changing needs of consumers by developing new products with tailor-made properties. Surprisingly, "engineered" foods are not only in high demand in industrialized countries but also in less developed areas, where nutritional needs must be satisfied with appealing but low-cost foods.

Foods are largely composed of polymers and water. The available polymers are a few simple macromolecules (proteins, polysaccharides, and lipids) made up of even simpler repeating units. Of capital importance is that they play nutritional roles as well. Water in food exists predominantly as an aqueous solution of small solutes (sugars and ions). Some very complex food structures are formed, not as a result of the abundance of elemental components, but as a result of the multiple interactions that proteins and polysaccharides can display under different conditions in an aqueous medium and structures derived thereof. This is what this chapter is all about.

As structures are formed, the level of complexity increases. If nutrients were consumed as dilute solutions, life would be simple for the food technologist. Figure 3–1 presents a schematic preview to this chapter and a simplistic attempt to dissect the structuring process of foods. At the lowest dimensional scale are macromolecules. Nowadays the primary structures of some proteins are well understood, and protein engineering can redesign almost any molecule to perform a desired task. Less is known about the intermolecular interactions between these biopolymers. Foremost among these interactions are those of food polymers—proteins and polysaccharides—in the aqueous milieu.

The microstructure of processed foods is the result of specific and nonspecific interactions spanning the molecular level to the supramolecular level. At the molecular level, specific interactions between molecules predominate. Specific interactions between distinct atoms or residues result in covalent bonds, hydrogen bonding, ion bridging, enzyme-substrate coupling, and so on. Hydrophobic interactions at the molecular level may also be considered as localized interactions.

At the macromolecular level (sometimes called the *ultrastructural* level when things are viewed down the dimensional scale), nonspecific interactions predominate. Structure formation at the colloidal level is driven by the presence of interfaces, hydrophilic-hydrophobic balance, net charge on surfaces, and so on. Interactions of macromolecular aggregates of colloidal dimensions result in

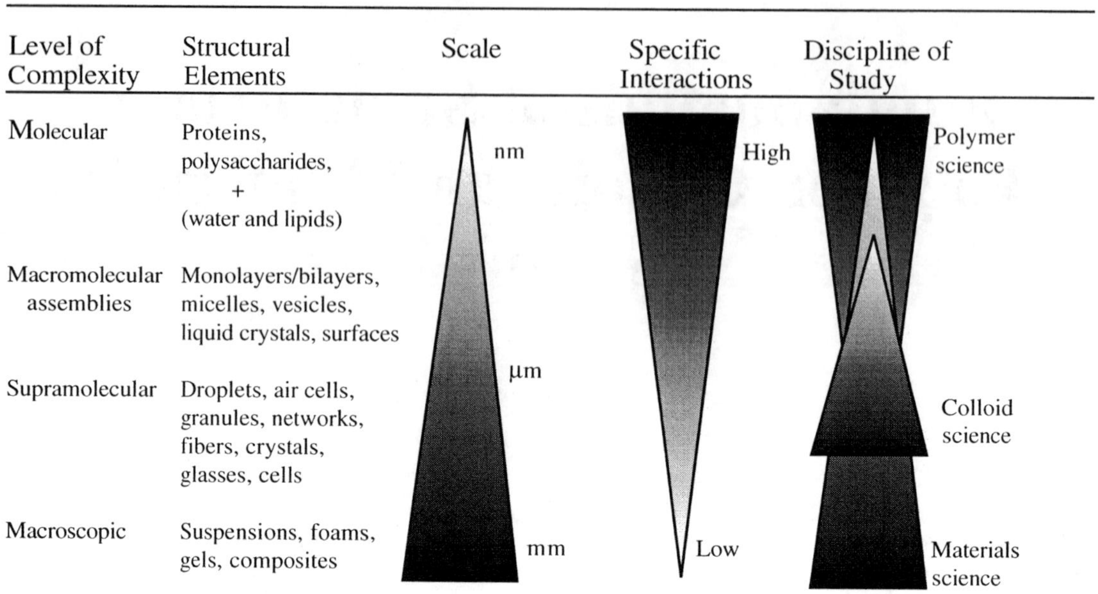


**Figure 3–1** Structural elements and the level of complexity of common food structures. Polymer, colloid, and materials sciences play fundamental roles in the understanding of food structuring.

formation of three-dimensional structural elements: networks in particulate gels and stabilizing interfaces in foams and emulsions, among others. Finally, macroscopic structures perform as food materials, and their properties are characterized (e.g., in rheological and mechanical terms) by materials science.

Most structural elements present in foods at the supramolecular (or microstructural) level are thermodynamically metastable and at nonequilibrium (e.g., amorphous phases), where the nature and kinetics of interactions between them are largely unknown and uncontrolled. Knowledge of the thermodynamics of simple mixtures provides a reference point to assess the potential behavior of the extremely complex multicomponent system that is a real food and the effect on it of variables such as temperature, pH, ionic strength, concentration, and so on (Tolstoguzov, 1996).

An understanding of polymer science principles is essential for following the evolution of food materials science, which started in the 1980s. The basic premise of this science is that since most foods are formed by polymers, they must comply with the principles and theories that apply to synthetic polymers. It tries to interpret physical and chemical phenomena in food systems through concepts such as thermodynamic incompatibility of polymer solutions, the glass transition, state diagrams, polymer rheology, and so on. Central to this approach is the notion that foods are generally metastable or kinetically constrained systems (Slade & Levine, 1991).

Materials science is a well-developed discipline that, building on chemistry and physics, covers such subjects as internal properties of materials, phase transformations and phase equilibria, strength and fracture of materials, and surface and transport properties. Other disciplines that focus on biological materials, such as botany, biology, and medicine, make use of basic principles and applied concepts of materials science. Microscopy has been extensively used in materials science as a tool to study metals, alloys, polymers, and ceramics, since many of their mechanical, thermal, optical, and electrical properties depend on their microstructure. Foods are also complex materials whose desirable characteristics and properties are frequently dependent on their microstructure or the spatial arrangement of structural elements at the micron and submicron level. However, the study of foods from the materials science perspective is yet to be fully developed.

This chapter is an attempt to present some of the scientific basis of structure development and of the physical characterization of foods. It deals mostly with polymers, polymer physics, and underlying physicochemical principles. Because these topics are difficult and somewhat dry, an effort has been made, whenever possible, to provide examples and data from food technology. In any case, the reader should try to understand these topics, since they are the basis of modern food materials science. This chapter's emphasis on relationships and equations is doubly justified. First, the equations show which are the relevant variables and parameters and indicate the importance of their effects (linear, exponential, directly or inversely proportional, etc.). Second, the basic relationships are applied to experimental numerical data obtained from real foods in order to model their physical or mechanical behavior. Key references for this chapter are the books by Sperling (1992) and Young and Lovell (1991) on polymers; the books by Adamson (1990) and Hiemenz (1986) on colloid science and the chapter by Walstra (1966); the book by Roos (1995) and the seminal paper by Slade and Levine (1991) on food polymer science and phase transitions; and the book by Vincent (1990) on the structural properties of biomaterials.

## 3.2 BASIC POLYMER SCIENCE CONCEPTS

### 3.2.1 Nomenclature

Some basic terminology is needed to understand topics presented later. The word *polymer* refers to a substance formed by molecules in large sequences of one or more species *(monomers)* linked together by covalent bonds. A macromolecule with only a few monomers is called an *oligomer*. The process that converts a monomer into a polymer is called *polymerization*. *Functionality* is the number of links that a monomer can exhibit during polymerization. There are two types of basic polymerization reactions: *condensation* and *addition*. Proteins are formed by condensation of two amino acids with the elimination of water, while ethylene is polymerized by addition through breakage of the double bond. A *homopolymer* such as starch consists of only one type of monomer (D-glucose), while a *copolymer* is formed by two or more monomers (glycoproteins and some algae polysaccharides are copolymers). Several types of copolymers are possible: *block*, *graft* (branched), *alternating*, and *statistical* copolymers. Polymers may be present in several types of molecular architecture (*linear*, *branched*, and *network*), which is relevant in phenomena such as crystallization and in the rheology of suspensions.

The term *configuration* refers to the permanent structure of a polymer. A change in configuration requires the rupture of covalent bonds involving large dissociation energies (300–500 kJ/mol). For example, different configurations exist for polymers based on the succession of monomers along the chain (*isotactic*, *syndiotactic*, and *atactic*). *Tacticity* is important in that an atactic polymer (a polymer with an irregular configuration) will never crystallize but rather will solidify into an amorphous mass. A change in shape of a molecule through torsion of a single $\sigma$ bond is called *conformation* and involves only small energy barriers (about 10 kJ/mol).

It is commonly assumed that the conformations of amorphous chains in space are random coils. Changes in conformation are responsible for the sudden extension of a rubber polymer on loading and for protein denaturation. Sometimes long chain molecules form entanglements that limit mobility and may raise viscosity or even produce gels. Melting results mainly in rupture of secondary bonds (e.g., van der Waals' forces and H-bonds) while degradation (e.g., hydrolysis) of polymers involves breaking of covalent bonds.

It is common to classify polymer materials into *thermoplastics, thermosets,* and *elastomers.* Thermoplastics are linear or branched polymer molecules that soften and flow (melt) on application of heat or pressure; they are only semi-crystalline and possess interspersed crystalline and amorphous regions. Elastomers are cross-linked rubbery polymers. Thermosets are irreversibly cross-linked polymers that do not melt but rather degrade upon application of heat.

*Biopolymers* have existed in nature since life began. They are abundant components of the Earth's biomass and play crucial roles in the lives of plants and animals, providing structure, transfer of information, and energy storage capabilities. Nature makes three basic types of biopolymers: *polypeptides* (proteins), *polysaccharides,* and *polynucleotides.* Miscellaneous biopolymers (e.g., rubber, lignin, and chitin) and copolymers (glycoproteins) also exist. From the earliest times, man has exploited biopolymers as materials for food, clothing, shelter, tools, and weapons.

### 3.2.2 Size and Shape of Polymers

The enormous size of polymers imparts some of their unique properties. These properties are more directly related to the conformations and shapes that the polymers adopt than to their primary structure. Polymerization results in polymers with different molecular sizes. Molar mass *(M)* is the mass of 1 mol of polymer (g $mol^{-1}$), and it is often called the molecular weight ($M_W$). The molar mass of a homopolymer is related to the degree of polymerization DP (or $x$) and to the molar mass of the monomer ($M_0$) as follows:

$$M = xM_0 \qquad \textbf{Equation 3–1}$$

Molecules in a pure polymer material (e.g., amylopectin or a protein hydrolysate) exist in different sizes, which are often expressed by a range of molecular weights (in daltons, Da) or by DP. Many times it is the distribution of molar masses that is the relevant parameter, and in polymers it may be determined by size exclusion chromatography (SEC). Molar mass averages are used to characterize molar mass distributions. The average molar mass based on number is defined as

$$\bar{M}_n = \frac{\sum_i N_i M_i}{\sum_i N_i} = \sum_i n_i M_i \qquad \textbf{Equation 3–2}$$

where $N_i$ is the number of molecules of molar mass $M_i$, and $n_i$ is the fraction by number of units. The average molar mass as based on weight is given by this expression:

$$\bar{M}_w = \frac{\sum_i N_i M_i^2}{\sum_i N_i M_i} = \sum_i w_i M_i \qquad \textbf{Equation 3–3}$$

where $w_i$ is the mass fraction of molecules of molar mass $M_i$. A full description of molecular weights requires information on their statistical distribution. The polydispersity index (*PDI*), a convenient measure of the molar mass distribution, is defined as

$$PDI = \frac{\bar{M}_w}{\bar{M}_n} \qquad \textbf{Equation 3–4}$$

The closer *PDI* is to unity, the more monodisperse or uniform is the sample. *PDI* may be an adequate parameter to follow changes in molecular size during acid or enzymatic hydrolysis of proteins (e.g., soy or fish hydrolysates).

The shape of a polymer molecule influences many of its properties. Shape is to a large extent determined by the effect of chemical structure on chain stiffness. A polymer molecule in solution, the molten state, and probably also in the amorphous glassy state, exhibits the shape of a random coil (also called a Gaussian chain). As the stiffness of the backbone increases, molecules may adopt wormlike and eventually rodlike shapes. The size of random coil chains depends on the solvent: a good solvent expands the coil while a poor one causes shrinkage. In between is a type of solvent called $\theta$-solvent, such that inter- and intramolecular interactions are similar in magnitude (see Section 3.4.2). The size of polymers in solution is expressed by the end-to-end distance *(r)* or distance between the chain ends, and the radius of gyration *(s)*. The following expression for $r$ is valid for polymers dissolved in $\theta$-solvents (represented by subscript o):

$$r_o^2 = Cnl^2 \qquad \textbf{Equation 3–5}$$

where $C$ is a constant that depends on the nature of the polymer and $n$ and $l$ are the number and length of bonds, respectively. Linear, branched, and cyclic macromolecules are sometimes characterized by the radius of gyration $s$, defined as the root-mean-square distance of the collection of atoms from their common center of gravity:

$$s^2 = \frac{\sum_{i=1}^{n} m_i \bar{r}_i^2}{\sum_{i=1}^{n} m_i} \qquad \textbf{Equation 3–6}$$

where $m_i$ is the mass of each atom and $\bar{r}_i$ is the vector from the center of gravity to atom $i$. For a random coil, the following relationship is often used:

$$s^2 = \frac{r^2}{6} \qquad \textbf{Equation 3–7}$$

where $r$ is the end-to-end distance. For polymer coils in solution, the following relation exists between the radius of gyration and the molecular weight (Sperling, 1992):

$$s = K\overline{M}_w^{1/2} \qquad \textbf{Equation 3–8}$$

Typical values for $s$ are 5 nm for globular proteins, 25 nm for flexible polysaccharides, and >100 nm for stiff (wormlike or rod) proteins and polysaccharides (Ross-Murphy, 1995).

## 3.3 FOOD POLYMERS

### 3.3.1 Comment

Any basic book on synthetic polymers has an introductory section describing the most important macromolecules. Proteins and polysaccharides are biopolymers that play important nutritional and functional roles in foods. It is unfortunate that treatments of proteins and polysaccharides as biopolymers are seldom found together (a drawback of specialization). The objective of this section is to describe briefly the main types of macromolecules that constitute the family of food polymers. The reader is urged to look at specialized texts for detailed information on the chemical aspects and properties of individual food polymers and biopolymers, such as Fennema (1996), Harris (1990), Imeson (1992), and Vincent (1990).

### 3.3.2 Proteins as Structural Polymers

Proteins, which are polymers of amino acids linked by peptide or amide bonds, are formed by condensation during synthesis in the ribosome. Conformation of proteins is largely dictated by the occurrence and position of specific amino acids. The *native conformation*—that assumed after synthesis—has a relatively low free energy. Side residues are free to interact with each other, and the strength of the interactions dictate some of the physicochemical and mechanical properties of proteinaceous materials. The position of an amino acid in the chain and its type partly determine the kind of interaction and the conformation derived thereof. Amino acids are classified as helix-breaker, helix-former, or indifferent, and also as polar/nonpolar, acidic/basic, or neutral (based on their chemical properties). Interactions between biological macromolecules are mainly effected through three types of noncovalent bonds (in parentheses are the approximate energy involved): ionic bonds (12.5 kJ/mole), hydrogen bonds or H-bonds (4 kJ/mole), and van der Waals' attraction forces (0.4 kJ/mole).

Several conformations can be assumed by the polypeptide chain, including antiparallel and parallel $\beta$-sheets, the $\alpha$-helix (as well as other helices), and coils. Certain combinations of $\alpha$-helices and $\beta$-sheets form folded, compacted globular units called *domains*. Protein conformation is discussed later in Section 4.4.1. A good description of the structure, shape, and interactions of macromolecules can be found in Alberts et al. (1989). Some examples follow to illustrate characteristics of protein that perform as structural elements in biological systems or as sources of amino acids in seeds.

#### *Collagen*

Collagen is the most common fibrous protein in the animal kingdom, a structure of importance in meat texture and the basis of the gelatin industry. Collagen almost entirely composes the connective tissue in tendons and muscle, where it functions by transmitting tensile stresses (Figure 5–2). More than half of collagen consists of the amino acids glycine (30%) and proline and hydroxyproline (25%), and a typical amino acid sequence would be:

-Ser-Gly-Pro-Arg-Gly-Leu-Hyp-Gly-Pro-Hyp-Gly-Ala-Hyp-Gly-

In this sequence, glycine occurs at every third residue, enabling the protein chains to approach closely and accommodating the hydrogen atom (side chain of Gly) in a small space via hydrogen bonding. The conformation of the polypeptide chains is that of extended left-handed helices. Three of these helices coil around each other to form a structure called *tropocollagen* in a right-handed triple helix. Five tropocollagen molecules are staggered longitudinally (overlapping about one-quarter of their molecular length) to form a microfibril with a diameter of 3.6 nm. Overlapping permits shear stresses to pass from one molecule to the next. Collagen microfibrils consist of five cross-linked tropocollagen molecules arranged in a longitudinal staggered form 280–300 nm in length and with a binding periodicity of 67 nm. In animal tendon, strength is derived from lateral cross-links between neighboring molecules (Vincent, 1990). The number of cross-links increases with animal age, resulting in a perceived reduced tenderness of meat. During heating, hydrogen bonds maintaining the collagen structure are weakened, and if heating is prolonged (as in stewing), some of the most labile cross-links are also broken. The solubilized and leached collagen causes formation of a "gelatinous" mixture as new H-bonds are formed upon cooling. Further discussion of collagen as part of the hierarchical structure of tendon occurs in the sections "Protein Confirmation and Function" (Chapter 4) and "How Does Nature Form Structures" (Chapter 5).

#### *Keratin*

Keratins are a group of proteins that contain significant amounts of sulfur or tyrosine-based cross-links that stabilize the material. In mammals, keratin occurs in skin, hair, wool, horn, and hoofs, and in avians, it occurs in feathers. Keratin exists as an $\alpha$-helix or extended as $\beta$-sheets when heated (40–60°C).

#### *Elastin, Fibroin, and Sericin*

Previously mentioned proteins play structural roles due to their high modulus. Elastin is the main elastic protein in vertebrates, and it is usually associated with collagen. It occurs as fibers formed by coils with an open helical structure and has water in the core that acts as a plasticizer. An interesting feature of elastin is its rubberyness, and its mechanics can be described by rubber elasticity theory (Vincent, 1990).

As a final example of proteins as structural elements, the liquid silk of the silkworm is a highly viscous aqueous solution of two proteins, fibroin and sericin (Magoshi & Nakamura, 1992).

### 3.3.3 Proteins as Storage Polymers

Proteins are also accumulated as storage deposits or protein bodies in seeds of cereals and legumes (see Figure 3–23). Seed proteins have been classified traditionally according to their solubility in different solvents (Osborne classification). The fraction extracted by water is defined as albumins, the fraction extracted by dilute salts as globulins, the fraction extracted by ethanol as prolamins, and the fraction extracted by acid or base as glutelins. The typical prolamin-type proteins are those of corn (zein), wheat, rye, and barley. Typical globular proteins (globulins) include the 7S proteins present in legume seeds and the 11S proteins in legumes and some cereals. The structure of plant storage proteins and their functionality is reviewed by Fukushima (1991). Milk proteins are treated in Section 7.2.3.

Plant proteins, most notably from soybean, are a main source of refined vegetable proteins for food and feed in the form of flours (50% protein), concentrates (70% protein), and isolates (>90% protein) (see Section 8.4.3). The relevant point is that the potential of these "amorphous" proteins to be structured into foods by extrusion or spinning, gelation, or baking appears to be dictated by the type and amount of certain amino acids, the secondary structures that the chains may adopt, and the reactivity of some residues.

Several chemical properties of vegetable proteins are important in extrusion. Plant storage proteins have glutamic acid and glutamine in large quantities compared to other proteins. Fibrous development in thermoplastic extrusion appears to be favored when the protein contains high enough levels of glutamic acid ($COO^-$) and free-lysine ($NH_3^+$) residues to participate in covalent linkages formed at high temperatures. Residues that promote disulfide linking (e.g., cysteine) further stabilize the structure during cooling (Ledward & Tester, 1994).

The chemical makeup of soybean proteins is essential for forming gels by heating. Gel formation of 11S globulins is favored by the presence of hydrophobic residues exposed at the surface of the protein by partial unfolding during heating. Intermolecular disulfide bonds are also formed through -SH/S-S interchange reactions, with participation of some of the 20 disulfide and 2 sulfhydryl groups present per mole of protein (Fukushima, 1991).

The high molecular weight subunit of glutenin (838 amino acid residues) in wheat contains a central domain of $\beta$-turn structures and extremes rich in $\alpha$-helices. The central domain consists of 670 amino acids characterized by repeats of two sequences (motifs) of 6 and 15 amino acids, mostly glycine, glutamine, proline, and tyrosine (Fukushima, 1991). The $\alpha$-helices and $\beta$-turn structures are responsible for the elasticity of glutenin and its stretchability. In addition, three types of groups and bonds are known to contribute to the viscoelastic properties of dough: amide groups, sulfhydryl groups and disulfide bonds, and hydrogen bonds, which stabilize the dough structure (Blanshard, 1988).

The complete amino acid sequence has been determined for most cereal and legume proteins. When the molecular and macromolecular bases of structure formation become resolved, specific modifications by application of genetic engineering techniques might be able to markedly improve not only the processability of plant proteins but their nutritional properties as well. For example, an increase in the hydrophobicity of the polypeptide chains would improve their emulsifying and oil-holding properties. Similarly, introduction of cystein residues (-SH groups) in the terminal regions of gliadin in wheat gluten would result in enhanced dough functionality.

### 3.3.4 Polysaccharides as Structural Polymers

Unlike proteins, the polysaccharides important to food scientists occur mostly in plants, where they have three major roles: energy reserve (starch), water stabilization (gums), and structure (cellulose). Also unlike proteins, polysaccharides can be linked to form branched molecules. The structure of some food polysaccharides is presented in Figure 3–2.

A major difficulty in discussing polysaccharides is the chemical nomenclature of monomers (sugars). For example, "D-glucose" is $\alpha$-D-glucopyranose. The suffix *-ose* implies that it has a carbonyl group; this carbonyl group reacts with the hydroxyl group at the end of the chain to form a ring structure characterized as *hemiacetal.* When the ring is six-membered, it is referred to as pyranose, and when it is five-membered, furanose. The $\alpha$ anomer is the one having the hydroxyl group in carbon number one (C1) above the plane of the ring (if the OH below the plane is $\beta$). The D indicates that a solution of this sugar will rotate the plane of polarized light to the right. Any good biochemistry text will list the multiple names of members of the family of aldoses (carbonyl group at the end of the chain) and ketoses (carbonyl group in any other position), which later become the monomers of polysaccharides chains. Fortunately, most polysaccharides of interest to us contain one type of residue (e.g., cellulose, starch) or two, arranged either periodically (agar) or in blocks (alginates). The fact that polysaccharides have sugars with free OH groups as monomers has important structural consequences: it creates opportunities for high hydration of individual molecules and/or formation of hydrogen bonds and ionic interactions over portions of the chains.

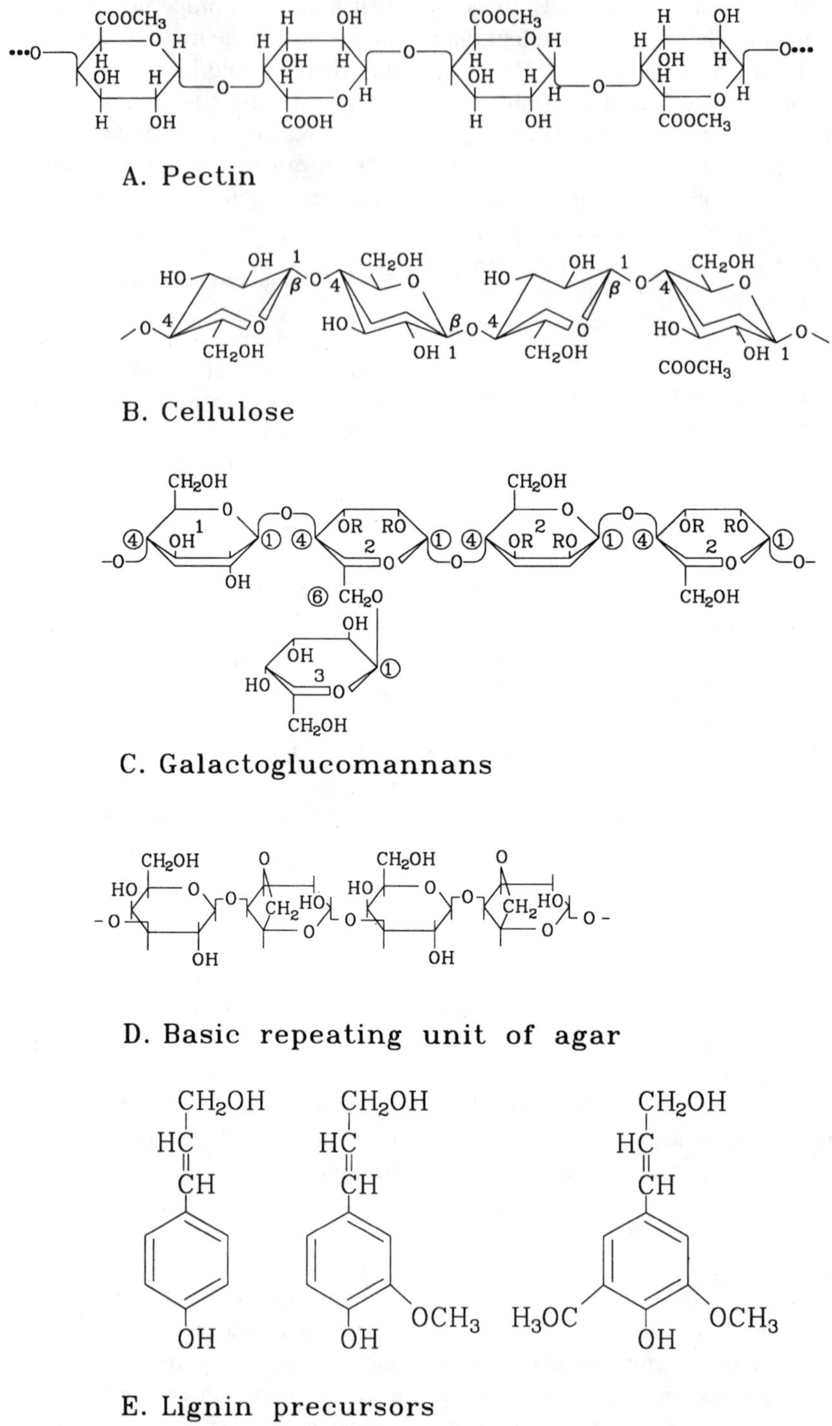


**Figure 3–2** Molecular structures of some polysaccharides present in foods and different forms in which their molecules are depicted in the literature.

***Cellulose***

Cellulose, said to be the most abundant organic polymer on earth, is an essential component of all cell walls of higher plants. It is a linear biopolymer consisting of at least 3,000 $\beta$-linked glucose units tightly held in a flat ribbon maintained by intermolecular hydrogen bonds. Cellulose is probably formed just outside the cell membrane in such a way that it can polymerize into highly H-bonded fibrils (see Section 4.5.2). Cellulose is found in the form of microfibrils several micrometers in length and about 20 nm in diameter. Cellulose molecules tend to form crystalline regions separated by less ordered regions. This ordered structure gives cellulose both its insolubility in almost all solvents and its high mechanical strength. Cellulose is insoluble in water and indigestible by the human body (as a part of nutritional fiber). Water solubility is accomplished by derivation (e.g., formation of cellulose ethers). Some cellulose derivatives of importance in foods are produced by the reaction of alkali cellulose with (1) methyl chloride to form methyl cellulose (MC); (2) propylene oxide, to form hydroxypropylcellulose (HPC); and (3) sodium chloroacetate, to form sodium carboxymethylcellulose (CMC). Microcrystalline cellulose (MCC) is produced by hydrolysis of the amorphous regions, which releases small crystals.

***Hemicelluloses***

Hemicelluloses in plants are closely associated in cell walls with cellulose, from which they can be extracted with alkaline solutions. Three types of hemicelluloses are recognized: the xylans, the mannans and glucomannans, and the galactans and arabinogalactans. Xylans are major components of the seed coats of cereal grains. As for their basic structure, xylans have a linear or occasionally branched backbone of $\beta(1\rightarrow4)$–linked xylopyranose residues, with a few single groups hanging from the chain.

***Pectins***

Pectic substances (pectin) are present in the middle lamella of cell walls. The main constituent of pectic substances is D-galacturonic acid, joined in chains by $\alpha(1\rightarrow4)$ glycosidic linkages. Inserted in the main chain are rhamnose units, which introduce a kink into the otherwise straight chain. The major part of all commercial pectin has the carboxyl groups esterified to various extents with methyl alcohol. The degree of esterification ranges from 60% to 90% and has a bearing on the firmness of plant tissue (a lower degree of esterification results in higher cohesion) and of the commercial gelled products (see Section 5.9.8). A comprehensive treatment of pectins can be found in Walter (1991).

***Seaweed Polysaccharides***

The place of pectins in higher plants is taken in algae by either the alginates (in the brown algae, the *Phaeophyceae*) or the agars and carrageenans (in the red algae, the *Rhodophyceae*). These biopolymers occur in the cell walls and intercellular spaces, providing flexibility and strength. Alginates (alginic acid) are linear polymers of two different monomers: $\beta$-D-mannuronic acid (M) and $\alpha$-L-guluronic acid (G). Both monomers occur together in the same chain, linked in different sequences by $\alpha$ or $\beta(1\rightarrow4)$ glycosidic links as blocks of only M or G or of alternating M and G. Alginates are high molecular weight polymers (DP 100–3,000 and MW 20 to 600 kDa), with flexible ribbonlike sections in M-block regions and buckled and stiff sections in G-block regions (Onsoyen, 1992). Although alginic acid is insoluble, the alkali-metal salts are freely soluble in water.

Agars and carrageenans have a linear galactose backbone composed of galactose (G) and 3,6-anhydro-$\alpha$-L-galactose (AG). The disaccharide agarobiose (G-AG) is the common structural unit. There are three basic types of carrageenan: kappa ($\kappa$), iota ($\iota$), and lambda ($\lambda$). Carrageenans are extracted from red algae: *Chondrus crispus,* also known as Irish Moss ($\kappa$ and $\lambda$ types), *Eucheuma* ($\kappa$ and $\iota$), and *Gargantina* species ($\kappa$ and $\lambda$). They are differentiated from agar and furcellan by the number and position of the ester sulfate groups and the amount of AG. Variations in these components influence gel strength, solubility, synergisms, and melting temperatures. For example, $\kappa$-

carrageenan forms a firm gel in the presence of $K^+$ ions. Carrageenan extracted from seaweed is not assimilated by the human body, providing only bulk but no nutrition.

#### *Plant Gums*

The term *gum* is a generic name for polysaccharides that show great affinity for water and high viscosity in solution without forming gels. These compounds are also termed hydrocolloids. Gum tragacanth is an exudate from the tree *Astralagus gummifer* while guar gum is the storage polysaccharide in the endosperm of seeds of the leguminous shrub *Cyamopsis tetragonoloba.* The essential structural feature of all gums is the extensive branching, which leaves no length of backbone to form junction zones and thus gels but permits intermolecular interactions and water trapping that leads to viscous solutions.

#### *Microbial Polysaccharides*

Several sources of microbial polysaccharides are used in foods. Dextran is a glucan that has contiguous $\alpha(1\rightarrow6)$–linked glucose residues with varying percentages of branched linkages [largely $\alpha(1\rightarrow3)$]. Gellan is the generic name for the extracellular polysaccharide secreted by the bacterium *Pseudomonas elodea.* It is a linear anionic heteropolysaccharide with a molecular weight of around 500 kDa, composed of a tetrasaccharide with repeat units of glucose, glucuronic acid, glucose, and rhamnose. Xanthan gum is produced by the bacterium *Xanthomonas campestris.* The backbone of the molecule is composed of $\beta(1\rightarrow4)$–linked glucose units (like cellulose), with side chains containing two mannose and one glucuronic acid molecules.

### 3.3.5 Polysaccharides as Energy Storage Polymers

Starch is quantitatively the largest nonwater component of human diets. A sizable fraction of the requirements for energy are provided by the starch of cereal grains and tubers. Starch is found in the form of granules of irregular rounded shapes ranging in size from 2 to 100 $\mu$m (see Section 4.5.3). Starch consists of two types of glucose polymer: one shorter and essentially linear, amylose ($10^5$–$10^7$ Da, DP 500–5,000), and a larger branched one, amylopectin ($10^7$–$10^9$ Da). Both are illustrated in Figure 3–3. Most starches contain 20–25% amylose.

Amylose consists of long chains of $\alpha(1\rightarrow4)$ D-anhydroglucose residues, with a few $\alpha(1\rightarrow6)$–linked units (9–20 per molecule). The main linkage produces a natural twist of the amylose molecule in a helical conformation, with six glucose residues per turn so that all the hydrophilic hydroxyl groups are on the external side of the helix. The hydrophobic cavity may accommodate many small molecules, such as the carbon chain of fatty acids (amylose-lipid complexes) and iodine (blue color). Amylose can be processed in a similar way to synthetic polymers. It can be formed into transparent films that are edible, as sponges or thermoplastically extruded. Fabrication and properties of starch plastics are reviewed by van Soest and Vliegenthart (1997).

Amylopectin is a branched polymer and one of nature's largest molecules. It is composed of three types of chains: A-chains, B-chains, and one C-chain. A-chains are unbranched and are attached to the molecule by a single linkage, B-chains are branched and are attached to other chains, and the C-chain contains the sole reducing group. All chains are assembled in a cluster structure within the granule, which shows between 15% and 45% crystallinity in several crystal types (denoted V, A, B, and C) depending on the packing density of single or double helices and the water content. Further details on starch polymers and granule structure can be found in Oates (1997), in Zobel (1992), and in Section 4.5.3.

### 3.3.6 Lignins

Lignins are a different type of natural polymer altogether. They form three-dimensional polymers of phenylpropane units that occur in cell walls of true vascular plants but not in algae or microorganisms. Lignins are linked by several different carbon-to-carbon and ether linkages that are not hydrolyzable by most microorganisms. It seems that in plant cell walls, a considerable part of the lignin molecules are linked to the primary alcohols and carboxyl groups of hemicelluloses

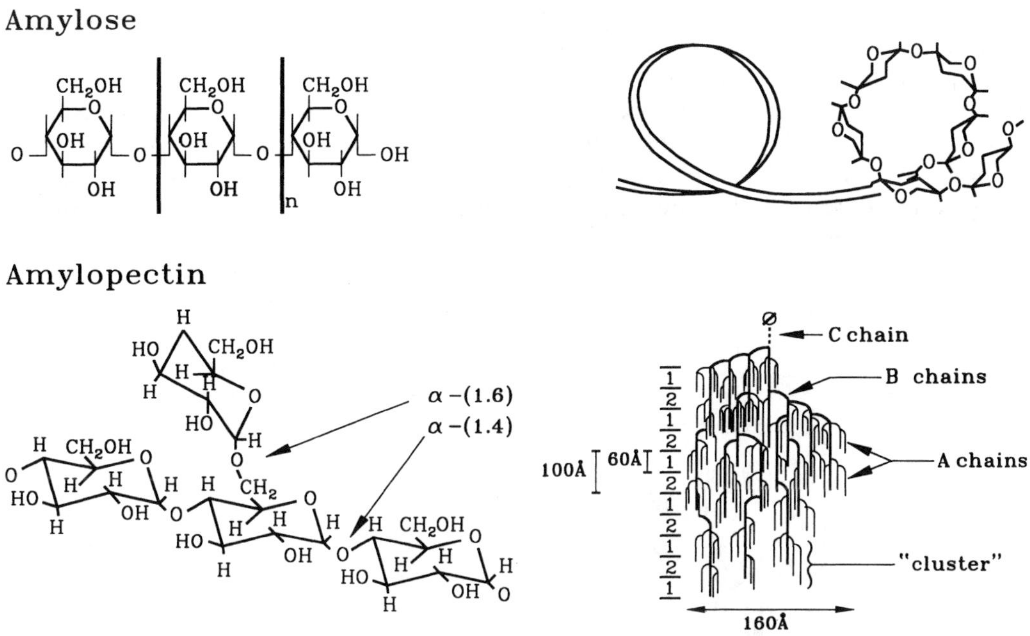


**Figure 3–3** Schemes of amylose and amylopectin in starch. Molecular bonding and macromolecular arrangement ($\phi$ is the reducing end). Note alternate crystalline (1) and amorphous (2) regions formed in amylopectin. *Source:* Reprinted from *TIBTECH*, Vol. 15, J.J.G. van Soest and J.F.G. Vliegenthart. Crystallinity in Starch Plastics: Consequences for Materials Properties, pp. 208–213, Copyright 1997, with permission from Elsevier Science.

and pectins (see Figure 6 in Higuchi, 1990). Lignins are generally distributed with hemicelluloses in the spaces of cellulose fibrils in primary and secondary walls, and in the middle lamellae as a cementing component to connect cells and harden the cell walls of xylem tissue. They also have a role in shielding against water.

## 3.4 POLYMER SOLUTIONS

### 3.4.1 A Little Thermodynamics of Solutions

Once the molecular makeup of polymers is understood, the microstructural engineer is interested in predicting and controlling the interactions that may take place and their consequences. Macromolecules in food are commonly found as mixtures in an aqueous milieu rather than as solventless polymer blends. A solution is any phase having more than one component, and in food technology aqueous solutions are of major importance. Food scientists are interested in what happens when macromolecules (of one type or in mixtures) are combined with water.

Thermodynamics provides valuable information as to the direction in which a system will move, what conditions will be reached at equilibrium, and what would be the effect of variables such as temperature, concentration, pH, ionic force, and so on. The Gibbs free energy $G$ is the key thermodynamic parameter for studying phases at equilibrium (Atkins, 1982). A necessary (but insufficient) condition for a homogeneous solution to be formed after mixing is given by this expression:

$$\Delta G_{mix} = \Delta H_{mix} - T\Delta S_{mix} < 0 \quad \textbf{Equation 3–9}$$

where $\Delta G_{mix}$ or $(G_{mixture} - G_{pure\ components})$ is the free energy of mixing, $\Delta H_{mix}$ is the enthalpy of mixing, $T$ is temperature, and $\Delta S_{mix}$ is the entropy of mixing. Thus, mixing generally involves changes in enthalpy and entropy.

An *ideal solution* is a fictitious model for mixtures of identical molecules in which molecular interactions are the same (or none) and the change in volume after mixing is zero. For an ideal solution of small molecules (e.g., those that follow Raoult's law), $\Delta H_{mix} = 0$ (athermal mixing), so the sign of $\Delta G_{mix}$ depends only on the entropic term. For the so-called *regular solution,* $\Delta H_{mix}$ is finite (e.g., equal to $Bx_1x_2$), and the free energy of mixing takes this form:

$$\frac{\Delta G_{mix}}{N} = \Delta H_{mix} + RT(x_1 \ln x_1 + x_2 \ln x_2)$$

**Equation 3–10**

where $x_1$ and $x_2$ are the molar fractions of solvent and solute, respectively, and $N$ is the total number of moles. Since $\ln x_1$ and $\ln x_2$ are always negative, components 1 and 2 will always mix if they behave as an ideal solution ($\Delta H_{mix} = 0$). The curve of $\Delta G_{mix}$ $N$ versus $x_2$ for ideal solutions is always concave, indicating full miscibility at all compositions (Figure 3–4).

When the enthalpic term is taken into account, as is the case for regular solutions, the graphical representation gets more complicated. If concavity exists, full miscibility is still present in the en-

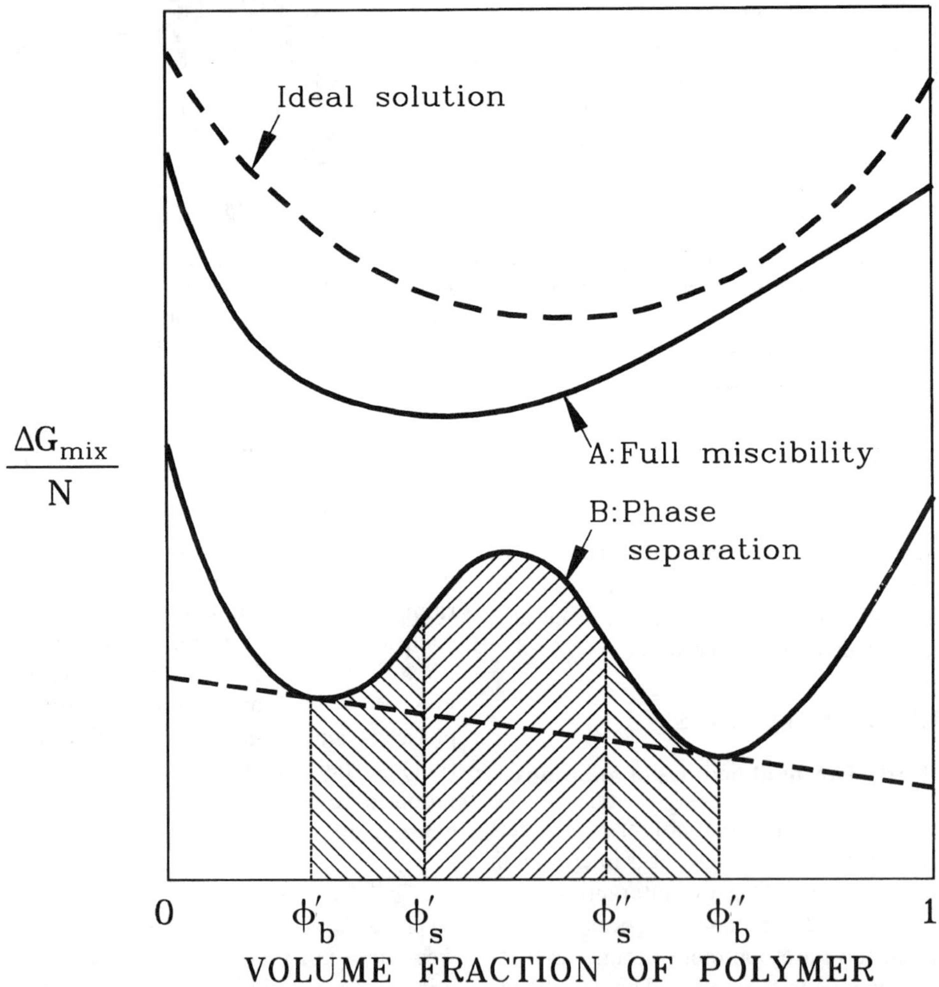


**Figure 3–4** Free energy of mixing for polymer solutions (at constant temperature). Curve A represents full miscibility and curve B phase separation (hatched area). $\phi_s$ and $\phi_b$ represent the spinodal and binodal compositions, respectively.

tire concentration range. However, under some conditions the curve for regular solutions (i.e., when B is large) shows two minima and two inflection points at intermediate concentrations. So even for small molecules, deviation from the ideal behavior makes things more complicated.

### 3.4.2 The Flory-Huggins Theory

A polymer solution behaves differently than a solution of small molecules, obviously because of the large size of the polymers in comparison to the solvent molecules. A polymer solution may be envisioned as a "watery" soup containing entangled, wriggling spaghetti with a length-to-diameter ratio of 10,000:1. However, the theoretical treatment of conditions for polymer-solvent miscibility is not very different from that used for dilute ideal solutions of small molecules. As was the case before, it involves calculating entropic and then enthalpic effects and determining their contribution to $\Delta G_{mix}$.

In the Flory-Huggins theory, the polymer solution is modeled as a lattice, where each lattice site (a volume) is occupied by either a solvent molecule or a polymer segment (the number of segments per polymer molecule is $x$). The change in free energy of mixing for a polymer solution is given by the Flory-Huggins equation (see Gedde, 1995, for a complete derivation):

$$\frac{\Delta G_{mix}}{N} = RT\left(\chi_{12}\phi_1\phi_2 + \phi_1 ln\phi_1 + \frac{\phi_2}{x} ln\phi_2\right)$$

**Equation 3–11**

Let us compare equations 3–10 and 3–11. The last two terms in the Flory-Huggins equation contain the entropic contribution arising from the different placements that polymer (component 2) and solvent (component 1) may have in the lattice. These terms are similar to the last two terms of equation 3–10, but the molar fractions of solvent and the small solute have been replaced by the volume fractions of solvent ($\phi_1$) and polymer ($\phi_2$). The first term in equation 3–11 represents the enthalpic contribution or interaction energy between the solvent molecules and the polymer segments and is equivalent to the $Bx_1x_2$ term of regular solutions (again volume fractions replace molar fractions). The coefficient $\chi_{12}$ is called the Flory-Huggins interaction parameter and is equal to

$$\chi_{12} = \frac{\Delta H_{mix}}{RTN_1\phi_2}$$

**Equation 3–12**

where $\Delta H_{mix}$ is the excess energy involved in neighbor interaction, $N_1$ is the number of moles of solvent, and $R$ is the gas constant. $RT$ is a sort of "thermal energy" that at normal temperatures is of the order of magnitude of the energies involved in intermolecular bonds such as hydrogen bonds or van der Waals' forces. Thus, $\chi_{12}$ is a kind of ratio between the energy involved in the interaction of neighboring molecules and the thermal energy, and it is positive for endothermic mixing, and negative for exothermic mixing. Negative values for $\chi_{12}$ indicate miscibility, while positive values indicate repulsion.

We are often interested in limiting values or critical values for an event. According to the Flory-Huggins theory, the critical value of the interaction parameter for phase separation of a polymer-solvent mixture is given by

$$\chi_{12c} = \frac{1}{2} + \frac{1}{2x} + \frac{1}{\sqrt{x}}$$

**Equation 3–13**

Therefore, for a monomeric mixture ($x = 1$), $\chi_{12c} = 2$, whereas for a large polymer ($x \rightarrow \infty$) in solution, it approaches 1/2. So if $\chi_{12} < 1/2$, the polymer should be soluble if amorphous and linear (Sperling, 1992). For a mixture of two long polymers, it can be shown that $\chi_{12c}$ approaches zero, which explains why binary polymer blends almost always phase separate. For further discussion along these lines, see Piculell, Bergfeldt, and Nilsson (1995).

Figure 3–4 depicts the two general forms that the Flory-Huggins equation may take when $\Delta G_{mix}/N$ is plotted against $\phi_2$, the volume fraction of polymer. The concave curve A represents miscibility in all proportions, while curve B corresponds to phase separation into two coexisting phases. Components in solution may separate by two mechanisms: (1) nucleation and growth and (2) spinodal decomposition. The first is associated with metastability, implying the existence of an energy barrier. Spinodal decomposition is a process by which a mixture separates, having no nucleation free energy barrier. In either case, *equilibrium* compositions after separation are represented by the minima or *binodal points,* which share a common tangent (Figure 3–4). The common tangent implies that the components have the same chemical potential at both minima (e.g., those compositions are at equilibrium). Binodal and spinodal curves are also depicted in graphs where $\phi_2$ is plotted against temperature. In this case, complete miscibility between polymer and solvent may exist either above an upper critical solution temperature (UCST) or below a lower critical solution temperature (LCST).

Another important parameter derived from the Flory-Huggins theory is the theta temperature ($\theta$). When $T = \theta$, polymer segments do not sense other segments of the same chain, and chains are said to be "unperturbed." The polymer is expanded and filled with solvent. Theta conditions are a sort of "ideal state" for dilute polymer solutions, and it can be demonstrated that they exist when $\chi_{12} = 0.5$ (Gedde, 1995).

In the original Flory-Huggins theory, the contact energy term ($RT\ \chi_{12}$) was considered purely enthalpic, but it has been shown to have an important entropic part as well. This explains phase separation in polymers at low as well as high temperatures and the existence of upper as well as lower critical solution temperatures. To summarize, using the Flory-Huggins theory, it is possible to account for equilibrium thermodynamic properties of polymer solutions such as deviations from Raoult's law, phase separation, melting point depression in crystalline polymers, and swelling of networks.

### 3.4.3 Phase Separation

So far we have seen interactions between one polymer and the solvent. Often we are more interested in the fate of mixtures of polymers in an aqueous solution. Phase separation in mixed polymer solutions is quite common and has important technological applications in foods and biotechnology. Two types of phase separation are commonly encountered in mixtures of two polymers: segregative and associative (Figure 3–5). *Segregative phase separation,* also known as *thermodynamic incompatibility,* is typical of ternary systems (protein-polysaccharide-water) and results in one phase containing most of the protein and the other phase nearly all of the polysaccharide. In contrast, *associative separation,* or *complex coacervation,* results in one of the two phases being enriched in both of the polymeric components. This latter behavior is typical of (but not unique to) oppositely charged polyelectrolytes (e.g., gelatin and gum arabic).

Incompatibility of mixed polymers in solution depends on interactions between polymers ($P_1$ and $P_2$), measured by the Flory-Huggins interaction parameter $\chi_{P_1-P_2}$, and interactions of each polymer with the solvent, characterized by $\chi_{P_1-S}$ and $\chi_{P_2-S}$. A positive value for $\chi_{P_1-P_2}$ is indicative of the exclusion of one polymer from the neighborhood of the other, whereas a negative value signifies miscibility. In simple terms, thermodynamic incompatibility is the result of each polymer preferring to be surrounded by its own molecules rather than by molecules of the other polymer. Larger molecular weights, lower electrostatic interactions (higher polymer concentration), and unfolding of molecules enhance incompatibility (Polyakov, Grinberg, & Tolstoguzov, 1997). Good solvents favor mixing while poor solvents lead to phase separation. As previously seen, the overall result is that the Gibbs free energy of the separated system is lower than that of the molecularly homogeneous mix (Grinberg & Tolstoguzov, 1997).

Part B of Figure 3–5 shows a typical phase diagram for a protein-polysaccharide-water system in rectangular coordinates. The black curve is the binodal, and tie-lines join points corresponding to equilibrium compositions (white circles). The dotted line is the rectlinear diameter representing the composition of systems breaking down into phases of the same volume. A mixture such as E, produced by mixing 2.4 parts of sodium alginate solution (A) and 1 part of casein solution (B), will separate into two solutions of composition $E_1$ (largely alginate) and $E_2$ (almost pure casein solu-

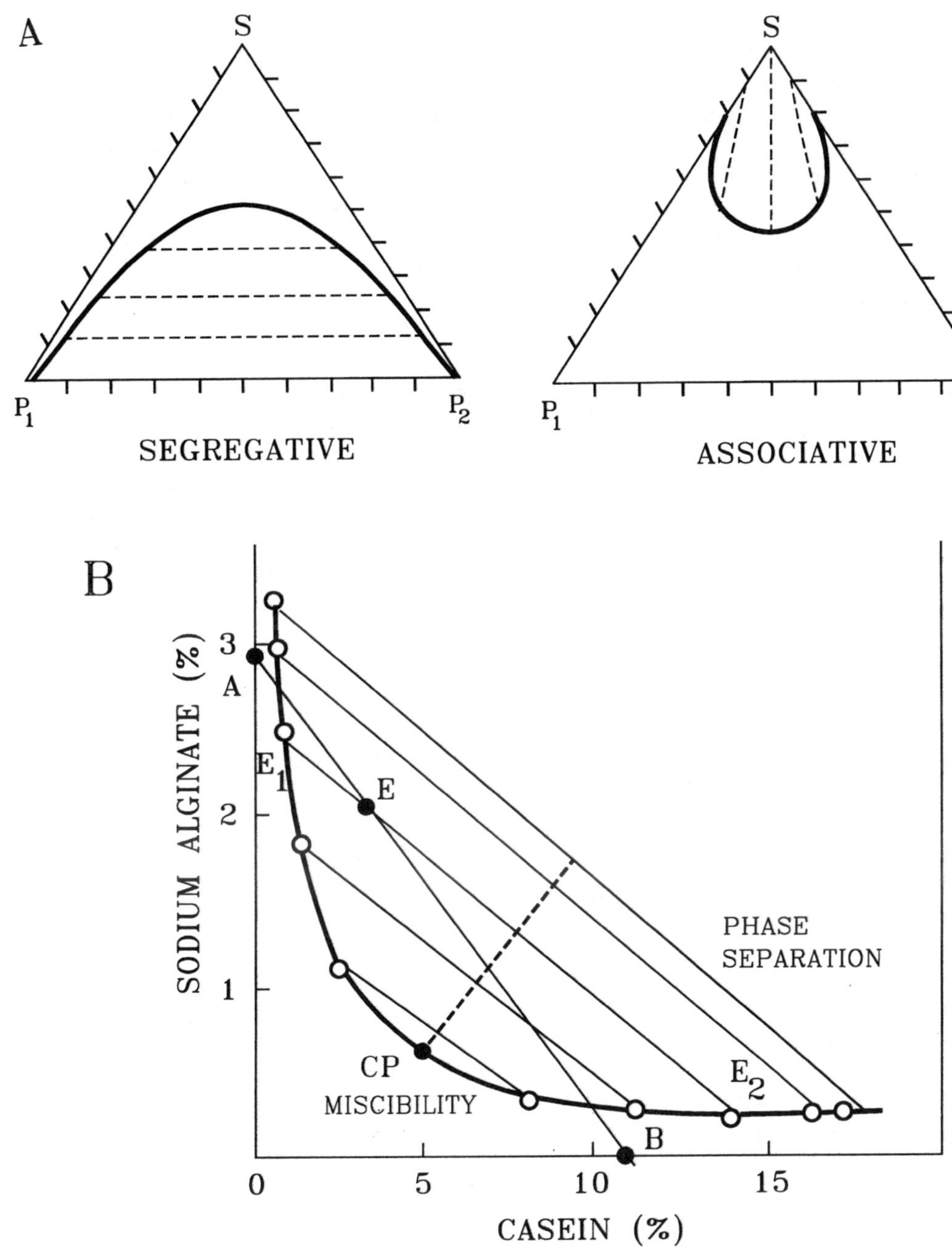


**Figure 3–5** (A) Phase diagrams representing phase separation in ternary polymer systems. *S* is the solvent and $P_1$ and $P_2$ the polymers. Segregative (thermodynamic incompatibility) phase separation and associative (complex coacervation) phase separation are shown. (B) Typical rectangular phase diagram of a protein-polysaccharide-water system. The black curve represents the binodal and the dotted line the rectilinear diameter. *Source:* Reprinted from *Food Hydrocolloids,* Vol. 11, V.B. Tolstoguzov, Thermodynamic Aspects of Dough Formation and Functionality, pp. 181–193, Copyright 1997, with permission from Elsevier Science.

tion). Data for several ternary systems involving food proteins, polysaccharides, and water can be found in Tolstoguzov (1988). Thermodynamic incompatibility occurs for the main types of proteins defined by the Osborne system (albumins, globulins, glutelins, and prolamins) as well as for neutral and anionic (including carboxyl and sulfated) polysaccharides, linear or branched. It is also present among proteins and polysaccharides and in mixtures of native and denatured proteins of the same class. Segregative phase separation is favored over complexing of biopolymers when the pH is far removed from the isoelectric point and at high ionic strength. Thermodynamic incompatibility has been postulated to play a key role in structure formation where food polymers are found in high concentration, such as in mixed gels (Zasypkin, Braudo, & Tolstoguzov, 1997), extruded food polymer mixtures (Yurjev et al., 1989), and dough (Tosltoguzov, 1997).

The preceding paragraphs describe the thermodynamic or equilibrium conditions for phase separation but say nothing about how fast this separation would take place. Since equilibrium may not be attained during processing of food polymers, the kinetics of demixing of incompatible polymers in solution are of primary importance, yet difficult to determine in practice. Different food structures may result if the demixing process is halted at a certain moment (e.g., by fixing the matrix) or if shear is applied to deform the mixture (Tolstoguzov, 1988). This subject as it applies to mixed gels has been addressed by Clark (1995).

In short, immiscibility of polymers in solution is a common phenomenon, and we should be aware of it, for it has structural consequences. The Flory-Higgins theory has been used to explain phenomena such as phase separation and the kinetics of demixing, water partition in mixed gels, adsorption of polymers at interfaces, and so on. The role of polymers in mixed solutions as applied to foods is reviewed in *Biopolymer Mixtures* (Harding, Hill, & Mitchell, 1995).

### 3.4.4 The Excluded Volume Effect

Another peculiar aspect of macromolecules in solution that makes their study different from volumeless or small molecules is that they occupy a certain volume of the solution and "exclude" other molecules from it. This volume has to be subtracted from the total volume, so that for concentrated solutions the excluded volume approaches the real volume of macromolecules. For example, a polymer chain existing as a random coil is filled with a mean or "statistical" volume of solvent (e.g., a sphere), which is then not available for other chains because it is hidden inside the polymer volume. Phase separation may be facilitated by this "excluded volume effect," which raises the effective concentration of polymers. According to Tolstoguzov (1997), the excluded volume has the following effects in food systems: (1) enhancement of association of biopolymers, including denatured globular proteins; (2) reduction in solubility and cosolubility of macromolecular solutes; and (3) destabilization of polymer suspensions (by depletion flocculation).

## 3.5 PHASE TRANSITIONS

### 3.5.1 Phases and Phase Diagrams

A phase is an homogeneous region of matter. Many food processes involve the transition of one phase into another, entailing changes in structure and properties (hence, the relevance for this book). Phases depend on temperature ($T$), pressure ($P$), and composition. Phase diagrams for binary systems show phases prevailing as a function of composition and temperature (at constant $P$). Ternary data are depicted in triangular diagrams in which each vertex represents a pure component (and $P$ and $T$ are constant). Phase diagrams show equilibrium conditions but say nothing about the kinetics of attainment of equilibrium. For example, although the crystalline form of many food components is thermodynamically stable (at certain $T$ and $P$), the material may remain temporarily amorphous or vitreous due to kinetic constraints (see Section 3.5.6).

Food technologists deal with products and components in all three fundamental states of matter: solid, liquid, and gas. The term *solid* describes a state of aggregation in which the substance possesses definite volume and shape. The ability of solids to withstand shearing forces and to regain their original shape after a small deformation distinguishes them from liquids and gases. The structure of solids may vary from simple crystalline forms to complex amorphous structures, depending on geometrical conditions imposed by the bonding and packing of atoms and molecules. Crystal structures are regular, three-dimensional patterns of atoms or molecules in space that are idealized by the geometric concept of a space lattice. Amorphous solids, representing metastable conditions, differ from crystalline solids in that they do not possess long-range order, and they differ from liquids in that they do not show short-time fluidity.

Liquid foods are important because they usually provide ample amounts of water, an essential nutrient. In fact, human life may be maintained by supplying all required nutrients in an aqueous milieu. The properties of liquids as foods are discussed in detail in Chapter 7, which is entirely devoted to the fluid food milk. Solids and liquids make up the vast majority of our daily foods, and food components existing in a gas or vapor phase are rare. However, many foods would lose much of their attractiveness if the molecules that cause their aroma were not able to enter the gas phase.

### 3.5.2 First- and Second-Order Transitions

As mentioned before, phase transitions are changes in the physical state of matter due to changes in $P$ or $T$. Thermodynamic transitions from a solid crystalline state to liquid and from liquid to gas are examples of first-order transitions. Other important first-order transitions in foods are those between polymorphic states in fats, starch gelatinization, and protein denaturation. Phase transitions as they pertain to foods are treated comprehensively by Roos (1995), which also is the source of some concepts and examples presented below.

Again, the Gibbs free energy is the thermodynamic parameter to observe, although it is now in the form of the chemical potential ($\mu$), which is the driving force for a phase transition:

$$\mu = \left(\frac{\partial G}{\partial n_i}\right)_{T,P,n_j} \qquad \textbf{Equation 3–14}$$

The prevailing phase after a change in $T$ or $P$ is the phase with the lower $\mu$. Since most transitions of importance occur after a change in $T$ at constant $P$, this case will be used for the purpose of discussion. The first derivative of the chemical potential with temperature is

$$\left(\frac{\partial \mu}{\partial T}\right)_P = -S_m = -\frac{H_m}{T} \qquad \textbf{Equation 3–15}$$

where $S_m$ and $H_m$ are the molar entropy and the molar enthalpy, respectively. It can be concluded that, as the temperature rises ($\Delta T > 0$), the chemical potential decreases, because $S_m$ is always positive. First-order transitions show a step change in enthalpy, entropy, and volume at the temperature of the transition. Melting, crystallization, vaporization, condensation, and sublimation are first-order transitions. At the transition, the specific heat or heat capacity of the substance ($c_p$) is defined by

$$\left(\frac{\partial H}{\partial T}\right)_P = c_p \qquad \textbf{Equation 3–16}$$

and has an infinite value, because the addition of energy is used for driving the transition rather than for increasing the temperature (Atkins, 1982). It follows that a first-order transition is characterized by an infinite heat capacity at the transition point.

Second-order transitions are those in which the second derivative of the Gibbs free energy shows a discontinuity at the transition:

$$\left(\frac{\partial^2 G}{\partial T^2}\right)_P = -\frac{c_p}{T} \qquad \textbf{Equation 3–17}$$

This equation suggests that a second-order transition can be located experimentally by finding a step change in heat capacity.

### 3.5.3 Differential Scanning Calorimetry (DSC)

There are several experimental ways in which phase transitions can be determined (see Sperling, 1992, p. 321), but differential scanning calorimetry (DSC) is a method traditionally preferred by food scientists. A differential scanning calorimeter is an instrument capable of keeping the temperature of a sample the same as that of a reference holder (the so-called "null balance principle") by continuous and automatic adjustment of the heater power. The unique feature of DSC is that the sample and reference pans (e.g., pans containing air) are heated independently. A signal proportional to the heat power difference *dH/dt* between sample *(s)* and reference *(r)* is recorded. The heat balance equation for DSC is

$$\frac{dH}{dt} = -\frac{dq}{dt} + (c_s - c_r)\frac{dT_r}{dt} - Rc_s\frac{d^2q}{dt^2}$$

**Equation 3–18**

where $dq/dt$ is the differential heat flow, $c_s$ and $c_r$ are the heat capacities of the sample and reference, $dT_r/dt$ is the scanning rate, and $R$ is the thermal resistance between the block and the sample cell (Gedde, 1995). $H$ (and $q$) have units of J (joule) and $dH\,dt$ of $Js^{-1}$ (or watts). Values of $\Delta H$ for a first-order transition are normalized dividing by the sample weight so that enthalpies of phase change are in J/g.

Figure 3–6 shows a typical DSC thermogram of an undercooled semi-crystalline polymer (Roos, 1995). Initially sample and reference are at the same temperature (balance). When a glass transition is reached, an increase in endothermal heat flow to the sample (due to a change in $c_s$) is required to keep the temperatures equal. The presence of a glass transition appears in the thermogram as a change in $c_s$. Glass transitions in foods are usually not sharp, and the glass transition temperature ($T_g$) may be reported as the onset, middle,

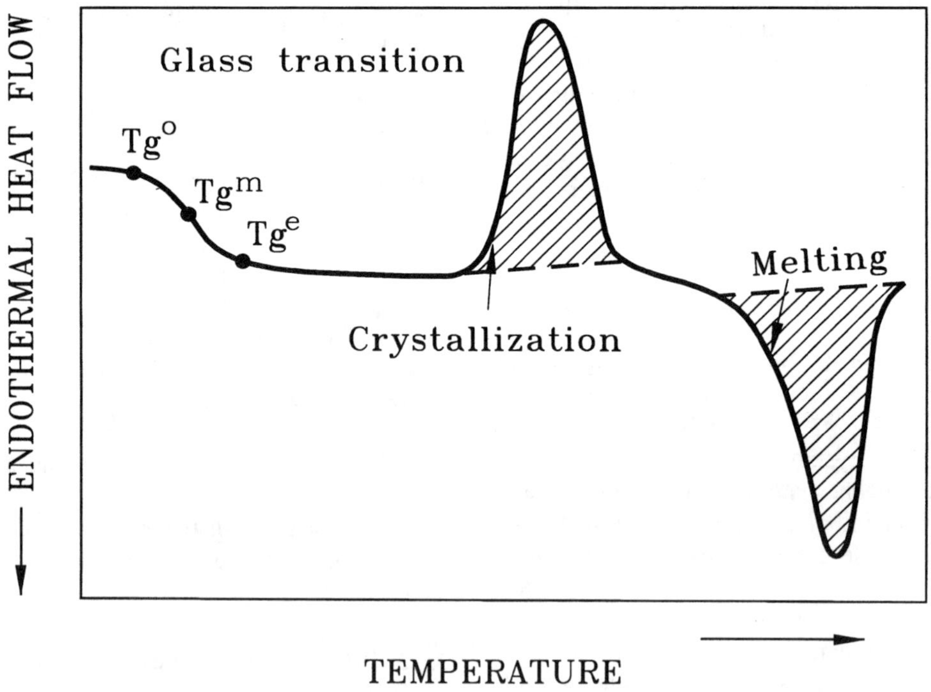


**Figure 3–6** Differential scanning calorimetry tracings of an undercooled crystallizable polymer during reheating. The initial amorphous state induced by fast cooling undergoes a glass transition (o = onset, m = middle, and e = end points), followed by crystallization and melting. Crystallization is possible only after the system has passed $T_g$ and exhibits enough mobility. Hatched areas represent the enthalpies of phase change (first-order transitions) involved. *Source:* Reprinted with permission from Y. Roos, *Phase Transitions in Foods*, p. 213, © 1995, Academic Press.

or end point. First-order transitions for crystallization and melting appear as peaks whose areas underneath (when divided by the sample weight) are equal to the crystallization and melting enthalpies. The calibration factor transforming peak area to enthalpy is independent of the thermal resistance; it is simply the electrical conversion factor, with no dependence on the temperature (Gedde, 1995).

### 3.5.4 The Glass Transition

Different physical states may be attained when a molten polymer is cooled down. For crystal-forming materials, the basic factor that determines whether the crystalline or the amorphous state will occur is the extent to which the molecules can attain equilibrium conditions. Metastable conditions arise when there are free energy barriers that hinder the attainment of the equilibrium state. Important barriers to crystallization are nucleation and hindrance of molecule rotation. Structural barriers such as high viscosity or hindered diffusion may also delay attainment of the equilibrium state. If a liquid polymer is cooled (Figure 3–7) and the rate of cooling is sufficiently rapid, the liquid will not crystallize at $T_m$, the melting temperature, but instead will become a metastable liquid—a glass. At the glass transition temperature, $T_g$, the material is no longer a liquid but a glass. The detected $T_g$ increases with increasing cooling rate, since the time available to the system is shorter and it starts to deviate from equilibrium at a higher temperature. The term *glass transition* refers to the temperature (or the temperature range) at which a glass begins to soften and flow. To the microstructural engineer, it would not be unexpected that the glass transitions in a heterogeneous food are usually not sharp but a diffuse change, as several $T_g$'s are known to exist for microregions in the material.

A number of structural factors are supposed to affect $T_g$. Among those related to the chemical structure of the polymer are the intrachain steric hindrance and bulk or stiff side groups. Intermolecular forces and high cohesive energy density increase the energy required for the onset of molecular movement and consequently increase $T_g$. Molecular weight $M_w$ is inversely related to $T_g$, and an expression proposed by Fox and Flory is frequently used to describe its effect:

$$T_g = T_{g\infty} - \frac{K}{M_w} \qquad \textbf{Equation 3–19}$$

where $T_{g\infty}$ is the hypothetical $T_g$ of an infinitely large chain and $K$ is a constant specific to the particular polymer. This effect has been corroborated for protein hydrolyzates, which exhibit a $T_g$ lower than the parent proteins. Levine and Slade (1988) showed a strong linear correlation between $M_w{}^{-1}$ and the $T_g$ of commercial starch hydrolyzates of different dextrose equivalents. The $T_g$ decreased more than 30°C as $M_w$ was reduced from 60 to 5 kDa.

The $T_g$ of a polymer blend is highly dependent on the microstructure. If the polymers are immiscible and amorphous, the mixture will exhibit two $T_g$'s corresponding to the two pure phases. In fact, this a way of checking phase separation in amorphous blends. The case of miscibility is treated under "plastization by water" in Section 3.5.6.

### 3.5.5 The Williams-Landel-Ferry Equation

The glass transition is also related to relaxation phenomena. Relaxation is the time-dependent response in a property following a perturbation. The temperature dependence of a relaxation process characterized by $\tau_o$ is usually represented by the familiar Arrhenius equation:

$$\frac{1}{\tau_o} = k_o \exp\left(\frac{-E_a}{RT}\right) \qquad \textbf{Equation 3–20}$$

where $E_a$ is the activation energy and $k_o$ is a constant.

In the case of amorphous materials in the rubbery state between $T_g$ and $T_m$ (or between $T_g$ and approximately $T_g + 100$ °K [see Section 3.5.8]), a totally different temperature dependence exists for phenomena influenced by restricted mobility. Above and below this temperature range, Arrhenius kinetics apply (Figure 3–8). Since most low-moisture foods have amorphous phases and $T_g$'s in the above mentioned range, the derivation of new relation for temperature dependence is presented in some detail.

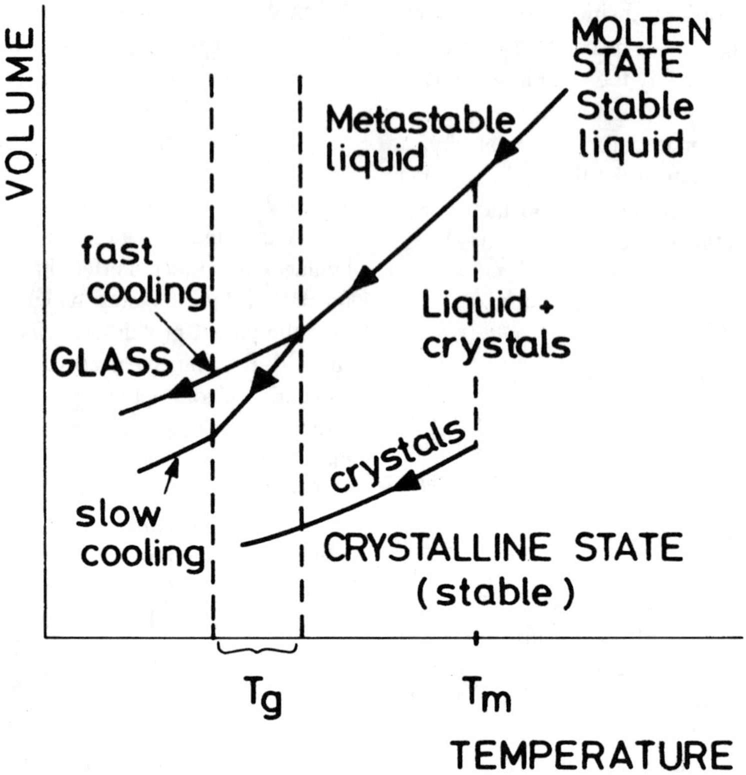


**Figure 3–7** Formation of glassy and crystalline states during cooling of a melt. The crystalline state will be formed at $T_m$ only if the sample is cooled slowly and nucleae are present. Faster cooling may lead to the glassy state, where the rate of cooling will also have an effect on $T_g$. Note that the amorphous state has a higher specific volume than the crystalline state.

Free-volume theory states that in the glassy condition polymers have a universal free volume related to the internal mobility of the system (e.g., viscosity). The empirical relationship between viscosity ($\eta$) and free volume is given by the Doolittle equation:

$$\eta = A\exp\left(\frac{B}{f}\right) \quad \textbf{Equation 3–21}$$

where $A$ and $B$ are constants and $f$ is the fractional free volume that at a temperature $T$ (such that the system is in the rubbery state) can be expressed as a linear function of the temperature difference $T - T_g$ and the fractional free volume at the glass transition $f_g$:

$$f = f_g + \alpha_f(T - T_g) \quad \textbf{Equation 3–22}$$

where $\alpha_f$ is the expansion coefficient of the free volume. From equations 3–21 and 3–22, the ratio of the viscosity at temperatures $T$ and $T_g$ can be derived. Further algebraic manipulation delivers the well-known Williams-Landel-Ferry equation (Sperling, 1992):

$$\log\left(\frac{\eta}{\eta_g}\right) = \frac{C_1(T - T_g)}{C_2 + (T - T_g)} \quad \textbf{Equation 3–23}$$

$C_1$ and $C_2$ are known as the universal constants (−17.44 and 51.6, respectively), because the fractional free volume at the glass transition and the thermal expansion coefficient of the fractional free volume are universal constants as well (0.025 and $4.8 \times 10^{-4}$ $K^{-1}$, respectively). The Williams-Landel-Ferry (WLF) equation has been found to ap-

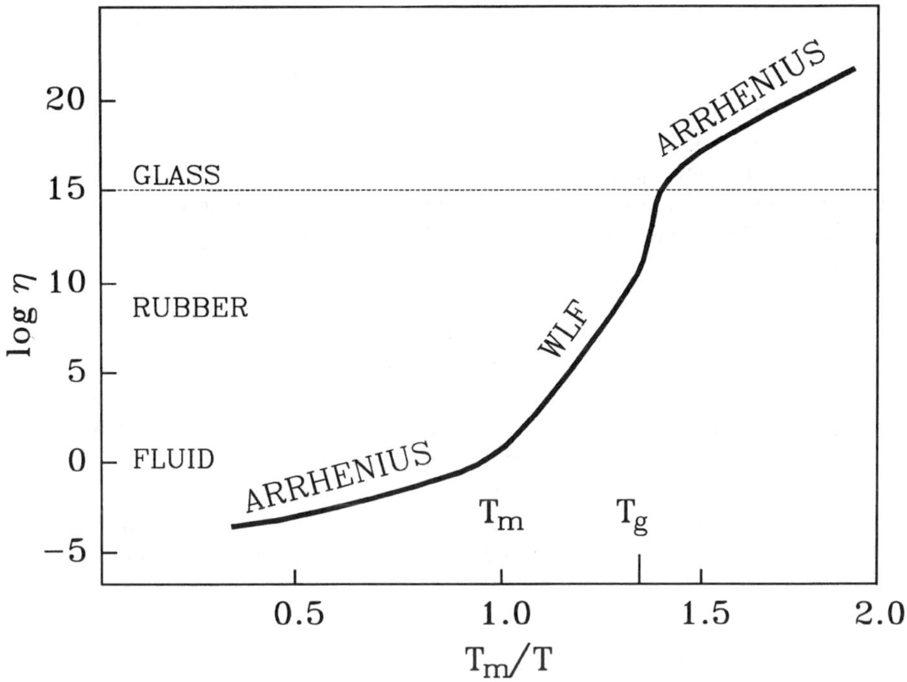


**Figure 3–8** Temperature dependence of viscosity as a function of reduced temperature for amorphous and semicrystalline polymers. Between $T_g$ and $T_m$, the temperature dependence is large and nonlinear, in accordance with the Williams-Landel-Ferry equation. Outside this range, an Arrhenius-type relationship occurs. *Source:* Reprinted with permission from L. Slade and H. Levine, Beyond Water Activity: Recent Advances Based on an Alternative Approach to the Assessment of Food Quality and Safety, *Critical Reviews in Food Science and Nutrition*, Vol. 30, pp. 115–359, © 1991, CRC Press, Boca Raton, Florida.

ply to amorphous foods with different values for the "constants $C_1$ and $C_2$". Figure 3–8 presents a diagram showing that the temperature dependence predicted by the WLF equation is greater than that stipulated by the Arrhenius equation. In practice, equation 3–23 states that WLF kinetics prevailing above $T_g$ in the rubbery region for diffusion-controlled relaxation phenomena depends exponentially on the magnitude of $\Delta T = (T - T_g)$.

### 3.5.6 The Glassy State

Almost any substance (even water) can be obtained in the glassy state provided it can be cooled sufficiently fast to prevent the orderly arrangement of atoms and molecules. The glassy state is characterized by the freezing out of long-range motion in the case of small molecules (sugars) and of the wriggling motion of chains in the case of polymers. In the latter case, molecular mobility is restricted to vibrations, rotations, and motions by relatively short segments of each chain (see Section 3.8.7). Structurally, glasses are isotropic solids showing no grain boundaries, and consequently most of them are transparent to light (e.g., clear, hard candy).

The most representative parameter of the glass transition is $T_g$. Restriction of movement at the glass transition temperature results in a very high apparent viscosity ($>10^{12}$ Pa·s) and shear modulus ($>10^9$ Pa). In frozen food systems $T_g'$ is defined as the $T_g$ of the maximally freeze-concentrated matrix (see Figure 5–29). $T_g'$ values for fresh fruits vary between −33 and −42°C, while those for fish and meat are fairly constant at around −12°C (Fennema, 1996). We will come back to $T_g'$ when we study stability of frozen foods in Section 5.7.3.

An interesting phenomenon occurs when a glass is held at constant temperature below $T_g$: physical aging. The specific volume and enthalpy continue to decrease with time, approaching equilibrium, and engineering properties change significantly. The tensile modulus and yield stress increase, while fracture toughness, impact strength, and permeability decrease. This is a physical manifestation of the fact that $T_g$ is a kinetic event.

Plasticizers are added to increase the plasticity and flexibility of polymeric materials by weakening the intermolecular forces between macromolecules. Water is the ubiquitous plasticizer in foods and in nature. The practical effect of plasticizers is to lower dramatically the $T_g$ of amorphous solids (Figure 3–9). As a rough estimate, the value of a 10°C decrease in $T_g$ per percentage point of water increase can be used below 10% moisture. A better prediction of the glass transition $T_g$ of a water-plasticized material in an extended moisture range can be made using mathematical mixing models such as the simple Gordon-Taylor equation:

$$T_g = \frac{w_1 T_{g1} + k w_2 T_{g2}}{w_1 + k w_2} \qquad \textbf{Equation 3–24}$$

where $T_{g1}$ and $T_{g2}$ are the $T_g$'s of components 1 (water) and 2 (polymer), $w_1$ and $w_2$ are the weight fractions, and $k$ is an empirical constant. $T_{g1}$ is usually taken to be −135°C for water. More elaborate equations for predicting the $T_g$ of mixed systems are found in Roos (1995).

Research on the glassy behavior of food materials has increased significantly over the past 20

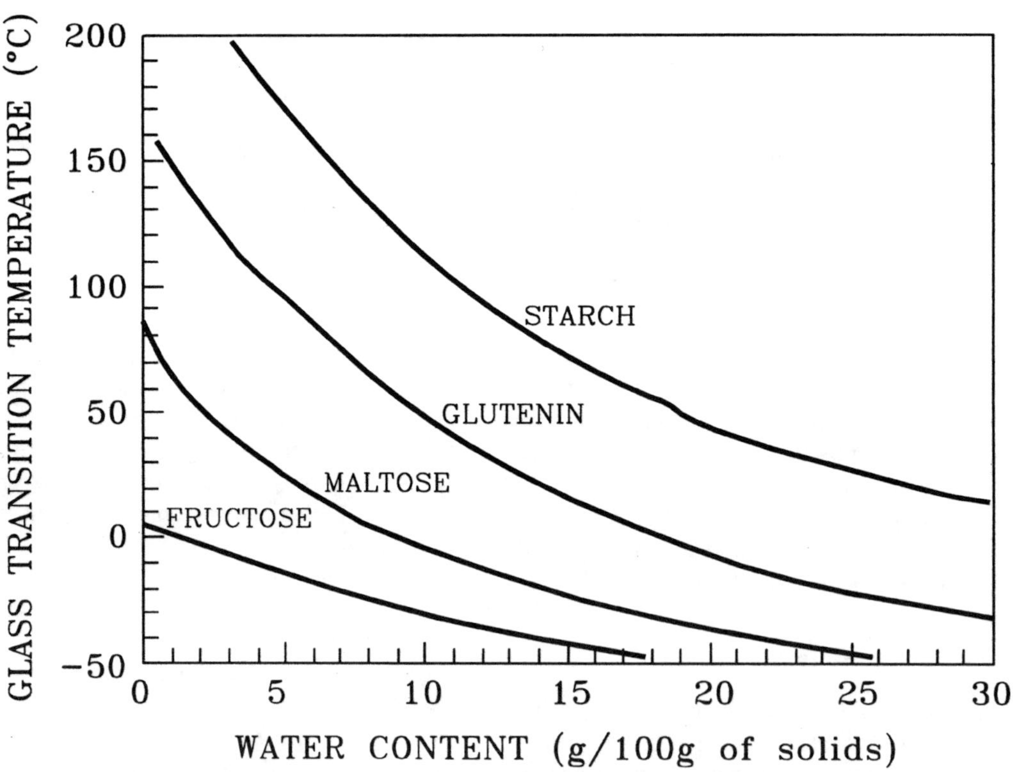


**Figure 3–9** Plasticization of amorphous biopolymers by water. $T_g$ decreases dramatically with increasing water content, particularly at low-moisture contents. Note that $T_g$ (anhydrous) for amorphous polymers and some sugars is above ambient temperature. *Source:* Reprinted with permission from Y. Roos, *Phase Transitions in Foods*, p. 222, © 1995, Academic Press.

years, and important technological advances have occurred in food processes such as freezing, dehydration, and baking, and in the development of state diagrams (Roos & Karel, 1991). Further details about glasses and their role in foods and food processing can be found in the book edited by Blanshard and Lillford (1993).

The curious reader may find interesting a recent review by Aguilera and Karel (1997) on how life is preserved under extremely low moisture and high temperature conditions (e.g., in a desert) and also how knowledge of life-preserving mechanisms might be applied in food technology and biotechnology. It appears that the protection mechanisms are provided by the glassy state as well as by the specific effects of some sugars on membranes and macromolecules.

### 3.5.7 The Crystalline State

As explained before, a crystal is a portion of matter in which atoms (or unit cells) are arranged in a regular three-dimensional pattern extending through the whole mass (long-range order). Crystals other than the classical examples of salt and sugar are ubiquitous in many foods, as becomes evident with microscopy. For example, crystals of calcium oxalate are common in the testa fraction of seeds, possibly to protect them during dormancy against physical injury or infection (Webb & Arnott, 1982). Several types of dense crystals that contained P, Ca, Mg, Si, S, Fe, Zn, and Al were detected by TEM and dispersive X-ray spectrometry in wheat flour (Geneves, Rutin, & Halpern, 1985).

Polymers in the solid or condensed state are only semi-crystalline as a consequence of their long chains and possible entanglements. Regularity in the chain structure and the presence of preferred chain conformations favor crystallization. Atactic polymers and polymers having bulky side groups almost never crystallize, while in branched polymers only the linear sections of branches may crystallize (e.g., amylopectin). Large arrangements of crystals in space are called lattices. The smallest repeating period in a lattice is called the unit cell. The most popular structure of crystalline polymers is the fringed micelle, in which a long polymer chain passes in and out many crystallites, thus interspersing amorphous and ordered regions. The degree of crystallinity has a profound effect on physical and mechanical properties of a polymer. Crystallinity can be determined by differential scanning microscopy (first-order transition), infrared absorption. Raman spectra, and X-ray or electron diffraction (Sperling, 1992).

Formation of a stable crystalline phase from the liquid state depends on two separate phenomena: nucleation and crystal growth. The first step in the growth of a crystal is the formation of a nucleus—a small and stable cluster of atoms or molecules that have achieved the right crystalline arrangement. Homogeneous nucleation involves the spontaneous formation of the new phase and usually requires a high degree of supersaturation. The total change in Gibbs free energy for the crystallization ($\Delta G$) of a spherical crystal of radius $r$ (embryonic nucleus) is the sum of a volume and a surface contribution:

$$\Delta G = \frac{4}{3}\pi r^3 \Delta G_v + 4\pi r^2 \gamma$$

**Equation 3–25**

where $\Delta G_v$ is the free energy change per unit volume associated with the phase change and $\gamma$ is the surface free energy of the formed nucleus (see Section 3.6.2). Since $\Delta G_v$ is negative and the second term is always positive, and since they depend differently on $r$, a plot of $\Delta G$ versus $r$ goes through a maximum at the critical radius $r_c$. If $r < r_c$, the embryo will not survive, and no nucleus will be formed. For freezing water, $r_c$ is approximately 10Å at $-40°C$ ($\approx$100 molecules of water) and 100Å at $-4°C$. Hence, homogeneous nucleation requires a high degree of supersaturation or supercooling. In heterogeneous nucleation, the nuclei are formed on foreign particles, surfaces, or imperfections, which lower the net energy associated with their formation.

The rate of nucleation is defined as the number of nuclei formed per unit time and unit volume of the parent phase. The rate of homogeneous nucleation can be expressed by an Arrhenius-type expression:

$$rate = \nu N \exp\left(\frac{-\Delta G}{kT}\right)$$

**Equation 3–26**

where $\nu$ is the frequency of collisions of the nucleus with atoms or molecules, $N$ is the concentration of atoms per unit volume, and $\Delta G$ is the overall free energy change for the formation of a nucleus. As seen in equation 3–26, the nucleation rate rises steeply with decreasing temperature. Once a stable nucleus has been formed, crystal growth proceeds at a velocity controlled by two distinct processes, the rate at which molecules are added to the crystal lattice and the diffusion of heat away from the interface. The typical dependence of nucleation and crystallization rates on the degree of supercooling is presented in Figure 3–10.

As usual we are interested in the rate at which crystallization proceeds. Any crystallization kinetics (for large or small molecules) can be modeled by the Avrami equation:

$$1 - \phi_c = \exp(-kt^n) \qquad \textbf{Equation 3–27}$$

where $\phi_c$ is the volume fraction of crystalline material formed and $k$ and $n$ are constants typical of the nucleation and growth mechanisms. Values of $n$ vary between 1 (for linearly growing systems) and 5 or 6 (for three-dimensional complex geometries). If the growth of a crystal is diffusionally controlled, the value of the exponent is reduced by half with respect to the "free" growth case. The exponent $n$ of a crystallization process can be determined from experimental data as the slope of a $\ln t$ versus $\ln[-\ln(1 - \phi_C/\phi_{C\infty})]$ (Gedde, 1995).

Crystalline polymers melt, but the melting behavior of polymers is different from that of low molar mass compounds. Melting depends upon specimen history and the rate of heating, and it usually occurs over a range of temperatures (Young & Lovell, 1991). The observed melting temperature ($T_m$) is always higher than the crystallization temperature and much higher than $T_g$. Factors that control $T_m$ are hydrogen bonding, polarity, and "impurities" (e.g., branching chain ends). The effect of a diluent on the melting temperature can be predicted from the equation

$$\frac{1}{T_m} - \frac{1}{T_m^o} = \frac{R}{\Delta H_m}\frac{V_u}{V_1}(\phi_1 - \chi_{12}\phi_2^2)$$

**Equation 3–28**

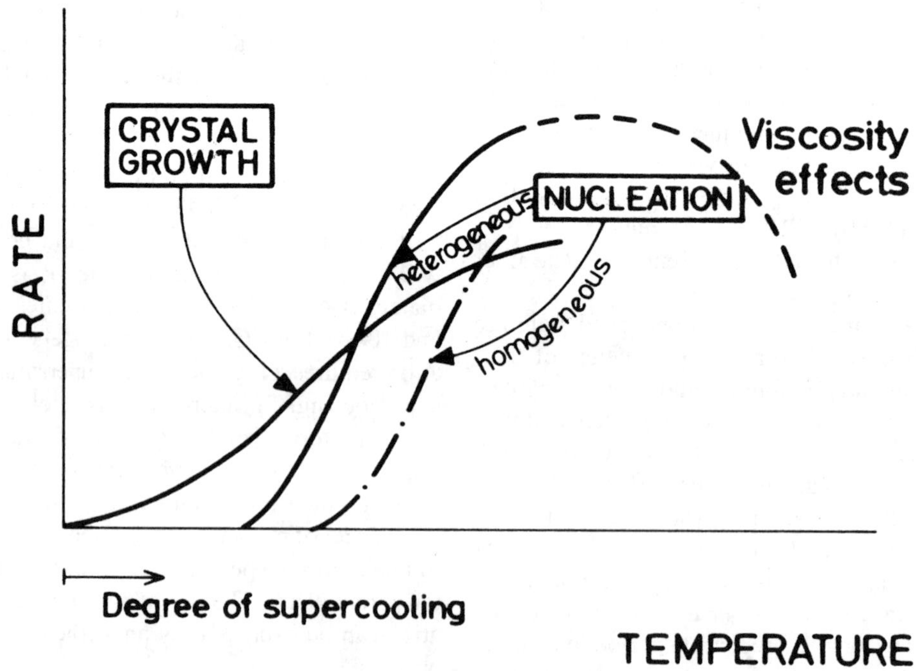


**Figure 3–10** Nucleation and crystal growth rates as influenced by supercooling.

where $T_m^o$ is the melting point of the pure polymer; $\Delta H_m$ is the enthalpy of fusion; $V_u$ and $V_1$ are the molar volumes of the repeating unit ($u$) and diluent (1), respectively; the $\phi$s are volume fractions, and $\chi_{12}$ is the Flory-Huggins interaction parameter. This expression has been used to study the effect of adding components to starch in foods (Zobel, 1992).

An interesting example of crystalline (actually semi-crystalline) behavior in foods is the behavior of starch. Native starch granules contain 15–45% crystallite material alternating in concentric rings with amorphous regions. Crystallinity occurs within the order domains of amylopectin and is created by the intertwining of chains to form double helices (Oates, 1997). Starch granules are insoluble, but when heated in the presence of excess water the granules swell, and if the temperature exceeds a value typically between 50°C and 80°C (depending on plant species), the crystal structure is disrupted (see Section 4.5.3). The process is known as gelatinization, and it is, in fact, an order-disorder transition. If water is reduced or solutes are added, the gelatinization temperature increases. Four different thermal transitions have been identified for potato starch (Willenbücher, Tomka, & Müller, 1993): gelatinization G (60–70°C), crystalline melting $S_1$ (110°C) and $S_2$ (155°C), and a high-temperature H transition (165–220°C). After the transition $S_2$, neither crystalline order nor helical order can be detected, and hence melting of crystallites is assumed to be completed. A thorough discussion of starch crystallinity is found in Zobel (1992). After cooling or aging, starch molecules can reassociate into crystalline segments (retrograde) to an extent that depends on factors such as chain length, linearity of the molecules, temperature of cooling, time, and concentration of starch. Recrystallized amylopectin, partly responsible of the staling of bread, can be rendered amorphous if heated to 55–95°C in the presence of water.

### 3.5.8 The Rubbery State

We have commented previously on how a "rubbery" state is attained in amorphous materials that have been heated above $T_g$. In a rubbery condition, materials have a viscosity (or modulus) several orders of magnitude smaller than in the glassy state (Figure 3–19).

Rubbers or elastomers subjected to a force deform in real time but return to their original state if the force is released. This phenomenon is called *rubberlike elasticity* and has attracted much attention from those interested in the structure-property relationships of materials. Rubberlike materials show very high deformability (they can be stretched to several times their original length) and on release of the stress exhibit essentially complete recovery (instantaneous return to their original length). To possess these two characteristics, molecules must be long and highly flexible chains and capable of forming a network by light cross-linking or entanglement. Permanent flow does not occur precisely because of the existence of these restraining points and of hard microdomains (crystalline or amorphous).

Rubberlike elasticity is of interest to food scientists because many foods are composed of long polymeric molecules (e.g., most gels) or elastic fibers (e.g., meat), and thus their mechanical behavior may be analyzed by analogy to rubberlike materials. As explained in the following sections, there is a fundamental difference in the way rubberlike solids and nonpolymeric solids behave under stress.

### 3.5.9 Rubberlike Elasticity

Chewing gum (chicle) is a typical rubbery material. Elastomers or rubbers exhibit some unusual properties: they shrink when heated and give out heat reversibly when stretched (which you can sense by putting your lips on a rubber band as it is stretched). Hookean elastic behavior shown by most solids (e.g., metals) is basically energy driven (see Section 3.9.1), which means that the displacement of atoms during deformation causes an increase in internal energy $U$. Elastomers, on the other hand, exhibit predominantly entropy-driven elasticity, which implies that upon stretching, segments in the macromolecule orient themselves in the direction of the stress into a more "ordered" state, thus decreasing the entropy of the material.

Molecularly, an elastomer is a lightly cross-liked amorphous polymer above its glass transition (Sperling, 1992) (see Section 3.8.8). A chain in its relaxed state has an end-to-end distance $r$ smaller than when stretched. Typical of rubbers is that the volume remains approximately constant on deformation. The elastic force $F$ is related to the internal energy (U) and entropic changes accompanying deformation $dL$ (at constant $T$ and $P$) as follows:

$$F = \left(\frac{\partial U}{\partial L}\right)_{T,P} - T\left(\frac{\partial S}{\partial L}\right)_{T,P} \quad \textbf{Equation 3–29}$$

This equation shows that the elastic force has two contributions. The first term is associated with the change in internal energy accompanying a deformation (as in a solid). The second term represents the change in entropy. The peculiarity of rubber behavior is that over 80% of the elastic force is due to the entropic contribution.

Below is a basic equation used in the mechanical characterization of foods showing some elastic behavior. It indicates how the uniaxial retractive stress ($\sigma$) is related to the extension ratio $\alpha$ (= $L/L_O$):

$$\sigma = \frac{\rho RT}{\overline{M}_c}\left(\alpha^2 - \frac{1}{\alpha^2}\right) \quad \textbf{Equation 3–30}$$

where $\overline{M}_c$ is the average molar mass of chains between cross-links and $\rho$ is the density. The modulus, $\rho RT/\overline{M}_c$, increases linearly with temperature. The mechanical properties of protein gels have been modeled using this equation.

We have commented previously on how nature deals with structure-property relationships. Elastomeric biopolymers are needed in tissues that must exhibit deformability and recoverability, such as those in skin, arteries and veins, and organs such as the lungs and heart. Some proteins show rubberlike elasticity: elastin, denatured collagen, resilin, abductin, and viscid liquid in spiders. Plasticization (and swelling) of elastin by body fluids is crucial to its functionality at room temperature, as the $T_g$ (anhydrous) is approximately 200°C (based on extrapolation data).

Another interesting concept derived from rubbers is the notion of a polymeric network. A network is a permanent structure formed by polymeric chains in the form of a three-dimensional structure. Fixed points in a network are called junctions, and the number of chains meeting at each junction equals the functionality of that junction. All synthetic and biological networks swell when exposed to low molecular weight solvents. The degree of swelling at equilibrium depends on factors such as temperature, the length of the network chains, the size of solvent molecules, and the strength of the thermodynamic interaction between polymer and solvent (e.g., the Flory-Huggins parameter). We will come back to networks when we look at gels and their formation in Section 5.8.

### 3.5.10 Polymers and the Liquid Crystalline State

While amorphous systems are mostly disordered and crystalline materials are ordered in all three directions, liquid crystals show order in one or two directions. A simple example is the natural alignment of toothpicks on the surface of water as they become more concentrated. As we will see, liquid crystals are more important in nature and foods than commonly thought.

Liquid crystals must satisfy some basic requirements (Sperling, 1992). First, they should exhibit a first-order transition between the true crystalline state and the liquid crystal state, and another first-order transition leading to the isotropic liquid state (or another liquid crystal state) at a higher temperature. Second, liquid crystals must exhibit long-range order in one or two directions but not in all three. Third, they must display some kind of "fluidity." Molecularly, liquid crystals are usually rod- or disk-shaped polymers. Liquid crystal polymers are classified as lyotropic, thermotropic, and mesogenic. Lyotropic polymers form ordered states only in concentrated solutions (they require the presence of solvents), similar to the toothpick analogy. Thermotropic liquid crystals exist as pure and ordered melts without decomposing. Mesogenic polymers, a subclass of the thermotropic liquid crystals, usually have random coil backbones with rod-shaped side groups, and it is the latter that form the liquid crystal structures. Since phases are usually clearly distinguishable, microscopy has been used to construct phase diagrams (particularly using polarized light), while differential scanning calorimetry is used to determine first-order transitions between phases (Sperling, 1992). Lyotropic phases of surfactants (surface active agents) or emulsifiers in the presence of water are important in "zero-fat" margarines (see Section 5.6.2).

Liquid crystal structures are organized into different mesomorphic phases (mesophases). The *lamellar* or *neat phase* is ordered in one direction only, with chains lying parallel to each other forming bilayers separated by water. Lamellar structures are usually less viscous and less transparent than other mesomorphic phases. *Hexagonal* or *smectic* liquid crystals form cylinders arranged two-dimensionally in a hexagonal array. The viscous *isotropic* mesophase consists of spherical micellar units in a face-centered cubic lattice.

Surfactants, also called amphiphiles, are chemical compounds constituted by a hydrocarbon tail (nonpolar) and a polar or ionic portion (Figure 4–10). The polar portion interacts strongly with water via dipole-dipole or ion-dipole interactions and is solvated while the hydrocarbon chain is usually squeezed out of the aqueous phase. A first effect of surfactants (at low concentrations) is to lower the surface tension of the liquid phase when positioned at the interface, the energetically most favorable position. In concentrated aqueous dispersions, these type of molecules give rise to mesomorphic phases such as micelles or multilamellar structures, some of which are schematically illustrated in Figure 4–10.

A polymer may exhibit multiple mesophases at different temperatures and pressures. As temperature is raised, the polymer goes through several first-order phase transitions from relatively ordered to more disordered states. Phase diagrams exist in which the various mesophasic first-order transitions are depicted. Figure 3–11 presents a phase diagram of an industrially distilled saturated monoglyceride (DGMS)–water system (Krog & Lauridsen, 1976). At low temperature, a single phase formed by a mixture of $\beta$-crystals and water exists. A crystal transformation from $\beta$ to $\alpha$ form and penetration of water into the polar region results in formation of an $\alpha$-crystalline gel before a dispersion or a lamellar phase is formed. The gel is composed by a lamellar structure and a water layer of about 10 Å. The lamellar or neat phase consists of a lipid bilayer of about 38 Å and a water layer thickness of around 16 Å. Under special conditions, a much thicker water layer can be occluded between the bilayers. At high water content, the neat phase transform into a liquid crystalline dispersion in which water not incorporated in the lamellar phase exists as free water in "pockets." When the temperature is increased, a cubic phase (viscous isotropic mesophase) is formed. The cross-hatched area in Figure 3–11 represents a mixture of neat and viscous isotropic phases. Emulsifiers like DGMS are used in emulsions and low-calorie spreads.

Let's again look at nature. Biological membranes are important ultrastructural elements often referred to in this book (see Section 4.6.2). A working model of a biological membrane structure adapted from the fluid mosaic model of S.J. Singer is schematically illustrated in Figure 4–10. The phospholipid bilayer is envisioned as a fluid matrix 60–100 Å thick in which a number of proteins are accommodated according to the affinity of the hydrophilic and hydrophobic sections.

## 3.6 COLLOIDS AND SURFACE CHEMISTRY

### 3.6.1 Interfaces and Colloids

Many times the microstructure of foods is formed by more than one phase. Think of a salad dressing (e.g., a dispersion of fine oil droplets in water), whipped cream (air bubbles dispersed in a viscous liquid), or spreading butter over a slice of bread (increase in the contact surface between butter and air). In all cases a very peculiar boundary, the interfacial region (the region that separates two phases), has to be formed and maintained. In an emulsion, the interfacial area separating the dispersed and continuous phases is much larger after emulsification. Consequently, there are now some molecules in the interface whose properties are different from those in the bulk of either phase.

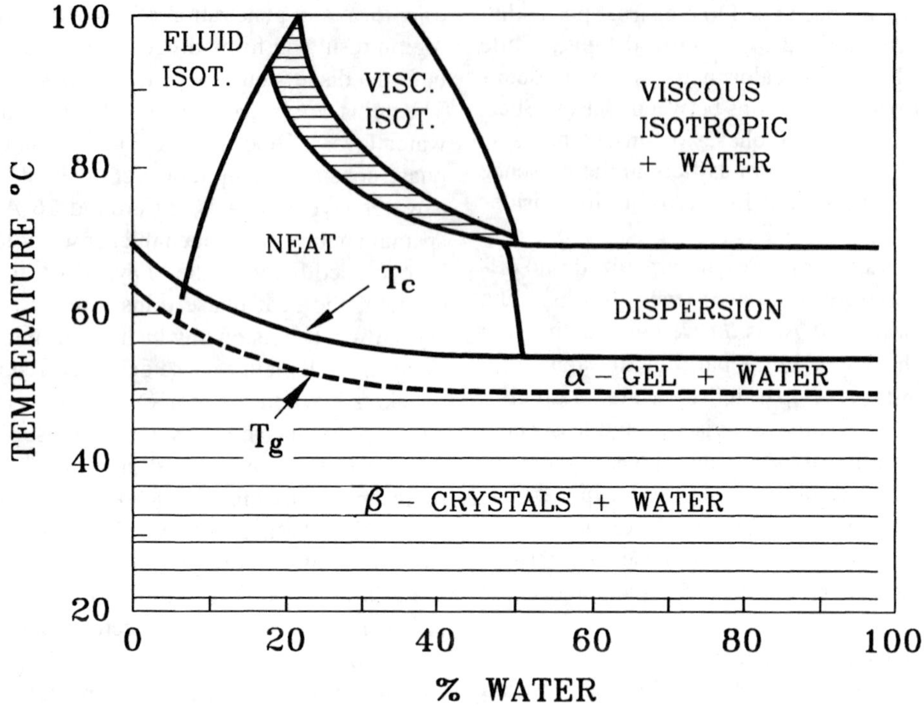


**Figure 3–11** Phase diagram of monoglyceride-water systems. *Source:* Reprinted with permission from N. Krog and J.B. Lauridsen, Food Emulsifiers and Their Association with Water, in *Food Emulsions*, S. Friberg, ed., pp. 67–139, by courtesy of Marcel Dekker, Inc.

Systems involving an interface are very often thermodynamically metastable.

Colloids are loosely but conveniently defined as particles or macromolecules larger than most molecules (e.g., solvent molecules) but definitely too small to be visible (Dickinson & Stainsby, 1982). For convenience, we will choose a size range between 1 nm and 1 $\mu$m, but obviously there will be exceptions. These dimensions are below the resolution of simple optical microscopes, so direct imaging of colloidal particles is best achieved by electron and atomic force microscopes.

Technologically, we will be interested in keeping colloidal particles apart (to stabilize emulsions) or bringing them together (as in coagulation and gelation). Proteins and polysaccharides will again play major roles as colloids in forming interfacial layers, aggregates, and networks; bridging particles together; structuring water; and modifying the rheological properties. We need to understand the physicochemical aspects ruling the behavior of structures in which surface play an important role (bubbles, droplets, capillaries, etc.) as well as related phenomena (wetting, coalescence, adhesion, etc.). Since there is a wide body of theory describing colloidal phenomena, only fundamental concepts and relationships will be presented here. Detailed and rigorous treatments of the subject, including explanatory schemes and figures, are contained in Adamson (1990), Everett (1993), and Hiemenz (1986).

### 3.6.2 Thermodynamics of Surfaces

The presence of an interface affects the thermodynamic properties of a system, and its study is also based on the Gibbs free energy. The extra free energy $dG$ of creating a new area $dA$ (at constant $T$ and $P$) is given by

$$dG = \gamma dA \qquad \textbf{Equation 3–31}$$

which constitutes the definition of $\gamma$, the interfacial or surface tension. The surface tension may also be expressed as free energy per unit area or as force per unit length (e.g., in the case of stretching a soap film). The SI units are J/m$^2$ or N/m (which are dimensionally identical), and in the cgs system, the units are erg/cm$^2$ or dyne/cm (the conversion factor from SI to cgs units is $10^{-3}$). Care must be taken because $\gamma$ is commonly defined for one interface; hence, $2\gamma$ should be used for two-sided films. The surface tension of most liquids decreases with temperature in a nearly linear fashion (Adamson, 1990).

An interesting result of equation 3–31 is that, anywhere interfacial effects are present, an extra term must be added to describe the total Gibbs free energy of a system. The new expression of $G$ (in differential form) for a multicomponent system takes this form:

$$dG = -SdT + Vdp + \sum_i \mu_i dn_i + \gamma dA$$

**Equation 3–32**

When $T$, $P$, and the number of moles remain constant, equation 3–32 reduces to equation 3–31. An important effect of the definition of $\gamma$ is that if $A$ decreases ($dA < 0$), then $dG$ is negative, which explains the spontaneous tendency of surfaces to contract (e.g., formation of drops) and the need to supply energy to create new surfaces (e.g., formation of emulsions).

When two phases $\alpha$ and $\beta$ are in contact (e.g., liquid and vapor), there is no sharp transition at the interface. In this region, the composition of certain species $j$ changes with the distance perpendicular to the interfacial surface. Thus, a mass balance for the number of moles of $j$ in each phase and at the interfacial diffuse zone can be written

$$n_j^I = n_j - (n_j^\alpha + n_j^\beta) \qquad \textbf{Equation 3–33}$$

where $n_j$ and $n_j^I$ are the number of moles of species $j$ in the whole system and at the interfacial zone, respectively, and $n_j^\alpha$ and $n_j^\beta$ are the number of moles in the bulk of each phase. The "excess of molecules of $j$" is defined as the surface excess (or surface concentration [in units mg/m$^2$]) through the relationship

$$\Gamma_j = n_j^I/A \qquad \textbf{Equation 3–34}$$

The real problem is that it is very difficult to measure surface excess (sometimes labelled isotopes are used). Changes in surface tension at constant $T$ can be related to the surface excess by an equation known as the Gibbs adsorption isotherm:

$$d\gamma = -\Gamma_j \, d\mu_j \qquad \textbf{Equation 3–35}$$

Remember that at equilibrium $\mu_j$ the chemical potential of species $j$ is the same everywhere in the system: It can be demonstrated that for dilute system (ideal) binary solutions in which only component 2 accumulates at the surface, the surface excess is given by the Gibbs equation (Everett, 1993):

$$\Gamma_2 = -\frac{1}{RT} \times \frac{\partial \gamma}{\partial \ln c_2} \qquad \textbf{Equation 3–36}$$

This equation states that when a solute accumulates in the interface ($\Gamma_2 > 0$) it causes a decrease in surface tension (i.e., $\partial\gamma/\partial c < 0$). This behavior is typical of most non-ionized organic compounds and amphiphylic species (surfactants). Surface excess is a preferred parameter for studying the kinetics of the adsorption-desorption of molecules (e.g., proteins) at interfaces, as in emulsions.

Since for small concentrations of component 2 the surface tension varies linearly with concentration, it can be demonstrated that the monolayer formed exerts a lateral pressure given by

$$\pi = \Gamma_2 RT \qquad \textbf{Equation 3–37}$$

where $\pi$ is the so-called film pressure (units of N/m). This expression is in fact, analogous to the ideal gas law, but for two dimensions. The lateral pressure can now be regarded as arising from the bombardment of adsorbing molecules in the surface against the walls of a container.

### 3.6.3 Contact Angle, Wetting, and Related Phenomena

There are several situations where high contact (wetting) is desired between a liquid, usually an aqueous solution, and an oily or waxy surface, for example, when treating fruits with an insecticide spray or when trying to remove the waxy layer of grapes to increase moisture diffusion during air drying.

The general situation of the contact angle in wetting is illustrated in Figure 3–12. The question is: At equilibrium, what is the angle between the liquid and the solid? A change in surface free energy occurs when there is a small displacement of the liquid, resulting in a change in the area of the solid covered by the liquid ($\Delta A$), resulting in the expression

$$\Delta G = \Delta A\,(\gamma_{SL} - \gamma_{SV}) + \Delta A\,\gamma_{LV}\,(\cos\theta - \Delta\theta)$$

**Equation 3–38**

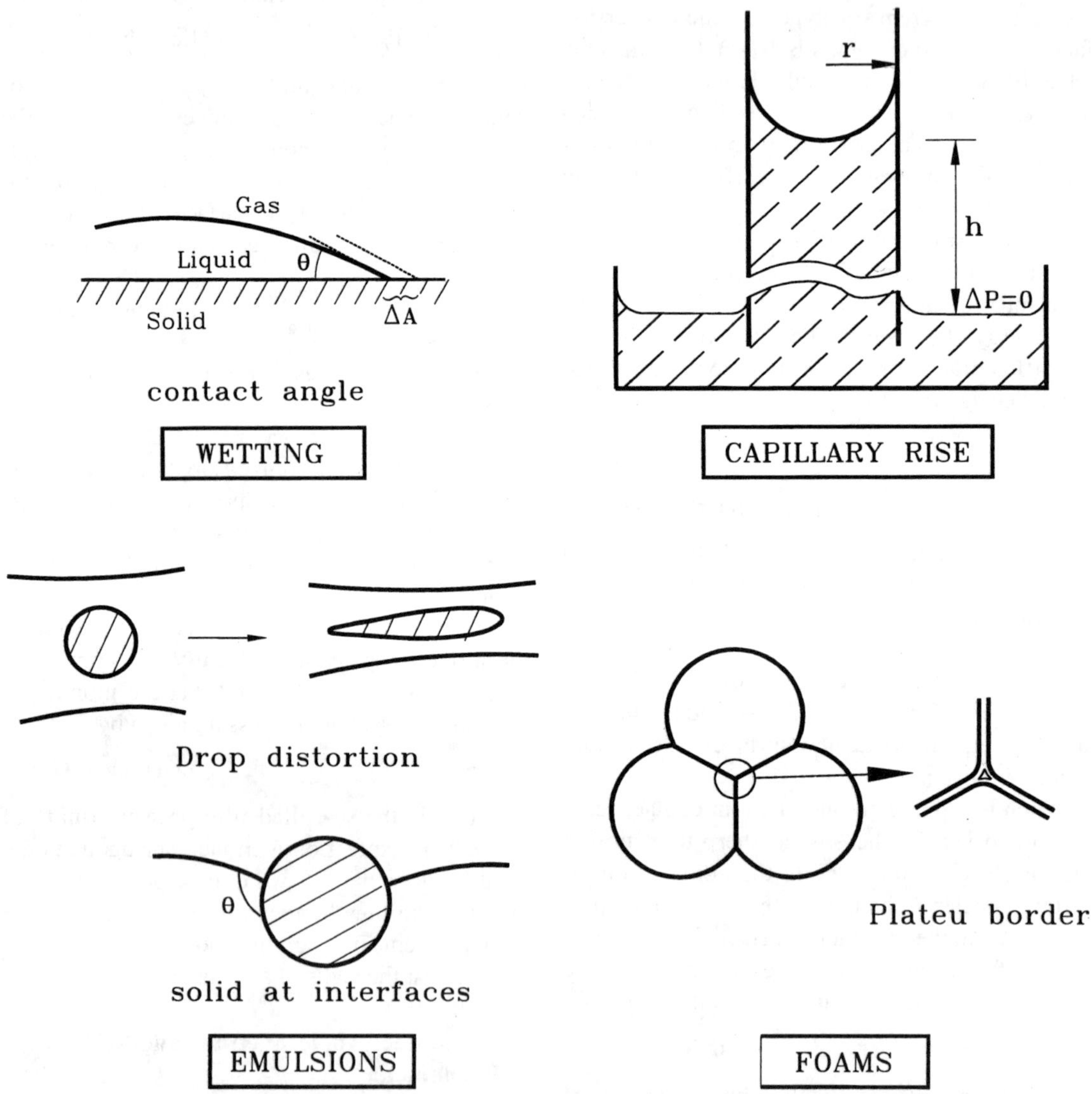


**Figure 3–12** Schemes of surface phenomena commonly found in foods: wetting, capillary rise, droplet deformation during homogenization, solids at interfaces, and foams.

where the subindices of the surface tension terms define the phases involved and the angle $\theta$ is conventionally measured in the densest fluid phase (liquid in this case).

At equilibrium, $\Delta G/\Delta A \rightarrow 0$ (since $\Delta\theta/\Delta A$ is a second-order differential, it also approaches 0), and the previous equation reduces to Young's equation (sometimes called the Young-Dupré equation):

$$\cos\theta = \frac{\gamma_{SV} - \gamma_{SL}}{\gamma_{LV}}$$ **Equation 3–39**

Wetting usually means that the contact angle between a liquid and a solid is zero and the liquid spreads over the solid (see below). Nonwetting implies that the contact angle is greater than 90° and the liquid tends to ball up and run off the surface easily (e.g., water over an hydrophobic surface). (Remember that cos 0° = 1; cos 90° = 0; cosine of an angle greater than 90° and less than 180° is less than 0; and cos 180° = −1).

If a structure has a rough rather than a smooth surface, Young's equation needs to be corrected. Roughness is empirically expressed as the ratio ($r > 1$) of the actual surface area to the projected area and is a weighting factor for cos $\theta$. Another interesting case is when the surface is heterogeneous or made of patches of various kinds of solids. If there are two kinds of patches occupying fractions $f_1$ and $f_2$ of the surface, equilibrium of forces leads to an equation from which cos $\theta$ can be obtained from the contact angles (Adamson, 1990):

$$(1 + \cos\theta)^2 = f_1(1 + \cos\theta_1)^2 + f_2(1 + \cos\theta_2)^2$$

**Equation 3–40**

Often the wetting or spreading of a liquid over a solid surface is required, but in other cases waterproofing is desired. A spreading coefficient $S_{L/S}$ is used to analyze whether a liquid would spread over a solid:

$$S_{L/S} = \gamma_{SV} - \gamma_{LV} - \gamma_{SL}$$ **Equation 3–41**

In general, the larger and the more positive $S_{L/S}$, the easier it is to spread a liquid film over the solid. Hence, for spreading to occur, $\gamma_{SL}$ and $\gamma_{LV}$ should be made as small as possible. From a practical standpoint, minimization is best achieved by adding to the liquid phase a surfactant that is adsorbed at the solid-liquid and the liquid-air interfaces and thus lowers the interfacial tensions.

### 3.6.4 Adhesion

Adhesion is a subject both of polymer and surface science. In food technology, it is desirable to have strong adhesion when binding pieces of meat in a restructured product but low adhesion between dough and muffin tins (so the dough will not stick). In general, low values for the surface energy of the solid mean low adhesion. Nonstick applications of Teflon® for frying are compatible with its low $\gamma$ (around 17 mJ/m$^2$ at 140°C). On the other hand, metal surfaces (e.g., processing equipment) show surface energies well exceeding 100 mJ/m$^2$.

The first property of a good adhesive is its capability to wet and spread over the surfaces of the pieces to be stuck together. Several mechanisms participate in good bonding between the adhesive and the substrate (Jolley & Purslow, 1988; Lee, 1989):

- diffusion of the molecules of the adhesive
- physical entanglement between polymer and substrate molecules
- physical adsorption (van der Waals' forces or acid-base interactions)
- chemical bonding (covalent bonds)
- electrostatic mechanisms (important in particle adhesion)

Adhesion is a particularly important problem in the baking industry. Wheat flour dough sticks to processing equipment by an adhesive bond. Tack is the ability of two materials (normally dissimilar) to resist separation after their surfaces come into contact. The adhesive bond strength formed at the interface depends directly on contact time and inversely on the number of monomers in the polymer chain (Lee, 1989). It may be acceptable for a food material to exhibit tack as long as separation (failure) is adhesive (separates cleanly from surface) rather than cohesive (leaves residue). In dough, the mode of failure depends on the separation rate, temperature, and moisture content (Saunders, Hamann, & Lineback, 1992)

### 3.6.5 Capillarity Phenomena

Capillaries are naturally present in many solid foods (even in fresh fruits and vegetables) and as remnants of passages for moisture diffusion in dried products (e.g., powders). Often they are purposely induced by processing, as in the agglomeration of dry powders to assist in the wetting and sinking of particles. The Laplace equation gives the pressure difference across the interface ($\Delta P$) between two phases $\alpha$ and $\beta$ when they are separated by a curved surface:

$$\Delta P = \gamma\left(\frac{1}{R_1} + \frac{1}{R_2}\right) \quad \text{Equation 3–42}$$

where $R_1$ and $R_2$ are the principal radii of curvature describing the curved surface and the term in parentheses is the curvature. If the center of curvature corresponding to $R_1$ lies in phase $\alpha$, then $\Delta P = (P_\alpha - P_\beta)$ is positive. If $R_1$ and $R_2$ are both positive, then the pressure in $\alpha$ is higher than in $\beta$. For a sphere of radius $r$ (e.g., a bubble or drop) $R_1 = R_2 = r$ and the curvature is 2 $r$. Laplace's equation has important consequences.

First, drops and bubbles tend to be spherical. If a drop or bubble is not, material will move from high-pressure areas (smaller curvature) to those with a lower pressure until a spherical shape is obtained.

Second, if a vertical capillary of circular cross-section (not too large in radius) contains a liquid, a concave meniscus will form. Because the pressure just below the meniscus is lower than the atmospheric pressure just above it, a liquid column will rise to a height $h$ above a flat surface (for which $\Delta$P is zero), as shown in Figure 3–12. If the liquid does not wet the walls of the solid, the capillary rise is given by the expression

$$h = \frac{2\gamma \cos\theta}{\Delta\rho r g} \quad \text{Equation 3–43}$$

where $r$ is the radius of the capillary, $\theta$ the contact angle, and $\Delta\rho$ is the difference in density between the liquid and the gas above it. If the liquid wets the solid, then equation 3–43 reduces to

$$h = \frac{2\gamma}{\Delta\rho r g} \quad \text{Equation 3–44}$$

Third, since the chemical potential depends on pressure, the properties of a liquid inside a drop will be different from those in the bulk. In particular, the effect of curvature on the vapor pressure of a spherical drop is given by the Kelvin equation:

$$RT\ln\frac{p}{p^o} = \frac{2\gamma V_m}{r} \quad \text{Equation 3–45}$$

where $p$ is the vapor pressure observed over the curved surface of mean radius of curvature $r$, $p^o$ is the normal vapor pressure, and $V_m$ is the molar volume of the liquid. Because a liquid drop has a positive curvature, $p/p^o > 1$ and the vapor pressure is higher than in a flat surface. In the case of a meniscus (e.g., liquid inside a capillary) or a bubble immersed in liquid, $p/p^o < 1$.

It is easily recognized that $p/p^o$ in the case of water is the water activity ($a_w$). For capillaries of radius $10^{-6}$, $10^{-7}$, $10^{-8}$, and $10^{-9}$ m filled with water at 20°C, the Kelvin equation predicts values for $p/p^o$ of 0.9989, 0.9893, 0.8976, and 0.339 (!), respectively (Hiemenz, 1986), which means that water in capillaries lowers water activity (see Section 9.2.2). It is questionable, however, whether the last value for capillary dimension (1 nm) has any real physical meaning.

Fourth, an expression similar to equation 3–45 holds for solubility (since the solubility of a gas is proportional to its pressure). The increased solubility of particles of small radius (e.g., less than 0.1 $\mu$m) promotes the growth of large particles at their expense, a phenomenon called *Ostwald ripening,* which is responsible of the enlargement of ice crystals during storage of ice cream and the growth of bubbles, among others.

### 3.6.6 Emulsions and Their Formation: Homogenization

Emulsions are dispersions of two immiscible liquid phases, one of which is the continuous phase while the other is the dispersed phase. Macroemulsions, described below, are emulsions in which the mean droplet diameter of the dispersed phase is between 0.1 and 50 $\mu$m. The commonest examples of food emulsions are dispersions of fine oil droplets in aqueous media (O/W) or of fine aqueous droplets in an oily phase (W/O) (Figure 3–13).

As is the case in all size reduction operations, formation of an emulsion involves an increase in interfacial area between both phases, which is accompanied by an increase in free energy (see equation 3–31). Emulsions are thermodynamically unstable: they need energy to be formed and tend to separate unless stabilized. The ease of formation of an emulsion is measured by the energy that must be supplied to form it. Also, the lower the interfacial tension, the easier the formation of the emulsion. Thus, emulsifying agents (surfactants) or molecules that adsorb at the interface are often added. Lowering the surface tension, however, is by no means sufficient to produce stable emulsions.

Several types of machines are used to supply the energy to form emulsions: stirring vessels (for production of coarse emulsions), colloid mills, toothed-disc dispersing machines, and homogenizers. In high-pressure homogenizers, the premix

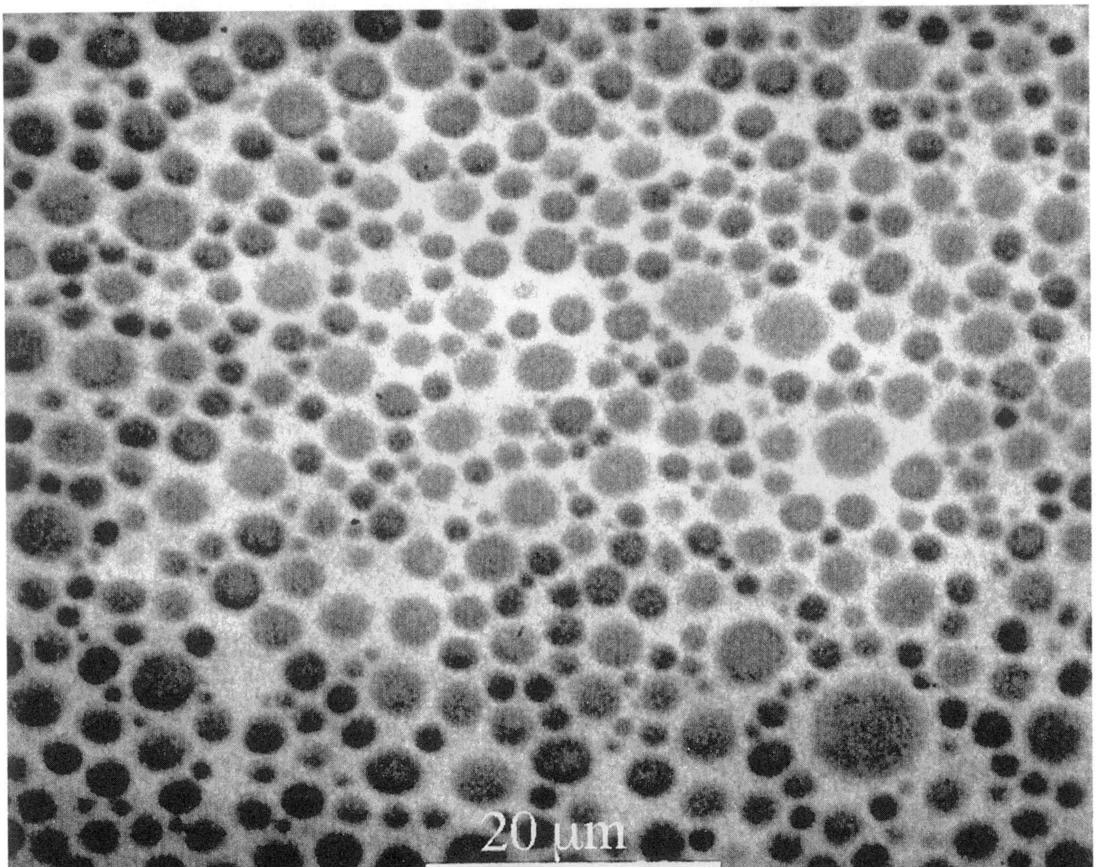


**Figure 3–13** ESEM micrograph of a vegetable oil-in-water emulsion in its liquid state, with no prior treatment. This entirely liquid sample shows remarkable contrast between both phases, with the oil phase being darker than the more intense continuous water phase. *Source:* Reprinted with permission from D.J. Stokes, B.L. Thiel, and A.M. Donald, Direct Observation of Water-Oil Emulsion Systems in the Liquid State by Environmental Scanning Electron Microscopy. *Langmuir*, Vol. 14, No. 16, pp. 4402–4408. Copyright 1998 American Chemical Society.

is forced to flow at high speed through a narrow gap under elevated pressure, resulting in high shear and cavitation. Droplets are effectively disrupted by colloid mills and toothed-disc dispersing machines if the viscosity is high but only homogenizers produce low-viscosity emulsions efficiently (e.g., milk). Homogenization of fat globules is reviewed in Section 7.3.2.

Emulsions are formed by a complex physical process of droplet deformation. The mechanism involves elongation of the droplet under shearing forces, necking, and finally separation into smaller drops (Figure 3–12). At the droplet level, any stress acting on a particle is opposed by the Laplace pressure (equation 3–42). Thus, a larger stress has to be applied if deformation and breakage is to occur. Notice that in Laplace's equation (equation 3–42) the stress is less if the interfacial tension is lower. In practice, shearing stresses arise from a velocity gradient or from pressure differences. As discussed by Walstra (1993), the actual energy supplied to make an emulsion is on the order of 1,000 times the theoretical energy needed to create the new surface, the difference being dissipated as heat in the bulk of the liquid (this dissipation occurs in all practical particle reduction processes).

In laminar flow, the Weber number (*We*) is defined as the ratio of the stress exerted by the surrounding fluid on a drop due to the velocity gradient $\nu_Z$ to the Laplace pressure:

$$We = \frac{\nu_z D_p}{4\gamma} \qquad \textbf{Equation 3–46}$$

Notice that in Laplace's equation [equation 3–42] the radius has been replaced by the droplet diameter Dp. The stress is given by Newton's law for viscosity and depends on the viscosity and the velocity gradient in the surrounding fluid. Correlations are made between the *We* and the viscosity ratio between the dispersed and the continuous phase ($\eta_d/\eta_c$) such that a critical *We* ($We_{cr}$) is defined for drop breakup. $We_{cr}$ varies between 1.3 and 4 for $\eta_d/\eta_c < 4$ but becomes extremely large above this value, meaning that no breakup occurs (Walstra, 1993).

In turbulent flow, where there are large fluctuations in local velocity, droplet disruption and breakup depend on the energy density per unit volume $E_v$ and the viscosity of the dispersed phase. $E_v$ is the main parameter characterizing turbulence, and it is related to specific power $P_v$ and the residence time of the emulsion in the dispersing zone $t_r$ as follows (Karbstein & Schubert, 1995a):

$$E_v = P_v t_r \qquad \textbf{Equation 3–47}$$

Under these conditions the largest droplet size that can remain in the turbulent field is given by this expression (Walstra, 1993):

$$D_p^{max} \approx E_v^{-2/5} \gamma^{3/5} \rho^{-1/5} \qquad \textbf{Equation 3–48}$$

The energy density may vary considerably, from $10^4$ W/m$^3$ for a simple stirrer to $10^{12}$ W/m$^3$ for a high-pressure homogenizer. Surface tensions for O/W systems typically range from 27 mN/m (for pure interfaces) to 13–15 mN/m (for systems containing proteins) and 3–10 mN/m (if surfactants are used) (Krog, Bradford, & Sanchez, 1989).

During emulsification, three processes occur simultaneously: (1) droplet deformation, (2) adsorption of surfactants onto deformed and newly created droplets, and (3) encounters or collisions between droplets and possibly coalescence. Droplets formed are unstable if emulsifier molecules cannot diffuse rapidly to the newly created surfaces. When droplets leave the dispersing zone of equipment, the contact time increases and the droplets tend to coalesce. Thus, multiple passes through an homogenizer have the effect of disrupting coalesced droplets rather than further reducing particle size. The rate of adsorption of surfactants to newly formed surfaces is of major importance for stabilizing droplets and avoiding coalescence (Karbstein & Schubert, 1995b). Globular proteins and egg yolk adsorb slowly, whereas some non-ionic emulsifiers stabilize interfaces in milliseconds.

### 3.6.7 Stabilization of Colloidal Particles

De-emulsification (a spontaneous process) occurs through the contact of droplets and their coalescence into larger droplets. Several mechanisms can lead to the stabilization of colloidal particles.

If the colloidal particles have a net charge, the interaction between two particles of the same sign can be explained by the so-called Deryagin-Landau-Vorwey-Overbeek (DLVO) theory, which takes into account the opposite contributions of electrostatic repulsion forces and attraction forces. In such a case, the stability of a colloidal suspension depends on the summation of the repulsion force due to the electrical double layer [proportional to $\ln(1 - e^x)$] and the van der Waals' attraction forces (proportional to $1/x$). Since the dependence of these two forces on interparticle distance $x$ is different, one result may be the appearance of two minima at which net attraction occurs, as shown in Figure 3–14. Outside these two zones, net repulsion prevails and charged particles become stabilized against aggregation or flocculation. Counter-ions in solution tend to decrease the width of the electrical double layer and thus allow closer proximity between particles. Important parameters to control are the ionic force and the concentration and valence of counter-ions in the solution. It has been proposed that the secondary minimum corresponds to for-

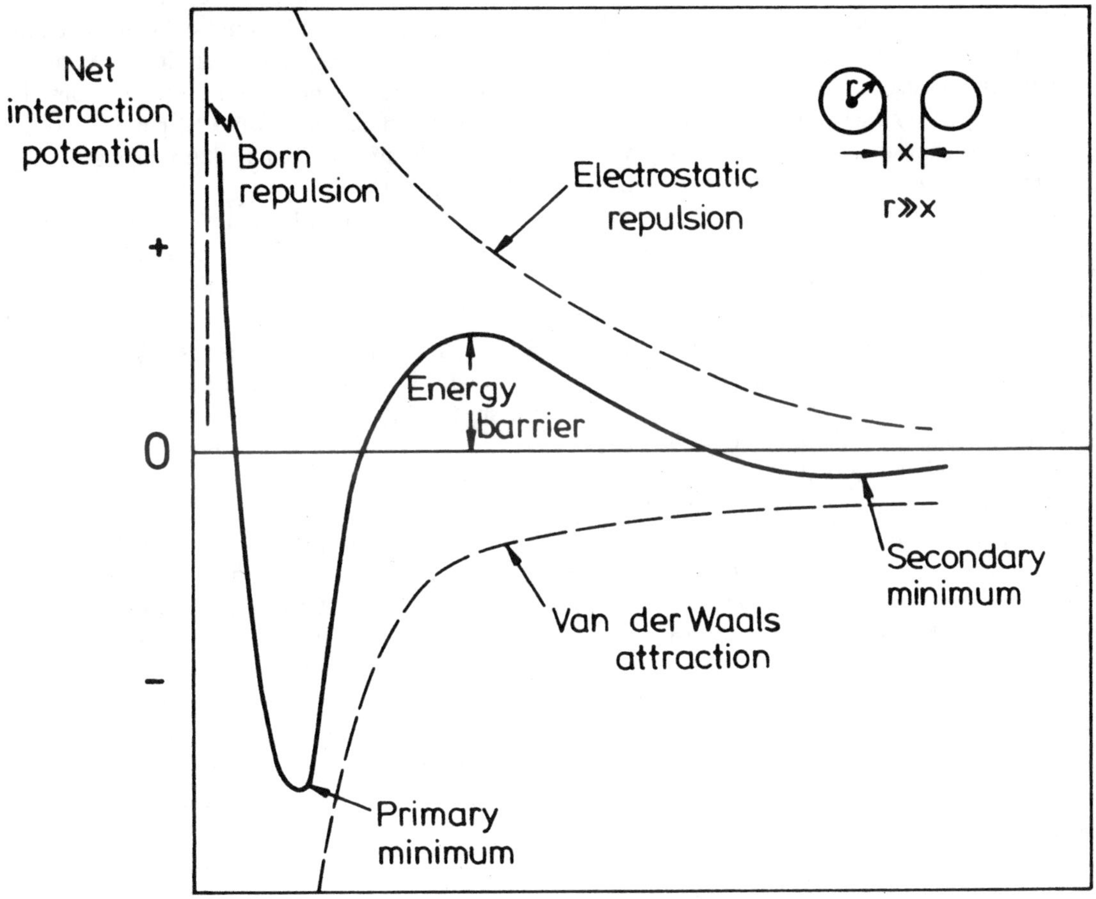


**Figure 3–14** Repulsion, attraction, and total interaction energies between two charged colloidal particles according to the DLVO theory ($x$ = interparticle distance). Minima in total interaction curve represent conditions for flocculation and precipitation. Polymers in solution and adsorbed on the surface of particles also affect stability of colloidal suspensions.

mation of loose flocs (flocculation) while the primary minimum, which occurs at closer interparticle distances, leads to precipitation.

Polymers may also stabilize (repel) or destabilize (attract) colloidal particles in solution, depending on whether they are adsorbed on the surface or remain in solution. Adsorbed polymers (e.g., denatured proteins) tend to position themselves at the interface and protrude into both phases. Upon collision between droplets, the polymer chains will interpenetrate, leading to two stabilizing effects (so-called *steric stabilization*): osmotic and entropic. The first is due to the diffusion of medium into the region between two surfaces, which reduces the concentration of polymer segments and drives the two particles apart. The second lowers the entropy of the polymer segments by reducing the number of configurations they can adopt; hence an increase in free energy drives the particles back. We will see in the next section that electrostatic charges also contribute to emulsion stabilization. Proteins are used to provide steric stabilization of emulsions and prevent coalescence. The best proteins are those that are both soluble and surface active. The amount of protein adsorbed at the interface—the surface excess (Section 3.6.2)—may be close to 3 $mg/m^2$ for a monolayer of globular proteins. Transmission electron microscopy has been used to qualitatively determine the relative adsorption of different proteins (Krog et al., 1989). This subject has been recently reviewed by Dagleish (1997).

The presence of nonadsorbed polymers in solution may cause aggregation by *depletion-flocculation*. On the close approach of two particles, there may be established a depletion layer from which large polymer molecules will be excluded (they become negatively adsorbed). Hence, the osmotic pressure of the bulk polymer solution will remove solvent from within the interparticle space and push particles together as if there were an attractive force between them. Flocculation may also be induced by the bridging of particles. The two ends of a long polymer chain may adsorb on separate particles and draw them together (e.g., an effect used in water purification).

Emulsions (like foams) may also be stabilized by fine solids that adsorb at the interface (Figure 3–12), providing steric hindrance to a close approach. The equilibrium condition for a solid sphere of radius $r$ positioned between two phases $\alpha$ and $\beta$ is (Adamson, 1990)

$$dG = \gamma_{S\alpha}(2\pi r dh) + \gamma_{S\beta}(-2\pi r dh) - \gamma_{\alpha\beta}\pi(2r - 2h)dh = 0$$

**Equation 3–49**

where $h$ is the distance the sphere protrudes into the $\alpha$ phase. Equation 3–49 leads to the Young-Dupré equation. Hence, at equilibrium a particle adopts a position in the interface such that the angle $\theta$ is the contact angle. If $\theta$ is finite, the particle will be stable at the surface. Theoretically, for a particle to become stabilized at an interface, it should have a small radius (roughly $<$ 1 mm; if larger, gravity may pull it from the interface). Thus, finely divided particles (e.g., as occur in mustard) may help in stabilizing emulsions.

The stability of solid multicomponent emulsions and foams is much more complex. In whipped cream, the crystalline fat fraction provides the rigidity that ensures the stability of the foam, while in ice cream the system is further stabilized by ice crystals and a glassy matrix (Dagleish, 1997).

### 3.6.8 Creaming and Phase Separation

Emulsions can also lead to phase separation, as in the phenomenon of "creaming." Creaming is basically a sedimentation process by which fat droplets migrate to the top of the vessel and separate out as a concentrated "phase," leaving a lean or defatted liquid or plasma underneath (see Figure 7–5). It is interesting to explain the relation between creaming and the microstructure of the emulsion using Stoke's equation:

$$v_t = \frac{\Delta\rho D_p^2}{18\eta} g \cdot \psi$$

**Equation 3–50**

where $v_t$ is the terminal velocity, $\Delta\rho$ is the density difference between particle and medium, $D_p$ is the particle diameter, $\eta$ is the viscosity of the continuous phase, and $\psi$ is a correction factor for deviations from idealities implied in the derivation of the equation. The terminal velocity is the constant speed at which a droplet or particle would move inside a vessel (up or down depending on the sign of $\Delta\rho$) under the action of gravity. We can retard the process (make $v_t$ small) by increasing $\eta$ (e.g., by modifying the structure of the dispersed phase through adding a polymer) or, much better, by changing the microstructure of the emulsion and reducing the size of the droplet (the term $D_p$ is raised to the square power). Manipulation of the term $\psi$ is less obvious, but it is also related to the microstructure of the system—for example, when interference between particles exists. Emulsion stability is discussed in relation to milk in Chapter 7.

### 3.6.9 Foams and Bubbles

A foam is a coarse dispersion of small gas bubbles separated from a liquid by thin films or lamellae. Just as a curiosity, the pattern of bands of rainbowlike colors seen in soap bubbles results from the interference of light reflected from the front and the back surfaces of the film. Foam formation (foaming) and stability have ample application in food technology, and they are "functional properties" of proteins ("emulsifying capacity" is another). Foam stability is desirable in beer (so the "head" will stay in the glass during drinking) but unwanted during the agitation of milk. It is also promoted in whipped egg white so the foam will not break down before it is solidified in the oven as a meringue sets.

The pressure difference across interfaces is usually very small, resulting in individual cells of the foam being of polyhedral shape bounded by plane faces (Figure 3–12). Under the influence of gravity and surface tension forces, a foam drains along lamellae to the curved menisci at the junction of thin lamellae (plateau borders), where the pressure is lower (see equation 3–42). Among other factors, high viscosity of the liquid phase increases the lifetime of a foam by reducing the draining rate.

Stable films present notable resistance to mechanical disturbances. Gibbs defined an "elasticity" for the film as

$$E = \frac{2d\gamma}{d\ln A} \qquad \textbf{Equation 3–51}$$

Qualitatively, $E$ is a measure of a film's ability to adjust its surface tension in a moment of stress, and a greater value tends to produce more stable foams. Typical values of $E$ are in the 10–40 mN/m range (Adamson, 1990). For pure liquids, $E$ should be zero, and this is in accord with the observation that pure liquids do not foam (or the foams are extremely unstable). An interesting stabilization mechanism is established when there is a surfactant in the liquid film. As the area is extended, the surface excess of the surfactant drops and the surface tension rises accordingly, opposing the stretching and thus protecting the film against rupture. If "surface healing" is not produced, the film thins and eventually collapses. Once a hole is made, it propagates spontaneously, reducing the surface area of the system.

Once in a while it is necessary to destroy unwanted foams. In some cases destruction may be achieved by blowing hot air across the surface of the foam. More often antifoaming agents are used (e.g., ether, n-butanol, and silicone). They seem to accomplish their effect by replacing the foaming agent at the surface, leading to the concomitant collapse of the foam.

## 3.7 MECHANICAL AND RHEOLOGICAL PROPERTIES

### 3.7.1 On Structures and Properties

The purpose of this section is to provide the reader with some basic concepts and models that characterize the behavior of foods as materials. The subject of physical properties of foods has been extensively investigated in the last 20 years, especially from the standpoint of methodology and data acquisition for engineering applications. Food engineers have advantageously adopted methods for measuring the mechanical and transport properties of prepared foods from chemical and mechanical engineers. For numerical data on the physical properties of whole foods, consult the computer database compiled by Singh (1993).

A property is a trait of a thing. Most physical properties of foods—including mechanical, rheological, electrical, and transport (heat and mass diffusivity) properties—are common to all industrial materials. Some food attributes are more specific, such as color and sound. Still others are even more difficult to define or measure, for example, flavor, texture, nutritional value, and shelf-life stability. All of them are important, and structure will usually play a role in their manifestation; therefore, a structure-property relationship needs to be found.

Vincent (1990) has correctly pointed out that the first task from an engineering perspective is to determine whether one is dealing with the property of a material or of a structure. A material is usually homogeneous and isotropic (properties do not vary with direction; for example, the stiffness is the same in tension, compression, and bending). Pure substances and alloys or uniform solid blends (which foods sometimes are) can also be regarded as materials. But, how do we define the structure of a food? Structures are composed of more than one material and normally exhibit some regularity or pattern that, when clearly discernible, is referred to as the "architecture." Thus, the microstructure of a food can only be understood when its architecture and elements (solid, liquid, and gaseous) are considered together. It is in the development of structural (or physical) models of the food that microscopy is irreplaceable.

Most foods are not simple, homogenous materials: they have complex structures, some of which are biochemically active and change with time (e.g., ripening, dough leavening). Many foods, are, at the microstructural level, highly structured tissues, composites, or complex suspensions. Consequently, determining a food's physical properties in bulk—in other words, considering the food as a "black box"—may be appropriate only when dealing with a standardized food. Alternatively, individual physical properties may be obtained for key architectural elements, and the overall property may be calculated from a physical model of the structural arrangements and interactions. The latter approach, which lies at the heart of engineering and materials science, provides flexibility for changes in formulation and processing. A good physical model is essential for uncovering the relationships between a material's structure and its mechanical, rheological, or transport properties. The differences between a material and a structure are considered in detail in later chapters.

In addition, the problem of discovering a food's properties does not end in mere numbers, regardless of how precisely they represents the properties. The ultimate challenge is to find the relationships between elements of the structure–property–sensorial effect triad. Understanding the microstructure of a food is not enough; we want to know how the food breaks down in the mouth and what are the eater's derived perceptions. Food science is just starting to peep into this complicated subject, and much more about how the brain functions will need to be learned before food properties can be fully explained.

### 3.7.2 Structure-Property Relationships

Structure-property relationships are crucial in materials science, for example, in the design of alloys and composites of great strength, light weight, and high resistance to heat for aeronautic applications. These relationships, in fact, determine the character of all the three-dimensional objects we come into contact with—artificial and natural. The shell that protects a macadamia nut is stronger than most metals yet is less than half as dense, thus minimizing the plant's weight burden (Niklas, 1992).

Nature constructs several structures to perform complex and multiple tasks. As in the case with foods, these structures are made up of a few rather simple macromolecules that themselves consist of even simpler repeating units. Thus, the complexity indigenous to biomaterials is firmly based on the intricacies of their molecular and supramolecular architecture. This complexity has been continually unraveled through the use of increasingly powerful analytical techniques, such as microscopy, thermal and mechanical analysis, and advanced spectroscopy (Baer, Cassidy, & Hiltner 1991). Fine-grained analysis of structure is also the direction in which food materials science is moving, and lessons from nature and biology may be applied advantageously.

The structure-property relationships of foods strongly affect their physicochemical, functional, technological, and even nutritional properties. Several examples are listed in Table 3–1.

Although advances in conceptualizing how food structure is formed have occurred and efforts are underway to relate structure to physicochemical properties (in particular mechanical and rheological properties), food science lags far behind materials science. Some of the interaction forces participating in structure formation and its stabilization are beginning to be understood qualitatively, but quantitative data and models are yet not available. Chapter 10, Section 2 returns to the topic of structure-property relationships in foods, including their importance and what to expect from them in the future.

### 3.7.3 The Ideal Solid-Liquid Spectrum

Scientists and engineers work with simple models representing the behavior of complex situations found in nature. When a force is applied to a material, the material may flow or deform. The simplest behavior for flow is that of the Newtonian or ideal liquid. On the opposite side of the spectrum is the Hookean or ideal solid, which does not flow but deforms elastically. In between these two ideal extremes lies viscoelasticity, or the behavior of soft solids under stress (Figure 3–15). A viscoelastic material has a response to stress that is partly viscous and partly elastic. The ideal model is the linear viscoelasticity model, in which simple relationships have been derived (although the linear viscoelastic range in real materials may be very small). The following sections revise some of the basic concepts and relationships of the rheology of fluids and suspensions and the viscoelasticity and mechanical properties of homogeneous and composite solids. A useful complementary reference is Stanley, Stone, & Tung (1996).

## 3.8 RHEOLOGY OF FOODS

### 3.8.1 Rheology: Empirical Models

Rheology is the study of the deformation and flow of materials under defined conditions. Rheometric

**Table 3–1** Examples of Structure-Property Relationships in Foods

| *Property Class* | *Measured or Observed Property* | *Example Related to Structure* |
|---|---|---|
| Sensory | Texture | Hardening of middle lamella in stored legumes |
| | Flavor release | Type of emulsion (W/O rather than O/W) |
| | Color | Size of pores in freeze-dried coffee powders |
| Mechanical | Strength | Reinforcement of protein gels with starch granules |
| | Structural stability | Caking of amorphous food powders |
| Rheological | Creaming | Reduction of fat globule size by homogenization |
| | Viscosity | Addition of gums in salad dressing formulation |
| Transport | Oil uptake | Surface treatments before frying of potatoes |
| | Diffusion of solutes | Rate of salt impregnation in cheeses |
| Functional | Water holding capacity | Water structuring in low-fat spreads |
| | Desiccation resistance | Formation of a glassy matrix |
| Nutritional | Retention of nutrients | Encapsulation of vitamins |
| | Nutrient availability | Cooking of starchy materials |

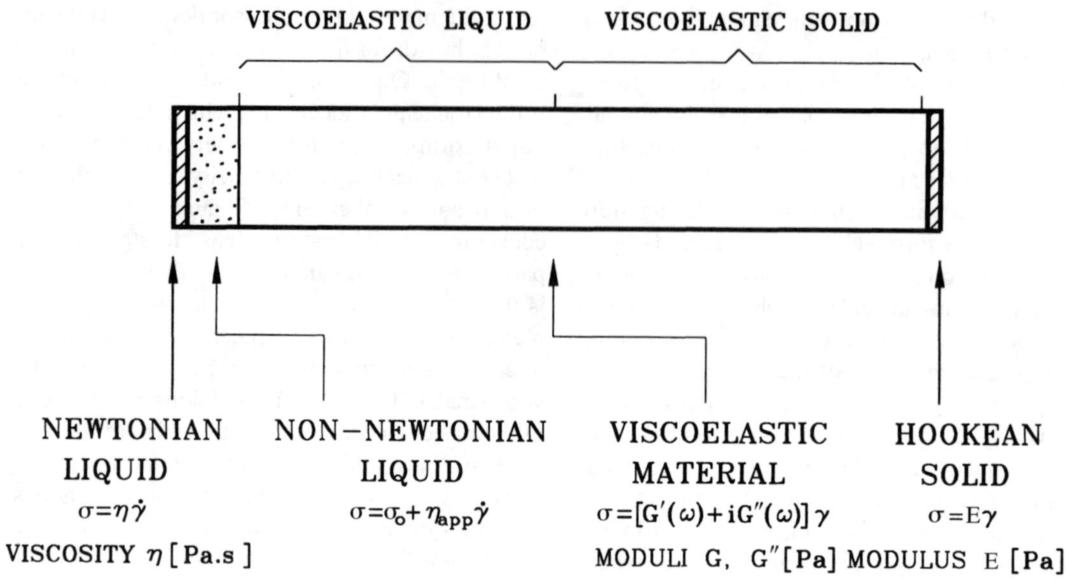


**Figure 3–15** Spectrum of the mechanical or rheological behavior of food materials from the ideal liquid to the ideal solid. Some foods exhibit viscoelasticity or simultaneous liquidlike and solidlike behavior. *Source:* Adapted from Ziegler and Foegeding, 1990.

flow can be divided into shear and elongational flow. Shear flow is best studied in a rheometer, in which a material moves between two surfaces, one of which rotates at a constant velocity. The rheometer basically measures either the torque $\sigma$ (e.g., force/area of contact between fluid and moving surface) or the shear rate $\dot{\gamma}$(e.g., velocity/gap occupied by the fluid). The parameter determined is the internal frictional resistance to flow or the *shear viscosity* $\eta$. In elongational flow, material is stretched and an *elongational viscosity* is determined. What follows is concerned only with the shear viscosity. Fundamental aspects of rheology are the subject of books such as those by Barnes, Hutton, and Walters (1989) and Macosko (1994). The rheology of food biopolymers is reviewed by Kokini (1992).

When stress ($\sigma$) is plotted against shear rate ($\dot{\gamma}$), all Newtonian fluids exhibit the following linear relationship, through the origin and with a slope equal to the *Newtonian viscosity* ($\eta$):

$$\eta = \sigma/\dot{\gamma} \qquad \textbf{Equation 3–52}$$

This equation may be regarded as the constitutive equation for an ideal liquid. In SI units, force is expressed as Newtons (N). The force exerted by gravity on a mass of 100 g (e.g., an apple) is 1 N. The unit of stress is the Pascal (Pa), or 1 N/m$^2$, and the unit of shear rate is s$^{-1}$. Viscosity is expressed as Pa·s in SI units and as poise in cgs units (0.1 Pa·s = 1 poise). One mPa·s or 1 centipoise approximately equals the viscosity of water at room temperature. Table 3–2 presents the order of magnitude of viscosity of some common food materials. Note that the range of viscosity for foods that flow readily at room temperatures may span 5 to 6 orders of magnitude and that a higher viscosity may not always imply non-Newtonian behavior. Table 3–3 shows the magnitude of shear rates found in some common food-processing operations. Shear rates can be estimated by dividing an average velocity for the fluid by a representative distance or by measuring a velocity gradient at a certain point.

In many food liquids and suspensions, $\sigma$ versus $\dot{\gamma}$ data do not fit a straight line (e.g., the viscosity does not remain constant), but rather $\sigma$ varies with $\dot{\gamma}$, so that an *apparent viscosity* (sometimes labeled $\eta_{app}$) has to be defined at each $\dot{\gamma}$. Moreover,

**Table 3–2** The Viscosity of Some Familiar Materials at Room Temperature

| *Liquid* | *Approximate Viscosity (Pa·s)* |
|---|---|
| Food glass | $10^{12-14}$ |
| Molten polymers | $10^3$ |
| Heavy syrup | $10^2$ |
| Honey | $10^1$ |
| Glycerol | $10^0$ |
| Milk fat | $10^{-1}$ |
| Light oil | $10^{-2}$ |
| Water | $10^{-3}$ |
| Air | $10^{-5}$ |

*Source:* Reprinted from H.A. Barnes, J.F. Hutton, and K. Walters. *An Introduction to Rheology*, Copyright 1989, p. 11, with permission from Elsevier Science.

the relationship may not start at the origin, which entails that some materials will not flow unless a critical stress or *yield value* $\sigma_o$ is reached, after which they will flow like liquids (Figure 3–16).

There are several empirical equations that fit nonlinear data obtained in a rheology experiment (Table 3–4). If data fitting is what one is after, they can be used, but they can also be abused.

The power law is the simplest and most popular empirical model applied to non-Newtonian flow. Parameters $K$ and $n$ are the consistency coefficient and the flow behavior index, respectively. They are determined from the intercept (at $\dot{\gamma} = 1$) and the slope when the equation is linearized in logarithmic form. The flow behavior index provides a convenient classification of flow into shear thinning ($n < 1$) and shear thickening ($n > 1$). Notice that $K$ has strange units ($Pa{\cdot}s^n$). For a power-law fluid, the viscosity shows the following dependence on shear rate:

$$\eta = K\dot{\gamma}^{n-1} \qquad \textbf{Equation 3–53}$$

A problem with this model is that, since most of the time $n < 1$, it predicts an infinite viscosity at low shear rates rather than the constant value observed experimentally. Models that overcome this deficiency are presented later.

The temperature dependence of the viscosity is most commonly expressed according to an Arrhenius relationship:

$$\eta = A\exp(-B/T) \qquad \textbf{Equation 3–54}$$

where $T$ is the absolute temperature and $A$ and $B$ are constants typical of each fluid.

### 3.8.2 The Effect of the Dispersed Phase

The rheology of multiphase fluid systems depends on the fraction of space occupied by the dispersed phase, termed the phase volume fraction ($\phi$). The

**Table 3–3** Typical Shear Rates Present in Food Processing

| *Situation* | *Typical Range of Shear Rates ($s^{-1}$)* | *Application* |
|---|---|---|
| Sedimentation of fine powders in a suspending liquid | $10^{-6}$–$10^{-4}$ | Food suspensions |
| Leveling due to surface tension | $10^{-2}$–$10^{-1}$ | Wetting of solids |
| Draining under gravity | $10^{-1}$–$10^{-1}$ | Dripping of liquids |
| Screw extruders | $10^0$–$10^2$ | Food extrusion |
| Chewing and swallowing | $10^1$–$10^2$ | Mastication |
| Dip coating | $10^1$–$10^2$ | Confectionary |
| Mixing and stirring | $10^1$–$10^3$ | Emulsification |
| Pipe flow | $10^0$–$10^3$ | Transport of liquids |
| Spraying and brushing | $10^3$–$10^4$ | Spray-drying |
| Rubbing | $10^4$–$10^5$ | Spreading of butter |
| Milling | $10^3$–$10^5$ | Size reduction |

*Source:* Reprinted from H.A. Barnes, J.F. Hutton, and K. Walters. *An Introduction to Rheology*, Copyright 1989, p. 13, with permission from Elsevier Science.

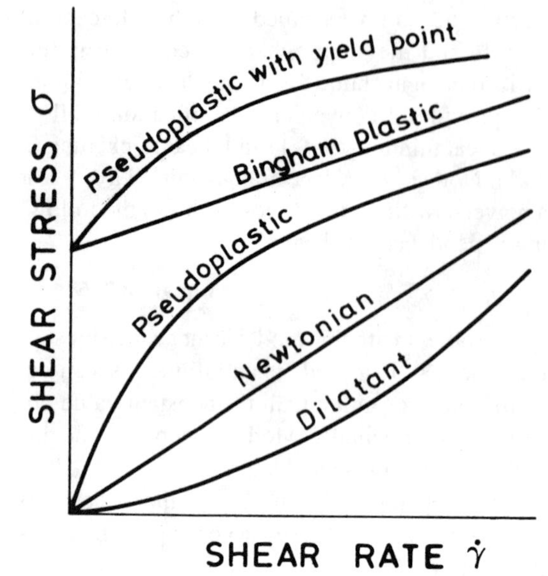


**Figure 3–16** Rheograms for the common behavior of Newtonian and non-Newtonian fluids. The apparent viscosity of non-Newtonian fluids is calculated as the slope of the line joining the origin and a curve at a preset value of $\dot{\gamma}$.

reason why $\phi$ is so important is that rheology depends largely on hydrodynamic forces acting on the surface of particles rather than on density. Three types of forces are relevant in flowing suspensions: (1) interaction or "structural" forces between particles; (2) those related to Brownian (thermal) movement, which are important for sizes less than 1 $\mu$m; and (3) viscous forces, which are proportional to the local velocity differences between the particle and the surrounding fluid. Thus, macroscopically measured rheological behavior is strongly dependent of microstructural aspects of the suspension. The rheology of suspensions is thorougly treated in Barnes et al. (1989) and Macosko (1994).

A large number of expressions exist that predict the viscosity of particle-filled liquids. The basic one is Einstein's equation for the viscosity $\eta$ of a dilute suspension of rigid spheres (generally <10% phase volume) in a solvent of viscosity $\eta_s$:

$$\eta = \eta_s(1 + 2.5\,\phi) \qquad \textbf{Equation 3–55}$$

where $\phi$ is the volume fraction of spherical particles. It is interesting to note that Einstein's equation is independent of the size of the particles. Strong deviations (i.e., higher viscosity) are observed even at a low concentration if the particles have colloidal dimensions. Many other relationships extend Einstein's equation to the moderate concentration range by forming a polynomial in $\phi$, such as

$$\eta = \eta_s(1 + 2.5\phi + \beta\phi^2 + \gamma\phi^3 + \ldots)$$

**Equation 3–56**

Except for providing some limiting conditions at high dilution, these equations are of little relevance for food suspensions. Normally viscosity curves upward (at constant shear rate) with increasing phase volume fraction. The Krieger-Dougherty relation takes into account the fact that the relative viscosity (see Section 3.8.3) diverges at a finite volume fraction when particles just touch each other, the so-called maximum packing condition $\phi_{max}$:

$$\eta_{rel} = \left(1 - \frac{\phi}{\phi_{max}}\right)^{-[\eta]\phi_{max}} \qquad \textbf{Equation 3–57}$$

**Table 3–4** Typical Mathematical Models for Non-Newtonian Behavior of Foods

| *Model* | *Equation* | *Example* |
|---|---|---|
| Ostwald or power law | $\sigma = K\,\dot{\gamma}^n$ | Tomato juice |
| Bingham plastic | $\sigma = \sigma_o + K\,\dot{\gamma}$ | Ketchup |
| Casson | $\sigma^{1/2} = K + m\,\dot{\gamma}^{1/2}$ | Molten chocolate |
| Hershel-Bulkley | $\sigma = \sigma_o + K\,\dot{\gamma}^n$ | Meat batters |

Maximum packing usually occurs at $\phi = 0.63$ to 0.71 for low and high shear rates (Macosko, 1994). Ball and Richmond (1980) (cited by Barnes et al., 1989) have proposed an alternative expression:

$$\eta = \eta_s(1 - K\phi)^{-5/(2K)} \qquad \textbf{Equation 3–58}$$

where $K$ is now a "crowding" factor due to packing.

For polymers, interpenetration of chains segments may occur at high concentrations, resulting in an apparent packing higher than if considered as solid spheres. A critical concentration $c^*$ is observed above which there is a marked increase in viscosity due to interpenetration of segments, implying a "structural" transition of the solution from one of isolated molecules (dilute regime) to another formed by entangled molecules (semi-dilute). From a practical viewpoint, if a viscosity increment is needed at low concentration, extended coils or rodlike-polymers will be more effective than random coils or globular polymers (Ross-Murphy, 1995). We will apply these concepts when dealing with stabilization of ice cream by hydrocolloids or gums (Section 7.4.7).

In the case of emulsions that are flowing (e.g., through a pump or during spreading), the liquid inside the droplets also flows, creating circulation. A *viscosity ratio* ($\eta_{dr}$), defined as the ratio of the viscosity of the liquid droplet ($\eta_d$) to that of the suspending medium, is introduced (Macosko, 1994):

$$\eta_{dr} = \eta_d / \eta_s \qquad \textbf{Equation 3–59}$$

and the steady shear viscosity becomes

$$\eta = \eta_s\left[1 + \left(\frac{1 + (5/2)\eta_{dr}}{1 + \eta_{dr}}\right)\phi\right]$$

**Equation 3–60**

Note that for large $\eta_{dr}$, the drop becomes rigid and the coefficient of $\phi$ goes to 5/2, as predicted by Einstein's equation.

### 3.8.3 A Few More Definitions of Viscosity

Most food liquids and pastes have an aqueous continuous phase that taken alone often behaves as a Newtonian fluid (even if it has some small solutes dissolved). A *relative* or *reduced viscosity* $\eta_{rel}$ has been be defined as

$$\eta_{rel} = \frac{\eta\,(multiphase\ system)}{\eta\,(continuous\ phase)} = \frac{\eta}{\eta_s}$$

**Equation 3–61**

assuming that the continuous phase is a solvent of viscosity $\eta_S$. Measurements of viscosity in dilute polymer solutions may give information on chain dimension, molecular size, degree of polymerization, and solute-solvent interactions (Young & Lovell, 1991). The *intrinsic viscosity* $[\eta]$ is the "intrinsic" ability of a polymer to increase viscosity. It is defined by this equation:

$$[\eta] = \lim_{c \to 0} [(\eta_{rel} - 1)/c] \qquad \textbf{Equation 3–62}$$

where $c$ is the concentration of polymer. Note that $[\eta]$ has units of reciprocal concentration. The product $[\eta]\,\overline{M}_w$ is proportional to the hydrodynamic volume of a polymer molecule, that is, the volume occupied by the polymer and the occluded solvent (Young & Lovell, 1991).

Molecular weight and its distribution affect the viscosity of polymeric solutions. The relation derived by Mark and Houwink for polymer-solvent systems relates intrinsic viscosity to average molar weight ($\overline{M}_w$):

$$[\eta] = K(\overline{M}_w)^{\alpha} \qquad \textbf{Equation 3–63}$$

where $K$ and $\alpha$ are empirical constants with typical values for flexible chains (in good solvents) of $10^{-3}$–$10^{-1}$ and 0.5–0.8, respectively. For pectin samples of molecular weight between 20 and 200 kDa, $K = 2.16 \times 10^{-2}$ and $\alpha = 0.79$ (Oakenfull, 1991). Thus, $K$ and $\alpha$ may give some information on how compact a polymer coil is in solution, with larger values indicating more expanded coils. Note that equation 3–63 can also be used to determine the molecular weight of polymers. Branched polymer molecules have a smaller hydrodynamic volume and a lower intrinsic viscosity than a similar linear polymer of equivalent mass, since the latter "sweeps" a higher hydrodynamic volume. Moreover, the shape of the molecule affects rheological behavior under varying shearing conditions. For example, due to the length and stiffness of hydrated alginate molecules, an aqueous solution of alginate has shear-thinning or pseudoplastic characteristics (Imeson, 1992). In low-shear conditions, alginate molecules are placed more or less randomly, while at high shear rates, they start to orient themselves in a more or less parallel fashion, opposing less resistance to flow, with lower apparent viscosity the result.

### 3.8.4 A Structural Rheological Model for Foods

At this point, we will ask a very fundamental question: What kind of mathematical description do we want for the behavior of a complex physical system? The answer is that we want a model that is able to bring order to our experience and observations as well as to make specific predictions about certain aspects of the world we experience (Bailey, 1998).

Previous empirical models (listed in Table 3–4) are able to express satisfactorily through one equation all data acquired in a rheogram. However, they do not provide insights into the basic mechanisms underlying the observed behavior. Since viscosity is a measure of the resistance to flow, it must be related to the structure of the fluid and the interactions among components. Many fluid foods are composed of discernible elements, such as macromolecules, colloidal aggregates, granules, particles, and droplets, that interact to form suspensions of hydrated particles, emulsions, creams, or pastes.

A better description of a system's behavior and the effect of structure may be acquired from data on apparent viscosity versus shear rate. Many polymer solutions and colloidal suspensions give a viscosity that decreases with increasing shear rate (shear-thinning behavior) at some intermediate region between two Newtonian plateaus (at low and high $\dot{\gamma}$). An equation often used to fit viscosity data for materials exhibiting Newtonian behavior at low ($\eta_o$) and high ($\eta_\infty$) shear rates is the Cross equation (Barnes et al., 1989):

$$\eta = \eta_\infty + \frac{(\eta_o - \eta_\infty)}{[1 + (K\dot{\gamma})^m]} \qquad \textbf{Equation 3–64}$$

where $K$ and $m$ are constants to be derived from experimental data. When $\dot{\gamma}\rightarrow 0$, then $\eta \rightarrow \eta_o$, while at high $\dot{\gamma}$, the viscosity tends toward $\eta_\infty$.

To account for shear effects on the shape and interactions between components, Windhab (1995) has proposed an interesting rheogram (based on classical suspension rheology; see Barnes et al., 1989, Chapter 10) relating qualitative structural aspects of a complex food material to its rheological behavior. The fluid is considered to be composed of a continuous phase (a dilute polymeric solution) and, as dispersed phases, solid particles and deformable droplets. The rheogram is divided in three regions (Figure 3–17).

Under low shear rates and below a first critical shear rate ($\dot{\gamma} < \dot{\gamma}_1^*$), two cases must be distinguished. If the concentration of the dispersed phase is high, structural forces (e.g., those keeping the structure together; see Section 3.6.7) predominate, and the interaction between particles generates an apparent yield stress $\sigma_o$. The apparent viscosity goes to infinity. The force that must be applied before flow occurs is best measured using a stress-controlled rheometer. For low concentrations of the dispersed phase, Brownian motion predominates and $\eta_o$ is independent of shear rate.

In the intermediate shear rate region ($\dot{\gamma}_1^* < \dot{\gamma} < \dot{\gamma}_2^*$), hydrodynamic forces (e.g., those associated with viscous drag) prevail over structural and Brownian motion forces. If the magnitude and time of application of shearing forces are sufficient, deformable particles or agglomerates become deformed and asymmetrical particles are oriented in flow to present minimal resistance, thus viscosity decreases. If enough time is allowed for the experiment, an "equilibrium structure" is attained.

Under high shear rates and above a second critical shear rate ($\dot{\gamma}_2^*$), instabilities may appear and the state of maximum orientation may be disturbed. For a particulate dispersion, microstructural phenomena may be related to particle-particle interactions or changes in particle structure (breakdown), leading to increased dispersion and a net increase in viscosity. This viscous behavior is called *shear-thickening* or *dilatancy*. Alternatively, some macromolecular

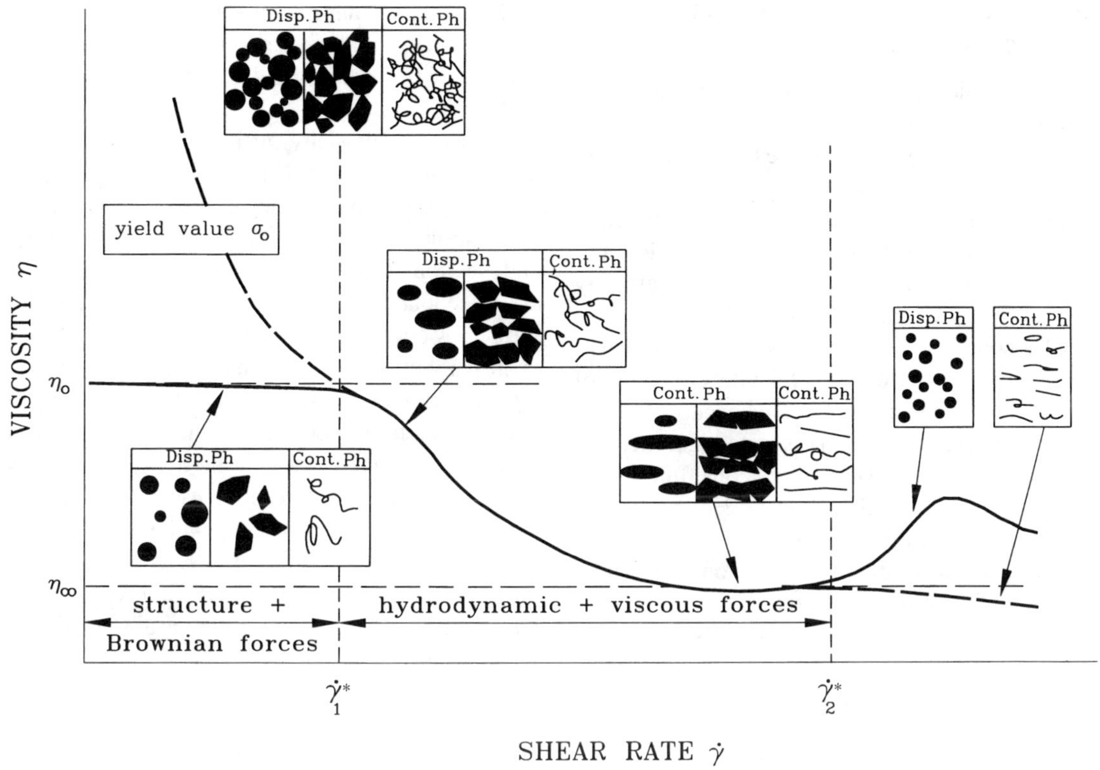


**Figure 3–17** Rheological behavior of a complex suspension and its relation to structure. The model suspension is formed by a polymer solution as continuous phase and by solid particles and deformable droplets as dispersed phase. See text for details. *Source:* Reprinted from E.J. Windhab, Rheology in Food Processing in *Physiochemical Aspects of Food Processing*, S.T. Beckett (ed.), p. 86, © 1995, Aspen Publishers, Inc.

solutions may behave differently than particulate systems owing to the fact that totally aligned macromolecular networks may break down because of localized shear and cause a further decrease in viscosity.

Windhab (1995) has proposed a structural model based on the superposition of the various effects leading to the shear-induced structure of the suspension:

$$\sigma = \sigma_o + \eta_\infty\dot{\gamma} + (\sigma_1 - \sigma_o)[1 - exp(-\dot{\gamma}/\dot{\gamma}^*)]$$

**Equation 3–65**

where $\sigma_o$ is the yield stress and $\sigma_1$ is the stress at the shear structuring limit (or when the final structure is attained). According to the Windhab model, the rheological pattern of concentrated emulsions (e.g., mayonnaise), suspensions (e.g., chocolate), and aerated foams (e.g., "mousse" products) can be described using equation 3–65.

Relating rheological behavior to microstructure demands observation and quantitation of structural changes with shear rate in real time. A "transparent rheometer," which allows the shape of oil droplets in an emulsion to be observed by microscopy and their sizes estimated by laser diffraction, has been referred to by Windhab (1995). Use of such equipment must obviously become a trend if adequate microstructure-rheological properties are to be derived and structural models developed for complex food fluids.

### 3.8.5 Intermission: The Rheology of Biological Fluids

One of the biological fluids whose rheology has been most studied is human blood, and the food rheologist can learn from the experience gained. Blood is a non-Newtonian suspension of cells in an aqueous solution of electrolytes and nonelectrolytes. It can be separated by centrifugation into plasma and cells. Plasma is about 90% water (by weight), 7% protein, and 2% small solutes. The cellular contents, which are essentially all disk-shaped erythrocytes or red cells with a diameter of 7.6 $\mu$m and a thickness 2.8 $\mu$m, occupy 50% of the blood volume. If plasma is tested in a rheometer, it behaves as a Newtonian viscous fluid with a viscosity of about 1.2 mPa·s (1.2 cP), slightly higher than water (Fung, 1981).

Since plasma is Newtonian, the non-Newtonian behavior of human blood is due to blood cells. Human blood cells form aggregates known as rouleaux whose existance depends on the proteins fibrinogen and globulin in plasma. The smaller the shear rate, the more prevalent are the aggregates, and at zero shear rate, blood may be regarded as a large aggregate exhibiting a yield stress. As $\dot{\gamma}$ increases from zero, the aggregates tend to break down and the apparent viscosity diminishes. If $\dot{\gamma}$ increases further, cells tend to become elongated and line up with the streamlines of flow. At intermediate values of $\dot{\gamma}$, the Cross equation (equation 3–64) fits data well with $\eta_o$ = 125 mPa·s, $\eta_\infty$ = 5 mPa·s, $K$ = 52.5, and $m$ = 0.715 (Barnes et al., 1989).

### 3.8.6 Linear Viscoelasticity

Some materials simulataneously exhibit viscous and elastic responses depending on the time scale of the experiment and its relation to a characteristic time of the material ($\tau$). The time $\tau$ is infinite for a Hookean solid and zero for a Newtonian liquid, but many materials exhibit intermediate responses that fit the definition of viscoelasticity. Viscoelastic phenomena in a polymeric liquid are due primarily to intramolecular forces that arise from changes in conformation caused by deformation of the liquid. The presence of other molecules, polymer or solvent, retard the reorientation process and give rise to the viscous component of the rheological effect (Barnes et al., 1989).

Linear viscoelasticity can be studied by stress relaxation or creep experiments and can be represented by mechanical models of springs and dashpots. Hookean elasticity is represented by a spring and Newtonian flow by a dashpot. The behavior of any viscoelastic material can be adequately described by connecting these basic elements in series or in parallel or in combination. Figure 7–15 presents a spring and dashpot model for ice cream. In stress relaxation experiments, the sample is subjected to constant strain and the decay in stress is monitored over time. Creep is slow deformation of a material under constant stress, while strain is measured over time (Stanley et al., 1996).

Dynamic tests are preferred for investigating viscoelastic behavior, since they are more versatile and cover a wider range of conditions. Oscillatory dynamic testing either applies a stress varying sinusoidally with time and measures the resulting strain or the reverse. The applied stress (or strain) must be small enough to stay within the limits of linear viscoelasticity. If the material is perfectly elastic, then the resulting stress will be exactly in phase with the strain wave. On the other hand, if the material is a viscous fluid, the stress wave will be exactly 90° out of phase with the deformation. Any viscoelastic material will lie between these two extremes, with a phase angle ($\delta$) between 0° and 90°. In the case of shear experiments, a complex shear modulus is defined as

$$G^*(\omega) = \frac{\textit{complex stress}}{\textit{complex strain}} = \frac{\sigma^*}{\gamma^*} = G'(\omega) + iG''(\omega) \qquad \textbf{Equation 3–66}$$

$$\sigma = \gamma_o[G' \sin \omega t + G'' \cos \omega t] \qquad \textbf{Equation 3–67}$$

where $G'$ is the *storage or elastic modulus*, $G''$ is the *loss or viscous modulus,* and $\omega$ is the angular frequency ($2\pi$ times the frequency in Hz). $G'$ and $G''$ represent the energy stored and the energy loss per cycle of deformation under frequency $\omega$, respectively, and they can be regarded as the "solidlike" and "liquidlike" viscoelastic behaviors of the material. $G'$ is also called the *dynamic rigidity*. The loss tangent (tan $\delta$) is the ratio $G''/G'$. For a predominantly solid material, $G' > G''$ and tan $\delta < 1$, whereas for a primarily liquid material, $G'' > G'$ and tan $\delta > 1$.

Alternatively, a *complex viscosity* can be defined as

$$\eta^* = G^*/\omega = \sqrt{(\eta')^2 + (\eta'')^2}$$

**Equation 3–68**

where $\eta' = G''/\omega$ is the *dynamic viscosity* and the parameter $\eta''$ has no special name but is related to the dynamic rigidity by $\eta'' = G'/\omega$.

This information on viscoelasticity will facilitate understanding of the material presented below and in the section on gels (Section 5.8). For further details on relationships and formulae for linear viscoelasticity as applied to foods, the reader is referred to the chapter by Stanley et al. (1996) and the review by Kokini (1987). A classic reference on viscoelasticity is Ferry (1980).

### 3.8.7 Dynamic Testing and Structure

The dynamic properties of viscoelastic materials are of considerable practical interest, as the processability and properties of materials can often be directly related to the "elastic" or "viscous" moduli ($G'$ and $G''$) derived from such measurements. A main advantage of performing dynamic tests at low strain is that the microstructure of the material is usually stable throughout the experiment. In particular, dynamic mechanical testing (or mechanical spectroscopy) can yield insight into the microstructure, transition temperatures, and the evolution of dynamic viscoelastic properties with time under constant or variable temperature, pressure, and moisture conditions. Care has to be taken in interpreting these diagrams, because they are normally plotted in double logarithmic axes, so lines with small slopes may represent large absolute variations in value. Four applications of dynamic testing of food materials follow.

Oscillatory rheology has been used to distinguish entanglement solutions from weak or strong gels (Ross-Murphy, 1995). Entangled solutions formed by the topological interaction of polymer chains at very low $\omega$ (long time scales) flow as high-viscosity liquids. Both $G'$ and $G''$ increase linearly with frequency, but $G'' > G'$ and $\eta^*$ is independent of frequency. On the contrary, for soft and strong gels, $G'$ and $G''$ run practically parallel (but $G' > G''$) and independently of frequency, while $\eta^*$ decreases continuously in the whole frequency range.

Dynamic measurements can also be performed as a function of temperature at constant frequency. When log $G'$ is plotted against temperature for an amorphous material, a large decrease in modulus is observed at the glass-rubber transition ($\alpha$ transition), while the loss modulus $G''$ and tan $\delta$ show a pronounced peak. In partially crystalline polymers, the drop in modulus at $T_g$ is much smaller, and more than one transition may be observed below $T_g$. These transitions (subglass) are labeled $\beta$, $\gamma$, and so on, and they are due to local relaxations of the main chain and rotation of terminal groups or side chains (Kalichevsky, Blanshard, & March, 1993). These subglass transitions suggest that a glassy polymer does have some segmental mobility below $T_g$.

Gelling kinetics can also be characterized by small amplitude dynamic testing, because of the gradual development of viscoelasticity. In the case of a protein solution's gelling by heat, a *curing* curve is obtained as the experiment proceeds at constant temperature and frequency, representing the change from solution (sol) to network (gel) (Aguilera, 1995). As expected for a liquid, $G''$ is initially larger than $G'$, but beyond a lag time $G'$ increases faster than $G''$, and there exists a crossover point that is sometimes taken as the gelling time or the sol-gel transition (Figure 3–18). $G'$ continues to increase rapidly as the system becomes more "solidlike" owing to formation of the gel network, until it finally enters a pseudo-plateau region.

Mechanical spectroscopy has also been applied to the study of structure formation and the breakdown in water-plasticized food polymer systems (gluten, soy protein, etc.) as a function of temperature and moisture content. Changes in $G'$ and $G''$ are associated with reactions within the system that

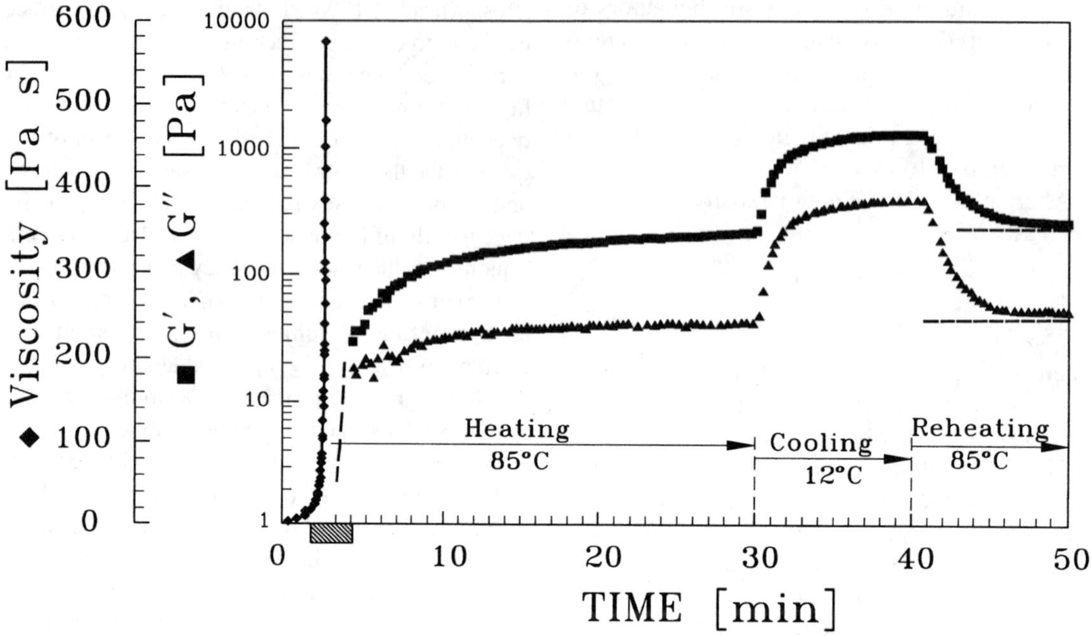


**Figure 3–18** A "curing" experiment for thermally induced protein gels. Development of viscoelasticity during gelation is followed by changes in viscosity, storage (G′) and loss (G″) moduli with heating time. The sol-gel transition (hatched area) may be observed at short times.

lead to structure building (e.g., cross-linking) or structure softening (e.g., debonding) and flow mechanisms (Kokini, Cocero, Madeka, & de Graaf, 1994). The first group of reactions result in a large increase in $G'$ (e.g., 100 times), while the second almost invariably implies a decrease in moduli. Little, Aguilera, Morales, and Kokini (1997), on the basis of data gathered by using high-pressure rheology and temperature sweeps, suggested that soybean protein under heat and high pressure undergoes first cross-linking (i.e., increase in $G'$), starting at 100°C, and then breakdown (decrease in $G'$), at temperatures above 150°C. These findings are in accord with the practical observation that an optimum temperature for the extrusion-texturization of vegetable proteins exists.

### 3.8.8 Regions of Viscoelastic Behavior

Amorphous polymers exhibit five regions of viscoelastic behavior depending on temperature (or time), as shown in Figure 3–19. For the time being, we will treat the vertical axis as viscosity, although it could be any modulus, as will be seen later.

At low temperatures in the *glassy state* (below $T_g$), amorphous polymers exhibit high and fairly constant modulus (e.g., $E = 3 \times 10^9$ Pa). As already explained, in the glassy state molecular motions are highly restricted and only vibrations and short-range rotational motions are allowed. The glass transition region is often called the *leathery region*, and it is recognized by a sizable drop in viscosity ($10^5$ times) or modulus ($10^3$ times) in a range of 20–30°C. Qualitatively, the glass transition region is where the onset of long-range, coordinated molecular motions occurs. The *rubbery plateau* is characterized by an almost constant modulus, with typical $E$ values of $2 \times 10^6$ Pa. If the polymer is amorphous, two cases have to be distinguished: for a linear polymer, the modulus will drop slowly, but if it is cross-linked, rubber-

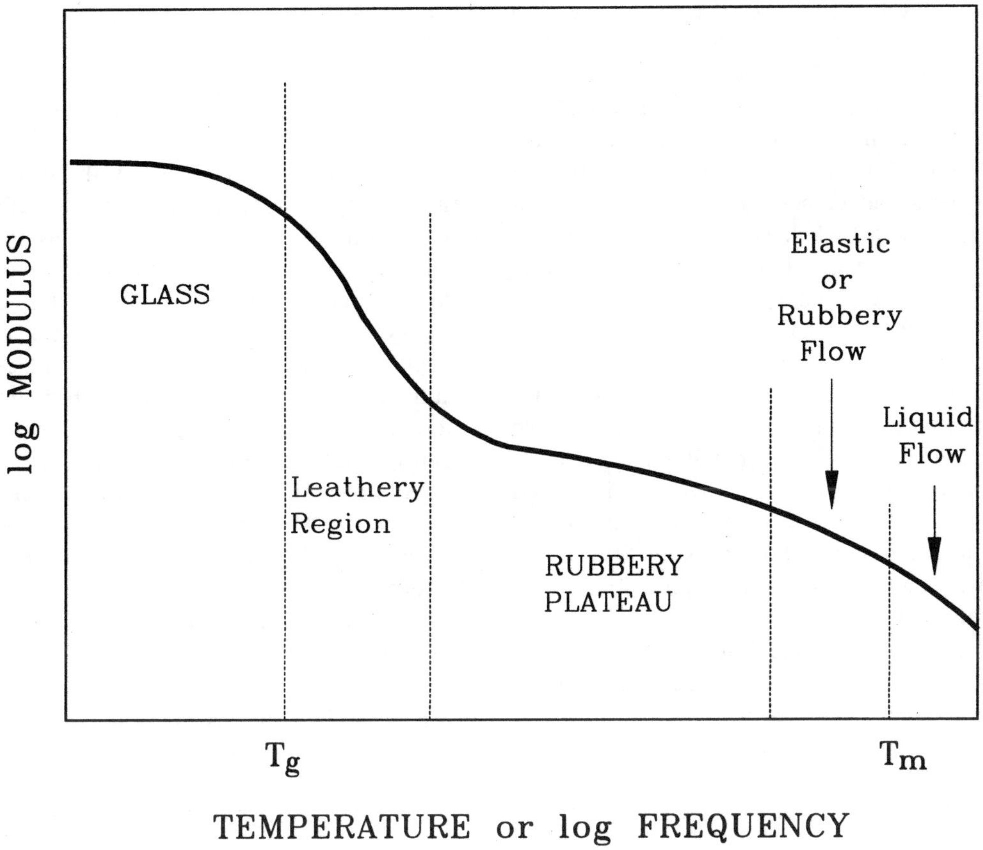


**Figure 3–19** Master curve for the modulus of an amorphous material showing the five regions of viscoelastic behavior. *Source:* Reprinted with permission from L. Slade and H. Levine, Beyond Water Activity: Recent Advances Based on an Alternative Approach to the Assessment of Food Quality and Safety, *Critical Reviews in Food Science and Nutrition,* Vol. 30, pp. 115–359, © 1991, CRC Press, Boca Raton, Florida.

like elasticity is exhibited. If the polymer is semicrystalline, the crystalline regions restrain movement and act as fillers of the amorphous regions, and thus the crystalline plateau extends until the melting temperature. On further increases of temperature, the *rubbery flow region* is reached for linear polymers. Entaglements may make the material look rubbery for short times or make it flow if longer times are allowed. At still higher temperatures, the polymer will flow readily in the *liquid flow region*.

A master curve of a modulus is useful, not only for determining the physical state as a function of temperature, time, or moisture content, but also for predicting the behavior of an amorphous food material during processing and storage (see Section 5.5.4).

## 3.9 MECHANICAL PROPERTIES OF FOOD SOLIDS

### 3.9.1 Simple Mechanical Relationships

The response of food materials when subjected to various forces is of the greatest importance to food scientists and engineers. The mechanical properties of foods are related to textural properties and influence the handling and processing of foods. In fact, this section sets the basic principles of instrumental procedures for texture measurements of foods, described in detail in Section 6.2.2. Works on the mechanical properties of foods and biomaterials include the book by Mohsenin (1970), which deals with the mechanical and physical properties of plant and animal materials, the book by Vincent (1990) on the structure of biomaterials, and a modern treatise on the biomechanics of plants by Niklas (1992).

Food solids are usually studied in compression, tension, or shear. Compression is the application of a uniaxial parallel force to cause flattening, whereas tension is the application of a uniaxial parallel force to cause extension. Shear is the application of uniaxial tangential forces to cause separation or cutting, while bending involves tension and compression, torque involves shear, and so on (see Figure 3–20). When a material is subjected to one or more of these forces, its original dimensions will change by some amount (i.e., deformation will occur).

The basic law for ideal solids assumes that when solids are subject to a force, they will deform elastically and that upon removal of the force causing the deformation they will return instantaneously to their original shape. Hooke's law for the ideal solid states that

$$\sigma = \frac{F}{A} = E\varepsilon \qquad \textbf{Equation 3–69}$$

where $\sigma$ is the force divided by the cross-sectional area of the specimen and $\varepsilon$ is the deformation or strain. This is a linear relationship only valid in the elastic or linear range. When used in cases of extension or compression, it defines the stiffness, or *Young's modulus E*, a characteristic of each solid material. Note that $\varepsilon$ is replaced by $\gamma$ in the case of shear ($\gamma = \tan\theta$, where $\theta$ is the deformation angle). An important difference between the behavior of liquids and solids when subject to a shear stress is that for the former $\sigma$ is proportional to $d\gamma/dt$ while for the latter it is proportional to $\gamma$. Another consequence of this simple formula is that if the cross-sectional area varies during testing, it must be determined at each particular strain in order for the actual stress to be known.

Strain can be expressed in a number of ways. The most common way is to calculate the *engi-*

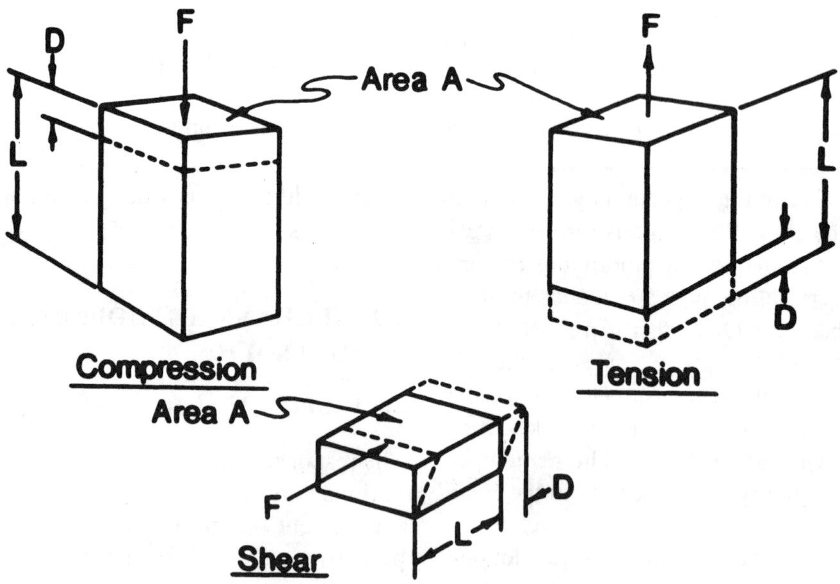


**Figure 3–20** Schematic diagram of the action of forces on solids: compression, tension, and shear. F = force, L = original dimension, D = displacement.

*neering* or *Cauchy's strain*, which is the change in length ($\Delta L$) per unit of original length ($L_o$):

$$\varepsilon = \frac{\Delta L}{L_o} \qquad \textbf{Equation 3–70}$$

Sometimes it is more convenient to use natural or *Hencky's strain* (e.g., when the deformation is small):

$$\varepsilon = \ln\left(\frac{L_o}{L_o - \Delta L}\right) \qquad \textbf{Equation 3–71}$$

A stress-strain relationship can also be defined for shear, and a *shear modulus* $G$ can be obtained:

$$\sigma = G\gamma \qquad \textbf{Equation 3–72}$$

In an isotropic material, Young's modulus and the shear modulus are related as follows:

$$G = \frac{E}{2(1 + \nu)} \qquad \textbf{Equation 3–73}$$

where $\nu$ is *Poisson's ratio* (the relationship between the lateral and the axial deformation). Typical values of $\nu$ are 0.5 when there is no change in volume during stretch, 0.0 when there is no lateral contraction, and 0.2–0.4 for plastics.

### 3.9.2 Stress-Strain Relationships and Fracture

What a rheogram is for a liquid, the stress-strain curve is for a solid. Although it is generated experimentally as a force (N) versus deformation (mm) curve, its correct representation is as stress-strain curve. Several types of solid behavior are depicted in Figure 3–21, and some definitions useful for describing the behavior of solids are presented in Exhibit 3–1.

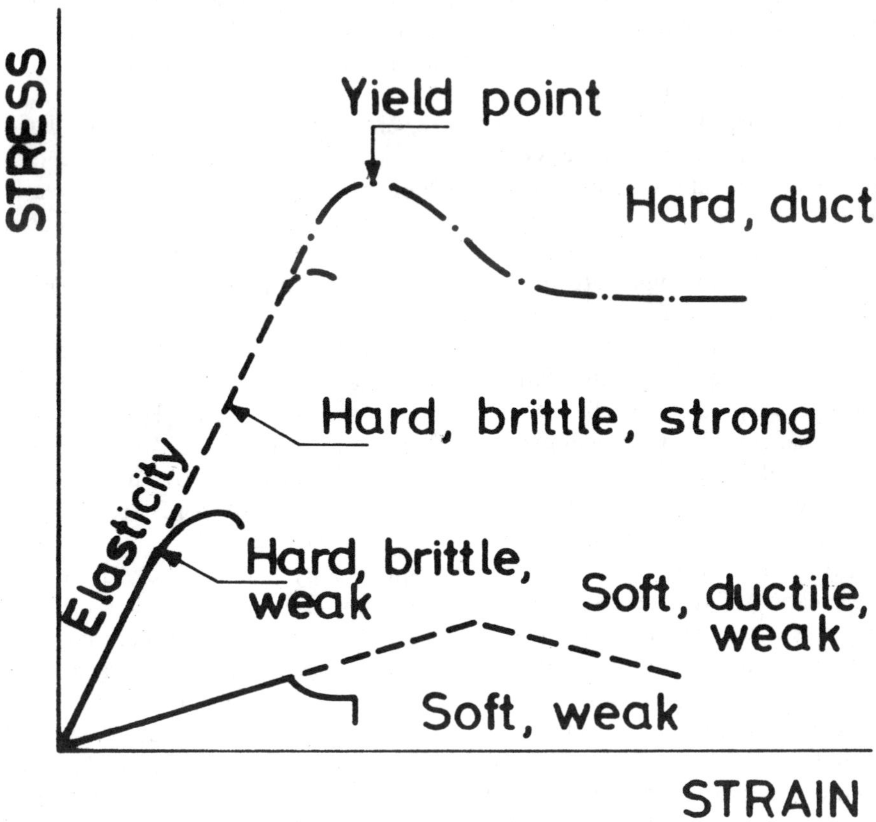


**Figure 3–21** Stress-strain curves for several types of solid materials.

**Exhibit 3–1** Glossary of Strength-Related Terms

**Breaking stress (strain).** The stress level at which a nonductile material breaks.
**Brittle.** The property of breaking at low strain with little or no plastic deformation.
**Ductile.** The property of being able to undergo plastic deformation before failure.
**Elastic limit.** The stress level at which elastic behavior is lost.
**Fracture.** Separation of a material by physical force into two or more parts.
**Resilience.** The capacity of a material to absorb energy in the elastic range. It is measured by the ratio of energy recovered to energy applied.
**Strength.** The ability of a material to resist applied forces without yielding or fracturing. Materials that are strong and ductile under a static load may appear weak and brittle under impact stresses.
**Toughness.** The ability of the material to absorb energy during plastic deformation. The area under the stress-strain curve provides a measure of the material's toughness.
**Yield point.** The point of strain and stress when plastic behavior is initiated. Beyond this point there is no elastic recovery.

Fracture properties influence the eating quality of solid foods (texture), their handling (e.g., slicing), and their storage capabilities (e.g., packaging). Fracture theory supposes that all materials are inhomogeneous and contain weak points or defects that are the source of small cracks. When a stress is applied (e.g., in tension), energy tends to concentrate at the tip of a crack, and the material fails if this local stress exceeds the breaking stress of the material. The crack may propagate spontaneously if during its growth the energy released exceeds that required to form the two new surfaces. It is important to realize that, because of inhomogeneities, structures fail at lower stresses than those predicted for the perfect material (e.g., the matrix). The traditional theory for brittle materials failing at low strains (e.g., a fresh cracker at low moisture content or chocolate at low temperature) is the so-called Griffith fracture theory. It simply considers fracture as a problem of the creation of two new surfaces, a process that requires energy to stretch and break bonds involved in the fracture. Since the energy necessary to break atomic bonds is similar for different materials, the energy per unit area that needs to be supplied varies between 1 and 10 $Jm^{-2}$. For ductile materials (semi-hard cheese or toffee), however, the energy supplied can be dissipated by deformation over a larger volume around the crack, reaching approximately $10^3$–$10^6$ $Jm^{-2}$. A Griffith critical crack length $L_G$—the minimum size of a crack needed for fracture—is the studied parameter in brittle fracture. It is defined by this formula:

$$L_G = \frac{2WE}{\pi\sigma^2}$$ **Equation 3–74**

where $W$ is the work of fracture (area under the stress-strain curve to the fracture point), $E$ is Young's modulus for the material, and $\sigma$ is the overall stress at the onset of fracture propagation. Because of energy dissipation mechanisms already mentioned, this formula largely underestimates the fracture of polymer materials that undergo plastic deformation. For deeper insight into the fracture of biological materials, see Niklas (1992) and Vincent (1990); and for deeper insight into the fracture of foods, see van Vliet and Luyten (1995).

Microscopy is extensively used in materials science to observe and characterize fracture surfaces. An SEM study of the deformation and flow of spun soybean fibers during fracture induced by different texture-measuring devices is presented by Aguilera, Kosikowski, and Hood (1975). Some fibers fractured in tension showed the presence of internal voids, probably caused by air bubbles entrapped in the spinning dope.

Last but not least, the mechanical behavior of biopolymers plasticized by water shows a strong dependence on moisture content. Low-moisture foods such as crackers, fried snacks, and extruded cereals exhibit a drastic reduction in mechanical parameters as moisture increases. At low moisture levels (<10%), the moduli are high and almost constant, possibly because they are in a glassy condition below $T_g$. As shown in Figure 3–9, under these moisture conditions starch and glutenin are glassy at room temperature. At higher moisture contents, there is a dramatic reduction in the moduli (and fracture stress), which correlates with increased plasticization of amorphous regions by water (Figure 3–22).

### 3.9.3 Concepts of Rubberlike Elasticity

Some foods may exhibit "elastic" behavior, especially those having a cross-linked polymeric net-

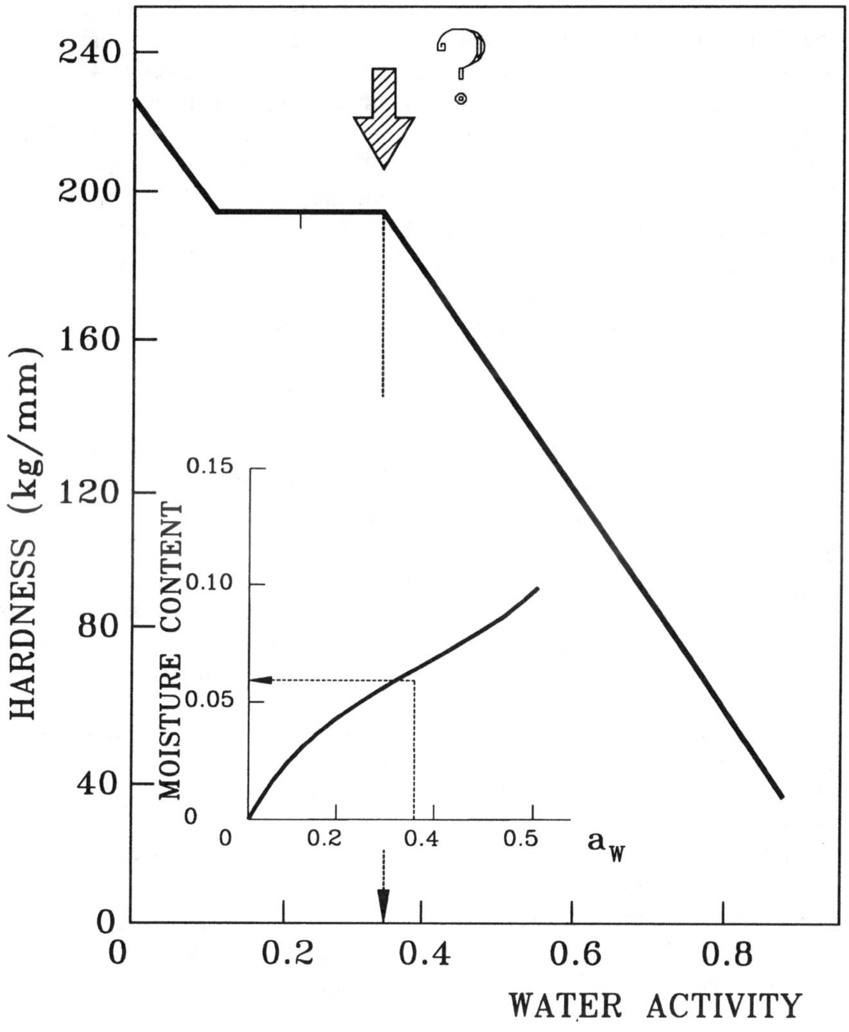


**Figure 3–22** Effect of moisture content on the hardness (measured as the initial slope of the force-deformation curve) of a saltine cracker. Inset: The moisture isotherm at 20°C. Arrow points at a break in the curve at a moisture content of 0.06 g/g, possibly related to a glass transition. *Source:* Reprinted with permission from E.E. Katz and T.P. Labuza, Effect of Water Activity on the Sensory Crispness and Mechanical Deformation of Snack Food Products, *Journal of Food Science*, Vol. 46, pp. 403–409, © 1981, Institute of Food Technologists.

work at the microstructural level. A few concepts of rubber elasticity are therefore presented in this section.

In accordance with the second law of thermodynamics, the reactive stress of an elastomer arises through a reduction in entropy rather than through changes in enthalpy. The basic relationship relating the reactive stress $\sigma$ of an elastomer to its extension ratio ($\alpha = L/L_o$) is given by (Sperling, 1992)

$$\sigma = \nu RT\left(\alpha - \frac{1}{\alpha^2}\right) \qquad \textbf{Equation 3–75}$$

where $\nu$ is the number of active network segments per unit volume. This quantity is equal to $\rho/M_w^c$, where $\rho$ is the density and $M_w^c$ is the molecular weight between cross-links. Note two things about this equation: it is similar to the ideal gas law and it expresses a nonlinear relation (i.e., the simple Hookean proportionality between stress and strain does not hold).

A similar equation, known as the Mooney-Rivlin equation, is derived from a phenomenological approach in which the elastomer is treated as a continuum:

$$\sigma = \left(2C_1 + \frac{2C_2}{\alpha}\right)\left(\alpha - \frac{1}{\alpha^2}\right)$$

**Equation 3–76**

The constants $C_1$ and $C_2$ depend on the material, and according to Flory, the constant $C_2$ is related to the looseness with which the cross-links are embedded in the structure.

### 3.9.4 Composites

The term *material* has been used rather loosely up to now. In engineering, the term is strictly utilized for a pure substance or for a homogeneous alloy. Composite materials are constructed out of at least two different types of materials in order to provide a property that any single material would lack. Modern structural composites are usually a combination of a matrix or binder material and some kind of reinforcement material, like fibers or particles. For instance, fiber-reinforced composites have superior tensile strength, mainly due to the fibers, and the weakness of the matrix becomes relatively unimportant. In vulcanized rubber, the presence of fine, hard particles increases the elastic limit because they perturb the flow pattern of stress deformations, causing rapid hardening. Larger particles (>1 $\mu$m) exhibit a strengthening effect by hydrostatically restraining the movement of the matrix.

At the microstructural level, several foods can be viewed as composites. Figure 3–23 shows a cross section of a bean cotyledon in which starch granules are perfectly embedded in a proteinaceous matrix. Similarly, the outside primary cell wall of plants may be regarded as a composite of cellulose microfibrils loosely woven together and embedded in a matrix of hemicelluloses and other polysaccharides (see Section 4.6.1 and Color Plate 7). In turn, a parenchymal cell may be considered as a liquid-filled foam. As suggested by Aguilera (1992), at a larger scale several high-moisture foods may be regarded as composite gels. Such foods include meat (a fibrous composite) and hard cheeses (a particle-filled composite). Specialized tissues that serve unique functions in plants and mammals possess a complex architecture and can be regarded as "composites" of composites, or supercomposites (Vincent, 1990). Thus, at the macroscale, muscle tissue can be regarded as a supercomposite of fibers cemented by connective tissue. Other interesting composite structures in the biological world include bone (crystals of hydroxyapatite embedded in a collagen matrix), wood (cellulose fibers in a matrix of lignin), and horn (keratin fibers in an amorphous matrix).

Models for moduli of polymer composites reinforced by particulate material have been reviewed by Ahmed and Jones (1990). When rigid inclusions are embedded in a nonrigid matrix, the equations describing the modulus are similar to Einstein's equation for suspensions, and they predict an increase in rigidity as the volume fraction of particulate inclusions increases. Again, a limitation of the model is that the stiffening action of the filler is independent of its size. The best of the improved equations, particularly for high-volume fractions of filler, is due to Mooney:

$$E_c = E_m \, exp\left(\frac{2.5\phi}{1 - S\phi}\right) \qquad \textbf{Equation 3–77}$$

where $E_c$ is the modulus of the composite, $\phi$ is the volume fraction of particles (spheres), and $S$ is the crowding factor (volume occupied by the filler or

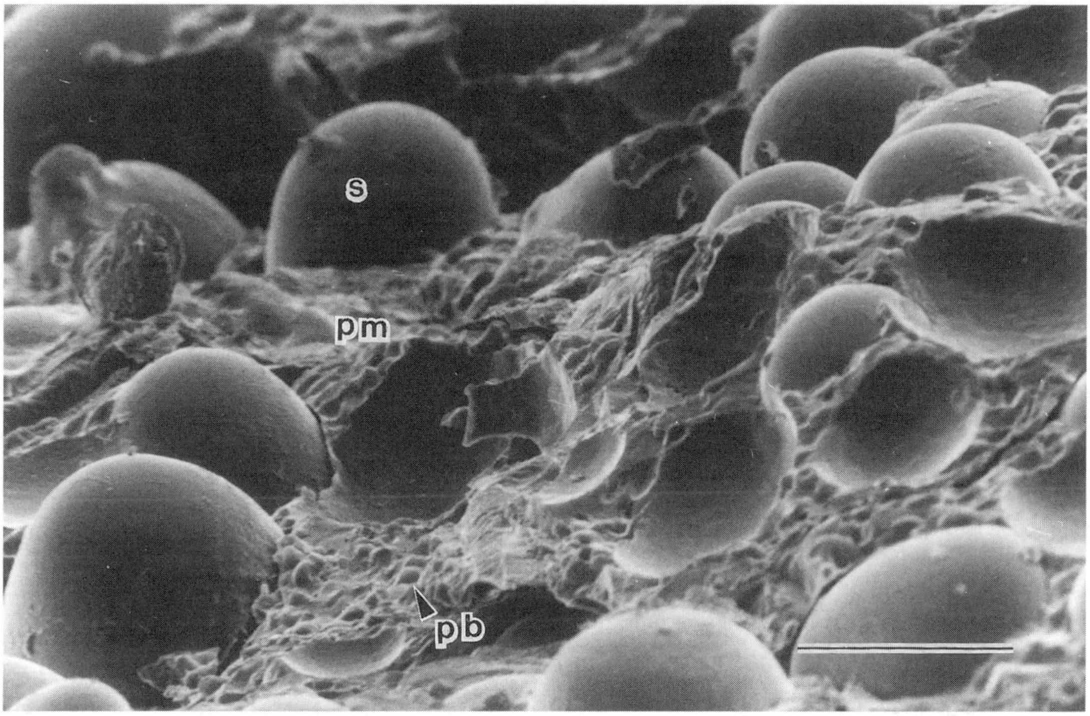


**Figure 3–23** SEM micrograph of the fracture surface of a bean cotyledon. Starch granules (s) are embedded in a protein matrix (pm) as if the interior of the bean was a composite structure of these two materials. pb = protein bodies. Marker = 10 $\mu$m.

true volume of the filler). For closed-packed spheres, $S = 1.35$.

When perfect adhesion exists between matrix and filler, the modulus for binary composites can be calculated using simple additivity laws. In the case of rigid inclusions in a rigid matrix, the most widely utilized expressions are for the series and parallel arrangements (sometimes called the Takayanagi model). In a parallel arrangement, isostrain conditions are assumed in the two phases, and the upper boundary is given by (Ahmed & Jones, 1990)

$$E_c = \phi_f E_f + \phi_m E_m \qquad \textbf{Equation 3–78}$$

where the subindices $f$ and $m$ stand for filler and matrix, respectively. Note that for a binary system $\phi_f = 1 - \phi_m$. In a series arrangement, the stress is assumed to be uniform in both phases, and the lower bound becomes

$$\frac{1}{E_c} = \frac{\phi_f}{E_f} + \frac{\phi_m}{E_m} \qquad \textbf{Equation 3–79}$$

For example, the modulus of fibrous food composites loaded in tension or compression and parallel to the fiber direction can be determined by equation 3–78, but if they are loaded normal to the fiber direction, equation 3–79 applies. Analogous expressions can be used for shear and brittle fracture of composites (Jeronimidis, 1991).

The Takayanagi model further combines the previous equations in a series-parallel model:

$$E_c = \left[\frac{\alpha}{(1-\beta)E_m + \beta E_f} + \frac{(1-\alpha)}{E_f}\right]^{-1}$$

**Equation 3–80**

where parameters $\alpha$ and $\beta$ represent the states of the parallel coupling and series coupling of the structure and are functions of the volume fraction of the series and of the parallel elements, respectively. Further modifications are needed if the filler consists of fibers of finite length not lying parallel to each other. In this case, equation 3–78 needs to be corrected by introducing parameters dependent on fiber length ($A$) and orientation ($B$) (Jeronimidis, 1991):

$$E_c = AB\phi_f E_f + \phi_m E_m \qquad \textbf{Equation 3–81}$$

It is obvious that in selecting the model to use as well as in estimating parameters such as $\alpha$, $\beta$, $A$, $B$, and $\phi$, the microstructure of the food must be examined first. The observed architecture of structural elements will dictate which model to use, and image analysis techniques may assist in quantifying the geometrical parameters of the model. (See Section 5.12.4 for application of Takayanagi's model to mixed gels.)

An important factor not considered by these simple models is the degree of interaction between the filler and the matrix. Evidently load transfer within the composite depends on the binding between the filler and the matrix. Figure 3–24 shows the case of a whey protein gel embedded with but poorly bound to starch granules. Narkis, Chen, and Pipes (1988) review methods for characterizing the fiber-matrix interaction.

The idea behind the previous expressions is that mathematical models should be used whenever possible to fit stress-versus-strain data for composite foods. They provide a more fundamental understanding of the role of structure and the effects of architectural elements than the graphical information alone or empirically fitted equations do.

### 3.9.5 Cellular Structures

We will now consider mechanical models for plant tissue (fruits and vegetables). The most prevalent structural units are closed cells surrounded by a semi-rigid cell wall and filled with a viscous liquid. Cell walls are distended under turgor pressure and adhere one to another. This structure is responsible for the crispness of fresh fruits and vegetables. In materials science, so-called "cellular materials" are (1) honeycombs with parallel prismatic cells and (2) solid foams with polyhedral cells. This terminology is rather confusing for food and biological engineers. We will come back to these materials later, but they may not be the most suitable models for cellular plant tissue.

A basic model for an isolated plant cell (e.g., a parenchymal cell) is that of the *hydrostat*, a thin-walled inflated structure (Niklas, 1992). The wall of plant cells is placed in tension by the internal (or turgor) pressure $P$ exerted by the fluid [which is on the order of 1 MPa (10 atm)]. The tensile stress $\sigma_t$ in the wall of the hydrostat is given by

$$\sigma_t = \left(\frac{P\alpha}{2}\right)\left(\frac{r_o}{\delta}\right) \qquad \textbf{Equation 3–82}$$

where $\alpha$ is the wall stretch ratio (the tensile strain), $r_o$ is the radius of the hydrostat, and $\delta$ is the wall thickness. Apple cells, for instance, are about 100 $\mu$m in diameter and the cell walls are about 2 $\mu$m thick. Interestingly, the tensile stress increases with the radius of the hydrostat, which means that actual cells have an optimum size. Nilsson, Hertz, and Falk (1958) derived a formula that predicts the elastic modulus of parenchymal tissue for any turgor pressure:

$$E = 3P\left[1 + \frac{(7 - 5\nu)}{20(1 + \nu)}\right] + \left(\frac{E_c\delta}{r_o}\right)\left[\frac{3(7 - 5\nu)}{10(1 + \nu^2)}\right]$$

**Equation 3–83**

where $\nu$ is the Poisson ratio and $E_c$ the elastic modulus of the cell wall (e.g., 1 GPa). The first term in equation 3–83 expresses the contribution to the elastic modulus of the tissue resulting from turgor pressure. The second term reflects the contribution of the material properties and the geometry of the cell walls. The maximum elastic modulus for potato tissue parenchyma with a turgor pressure of 0.67 MPa is 19 MNm$^{-2}$ (Niklas, 1992). It must be kept in mind that the elastic modulus for a piece of turgid tissue is different (normally higher) than that of a single cell, because the inflated protoplasts reduce the freedom of cell walls to buckle under a compressive stress.

The hydrostat model should be used only as a first approximation for vegetable tissue. Stress-strain curves are not linear, and parenchyma exhibits short-term elasticity recovery, long-term plasticity, stress relaxation, and creep. Moreover, cells flatten in the direction perpendicular to the applied load, and water may escape from the cells across the plasmalemma membrane (Niklas, 1992).

It is also possible to treat plant tissue as a sponge filled with liquid (Gibson & Ashby,

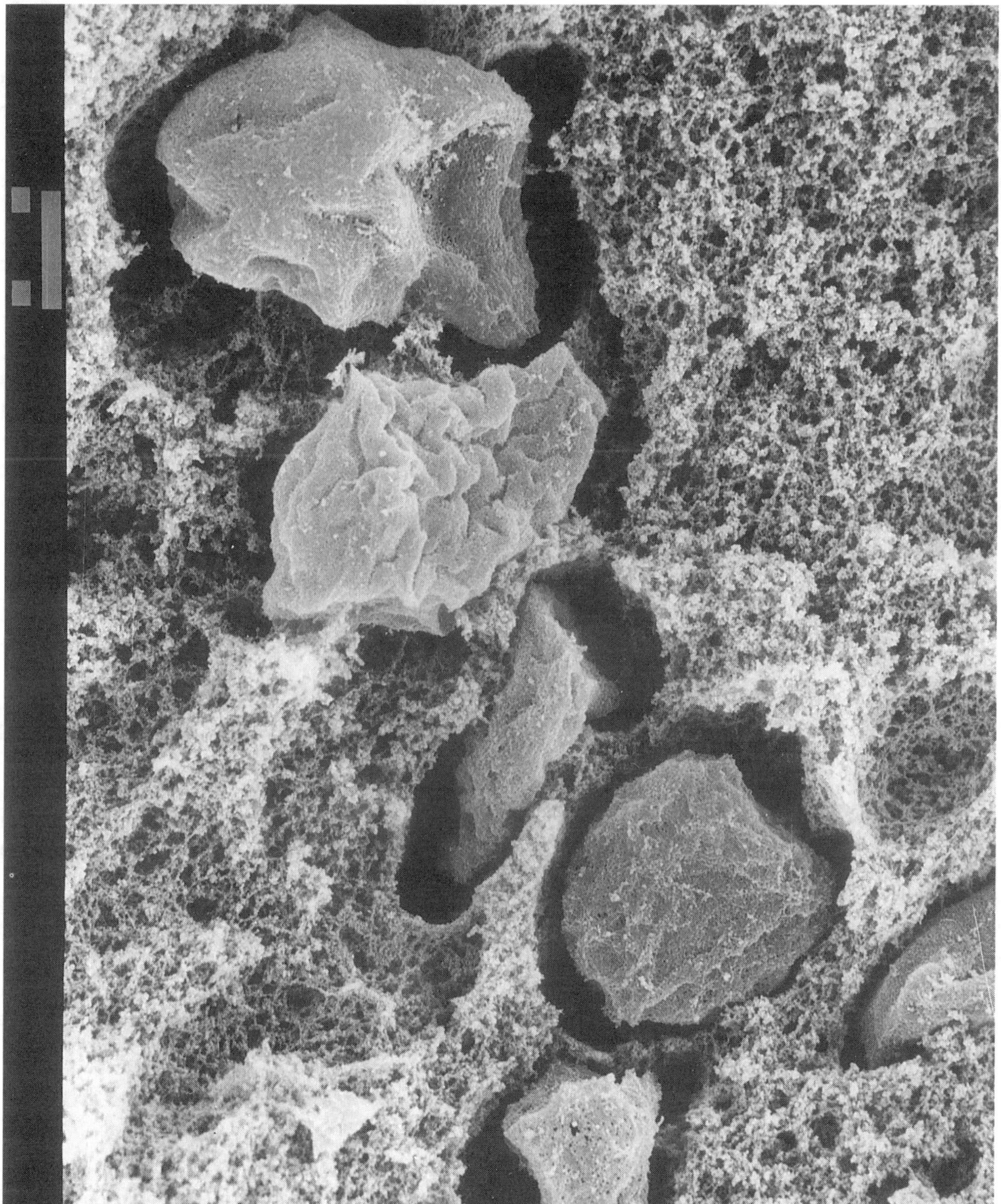

**Figure 3–24** SEM micrograph of a whey protein gel with embedded gelatinized starch granules. Note poor bonding between the matrix and the filler. Marker 10 $\mu$m.

1988). Although cells in sponges are open, as a first approximation they can be viewed as representing tissue that leaks liquid during compression. We are now dealing with a piece of open-cell foam (see next section) of base $L$ and cell edge length $l$, whose strength $\sigma$ is given by

$$\sigma = \frac{K\eta\dot{\varepsilon}}{1-\varepsilon}\left(\frac{L}{l}\right)^2 \qquad \textbf{Equation 3–84}$$

where $\eta$ is the liquid viscosity, $\varepsilon$ is the strain, $\dot{\varepsilon}$ is the strain rate, and $K$ is a constant that includes the various proportionality constants involved in the derivation of the formula. Note that the strength of the filled foam is directly proportional to the viscosity of the filling liquid and to the rate of deformation.

A further complication in modeling the mechanical behavior of plant cellular tissue is that the type of fracture (Section 3.9.2) depends on processing and/or physiological conditions. In mechanical terms, "crispness" is a combination of high turgor and strong bonding between cells (Vincent, 1990). As parenchymatous fruits ripen or legumes are cooked, the mode of failure at fractures changes from cell wall rupture to cell debonding (see Sections 6.5.5 and 6.5.7).

### 3.9.6 Solid Foams

Technically, a solid foam is a three-dimensional structure made up of polyhedral cells. Open-cell foams consist only of the edges of the cells whereas closed-cell foams have the faces covered by a solid membrane. The single most important structural characteristic of a foam is its relative density $\rho_c/\rho_s$ (the density of the foam divided by the density of the solid from which the walls are made). Mechanical properties of foams depend on the relative density, cell wall properties, and cell geometry.

The compressive stress-strain curve of a foam has a characteristic sigmoid shape that represents the prevailing deformation mechanism (Figure 3–25). The following regions can be distinguished: (1) an initial linear elastic section (AB) due to wall bending; (2) a stress plateau (BC) produced by elastic buckling; and (3) a final, steeply rising stress corresponding to densification after cell collapse (D) (Gibson, 1989).

The Young's modulus $E_c$ of an isotropic foam undergoing elastic collapse in tension or compression is given by

$$E_c = k_i E_s \left(\frac{\rho_c}{\rho_s}\right)^n \qquad \textbf{Equation 3–85}$$

where $E_s$ is the modulus of the cell membrane material and $k_i$ and $n$ are constants. The buckling stress presents a similar relationship with $E_s$ but different constants. For the crushing of brittle foams, an exponent $n$ of 2 for open interconnected cells and an exponent of 3 for closed-cell foams are predicted. Further theoretical aspects of the mechanical properties of foam materials are presented in the review by Gibson (1989). The topic of foam in relation to food materials is covered in Jeronimidis (1991) and Peleg (1997).

Cellular solids abound in foods: bread, meringue, extruded snacks, puffed products, and so on. Expanded starchy products made by extrusion, such as flat breads and corn snacks, can be regarded as closed foams with bulk densities down to 0.03 g/cm$^3$. Hayter, Smith, and Richmond (1986) found that the relationship between stress and bulk density for expanded starchy products produced by extrusion was similar that expressed by equation 3–85, with an exponent equal to 1.1. General agreement between the theoretical equation and the actual behavior of several food foams (e.g., sponge cakes) has also been reported. The main deviations are caused by the large size and nonuniformity of pores as well as the effects of moisture and composition (Peleg, 1997).

## 3.10 FOOD STRUCTURE IN THE MOUTH AND BEYOND

The ultimate goal of the microstructural food engineer is to find and/or develop "appropriate" structure-property relationships in foods. But do we know enough about how food structure is sensed in the mouth? This is a legitimate concern of food technologists, since "a food is not a food until it is eaten."

The topic is quite complex, as it encompasses the flow and deformation of food pieces in the

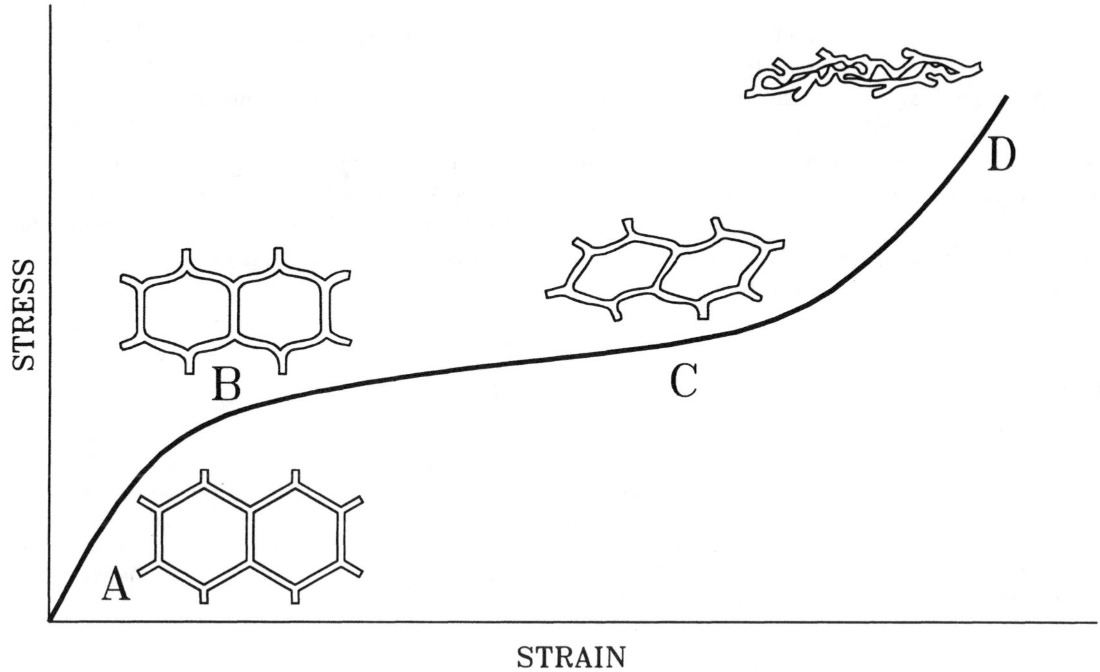


**Figure 3–25** Stress-strain behavior of foams and its relationship to structural changes.

mouth and sensorial evaluation during structure breakdown (Section 6.3). Three phases have been identified during oral processing of a food (Heath & Lucas, 1988): an initial ingestion phase, a repetitive chewing phase (most people chew at a rate of 40–80 masticatory strokes per minute), and swallowing. Ingestion demands an assessment of the overall quality of the food, and some properties related to structure, such as surface appearance and color, may be important. It is during incision and chewing that the internal structural properties are displayed and predominate and that the perception of physical parameters such as melting, viscosity, sound, and firmness of foods are realized.

Viscosity is recognized in the mouth by forcing liquid foods to flow in the space between the surface of the tongue and the roof of the mouth (Sherman, 1988). The lingual pressures associated with viscosity of undiluted Newtonian fluids (e.g., glucose syrup) are as high as $3 \times 10^4$ Pa, and shear rates range from 10 to $10^3$ $s^{-1}$ as liquids became less viscous (compare these to those Table 3–3). For Newtonian liquids such as wine, grape juice, and oils, the stimulus perceived in the mouth correlates well with their viscosity. A master curve has been proposed by Sherman (1988) to determine the conditions under which rheological measurements should be made if those during sensory assessment were to be simulated.

Solid foods in the mouth undergo three major processes: (1) the reduction of particle size by mastication; (2) the lubrication of pieces by saliva, juices released from the particles, and molten fats; and (3) reassembly before swallowing (Lillford, 1991). Evidently, all three processes are related to structure and its breakdown. Lillford proposed that the "mouth processing" of foods may be represented in a three-dimensional diagram having as axes "degree of structure," "degree of lubrication," and time. The resulting pathways followed by several foods to the swallowable state were significantly different. Lillford suggested that for each food, this sequence of events is inherently "engraved" in people, and when it deviates from the expected pattern a conscious response is triggered. A more mechanistic model has been proposed by Prinz and Lucas (1997), who suggest that the swallowing of a food requires the formation of a bolus in which food particles are tightly bound by viscous forces. A crucial element for bolus formation is structure breakdown and particle size reduction. Research efforts such as these are of importance for understanding the relation between the structure of foods, its breakdown and transport phenomena in the mouth, and sensorial properties.

Thus far, the chapters have been devoted to examining food microstructure and the basics of food materials science. The next step is to examine imitative instrumental methods that are expected to correlate better with the sensory evaluation of structure than pure mechanical tests. Note that we know even less about the effect of structure on nutrition, except that certain structures are unavailable for hydrolysis (e.g., "fiber") and not all starch may be digested. This is another topic that deserves to be researched thoroughly.

## BIBLIOGRAPHY


Adamson, A.W. (1990). *Physical chemistry of surfaces* (5th ed.). New York: Wiley.

Aguilera, J.M. (1992). Generation of engineered structures in gels. In H.G. Schwartzberg & R.W. Hartel (Eds.), *Physical chemistry of foods* (pp. 387–421). New York: Marcel Dekker.

Aguilera, J.M. (1995). Gelation of whey proteins. *Food Technology, 49*(10), 83–86, 88–89.

Aguilera, J.M., & Karel, M. (1997). Preservation of biological materials under desiccation. *Critical Reviews in Food Science and Nutrition, 37,* 287–309.

Aguilera, J.M., Kosikowski, F.V., & Hood, L.F. (1975). Ultrastructure of soy protein fibers fractured by various texture measuring devices. *Journal of Texture Studies, 6,* 549–554.

Ahmed, S., & Jones, F.R. (1990). A review of particulate reinforcement theories for polymer composites. *Journal of Materials Science, 25,* 4933–4942.

Alberts, B., Bray, D., Lewis, J., Raff, M., Roberts, K., & Watson, J.D. (1989). *Molecular biology of the cell* (2nd ed.). New York: Garland Publishing.

Atkins, P.W. (1982). *Physical chemistry* (2nd ed.). San Francisco: W.H. Freeman & Co.

Baer, E., Cassidy, J.J., & Hiltner, A. (1992). Hierarchical structure of collagen composite systems. In W. Glasser & H. Hatakeyama (Eds.), *Viscoelasticity of biomaterials* (pp. 2–23). ACS Symposium Series 489. Washington, DC: American Chemical Society.

Bailey, J.E. (1998). Mathematical modeling and analysis in biochemical engineering: Past accomplishments and future opportunities. *Biotechnology Progress, 14,* 8–20.

Barnes, H.A., Hutton, J.F., & Walters, K. (1989). *An introduction to rheology.* Amsterdam: Elsevier.

Blanshard, J.M.V. (1988). Elements of cereal product structure. In J.M.V. Blanshard & J.R. Mitchell (Eds.), *Food Structure: Its creation and evaluation* (pp. 313–330). London: Butterworths.

Blanshard, J.M.V., & Lillford, P.J. (1993). *The glassy state in foods.* Nottingham, England: Nottingham University Press.

Clark, A.H. (1987). The application of network theory to food systems. In J.M.V. Blanshard & P. Lillford (Eds.), *Food structure and behavior* (pp. 13–34). London: Academic Press.

Clark, A.H. (1995). Kinetics of demixing. In S.E. Harding, S.E. Hill, & J.R. Mitchell (Eds.), *Biopolymer mixtures* (pp. 37–64). Nottingham, England: Nottingham University Press.

Dagleish, D.G. (1997). Adsorption of protein and the stability of emulsions. *Trends in Food Science and Technology, 8,* 1–6.

Dickinson, E., & Stainsby, G. (1982). *Colloids in foods.* London: Applied Science Publishers.

Everett, D.H. (1993). *Basic principles of colloid science.* London: Royal Society of Chemistry.

Fennema, O.R. (Ed.). (1996). *Food chemistry* (3rd ed.) New York: Marcel Dekker.

Ferry, J.D. (1980). *Viscoelastic properties of polymers.* New York: Wiley.

Friberg, S. (1976). *Food emulsions.* New York: Marcel Dekker.

Fukushima, D. (1991). Structures of plant storage proteins and their functions. *Food Reviews International, 7,* 353–381.

Fung, Y.C. (1981). *Biomechanics.* New York: Springer-Verlag.

Gedde, U.W. (1995). *Polymer physics.* London: Chapman & Hall.

Gekas, V. (1992). *Transport phenomena of foods and biological material.* Boca Raton, FL: CRC Press.

Geneves, L., Rutin, J., & Halpern, S. (1985). Formes et composition elementaire de cristaux observes dans une farine de blé. *Comptes Rendus Academie Scientifique, Paris 301.* Serie III. No. 16, pp. 717–720.

Gibson, L.J. (1989). Modelling the mechanical behavior of cellular materials. *Materials Science Engineering, A110,* 1–36.

Gibson, L.J., & Ashby, M.F. (1988). *Cellular solids: Structure and properties* (pp. 178–180). Oxford: Pergamon Press.

Grinberg, V.Y., & Tolstoguzov, V.B. (1997). Thermodynamic incompatibility of proteins and polysaccharides in solution. *Food Hydrocolloids, 11,* 145–158.

Harding, S.E., Hill, S.E., & Mitchell, J.R. (1995). *Biopolymer mixtures.* Nottingham, England: Nottingham University Press.

Harris, P. (1990). *Food gels.* London: Elsevier Applied Science.

Hayter, A.L., Smith, A.C., & Richmond, P. (1986). The physical properties of extruded food foams. *Journal of Materials Science, 21,* 3729–3736.

Heath, M.R., & Lucas, P.W. (1988). Oral perception of texture. In J.M.V. Blanshard & J.R. Mitchell (Eds.), *Food structure: Its creation and evaluation* (pp. 465–481). London: Butterworths.

Hiemenz, P.C. (1986). *Principles of colloid and surface chemistry.* New York: Marcel Dekker.

Higuchi, T. (1990). Lignin biochemistry: Biosynthesis and biodegradation. *Food Science and Technology, 24,* 23–63.

Imeson, A. (1992). *Thickening and gelling agents for food.* London: Blackie Academic and Professional.

Jeronimidis, G. (1991). Mechanical and fracture properties of cellular and fibrous materials. In J.F.V. Vincent & P.J. Lillford (Eds.), *Feeding and the texture of food* (pp. 1–17). Cambridge: Cambridge University Press.

Jolley, P.D., & Purslow, P.P. (1988). Reformed meat products: Fundamental concepts and new developments. In J.M.V. Blanshard & J.R. Mitchell (Eds.), *Food structure: Its creation and evaluation* (pp. 231–264). London: Butterworths.

Kalichevsky, M.T., Blanshard, J.M.V., & March, R.D.L. (1993). Applications of mechanical spectroscopy to the study of glass biopolymers and related systems. In J.M.V. Blanshard & P.J. Lillford (Eds.), *The glassy state in foods* (pp. 133–156). Nottingham, England: Nottingham University Press.

Karbstein, H., & Schubert, H. (1995a). Basic criteria for the selection of emulsifying machines. *Food Ingredients Europe Conference proceedings (Frankfurt) 1995,* pp. 93–100.

Karbstein, H., & Schubert, H. (1995b). Developments in the continuous mechanical production of oil-in-water macroemulsion. *Chemical Engineering Processing, 34,* 205–211.

Katz, E.E., & Labuza, T.P. (1981). Effect of water activity on the sensory crispness and mechanical deformation of snack food products. *Journal of Food Science, 46,* 403–409.

Kokini, J.L. (1987). Viscoelastic properties of semisolid foods and their biopolymeric components. *Food Technology, 41*(3), 89–95.

Kokini, J.L. (1992). Rheological properties of foods. In D.R. Heldman & D.B. Lund (Eds.), *Handbook of food engineering* (pp. 1–38). New York: Marcel Dekker.

Kokini, J.L., Cocero, A.M., Madeka, H., & de Graaf, E. (1994). The development of state diagrams. *Trends in Food Science and Technology, 5,* 281–288.

Krog, N., Bradford, N.M., & Sanchez, R.M. (1989). Interfacial phenomena in food emulsions. *Journal of Dispersion Science and Technology, 10,* 483–504.

Krog, N., & Lauridsen, J.B. (1976). Food emulsifiers and their association with water. In S. Friberg (Ed.), *Food emulsions* (pp. 67–139). New York: Marcel Dekker.

Ledward, D.A., & Tester, R.F. (1994). Molecular transformations of proteinaceous foods during extrusion processing. *Trends in Food Science and Technology, 5,* 117–120.

Lee, L.H. (1989). New perspectives on polymer adhesion mechanisms. In L.H. Lee (Ed.), *New trends in physics and physical chemistry of polymers* (pp. 95–127). New York: Plenum Press.

Levine, H., & Slade, L. (1988). "Collapse" phenomena: A unifying concept for interpreting the behavior of low moisture foods. In J.M.V. Blanshard & J.R. Mitchell (Eds.), *Food structure: Its creation and evaluation* (pp. 149–180). London: Butterworths.

Lillford, P.J. (1991). Texture and acceptability of human foods. In J.F.V. Vincent & P.J. Lillford (Eds.), *Feeding and the texture of foods* (pp. 231–243). Cambridge: Cambridge University Press.

Little, C., Aguilera, J.M., Morales, A., & Kokini, J. (1997). Changes in soy protein during heating analyzed by pressure rheometry. *Food Science and Technology International (Tokyo), 3,* 130–133

Macosko, C.W. (1994). *Rheology: Principles, measurements and applications.* New York: VCH Publishers.

Magoshi, J., & Nakamura, S. (1992). Gelation and subsequent molecular orientation of silk fibroin. In W.G. Glasser & H. Hatakeyama (Eds.), *Viscoelasticity of biomaterials.* (pp. 231–248). Washington, DC: American Chemical Society.

Mohsenin, N.N. (1970). *Physical properties of plant and animal materials.* New York: Gordon & Breach Science Publishers.

Narkis, M., Chen, E.J.H., & Pipes, R.B. (1988). Review of methods for characterization of interfacial fiber-matrix interactions. *Polymer Composites, 9,* 245–251.

Niklas, K.J. (1992). *Plant biomechanics.* Chicago: University of Chicago Press.

Nilsson, S.B., Hertz, S.H., & Falk, S. (1958). On the relation between turgor pressure and tissue rigidity. 2. Theoretical calculations on model systems. *Physiological Plantarum, 11,* 818–837.

Oakenfull, D.G. (1991). The chemistry of high-methoxyl pectins. In R.H. Walter (Ed.), *The chemistry and technology of pectins* (pp. 87–108). New York: Academic Press.

Oates, C.G. (1997). Towards an understanding of starch granule structure and hydrolysis. *Trends in Food Science and Technology, 8,* 375–382.

Onsoyen, E. (1992). Alginates. In A. Imeson (Ed.), *Thickening and gelling agents for food* (pp. 2–24). London: Blackie Academic and Professional.

Peleg, M. (1997). Review: Mechanical properties of dry cellular solid foods. *Food Science and Technology International, 3,* 227–240.

Piculell, L., Bergfeldt, K., & Nilsson, S. (1995). Factors determining phase behavior of multicomponent polymer systems. In S.E. Harding, S.E. Hill, & J.R. Mitchell (Eds.), *Biopolymer mixtures* (pp. 13–35). Nottingham, England: Nottingham University Press.

Polyakov, V.I., Grinberg, V.Y., & Tolstoguzov, V.B. (1997). Thermodynamic incompatibility of proteins. *Food Hydrocolloids, 11,* 171–180.

Prinz, J.F., & Lucas, P.W. (1997). An optimization model for mastication and swallowing in mammals. *Proceedings of the Royal Society of London B 264,* 1715–1721.

Roos, Y. (1995). *Phase transitions in foods.* New York: Academic Press.

Roos, Y., & Karel, M. (1991). Applying state diagrams to food processing and development. *Food Technology, 45,* 68–71, 107.

Ross-Murphy, S.B. (1995). Rheology of biopolymer solutions and gels. In E. Dickinson (Ed.), *New physico-chemical techniques for the characterization of complex food systems* (pp. 139–156). London: Blackie Academic and Professional.

Ruben, C., & Larsson, K. (1985). Relations between antioxidant effect of a-tocopherol and emulsion structure. *Journal of Dispersion Science and Technology, 6,* 213–221.

Saunders, S.R., Hamann, D.D., & Lineback, D.R. (1992). A systems approach to food material adhesion. *Lebensmittel-Wissenschaft und Technologie 25,* 309–315.

Sherman, P. (1988). The sensory-rheological interface. In J.M.V. Blanshard & P. Lillford (Eds.), *Food structure and behavior* (pp. 417–432). London: Academic Press.

Singh, R.P. (1993). *Food properties database.* Version 1.1. Boca Raton, FL: CRC Press.

Slade, L., & Levine, H. (1991). Beyond water activity: Recent advances based on an alternative approach to the assessment of food quality and safety. *CRC Reviews in Food Science and Nutrition, 30,* 115–359.

Sperling, L.H. (1992). *Introduction to physical polymer science* (2nd ed.). New York: Wiley-Interscience.

Stanley, D.W., Stone, A.P., & Tung, M.A. (1996). Mechanical properties of foods. In L.M.L. Nolett (Ed.), *Handbook of food analysis* (pp. 93–137). New York: Marcel Dekker.

Tolstoguzov, V.B. (1988). Concentration and purification of proteins by means of two-phase systems: Membraneless osmosis process. *Food Hydrocolloids, 2,* 195–207.

Tolstoguzov, V.B. (1996). Structure-property relationships in foods. In N. Parris, A. Kato, & L.K. Creamer (Eds.), *Macromolecular interactions in food technology* (pp. 1–14). ACS Symposium Series 650. Washington, DC: American Chemical Society.

Tolstoguzov, V.B. (1997) Thermodynamic aspects of dough formation and functionality. *Food Hydrocolloids, 11,* 181–193.

van Soest, J.J.G., & Vliegenthart, J.F.G. (1997). Crystallinity in starch plastics: Consequences for materials properties. *TIBTECH, 15,* 208–213.

van Vliet, T., & Luyten, H. (1995). Fracture mechanics of solid foods. In E. Dickinson (Ed.), *New physico-chemical techniques for the characterization of complex food systems* (pp. 157–176). London: Blackie Academic & Professional.

Vincent, J. (1990). *Structural biomaterials* (revised ed.). Princeton, NJ: Princeton University Press.

Walstra, P. (1993). Principles of emulsion formation. *Chemical Engineering Science, 48,* 333–349.

Walstra, P. (1996). Dispersed systems: Basic considerations. In O.R. Fennema, (Ed.), *Food chemistry* (pp. 95–135). New York: Marcel Dekker.

Walter, R.H. (1991). *The chemistry and technology of pectin.* New York: Academic Press.

Webb, M.A., & Arnott, H.J. (1982). A survey of calcium oxalate crystals and other mineral inclusions in seeds. *Scanning Electron Microscopy, III,* 1109–1131.

Willenbücher, R.H., Tomka, I., & Müller, R. (1993). Thermally structural transitions in the starch-water system. In J.M.V. Blanshard & P.J. Lillford (Eds.), *The glassy state in foods* (pp. 491–497). Nottingham, England: Nottingham University Press.

Windhab, E.J. (1995). Rheology in food processing. In S.T. Beckett (Ed.), *Physicochemical aspects of food processing* (pp. 80–116). Gaithersburg, MD: Aspen Publishers.

Young, R.J., & Lovell, P.A. (1991). *Introduction to polymers.* London: Chapman & Hall.

Yurjev, V.P., Likhodzievskaya, I.B., Zasypkin, D.V., Alekseev, V.V., Grinberg, V.Y., Polyakov, V.I., & Tolstoguzov, V.B. (1989). Investigation of the microstructure of textured proteins produced by thermoplastic extrusion. *Die Nahrung, 33,* 823–830.

Zasypkin, D.V., Braudo, E.E., & Tolstoguzov, V.B. (1997). Multicomponent biopolymer gels. *Food Hydrocolloids, 11,* 159–170.

Ziegler, G.R., & Foegeding, E.A. (1990). The gelation of proteins. *Advances in Food Science Nutrition, 34,* 203–298.

Zobel, H.F. (1992). Starch granule structure. In R.J. Alexander & H.F. Zobel (Eds.), *Developments in carbohydrate chemistry* (pp. 1–36). St. Paul, MN: American Association of Cereal Chemists.

## CHAPTER 4

# Microstructural Components and Food Assemblies

### 4.1 INTRODUCTION

Chemically, foods are admixtures composed of organic compounds, inorganic compounds, and, mainly, water. This description, however, fails to recognize the complexity of structures associated with foods. As an example, many fruits and vegetables resemble cow's milk in containing about 9 parts in 10 of water. Yet, the former are solids and the latter is a liquid. Such a difference in physical state can only be attributed to microstructure.

It is now becoming appreciated that, in order to completely deal with the components of foods, it is necessary to have a basic understanding of how they are assembled and disassembled. Chapter 3 dealt with the chemical and molecular structures of food biopolymers. However, most macromolecules in nature are found as assemblies or elements of three-dimensional structures. Some biological macromolecular assemblies (e.g., membranes) are extremely complex because they are multicomponent systems; others are massive yet monocomponent (e.g., starch). These macromolecular systems are formed and interact in an aqueous medium, so the properties (and structure) of water need to be studied as well.

This chapter discusses the structural features of food components and gives examples of ways in which they can combine to form foods. Further, it examines the relationship between the structural aspects of food and the transformations that foods undergo, because, during processing, the former dictates the latter. This chapter is not intended to be a complete atlas of food assemblies; rather its goal is to explore some structures found in food materials and relate these to their properties.

### 4.2 WATER AND ICE

On our planet, water is the only substance that occurs abundantly in all three physical states. In foods, water is seen to be a major component when considered on the basis of weight, but it appears even more important when composition is expressed on a molar basis. Even in dry foods, water molecules are comparable in number to the molecules representing the solids. At the microstructural level, the location and interaction of water with other components is probably more relevant than the actual amount.

Although a simple and small molecule (main diameter = 3.3Å), water exhibits unusual properties owing to the existence of strong attractive forces among molecules. The V-like form and the polarized nature of the O-H bond results in an asymmetrical charge distribution that is the basis of the intermolecular attractive forces. Water can engage in multiple three-dimensional hydrogen bonding with bond energies varying between 2 and 40 kJ/mol. The bonds explain some of the unusual properties of water such as the large heat capacity, the melting and boiling points, the surface tension, and the enthalpies of phase transitions. They are also responsible for the basic structure of water—a tetrahedral arrangement in which each water molecule is coordinated through hydrogen bonds to four other molecules. This structure seems to be retained to a certain extent in the liquid state, influencing the orientation and mobility of molecules and suggesting that liquid water is "structured." In the same sense, solutes may enhance or disrupt the normal structure of water. Basic aspects of the water molecule and the structure of water and ice are described in Fennema (1996).

One would think that, given water's universality and the simplicity of its composition, our knowledge of water would, by now, be complete. Despite the enormous amount of research already done on water, the experimental evidence is often contradictory. Thus, not surprisingly, discrepancies exist concerning the role of water in food. For example, it has been common to represent the role of water in controlling changes in the processing and storage of foods through the use of water activity ($a_w$) as a measure of food stability. The decrease in reaction rates is attributed to reduced availability of water, since it is necessary for the mobility of the reactants. The mobility of water is important, because the reactions can occur to the fullest extent in solution, and at a lower $a_w$ an increasing fraction of the reactants are not in solution. Another way of looking at this is by invoking diffusional limitations. At low $a_w$, the observed rate constant is lower than the true constant by a function of the diffusivity of the critical reactant. The importance of this theory is that it suggests why water is not available at low $a_w$: it is "bound" chemically or physically (e.g., associated with biopolymers such as proteins and starch or trapped in micropores) and unavailable for reaction. Although the term *bound water* is controversial, it should be understood to mean water that does not exhibit the same properties as bulk water. The topic of water in foods is taken up again in Chapter 9.

Freezing is one of the most significant phase changes that water undergoes, and it has important consequences in food processing and also in microscopy, as mentioned in the discussion of cryomicroscopy (Section 1.5.4). For the transition from water to ice to occur, water molecules must link together through persistent hydrogen bonds to form tetrahedral frameworks that can then be arranged in a variety of lattice structures. The most common form of ice consists of the orderly array of hexagonal cells shown in Figure 4–1.

As the temperature is lowered during the freezing of foods, the molecular motion of the water molecules is slowed. Ice is formed as pure water goes through a two-step (nucleation and propagation) crystallization process. Extremely pure water may be supercooled to fairly low temperatures before nucleation occurs. Solid particulates act as catalysts for the appearance of water clusters upon which ice crystals can grow. Hexagonal ice crystals will form as these nuclei increase in size. The final size of the ice crystals, of great importance in frozen foods, is determined by the rate of cooling. Slow rates of cooling lead to a small number of large crystals, resulting in ruptured cells and poor quality, while faster cooling yields many small ice crystals, a less harmful configuration. Water may also solidify in an amorphous or vitreous state having a glass transition (onset) of $-138°C$. More on freezing as related to foods is found in Sections 5.3.2 and 5.3.3.

## 4.3 PROTEINS

Proteins provide the structural elements for many foods. The twenty-some amino acid building blocks of proteins can be arranged in many ways, leading to a broad spectrum of possible protein conformations and resulting structures. It has been estimated that there are around $10^{38}$ nonrelated ways to construct a protein containing 150 amino acids and that there are a total of $10^{10}$–$10^{12}$ different kinds of protein molecules existing in the spectrum of living species. Obviously, biochemists have examined only a handful of these in any depth. Thus, it seems a bit pretentious to assume that all of the large number of food proteins must fit into the mold generated from studying .000001% of the population (Branden & Tooze, 1991). However, certain apparently universal principles have been extracted that seem to govern the structure and function of proteins. This section focuses on how these molecules are configured into natural protein assemblies and the ways that food scientists can modify both their structure and behavior to fabricate protein-based foods.

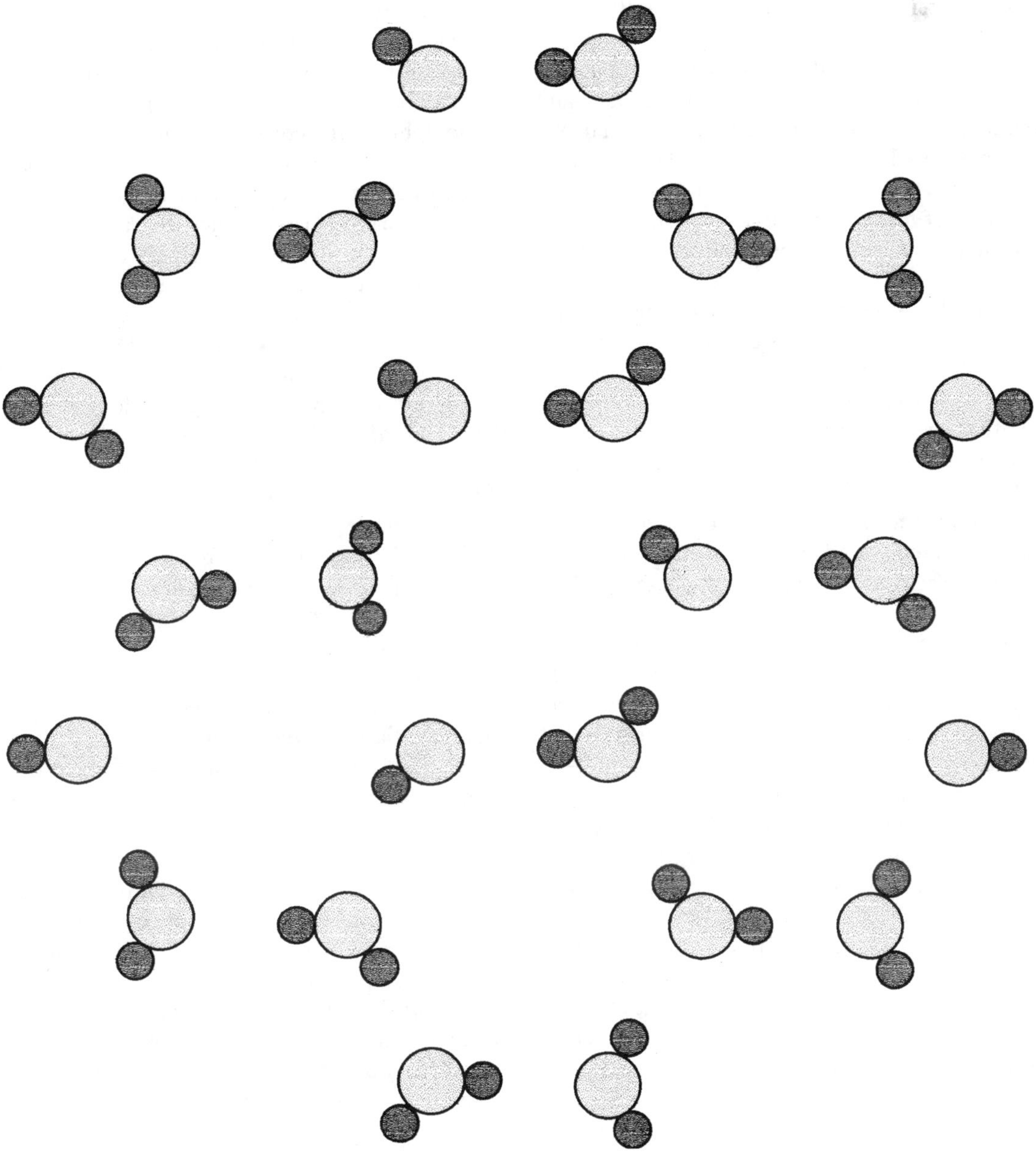

**Figure 4–1** Crystal structure of hexagonal ice. View along the c-axis showing the open structure. *Source:* Reprinted with permission from P. Echlin, *Low-Temperature Microscopy and Analysis*, © 1992, Kluwer Academic/Plenum Publisher.

### 4.3.1 Protein Conformation and Function

Over 3 billion years of evolution have produced a huge variety of different protein molecules, and during the past few thousand years plant and animal breeders have contributed to the variety as well. But the development, in the past few decades, of molecular genetics, in particular, the techniques of gene cloning and gene insertion, means that we are no longer restricted to those proteins that arise in nature through mutation and natural selection. Scientists now have the capability to alter proteins to achieve a predictable outcome or desired function. The implications for food science in general and food microstructure in particular are vast and only partially discernible at this time.

It has proven useful, when discussing the structure of proteins, to use the convenient terms of *primary, secondary, tertiary,* and *quaternary* levels of organization, since this nomenclature recognizes the hierarchical nature of the protein complex. The conformation of a protein molecule is determined first by the genetic code, which sets the primary sequence, and then by parameters that produce the transition from a linear polypeptide to a three-dimensional folded structure. The primary sequence is important, because it determines the type of secondary structures formed. These include conformations such as several $\alpha$-type helices, several $\beta$-type structures (pleated sheets and bends or turns), and the often invoked random coil. How these secondary structures organize into a three-dimensional form determines the tertiary structure of the protein. The tertiary structure is made up of domains or regions of the polypeptide chain that fold up into the tertiary form independently.

Many proteins of interest to food scientists result from the assembly of different peptide chains, and the supramolecular structures resulting from the combination of different chains are the referents of the term *quaternary.* Formation of a quaternary structure is driven by the thermodynamic requirement to bury exposed hydrophobic regions of subunits. Functionally, associated proteins displaying quaternary structure are composed of oligomers; that is, they are built up from subunits or noncovalently linked modules and possess the property of disassociation. The smallest subunit that can be isolated from a quaternary structure without breaking covalent bonds is termed a *monomer,* and the whole molecule may be composed of identical or different monomers. The size of these protein assemblies can vary from just two subunits that together possess a molecular weight of about 6 kD (bovine insulin proteins) to 2,130 subunits having a weight of 10,000 kD (tobacco mosaic virus coat proteins). Of course, it may be more useful for food scientists to view *in situ* protein assemblies as a single biological functional unit and disregard chemical and physical distinctions, such as molecular weight, that seem of little consequence in the case of such complex organizations.

The native structure of a protein is the result of intramolecular forces as well as the interactions of protein groups with the surrounding aqueous milieu. The native state is thermodynamically the most stable state with the lowest free energy at physiological conditions. Changes in the secondary, tertiary, and quaternary structures without cleavage of backbone peptide bonds constitute "denaturation." In practical terms, denaturation means the unfolding of the native structure, and there are in fact several "denatured states," each differing slightly in free energy. Fully denatured globular proteins resemble random coils. The difference in free energy between the native and the denatured state of food proteins is on the order of 20–50 kJ/mol. Protein denaturation usually has a negative connotation, since it is associated with loss of functionality in foods. However, it is often a prerequisite for improved digestibility, biological availability, and performance (e.g., emulsification or gelation).

For many food scientists, the quaternary structure of proteins is of greatest interest. The reasons for this include the fact that many food proteins are large and oligomeric and that the influence of processing on these structures and the subsequent structure formation is of major importance in food systems. It is indeed worthwhile to examine how the subunits of these proteins are held together, since the strength of these bonds and how they react to perturbations determine their structural conformation and functional properties. Of the noncovalent interactions possible, including hydrogen bonding, dipolar interactions, ionic interactions, and hydrophobic interactions, the latter seem of the greatest importance in relation to food proteins such as muscle contractile proteins, casein micelles, and plant storage proteins. The structure of a plant storage protein is shown in Figure 1–19.

It is a truism that protein conformation dictates function. For example, while globular and random coil protein molecules are frequently characterized by their solubility, rod-shaped proteins are often insoluble and can self-associate to form structural elements through the interaction of subunits. The ability of a protein to form structures depends mainly upon protein-protein and protein-water interactions. The structure achieved in turn allows certain processes to occur, such as gelation, texturization, dough formation, emulsification, and foaming, all of which can lead to stable food structures. These processes are complicated in foods because of the intentional or unintentional modification of proteins resulting from processing steps such as heating, which can lead to unfolding and association with other components, including carbohydrates and lipids.

### 4.3.2 Natural Protein Assemblies

Nature forms supramolecular structures (e.g., ribosomes, protein filaments, and membranes) by the noncovalent assembly of preformed macromolecular subunits. These structures are then responsible for complex functions, such as transport and molecular or cell recognition. Food scientists are mostly concerned with nonliving systems. For those interested in proteins, this means that rarely are natural protein assemblies encountered. In other words, *in situ* arrangements of proteins are susceptible to both enzymatic and microbiological attack, which, while perhaps not degrading the large structural proteins, may destroy the couplings that connect these units. The first step in this process is often the deterioration of membranes, which leads to the peroxidative attack of polyunsaturated fatty acids and the concomitant production of free radicals, the activation of certain membrane-bound enzymes, and associated cellular damage (Stanley, 1991). Since natural protein assemblies are not often found in post-mortem or postharvest food systems, it is reasonable to ask of what interest they are to food scientists.

There are several important reasons to study protein assemblies. First, it is important to understand how protein association aids in studying reactions involving their disassociation. These reactions lead ultimately to major losses in food quality and thus are significant. Second, impressive progress is being made by food scientists interested in how proteins interact and how the interactions can be modified to achieve greater functionality. Knowledge of natural protein structures helps researchers to determine which configurations and bonding types lead to stable designs. For example, one of the protein's characteristics most important for dictating functional characteristics is its thermal stability. This characteristic appears to result from certain structural parameters, including amino acid composition, compact packing or protein-protein contacts, binding of metals and other prosthetic groups, and intramolecular interactions and linkages. Thermal stability can be altered through chemical modification. For example, one strategy for adding rigidity to the structure is to introduce more noncovalent bonds (Stanley & Yada, 1992).

Naturally occurring protein assemblies exhibit a delightful structural elegance, and food scientists can convert them into a wide range of important food products. Table 4–1 lists examples of important protein structures in foods, and some of these structures are shown in Figure 4–2.

Because of their structural importance and abundance, we will use the natural protein assemblies associated with myosystems as an example. Food scientists wishing to investigate the structure of skeletal muscle and how muscle proteins are assembled are faced with the problem that this tissue is the most structurally complex material used as a food. In fact, it is probably the most elaborate structure found in nature. Fortunately, the

**Table 4–1** Examples of Important Protein Structures in Food

| *Protein Source* | *Protein* | *Structure* |
|---|---|---|
| Meat | Actin, myosin | Interact to form actomyosin, texturally important<br>Form emulsified, stable heat-set gels<br>Form binding system for restructured meats |
| | Collagen | Forms stable cross-links, texturally important |
| Fish | Actin, myosin | TMAOase forms HCHO, resulting in cross-linking<br>Form stable gels (kamaboko) |
| Legumes | Conglycinin, glycinin | Form stable gels with calcium and heat (tofu)<br>Form stable fibers with freezing (kori-tofu)<br>Form stable lipid-protein films (yuba)<br>Form stable fibers by isoelectric precipitation (spun soy fibers)<br>Form stable fibrous structures by thermoplastic extrusion (texturized soy protein) |
| Wheat | Gluten | Forms viscoelastic doughs (bread) |
| Milk | Caseins | Form stable emulsions (ice cream, butter, processed cheeses)<br>Form stable foams (whipped cream, milkshakes, mousses) |
| | Whey proteins | Form stable gels |
| Eggs | Yolk lipoproteins | Form stable emulsions (mayonnaise, salad dressings) |
| | Albumin | Forms stable foams (meringues, souffles, omelets) |

protein content is high, about 20% on a wet weight basis, and the contractile proteins are large and arranged in a highly ordered fashion, making it possible to observe the structural organization directly with light and electron microscopy.

Muscle tissue is so complex and sophisticated because of its biological purpose—to produce harmonized, controlled movement. Muscle contraction is the source of locomotion in animals used as food. As with all tissue, skeletal muscle is constructed of cells—termed *fibers* because of their threadlike appearance. These cells contain not only contractile machinery but also structures to regulate contraction and to supply energy. Cylindrical muscle fibers from 10 to 100 $\mu$m in diameter and from 20 $\mu$m to several centimeters in length are arranged in parallel bundles to form a whole muscle. The sarcolemma, a true biological membrane with an associated connective tissue component, surrounds the cell, and it is to this surface that nerve impulses are conducted initially. The fine structure of muscle tissue is examined in subsequent sections.

What is most important to food scientists about muscle tissue is that it provides a source of actin and myosin, the two major muscle proteins. Myosin, like most fibrous protein molecules, contains a mixture of elongated helical chains and compact globular areas. The structure of myosin is represented by a long twin-chain rod terminating in two large globular heads that are the location of adenosine triphosphate enzyme activity (Figure 4–3). Actin *in situ* consists of a double-stranded "string of pearls" helical fibrous structure formed from polymerized individual actin monomers. Both these proteins, either singularly or combined together as actomyosin, are prized food components because of their functional properties, including binding and emulsification ability and the ability to form heat-set gels.

As regards the physical properties of meat, collagen is the most significant of the connective tis-

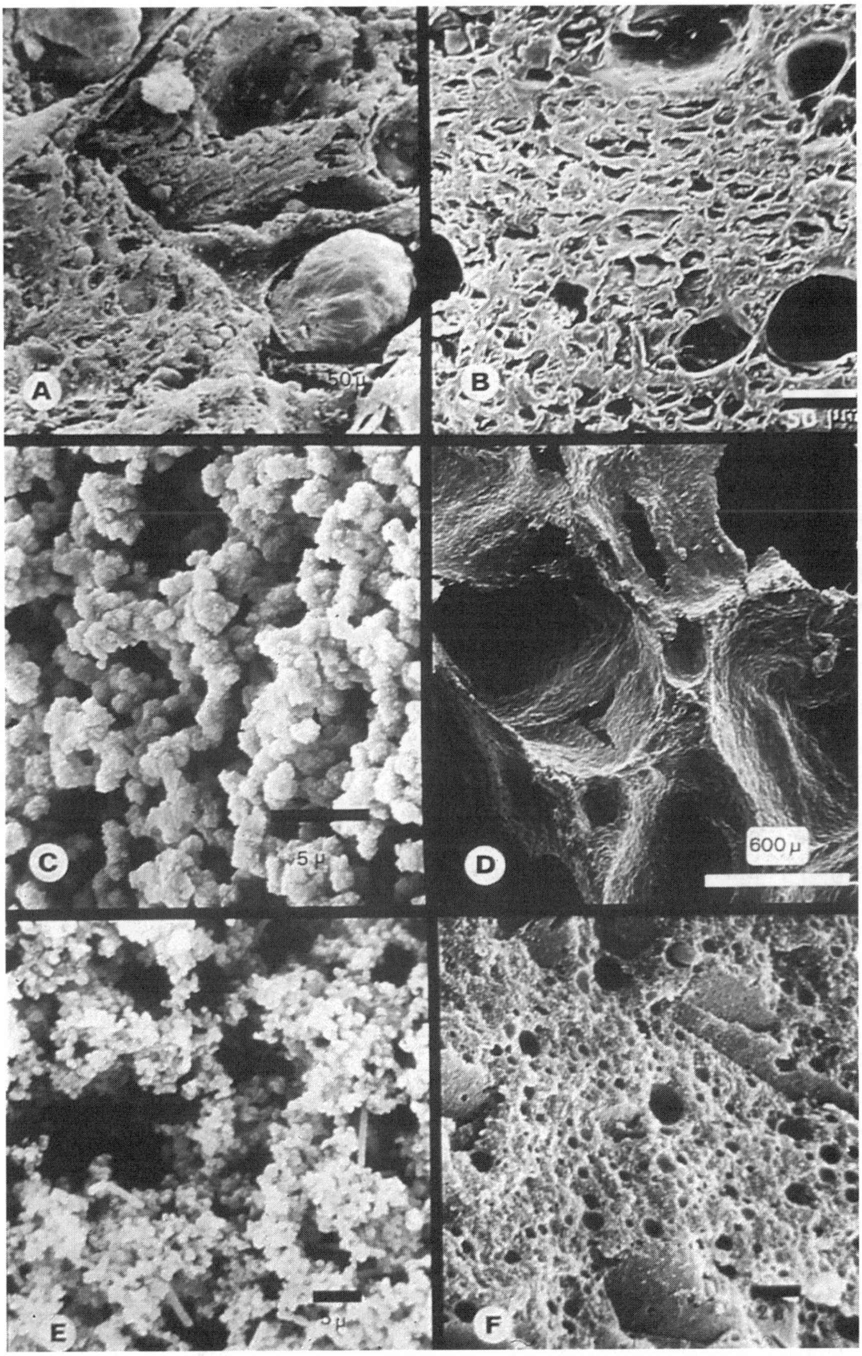


**Figure 4–2** Examples of protein structures in foods. (A) Scanning electron micrograph of commercial frankfurter prepared from mechanically deboned chicken meat, 2.5% salt, 18% fat. *Source:* Courtesy of S. Barbut. (B) Scanning electron micrograph of commercial surimi. *Source:* Courtesy of A.K. Smith. (C) Scanning electron micrograph of scrambled egg. *Source:* Courtesy of S. Barbut. (D) Scanning electron micrograph of bread. *Source:* Courtesy of J.L. Smith. (E) Scanning electron micrograph of yogurt. *Source:* Courtesy of E.M. Parnell-Clunies. (F) Scanning electron micrograph of processed Colby cheese. *Source:* Courtesy of K.W. Baker.

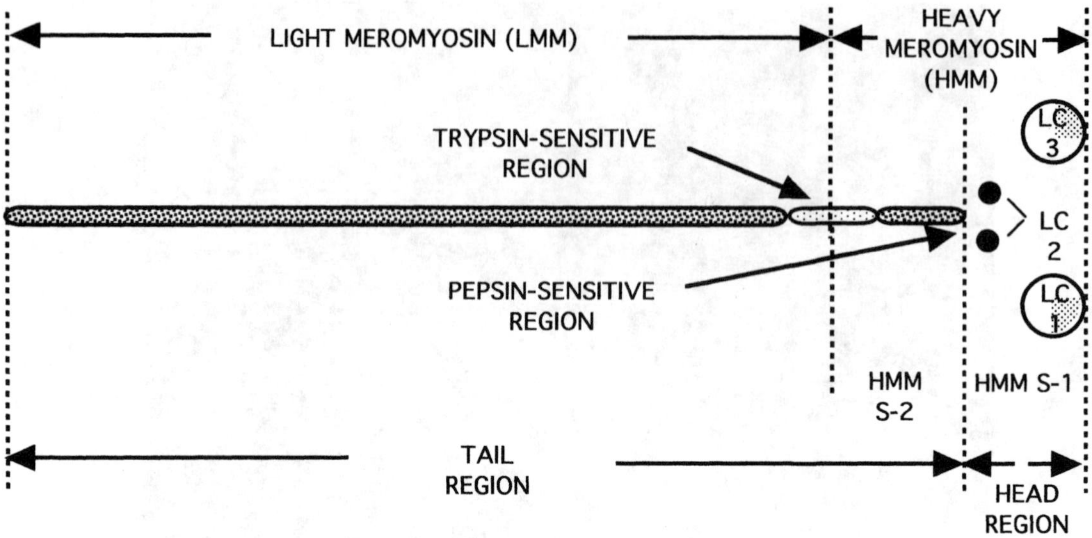


**Figure 4–3** Diagram of the myosin molecule. *Source:* Reprinted from *Food Research International*, Vol. 25, A.P. Stone and D.W. Stanley, Mechanisms of Fish Muscle Gelation, pp. 381–388. Copyright 1992, with permission from Elsevier Science.

sue proteins. It is a rod-shaped molecule consisting of three subunits that interact to form a compact triple helix. Collagen fibrils are formed by the association of a number of these helices so as to form characteristic banding with a 68 nm repeat pattern, and bundles of fibrils form a collagen fiber. Figure 4–4 shows a collagen fibril as viewed in the transmission electron miscroscope (TEM) and also how collagen is distributed in muscle tissue. Collagen possesses high tensile strength, about two orders of magnitude greater than the muscle fibers it surrounds, and this strength is its major contribution to meat texture. Several other factors, however, must be taken into consideration. One is the orientation of collagen fibrils. Post-mortem events dictate the alignment of endomysial collagen fibers (Figure 4–5) that contributes to the toughness associated with contracted muscle. Also, collagen can form stable, covalent, intermolecular cross-links that increase in number and stability with animal age, leading to enhanced toughness. Cooking has a dramatic effect on collagen properties, since it sequentially produces softening, shrinkage, and conversion to gelatin, composed of much less structured molecules and possessing concomitantly reduced physical properties.

### 4.3.3 Engineered Protein Assemblies

Proteins can produce a wide range of important food structures. As food scientists learn to manipulate and control their formation, "engineered" structures will become available that are specifically designed for particular end uses. Biomolecular engineering is an emerging field that integrates the structural and physical properties of macromolecules to optimize function. Optimization can occur by either modification or synthesis, and it might be useful to make some distinctions among activities often grouped together as "protein engineering." These have been divided as follows (Feeney & Whitaker, 1986):

- Protein modification consists of any physical or chemical change caused by treatment of the protein by chemical, enzymatic, or physical means.
- Protein tailoring involves specific protein modification for a specific purpose.
- Protein engineering comprises those modifications caused by changes in the genetic code.
- Recombinant proteins are proteins made by *in vitro* mutagenesis.

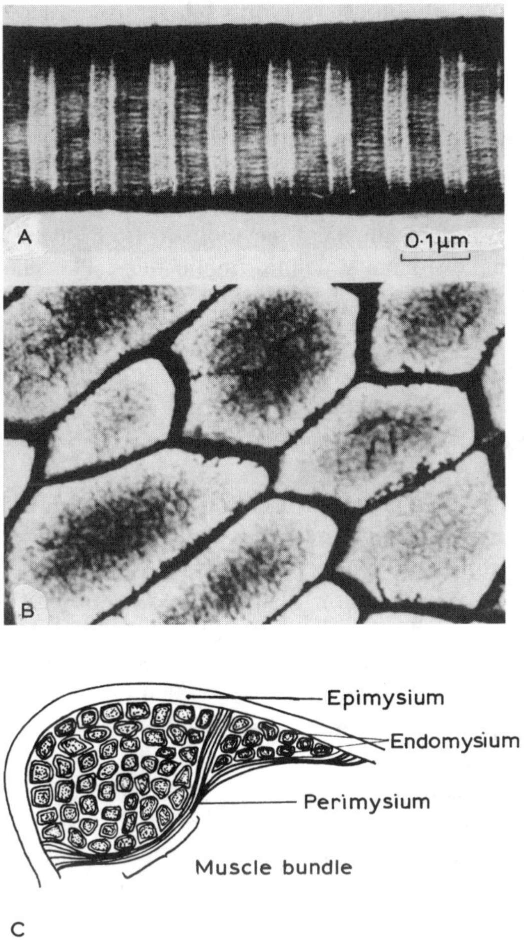


**Figure 4–4** Muscle connective tissue. (A) Transmission electron micrograph of collagen fibril showing 68 nm periodicity. (B) Cross section of porcine muscle stained with silver to demonstrate reticular fibers of the endomysium. (C) Diagram of muscle cross section showing major connective tissue components. *Source:* Stanley, 1983a.

Much work has been done in the area of tailoring food proteins by chemical means to achieve improved functionality, and numerous studies can be found in the literature of dairy and plant proteins. The use of various plant proteolytic enzymes to achieve tenderization of meat may be cited as an example of chemical modification, but generally food scientists wishing to refashion muscle foods take a physical or mechanical approach.

Engineering proteins to create desirable physical properties and to facilitate materials applications is certainly within the realm of current knowledge. Synthesis of a certain class of proteins characterized by repetitive amino acid sequences—including silks, collagens, elastins, and bioadhesives—has been achieved, and it seems likely that in the near future protein engineers will be able to design and produce novel proteins using amino acid sequences not found in nature. *De novo* protein design will undoubtedly provide new food materials.

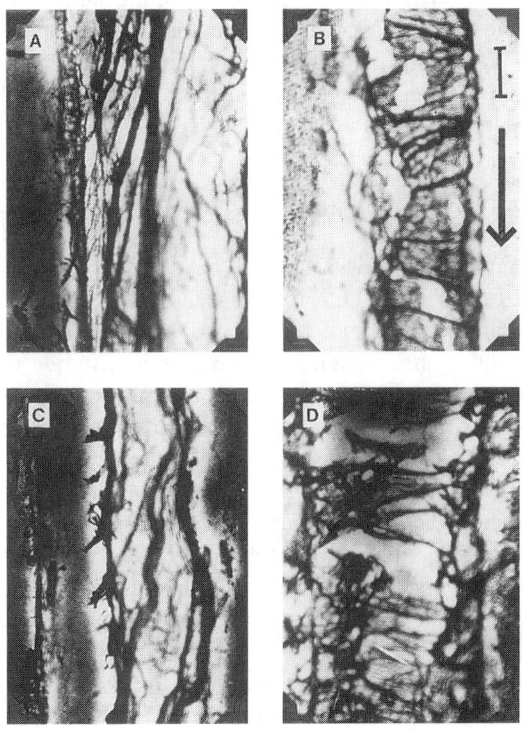


**Figure 4–5** Light microscopy of porcine muscle stained with silver to demonstrate reticular fibers of the endomysium. (A, C) Restrained during rigor at approximately 150% of rest length. (B, D) Unrestrained during rigor. Bar = 20 $\mu$m, arrow indicates orientation of fiber axis. *Source:* Stanley and Swatland, 1976.

## 4.4 LIPIDS

Natural triglycerides, the most common storage lipids, are usually not considered a major structural element in food tissue. On the other hand, structured lipids, assembled triglycerols containing mixtures of short-, medium-, and long-chain fatty acids, are becoming widely used for their specific functionality. These food materials can be produced chemically, enzymatically, or by genetic modification, although only the latter two methods can result in specific placement of fatty acids on the glycerol molecule. An example of their use, structured triglycerides are employed as fat replacers in food products (see Section 5.6.3). By synthesizing triglycerides that incorporate fatty acids such as behenic, a C22 molecule that is poorly absorbed, food scientists can significantly reduce their caloric content. There are numerous potential applications for structured lipids in the food industry, but their acceptance by consumers will dictate how widespread their utilization will eventually become.

### 4.4.1 Phospholipids

Some forms of derived lipids have quite important structural roles. One of these is the family of phospholipids, essential for the formation of various cellular and subcellular membranes. Studies of how triglycerides, fatty acids, and phospholipids react when placed in aqueous environments have revealed how the hydrophobic hydrocarbon tails of these molecules tend to interact—either as a monolayer on the surface or as spherical micelles, depending upon the concentration. A micelle is an assembly of polar and nonpolar constituents in aqueous solution. Above a critical concentration, spherical micelles are formed that consist of a hydrocarbon kernel surrounded by the hydrophilic headgroups. These configurations allow the hydrophobic moieties to avoid, as much as possible, contact with the polar medium, while the polar or hydrophilic ends of the lipid molecules orient themselves toward the aqueous phase. This led to the concept of the lipid bilayer, which has since dominated thinking on membrane structure.

Phospholipids will form bilayers in aqueous media in which their ionic and other polar groups are exposed to water and are free to form close associations with proteins. One result is the formation of biomembranes that possess the important property of selective permeability and also have the ability to act as a platform or site for certain aspects of biochemical metabolism. Both of these functions have structural implications. Phospholipids such as soybean lecithin and those found in egg yolk also serve as emulsifiers that promote oil-in-water emulsions.

### 4.4.2 Triacylglycerols (Triglycerides)

*Fat* and *oil* are terms used for lipid materials that are solid or liquid at room temperature, respectively. Fats are esters of fatty acids and glycerol, mostly in the form of triacylglycerols (triglycerides). When a melted fat cools, it solidifies into a solid crystalline material in which triglycerides are shaped like an elongated h. Fat crystals usually take the form of needles or platelets. Polymorphic forms are crystalline phases of the same chemical composition; they differ among themselves in structure but yield identical liquid phases upon melting. Each polymorphic form is characterized by specific properties (density, melting point, etc.), and changes from one form into another may occur in the solid state without melting. Although differential scanning calorimetry is a preferred method of determining phase transitions in fats, successive crystallization and melting between different forms may cause difficulties in the interpretation of thermograms.

A problem with the different polymorphic forms of triacylglycerols is that the nomenclature in the earlier literature is extremely confusing. If a melted fat is cooled rapidly, the least-ordered form ($\alpha$-crystals) crystallizes at a temperature just below the melting point. If cooled extremely slowly, the high-melting triglycerides have time to form stable $\beta$ crystals. At intermediate cooling rates, the fat first forms $\alpha$ crystals, which melt and recrystallize into the metastable $\beta'$ form. Properties of the different crystal forms are shown in Table 4–2.

**Table 4–2** Characteristics That Control the Type of Crystal in Fats

| *Characteristic* | *α-form* | *β′-form* | *β-form* |
|---|---|---|---|
| Molecular packing | Lowest | Intermediate | Highest |
| Molecular tilting | None | Tilted (70°) | Tilted (60°) |
| Growth rate | Fastest | Lower | Lowest |
| Stability | Least stable | Intermediate | Most stable |
| Melting point | Lowest | Intermediate | Highest |

*Source*: Reprinted with permission from G.G. Jewell and J.F. Heathcock, Structured Fat Systems, in *Food Structure and Behavior,* J.M.V. Blanshard and P. Lillford, eds., pp. 279–295, © 1988, Academic Press.

Crystals in turn may associate to form a network. The nature of the network structure depends on both composition and processing. Fat in margarine (80% fat) consists of a mixture of liquid oil and crystallized fat. The structure is stabilized by "shells" of fat crystals interconnected into a three-dimensional network surrounding water droplets and oil. A finer emulsion results in a more dense structure, but it is possible to incorporate up to 80% water into margarines. Obviously, high-water-content margarines are unsuitable for baking or frying (because of spattering due to the rapid evaporation of water).

## 4.5 CARBOHYDRATES

Carbohydrates constitute the most heterogeneous group of the major food elements, ranging widely in size, shape, and function. While the smaller, simpler molecules are only rarely involved, larger carbohydrates often play significant roles in the microstructure formation of plant foods. Polysaccharides such as starch, cellulose, hemicellulose, pectic substances, and plant gums provide textural attributes such as crispness, hardness, and mouthfeel to many foods. Many can form gels that will provide microstructure and also enhance viscosity of solutions owing to their high molecular weight.

### 4.5.1 Polysaccharides (Gums)

One of the major achievements in food science in the past few decades has been an increased appreciation of the role of polysaccharides in food structuring. Polysaccharides are natural macromolecules present in almost all living organisms, and they function either as a source of energy or as structural units in the morphology of the living material. These high-molecular-weight polymers are formed by simple sugars covalently linked through glycosidic bonds. The structure and conformation of polysaccharides and their intermolecular associations give polysaccharide dispersions, solutions, and gels their distinctive properties. Polysaccharide primary structures frequently show simple repeating sequences; the geometry of the individual monosaccharide ring is essentially rigid but the relative alignment of component residues about the glycosidic linkage determines the overall conformation of the polysaccharide secondary structure. The introduction of 1.6 linkages imparts rotational angles and gives an additional element of flexibility. The overall dimensions of these molecules are affected by the extent and type of branching between individual monosaccharides. Tertiary structures observed in polysaccharides arise from the folding pattern due to the secondary and primary structures. These include extended ribbon, flexible helix, crumpled ribbon, and flexible coil geometries. Some have more complicated conformations, such as double helices and fivefold helices. Quaternary structures involve subunit aggregations of like molecules or unlike molecules through noncovalent bonding. Polysaccharides in solution may develop quaternary structures from the cross-linking of tertiary structures.

The wide range of rheological behavior demonstrated by polysaccharides in solution is due to the variety of possible conformations and chain flexibility. When a soluble polysaccharide is placed in water, the water molecules quickly penetrate the amorphous regions and surround available polymer sites. During hydration, segments of the polysaccharide chain become fully solvated: as hydration continues, the polysaccharide molecules become completely surrounded by partially immobilized water, thus the alternative name of hydrocolloid. The majority of polysaccharide thickeners exist in solution as conformationally disordered random coils. These can produce an enhanced viscous effect by virtue of entanglements. Polysaccharides that form gels do so by specific, permanent chain-chain polymer interactions. The gelation mechanism depends on the polysaccharide but invariably involves the physical entrapment of water in a three-dimensional network of ordered polysaccharide chain segments.

The majority of polysaccharides or gums used in the food industry are derived from plant materials such as seaweeds, seeds, and tree exudates. Their commercial usefulness is based on their ability to alter the basic properties of water. Polysaccharides are used primarily to modify texture through thickening or gelling. Related properties include the ability to cause suspension of particulates, inhibition of syneresis, film formation, and encapsulation as well as the ability to control crystallization. The suitability of a polysaccharide for a particular application is primarily contingent on the functional behavior of the polysaccharide in the presence of other food ingredients and its response to time and temperature conditions, pH changes, and mechanical treatment. Polysaccharides are heterodisperse and impart unusually high solution viscosities at low concentrations; they are normally used at concentrations ranging from 0.05% to about 5%, but concentrations less than 1% are typical.

### 4.5.2 Cellulose

Cellulose is both the most abundant carbohydrate and also the principal structural component of plant tissue. Although it is insoluble and indigestible, cellulose is of prime concern to the food scientist because of its contribution to the texture of plant foods. It exists in nature as linear chains of glucopyranose joined by linkages that are impervious to attack by enzymes found in the human gut. The major function of cellulose is in plant cell walls, where it combines with hemicellulose, proteins, pectins, and lignin to provide necessary structural integrity. Cell walls have been described as consisting of interlaced cellulose microfibrils embedded in an amorphous matrix composed mainly of pectic substances and hemicelluloses and in which the cellulose serves a structural role similar to the steel rods in reinforced concrete. The introduction to this chapter mentioned that plant tissue, although having a water content similar to that of milk, exhibits a characteristic solid microstructure. Its solidity is due to the small amount of polysaccharides found in the cell walls. Typical plant cell walls from potato tissue are shown in Figure 4–6. Thus, milk is an example of a solid-in-liquid food while fruits and vegetables are liquid-in-solid foods. It is of interest to note that food scientists often wish to reverse this order; apples become apple juice and milk becomes cheese through structural inversion.

### 4.5.3 Starch

Starch is composed of two polymeric units: linear amylose and highly branched amylopectin (Section 3.3.5). Plants lay down starch as granules (normally 10–50 $\mu$m in size) in which molecules are organized into a radially anisotropic, semicrystalline unit (e.g., between 15% and 45% crystalline). Radial anisotropy is responsible for the Maltese cross (birefringence) appearing when the native granules are seen under polarized light. The center of the cross is at the hilum, the origin of growth of the granule. The semi-crystalline state becomes apparent when studied under X-ray diffraction.

Native starch granules of different origin are shown in Figure 4–7. There are several levels of structural complexity in starch granules. The first level is the "cluster arrangement" of the amylopectin branches, in which arrangement alternate regions of ordered, tightly packed parallel glucan chains alternate with less ordered regions corresponding to branch points. Thus, the starch granule appears to be formed by alternating concentric

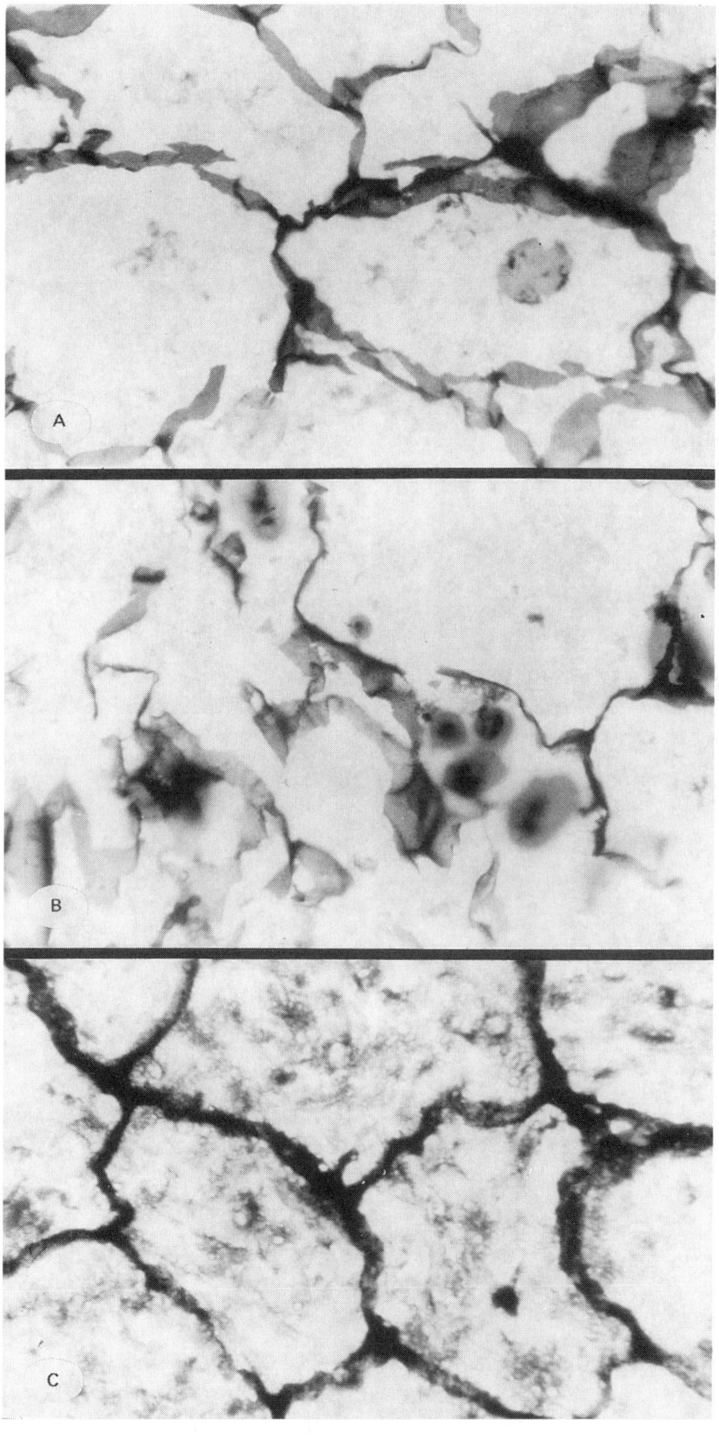


**Figure 4–6** Light micrographs of cell walls from potato tissue. (A) Raw. (B) Frozen. (C) Cooked. *Source:* Stanley and Tung, 1976.

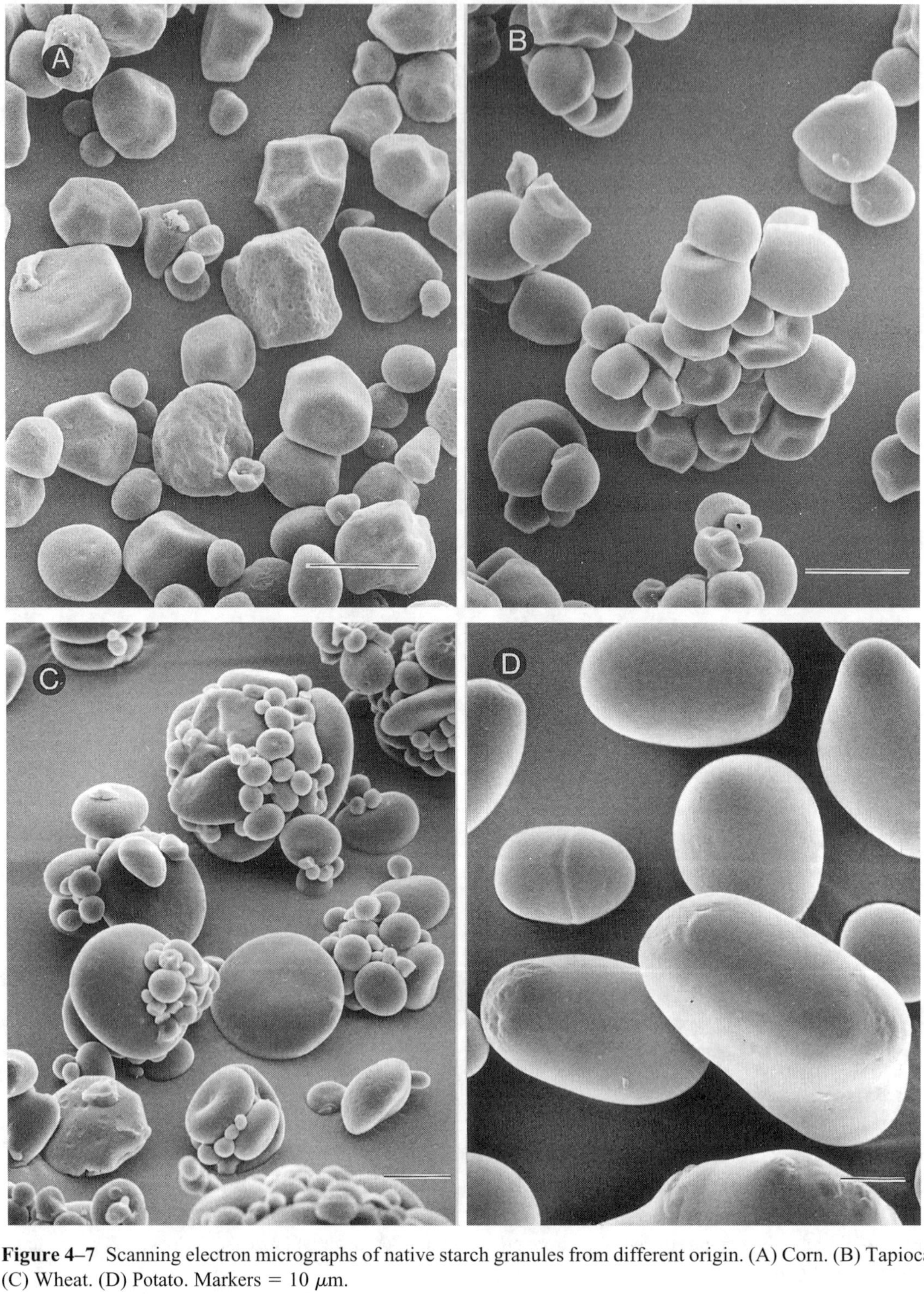


**Figure 4–7** Scanning electron micrographs of native starch granules from different origin. (A) Corn. (B) Tapioca. (C) Wheat. (D) Potato. Markers = 10 $\mu$m.

rings or clusters of amorphous (branching points) and crystalline (glucan chains) lamellae (see Figure 3–3). The size of each cluster is about 9–10 nm. Crystallinity in amylopectin is created by the intertwining of long chains in adjacent branches as double helices (Oates, 1997). Three types of polymorphs exist in starches. The differences between them are related to packing differences, and which type is present can be determined by X-ray diffraction. The A-type pattern is exhibited by cereals, whereas the B-type is found in tuber and fruit starches as well as in retrograded starch. C-type starch is typical of legumes. The unit crystal structures of A- and B-type polymorphs contain 4 and 12 water molecules between the helices, respectively.

The amylose fraction is assumed to exist in the native granule in the amorphous state as individual molecules randomly interspersed among the amylopectin molecules (in both the amorphous and crystalline regions). Thus, small amylose molecules located at the periphery are free to leach out of the granule during gelatinization (Oates, 1997).

Starch granules are insoluble in cold water. When heated in the presence of excess water, the amorphous regions that pervade the whole granule swell tangentially, forming a continuous gel phase. As the temperature exceeds a value typical for each plant species (roughly between 50 and 80°C), the crystal structure is also disrupted by gelatinization. If water is reduced or solutes are added, the gelatinization temperature is increased. Because starch can be considered a polymer spherulite, *gelatinization* may be viewed as a melting process and as corresponding to an order-disorder transition. Hence, gelatinization of starch having a high water content is most commonly characterized in a differential scanning calorimetry curve as a relatively sharp endothermic peak that spreads over the baseline for 10 to 15°C and involves enthalpies on the order of 10–20 J/g. Under conditions of limited water availability, the transition shifts to higher temperatures, and a second peak may appear. The phase transitions of starches are discussed in Roos (1995). Microstructurally, as gelatinization proceeds, the granule's network is obliterated and individual linear macromolecules (amylose) diffuse into the aqueous medium, increasing its viscosity. Further heating and/or shear disrupts the granule, and a starch paste consisting of a continuous phase of amylose (and amylopectin) and a dispersed phase of granule remnants is formed.

After cooling, dilute solutions of starch will precipitate, but concentrated dispersions may form a firm, viscoelastic gel having crystallites as junction zones. During aging, starch molecules can reassociate into crystalline segments (retrograde) to an extent that depends on factors such as the source of the starch, the amylose:amylopectin ratio, the molecular weight and linearity of the starch molecules, the time and temperature of cooling, the moisture content, and the concentration of starch (Roos, 1995). This recrystallization phenomenon is known as *retrogradation* and has been detected by X-ray diffraction and by an increasing endothermic peak of melting with storage time. Amylose crystallization occurs rapidly after cooling of gelatinized starch, while amylopectin crystallization is a slower process. Recrystallized amylopectin, partly responsible for the staling of bread, can be rendered amorphous if heated in the presence of water to 55–95°C. Recrystallization of amylose is known to be an inhibiting factor in the enzymatic degradation of starch and consequently impairs its digestion (Morris, 1990). A schematic representation of the most important transformations of starch during food processing is presented in Figure 4–8.

## 4.6 CELLS AND CELL MEMBRANES

The primary structural feature of all biological tissue is its cellular organization. Early light microscopists quickly found this uniting element in both plants and animals. Although their primitive instruments did not allow a detailed examination of the fine microstructure existing within cells, they soon realized that cells contained myriad organelles and were surrounded by barriers that have important structural roles. A generalized diagram of a cell containing both animal and plant organelles is shown in Figure 4–9.

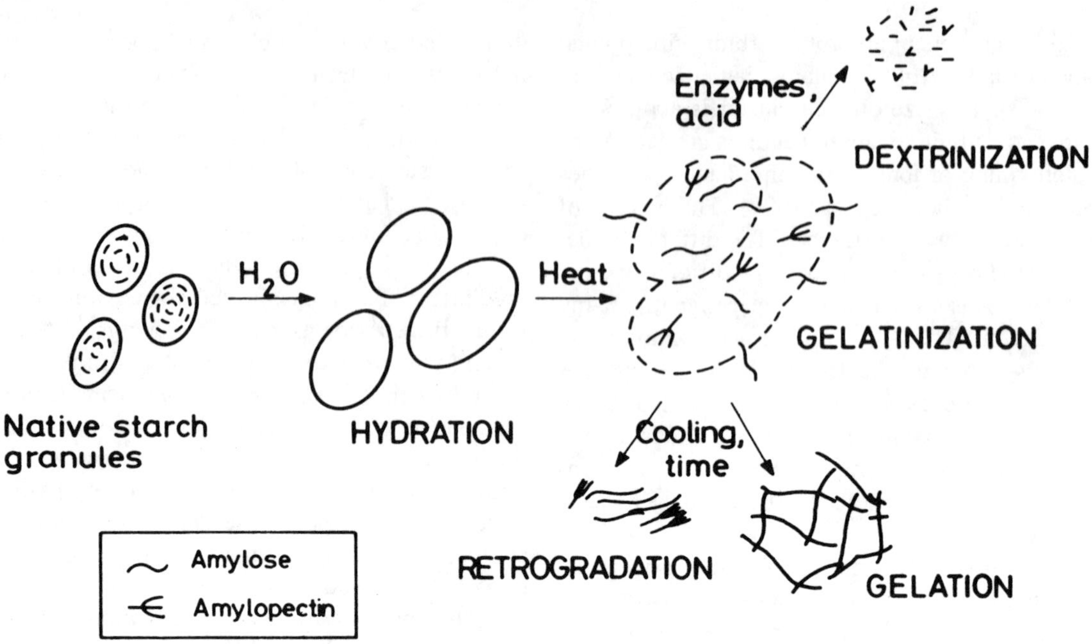


**Figure 4–8** Diagrammatic representation of main starch transformations during food processing.

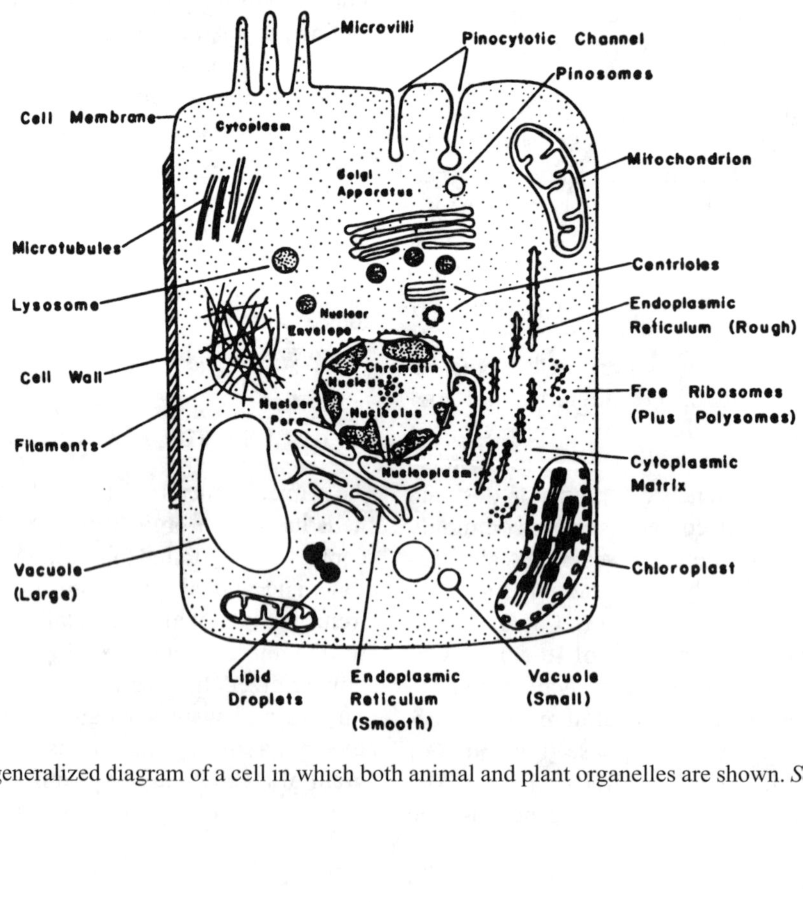


**Figure 4–9** A generalized diagram of a cell in which both animal and plant organelles are shown. *Source:* Moner, 1972.

In a living plant cell, the medium in which physiological activity takes place is the cytoplasm, a gel-like material. It contains the nucleus, the plastids, and a variety of other organelles, including mitochondria and the endoplasmic reticulum, a complex system of internal and external membranes. Major food components such as starch, storage proteins, and most lipids are contained in discrete, homogeneous packets embedded in the cytoplasm. Compartmentalization is an important means of regulating plant metabolism and also contributes to the "architecture" of the cell. Central vacuoles contain water and considerable amounts of sugars and salts in solution, which largely accounts for the osmotic potential. The cytoplasm (sometimes called protoplasm) is bound by the vacuolar membrane or tonoplast and the plasma membrane or plasmalemma. These limiting membranes are semipermeable and control the movement of water and solutes between and inside cells (see Figure 8–3). Further description of animal and plant cells is presented later in this chapter. The topic of plant cells is revisited in connection with extraction of solutes in Section 8.2.5.

Biological materials exhibit a structural hierarchy. This is not to say that, like the fractal objects discussed previously, they are scale invariant. Rather, the structure and properties manifested at each successive level are dependent on the attributes of elements in the preceding level, the elements' relative concentrations, the physical forces involved in their interaction, and the manner in which the elements are spatially arranged. (See Section 5.1.2 for a discussion on hierarchy in nature and also see Figure 5–2.)

### 4.6.1 Cell Walls

Higher magnification, when it became available, revealed that cell walls are composed of two main elements—an outer framework that supplies structural rigidity and an inner lipid-protein membrane responsible for boundary formation and active transport. Food technologists and engineers ought to have a basic idea of how these elements are organized, their architecture and structural features, and the transport processes taking place through them.

The cell walls of both plant and animal tissue are of particular interest to food scientists because of their contribution to microstructure. Also, the destruction of cell walls to extract their contents, which is occasionally necessary, has important economic consequences. The two major structural elements involved in cell walls, cellulose in plants and collagen in animals, have already been described (see Section 3.3), but it is curious to see the structural convergence between these two chemically dissimilar molecules (see Exhibit 4–1). Of course, they do have similar functions, yet there are some obvious chemical differences between the animal and plant system: the plant cell utilizes carbohydrate building blocks in its cell walls whereas the animal cell uses protein. The use of protein may be explained by the fact that animals are motile, and motion places significant levels of stress on the cell. Thus, the structural element of the animal cell must have flexibility, a property associated with linear protein polymers but not linear carbohydrate polymers. The cell walls of plant tissue, which are especially important to food scientists, are described more fully later in this chapter.

### 4.6.2 Cell Membranes

The biological imperative to possess a membrane is obvious: every living cell has at least one membrane to differentiate that cell from its environment. Indeed, the most fundamental and primitive role of membranes is to provide a boundary. The simplest of cells, the prokaryotic cell (found in bacteria), is identifiable by its singular extracellular or plasma membrane. Eukaryotic cells, on the other hand, are more complex and possess membrane-bound compartments or organelles that allow life processes to be segregated. Common organelles include structures such as the nucleus, lysosomes, Golgi bodies, the endoplasmic reticulum, mitochondria, and chloroplasts. Studies of these various membranes show that extracellular membranes differ from intracellular ones in function, structure, and composition.

**Exhibit 4–1** Structural Convergence of Cellulose and Collagen

- Found in small amounts (approximately 5% wet basis) but ubiquitous
- Linear molecules formed into polymers
- High functional molecular weight
- Possess crystalline regions
- High tensile strength, which increases with age
- Soften upon heating
- Low nutritional value
- Water insoluble in native state
- Resistant to enzyme attack, stable
- Provide structural integrity to cells
- Found extracellularly, part of intercellular network
- Major factor in food texture

The two major nonaqueous components of membranes are lipids (mainly phospholipids and cholesterol) and proteins (both structural and catalytic). It is the way in which these molecules arrange themselves that determines membrane structure. Early experimentation led to the conclusion that membrane lipids exist in a bilayer conformation. The lipid bilayer has since dominated thinking on membrane structure. Regarding proteins, an appreciation of the hydrophobic nature of the constituent amino acids in membrane proteins and its influence on structure led to the realization that lipids and proteins interact so as to segregate hydrophobic areas as much as possible. This and the discovery that membrane components can move laterally prompted the development of a general model of membrane structure that has been accorded substantial agreement by researchers. The "fluid mosaic" model (Figure 4–10) is generally considered to be the most realistic membrane model yet developed. This model depicts membranes as *fluid,* meaning that components can diffuse laterally in the plane of the membrane, and as *mosaic,* meaning that the membrane proteins are not spread uniformly over the outer polar surfaces but some of them are buried deeply and discretely into the membrane interior, in certain cases traversing the entire structure.

Cell membranes and their deterioration play a major role in food quality. Besides regulating transportation of molecules across their boundaries and acting as sites for enzyme attachment, they allow, through their breakdown in necrotic tissue, enzymes to access previously unavailable substrates. These processes, coupled with free radical production from oxidation of membrane lipids, initiate many of the deleterious reactions that lead to quality loss in food. Despite all of this, biological membranes are rarely considered by food scientists when studying the deteriorative reactions that take place during the processing or storage of food tissues. Yet membranes and their deterioration play a major role in food losses, and recent biochemical information indicates that at least some deteriorative reactions can be controlled by procedures suited to food materials (Stanley, 1991).

Much of what passes for information about membrane degradation in food systems is incomplete and speculative. It is known, however, that in order to perform their many indispensable functions in cells, membranes are constituted mainly of phospholipids, proteins, and some carbohydrates arranged in thin, bimolecular sheetlike structures that serve to compartmentalize cells and their organelles. Membranes have imbedded

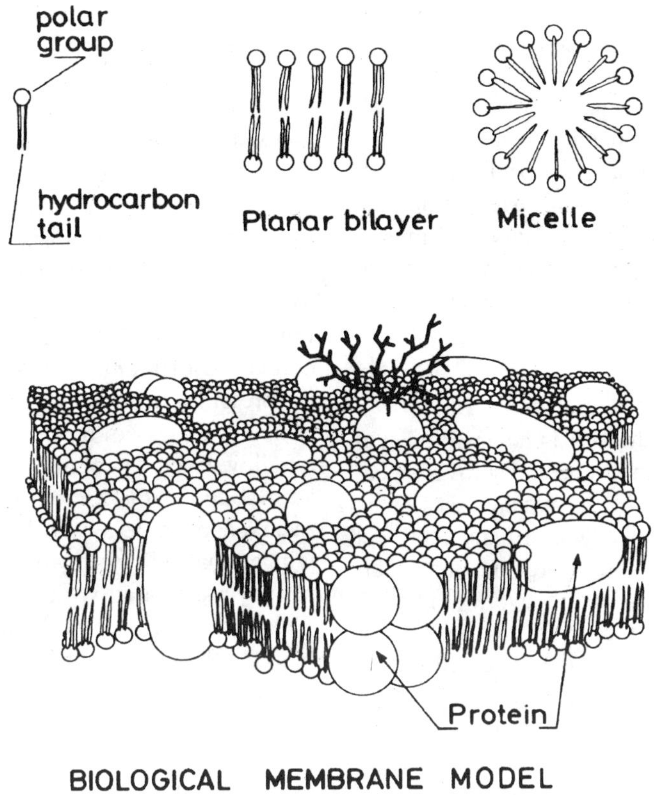


**Figure 4–10** Above: Fatty acid molecule and self-assembled structure (bilayers and micelles). Below: Fluid mosaic model of biological membrane structure.

in their asymmetric surfaces complements of catalytic and cytoskeletal proteins that serve permeability and structural functions. Membrane surfaces exhibit fluidity, due partially to the continuous lateral diffusion of lipids and some proteins. Two important consequences of fluidity are the ability of membrane phospholipids to exist in different interconvertible conformational phase structures and the formation of heterogeneous lipid domains on the membrane surface. While all lipid components appear capable of lateral diffusion, some proteins are immobile, presumably because of their linkage to the cytoskeleton.

Cellular death leads unavoidably to the initiation of membrane deterioration. Although the time course of this series of reactions differs in animal and plant tissue, both types of tissue are damaged by generally similar mechanisms. These include an initial peroxidative attack on polyunsaturated fatty acids and the concomitant production of free radicals. Many biological agents can act as accelerating agents in these reactions, including transition metal ions, heme compounds, radiation, illuminated chlorophyll, calcium, and ethylene. Once formed, free radicals catalyze further reactions that can affect all aspects of membrane function and cellular metabolism and can lead ultimately to significant losses in food quality through defects such as chilling injury and cold shortening. These defects are aggravated by many food processing steps, especially those that involve tissue disruption.

Control of membrane breakdown by exogenous chemical intervention has been practiced, but, at best, such intervention only slows the rate of the reactions. Newer approaches to solving the breakdown problem include dietary treatment of meat animals, modifying storage and packaging conditions, and genetic interventions. Membrane deterioration can be considered a universal mechanism that leads to significant quality loss, and food scientists need to become more aware of its causes and prevention.

### 4.6.3 Cell Cytoskeletons

The early view that the cell is a sac filled with structureless cytoplasm has been disproved: a cell in fact contains a number of smaller sacs or organelles. It is now known that virtually every cell is exquisitely ordered, and its organization is conferred by a three-dimensional network of microfibrous structures. This meshwork, termed the cytoskeleton, is found in all eukaryotic cells (those having a true nucleus with a nuclear membrane). It is a dynamic three-dimensional structure that fills the cytoplasm. It consists of three components—microfilaments, intermediate filaments, and microtubules—and most elements of the cytoskeleton are capable of assembly and disassembly (i.e., the insoluble filaments assemble from soluble subunits and can disassemble back into subunits).

Many functions have been attributed to the cytoskeleton, but overall it is responsible for mediating the interaction and roles of cytoplasmic organelles. It would seem that the cytoskeleton of the muscle cell is particularly well developed and is metabolically quite active, perhaps as a result of the function of these cells in motion (Figure 4–11). The cytoskeleton is also an internal "scaffolding" that helps give a cell its own unique shape. In addition, it is a transport mechanism responsible, for example, for the migration of vesicles in protein secretion, taking vesicles from the endoplasmic reticulum to the Golgi bodies and out to the plasma membrane. Finally, it helps cells change shape and is the driving force in cell motility processes.

Plant cells, in contrast to animal cells, consist of a semi-rigid cell wall encasing a plasma membrane that surrounds a thin layer of cytoplasm. This, in turn, encloses a large vacuole. Thus, the cytoplasm and its organelles, including the cytoskeleton, compose only about 10% of the total cell volume. The wall occupies another 10% and the vacuole as much as 80%. This cytoarchitecture has severely hampered studies on the plant cytoskeleton. The cytoskeleton of plants is an elaborate and highly dynamic network of filamentous and tubular structures; the major cytoskeleton elements include microfilaments, which are composed of a double-stranded helical array of the protein actin, and microtubules, which consist of a helical array of the protein tubulin that forms a hollow fiber. The third major component of the animal cell cytoskeleton, the intermediate filaments, appears less well developed in plants. As with animal cells, the cytoskeleton in plant cells is closely associated with the plasma membrane. Such interactions control many fundamental cellular processes (e.g., cell-cell adhesion and signal transduction). Not only does a close association exist between the microfilaments and microtubules, but a close association also exists between both of these structures and the endomembranes. These cytoskeleton-membrane complexes may be involved in the cytoarchitectural coherence of organelles, such as protein bodies and mitochondria. The role of the cytoskeleton in plant cells has not been as fully explored as in the case of animal cells, but it does appear to be smaller than its role in animal cells, conceivably because the plant cell wall causes some of the rigidifying that is the responsibility of the animal cell cytoskeleton. The cytoskeleton is transparent in standard light and electron microscope preparations and is therefore "invisible." It is usually left out of drawings of the cell, but it is an important, complex, and dynamic cell component that has implications for food scientists.

## 4.7 STRUCTURAL ASPECTS OF ANIMAL TISSUE

Animal tissue of interest to food scientists—primarily voluntary muscle—is an extremely complicated system because it is formed from many

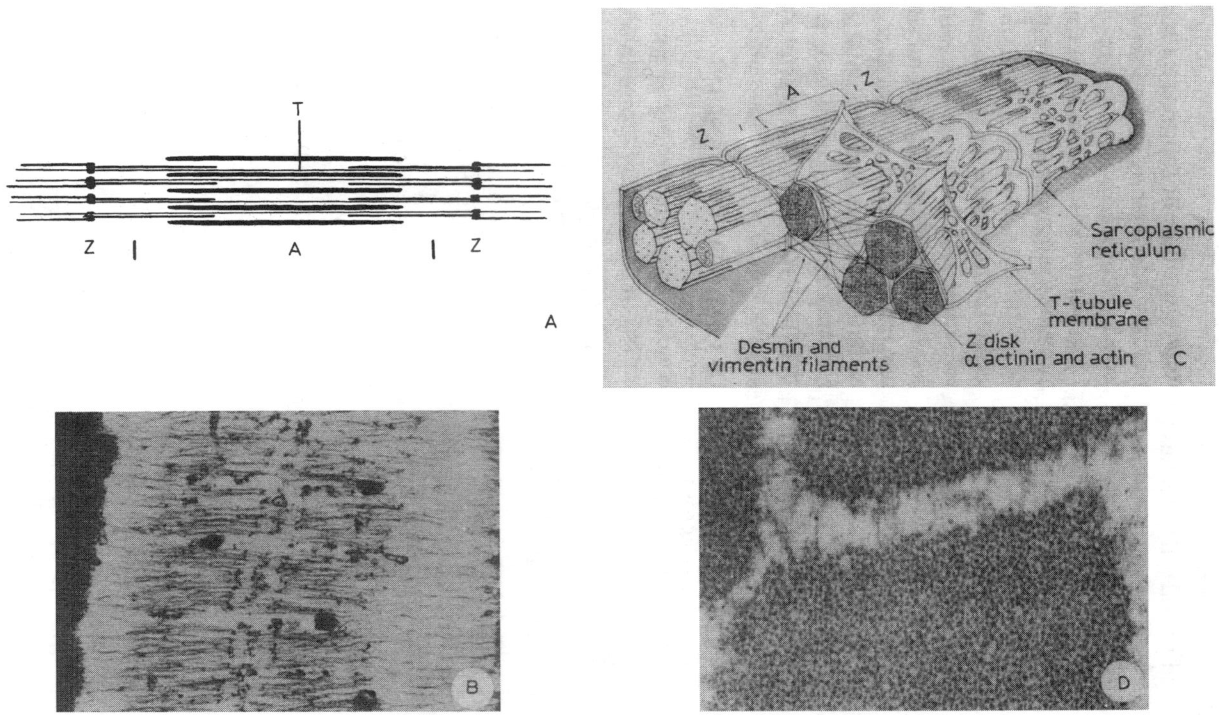


**Figure 4–11** Diagrams showing muscle cytoskeleton. (A) Longitudinal element (T filaments, gap filaments, connectin). (B) TEM of gap filaments. (C) Transverse elements (intermediate filaments, desmin). *Source:* Stanley, 1983b. (D) Transverse connections between adjacent myofibrils. *Source:* Swatland and Belfry, 1985.

individual elements that interact over a wide range of dimensions to form a structural hierarchy (Table 4–3).

Although meat includes a number of different tissues, food scientists have focused their research on muscle cells, since they make up the preponderance of edible tissue and provide the proteins necessary to form the continuous phase of comminuted meat emulsions required for stabilizing fat particles. Muscle cells or fibers have a threadlike appearance, being long (1–40 mm), thin (10–100 μm), and polyhedral. They are polynucleated and surrounded by a sarcolemma that combines a cell membrane overlaid with endomysial connective tissue.

The most striking feature of muscle fibers is their transversely parallel bands or striations. As noted in Chapter 1, under polarized light the bands that seem darker with ordinary illumination are anisotropic or birefringent (A bands) whereas those that appear light are isotropic (I bands). Although these bands seem continuous across the fiber, this is only because the underlying elements, myofibrils, usually appear in register with one another, a consequence of cytoskeletal organization.

Myofibrillar architecture has been shown to be based on an even more slender thread (1–2 μm) consisting of repeating units called sarcomeres that are generally 1.5 to 2.5 μm in length, depending upon their degree of contraction. In the individual sarcomere lies the origin of the A and I bands, now known to be associated primarily with the proteins myosin and actin, respectively, and the Z-disc that delineates the contractile unit. Contraction occurs by a mechanism of interdigitation of the set length A and I bands. Figure 4–12 provides views of muscle at different magnification. Because of the vast complexity of this topic, serious students should augment their understanding by consulting a modern physiology text.

The components mentioned above join to form numerous features necessary to fulfill the function of motion, but the major structural factors affecting meat quality are connective tissue, myofibrillar proteins (the structural proteins directly responsible for contraction), and the cytoskeletal system.

As stated, connective tissue is an ubiquitous component of the animal body. In muscle it is present in several forms but is composed mainly of the protein collagen. While it was thought formerly that the only factor influencing the role of collagen was its gross amount, now it is known that other factors are important, such as age-related cross-links, degree of contraction, postmortem breakdown, and differential effects of heat. Thus, whereas connective tissue was once viewed as a constant background factor in meat quality, its dynamic and changeable nature is now becoming known.

The muscle myofibril is an extraordinary biochemical mechanism endowed with the ability to shorten and relax quickly, uniformly, and repetitively. It does this through the interaction of numerous proteins, salts, membranes, and metabolites in an aqueous milieu. The interaction of the two major protein components, actin and myosin,

**Table 4–3** Structural Hierarchy of Muscle Tissue

| *Component* | *Structure Type* | *Instrument* | *Approximate Size* | *Relative Order of Magnitude of Diameter* |
|---|---|---|---|---|
| Beef carcass | Macrostructure | Human eye | 2 m × 1 m dia | $10^9$ |
| *L. dorsi* muscle | Macrostructure | Eye, hand lens | 1 m × 10 cm dia | $10^7$ |
| Muscle fiber (cell) | Microstructure | LM, SEM | 1–40 mm × 10–100 μm dia | $10^3$ |
| Myofiber | Ultrastructure | SEM, TEM | 1–40 mm × 1–2 μm dia | $10^2$ |
| Myosin filament | Ultrastructure | TEM | 1.5 μm × 15 nm dia | 1 |

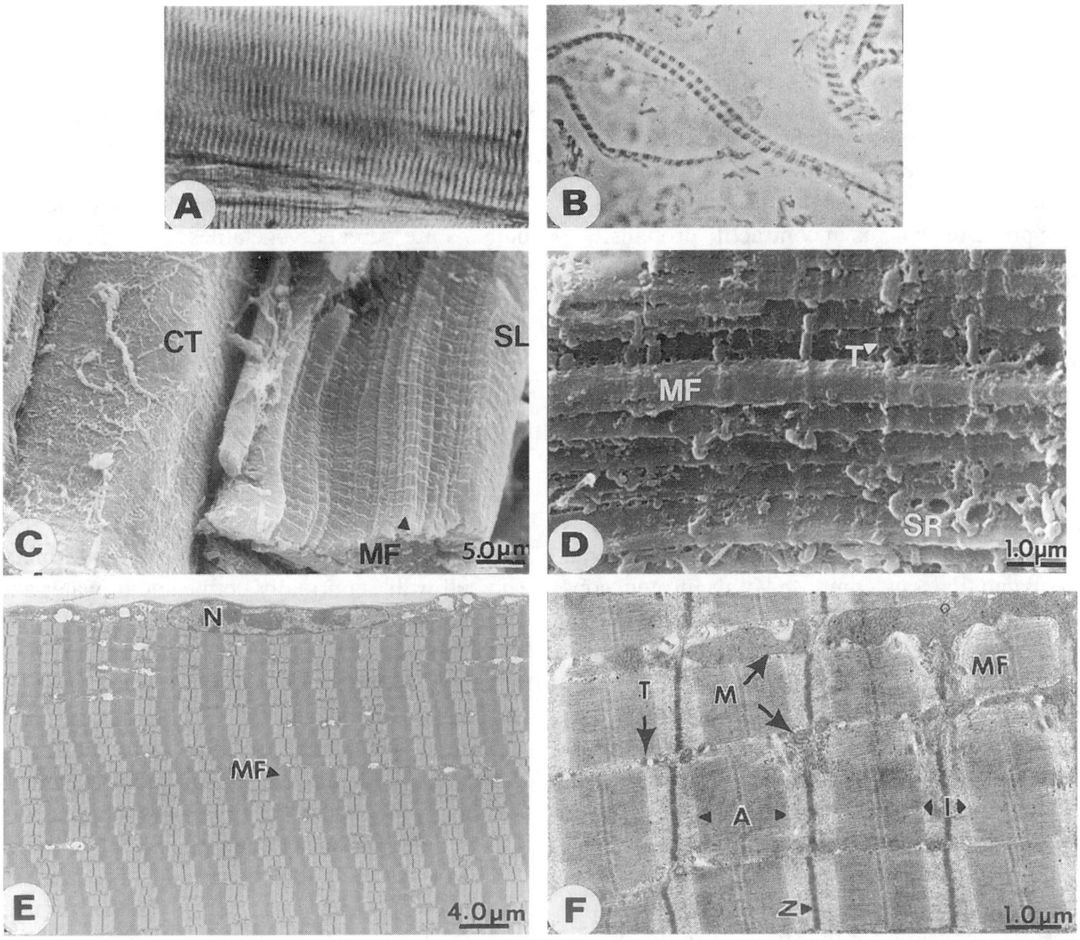


**Figure 4–12** Microstructure of muscle tissue. (A) Light micrograph (phase contrast) of muscle fiber showing transverse striations. Several nuclei are visible. (B) Light micrograph (polarized light) of muscle myofibrils. (C) Scanning electron micrograph of muscle fiber showing endomysium (CT), and myofibrils (MF). (D) Scanning electron micrograph of freeze-fractured muscle showing myofibrils (MF), T tubules (T), and sarcoplasmic reticulum (SR). (E) Transmission electron micrograph of muscle longitudinal section, low magnification. (F) Transmission electron micrograph of muscle longitudinal section showing myofibrils (MF), mitochondria (M), Z-discs (Z), A and I bands (A, I), and sarcoplasmic triads at A-I level (T). *Source:* Stanley, 1983a.

play the main role in determining sarcomere length during contraction. These two proteins overlap as a result of contraction, and the degree to which overlapping occurs is reflected in sarcomere length or the distance from one Z-disc to the next. This distance is often used as an index of myofibrillar contribution to toughness, since the more post-mortem sarcomeres are contracted, the tougher the muscle becomes, owing, to inextensibility, rigidity, and filament packing density. This is one reason why, in the case of species where toughness of meat can be a consumer problem (e.g., beef), muscles are left on the carcass at least until the rigor mortis process is completed, thus eliminating unwanted contraction during butchering.

It has proved difficult to differentiate completely the effects of connective tissue and myofibrillar proteins, since the two are, in fact, not independent. As has been shown (Figure 4–5), contraction affects both sarcomere length and connective tissue configuration. Also, muscles low in connective tissue would be expected to be more prone to changes in contractile proteins.

Because of the unique design of the muscle fiber, the function of its cytoskeleton would be expected to be different from that of nonmotile cells. Just as connective tissue serves as an extracellular support for the fiber, so would the cytoskeleton seem likely to hold myofibrils in place and provide an ordering of the contractile apparatus. The cytoskeleton (see the diagrams in Figure 4–11) of adult muscle fibers contains at least two components related to its physical properties. First, a thin (2 nm) gap filament, composed of the protein connectin, runs parallel to the fiber axis and either connects adjacent Z-discs or an A band to a Z-disc, perhaps extending to the sarcolemma. This element provides intracellular elasticity and tensile strength. The second link consists of the intermediate filaments, composed of the protein desmin, and they help to make up the Z-discs themselves and also interconnect the Z-discs and connect them to the sarcolemma. These filaments provide lateral organization and cause the axial register responsible for the striated appearance of muscle tissue. It is tempting to postulate a three-dimensional framework made up of lateral components (intermediate filaments) linked to axial components (gap filaments) at the level of the Z-discs. The statements above must be taken as tentative, however; new muscle proteins continue to be reported, and other cytoskeletal elements may be discovered in the future.

The structural nature of the cytoskeleton suggests that it probably plays a role in determining the physical properties of meat (Stanley, 1983a). Solid evidence for this thesis is currently lacking, but if endogenous proteolytic enzymes are activated post-mortem and their action is to disconnect the previously integrated cytoskeletal units, then changes could occur—loss of elasticity, decreased tensile strength, and increased tenderness—that coincide with what is presently known to happen to meat texture.

## 4.8 STRUCTURAL ASPECTS OF PLANT TISSUE

The three main divisions of plants consumed by humans are cereals, vegetables, and fruits. Cells of all these edible tissues are bounded by a more or less rigid cell wall composed of cellulose fibers and other polymers, including pectic substances, hemicellulose, lignin, and protein. A layer of pectic substances forms the middle lamella that acts to bind adjacent cells together.

Although the cytoplasm of neighboring cells are separated from one another by their cell walls, there is evidence that they are connected by thin strands of cytoplasm known as plasmodesmata. The cell wall is normally permeable to water and some solutes, and it works to contain cell contents by sustaining the outer cell membrane or plasmalemma against hydrostatic pressures and also provides structural support.

Plant tissues used for food, whether fruits, leaves, nuts, stems, seeds, or tubers, contain various cell types (mainly parenchyma but also collenchyma, sclerenchyma, and vascular bundles). Mature parenchyma cells are approximately 50–500 $\mu$m across and polyhedral in shape (Figures 4–13 and 8–3). These cells are associated with one another via mutual pressure arising from their confinement within a limiting epidermis or skin. Intercellular air space is common in parenchymous tissue and has been estimated at 20–25% of the total volume in apples, 15% in peaches, and 1% in potatoes. Parenchyma cells usually contain a single large vacuole that accounts for most of the cell volume. It is surrounded by a membrane and is filled with a watery solution of organic and inorganic solutes. Of importance to food scientists is the ability of the membrane to generate hydrostatic pressure because of its semipermeable nature. The membrane allows small molecules such as water to pass but restricts the transmission of larger molecules such as sugar, thus producing the phenomenon of turgor pressure. Turgor effects cause the vacuoles to

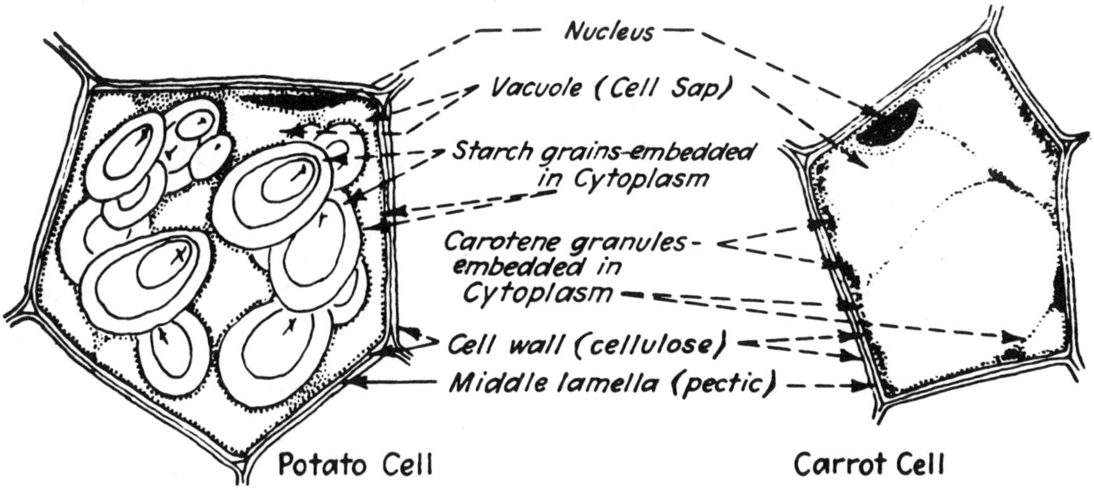


**Figure 4–13** Microstructure of parenchyma cells of carrot and potato. *Source:* Bourne, 1983.

enlarge and press against one another, imparting to the plant tissue turgidity, rigidity, crispness, and a fresh appearance.

Turgor is lost when fruits or vegetables are deprived of water or when they cease to respire. Perhaps the most important change induced by harvest is the deprivation of a water source to replace that lost by transpiration. Some important consequences of loss of turgor include wilting, a decline in transpiration leading to a lack of cooling vaporization, invasion of pathogen, and, of great economic importance, a dry appearance accompanied by a loss of gloss and color. Techniques such as heating or freezing destroy the water transport mechanism, and thus turgor is lost in processed fruits and vegetables. In some cases (e.g., canned potatoes and tomatoes), the addition of calcium salts will aid in maintaining textural integrity by cross-linking pectins and preventing cell sloughing during the heating process.

Plasmodesmata are cytoplasmic connections that link adjacent plant cells through their common cell wall (see Section 8.2.5). They create an intercellular continuum and regulate the transport of water, small molecules, and ions between cells. These organelles have diameters in the range of 30–60 nm, and each cell may contain 1–15 of these conduits per $\mu m^3$. Plasmodesmata are lined with plasmalemma membrane, and through the core of each structure runs a membrane tube thought to be continuous with the endoplasmic reticulum of the overlying cells. The upper size limit for transport through plasmodesmata is around 700–900 Da.

In addition to the watery vacuolized parenchyma cells, some edible plant parts are composed of storage tissue in which starch granules, protein bodies, and/or oil droplets are packed closely within cells that contain no vacuoles and little free water. These are the seeds of cereals and legumes. The first group is composed of members of the grass family. Wheat, corn, rice, oat, barley, rye, triticale, sorghum, and millet are the major cultivated cereals. The second group consists of the legume family, of which peas, beans, peanuts, lentils, and soybeans are the most common members. A third group should be mentioned. The so-called tree nuts—hickory nuts, almonds, hazel nuts, walnuts, pecans, chestnuts, and coffee beans—can be included here. The first two groups can be differentiated structurally on the basis of their endosperm content. In legumes, the endosperm is usually completely utilized by the developing embryo, especially by the cotyledons, while in cereals there is still a conspicuous amount of endosperm remaining in the mature seeds, for use following germination. Further, food legumes may be divided into pulses and oilseeds. A pulse is the dried edible seed of a cultivated legume; they are important in human nutrition because they contain a higher percentage of protein than any other natural plant source. Oilseeds are legumes used primarily for their oil content. Usually, they are arbitrarily defined as those legumes with a lipid content greater than 20%. Economically important oilseeds include soybean, rapeseed (canola), and sunflower. In addition to their high lipid content, oilseeds often have substantial amounts of protein stored in protein bodies.

Parenchyma cells form the bulk of the softer parts of plants, and frequently the nutrients of importance to humans are stored in these thin-walled living cells. The mature plant body contains many kinds of cells, however, differing in size, shape, and other characteristics. Collenchyma cells are also living, but in contrast to parenchyma cells their walls are thickened in the corners and they are often elongated. They occur frequently toward the outside of a stem, where they furnish support against bending. Their thickened cell walls, rich in pectic substances and hemicellulose, resist softening during cooking. Sclerenchyma cells have heavily lignified cell walls and are nonliving. These cells also furnish strength, since at maturity the walls are thickened to a point where only a small cell cavity remains. Sclerenchyma cells encountered in foods include the stringy fibers of green beans and asparagus and the spherical gritty stone cells in pears.

Also present within the cell is a matrix of organized polymeric proteins collectively referred to as the cytoskeleton. This matrix may function to provide a structural framework for cytoplasmic processes. The cytoskeleton is continuous with the plasmalemma and, it is thought, the cell wall. Like the subcellular membranes surrounding the vacuole, plastids, and organelles, the plasmalemma surrounding the cell controls the translocation of water and solutes.

The cellular structure of most consequence to food scientists is undoubtedly the cell wall–middle lamella complex. This structure is a major contributor to the texture of plant foods, and food processors often must disrupt cell walls in order to extract desirable cell components. The primary cell wall and middle lamella contain polysaccharides and smaller amounts of glycoproteins and phenolic compounds. The cell wall of the average parenchyma cell is thin (0.1–10 $\mu$m) but strong, and it is able to limit expansion due to the intracellular fluid and thereby generate turgor pressures of roughly 0.3–1 MPa and associated wall stresses of roughly 100–250 MPa. This internal pressure must be borne mainly by the wall if bursting of the cell is to be avoided. The middle lamellae between adjacent cells act much like adhesives; they are heat labile and in their absence plant cells separate easily. All cells possess primary cell walls. Secondary cell walls, if present, are deposited after the cessation of cell growth outside the plasmalemma but inside the primary cell wall. Their presence in plant tissues is associated with the development of "woodiness," such as found in stringy asparagus.

Cell walls contain numerous polymeric compounds (Table 4–4). With the exception of cellulose, these compounds, when extracted, are water soluble. Yet, in the wall they are organized into a water-insoluble matrix capable of bearing considerable stresses while simultaneously permitting growth. Cellulose, $\beta$-1,4-polyglucan, forms the skeletal scaffolding of the wall through formation of microfibrils about 5–15 nm in diameter and several thousands units long. Hemicellulose consists of rigid, highly branched, rod-shaped polymers of neutral sugars such as xylan, xyloglucan, and $\beta$-1,3 or $\beta$-1,4 mixed glucans (about 200 nm in length) that link with cellulose, pectin, and lignin by means of hydrogen bonding. Pectin, found in highest concentrations in the middle lamella, contains both "smooth" zones of partially esterified, PG-labile $\alpha$-galacturonic acid residues (homogalacturonan, $\approx$100 nm in length) in addi-

**Table 4–4** Major Polymers of the Plant Cell Wall

| *Polymer* | *$H_2O$ Solubility after Extraction* |
|---|---|
| Polysaccharides | |
| Cellulose | Insoluble |
| Hemicelluloses | |
| Xyloglucan | Soluble |
| Xylan | Soluble |
| Mixed glucans | Soluble |
| Pectins | |
| Homogalacturonan | Soluble |
| RG I | Soluble |
| RG II | Soluble |
| Glycoproteins | |
| Arabinogalactan proteins | Soluble |
| Extensin | Soluble |

*Source:* Reprinted from *Trends in Food Science Technology,* Vol. 6, R.L. Jackman and D.W. Stanley, Perspectives in the Textural Evaluation of Plant Foods, pp. 187–194, Copyright 1995, with permission from Elsevier Science.

tion to PG-resistant "hairy" zones (rhamnogalacturonan) that vary in degree of polymerization and neutral sugar content and that can contain phenolic acid or other side chains that facilitate cross-linking.

Glycoproteins are also present in roughly 5–10% of the dry weight of the walls of dicotyledon cells, with carbohydrate constituting as much as two-thirds of the glycoprotein mass. Several classes of these cell wall proteins are recognized, the extensins being the most well known. These constitute a group of glycoproteins rich in hydroxyproline; they possess the repeating pentapeptide sequence Ser-$(Hyp)_4$ and an extended, rodlike polyproline II helical structure around 80 nm long. Extensins are uniformly distributed across the cell wall but do not occur in the middle lamella. They form cross-links with other cell wall polymers, perhaps also pectin, and are therefore thought to contribute to the physicomechanical integrity of the wall. For mature cells, the list of cell wall polymers must be extended to include lignin, the most abundant material on earth next to cellulose. Lignins, which are phenyl propanoid polymers of varying molecular weight, can account for as much as 20–30% of the dry weight of plant tissue. Lignin formation begins in the primary walls or middle lamellae, but concentration is greatest in secondary walls, where polymerization and formation of a composite material (in which the linear wall polysaccharides are encased in a lignin "cage") occurs at the expense of water. The result is a strong hydrophobic lattice that surrounds other cell components. This network is incapable of extension, so cell growth ceases.

The cell wall is not a static structure. Rather, it is a dynamic organelle vital to cell growth, metabolism, attachment, shape, and disease and stress resistance. A recent model proposes that three structurally independent but interacting domains constitute a single layer of the growing cell wall and that several layers condense to form the complete wall (Carpita & Gibeaut, 1993). In this model (Color Plate 7), hemicelluloses constitute the main interlocking component. Their highly branched but linear conformation is conducive to orientation between cellulose microfibrils to which they bind. The resulting cellulose-hemicellulose domain constitutes 50–65% of the dry weight of the wall. It is embedded in a second domain consisting of pectic substances that account for an additional 30% of the wall mass. Pectin cross-linking can also occur via oxidative coupling of phenolic constituents such as ferulate, a mechanism gaining favor as a way to explain cell-cell adhesion. However, more often the cross-linking of the helical homogalacturonan chains of de-esterified pectin occurs via $Ca^{+2}$-bridging to form junction zones. De-esterification is mediated by the enzyme pectinesterase; however, not all sites of de-esterification become cross-linked. RG I represents a portion of the pectin polymer rich in arabinogalactan side chains that can interrupt the $Ca^{+2}$ junctions. A third structural domain consists of extensin units covalently cross-linked and oriented radially within the wall matrix. Extensin cross-linking is thought to be involved in locking the cell wall in a fixed shape once cell growth is complete.

During growth, the cell must expand, deforming the wall while retaining the strength to withstand turgor pressure. Cellulose microfibrils are deposited in a directed orientation as the hemicellulose network is enzymatically cleaved, leading to stretching of the wall to the extent that cellulose-hemicellulose interactions will allow. The alignment of the cellulose microfibrils transversely in a shallow helix permits wall extension to occur longitudinally; the noncellulosic polysaccharide matrix in which the microfibrils are embedded dictates the degree to which they are pulled apart during extension. As the wall stretches, chemical bonds or associations are broken and stress relaxation takes place, resulting in a reduction in turgor pressure. Expansion of the cell follows as it absorbs water in response to the reduction in turgor pressure.

It is interesting to note that this model of the cell wall resembles a lamellar composite such as plywood. Such a model can aid in the study and understanding of plant food structure, but it must be regarded as somewhat incomplete, if only because it is missing the structural component lignin that would be expected to play a major role in the texture of plant tissues in which a secondary wall is developed.

**BIBLIOGRAPHY**


Bourne, M.C. (1983). Physical properties and structure of horticultural crops. In M. Peley & E.B. Bagley (Eds.), *Physical properties of foods* (p. 207). Westport, CT: AVI Publishing Co.

Branden, C., & Tooze, J. (1991). *Introduction to protein structure*. New York: Garland Publishing.

Carpita, N.C., & Gibeaut, D.M. (1993). Structural models of primary cell walls: Consistency of molecular structure with the physical properties of the walls during growth. *Plant Journal, 3,* 1–30.

Echlin, P. (1992). *Low-temperature microscopy and analysis.* New York: Plenum Press.

Feeney, R.E., & Whitaker, J.R. (Eds.). (1986). *Protein tailoring for food and medical uses.* New York: Marcel Dekker.

Fennema, O.R. (1996). Water and ice. In O.R. Fennema (Ed.), *Food chemistry* (pp. 17–94). New York: Marcel Dekker.

Haard, N.F. (1995). Foods as cellular systems: Impact on quality and preservation. A review. *Journal of Food Biochemistry, 19,* 191–238.

Jackman, R.L., & Stanley D.W. (1995). Perspectives in the textural evaluation of plant foods. *Trends in Food Science and Technology, 6,* 187–194.

Jewell, G.G., & Heathcock, J.F. (1988). Structured fat systems. In J.M.V. Blanshard & P. Lillford (Eds.), *Food structure and behavior* (pp. 279–295). London: Academic Press.

Moner, J.G. (1972). *Cells: Their structure and function.* Dubuque, IA: Wm. C. Brown Co.

Morris, V.J. (1990). Starch gelation and retrogradation. *Trends in Food Science and Technology, 1,* 2–6.

Oates, C.G. (1997). Towards an understanding of starch granule structure and hydrolysis. *Trends in Food Science and Technology, 8,* 375–382.

Roos, Y.H. (1995). *Phase transitions in foods.* San Diego: Academic Press.

Stanley, D.W. (1983a). Relation of structure to physical properties of animal material. In M. Peleg & E.B. Bagley (Eds.), *Physical properties of foods* (pp. 157–206). Westport, CT: AVI Publishing Co.

Stanley, D.W. (1983b). A review of the muscle cell cytoskeleton and its possible relation to meat texture and sarcolemma emptying. *Food Microstructure, 2,* 99–109.

Stanley, D.W. (1991). Biological membrane deterioration and associated quality losses in food tissues. *C.R.C. Critical Reviews in Food Science, 30*(5), 487–553.

Stanley, D.W., & Swatland, H.J. (1976). The microstructure of muscle tissue—A basis for meat texture measurement. *Journal of Texture Studies, 7,* 65–75.

Stanley, D.W., & Tung, M.A. (1976). Microstructure of food and its relation to texture. In J.M. deMan, P.W. Voisey, V.F. Rasper, & D.W. Stanley (Eds.), *Rheology and texture in food quality* (pp. 28–78). Westport, CT: AVI Publishing Co.

Stanley, D.W., & Yada, R.Y. (1992). Physical consequences of thermal reactions in food protein systems. In H.G. Schwartzberg & R.W. Hartel (Eds.), *Physical chemistry of foods* (pp. 669–733). New York: Marcel Dekker.

Stone, A.P., & Stanley, D.W. (1992). Mechanisms of fish muscle gelation. *Food Research International, 25,* 381–388.

Swatland, H.J., & Belfry, S. (1985). Postmortem changes in the shape and size of myofibrils from skeletal muscle of pigs. *Mikroscopie, 42,* 26–34.

## SUGGESTED READING


Brett, C.T., & Waldron, K.W. (1996). *Physiology and biochemistry of plant cell walls.* (2nd ed.). London: Chapman & Hall.

Damodaran, S., & Paraf, A. (Eds.). (1997). *Food proteins and their applications.* New York: Marcel Dekker.

Eliasson, A.-C. (Ed.). (1996). *Carbohydrates in food.* New York: Marcel Dekker.

Franks, F. (1983). *Water.* London: Royal Society of Chemistry.

Nicklas, K.J. (1992). *Plant biomechanics.* Chicago: University of Chicago Press.

Phillips, L.G., Whitehead, D.M., & Kinsella, J. (1994). *Structure-function properties of food proteins.* New York: Academic Press.

Robards, A.W., & Lucas, W.J. (1990). Plasmodesmata. *Annual Review of Plant Physiology and Plant Molecular Biology, 41,* 369–419.

Stanley, D.W. (1994). Understanding the materials used in foods: Food materials science. *Food Research International, 27,* 135–144.

Stephen, A.M. (Ed.). (1995). *Food polysaccharides and their applications.* New York: Marcel Dekker.

Swatland, H.J. (1994). *Structure and development of meat animals and poultry.* Lancaster, PA: Technomic Publishing Co.

# CHAPTER 5

# Food Structuring

## 5.1 INTRODUCTION

### 5.1.1 Structure and Food Technology

Let us recapitulate our progress to this point. We began by reviewing a variety of available techniques for "seeing" the microstructure of foods, including some methods that probe into it and provide physical, chemical, and structural information down to the atomic level. We next discussed means of retrieving, processing, and generating quantitative information from images. We then considered some principles of polymer, colloid, and materials science that are essential for understanding the interactions between food components at different scales and the architecture of microstructures that may be formed. The concept of structure-property relationships was introduced, as were models for interpreting the mechanical and rheological behavior of foods based on microstructural information. Finally, we took a detailed look at the basic components and assemblies present naturally in foods, since they are the actual building blocks for structure formation. To understand the rationale behind this "microstructural approach" to foods, consider the following statement (and see Figure 5–1).

> *For the microstructural engineer, food processing may be understood as a controlled effort to preserve, transform, create or destroy structure.*

Preserving structure is a major objective of the postharvest processing of fruits and vegetables, as changes in structure lead to detrimental changes in texture, flavor, and even nutritional properties. The same objective is pursued by scientists interested in preserving the quality of meats after slaughter, fish and crustaceans after capture, and cereals and legumes after harvest. Success in preserving structure and concomitant quality is the reason why the freezing of foods has developed as a prominent technology. In summary, preserving structure is of major concern after the desired structure of a food has been attained, since changes in structure during storage and distribution can lower the quality of finished products.

Destroying structure during food processing is not trivial, since often the microstructural characteristics dictate the type of breakdown. Controlled destruction of structure in food processing is needed to release valuable components, facilitate handling of materials, and prepare refined ingredients. Milling of grains is based on the properties of microstructural elements, as are other unit operations related to size reduction (homogenization, grinding, crushing, slicing, etc.). Control must be maintained, since structural breakdown is accompanied by increased instability due to loss of natural preservation systems. Food technologists also recognize that extensive destruction of the food structure is achieved in the mouth prior to swallowing.

Structural transformations are a vast part of the modern food industry. Raw materials are changed into refined food materials by primary processing of agricultural output (e.g., oils and fats, milk, ce-

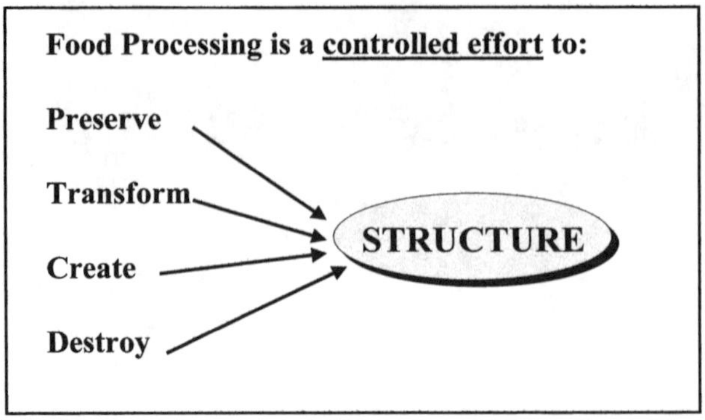


**Figure 5–1** Microstructural definition of food processing.

real and grain flours, sugar and starches, among others), and the refined materials are then mixed and assembled into traditional products representing the majority of processed foods consumed around the world, such as baked products, processed meats, dairy products, confectionery products, and many others.

Product development and product improvement are largely based on creating structures in which nutrients are conveyed in desirable textures and forms. Extrusion is an example of how cellular and fibrous structures are derived from starchy and proteinaceous flours, respectively. This chapter attempts to explain why and how these new structures are created.

### 5.1.2 How Does Nature Form Structures?

Nature compensates for the limited types of molecules available by utilizing the same macromolecular design and only varying the *hierarchical* structure. In other words, structures of higher and higher levels of organization are progressively assembled from the molecular to the macro scale until the desired properties and functions are achieved (Baer, Cassidy, & Hiltner, 1991). Hierarchical structures are found in practically all complex systems in nature, including cartilage, skin, wood, nacre, and other natural materials.

As an example, a tendon is a uniaxial hierarchical structure that serves as the primary linkage between muscle and bone. Tendons are subjected almost exclusively to uniaxial tensile loading directed along their length. They must be elastic to transmit muscular force while remaining capable of absorbing large amounts of energy without fracturing. In tendons, collagen microfibrils are primarily arranged in a lattice form consisting of microscopic subfibrils about 30 nm in diameter. These subfibrils are then assembled into the collagen fibril, which varies in diameter from 50 to 500 nm. Fibrils are subsequently surrounded by an extrafibrillar matrix and aligned parallel to one another between bone and muscle, thus forming the soft tissue (Hiltner, Cassidy, & Baer, 1985). The primary macromolecular component of the extrafibrillar matrix is a highly hydrated and swollen proteoglycan consisting of a core protein and numerous pendant mucopolysaccharide units. This proteoglycan aggregate forms a network that connects and maintains the hierarchical architecture of collagen fibrils. Finally, the tendon is surrounded by a reticular membrane. Figure 5–2 shows the hierarchical structure of a tendon as well as the microscopy techniques that have been used to resolve its structure at different scales. Collagen fibers embedded in a gel matrix of protein-polysaccharide are also used by nature to accomplish other specific functions: they are found in intestines (allowing the intestines to perform as

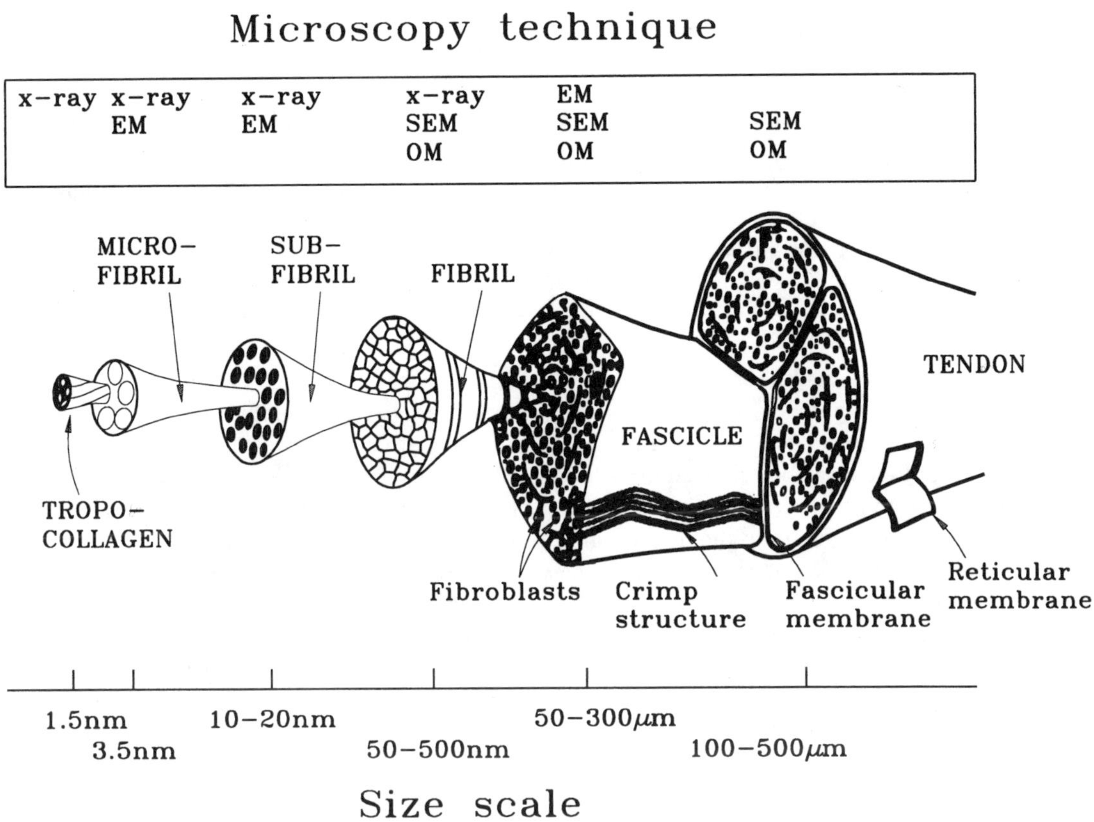


**Figure 5–2** The hierarchical structure of the tendon. *Source:* Reprinted with permission from E. Baer, J.J. Cassidy, and A. Hiltner, Hierarchical Structures of Collagen Composite Systems: Lessons from Biology, in *Viscoelasticity of Biomaterials*, W. Glaser and H. Hatakeyama, eds., ACS Symp. Series 489, pp. 2–23, © 1991, American Chemical Society.

tubes under multiaxial tension) and in intervertebral discs forming soft pads between rigid bones during compression.

The study of the hierarchical architectures of biological materials is rewarding and inspirational in several ways. It provides a means of understanding how natural food materials are assembled at different size scales and thus indicates alternatives for disassembling them. It suggests an approach to unveiling levels of organization in fabricated foods and to understanding interactions at each level. Finally, it reveals extraordinary combinations of performance properties and may hint at routes for new structuring techniques (National Research Council, 1994). Whenever possible, the microstructural engineer should try to dissect macrostructures into their hierarchical components as one way of analyzing structure.

### 5.1.3 Recognizing Hierarchical Structures in Foods

Hierarchical structures exist not only in foods of natural origin but in fabricated foods (Table 5–1). Of further significance is the possibility of identifying key parameters and substructures that influence selected properties of the food at different levels. Note that, in Figure 5–2, each level of the structure has to be probed by a specific analytical technique, which leads to another important con-

**Table 5–1** Examples of Hierarchical Structures in Foods

| | *Type of Food* | | | |
|---|---|---|---|---|
| *Size Range* | *Cellular* | *Fibrous* | *Gel* | *Crystalline* |
| Angstroms | Glucose | Amino acid | Monomer | Fatty acid |
| nm | Stem | Helix | Polymer | Lamellae |
| μm | Microfibril | Protofibril | Strand | Spherulite |
| 100 μm | Composite wall | Fiber | Gel network | Fat crystal |
| Macro | Fruit | Meat | Yogurt | Chocolate |

cept in the study of microstructure of foods—*the relevant scale*.

It is conceivable that added functionality and novel textures of foods may be achieved in the future by proper assembly of hierarchical structures from the microscopic level up to the macroscale. These new fabrication techniques will require understanding and precise control of the assembly process at all scales. Food technology has at its disposal variables not available to nature, such as temperature and pressure. Of particular importance will be the adhesion mechanisms in interfaces that link structural elements of dissimilar scales.

### 5.1.4 The Relevant Scale

Earthquake engineers can design complex structures and analyze their behavior under different controlled testing conditions partly because they can observe fracture and collapse mechanisms in real time. Unfortunately, thus far microstructural food engineers cannot see internal cracks developing during drying of spaghetti or see stabilizers diffusing into interfaces in emulsions, yet these phenomena are decisive for the properties of the finished products. The problem is that the relevant scale at which such events occur in foods is beyond the capacity of the naked eye. Microscopy and other imaging techniques allow us to probe into the scale that is relevant to the process and property under study, as described in Chapter 1.

Relevant scale is the dimensional level at which the effects of certain phenomena are realized or explained. It is usually found by direct observation, by inference using analytical techniques, or through theoretical analysis. For instance, the softening of dry legumes during cooking can be followed by mechanical analyses and is related primarily to debonding of the middle lamella between cells (Aguilera & Stanley, 1985), as shown in Figure 6–13. In the case of emulsion stabilization, physical chemistry theory predicts that the dynamics of adsorption-desorption of macromolecules would be the key phenomenon and thus defines the relevant scale. Figure 5–3 shows typical sizes of major structural components—organelles and molecules found in plant, meat, and dairy products—and indicates standard microscopy techniques for viewing them.

### 5.1.5 Objectives of Food Structuring

It is now time to discuss how engineered food structures are formed or transformed by processing. A major objective of food structuring is to recombine components derived from raw agricultural materials into acceptable human foods. Several benefits accrue from this approach, one being the improved utilization of food resources—previously unusable stocks can be rendered serviceable and underutilized ingredients can be upgraded to be more valuable. Also, texture, often the limiting factor in food acceptability, becomes an attribute of the resulting product. Thus, components that meet the criteria of nutritional adequacy, safety, and other quality tests can be restructured to yield foods that are attractive to consumers.

Even in a world that is critically short of essential nutrients, it has become evident that nourishment alone does not ensure acceptance. Restructuring is a critical step in making the final

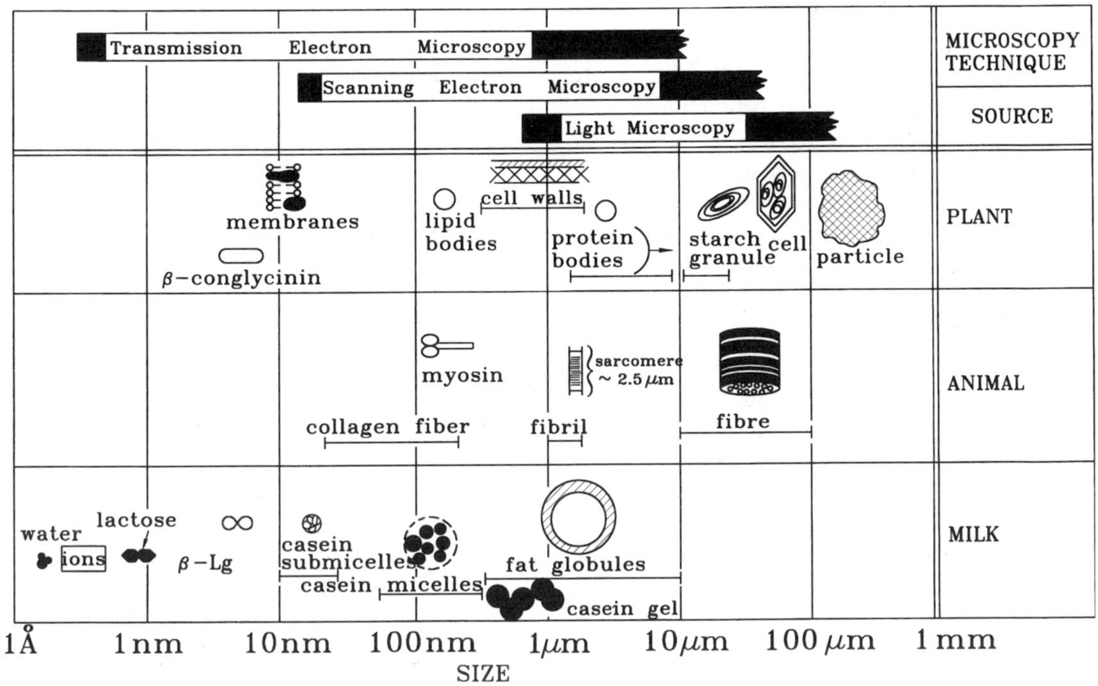


**Figure 5–3** Typical sizes for some food components.

connection in the food chain. Ironically, sometimes those members of the world's population that are at greatest risk are often the least willing to experiment with unfamiliar textures, colors, or flavors. As an example, the rural population of Guatemala derives significant amounts of protein, calories, vitamins, and minerals from cooked dried beans, but local populations traditionally consume only certain acceptable colors and sizes of this staple and are loath to try beans deviating in these characteristics, even if the nutritional value and cost are comparable (Watts, Elias, & Rios, 1987). Reluctance to eat unfamiliar or appealing foods, probably genetically ingrained in all humans as a survival mechanism, means that restructuring efforts are most often directed at mimicking known foods.

As emphasized earlier, the food scientist wishing to utilize the structuring option has four basic components with which to work: water, proteins, lipids, and carbohydrates (ranging from simple hexoses to complex polysaccharides). Micronutrients (vitamins and minerals) and other microcomponents such as flavors, colors, preservatives, and important functional additives (emulsifiers, stabilizers, etc.) can usually be added later to the formulation, although sometimes their incorporation causes further problems. In order to be successful at mixing components, it is necessary to know the basis of the chemical and physical interactions among the components that lead to the complex sensation of texture. If texture is to be controlled and improved to meet the requirements of the final product, then the effect of individual ingredients on formulations should be known. At the outset of a study of this type, the researcher faces two inherent limitations; first, a chemical one, for the product must be fabricated around the interactions possible given the raw materials that are to be used, and second, a physical constraint, for the form of the ingredients must be related to the final product.

Of the macrocomponents available for restructuring foods, proteins have been utilized more than the others. The reasons include the higher cost of protein-based foods (allowing more profit), the physicochemical and functional properties of proteins, and the worldwide demand for dietary protein. Starch-based snack foods restructured through thermal extrusion are a popular example of carbohydrate utilization (Guy & Horne, 1988). These products are characterized by porous structures composed of air pockets surrounded by laminar sheets of gelatinized starch (Figure 1–28). Baking can also be considered a restructuring process (Blanshard, 1988). As for lipid-based fabricated foods, margarine, whose crystal characteristics and liquid composition can be altered by blending raw materials and altering processing steps, meets the definition of a restructured food. Structured lipid systems are reviewed by Jewell and Heathcock (1988).

## 5.2 TRADITIONAL FOOD STRUCTURING AND TEXTURE IMPROVEMENT

### 5.2.1 Texture Improvement through History

Intentional food structuring can be traced back to several traditional technologies aimed at texture improvement rather than the creation of new or unique foods. It is instructional to examine some of these, since they provide the foundation for current efforts.

Much of our ancestors' time was spent in gathering food for immediate consumption. Gaining the ability to first store and later preserve foods marked an important shift that led, in time, to the development of civilization. Viviculture (keeping animals and plants alive until needed) was probably the first technique practiced to extend the harvest. It was followed by salting, drying, and fermentation, among preservation techniques. At the same time, it became evident that particle size control enabled certain foods to be more completely utilized. For example, meat was a prized food item, both for its nutrition and quality attributes, but certain cuts or muscles are quite high in connective tissue, a component that increases in toughness with animal age. One way to overcome this problem is particle size reduction or comminution. Chopping tough cuts of meat into coarse particles and mixing them with similarly prepared fat fragments, which provide a smooth coating to meat during chewing and aid swallowing, is the basis of sausage manufacture, a technology in use since prehistoric times (Borgstrom, 1968). Another consequence of meat comminution is the release and solubilization of muscle proteins that are capable, upon heating, of forming a stable matrix that serves to entrap the fat particles. Addition of salt to the formulation fosters protein extraction and helps explain its presence in sausage products, since the level at which it is presently used is below that required for effective preservation. These structures represent a type of coarse oil-in-water emulsion and are perhaps the first example of a manufactured food based on the principles of particle size control and emulsion formation.

Plant foods also offer examples of particle size control. Cereals and legumes yield their nutrients in the form of grain or seeds that can be naturally or artificially preserved by drying to reduce water activity. Many (perhaps all) of these potential food materials suffer from storage-induced hardening, that is, an inability to imbibe water and soften sufficiently during poststorage soaking and cooking. It is common practice, then, for grinding and milling to be used as unit operations to, among other benefits, break down hard structures and increase surface area, leading to enhanced water absorption and softening.

### 5.2.2 An Example of Food Structuring: Soybeans

The procedures described previously are examples of approaches to texture improvement. To gain an understanding of several interrelated traditional food structuring techniques that continue to be of importance in many parts of the world, let us examine the physical and chemical properties of the soybean. Oilseeds, in particular soybeans, are a major source of protein, lipid, energy, and micronutrients for much of the world's population and have served as such for centuries. They also represent an important contribution to animal feed. An approximate gross compositional analysis of mature soybeans indicates they contain about 40% protein, 25% lipid, 20% carbohydrate (simple and complex), and 10% water; the remainder of their contents consist of variable quantities of microcomponents. Structurally, the soybean is typical of oilseed legumes (Figure 5–4) in that the cotyledons store protein in the form of separate globoid protein bodies or aleurone grains that average 5–10 $\mu$m in diameter and contain up to 90% protein; these protein bodies account for 60–70% of the soybean's total protein. Lipid is stored in smaller ($<1 \mu$m) spherosomes interspersed with the protein bodies in the cell. The carbohydrate fraction consists mainly of soluble oligosaccharides and insoluble polysaccharides but not starch.

It may have been an effort to destroy components such as trypsin inhibitors and lipoxygenase enzymes that first led to the preparation of "soy milk," a suspension of ground soybeans in water that is boiled and then filtered. The resulting liquid has several end uses: it can be consumed as a beverage, fermented, or used as the starting material for several structured foods having characteristic textures. It seems to have been as apparent to earlier generations as it is to modern humans that, no matter how potentially nutritious, bland, uninteresting products have little appeal as foods.

Soy milk is also the basis for a series of structured products called *tofu* and *kori-tofu,* essentially curds that are used to provide textural variation in meat and vegetable dishes. Bean curd was invented in China between AD 200 and AD 900 (McGee, 1984). Tofu is made by precipitating the proteins in soy milk with a calcium salt to form a coagulum that is then drained, pressed, and washed to yield a cheeselike product containing about 6% protein and 88% water. A further process was used to produce a stronger, more stable texture, which led to the development of kori-tofu. In this case, tofu is frozen, causing partial protein denaturation and insolubilization that strengthens the structure (Smith & Circle, 1972; Wolf & Cowan, 1975). Figure 5–5 shows an example of soybean tofu structure.

Another traditional soy-based food popular in the Orient for centuries and meeting the definition of a structured product is *yuba.* Soy milk is heated in shallow pans and a lipid-protein film (yuba) results. After drying, yuba is used as wrapper for various vegetable and meat mixtures, not unlike the way tortillas are used in Latin America to form tacos and impart a crisp texture to the filling. High yields (80%) of protein from the original bean are obtainable, and, although labor intensive, the product is widely accepted. Studies of the structuring process (Farnum, Stanley & Gray, 1976) found that, as water is driven from the heated soy milk by evaporation, a protein matrix is formed in which lipid droplets are dispersed (Figure 5–6). The process is not unique to soy, and other oilseed milks, such as peanut milk, may be employed as well (Aboagye & Stanley, 1985).

## 5.3 FOOD TECHNOLOGY AND STRUCTURE

### 5.3.1 Microstructure and Processing

As noted, the microstructural engineer views food processing as a controlled effort to preserve, destroy, or transform the original structure of foods or create new structures from basic building blocks. Food processing is also a means of providing good nutrition by creating attractive food shapes and textures and using novel ingredients. In fact, food technology can be elegantly explained as a sequence of microstructural changes, as this book tries to document. The following sections present a few examples of how controlled effort is achieved and what its effect on microstructure is.

### 5.3.2 Freezing of Foods

To the food technologist, water is the medium that causes the most deleterious reactions, and its removal or immobilization, when possible, is highly desirable. Moreover, it is known that the rate of deleterious reactions decreases exponentially with temperature. Freezing, involving simultaneous immobilization of water as ice and temperature reduction, would seem the perfect preservation technique for foods. This would be true were it not for the extensive microstructural changes inside organized tissues caused by ice crystallization. Nature also suffers the detrimental effects of

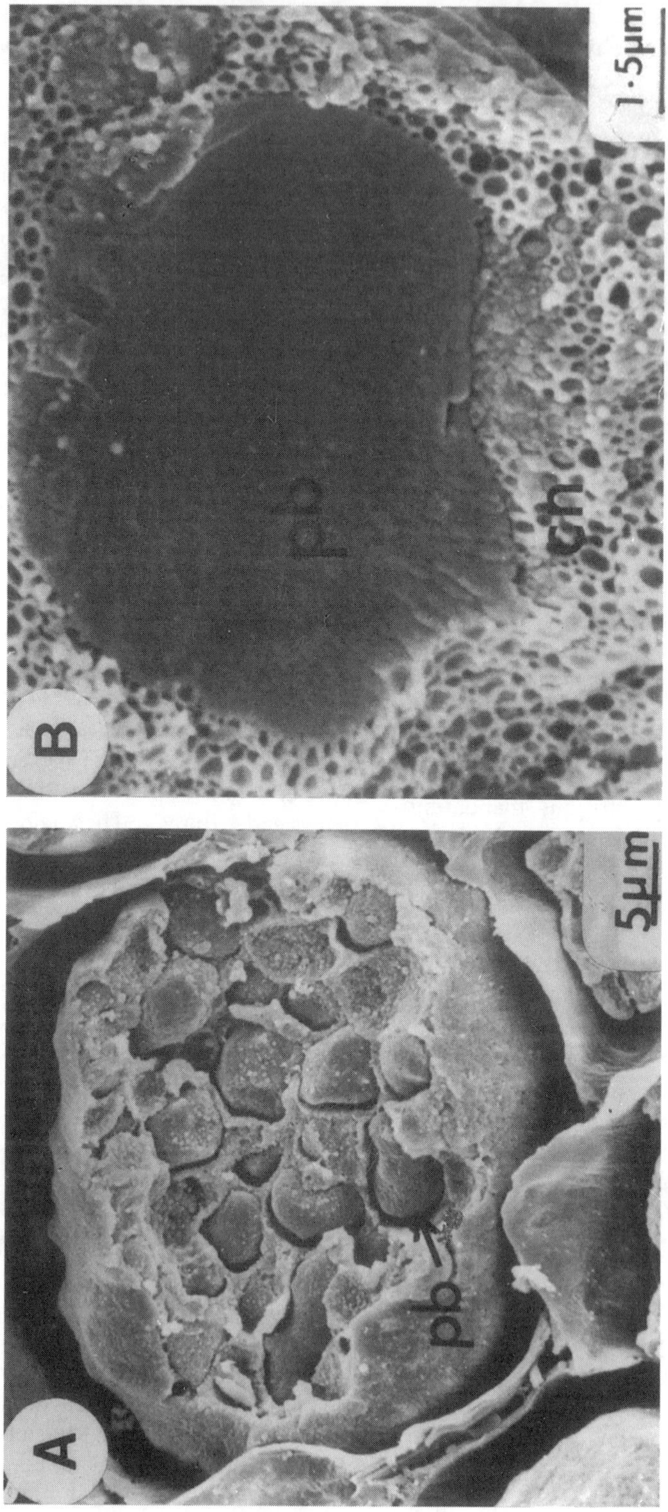


**Figure 5–4** Scanning electron micrographs. (A) Soybean cotyledon cell cross section. Note protein body (pb). (B) Cross section of pb surrounded by cytoplasmic network (cn) showing spherosome sockets. Source: Prattley and Stanley, 1982.

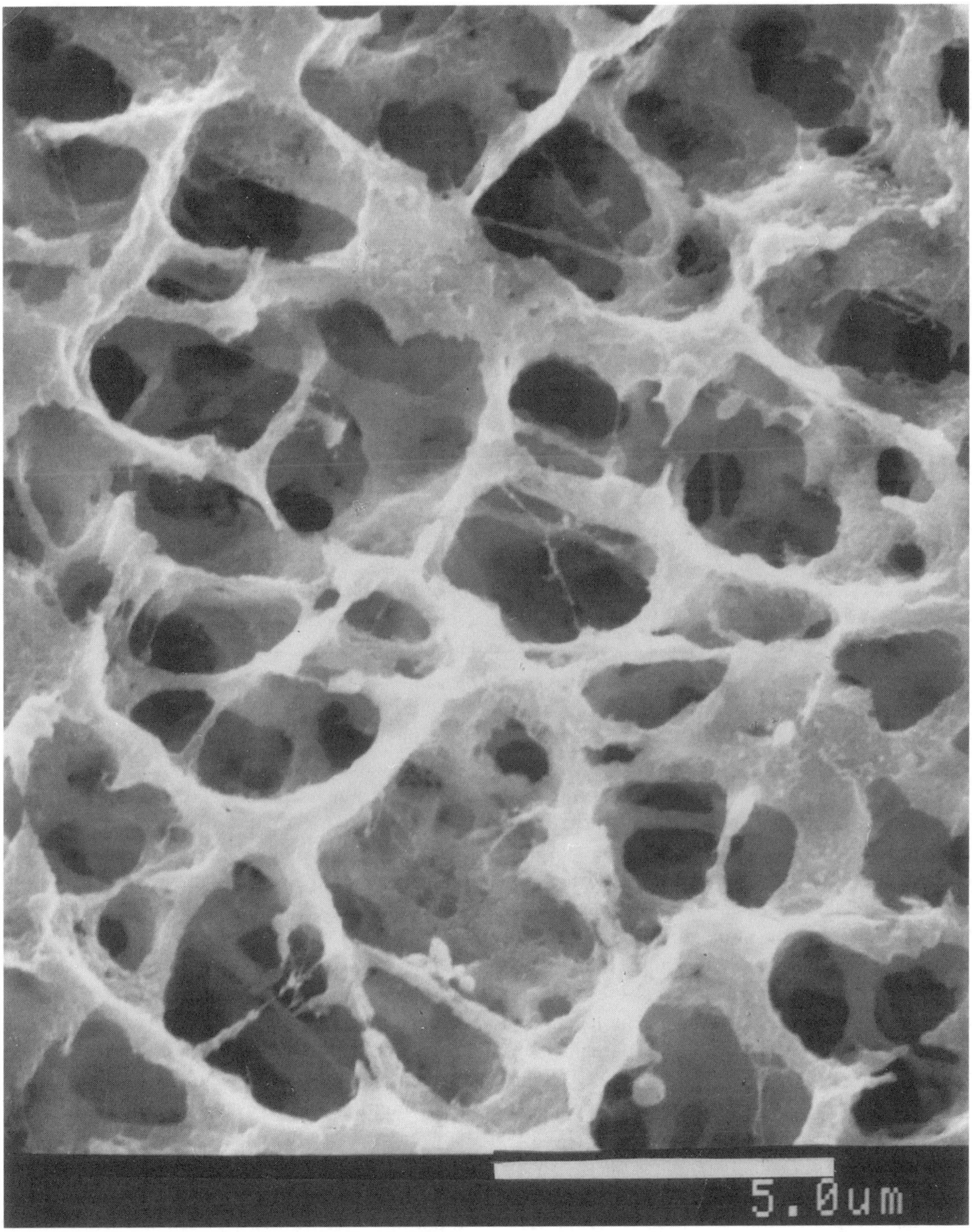


**Figure 5–5** Scanning electron micrograph of soybean tofu (cryopreparation). *Source:* Courtesy of B.T. Lim.

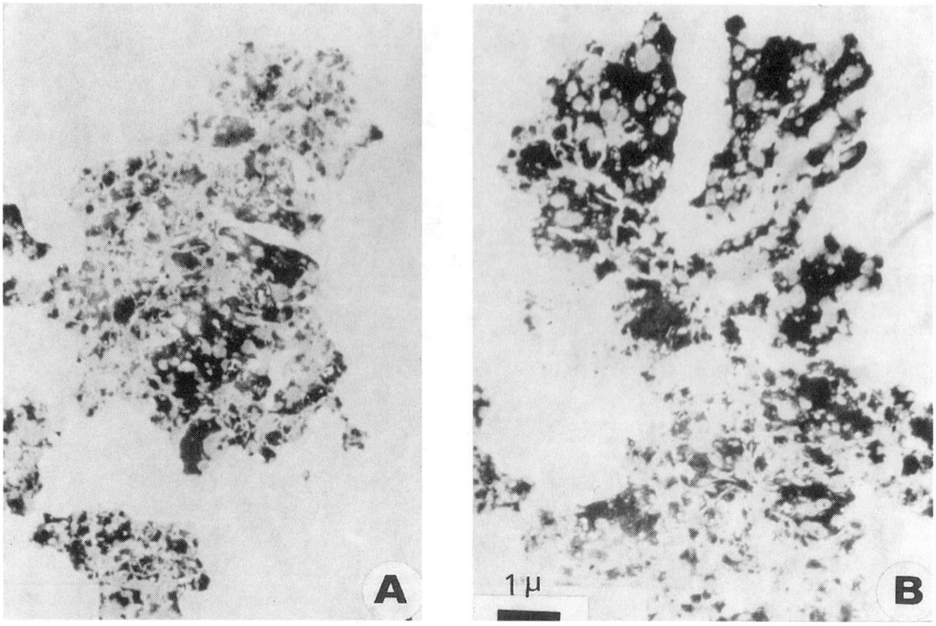


**Figure 5–6** Transmission electron micrographs of soy protein-lipid film (A, B). Dark areas—protein; light areas—lipids. *Source:* Farnum et al. (1976).

freezing and has learned how to cope with them (see Section 5.7.2).

When freezing occurs at a moderately fast rate, water molecules arrange themselves into hexagonal crystallization units, the only form of ice of importance in foods (Fennema, 1996). Other possible crystallization forms are irregular dendrites and coarse and evanescent spherulites (Luyet, 1968). Solutes also affect the structure of ice, as can be seen using light microscopy (Tanner, 1975).

More important than the actual shape of the ice crystals is where they are formed. The location of ice crystals depends mainly on the freezing rate. Slow freezing (at a rate of less than about 1°C/min) generally causes ice crystals to grow in extracellular locations. This results in large crystals, maximum dislocation of water, shrunken appearance of cells in the frozen state, and reduced food quality. On the other hand, rapid cooling produces small ice crystals, uniform crystallization, and product quality superior to that of slowly frozen food. Differences in tissue damage between strawberries slowly frozen (in a home freezer) and rapidly frozen (cryogenic freezing) are shown in Figure 5–7.

Recrystallization or "Ostwald ripening" during frozen storage is probably the most important change leading to quality losses in frozen foods. Recrystallization involves the enlargement of large crystals at the expense of smaller ones (see Section 3.6.5). As the temperature increases above freezing, the proportion of frozen water decreases and smaller crystals melt, since they have a lower melting temperature. Conversely, as the temperature decreases, water being frozen is deposited on the surface of the larger crystals, which have a more stable energy state. As a result, the total number of crystals diminish and the mean crystal size increases. This type of recrystallization in meat leads to physical disruption of the tissue and denaturation of proteins due to dehydration and increases in ionic strength (Martino & Zaritzky, 1988). Figure 5–8 shows changes in the tissue of fish after several thawing and freezing cycles.

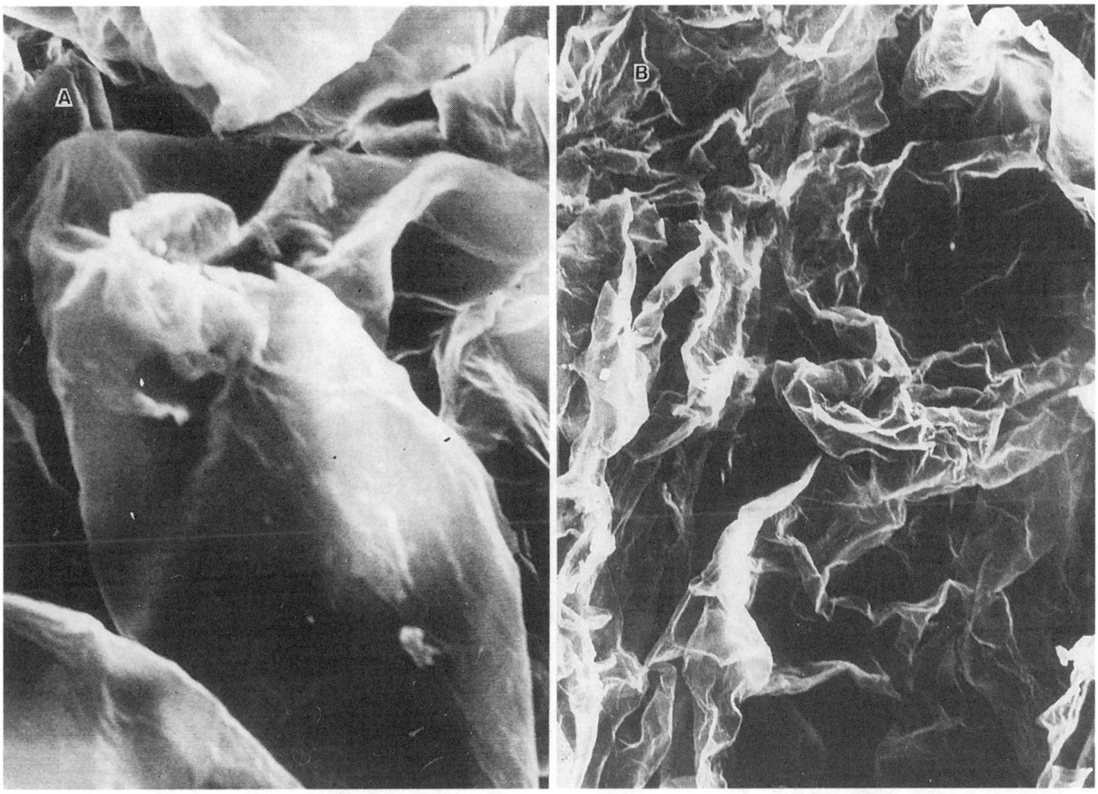


**Figure 5–7** Scanning electron micrographs of frozen strawberries. (A) Intact cell in a commercially individual quick frozen (IQF) berry; (B) tissue from a strawberry frozen in a home freezer, denoting ample cellular destruction. Magnification: approximately 1000×.

Microscopic techniques, including light microscopy (Bello, Luft, & Pigott, 1982; Love, 1968) and transmission electron microscopy (Järenbeck & Liljemark, 1975), have proven helpful in comparing microstructural changes accompanying the freezing process. However, most examinations have been performed on frozen or thawed tissue after "fixation" in a state suitable for microscopic observation, and the presence of artifacts cannot be ruled out. Direct observation can be done using a cryoscopic stage (or cryostage) attached to a light microscope (LM), but the materials have to be transparent. A comprehensive review of cryoscopic techniques applied to the study of freezing of biological systems is presented in Diller (1981). Reid (1978) discusses use of a microscopy cryostage in which the freezing rate could be controlled by adjusting the temperature of the circulating freezing liquid. Important changes during freezing, such as supercooling of cell contents, shrinkage due to osmotic loss of water, sequential freezing of different cell compartments, and cell membrane damage, were observed in vegetable tissue sections mounted in this cryomicroscope (Reid, 1980). A freezing stage attached to a scanning electron microscope (SEM) was used to identify product structure, evaluate the suitability of different fruit and vegetable varieties for freezing, and assess the effect of various blanching treatments on the structure and texture of green beans

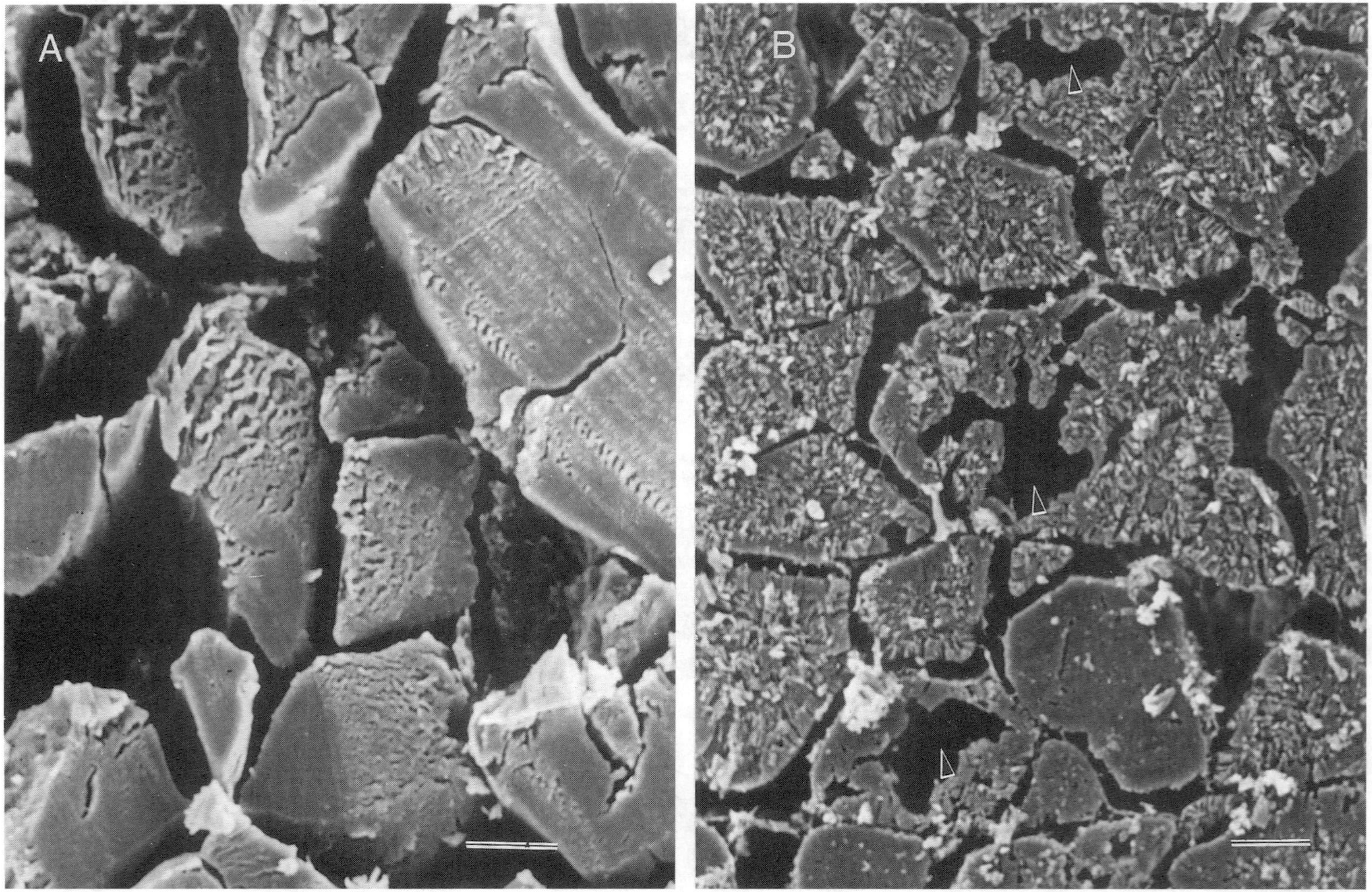


**Figure 5–8** Scanning electron micrographs of fresh and frozen fish. (A) Cross-section of cells from raw fish. (B) Tissue after repeated freezing-thawing cycles in a home freezer showing increased separation between individual fibers, shrunken cells, and internal voids (arrows). Marker 20 $\mu$m.

(Howlett & Hall, 1985). Bomben and King (1982) used a cold stage attached to an SEM to study heat and mass transfer during freezing. A sequence of prints obtained during the freezing of onion tissue in a cryostage attached to an LM is shown in Figure 5–9.

### 5.3.3 Structuring Water at Subfreezing Temperatures

It may appear incongruous to present ice cream as an example of a new restructured lactosystem, since a rudimentary form of ice cream was eaten

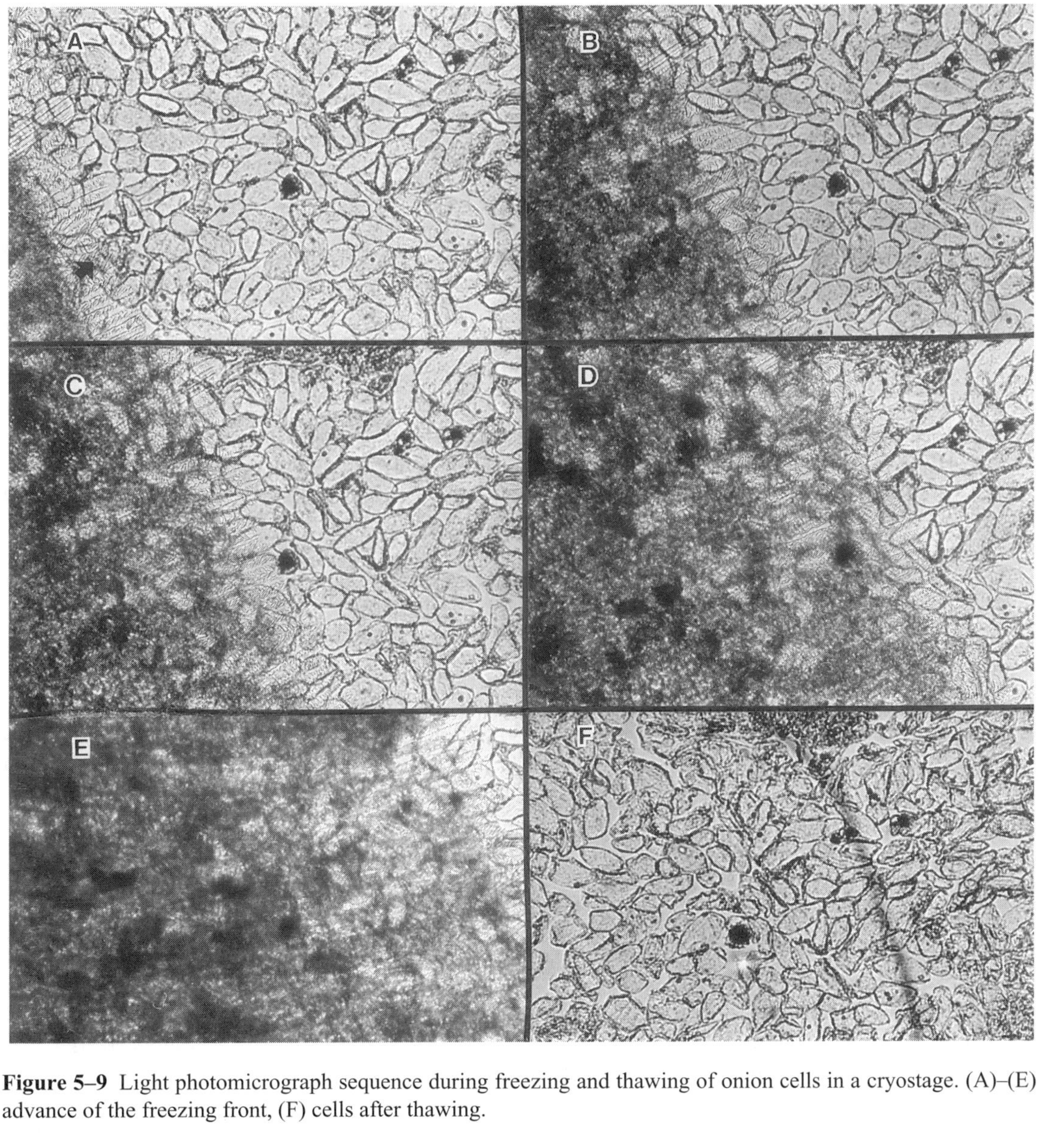


**Figure 5–9** Light photomicrograph sequence during freezing and thawing of onion cells in a cryostage. (A)–(E) advance of the freezing front, (F) cells after thawing.

in early Roman times. However, it is doubtful if contemporary ice cream would be recognizable by the Romans or even by people several generations ago. Ice cream does fit the accepted definition of a restructured food, because the processing it undergoes is directed primarily not at preservation but at producing an edible texture and because this texture would not be achieved or maintained without the addition of certain nondairy ingredients. Ice cream can be characterized as a partly frozen foam containing 40–50% air by volume (referred to as the overrun) and by weight, 10–14% fat, 12–15% sucrose or other sugars, 10–14% milk solids-not-fat, up to 0.5% stabilizers and emulsifiers, and 45–55% water (see also Section 7.4.7). The continuous phase of the foam is composed of a highly concentrated unfrozen aqueous solution or solid amorphous glass of soluble milk salts, lactose, and added sugars. The continuous phase also incorporates dispersed colloidal solids (casein, salts, and stabilizers) and a lipid phase (a partially coalesced emulsion made up of butter fat globules). Ice crystals, which form another coarsely dispersed phase, occupy a major portion of the space between the air cells (Arbuckle, 1986; Berger, 1990). Figure 5–10 shows an SEM micrograph of ice cream after cryopreparation.

Two of the main determinants of structure and texture are the freezing process, particularly the rate of freezing, and the addition of two categories of additives to the mix, *stabilizers* and *emulsifiers*. Because of their importance to ice cream manufacture and structuring in general, each of these will be discussed in detail.

Three methods can be employed in the freezing of ice cream: the traditional process, which employs ice and salt as the refrigerant, of interest only from an historical point of view; the batch freezing process, which employs an industrial refrigerant but takes 10–30 minutes to reach a satisfactory state for packaging and hardening; and the commercially important continuous freezing process, which also employs an industrial refrigerant but completes the dynamic freezing process in approximately 30 seconds (Arbuckle, 1986; Hartel, 1998a). Ice crystal nucleation and growth are time-dependent phenomena, and, as is shown in Figure 3–10, rapid freezing (high supercooling) promotes the creation of many ice crystal nuclei of small dimension (Karel, Fennema, & Lund, 1975). This has a great impact on the organoleptic evaluation of the ice cream. The traditional freezing and batch freezing processes produce, on average, ice crystals in the 60–70 $\mu$m range or greater, while continuously frozen and rapidly hardened ice cream samples contain ice crystals in the 25–30 $\mu$m range (Berger, 1990). The threshold for organoleptic detection of ice crystals in ice cream (grainy texture) is probably between 65 and 70 $\mu$m. Bearing in mind that ice crystals continue to grow throughout the storage and distribution of the ice cream, particularly if it suffers significant freeze-thaw or heat shock cycles, it is evident that continuously frozen samples will also be smoother to the palate than batch frozen samples or samples that have undergone severe heat shock.

Many high molecular weight polysaccharide gums are used in ice cream formulations, including locust bean gum, guar gum, carboxymethyl cellulose, and carrageenan. The primary purpose of these polysaccharides is to promote and maintain the smooth texture produced by rapid freezing of the product by controlling ice crystallization during storage (Arbuckle, 1986). The ability of these gums to hydrate and hold within their gel networks large quantities of water is said to be responsible for their ability to protect ice cream from heat shock (Wielinga, 1983).

Research by Levine and Slade (1988a) suggests that the mechanism of structural preservation in frozen systems may be related to the mobility of free water. During the freezing of a sugar solution, a glass transition temperature is reached at which the solution becomes maximally freeze-concentrated ($T_g'$) and below which no more water will freeze in the time frame of normal storage (Franks, 1985). In the glassy region of Figure 5–11, the serum phase exists as a high viscosity amorphous solid, and no ice crystal growth can occur because of kinetic constraints. In the rubbery state, above the glass transition line, the

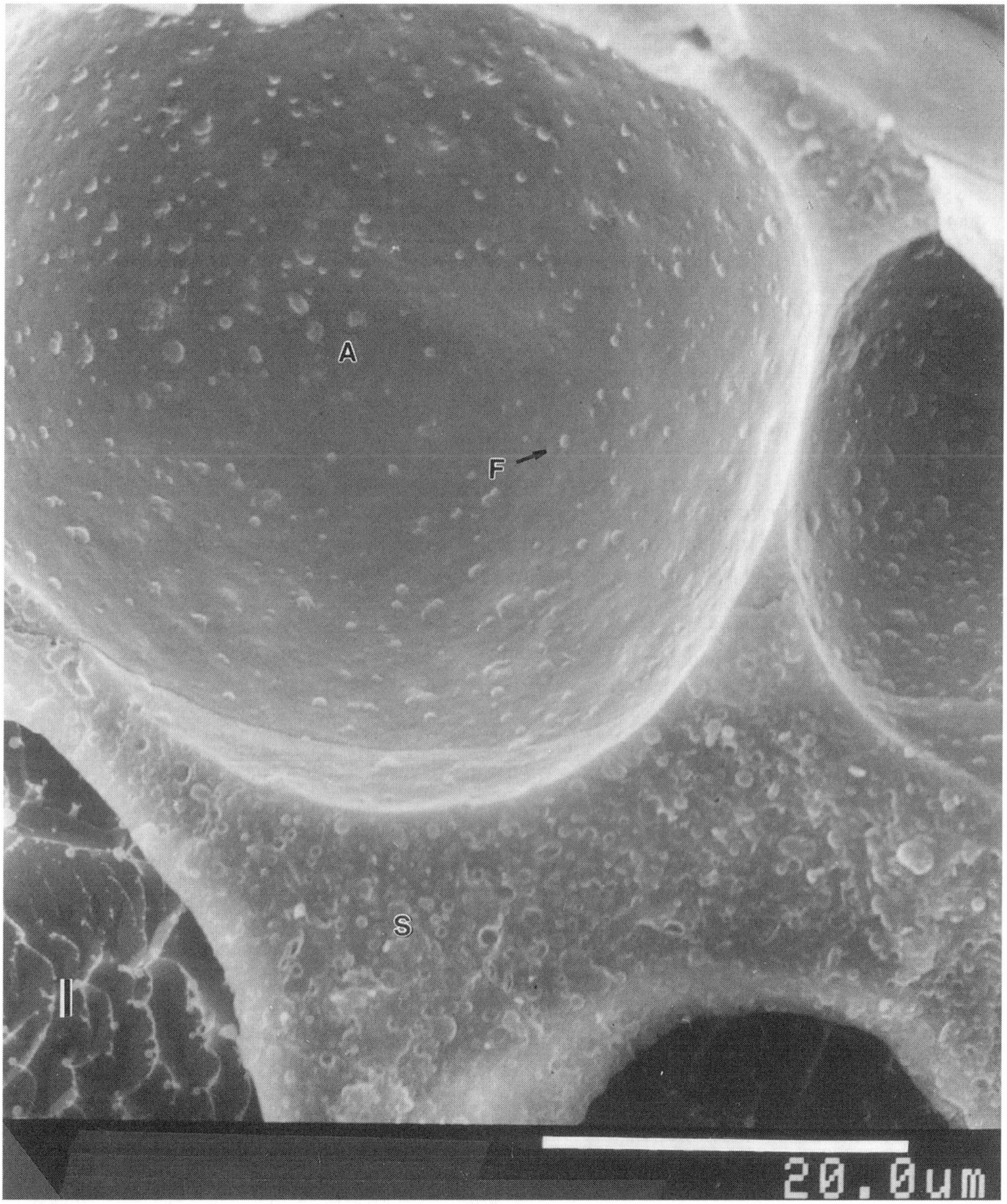


**Figure 5–10** Scanning electron micrograph of ice cream (cryopreparation). A = air bubble; I = ice crystal socket; S = serum phase, F = fat globule. *Source:* Courtesy of K.B. Caldwell.

serum phase becomes mobile and reactive and ice crystal growth can occur in time frames significant for food storage. The preservation of ice cream structure is thus dependent on its temperature of storage relative to its $T_g'$ (Blanshard & Franks, 1987; Levine & Slade, 1988a, 1988b).

The second category of additives in ice cream systems consists of emulsifiers (e.g., mono- and diglycerides and polysorbate esters). These contribute to structure in a different way than stabilizers. Emulsifiers are responsible for the production of a smooth texture in ice cream, for dryness of the product at the time of extrusion or drawing from the swept surface freezer, and for the production of slower rate of meltdown in the frozen product (Arbuckle, 1986). Their usefulness results from their effect on the fat emulsion at the time of homogenization and during the dynamic freezing process, when both agitation and air incorporation are occurring. The emulsifiers reduce the amount of protein adsorbed per surface area of fat (Figure 7–3) at the time of homogenization because of their ability to lower the fat-serum interfacial tension further than that of the proteins (Goff & Jordan, 1989; Goff, Liboff, Jordan, & Kinsella, 1987). As a result, when the mix is subjected to the high rate of shear encountered in dynamic freezing, fat globules have less physical protection from coalescence, and the emulsion begins to partially destabilize. This causes an internal matrix of fat to be produced that offers to the product many of the beneficial attributes mentioned above and is similar to what occurs in the whipping of heavy cream—essentially the forming of a stable foam (Brooker, Anderson, & Andrews, 1986; Stanley, Goff, & Smith, 1996). Fat destabilization in ice cream needs to be optimized; if the process proceeds too far, the partially churned fat can begin to depress overrun, give the product a "greasy" mouthfeel, and become visibly evident in the form of butter particles (Berger, 1990).

### 5.3.4 Milling and Crushing

A notable example of controlled destruction of the microstructure and exploitation of the mechanical properties of a food material for technological applications is flour milling. Milling of wheat to white flour attempts to achieve as completely as possible a separation of endosperm, bran, and germ. The process is difficult, owing to the presence of a crease in the wheat grain and the strong adhesion between the aleurone layer in the bran and the endosperm. Flour milling involves many grinding steps that combine shearing, scraping, and compression. Fluted rolls are used to break open the grain and scrape the endosperm from the bran by shear forces, while smooth rolls compress the endosperm, reducing the particle size, as illustrated in Figure 5–12. The process has been studied using SEM by Moss, Stenvert, Kingswood, and Pointing (1980).

As the moisture content of the wheat grain increases to 15–17% by conditioning or tempering, the bran becomes tougher and more ductile while the endosperm becomes mellower and more brittle (Kent, 1975). The germ is a separate structure that is easily segregated almost intact from the rest of the grain when the cementing material linking germ and endosperm is softened by conditioning. Figure 5–13 shows the microstructure of the starchy endosperm of a wheat kernel and a piece of bran from which flour has been scraped off by milling.

### 5.3.5 Dough Formation and Baking

Forming a foam or porous structure in plastics technology is simply done by selecting an appropriate polymer (e.g., polyurethane), generating gas bubbles by an internal reaction, and waiting until the final structure sets in. A large tonnage of a cereal foam—bread—is produced daily all around the world using the same basic sequence, but the source of polymers is largely restricted to those in one cereal: wheat. Dough formation and baking epitomize a polymer-based restructuring operation. It involves release of participating polymers from the native structure and modification of them.

Dough formation starts by mixing basic ingredients—wheat flour, baker's yeast, fat and emulsifiers, sugar, salt, and water—to form a viscoelastic mass. The dough is fermented at about 30°C and then divided into pieces that are shaped and inserted into baking pans for proofing. Proofing provides carbon dioxide for leavening, modifies dough proteins, and imparts desirable rheo-

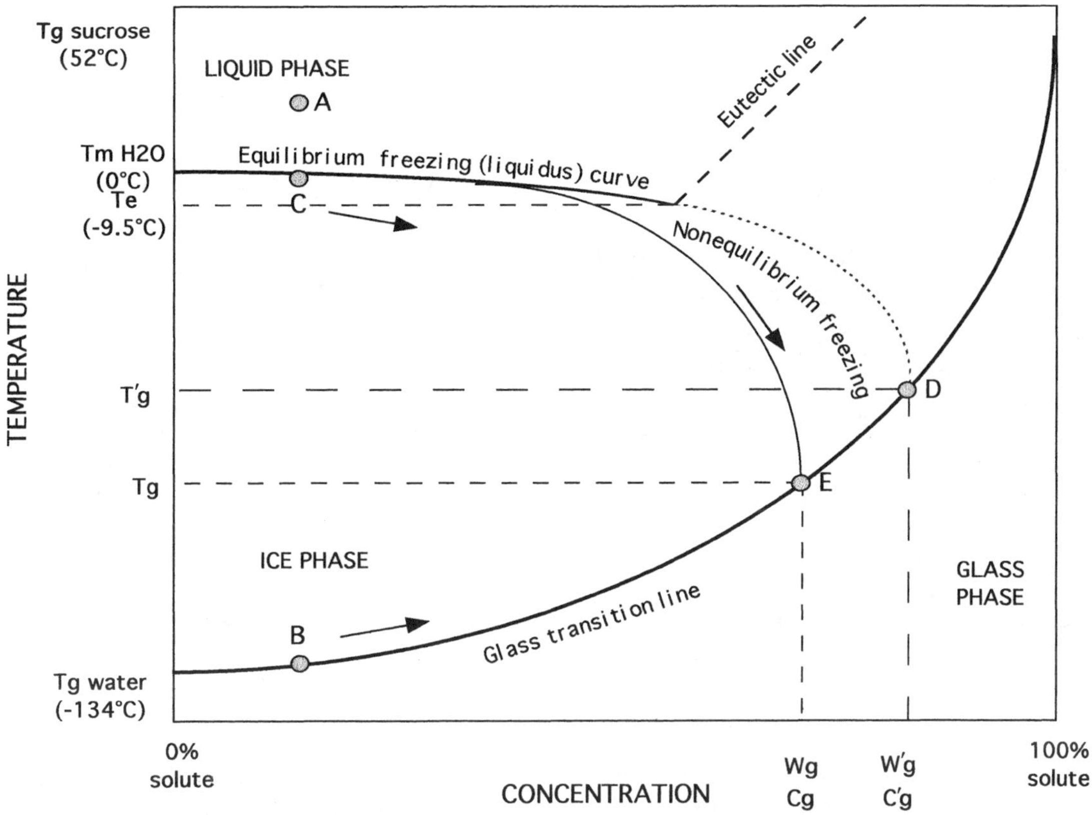


**Figure 5–11** Supplemented state diagram showing the solid-liquid coexistence boundaries and glass transition profile for a binary sucrose-water system. Below and to the right of the glass transition line, the solution is in the amorphous glass state, with or without ice present, depending on the temperature and freezing path followed. Above and to the left of the glass transition line, the solution is in the liquid state, with or without ice, depending on the temperature. Consider a sucrose solution with an initial concentration of 20% at room temperature (A). The initial $T_g$ of this solution at room temperature before phase separation is shown as B. However, upon slow cooling of the system somewhat below its equilibrium freezing point due to undercooling, nucleation and subsequent crystallization begin at C and initiate the freeze-concentration process, removing water in its pure form as ice. As ice crystallization proceeds, the continual increase in solute concentration and removal of water further depresses the equilibrium freezing point of the unfrozen fraction in a manner that follows the liquidus curve (shown as the path C→D), while the $T_g$ of the unfrozen fraction moves up the glass transition line (shown as the path B→D) due to increased concentration. This is accompanied by a rapid increase in viscosity, particularly in late stages of the freezing process. Co-crystallization of solute at the $T_e$ is unlikely, and thus freeze-concentration continues past this point into a nonequilibrium state, since the solute becomes supersaturated. When a critical, solute-dependent concentration is reached, the unfrozen liquid exhibits very reduced mobility, and the physical state of the unfrozen fraction changes from a viscoelastic rubbery liquid to a brittle, amorphous solid glass. The intersection of the nonequilibrium extension of the liquidus curve beyond $T_e$ and the kinetically determined glass transition curve at *D* represents the solute-specific, maximally freeze-concentrated $T_g$ of the frozen system $T_g'$, where ice formation ceases within the time scale of measurement. The corresponding practical maximum concentrations of water and sucrose trapped within the glass at $T_g'$ and unable to crystallize are denoted $W_g'$ and $C_g'$, respectively. Freezing becomes progressively slower as ice crystallization is hindered, and consequently more time is required for lattice growth at each temperature. Therefore, the kinetic restriction imposed on the system can lead to a situation in which nonequilibrium freezing can occur. The pathway followed during this nonequilibrium freezing (shown as C→E) leads to a lower $T_g$ than $T_g'$, with a corresponding lower sucrose concentration in the glass ($C_g$) and higher water content in the glass ($W_g$) due to excess undercooled water plasticized within the glass. This is often referred to as a dilute glass. The magnitude of deviation from the equilibrium curve, and hence the actual path followed, may be regarded as a function of the degree of departure from equilibrium. *Source:* Redrawn from H.D. Goff and M.E. Sahagian (1996).

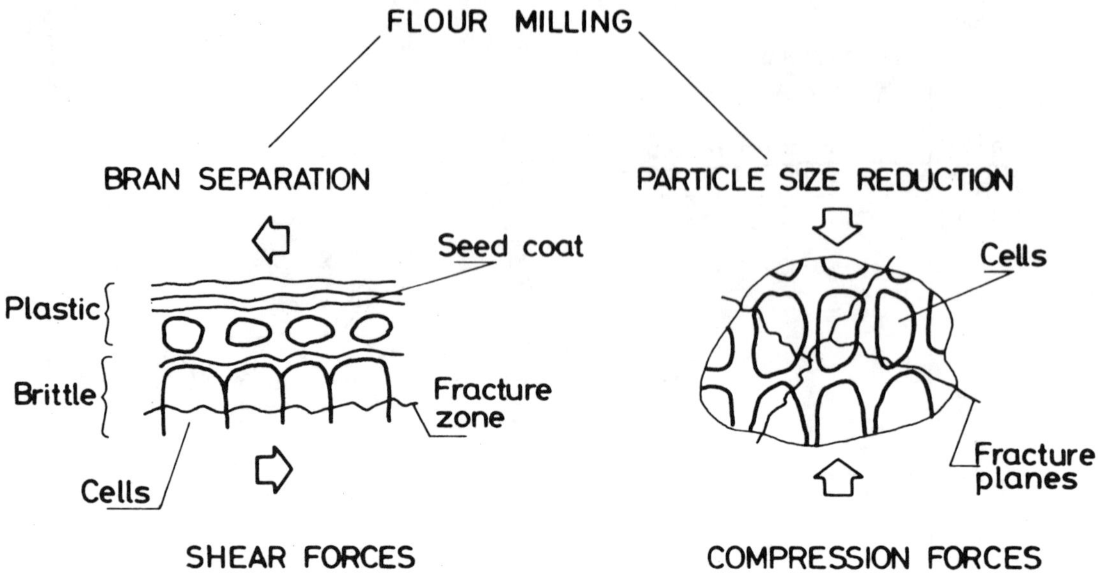


**Figure 5–12** Schematic representation of the wheat flour milling process aimed at simultaneous bran separation and particle size reduction. Shear and compression forces acting on the seed coat and endosperm respectively, induce fracture leading to particle size reduction. Arrows indicate the direction of the forces.

logical properties. Heating in the oven produces gelatinization of starch, coagulation of proteins, and the desirable permanent structure of crumb and crust (Pomeranz, 1970).

When water is added to wheat flour during dough mixing, the water-insoluble proteins hydrate and form gluten, an elastic and cohesive mixture of two types of proteins existing in almost equal quantities: glutenins ($\overline{M}_w = 40 - 150$ kDa) and gliadins ($\overline{M}_w = 40$ kDa). The glutenin fraction is tougher, is less easily stretched, and behaves as a cohesive elastic solid, whereas the gliadin fraction has less cohesiveness and elasticity, performs like a viscous liquid, and is responsible for the extensible properties (Blanshard, 1988). A liquid aqueous phase—formed by albumins, globulins, soluble starch, and pentosans—separates from the gluten phase (Tolstoguzov, 1997a). An appropriate amount of mixing is needed to develop the right structure of gluten; overmixing usually results in a sticky dough through shear degradation of proteins. Mixing is also determinant in gas cell formation, with fast shearing resulting in smaller bubbles (10–100 $\mu$m). Microstructural studies have been used to follow the development of the gluten network structure, gas cell architecture, location of fat, and starch gelatinization (Autio & Laurikainen, 1997). Some of these studies demonstrated that sheeting further contributes to the organization of the dough structure and to the reduction of bubble size. Figure 5–14 is a diagram of the main structural elements in an idealized baked product.

In few processes does the presence of vapor and gas have more effect on the final microstructure than in baking. The rate of gas production, via fermentation or the decomposition of baking powder, affects the growth and size of gas cells. Heating in an oven is accomplished mainly by convection, but the predominant heat transfer mode inside the food is conduction, which leads to the following phenomena and microstructural changes:

- vaporization and transfer of water vapor from the interior of the product to the outside
- denaturation of proteins and starch gelatinization, which strongly affects water partition between phases

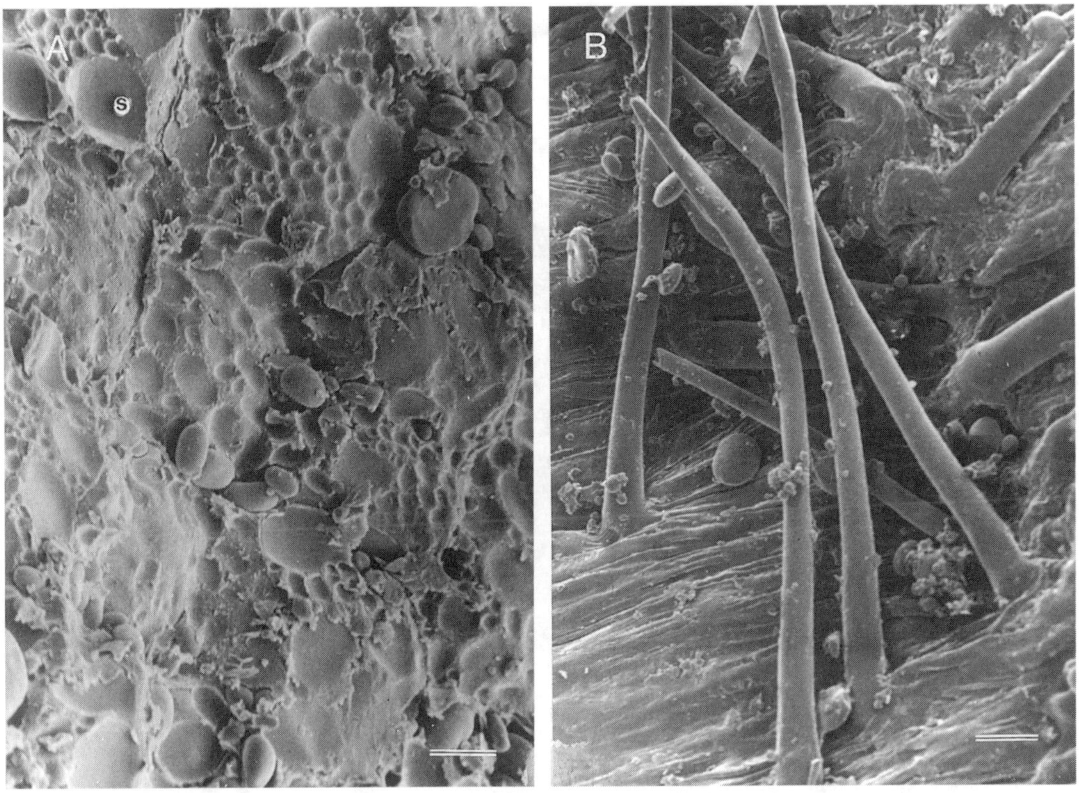


**Figure 5–13** Scanning electron micrographs of wheat flour and bran. (A) Starch granules (S) embedded in the compact protein matrix of the endosperm marker = 10 $\mu$m. (B) Piece of bran from the stigmatic end of the grain showing brush hairs. Marker = 20 $\mu$m.

- extension of the elastic gluten-starch matrix by $CO_2$ formed during fermentation or decomposition of baking powder, expansion of air bubbles, and/or vaporization of water, all of which generate a porous inner microstructure
- melting of fat crystals, which results in "lubrication" and expansion of the dough, allowing bubbles to grow without rupture
- formation of the crust

The structure of fresh bread is altered after baking by *staling*, a process almost completely dominated by starch. Initially it is the amylose fraction and amylose-lipid complexes that tend to crystallize. The later stages of staling and aging of bread result from the recrystallization of amylopectin, which involves moisture migration from amorphous matrix into the crystalline regions (Slade & Levine, 1991). Microstructural aspects of gelatinized starch in bread crumb are presented in Figure 5–15.

### 5.3.6 Crystallization and Tempering of Chocolate

Crystallization of fats has significant technological importance. These very complex molecules can exist in more than one crystalline form that is stable over a certain range of temperature, a phenomenon called *polymorphism*. Several factors determine the polymorphic form assumed after crystallization: purity, temperature, rate of cooling, presence of nuclei, and type of solvent. Transformations of one polymorphic form to another

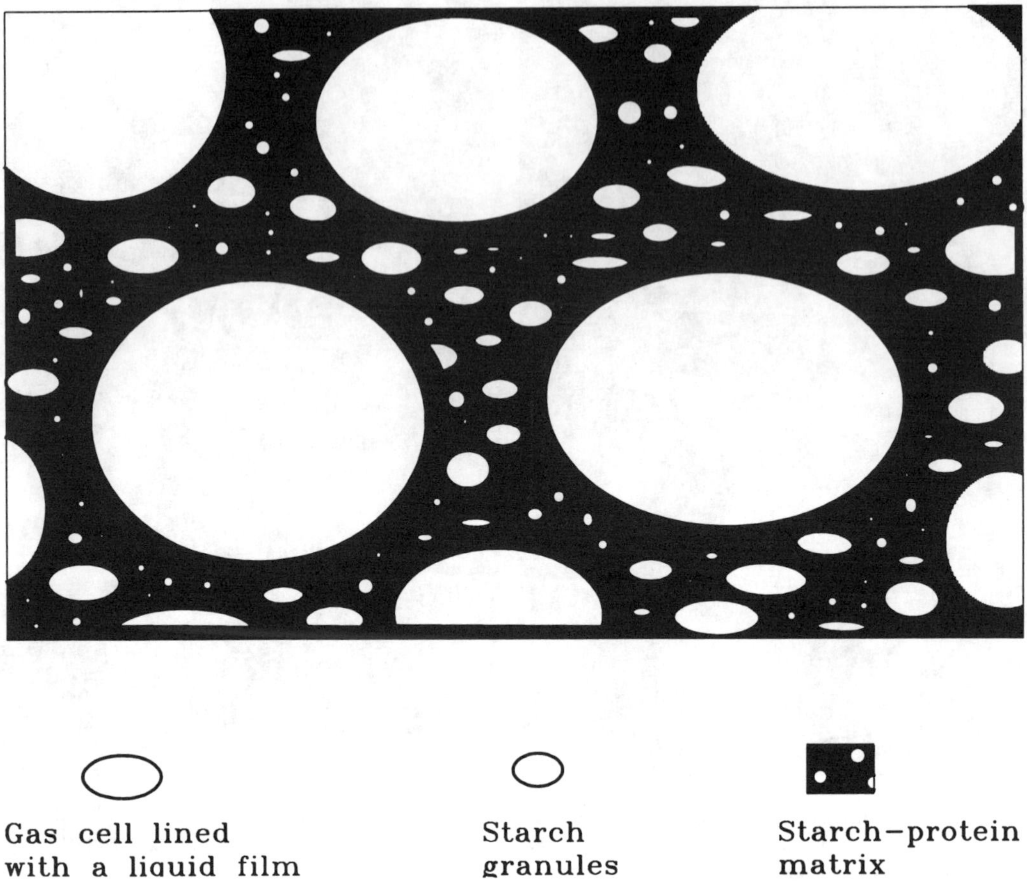


**Figure 5–14** A model of dough expansion. *Source:* Reprinted from *Trends in Food Science Technology*, Vol. 8, K. Autio and T. Laurikainen, Relationship between Flour/Dough Microstructure and Dough Handling and Baking Properties, pp. 181–185. Copyright 1997, with permission from Elsevier Science.

can take place even in the solid state without melting.

Natural fats having long nonpolar chains of various lengths interact with each other to form crystals after cooling from the molten state. Three basic crystal forms are usually distinguished: $\alpha$, which has the lowest molecular packing, has the fastest growth rate, is the least stable, and has the lowest melting point; $\beta'$ which is more densely packed than the $\alpha$ form, is more stable, and has a higher melting point; and $\beta$, which has the highest molecular packing, is the most stable, and has the highest melting point.

Chocolate is formed by a continuous phase of cocoa butter in which crystalline sugar, milk, and cocoa solids are dispersed in the presence of an emulsifier—lecithin. The fat phase in turn is made of liquid fat (approximately 15–20% for cocoa fat at room temperature) having fat crystal inclusions (Figure 5–16). Cocoa butter is 94–96% triglycerides, formed mainly by palmitic, stearic, and oleic acids. Triglycerides are believed to exist in tuning fork configurations, both in the liquid and crystalline states, and to interlock laterally to form lamellar arrangements. Cocoa butter has been shown to form six crystal types that are numbered from I to VI according to their increasing melting points: type I = 21.3°C, II = 23.3°C, III = 22.5°C, IV = 27.5°C, V = 33.8°C, and VI =

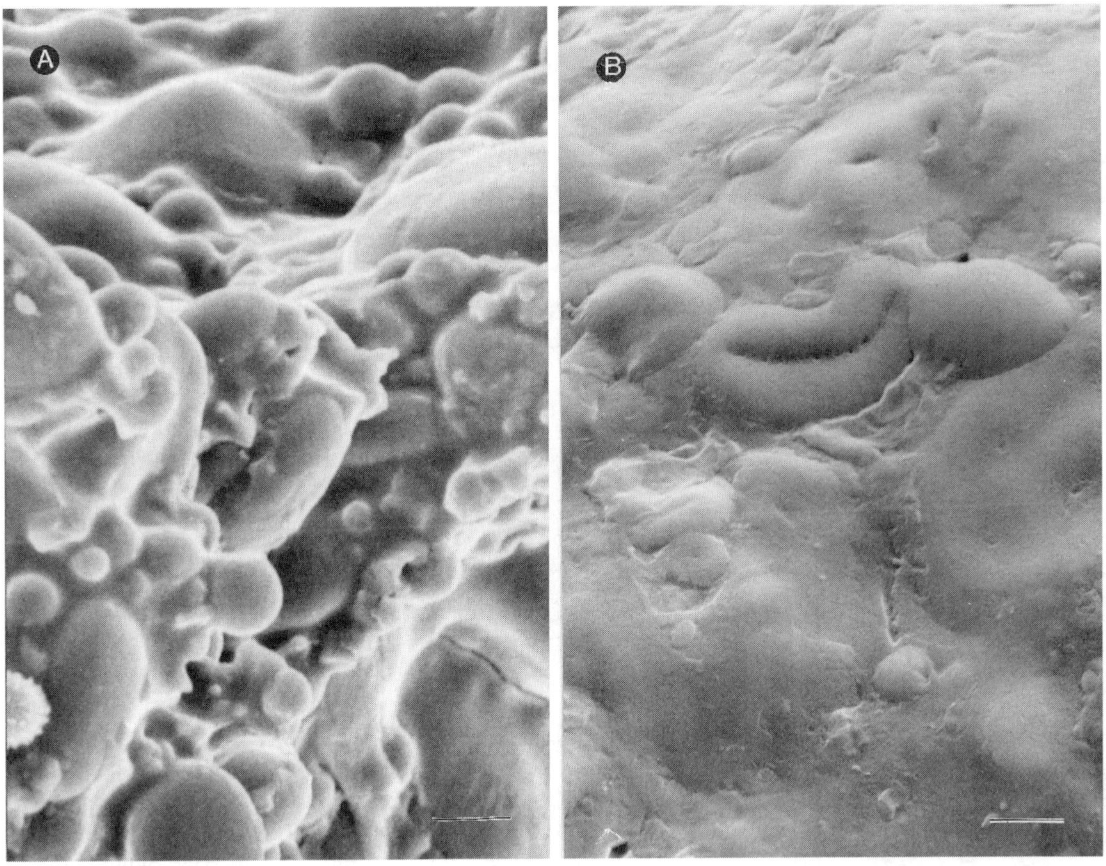


**Figure 5–15** Scanning electron micrograph of bread crumb. (A) Conventional SEM. (B) Quick-frozen and examined in cold stage. Markers = 5 $\mu$m.

36.4°C (Jewell & Heathcock, 1988). This cumbersome nomenclature is of course related to the basic polymorphic forms; for example, type II corresponds to $\alpha$ and V and VI are $\beta$-types. Details of cocoa butter crystallization are presented in Dimick (1991) and Hartel (1998b).

In chocolate, sugar has to be ground to a particle size smaller than 25 $\mu$m so that it does not feel gritty in the mouth, with most particles smaller than 5 $\mu$m. Two problems arise if sugar is ground too fine: more fat is needed to coat the individual particles during conching and sugar crystals are transformed into an amorphous phase. Amorphous sugar is more hygroscopic than crystalline sugar and picks up moisture, accompanying flavors, and odors. During storage, the amorphous phase slowly converts back to crystals and releases the moisture to neighboring particles (Beckett, 1995). The microstructure of chocolate has been studied by LM, transmission electron miscroscope (TEM) (Lewis, 1988), and confocal laser scanning microscope (CLSM) (Brooker, 1995).

Tempering of chocolate is a process that aims at formation of a large number of the smallest possible fat crystals and the right high melting point polymorphic forms (more stable). Properly tempered chocolate possesses good flow properties, sets rapidly on cooling, and presents a high gloss. Fat bloom, a well-known defect in chocolate, imparts a white or gray appearance to the surface. The

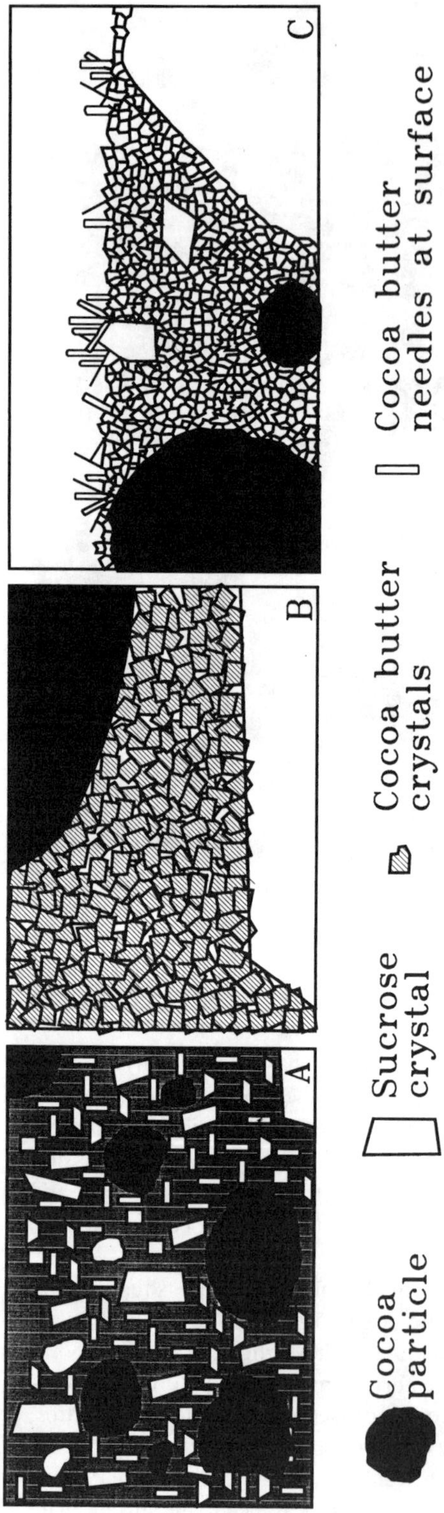


**Figure 5–16** Schematic representations. (A) The structure of chocolate. (B) Higher magnification showing discrete crystals of cocoa butter. (C) Bloom in chocolate. *Source:* Reprinted from R.W. Hartel, Phase Transitions in Chocolate and Coatings, in *Phase/State Transitions in Foods*, M.A. Rao and R.W. Hartel, eds., pp. 217–251, by courtesy of Marcel Dekker, Inc.

bloom corresponds to the VI polymorphic form (which is the most stable) and is characterized by large needle-shaped fat crystals (Figure 5–16, part C). Two kinds of fat bloom can be distinguished. One occurs when tempering is not done properly. Proper tempering involves warming a partially crystallized blend to about 32°C (or holding it at this temperature), followed by rapid chilling and storage at approximately 16°C. Improper tempering causes contraction and formation of crevices that scatter the incident light, giving rise to a whitish appearance (Vaeck, 1960). Bloom can also develop in perfectly tempered chocolate by two mechanisms: continuous temperature fluctuation, with an occasional rise above 15–20°C, or partial melting during storage and rapid crystallization (≈15 hr) into the VI form. Figure 5–17 shows the surface of a well-tempered chocolate and the surface of one where bloom has set in.

### 5.3.7 Comminuted Meat Products

Comminuted meat products are composite foods in which an oil and water (O/W) emulsion is entrapped in a gel formed by insoluble proteins and muscle fibers (Aguilera, 1992). Fat, which usually ranges between 20% and 45% of the total weight, is the dispersed phase in the emulsion, and fat droplets are surrounded by a protein film of myofibrillar proteins extracted by salt. Myofibrillar proteins also contribute to the formation of the gel matrix. The stability of so-called meat emulsions is different from that of normal emulsions, in that coalescence may occur but the gel matrix confines movement of the coalesced fat globules. Also, structural breakdown encompasses not only fat separation but also exudation of water from the matrix. Instability is induced by heating and occurs when the fibrous meat proteins denature and coagulate, losing some of their capacity to hold water and debilitating the matrix. Figure 5–18 shows the main microstructural elements of a meat emulsion.

The microstructure of meat emulsions has been studied using most types of microscopy. Early work done with TEMs showed a porous membrane surrounding the fat globules. Heating in-

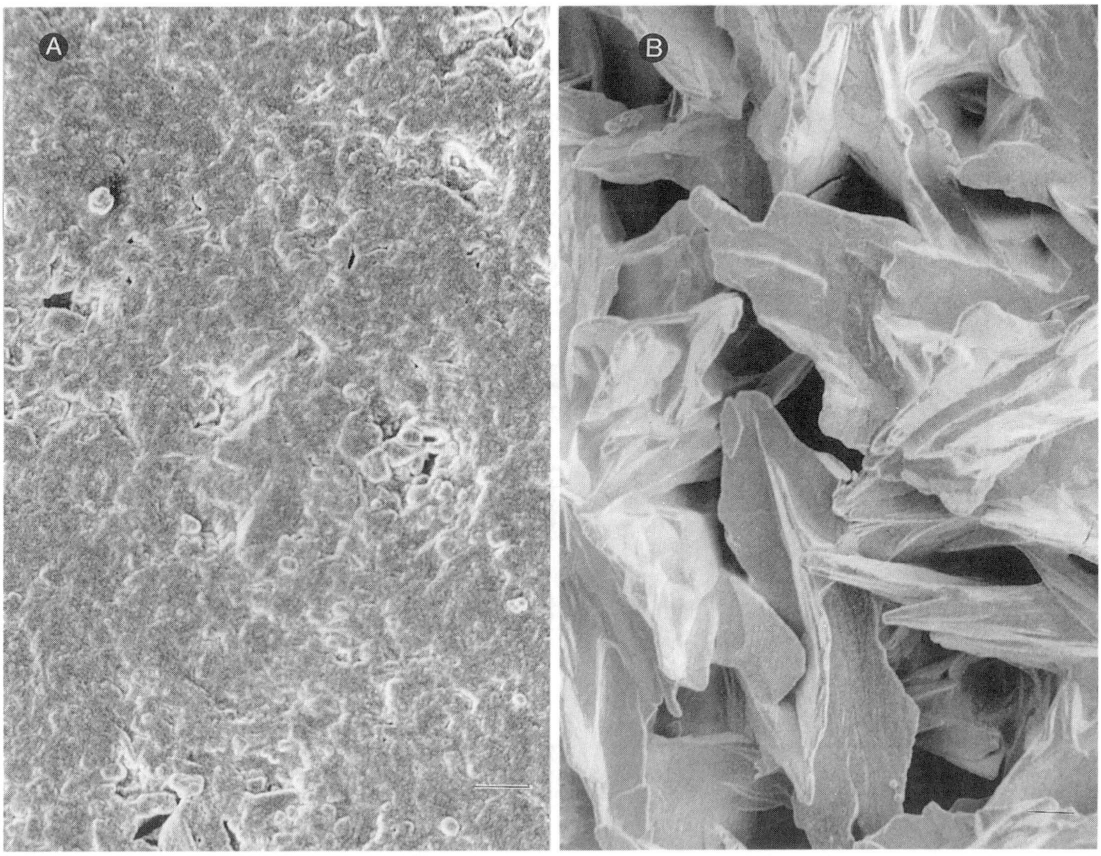


**Figure 5–17** Scanning electron micrograph of the surface of milk chocolate. (A) Properly tempered fat. (B) Large needlelike fat crystals present in a defect known as "fat bloom." Markers = 5 μm.

duced disruption of the protein matrix and coagulation into dense zones (Borchet, Greaser, Bard, Cassen, & Briskey, 1967). The important role of the protein matrix was demonstrated by Lee, Carroll, and Abdollahi (1981) through examining meat emulsions with LMs. Structure stabilization was favored by fat droplets of appropriate hardness and uniform distribution and the presence of a continuous protein matrix. Using SEM photomicrographs as evidence, Jones and Mandigo (1982) were able to postulate a mechanism of fat stabilization and breakdown in frankfurters. A critical maximum chopping temperature for emulsion stabilization was found to occur at 16°C; at this temperature, the protein coating surrounding the fat droplets was sufficiently thin and elastic to accommodate volume changes and the protein matrix formed was dense enough to retain its integrity during heat treatment. TEM studies showed that the dispersed fat particles in frankfurter-type sausages are surrounded not only by large protein molecules but by filaments radiating from the surface of the fat particle. Filamentous coverage results in a more stable emulsion than does the molecular coating (Oelker, 1987). Barbut (1988) used scanning electron microscopy to study the effect of reductions in salt content on poultry meat batters, and a minimum of 2.5% was found necessary to extract sufficient protein to stabilize fat droplets. A scanning electron micrography of a meat emulsion is presented in Figure 5–19.

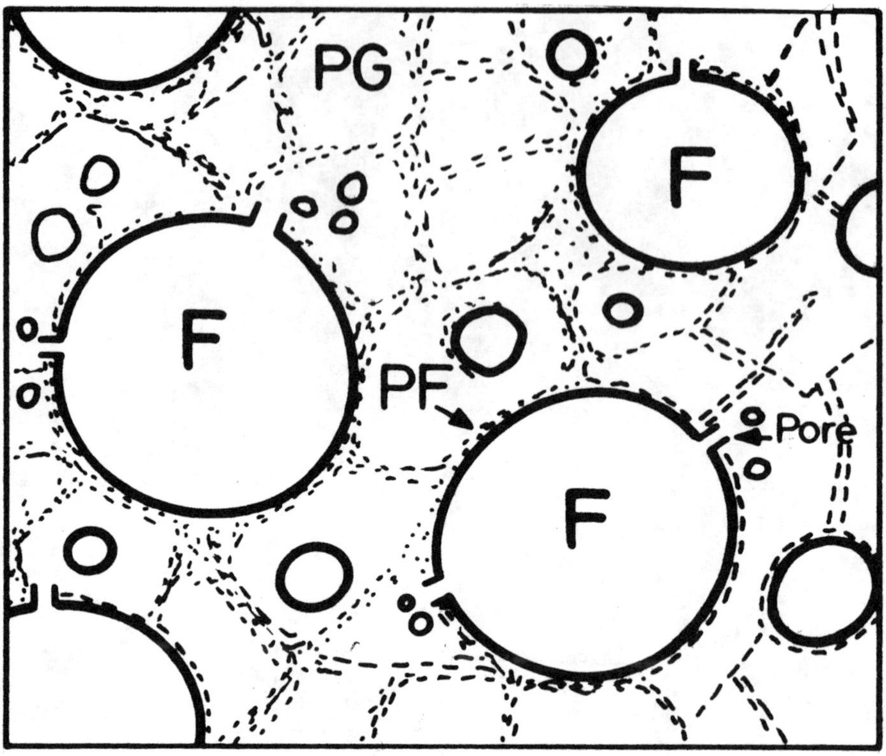


**Figure 5–18** Schematic representation of main components in a meat emulsion. F = fat globules, PG = proteinaceous gel matrix, PF = protein film surrounding fat globules.

### 5.3.8 Mayonnaise

Mayonnaise and salad dressings are low-pH O/W emulsions. In mayonnaise, egg yolk acts normally as the emulsifier. Gums, such as carboxymethylcellulose (CMC), guar gum, and xanthan gum, are used as emulsion stabilizers and thickeners of the aqueous phase. Mustard provides flavor and helps stabilize the emulsion. Fabrication usually involves two main stages: mixing of ingredients (oil, water, vinegar, salt, sugar, mustard, and other spices) and formation of emulsion in the presence of egg protein. It is important in the latter stage that the egg protein be undenatured.

Structuring in mayonnaise is achieved by finely dispersing the oil in the system, as demonstrated by the strong correlation between the rheological parameters ($G'$, $\eta$) and the size of the oil droplets. Dispersion is achieved by introducing energy and shearing effects into the system in a colloidal mill or homogenizer. The mean diameter of oil droplets in commercial mayonnaise was found to be 2.2 $\mu$m, with a major proportion being less than 1.5 $\mu$m. At the same time, a "structure-stabilizing" macromolecular network must develop in the continuous aqueous phase to inhibit droplet coalescence (Windhab, 1995). An SEM photomicrograph of a model emulsion (O/W) is presented in Figure 5–20.

Since the O/W phases in mayonnaise are separated, microbial stability concerns only the aqueous phase. Acid (pH), preservatives (sorbic or benzoic acids), and solutes should have a high effective concentration only in this phase (which represents 20% of the volume), which shows that compartmentalizing can be an effective way of

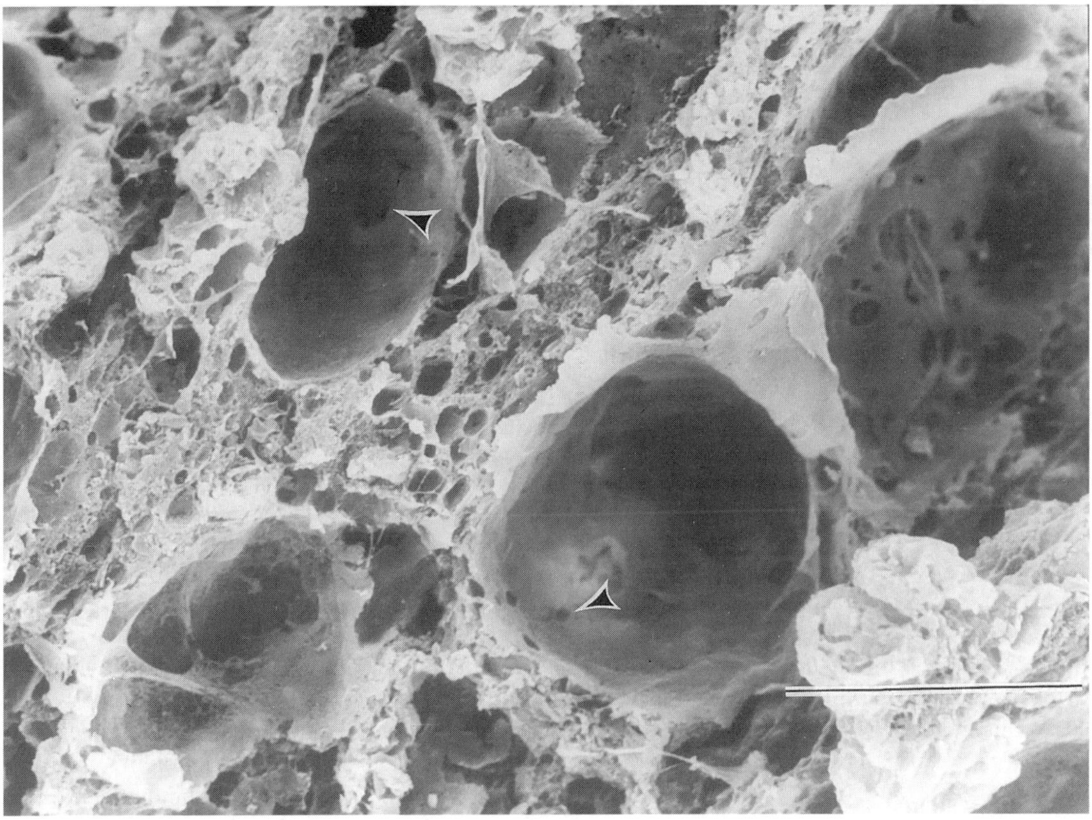

**Figure 5–19** Scanning electron micrograph of a meat emulsion. Arrows point at pores in the internal surface left by fat globules. Marker = 100 $\mu$m.

controlling the stability of food systems. The "measured" water activity in mayonnaise is around 0.80.

## 5.4 APPROACHES TO FOOD STRUCTURING

### 5.4.1 Rationale

The impetus to discover novel food structuring techniques is the desire of food scientists to improve methods of food production, respond to market opportunities, introduce new functional ingredients, and upgrade the utilization of byproducts. Engineers and scientists have developed the concept of fabricated foods—products that contain macromolecules (proteins, lipids, and carbohydrates) derived from various sources and that meet a set of nutritional, quality, and cost specifications. The major challenge facing the food industry is to achieve an acceptable texture for these products (Stanley, 1986). The goal is to create specific mechanical and sensory properties by controlling the composition of the product, interactions among components, and the structure-building process. Efforts in this direction, some of them successful, have been undertaken in the past few decades, despite the lack of a fundamental understanding of the relationship between properties and structure. Several basic

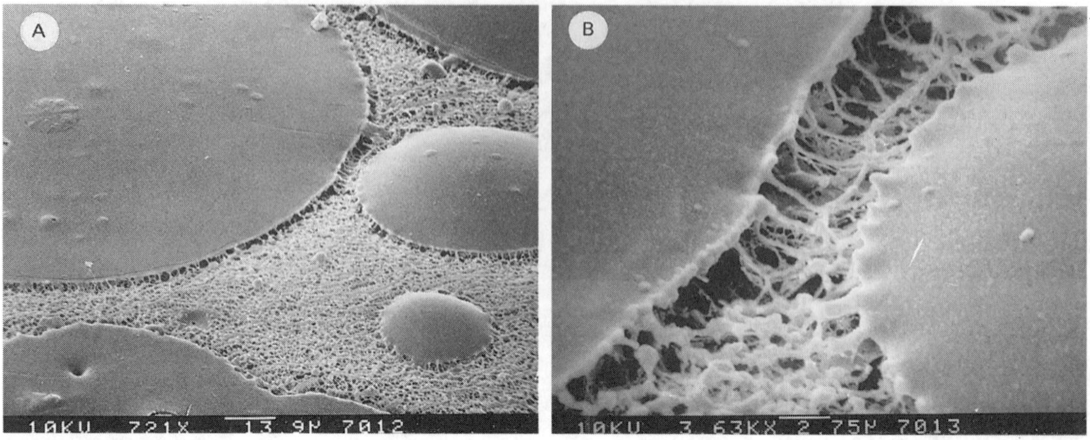


**Figure 5–20** Scanning electron micrograph of an emulsion using cryopreparative techniques. (A) SEM micrographs of freeze-etched O/W emulsion formed by corn oil (dispersed phase) in an aqueous solution containing pea protein (continuous phase) as stabilizer. (B) Higher magnification of interfacial area. *Source:* Courtesy of R.L. Jackman.

approaches to food structuring of proteins from various commodities will be presented as examples of this trend.

### 5.4.2 Myosystems

The most prized quality attribute of meat is its texture. Much research has been aimed at successfully fabricating products from underutilized raw materials that exhibit such texture. The strategy that has so far proven most successful is restructuring. Restructuring encompasses a group of manufacturing techniques that first reduce the particle size of low-value, high–connective tissue meat cuts or trimmings and then recombine them into more valuable products of uniform size and shape and desirable texture. Upgrading these materials, traditionally used for comminuted products such as ground beef and sausage, provides the manufacturer with increased profit and aids in satisfying consumer demand for whole muscle. Restructuring is reviewed by Jolley and Purslow (1988).

Crucial for the final texture of restructured products is the recombining or binding step. Following particle reduction by flaking, the meat is mixed or blended. This step extracts myofibrillar proteins in the form of a sticky exudate that will bind the pieces together during cooking. Salts, phosphates, and other ingredients are added during the mixing step to promote protein extraction. Since the muscle proteins only bind under the influence of heat, after the pieces are molded and pressed into a suitable shape, the product is frozen, sliced into individual portions, and stored and distributed (still in the frozen state). The product is cooked without thawing, and the bind develops during this step. The need to keep such products frozen tends to limit their use to institutional and food service outlets, since the distribution channels used for chilled fresh meats are not suitable. Development of a nonmeat cold-setting binder could be expected to foster the retail marketing of restructured products by giving these products textural characteristics closer to those of intact muscle and meeting consumer desire for fresh products. Several nonmeat binders such as alginates and methylcellulose have been tried (Bernal, Bernal, & Stanley, 1987; Bernal, Smajda, Smith, & Stanley, 1987; Bernal & Stanley, 1989; Means & Schmidt, 1986).

Another example of a structured myosystem is *surimi,* which is deboned fish muscle that has been washed and had cryoprotectants added to

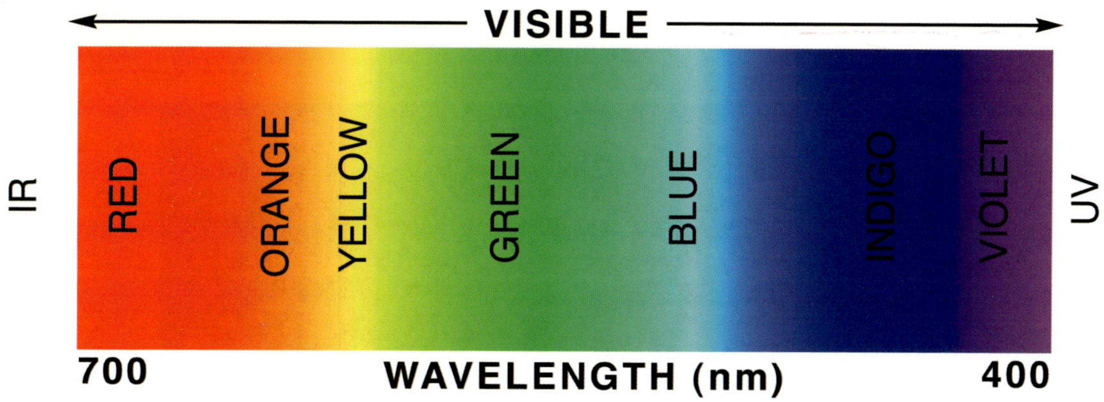


**Color Plate 1** Diagram of visible spectrum.

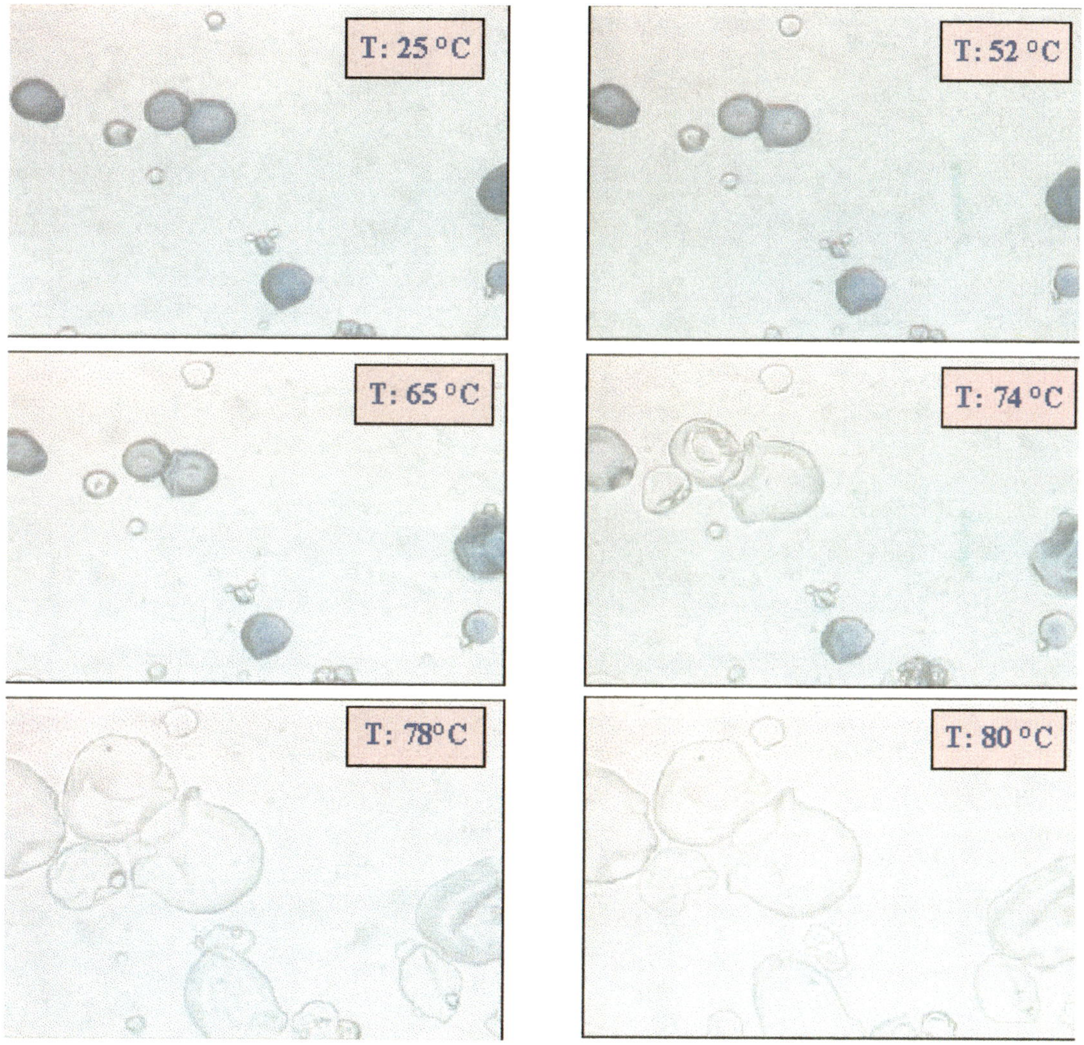


**Color Plate 2** Hot stage video microscopy of a-real-time sequence of swelling of cassava starch granules in water (heating rate 40°C/min, holding temperature 80°C). *Source:* Stanley et al., 1998.

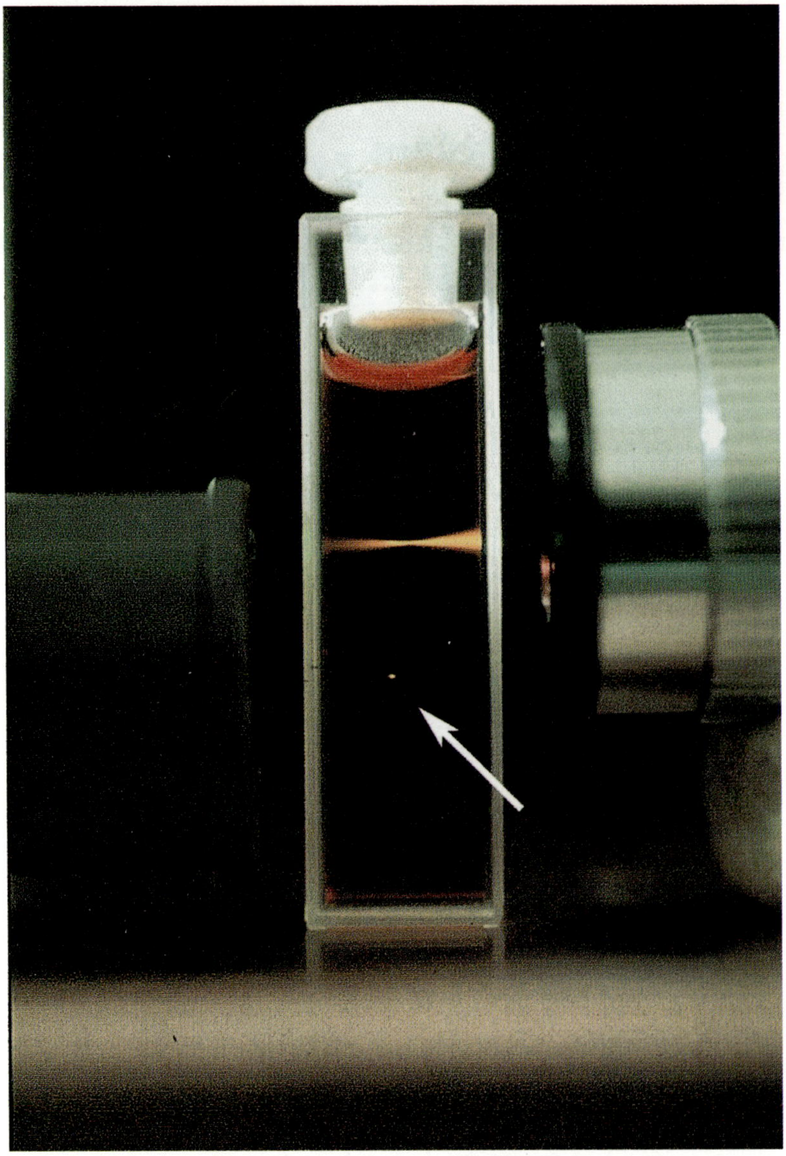

**Color Plate 3** Cuvette containing safranin dye, which normally needs green light to fluoresce. Single-photon laser excitation (top center) showing cone of fluorescence leading to and away from focal plane that can create photobleaching. Lens-focused two-photon excitation (lower center, arrow) showing pinpoint of fluorescence produced, which can eliminate photobleaching. Photons in this beam can be absorbed at the same time by the same molecule, adding their energies. This two-photon absorption only occurs at high intensities at the focal point. *Source:* Courtesy of W.B. Amos. Medical Research Council Laboratory of Molecular Biology, Cambridge, UK, and Science Photo Library, London, UK.

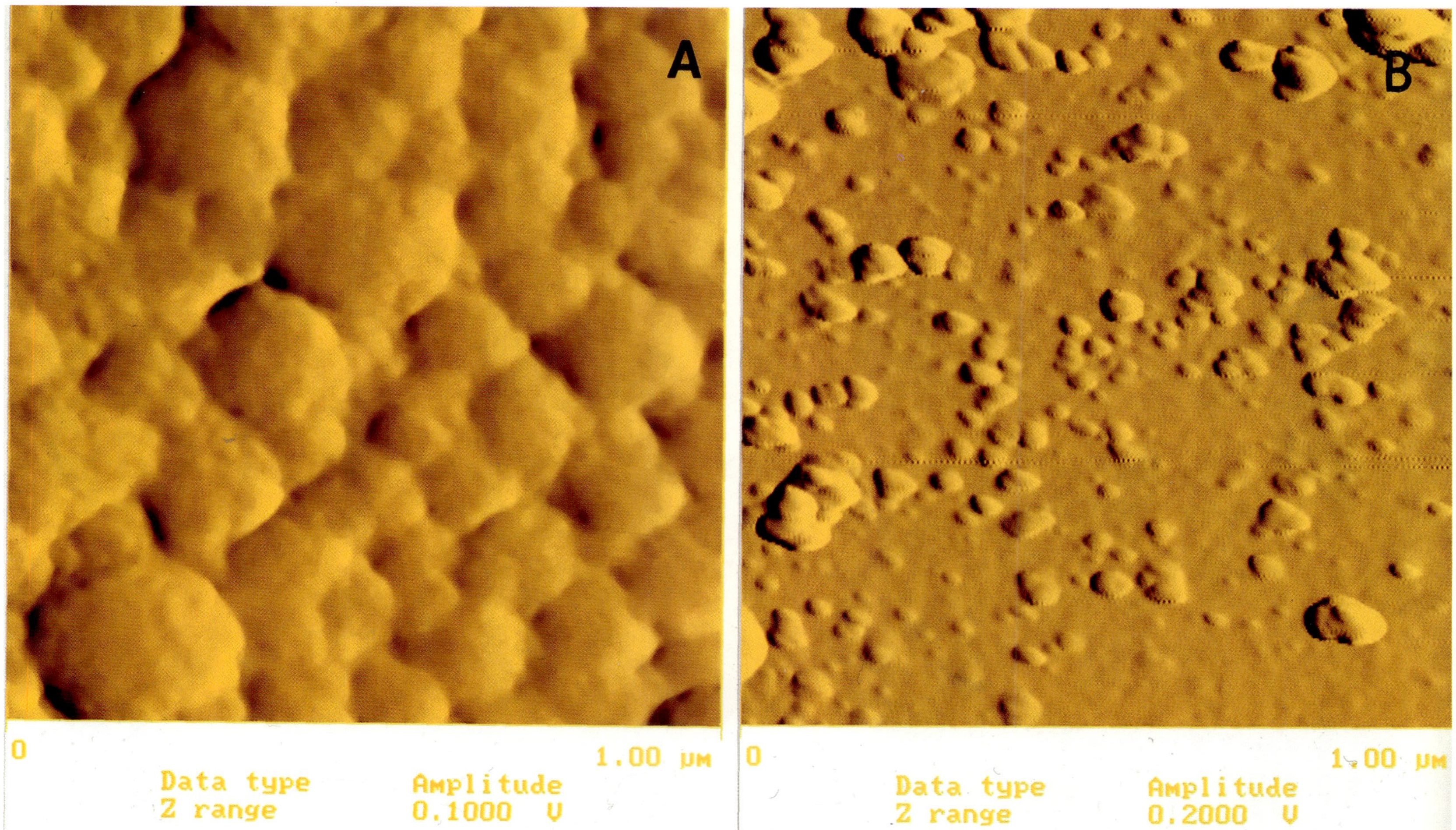


**Color Plate 4** Tapping mode AFM of (A) skimmed milk powder showing casein micelles and (B) emulsion of 10% skimmed milk powder and 10% soy oil. *Source:* Courtesy of K. W. Baker.

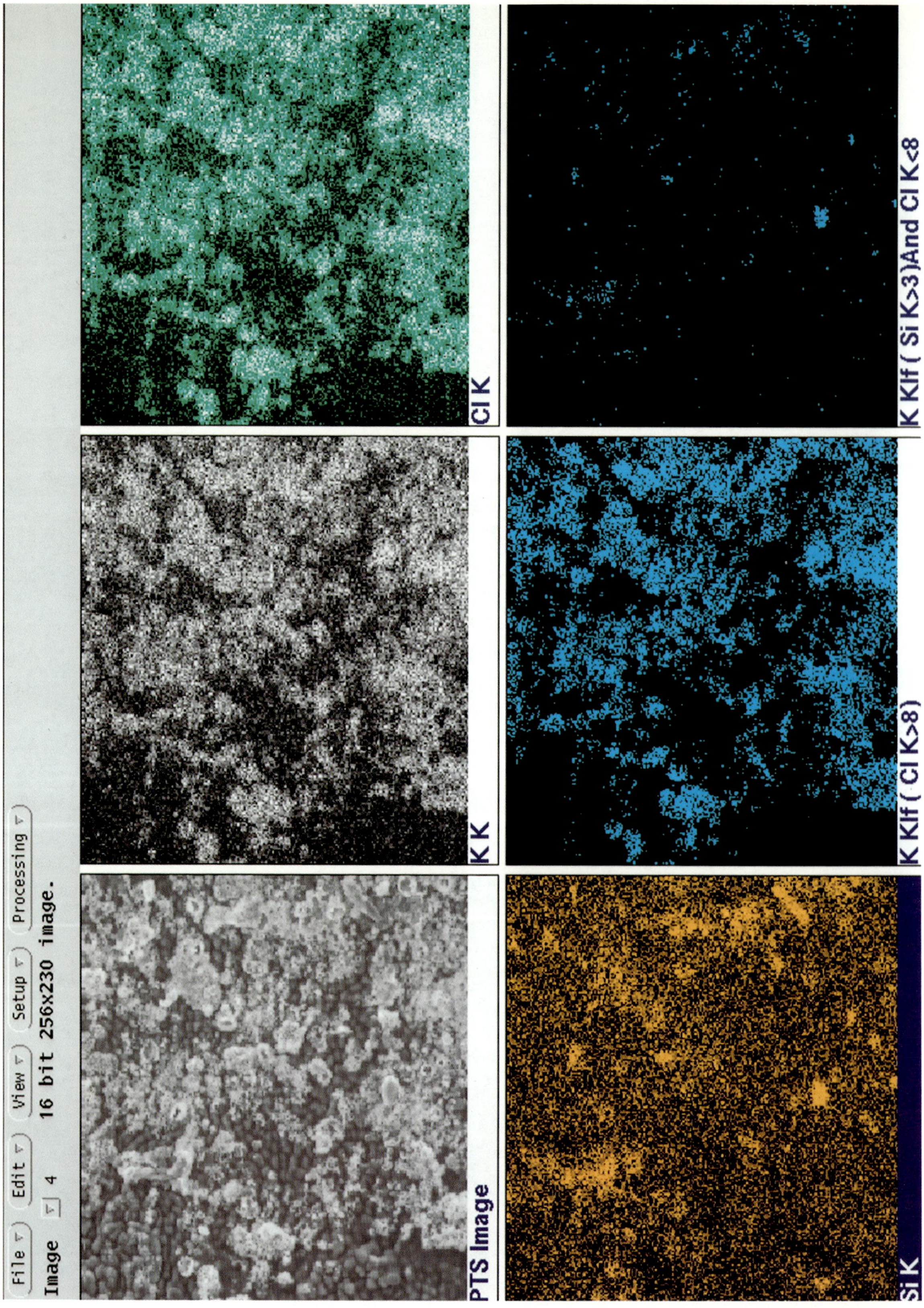
File
Edit
View
Setup
Processing
Image 4 16 bit 256x230 image.
PTS Image
K K
Cl K
Si K
K Klf ( Cl K>8)
K Klf ( Si K>3)And Cl K<8

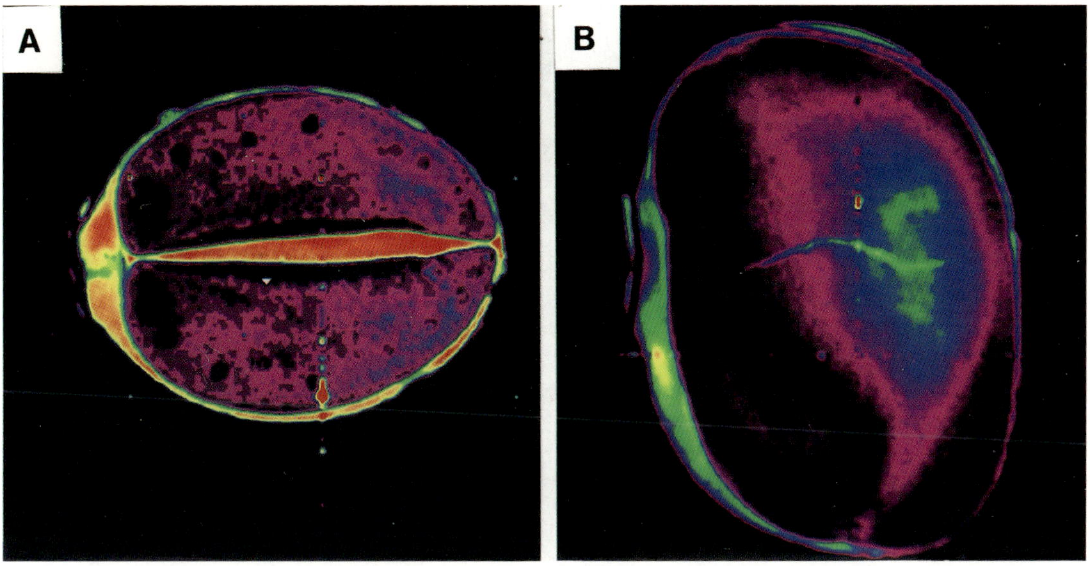


**Color Plate 6** MRI diffusion attenuated proton microimaging of bean following 48 h soaking in distilled water at ambient temperature. Each voxel is ≈60 $\mu$m × 60 $\mu$m × 500 $\mu$m (slice thickness). (A) Cross section. (B) Longitudinal section. Note crack in cotyledon. *Source:* Reprinted with permission from E.J. Kendall, D.W. Stanley, and K.B. Chatson, A Comparison of Water Self-Diffusion in Fresh and Stored Beans, in *Proceedings of the 77th Chemistry Society of Canada Conference*, Winnipeg, Man., May 29, 1994. Abs. # 510.

---

**Color Plate 5** Position-tagged spectrometry X-ray image of the surface of sun-dried dulce, an edible red seaweed. Imaging was done in an environmental scanning electron microscope so that the sample could be rehydrated. A complete EDX analysis was performed on the hydrated, uncoated material. Maps were then extracted from the original data for various elements, as shown. In the K, Si, and Cl maps, the brighter regions are the highest concentrations of the specified element. The two logical maps show K where Cl is greater than 8 counts, and K where Si is greater than 8 counts and Cl is less than 8 counts. These types of logical maps allow the analyst to look for possible phases in a sample. *Source:* Courtesy of K. W. Baker.

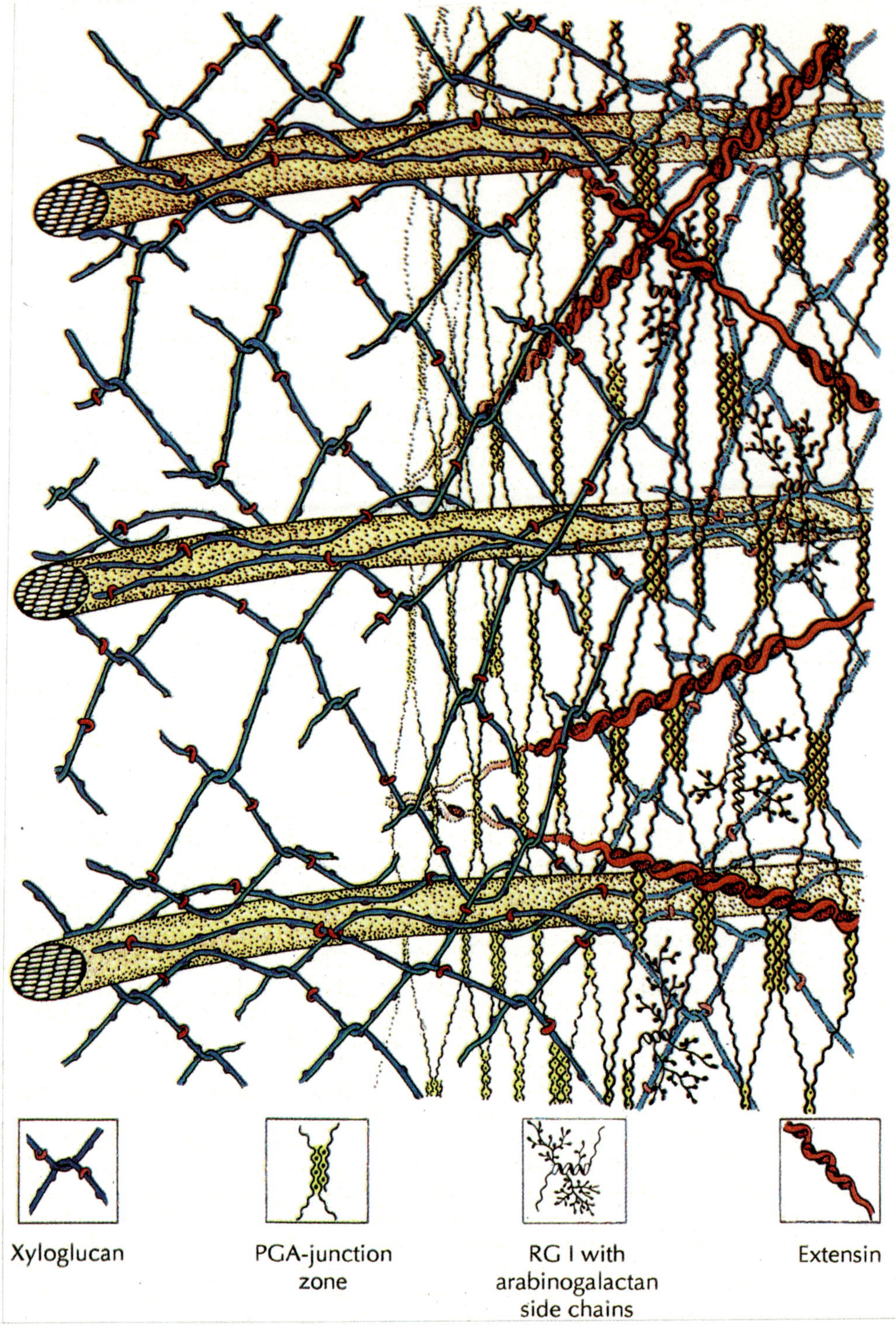
Xyloglucan
PGA-junction zone
RG I with arabinogalactan side chains
Extensin

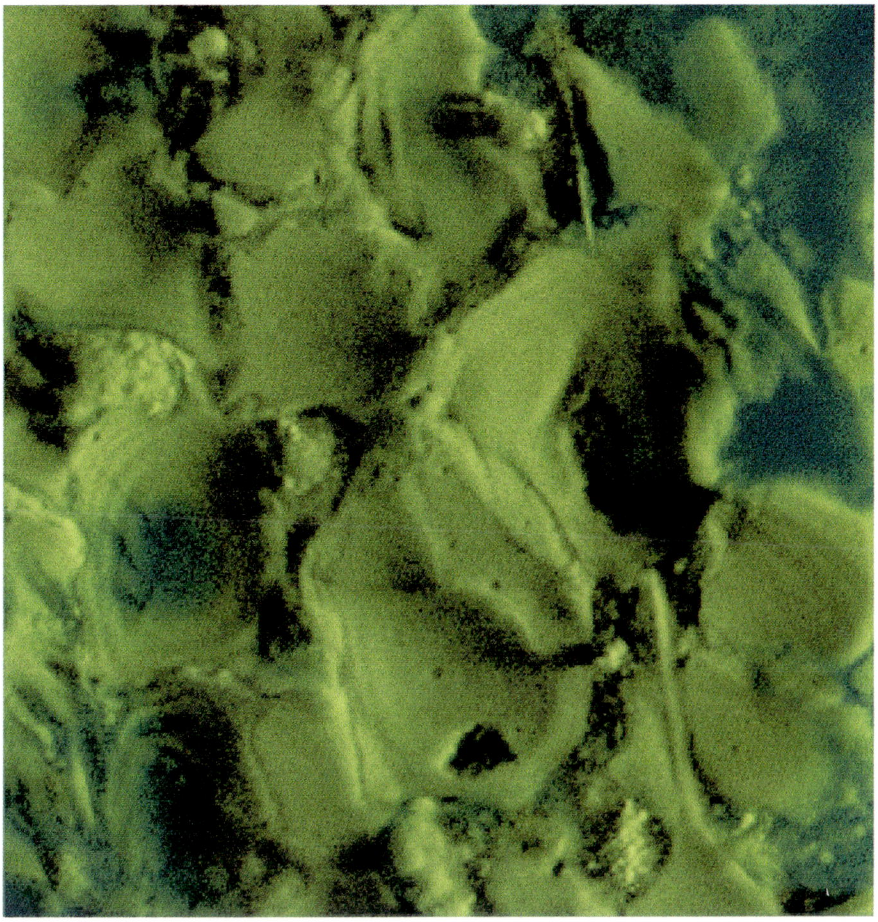

**Color Plate 8** Oil in the crust of a fried potato. Three-dimensional reconstruction of optical sections obtained by CLSM using the Software Imaris. Oil forms interconnected concave shells in intercellular spaces (egg-box structure). Each side of the square is 1,280 $\mu$m. *Source:* Copyright © Franco Pedreschi.

**Color Plate 7** A model of the expanding primary cell wall of flowering plants except grasses. A single layer is represented: several such layers condense to form the wall. Three thick Cellulosic microfibrils are aligned in parallel but in a helical formation around elongating cells. They are cross-linked with hemicellulosic xyloglucan polymers that have been partially cleaved to permit microfibril separation. This domain is embedded in a second consisting of a matrix of pectic polygalacturonic acid (PGA), which forms junction zones in the presence of $Ca^{2+}$ and rhamnogalacturonan I (RGI), with attendant arabinogalactan side chains. A third domain contains extensin molecules, which are inserted radially to stabilize the separated microfibrils and limit further stretching upon the cessation of growth. *Source:* Carpita and Gibeaut, 1993.

give it an acceptable frozen shelf-life (Lee, 1984). Mechanical deboners, originally developed for poultry, are used to separate fish flesh from the bones, skin, and scales of underutilized species and from racks remaining after filleting. The texture-forming ability of raw fish proteins is known to decrease during frozen storage because of denaturation and the cross-linking of proteins.

Surimi is essentially a gelled myofibrillar concentrate. Gelation of fish proteins is reviewed by Stone and Stanley (1992). When fish mince paste (sol) is heated to about 50°C, a loose network (*suwari*) is formed from actomyosin and myosin molecules; this process is referred to as setting. Setting is species dependent and occurs over a range of temperatures (up to 50°C) and to a varying extent. As the temperature is increased to around 70°C, suwari is partially disrupted to form a broken net structure (*modori*), a phenomenon attributed to the dissociation of myosin from actin and the possible fragmentation of the actin filament. Although initially alkaline proteases were cited as instigating the gel weakening, more recent evidence favors a thermally driven mechanism, the precise nature of which has yet to be determined.

Further heating at a temperature above 60–70°C produces an increase in strength (firmness) as the gel structure sets into *kamaboko*. Whereas the tail region of myosin has been involved in cross-linking interactions at lower temperatures, the globular head portion assumes a role above 60–70°C. Hydrophobic interactions between adjacent tail regions form the basis for the initial structure. The aggregation of myosin head regions is almost universally accepted as the primary mechanism promoting gel strength. Changes in the rheological properties of fish paste, actin, and myosin during heating are easily detected by following changes in $G'$ as a function of temperature.

Traditionally, surimi was prepared from fresh fish and processed immediately into various kamaboko (sausagelike) products. Around 1960, Japanese food scientists discovered that cryoprotectants—such as a 4% sucrose, 4–5% sorbitol, 0.2–0.3% polyphosphate mixture—stabilize frozen surimi and allow its production aboard processing ships rather than on shore. Although the exact mechanism of cryostabilization by these solutes is still unknown, it is obviously related to formation of a glassy state that protects proteins from denaturation and freeze-concentration effects (Ohshima, Suzuki, & Koizuma, 1993). This technology yields a food material with acceptable firmness, springiness, and water-binding properties and overcomes the limitations mentioned above. Many final products, mainly of a seafood analog type, can be produced from surimi. For example, to imitate the muscle structure of crab, thawed surimi is extruded in a continuous thin strip, heated to thermally set the proteins, and shredded into fine fibers; the fibers are then united using binders such as egg or starch, and the material is sliced and packaged. Skillful applications of colors and flavors produce a realistic product simulating intact muscle. Part B of Figure 4–2 shows the microstructure of a commercial surimi material, and Table 5–2 compares the sensory and instrumental parameters for scallop tissue and a surimi analog. Note the influence of anisotropy on this system.

### 5.4.3 Phytosystems

Several social and scientific forces have converged to make improved utilization of plant proteins a major challenge for food scientists. These include the growth in world population, upgraded expectations regarding food quality, a growing demand for moderately priced protein, the increased cost of animal protein, and a growing awareness of nutrition. The usual process of isolating and purifying plant proteins results in a bland powder, reasonably nutritious and functional but having little appeal as food. Successful structuring of these proteins could easily be ranked as one of the most significant technological developments in the food area. In this section, several newer approaches to structuring plant foods are examined.

It is interesting that in these cases the goal of structuring is the formation of a fibrous final product, one that imitates meat. Structured plant pro-

**Table 5–2** Sensory and Instrumental Parameters for Scallop and Surimi Analog

| *Parameter* | *Scallop* | *Analog* | *Significance* |
|---|---|---|---|
| Tenderness[a] | 12·7 | 10·4 | $p < 0.05$ |
| Juiciness | 8·5 | 7·6 | NS[b] |
| Residual connective tissue[a] | 12·9 | 11·5 | NS |
| Overall preference[a] | 11·9 | 9·0 | $p < 0.05$ |
| Compression (N/unit)[e] | 1·5 | 4·5 | $p < 0.05$ |
| Shear (N/g)[d] | 1·60 | 1·32 | NS |
| Anisotropy (C/S)[c] | 0·94 | 3·48 | |

[a]Sensory measurements using 8-member trained panel; data are from a 15-cm unstructured scale; larger number indicates higher attribute value.
[b]Not statistically significant
[c]Ratio of compression to shear.
[d]Warner-Bratzier shear across fiber axis.
[e]Flat plate compression to sheer.
*Source:* Stanley, 1987.

teins have not yet fulfilled their initial marketing projections. Although many factors are responsible for this shortfall, such as low meat prices, defects in commercial analogs, and protein overnutrition (Visser, 1988), one possible factor is the commercial decision to aim research efforts at mimicking the texture of the most expensive protein source, meat, rather than to develop novel foods. Mimicking has meant that consumers in developed countries often suspect they are purchasing inferior imitation products while, on the other hand, the populations of developing countries are reluctant to purchase the products because they are unfamiliar or unaffordable.

Freeze alignment is a method of protein texturization based on a much older product, *kori-tofu*. This traditional Oriental food is based on soybean curd that has been frozen and dried so that it becomes porous and spongy. The newer texturization process is aimed at developing a stable, fibrous structure that can be rehydrated to produce a meat analog. The freeze alignment process, developed by Lugay and Kim at General Foods (Lugay & Kim, 1981), can utilize as protein sources any edible protein or combination of proteins provided that a fraction of the protein mixture has heat-setting properties. Protein texturization is achieved in several steps: an aqueous protein dispersion is frozen so that elongated ice crystals are generated perpendicular to the cooling surface. This goal is achieved by preferentially cooling the bottom of the container holding the protein solution. The formation of ice crystals separates the protein material into distinct aligned parallel zones. Ice crystals form in a latticework, entrapping protein in an orderly fiberlike region between the elongated ice crystals. When freezing is completed, protein is distributed throughout the frozen mass in an aligned, fibrous arrangement. Water is then removed by freeze-drying, and the resulting dry mass is stabilized using pressurized steam to immobilize the protein in fibrous form. The resulting protein material can then be safely rehydrated with a solution containing the microconstituents (flavors, colors, nutrients, etc.) and a meat analog can be constructed. Figure 5–21 shows the freeze alignment process.

A totally different approach is taken in the production of the next example, *protein micellar mass.* The association between protein molecules mainly results from hydrophobic interactions in which large subunited storage proteins (of cereals and legumes) with nonpolar patches on their surface tend to come together to avoid aqueous environments. Protein hydrophobicity was exploited by Murray, Myers, Barker, & Maurice (1981), yet another group of General Foods scientists who developed a protein-structuring process. First, it was found that most proteins (>90%) from many plant species and other sources such as yeast

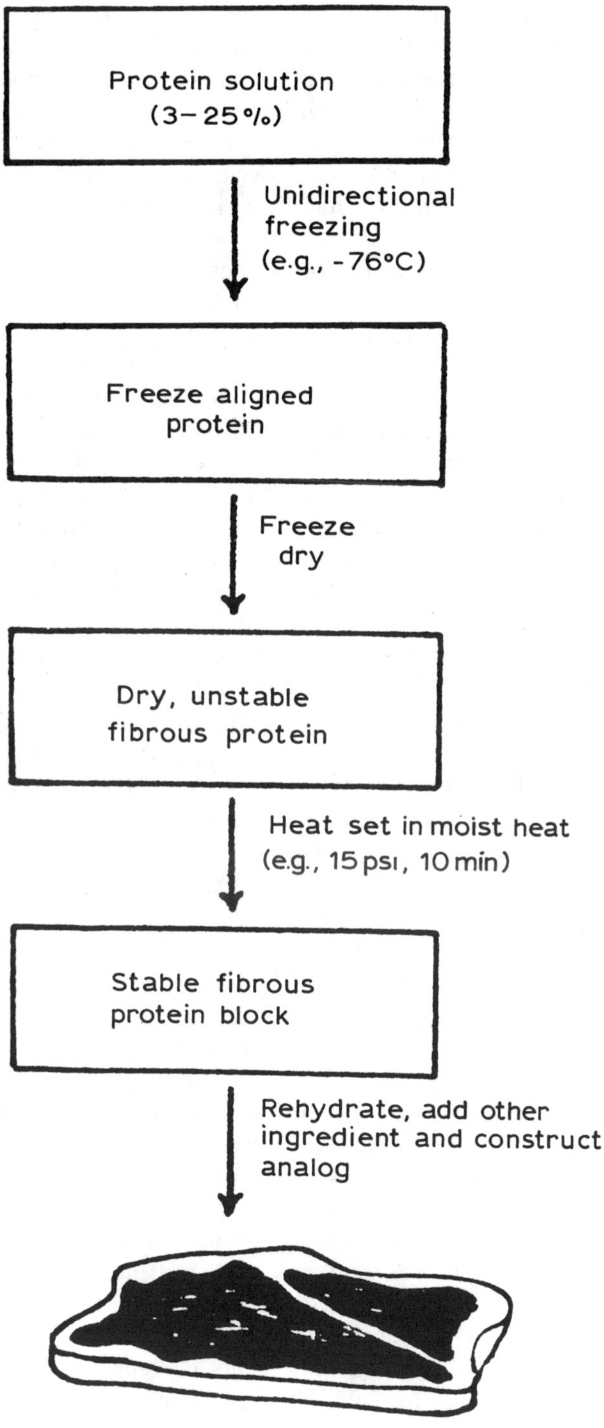


**Figure 5–21** Protein texturization by the freeze-alignment process. *Source:* Lugay and Kim, 1981.

could be solubilized by extraction with 0.3 M sodium chloride. This represented a classical salting-in of the proteins. That is, the surface of species having high charge densities interacts ionically with water. If this salt-protein extract is then diluted with excess water, the proteins self-associate as a result of hydrophobic interactions to form a viscous material called *protein micellar mass* because it consists of almost pure protein in the form of spherical particles 1–10 $\mu$m in diameter (Figure 5–22). The protein micelles can be made from a wide variety of proteins as long as they have some external hydrophobicity.

These protein micelles have several interesting properties: when injected into hot water, they will form fibers that are a potential starting material for meat analogs; upon settling or centrifuging, they form a gelatinous precipitate exhibiting binding and heat-setting properties similar to egg white; and they have sufficient gas trapping functionality to either extend or replace wheat gluten in leavened bakery products. These functional properties are attributed to the mild processing conditions; differential scanning calorimetry studies led to the conclusion that the proteins underwent little or no denaturation during their isolation. Structurally, each micelle is packed with protein subunits in an amorphous manner (i.e., it is not crystalline). Thus, micelles are a functional intermediate between individual (soluble) protein molecules and extensive protein networks (gels).

### 5.4.4 Lactosystems

Increasing the utilization of milk protein is a challenge facing the world's dairy industry. Only part of the whey protein produced from cheesemaking is used for food and feed; the rest is discarded, often to the detriment of the environment. On the other hand, milk proteins are popular as food ingredients, since they possess exceptional functional properties. These include bland flavor, micelle-forming ability, water-binding capacity, heat and enzymatic coagulatability, foamability, and ability to stabilize emulsions.

Considering these functional properties and the favorable regard dairy products are held in by consumers, it is surprising that more development work has not been directed toward restructuring dairy proteins. Two probable contributing factors include cost (the cost of caseinate, an acid- or enzyme-produced isolate, is roughly 1.5–2 times the cost of soy products) and the existence of a nutritional problem resulting from the consumption of lactose associated with some forms of milk proteins. Lactose intolerance is widespread among adults in nonwhite populations.

One approach that has been investigated is to coextrude casein and a starch source. This approach takes advantage of the fact that using more than one raw material can optimize product characteristics and reduce cost. In one study (van de Voort, Stanley, & Edamura, 1984), casein (83% protein) and wheat flour (13% protein) were mixed in various ratios and then adjusted to different water contents. The combinations were used as feed material for a thermoplastic extruder. It was found that casein did not alter the bland flavor characteristics of extruded flour. In fact, casein had little impact on any of the product characteristics, although it constituted about 10% to 30% of the nonwater ingredients. In other words, casein could be added without affecting product quality.

Another approach to structuring lactosystems is spinning. In dry spinning milk proteins, use is made of the ability of dairy products such as cheese and caseinate solutions to form fibers when heated moderately (Visser, 1988). A spinning "dope" is made from mixtures of casein and other ingredients, such as skim milk powder, soy protein, starch, and gluten. The ingredients are mixed with water to a final moisture content of 25–45% and heated to about 80°C. The dope, formed in an extruder, is then pumped through the narrow (0.25 mm) holes in a spinnerette into a current of hot air that evaporates surface water. Fibers are pulled away from the spinnerette and stretched over rollers prior to final air drying. This pilot operation, described in Visser (1988), was scaled up through the application of pasta extruders fitted with dies having smaller (0.25–0.65 mm) than normal orifices. Visser concludes that the simplicity of the process and low production

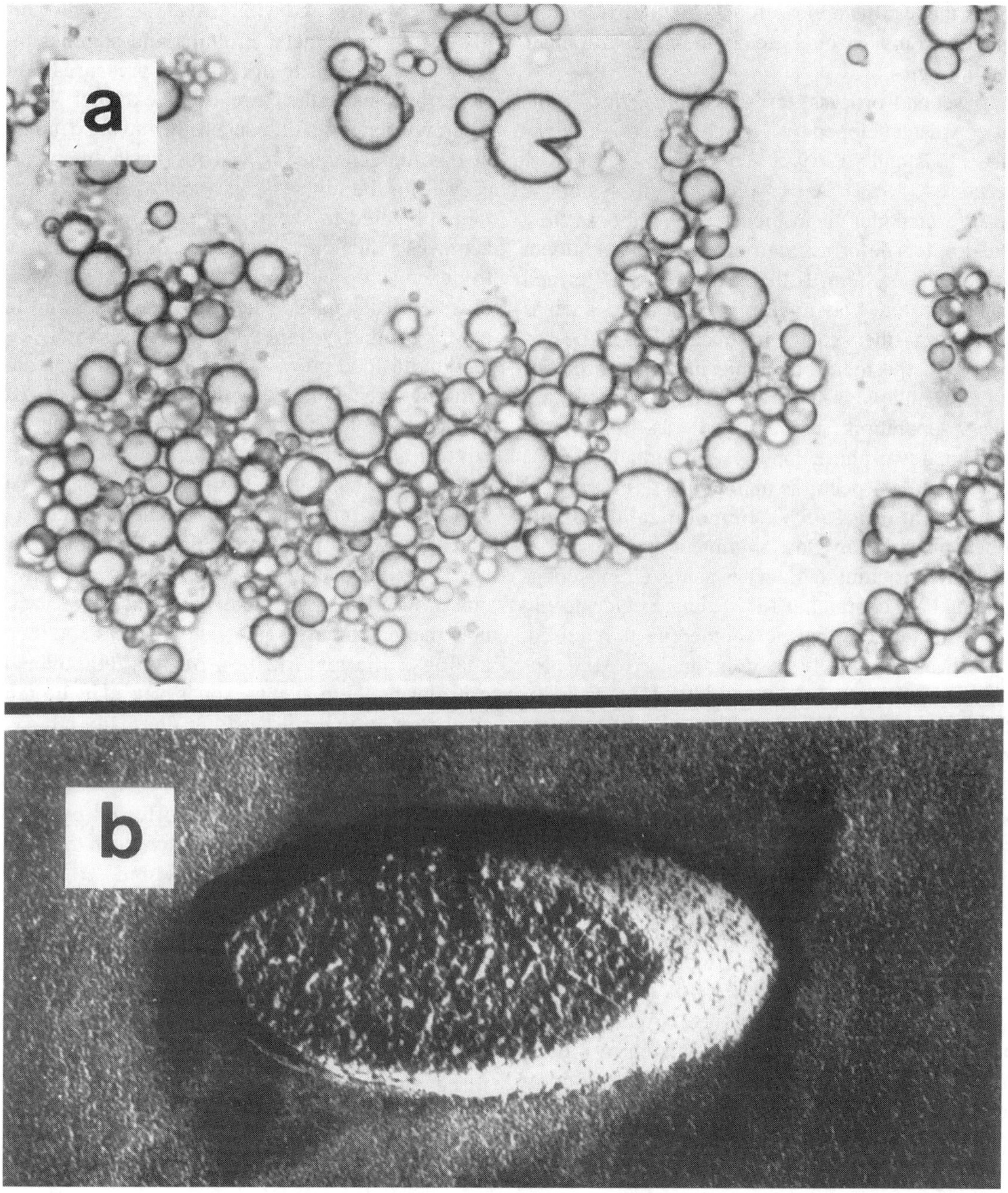


**Figure 5–22** Protein micelles. (a) Micelles of oat protein. Light micrograph showing size varieties as micelles coalesce (×320). (b) Micelle of fava bean. Transmission electron microscopy showing randomly packed, amorphous interior (×75,000). *Source:* Courtesy of Murray et al., 1981.

costs makes this approach to structural fiber formation from a casein base competitive with other technologies.

A second process, termed *spinneretless spinning,* was developed by a group of Russian scientists (Tolstoguzov, 1988; Tolstoguzov, Grinberg, & Gurov, 1985). A two-phase liquid system is made, and under the influence of an imposed flow the droplets deform and orient themselves into an anisotropic system. If the structures thus formed can be in some way immobilized or fixed, such as by gelling, fibers can be produced. The process is based on the thermodynamic incompatibility of proteins and polysaccharides and the concomitant phase separation (see Section 3.4.3). In a typical system, a two-phase dope system containing 20% casein and 2% pectin is maintained at pH 6.7 and a temperature of 45°C. Immobilization of the fibers produced by flow alignment through a nozzle onto a rotating cylinder is achieved in a coagulating bath containing 16% calcium chloride and 0.8% acetic acid, but no spinnerette is required. After washing, the fibers contained 26–31% protein and 0.3–0.5% calcium and had a pH value of 6.0. When the 0.2–1.0 mm diameter fibers were dried, their solubility was under 3.5% and they exhibited a water-holding capacity of 350%. They could be used at the 30% replacement level in meat products. Microscopic examination of the material showed that it was composed of oriented, parallel microfibers with a diameter of 0.1–0.3 $\mu$m. Many mixtures of proteins and polysaccharides may be used as starting materials, and various versions of the basic process have been developed. Several advantages are claimed for this technology over the classical spinning of protein fibers. Among these are that a wide variety of proteins can be used, mixtures of proteins can be prepared to optimize nutritional value and minimize cost, and the process is simpler and less expensive than wet spinning.

## 5.5 EXTRUSION AND SPINNING

### 5.5.1 Texturization of Proteins

The principles involved in cooking extrusion are basically the same as for the thermoplastic extrusion of polymers and will be explained as they apply to food polymers. Proteins and starches are subjected to high temperatures, pressures, and shear rates inside the barrel of the extruder, where a screw rotates at high speed. In most extruders, heat is autogenerated by viscous dissipation of energy from the high-viscosity polymer-water system subjected to shear. Once the food polymer achieves a rheological condition of flow in the extruder, macro- and microstructure are formed by diverse mechanisms. The use of extruders in the pet food industry started in the 1960s, rapidly extending to food processing. Today, extrusion and cooking-extrusion are widely used to texturize vegetable proteins, precook and form starches, in confectionery, and other applications.

A preliminary discussion of natural protein structures is found in Chapter 4. The restructuring of concentrated plant proteins by extrusion has been employed since the late 1960s as a commercial process for manufacturing plant-based products that can be used as meat extenders and meat analogs. The texturization of defatted soybean grits during thermal extrusion is caused by protein fiber formation resulting from thermally induced intermolecular cross-links. It must be remembered, however, that this is a multicomponent system, and extrusion is affected by other components, in this case, carbohydrates (soluble, insoluble, and fiber) and residual lipid. At the subcellular level, extruded soy meal is an aggregation of insoluble carbohydrates within a continuous protein matrix. While insoluble carbohydrates are not thought to play a major role in the development of texture or structural stabilization, the embedded, insoluble soy carbohydrates may reinforce weak hydrophobic interactions and engender additional stabilizing forces. The relatively low level of lipid remaining in the feed material after defatting provides some lubricating action in the extruder but does not seem to interfere significantly with fiber formation.

Of the restructuring systems examined in this chapter, thermal extrusion, although widely used, is, paradoxically, the least understood. The textural and physical characteristics of protein extrudates as well as the factors that dictate their utility to the consumer depend upon microstructure, which in turn stems from chemical interactions. Obviously, more information and understanding are required to increase our control of the extrusion process (Ledward & Mitchell, 1988; Ledward & Tester, 1994).

### 5.5.2 Wet Spinning of Protein Fibers

The *spinning* of protein fibers from an alkaline extract of soy isolate (>90% protein) to form meat analogs was described in the 1950s (Smith & Circle, 1972; Visser, 1988). The technology used in this process is quite similar to the manufacture of human-made textile fibers. Fibers are formed when the protein dispersion (or dope) is forced through a spinnerette containing many small holes (roughly 0.2 mm in diameter) into an acidic bath. Coagulation results as the pH is lowered toward the isoelectric point of the soy protein. The fibers are thin filaments (Figure 5–23) and composed predominantly of protein (Stanley, Cumming, & deMan, 1972). Research from several groups has demonstrated that intermolecular disulfide bonds are responsible for forming the structure of spun soy fibers (Aguilera & Stanley, 1986).

The fibers produced by spinning processing can be combined with fats, flavors, colors, and stabilizers to produce simulated meat. Spun protein fiber products are used by vegetarians, but they have not found wide consumer acceptance, perhaps because of the high cost associated with their production.

### 5.5.3 The Extrusion, Cooking, and Forming of Starches

Extrusion is also widely used in food processing to precook starchy flours and to produce expanded shapes that are then further processed into snacks and breakfast cereals. During extrusion, starches are subjected for short times (20–200 seconds) to high pressures (up to 7 MPa) at elevated temperatures (120–180°C) in the presence of mechanical shear. Depending on the moisture content, starches may undergo gelatinization, melting, and fragmentation, as described by Lai and Kokini (1991). Normally the onset of gelatinization of starches in excess water occurs between 55 and 70°C. Below 30% moisture, gelatinization is truly a water-assisted melting process (order-disorder transition) occurring at much higher temperatures and involving larger enthalpies. High pressure (e.g., 1.4 MPa) also increases the temperature of the transition, and two thermally induced enthalpy peaks are observed. Shear, on the other hand, results in fragmentation of the starch granule during extrusion, as demonstrated by microscopy, gel permeation chromatography, and viscosity measurements (Colonna, Tayeb, & Mercier, 1989; Gómez & Aguilera, 1984). Distinctive shapes, functional properties, and textures can be controlled by the die design, feed formulation, and operating conditions of the extruder.

Second-generation snacks are produced by direct expansion of the hot starchy dough exiting the die of the extruder. Once the molten starch emerges from the die, superheated steam under high pressure is flashed off and, together with the die-swelling effect, contributes to expansion and the porous structure of products. Third-generation snacks, however, are unexpanded half-products with moisture contents of 5–10% (below $T_g$). Expansion occurs later by frying, microwave heating, or oven heating at 160–190°C for 10–20 seconds. Final structure in this case is based on the transition from the amorphous to the rubbery and flow states of the starchy phase as temperature increases and water is transformed into steam bubbles encapsulated in a deformable matrix. From the fabrication standpoint, these products are closed-cell foams.

Gómez and Aguilera (1984) derived a model for microstructural events occurring during low-moisture extrusion of starch by comparing the physicochemical properties of raw, gelatinized, and dextrinized corn starch with those present after extrusion at varying moisture contents (Figure 5–24). In their study, they concluded that extruded corn products behave as a blend of gelatinized and dextrinized starch, with dextrinization becoming the predominant mechanism during low-moisture, high-shear extrusion. Extruded

**Figure 5–23** Scanning electron micrograph of spun soy fibers. (a) Low magnification (×235). (b) High magnification (×1150). *Source:* Stanley et al., 1972.

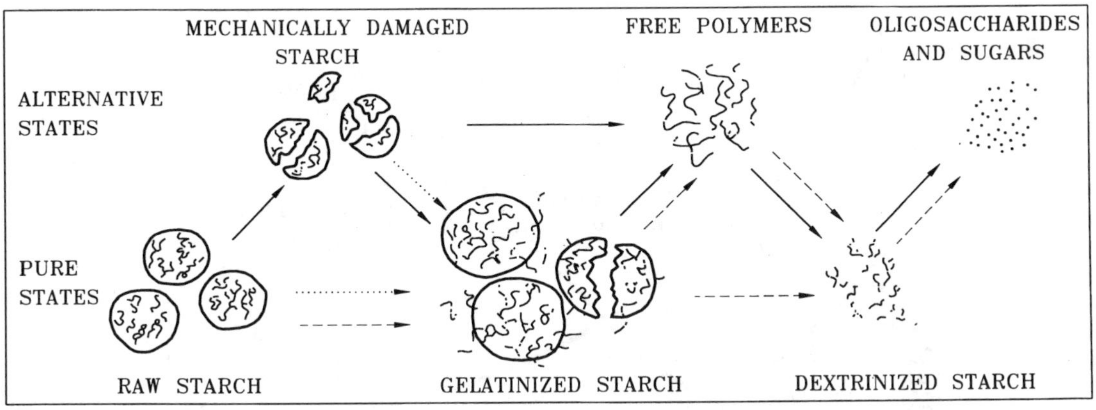


**Figure 5–24** Proposed model of changes in the starch granule during extrusion cooking. The effects of mechanical shear, heat, and moisture transform the native starch granule into degraded forms ranging from gelatinized starch to dextrinized material. *Source:* Reprinted with permission from M. Gómez and J.M. Aguilera, A Physicochemical Model for Extrusion of Corn Starch, *Journal of Food Science*, Vol. 49, pp. 40–43, 63, © 1984, Institute of Food Technologists.

corn products have high water solubility and low water absorption and yield lower viscosity than gelatinized starches. Fragmentation has also been correlated with the specific mechanical energy (kJ/kg) transferred to starch during extrusion, which increases as moisture content decreases.

### 5.5.4 The Use of State Diagrams in Extrusion

Proteins may exist in many physical states, and which particular state is obtained depends on moisture and temperature (Section 3.8.8). State diagrams depict the states of a system under specific conditions (generally temperature, time, and concentration) during processing. Transitions between states and chemical reactions can be identified and characterized using analytical methods such as differential scanning calorimetry, small-amplitude oscillatory rheometry, dilatometry, and dielectric constant measurement (Kokini, Cocero, Madeka, & de Graaf, 1994). The use of state diagrams was introduced by Slade and Levine (1991) to assess transformations in starch during baking, cooking, and puffing as a function of moisture and temperature.

Figure 5–25 shows a hypothetical state diagram for vegetable proteins during extrusion cooking as they undergo wetting and heating inside the extruder barrel, exit through the die, and expand and cool into a glassy product (Kokini et al., 1994). The two relevant lines in this state diagram are the glass transition curve and the rubber to free-flow transition line (determined by pressure rheometry). As the flour is wetted and warmed inside the barrel, the protein mass changes from glassy to rubbery. Further heating in the front part of the extruder induces a transition to the free-flow region, where the protein moves as a continuous mass (melt) with further increase in temperature (e.g., above 50°C) and pressure. At even higher temperatures (>70°C), several reactions take place (e.g., unfolding of the protein) that affect the rheological properties of the mass. It is believed that even higher temperatures (>120°C) are needed to favor intermolecular arrangements and induce "texturization." As the product exits the die, there is a fast release of steam and simultaneous expansion of the matrix. Evaporation-cooling brings the temperature and moisture to the lower left-hand corner of the diagram, where glassy conditions prevail again. Little, Aguilera, Morales, and Kokini (1997) used pressure rheometry to discover the conditions that optimize structure formation during texturization of soy proteins

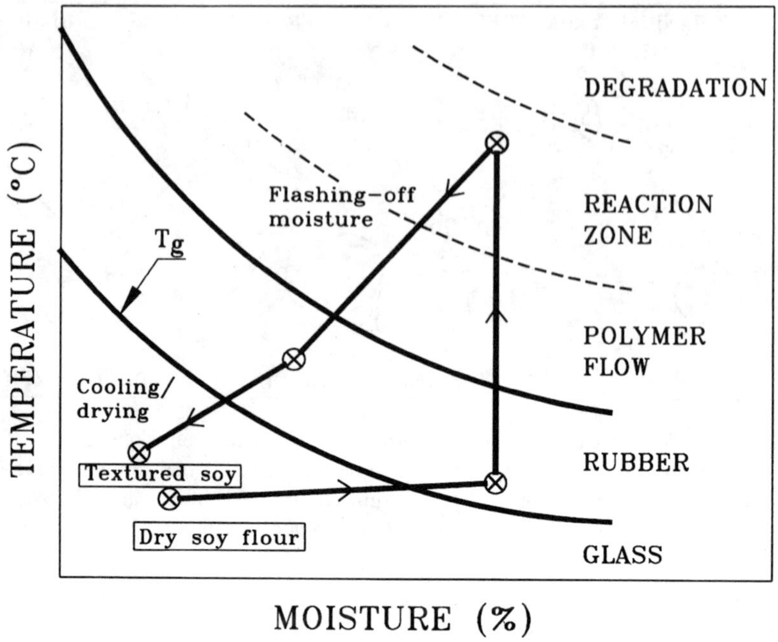


**Figure 5–25** State diagram showing transformations of proteins during wetting, heating, cooling, and drying stages of extrusion cooking. *Source:* Reprinted from *Trends in Food Science Technology*, Vol. 5, J.L. Kokini, A.M. Cocero, H. Madeka, and E. de Graaf. The Development of State Diagrams, pp. 281–288, Copyright 1994, with permission from Elsevier Science.

(measured as the increase in G′). Practical experience suggests that around 180°C is an optimum temperature for reactions leading to texturization, and beyond this temperature the modulus starts to decay.

## 5.6 STRUCTURING FAT PRODUCTS

### 5.6.1 Margarine

In the late 1860s, the food technologist Hyppolite Mège-Mouries was commissioned by the French navy to find a substitute for butter. He extracted some of the soft fat from animal byproducts, mixed it with alkaline water, and added some chopped cow's udder and milk for flavor, thus creating margarine, probably the first substitute food with commercial applications. Napoleon III awarded Mège-Mouries a prize, and a factory began commercial production in 1873, not without strong opposition from the producers of real butter (Tannahill, 1988). Butter and its 19th-century analog, margarine, contain 80% fat and contribute significant quantities of lipids to the human diet. Low-calorie products are important targets for the food industry in its effort to reduce the total intake of calories from fats in the diet.

Although margarine is also defined as a W/O emulsion, its microstructure is far more complex than thought and different from that of butter. In butter, a limited number of fat globules are present in the final product (most of them are destroyed during the intensive working). An interglobular phase is formed by a mixture of liquid oil, crystal aggregates, and membrane residues (Juriaanse & Heertje, 1988).

In margarine, crystallized fat forms a fine network of interconnected platelets composed of single crystals and sheetlike crystal aggregates that serve as "containers" for an emulsion of water in liquid oil (Heertje, 1993). Butter in turn, has a discontinuous structure composed mostly of fat globules. It is possible to make margarine that has this basic structure and contains up to 80% water, but it would be unsuitable for baking or frying. The next section describes an industrial development to make low-fat or diet margarine by using several physicochemical concepts and their structuring capabilities.

### 5.6.2 Low-Fat Spreads

Making margarine with lower fat content (e.g., 5%) requires a totally different approach, one in which water is structured rather than fat. The manufacture of low-fat margarine is described in a Unilever publication (Zeelenberg-Miltenburg, 1995). Structure development starts when surface-active agents or emulsifiers such as lecithins and monoglycerides form bilayers in response to exposure to an aqueous phase. These amphiphilic molecules, when dispersed in water, display a specific aggregation pattern that depends on their geometry. In particular, the lamellar assembly is composed of bilayers of amphiphilic molecules with alternating water layers that can be quite thick—thick enough for the total water content to be around 50%. Substances possessing these types of lamellar structures are sometimes referred to as "liquid-crystalline phases," and, depending on the surfactant used, they can be reasonably rigid at room temperature (see Section 3.5.10). Lamellae then become the rigid network structure provided in margarine by solid fat.

In order to include even more water, a second electrically charged surface-active agent is placed so as to "stick out" from the bilayers and thus cause an electric repulsion that further increases the gap between the bilayers. Finally, additional water structuring and spreadability are achieved by immobilizing free water through the formation of gelatine microgels. The structural components of this type of low-fat spread are shown in Figure 5–26.

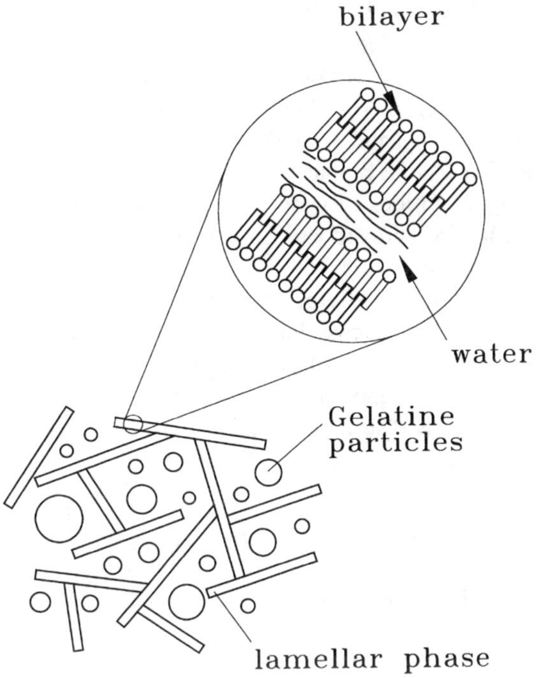


**Figure 5–26** Scheme showing elements contributing to the final structure of low-fat spreadable products.

### 5.6.3 Fat Replacers

Low-fat and fat-free products are given high priority in the food industry. Obvious ways of reducing the fat content of these products include removing fat through processing and formulation and adding more water. However, the removal or reduction of fat adversely affects both flavor and texture as well as the availability of fat-soluble vitamins. The term *fat mimetics* is used to characterize products that mimic the creamy mouthfeel and creaminess of fat. Fat replacers can be divided into those that are protein based, those that are carbohydrate based, and those that are fat based (Lucca & Tepper, 1994). Many commercial fat replacers are actually combinations of two or more ingredients.

Proteins may be induced to form small soluble aggregates or weak gels providing "body" and a soft texture. Microparticulated protein is produced by applying high shear after the protein has been coagulated and structured into a gel, thus forming small particles 0.1–2.0 $\mu$m in diameter. Dispersions of such small particles are perceived as a creamy and smooth fluid. Polysaccharides perform as fat replacers mainly by immobilizing large quantities of water in a gel-like matrix, resulting in lubricant and flow properties similar to those of fat. Starches, maltodextrins and dextrins, polydextrose, cellulose gel, and gums are often used as fat replacers. Fat-based replacers perform by sparing the use of fat (emulsifiers), being less absorbable or more inefficiently metabolized than natural fats ("structured lipids"), or simply being resistant to hydrolysis by digestive enzymes.

## 5.7 STRUCTURE AND STABILITY

### 5.7.1 Looking at Nature Again

Since food technologists often deal with intact biological systems (e.g., fruits, vegetables, and muscle tissue), their methods of providing stability usually involve control of external variables: low temperature, modified or controlled atmospheres, use of preservatives, and so on. Where permitted, formulations might be changed to include less reactive reactants (e.g., nonbrowning precursors), more inhibitors of deleterious reactions (e.g., antioxidants), and control of the aqueous media (e.g., pH). Other methods of preservation involve severe changes to the food structure such as heating (inactivation), removing water (drying and freezing), and addition of solutes to control water activity. Nevertheless, there is ample consciousness among food professionals that destroying the structure of natural foods results in rapid and uncontrolled deterioration. In the case of fabricated foods, there is an opportunity to engineer structures that will increase the stability of a food. Phase separation and segregation, induction of a glassy state, encapsulation, and formation of artificial membranes or coatings are but a few of the alternatives.

Nature has less flexibility but more wisdom in stabilizing structures against unwanted reactions. It does so mainly by three means:

- *Complexing key reactants into passive forms.* This mechanism operates at the molecular level. Typical examples include some enzymes that are maintained inactive until required to participate in reactions.
- *Restricting the mobility of the reactants.* The best example here is "anhydrobiosis," the maintenance of life in seeds and microorganisms by immobilization of the system and protection of key reactants (DNA and proteins) and organelles (membranes) under desiccation conditions (e.g., $<5\%$ moisture). Stability is provided by a diffusion-limited condition in which the system is "frozen" at ambient temperatures by the existence of a glassy state and the presence of compatible solutes (Aguilera & Karel, 1997).
- *Compartmentalizing and communicating.* Unlike a bacterium, which typically consists of a single compartment surrounded by a plasma membrane, a eukaryotic cell is divided into functionally distinct, membrane-bound compartments or organelles. The main compartments in a cell are (1) the nucleus and the cytosol (6% and 54% of total volume, respectively), which communicate through the nuclear pores; (2) the mitochondria and (in plants) the chloroplast; and (3) peroxisomes and other membrane-bound organelles (ER, Golgi apparatus, endosomes, and lysosomes), whose interiors communicate with one another and with the outside of the cell via transport vesicles.

Mention should be made of the enhanced activity of some enzymes in nonaqueous media, which may seem a "natural" oddity. It has been demonstrated that the water bound to an enzyme determines its stability and the catalytic activity, and thus as long as this "conformational water" is present its catalytic activity may not be impaired and the kinetics of organic solvents are favored by other mechanisms (Bell, Halling, Moore, Partridge, & Rees, 1995). Another interesting phenomenon is "molecularly imprinting" of enzymes with a ligand in anhydrous media (or by lyophilization) and the maintainance of "molecular memory," which also results in superior catalytic activity (Klibanov, 1995). Here we are discussing "nanotechnologies" that may be used to advantage when structuring a new food from its parts.

### 5.7.2 Freezing in Nature

The formation of ice within a tissue is potentially detrimental to its survival. The rigid ice lattice structure may penetrate cell walls and cell membranes beyond the point of reparation by normal cell processes. It is generally accepted that intracellular ice formation is lethal, and most plants can acclimate to cold weather by forming ice outside the cells as temperature drops. Extracellular ice is partially benign but acts as a nucleating site for ice crystal growth promoted by vapor drawn out of the cell, thereby desiccating the cell and causing shrinkage. As long as the cell wall remains attached to the plasma membrane, ice-driven desiccation does not involve cellular damage (Andrews, 1996). Otherwise, lethality from freezing injury is really caused by dehydration.

Insects survive subfreezing temperatures through preventing ice formation by a variety of mechanisms (Ring & Danks, 1994). First, they reduce water content so that a large proportion of water becomes "bound" and thus unfreezable by concentration effects or association to structural components. Second, they employ supercooling (even to $-20$ to $-40°C$); that is, they remain unfrozen below their thermodynamic freezing points. For an insect to achieve supercooling, all nucleators that provide a surface on which ice crystals growth can be initiated have to be masked or removed. Third, they depress the freezing and supercooling points of body fluids by accumulating low molecular weight solutes such as polyhydric alcohols (glycerol, mannitol, sorbitol, threitol, erythritol, and others) and some sugars (most notably trehalose). Fourth, they have molecules that cause specific noncolligative effects and increase cold hardiness. Finally, they have large cryoprotective proteins (polypeptide or lipoprotein molecules) that appear to prevent ice growth at any ice-water interface. Interestingly enough, Arctic insects are freezing-tolerant, that is, they survive extracellular freezing of the body by manufacturing ice nucleators that cause ice to form at relatively high subfreezing temperatures. This adaptation mechanism prevents rapid and injurious ice crystal growth from taking place at high supercooling.

At the molecular level, *in vivo* proteins, nucleic acids, and membranes contain water (about 0.25 g/g DW) that does not freeze readily (e.g., does not present a freezing exotherm on cooling). Removing this water results in profound changes in the biophysical properties of biomolecules such as proteins and lipid bilayers. Dehydration (i.e., desiccation) usually involves removal of the unfreezable water whereas freezing does not; hence, the two protective mechanisms are based on different principles. Any solute that is preferentially excluded from the hydration shell is a *cryoprotectant*. By contrast, stabilization of proteins against dehydration requires direct interaction between the solute and the biomolecule (water replacement hypothesis). Readers who are interested in looking to nature for new insights regarding food preservation and quality may benefit from reading Aguilera and Karel (1997); Crowe, Carpenter, Crowe, and Anchordoguy (1990); Potts (1994); and Ring and Danks (1994).

### 5.7.3 Stability of Frozen Foods

A good illustration of the relationship between structure and stability is provided by frozen foods. Freezing is one of the most significant phase changes that water undergoes and has important consequences in food processing and also in microscopy, as mentioned in the discussion of cryomicroscopy (Section 1.5.4). For the transition from water to ice to occur, water molecules must link together through persistent hydrogen bonds to form tetrahedral frameworks that can then be arranged into a variety of latticework structures. The most common form of ice consists of the orderly array of hexagonal cells.

As the temperature is lowered during the freezing of foods, the molecular motion of the water molecules is slowed. Ice formed from pure water goes through a two-step (nucleation and propagation) crystallization process. Solid particulates act as nuclei for water clusters upon which ice crystals can grow. Hexagonal ice crystals will form as these nuclei increase in size. The final size of the ice crystals, of great importance in frozen foods, is determined by the rate of cooling. Slow rates of cooling result in a small number of large crystals, causing ruptured cells and reducing quality, while faster cooling yields many small ice crystals, a less harmful configuration.

As temperature decreases and water is "removed" from a food in the form of ice, the solutes present in the unfrozen fraction are freeze concentrated. This equilibrium thermodynamic process can be modeled in a phase diagram as an equilibrium freezing (liquidus) curve that extends from the melting temperature ($T_m$) of pure water (0°C) to the eutectic temperature ($T_e$) of the solute, the point at which the solute has been freeze-concentrated to its saturation concentration (Figure 5–11). As the temperature is lowered in a food, the solute(s) will likely not crystallize at $T_e$ because of the high viscosity resulting from concentration of the solute and the low temperature; consequently, the freeze-concentration proceeds beyond $T_e$ in a nonequilibrium state. The highly concentrated unfrozen fraction can then go through a viscous liquid-glass state transition, driven by the reduction in molecular motion and diffusion kinetics resulting from the very high concentration and low temperature. Thus, the unfrozen liquid changes from a viscoelastic rubber to a brittle amorphous glass.

Below $T_g$, the water molecules are capable of vibrational movement only, which explains the high viscosity. The glass transition curve extends from the $T_g$ of pure amorphous water (−134°C) to the $T_g$ of pure solute (in the case of sucrose, 52°C). The equilibrium phase diagram and the kinetically derived state diagram can be modeled together on a supplemented state diagram (Figure 5–11). At $T_g'$, the maximally freeze-concentrated $T_g$ of the frozen system, the unfrozen water is not "bound" in an energetic sense; rather, it is unable to freeze within practical time frames. At $T_g'$, the supersaturated solute takes on solid properties because of reduced molecular motion, which is responsible for the tremendous reduction in translational, not rotational, mobility. Thus, the diffusion of a water molecule in such a system is reduced to around 1 cm/3 × $10^5$ yr! It is the intrinsic slowness of molecular reorganization below $T_g'$ that the food scientist seeks to create within the concentrated phase surrounding constituents of frozen food materials. However, warming from the glassy state to temperatures above $T_g'$ results in a tremendous increase in diffusion rates, not only because of the effects of the amorphous-to-viscous liquid transition but also because of increased dilution, for the melting of small ice crystals occurs almost simultaneously. The time scale of molecular rearrangement continually changes as the glass transition is approached, so that food scientists can also cryostabilize products at temperatures above $T_g'$ by minimizing $\Delta T'$ between the storage temperature and $T_g'$, either by reducing the storage temperature or enhancing $T_g'$ through freezing methods or formulation. Hence, knowledge of the glass transition provides a clear indication of molecular diffusion and reactivity and therefore shelf stability.

Cryostabilization of frozen foods is based on the idea that molecular mobility is the overriding mechanism in achieving stability (Slade & Levine, 1991) and that the quality of frozen foods will deteriorate if unfrozen water is allowed to migrate to existing ice crystals and recrystallize. Also, solute mobility and various chemical or enzymatic reactions in the unfrozen phase lead to a loss of quality. As foods freeze, pure water is separated from the surrounding milieu by ice crystal formation. However, the temperature of commercial food freezing equipment (normally −18°C) is not low enough to prevent the formation of an unfrozen aqueous phase containing freeze-concentrated solutes (e.g., −25°C). Within this phase, deleterious reactions, such as salt-induced protein denaturation, can occur that are controlled by diffusion, and thus temperature. As well, defects can occur in the microstructure of tissue foods during freezing as a result of ice crystallization that damages cells, releasing water and leading to sublimation or "freezer burn" during storage and excessive "drip" during thawing. Even when foods are in the frozen state, ice crystals can recrystallize to form large, sensorially detectable particles, further impairing quality (Goff, 1996). In food processing, it may be useful to add polysaccharide stabilizers (hydrocolloids or gums); these are thought to slow the diffusion process (mobility) and thus inhibit recrystallization.

The freezing process in real foods involves many steps. Because of the complexity of the process, researchers have turned to the study of model systems, the most common of which are aqueous carbohydrate solutions. These have been used extensively as model systems because in most the low molecular weight carbohydrates, the sugars, dictate the unfrozen water:ice ratio and the high molecular weight carbohydrates, the polysaccharides, have the greatest impact on ice recrystallization. Model systems are meant to mimic real freezing behavior yet only use those components that most influence the freezing process. Also, it is relatively easy to determine the $T_g$ of these systems. Determining the $T_g$ is especially important because at temperatures equal to or greater than $T_g$, ice recrystallization may occur, since the material is in the so-called rubbery phase, which is characterized by much more rapid diffusion.

Commercially, care is given to the freezing process. The goals include the creation of small ice crystal size to minimize tissue disruption and the development of the maximal freeze concentration to obtain the highest $T_g'$ possible, thus slowing ice recrystallization and promoting stability.

One might think that freezing temperatures should be as low as possible, but this is not always the case. Carrington, Goff, and Stanley (1996) investigated the structure and stability of rapidly frozen and slowly frozen carbohydrate solutions using cryo-SEM to examine structure and differential scanning calorimetry (DSC) to measure $T_g$. Solutions of 30% aqueous fructose ($T_g \approx -50°C$) with and without a stabilizer (CMC) were studied. The frozen samples were stored at either low temperatures ($-75°C$, $< T_g$) or high temperatures ($-25°C$, $> T_g$) and then examined for recrystallization by sublimating a freshly fractured plane to remove the exterior ice. Figure 5–27 shows the structure of the amorphous glass and crystalline phases in samples frozen rapidly without CMC. In this case, rapid freezing trapped in the glassy state a considerable quantity of unfrozen solution not maximally freeze-concentrated. The influence of the stabilizer and storage temperature may be seen in Figure 5–28. In all samples, the initial amorphous matrix had undergone a glass-rubber transition, and ice recrystallization had occurred. However, the extent of recrystallization varied considerably.

The use of CMC (Figure 5–28, part B) enhanced ice nucleation and/or reduced ice crystallization, since more but smaller ice crystals were found. This stabilizer may function by acting as a catalytic site for nucleation or by enhancing viscosity in the unfrozen phase to limit diffusion of water to the surface of a growing ice crystal. The second mechanism is more likely, for the determination of $T_g$ values by DSC indicated a statistically higher result for slowly frozen samples containing CMC ($T_g = -54.4°C$ versus $-48.1°C$), suggesting that the stabilizer enhanced viscosity and hindered diffusion and thus made it more difficult to achieve maximum freeze-concentration. Samples stored at higher temperatures underwent enormous recrystallization (Figure 5–28, part C) and had formed ice crystals about 25 times larger than those seen previously; again, the samples containing CMC had smaller crystals (Part D).

Freezing rate also had a significant effect on this system: solutions that were frozen slowly (Figure 5–29, part B) exhibited crystals several times larger than those produced by rapid freezing (Part A). However, remember that the rapidly frozen solution initially formed an amorphous matrix with very small crystals, but the samples in Figure 5–29 were stored at $-25°C$ and underwent extensive recrystallization. Specifically, the unfrozen water trapped in the fructose glass during freezing recrystallized as the sample was warmed above $T_g$. Thus, the initial freezing process, no matter to how low a temperature, failed to yield a matrix that could resist recrystallization at storage temperatures greater than $T_g$. On the other hand, the slow freezing conditions approached maximum ice crystallization at $-25°C$, and the slowly frozen sample did not change much during storage.

In this study, the DSC and cryo-SEM results showed that the presence of large amounts of an amorphous matrix produced by rapid freezing does not ensure increased stability if the material is subsequently stored at temperatures above $T_g$, since extensive ice recrystallization can occur. Thus, even if it were economically feasible to freeze foods cryogenically, storage and distribution at temperatures above $T_g$ would cancel initial benefit because of the development of coarse ice crystals. Successful commercial freezing of food requires as complete a crystallization as possible with as small a crystal size as possible, but time (i.e., conventional versus cryogenic freezing rates) must be allowed for the crystallization process to occur, a process that would undoubtedly take longer in a real food than in the model systems used in this research.

### 5.7.4 Control of Lipid Oxidation

Reference has been made to the formation of interfacial layers with surfactants or proteins to stabilize emulsions. These same "membranes" may also protect lipids inside droplets by acting as a barrier against the diffusion of molecules that initiates the oxidation reaction. The barrier effect, however, may be highly specific. For example, ascorbic acid is around 1,000 times less effective as an antioxidant when the lipid is contained inside negatively charged micelles and EDTA performs as a "pro-oxidant" than in the presence of other oxidizing agents. Because of the ability of

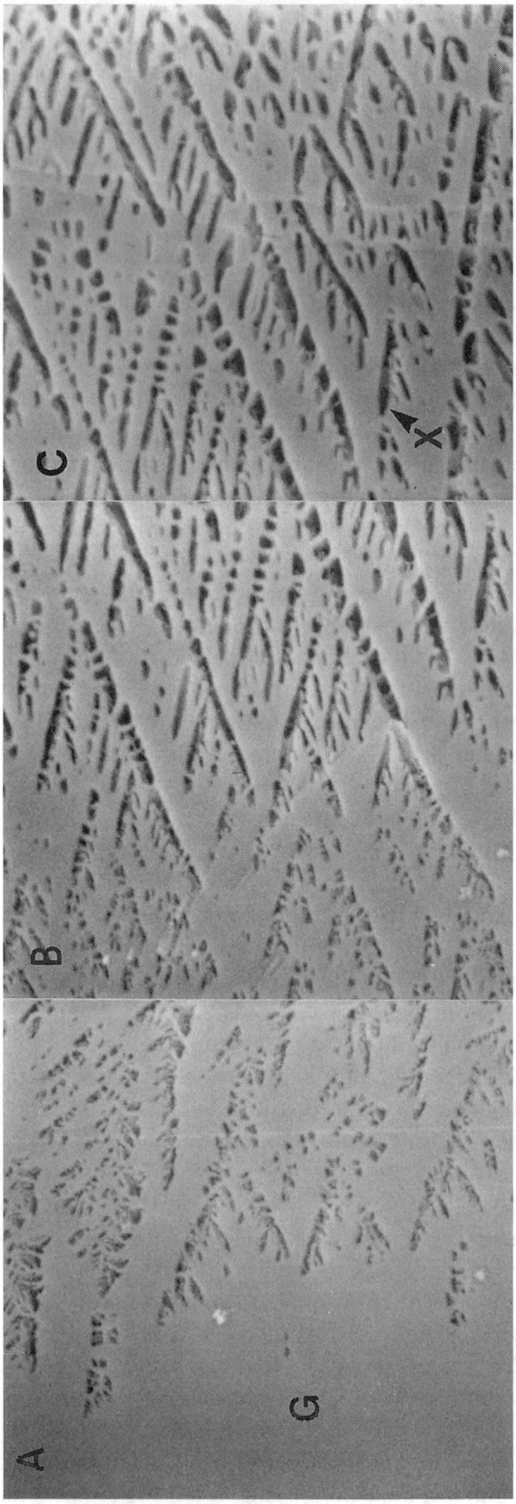


**Figure 5–27** Cryo-SEM micrographs showing the surface of amorphous glass (G) and crystalline (X) structures of 30% fructose rapidly frozen in Freon 22™. The sequence (A→C) proceeds from the outermost edge of the sample toward the center. Bar = 6 $\mu$m. *Source:* Reprinted from *Food Research International*, Vol. 29, A.K. Carrington, H.D. Goff, and D.W. Stanley, Structure and Stability of the Glassy State in Rapidly and Slowly Cooled Carbohydrate Solutions, pp. 207–213. Copyright 1996, with permission from Elsevier Science.

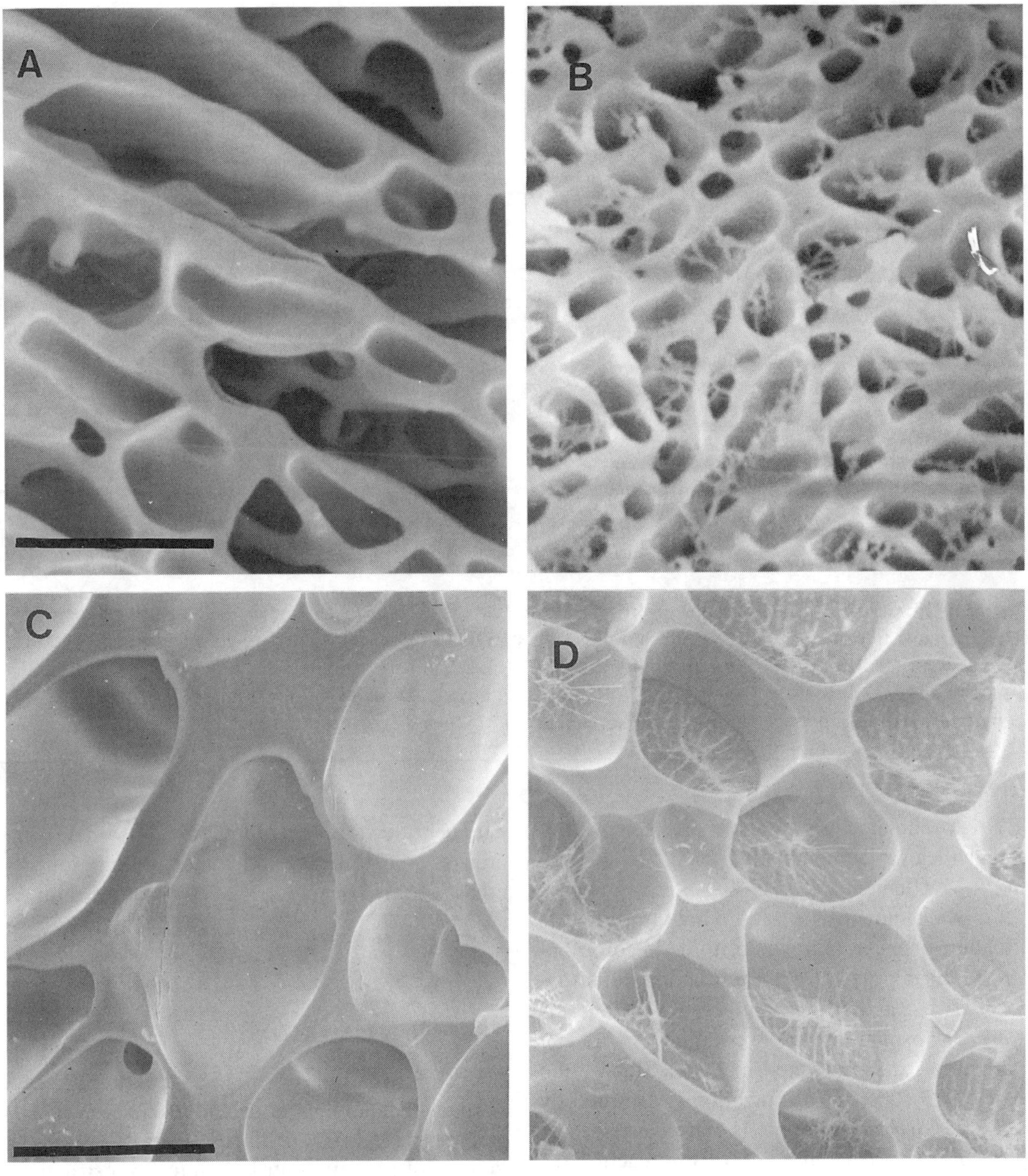


**Figure 5–28** Cryo-SEM micrographs. (A) 30% fructose. (B) 30% fructose + 1% carboxymethyl cellulose (CMC) after rapid freezing in Freon 22™ and storage of the samples for 2 weeks at −75°C (edge sections, bar = 3 μm). (C) 30% fructose. (D) 30% fructose + 1% CMC after rapid freezing in Freon 22™ and storage of the samples for 2 weeks at −25°C. Edge sections, bar = 30 μm. *Source:* Reprinted from *Food Research International*, Vol. 29, A.K. Carrington, H.D. Goff, and D.W. Stanley, Structure and Stability of the Glassy State in Rapidly and Slowly Cooled Carbohydrate Solutions, pp. 207–213, Copyright 1996, with permission from Elsevier Science.

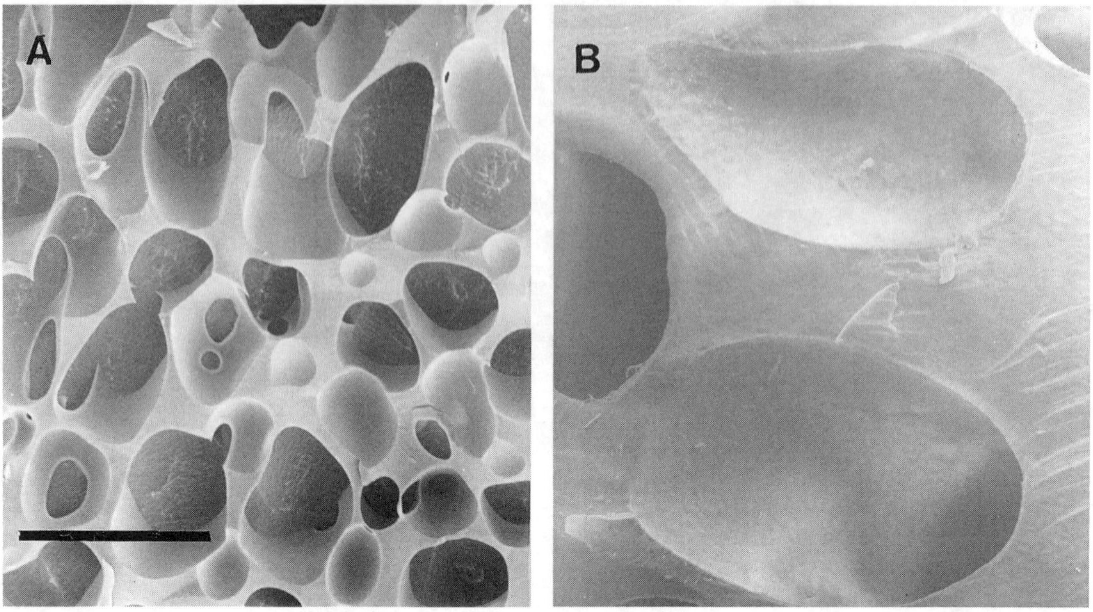


**Figure 5–29** Cryo-SEM micrographs of 30% fructose + 0.25% CMC. (A) After rapid freezing in Freon 22™. (B) After slow freezing at −18°C. Both samples were stored at −25°C for 3 days. Edge sections, bar = 60 μm. *Source:* Reprinted from *Food Research International*, Vol. 29, A.K. Carrington, H.D. Goff, and D.W. Stanley, Structure and Stability of the Glassy State in Rapidly and Slowly Cooled Carbohydrate Solutions, pp. 207–213, Copyright 1996, with permission from Elsevier Science.

EDTA to chelate iron catalysts, its penetration through the droplet membrane is facilitated (Coupland & McClements, 1996).

It is well known that dried whole milk resists oxidation because oxidizable lipids become entrapped during drying within an amorphous matrix of lactose and protein. In fact, this is the basis of dry encapsulation, widely used for flavors and pharmaceuticals. As stated in Karel, Buera, and Levy (1993), increased protection is afforded by the formation of an amorphous matrix and storage below $T_g$. Above $T_g$, the rate of the reaction increases more rapidly as $(T - T_g)$ increases. The variable to control is moisture content and its plasticizing effect, which impacts not only the diffusivity of reactants but also the amorphous-crystalline transition of the matrix (see Section 3.5.6). This latter phenomenon has been studied as it occurs in amorphous lactose, which crystallizes readily at 25°C when $a_w > 0.37$. As the difference between the storage temperature and $T_g$ ($\Delta T$) increases, the time needed for crystallization decreases from 5 years ($\Delta T = 5$°C) to a few minutes ($\Delta T = 40$°C).

## 5.8 GELS

### 5.8.1 What Is a Gel?

There are two kinds of solidlike structures present in foods that contain large amounts of water: cells and gels. Here gels are discussed, mostly from the viewpoint of their contribution to form structures (see Section 4.6 for a discussion of cells). A gel has been defined as "a form of matter intermediate between a solid and a liquid" (Tanaka, 1981). Gels are "soft solids" and ubiquitously present in high-moisture processed foods from all origins (Aguilera, 1992): jellies, jam, confectionery and dairy products (yogurt), processed meats (frankfurters), and fish (surimi), among others.

Gels have the ability to structure water into semisolid structures, which obviously has immense technological importance in food processing. In a sense, gels are a form of "solid water" at room temperature. For example, gelled agar (the material used to plate microorganisms in the laboratory) may contain as much as 998 parts water and only 2 parts polymer and yet stand up against gravity, at least for a time. Gels exhibit viscoelastic behavior and a moderate modulus (e.g., $10^6$–$10^8$ Pa).

Gels are formed from a polymer solution (sol) by diverse mechanisms (which will be explained later). A continuous network of interconnected material (molecules or aggregates) spanning the whole volume becomes swollen with a high proportion of liquid. Network-forming food polymers are treated in Section 3.3 as part of a group of compounds generically called "hydrocolloids." Swelling or uptake of a liquid by a gel is effected in the interstices (pores) by affinity due to interaction energy and polymer entropy (Tanaka, 1981). The reverse phenomenon, *syneresis*, is the expulsion of liquid from the gel and is generally regarded as a defect in food gels. Gels in which the liquid phase is an aqueous solution are called *hydrogels* and those where it has been removed are called *aerogels*.

A short introduction to food gels is provided by Walstra (1996). Clark and Ross-Murphy (1987) have written an excellent review of the structural techniques, mechanical characterization, and properties of biopolymer gels. Gelation mechanisms and theories and the mechanical properties of gels are discussed by Clark (1992). Specialized books on food gels and commercial applications include those edited by Harris (1990) and by Imeson (1992) and the book series *Gums and Stabilizers for the Food Industry*. The journal *Polymer Gels and Networks* is also useful.

Gels can be concocted but they also occur in nature, where their unusual properties are often exploited. Hidden in the secretory mechanism of many cells are gels that can swell or shrink in milliseconds in response to certain ions. During the past decade, researchers around the world have developed synthetic gels ("intelligent gels") that absorb or expel water in response to temperature, pH, or electric fields (none of these gels can yet be eaten). An interesting article on futuristic applications of gels in biomimetics was written by Osada and Ross-Murphy (1993). Food scientists should be alert to future developments in this field.

### 5.8.2 Classification of Gels

Gels are usually classified according to the cross-linking mechanisms intervening in formation of the polymer network and to the type of network structure. Cross-links in gels may be strong covalent bonds, found mostly in synthetic networks. In foods, junction zones are formed by weaker bonds such as electrostatic, ion-bridging, hydrophobic, and hydrogen bonds and those derived from van der Waals' forces. Physical cross-links may also be entanglements providing temporal barriers to chain movement.

Djabourov (1991) classified gels according to their formation mechanisms and came up with four main types:

- Fishing nets or branched three-dimensional networks built from linear flexible chains linked by covalent bonds. Usually these are of a synthetic nature and exhibit a rubbery consistency (e.g., acrylamide gels).
- Thermoreversible physical gels formed by partial crystallization of chains or by conformational coil-to-helix transitions. These substances switch from sol to gel and back again upon temperature changes. Their consistency varies widely, ranging from soft and highly deformable (e.g., gelatin) to hard and brittle (e.g., agarose gels).
- Egg-box structures formed by junction zones linked by ionic complexation, in which a divalent cation (e.g., $Ca^{++}$) bridges two strands of the polymer. Examples include alginate and pectin gels.
- Particle or colloidal gels consisting of strands of more or less spherical aggregates ordered into a string-of-beads or cluster arrangement. Casein and whey proteins form particle gels (see Section 7.4.2).

Other classificatory schemes group gels by the composition of the network or phases present (single or mixed), the role of heat in their formation (thermotropic) or stability (thermoreversible), the transmittance of light (translucent or opaque), and so on.

## 5.9 GELATION MECHANISMS

Following is a brief introduction to the proposed mechanisms by which food polymers form gels. The reader should refer to the publications cited in the text for details. The junction zones and supramolecular structures of some gels discussed below are represented in Figure 5–30.

### 5.9.1 Alginates

Alginates form heat-stable, irreversible gels in cold systems. In food applications, it is primarily the $Ca^{2+}$–mediated mechanism of gelation that is important (although acidic gels are also formed at pH $< 4$). Regions of polyguluronic acid are linked to similar regions in another polymer chain by calcium ions in a so-called *eggbox* structure (Figure 5–30, part C). Alginates gels have been used in the production of fruit analogs using

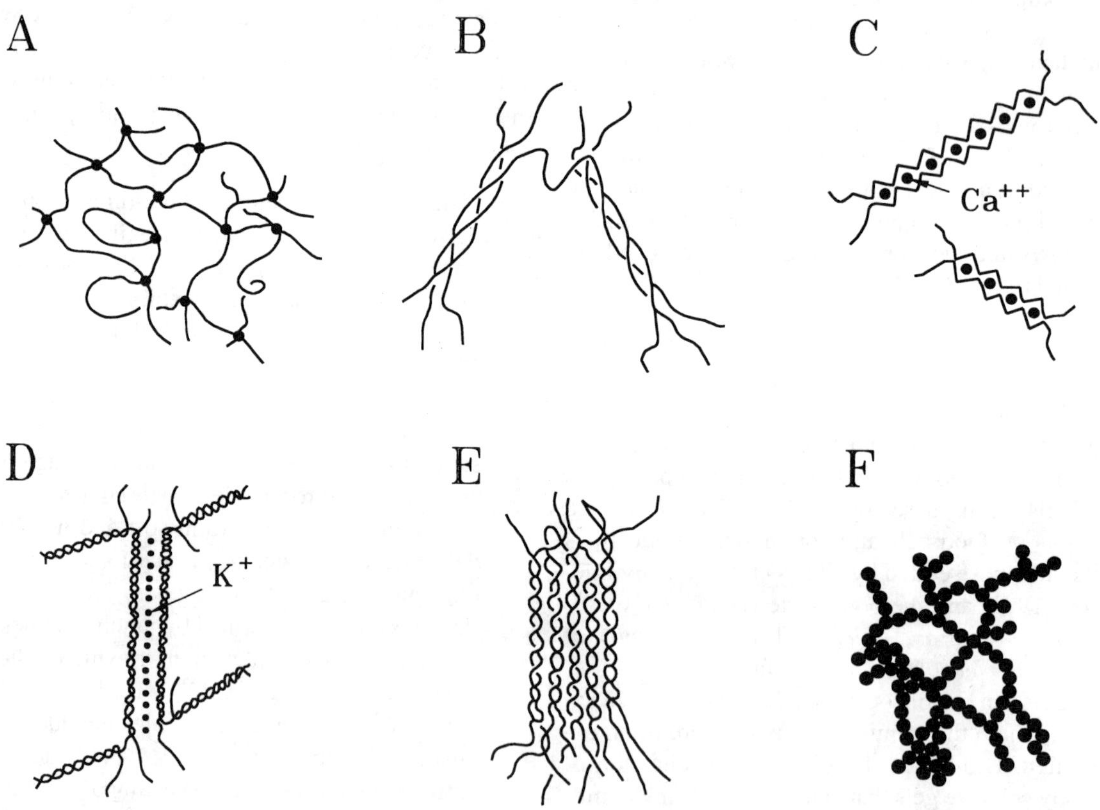


**Figure 5–30** Schematic representation of some supramolecular structures of pure gels. (A) Cross-linked or fishing net type (chemical gel). (B) Triple-helices of gelatin gels. (C) Egg-box structures of pectin and alginate gels in the presence of calcium ions. (D) Aggregated domains after carrageenan gelation. (E) Bundles of double helices in agarose gels. (F) Particulate gels formed by globular proteins. (Not at the same scale.)

fruit purees. Details on the properties of alginate gels can be found in Onsoyen (1992) and Sime (1990).

### 5.9.2 Agar

The process of network formation in agar gels is quite unlike that of alginates and does not require the presence of cations. In hot solutions, agarose molecules tend to behave as stiffened coils. Agarose forms stiff, turbid, and brittle gels reversible when hot solutions are cooled below around 40°C. The process involves first the formation of bundles of double or single helices in an order-disorder phenomenon involving polymer association (Clark & Ross-Murphy, 1987). The presence of these helices, however, is not enough to form a gel; they are the precursors of "super-junctions" of helices which are groupings of multiple helices (Figure 5–30, part E).

### 5.9.3 Carrageenans

Structurally, carrageenans are closely related to agar. The linkage pattern of the chain introduces a twist in the molecule, giving rise to helical structures. Their ability to gel is based on the association of the helical chains into double helices. Because of the anionic nature of the molecule (presence of sulfate), carrageenans require counter-ions to gel, preferentially calcium (for $\kappa$-carrageenan) and potassium (for $\iota$-carrageenan) (Figure 5–30, parts C and D). The gelling types of carrageenan ($\kappa$- and $\iota$-carrageenans and furcellaran) require heat to bring them into solution, and rigid gels are formed on cooling, often at very low concentration ($\approx$1% w/w).

### 5.9.4 Gelatin

Gelatin is obtained from collagen by controlled acid or alkaline hydrolysis. Its gelling properties are derived from the triple helical molecular structure of the tropocollagen rod (Figure 5–30, part B). Gelatin swells in contact with cold or warm water, and when heated to temperatures above its melting point, the swollen gelatin dissolves. Upon cooling, the molecules reform into triple helices like those in collagen and give rise to a transparent and elastic gel (this process can be reversed by heating). Groups of triple helices align themselves in a parallel array forming microcrystallites that act as cross-links. Gelatin gels melt between 27°C and 34°C and do so in the mouth (Poppe, 1992). The minimum concentration for gelation is 0.5%. The rheological properties of gelatin gels are discussed by Clark and Ross-Murphy (1987).

### 5.9.5 Whey Protein

Whey protein gels are typical thermotropic gels—gelation is induced by heating and the process is irreversible. Structurally, most whey protein gels are particle gels in which the units forming the network are protein aggregates (0.5–2 $\mu$m in diameter) associated as a string of beads or in clusters. Gelation of whey proteins (and of other globular proteins, such as those from soybean or egg albumin) appears to involve a series of transitions from the native to the gel state (Aguilera, 1995). These include unfolding of the native proteins, aggregation of the unfolded molecules, strand formation from aggregates, and association of strands into a network. A gel of globular proteins consists of fine strands or aggregates depending on the environmental conditions under which the gel was prepared (presence of cations, pH, etc.). Several stabilizing forces are thought to operate at the molecular level: electrostatic forces, covalent bonds, and hydrophobic forces. A detailed discussion of the gelation of food proteins is found in Ziegler and Foegeding (1990).

### 5.9.6 Myosin

The gelling of myosin is induced by heating (>60°C), as is the case for globular proteins, and it strongly depends on pH and ionic strength. Myosin can form stranded as well as aggregated gels. Nonfilamentous myosin gels made by heating solutions at temperatures above 60°C are strong and elastic. The mechanism is predominantly tail-tail association via noncovalent bonding. Gels formed from myosin filaments at low ionic strength result in very large aggregates cross-linked via the head groups (Clark & Ross-Murphy, 1987).

### 5.9.7 Starch

The gelling of starch dispersions depends on the source, concentration, and presence of starch components in the granular and aqueous phase. When heated in excess water, starch granules rapidly swell and imbibe enough liquid to reach several times their dry weight, a phenomenon called gelatinization (see Section 4.5.3). As a result, the crystalline portion of the granules melts, and amylose leaches out of the granules, which remain intact unless vigorously agitated. On cooling, the free amylose present as single helices become ordered into microcrystalline regions surrounding swollen granules, thus forming a gel. In fact, this type of starch gel is a composite gel of amylose matrices filled with swollen granules. Apparently, gels formed from potato starch differ: amylopectin leaches out of the granules, and gel formation does not take place immediately after cooling (Hermansson & Kidman, 1995). The gelling mechanism and microstructure are more complicated if both amylose and amylopectin are present in the aqueous phase, since they undergo phase separation at concentrations of each component above 20–30%.

### 5.9.8 Pectins

Pectins are usually classified as low methoxyl (LMP) and high methoxyl (HMP), and each type has its own gelation mechanism. In HMP gels, the junction zones are aggregates of chains of various sizes promoted by hydrogen bonding and hydrophobic interactions (Oakenfull, 1991). Gelation of HMPs is favored by the presence of sugar (minimum of 55% soluble solids) and a low pH ($<3.5$), conditions usually present in jams. Since LMPs are structurally similar to alginates (galacturonic acid blocks in pectins differ in configuration at only one carbon from guluronic acid blocks of alginates), it is not surprising that they gel by a calcium-bridging mechanism (i.e., egg-box arrangement).

### 5.9.9 Gellan Gum

Gellan produces gels at low concentration when hot solutions of the gum are cooled. The substituted form of gellan produces soft, elastic gels whereas the unsubstituted types yield hard, brittle gels. It is thought that the mechanism of gelation of gellan gum starts by formation of double helices, followed by ion-induced association of the double helices (Gibson, 1992).

## 5.10 MIXED GELS

Several food gels (as well as cosmetic and pharmaceutical gels) include in their formulation more than one type of gel-forming material. Many times a blend of gelling polymers provides superior properties than a single component, so it may be economically advantageous to replace an expensive gelling agent by a mixture of cheaper ones.

Aguilera (1992) has pointed out that most high-moisture structured foods may be regarded as combinations of gel matrices (hydrated or collapsed) with embedded microstructural elements such as fibers, globules, occluded solutions, and encapsulating walls. This section presents fundamental concepts and applications of mixed gels as well as illustrations of the connections between morphological aspects of mixed gels, processing, and product characteristics.

There are many types of mixed food gels and consequently several ways to classify them (Aguilera, 1992; Brownsey & Morris, 1988; Morris, 1990; Tolstoguzov, 1997b). Although not all mixed gels are phase-separated gels, Tolstoguzov (1997b) has succeeded in linking thermodynamic phase diagrams with possible structures of mixed gels (Figure 5–31). Mixtures of gelling polymers A and B form mixed gels if both are present in solution in amounts exceeding their critical concentration for gelation ($C^*_{gel}$); otherwise they form a mixed solution (zone I). Critical concentrations for anionic polysaccharides are usually greater than 0.1–0.3% and for proteins may range from 10% to 13%. Zones II and III represent gels of a

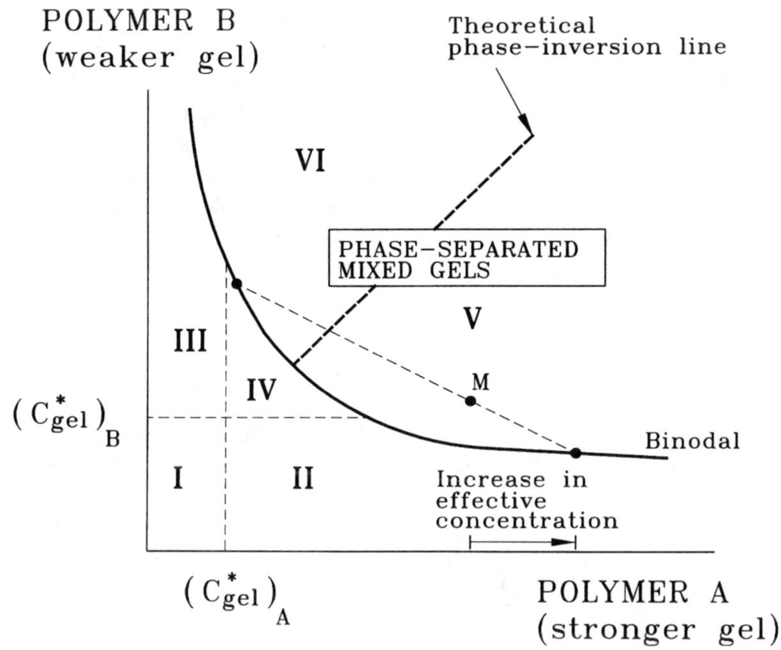


**Figure 5–31** Phase diagram for gelation of a mixed solution of polymers. *Source:* Reprinted from V.B. Tolstoguzov, Some Physico-Chemical Aspects of Protein Processing in Foods: Multicomponent Gels, *Food Hydrocolloids*, Vol. 9, pp. 317–332, Copyright 1995, with permission from Elsevier Science.

single polymer network (either A or B, respectively) in which the other polymer is present in the solution filling the pores. A type of "synergism" due to the excluded volume effect is observed. Since under these conditions each polymer appears as if it was more concentrated (because the other polymer takes up solvent volume that cannot be occupied), there is a decrease in the effective concentration for gelation.

Phase separation occurs when the overall composition of the mixture falls inside the binodal. In zone IV of Figure 5–31, both polymers are above their critical concentrations for gelation but not yet inside the binodal, so a homogeneous mixed gel would be formed. Inside the binodal, two phases are formed at equilibrium, one rich in polymer A and the other rich in polymer B, implying that there is a redistribution of water between the system phases. Conceivably, one phase may become so diluted that it is not able to gel and turns into a liquid filler (negative synergism). Synergistic effects in two-phase systems occur when the mixed gel is formed by a continuous phase containing the stronger gelling agent (which has become significantly concentrated) and the other polymer is present as a gelled filler or dispersed phase (zone V). Phase inversion results in the formation of a continuous phase of the weaker-gelling polymer and will decrease the physical and rheological properties of the mixed gel (zone VI).

As explained by Aguilera (1992), the gelling of mixed solutions is a dynamic, nonequilibrium process in which the physicochemical properties of the ungelled phase change with time. In the case of gels formed by cooling or heating, the binodal shifts as temperature varies during gelation, and so does the composition of the gelling phases. Tolstoguzov (1997b) has also pointed out that, since gelation always results in a transfer of molecules from the solution to the forming network, the excluded volume effect of the macromolecules decreases. This makes the dispersion medium a better solvent, indicating that the structure (and properties) of the mixed gel depends on the kinetics of gelation of individual molecules in the mixture.

Hypothetical network arrangements for mixed gels are presented in Figure 5–32. Some of these are more likely to occur in foods than others. Following are summaries of how the most common types of mixed gels are formed in foods.

### 5.10.1 Phase-Separated Gels

As discussed above, true phase-separated gels require thermodynamic incompatibility between polymers in solution (i.e., the development of turbidity upon mixing). As previously discussed, concentrated solutions of polymers almost always lead to phase separation. Hydrocolloids (proteins and polysaccharides) can form mixed gels when the concentration of both in the mixed solution exceeds the minimum concentration for gelation (Tolstoguzov, 1997b). Gelation of the separated mixed solution leads to mixed gels above the binodal and to single-phase gels below the binodal.

### 5.10.2 Copolymer Gels

Strictly speaking, a copolymer consists of two or more monomeric units in the chain. By extension,

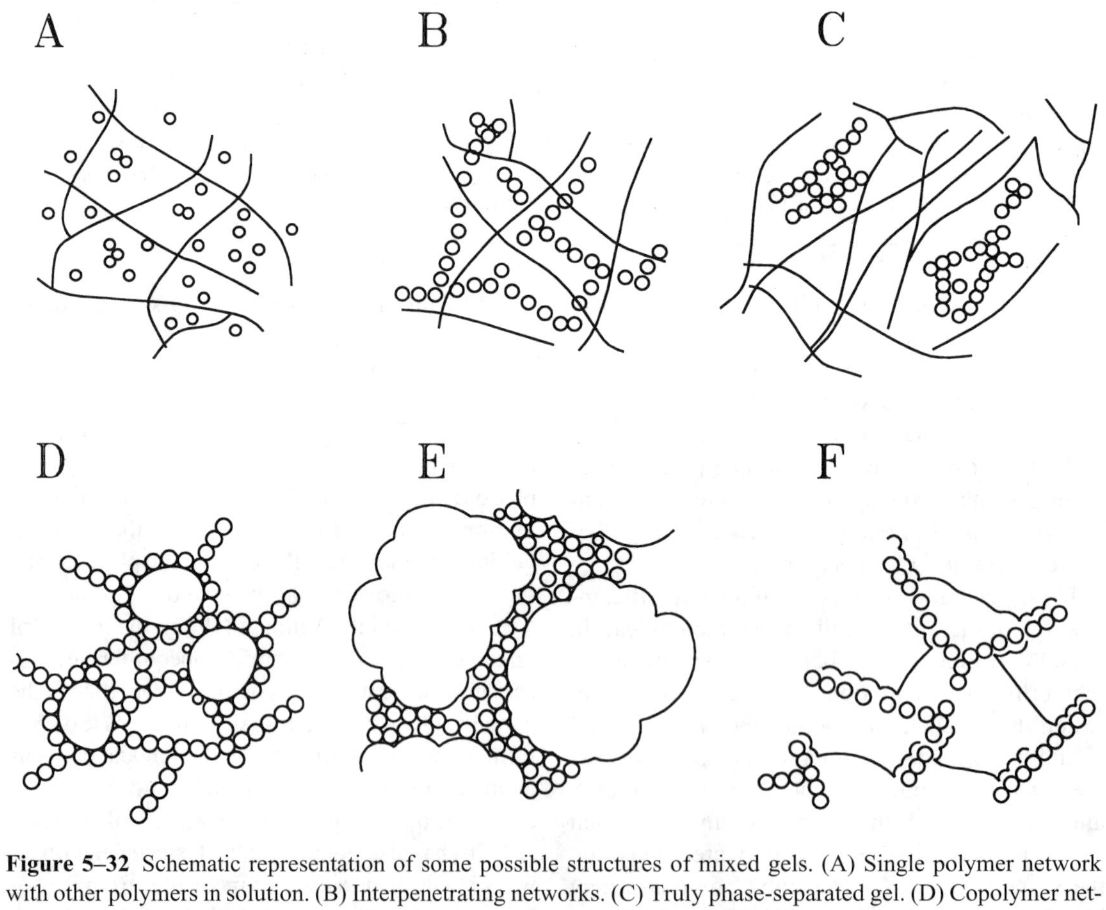


**Figure 5–32** Schematic representation of some possible structures of mixed gels. (A) Single polymer network with other polymers in solution. (B) Interpenetrating networks. (C) Truly phase-separated gel. (D) Copolymer network (containing fat globules covered with polymer; see text). (E) Starch-filled gel (active phase-separated gel). (F) Coupled or complex gel.

copolymer gels may be defined as those in which the network chain is formed by association of at least two species. Milk fat homogenized in the presence of protein (e.g., casein or whey protein isolate) results in fat globules with adsorbed protein membranes. If these small fat globules (e.g., size $<1$ $\mu$m), which may be considered monomer B, are dispersed in a solution of whey protein (monomer A) and heated under conditions similar to those needed to form a whey protein gel, they will become part of the protein network (Aguilera & Kessler, 1989). Such "copolymer" gels show textural properties (e.g., firmness) higher than whey protein gels of the same total concentration, even though they contain fat. If fat globules of the same size but without the protein membrane are subject to gelation, the resulting gel is significantly weaker. Thus, during the gelation process protein-covered fat globules are "recognized" by their proteinaceous surface and incorporated into the network as if they were whey protein aggregates (Xiong, Aguilera, & Kinsella, 1991). Fat in engineered foods modifies texture and acts a carrier for flavors, colors, and nutrients (e.g., fatty acids and vitamins). Many gelled dairy products, like cheeses and yogurt, contain fat, and it is important that fabricated gels accommodate the lipid phase without weakening the microstructure.

### 5.10.3 Complex or Coupled Gels

Complex gels may be defined as those in which chains of different polymers interact directly to form the network. For example, an anionic polysaccharide may associate to a charged protein molecule below its isoelectric point by charge-charge attraction. Mixed gels (pH 3.9) of gelatin and sodium alginate (or low methoxyl pectins) are formed under conditions where the polysaccharide does not gel (Morris, 1990). HMPs form stronger gels when mixed with alginates at low pH ($<4.0$) without the presence of sugar (required for the gelation of pectin) and without calcium (required for the gelation of alginate). In this case, gelation occurs through direct interactions of the two polymers, which form close-packing structures between chains that act as junction zones (Oakenfull, 1991).

Alternative interactions are also possible. All galactomannans (guar gum, locust bean gum, etc.) interact with xanthan gum. Galactomannans have an uneven distribution of galactose side units, allowing portions of the molecules with unsubstituted mannose to interact with xanthan gum. Long galactomannan molecules attached by this interaction to xanthan gum actually link and reinforce the xanthan gum gel network (Urlacher & Dalbe, 1992). As another example of such synergistic effect, locust bean gum, when combined with $\kappa$-carrageenan in desserts, decreases syneresis and provides a more gelatinlike texture. It is also known that $\kappa$-carrageenan interacts even at very low concentration with the surface of casein micelles, producing stronger milk gels.

### 5.10.4 Kinetically Induced Co-Gels

Regretfully, casein and whey proteins do not form mixed gels. During cheese manufacture, casein sets into a curd while whey proteins remain in solution. Since whey proteins gel by heating and casein gels by acidification, co-gels can be formed if an acid-producing compound is decomposed in the mixture during heating, triggering casein aggregation. Glucono-$\delta$-lactone, which yields gluconic acid when hydrolyzed by heat, has been used to form mixed gels of whey protein and casein (Aguilera & Kinsella, 1991). The kinetics of gel formation and properties are strongly dependent on temperature, as it affects the release of acid and the final pH. Co-gels from other binary polymer solutions with different gelling mechanisms are also known, such as those exploiting gelation of alginate by controlled release of Ca.

### 5.10.5 Active Phase-Separated Starch Gels

Mixed gels in which starch granules undergo hydration and swelling in the presence of a gelling agent (e.g., a globular protein) constitute a special class of phase-separated gels, for the two phases are present in the starting blend (Aguilera & Baffico, 1997). These mixed gels show enhanced mechanical and rheological properties at low concentrations of starch, permitting a strong continuous network of the gelling polymer to develop. This synergy is explained by the early removal of water from the system during swelling of the starch, which concentrates the polymer before gelation. Thus, the separate phase (consisting of the starch granules) is active in the sense that it introduces to gelation another kinetic mechanism of water removal. The formed structure of the mixed gel is that of a filled gel. This case is discussed in Section 5.13.

## 5.11 THE MICROSTRUCTURE OF GELS

Gels are difficult to study by microscopy because of the risks of artifacts during sample preparation. The high proportion of water in gels makes dehydration without collapse difficult, while the molecular nature of some networks requires a high resolution. Even a technique such as cryo-SEM, which involves freezing to immobilize the water in place (rather than removing it), is subject to artifact formation owing to the segregation of soluble solids in the aqueous phase during freezing and the crystallization of ice (even for fast freezing rates). In spite of such problems, most food gels are usually examined by both light and electron microscopy techniques (Colombo & Späth, 1981). Some of the best microscopy work on gels, protein and starch gels in particular, has been done by Hermansson and coworkers at SIK in Sweden (Hermansson, 1988; Hermansson & Kidman, 1995; Stading, Langton, & Hermansson, 1993).

Solutions of globular proteins form transparent or opaque heat-set gels, depending on the ionic strength and pH (Doi, 1993). Gels formed at low ionic strength or removed from the isoelectric point tend to be transparent, which is explained by the fact that thinner strands form the network (e.g., $<15$ nm); opaque gels are of a particulate (aggregate) nature. The size of pores in fine-stranded gels may be on the order of nm, while in particle gels their size is on the order of 10–100 $\mu$m. The size of the aggregates forming the particulated network of $\beta$-lactoglobulin gels varies with heating rate; it ranges from less than 1 to 5 $\mu$m (Stading et al., 1993). The microstructure of transparent and opaque gels made of globular proteins is best studied using transmission electron microscopy. Clark, Judge, Richards, Stubbs, and Suggett (1981) report the value of this type of microscopy for some enzymes and Aguilera (1995) reports its value for a whey protein isolate.

Hermansson and Kidman (1995), studying amylose gels by transmission electron microscopy, confirmed that they are formed by strands 10–40 nm in diameter and that double helices are placed contiguously and perpendicular to the direction of each strand. When amylopectin was present, it seemed to be partly adsorbed to the amylose strands. Kanzawa, Koreeda, Harada, Okuyama, and Harada (1990) studied gels of $\kappa$-carrageenan, alginate, gellan gum, and low methoxyl pectin by transmission electron microscopy. The microfibrillar nature of the pure gels was revealed, as were the thicker microfibrils (ca. 400 Å), when calcium cations were present. Carrageenan gels studied by cryo-TEM show an entanglement of thin threads of 6–10 nm in diameter. In agarose gels, the threads are about 10 nm in diameter (Guenet, 1992).

Microscopy is potentially a powerful tool for studying phase-separated gels, as the presence of phases and even their relative proportions (volumes) may be resolved. Abeysekera and Robards (1995) used several microscopy techniques (bright field, phase contrast, confocal laser scanning, and transmission electron microscopy) to study phase-separated gels of starch and gelatin or maltodextrin. Kalab, Allan-Wojtas, and Shea Miller (1995) have suggested using immunolabeling techniques to identify (through attaching colloidal gold markers to antibodies) the presence of specific macromolecules. Such techniques have already been used to locate the fine network of maltodextrin. The microstructure of fat-rein-

→

**Figure 5–33** TEM and SEM micrographs of whey protein isolate (WPI) gels. (A) TEM of a pure WPI gel using cryo-fracture. Courtesy of Mr. Tony Weaver. Unilever Research Laboratory. (B) TEM of a WPI gel having incorporated fat globules with reformed protein membranes. (C) SEM of a pure WPI gel.

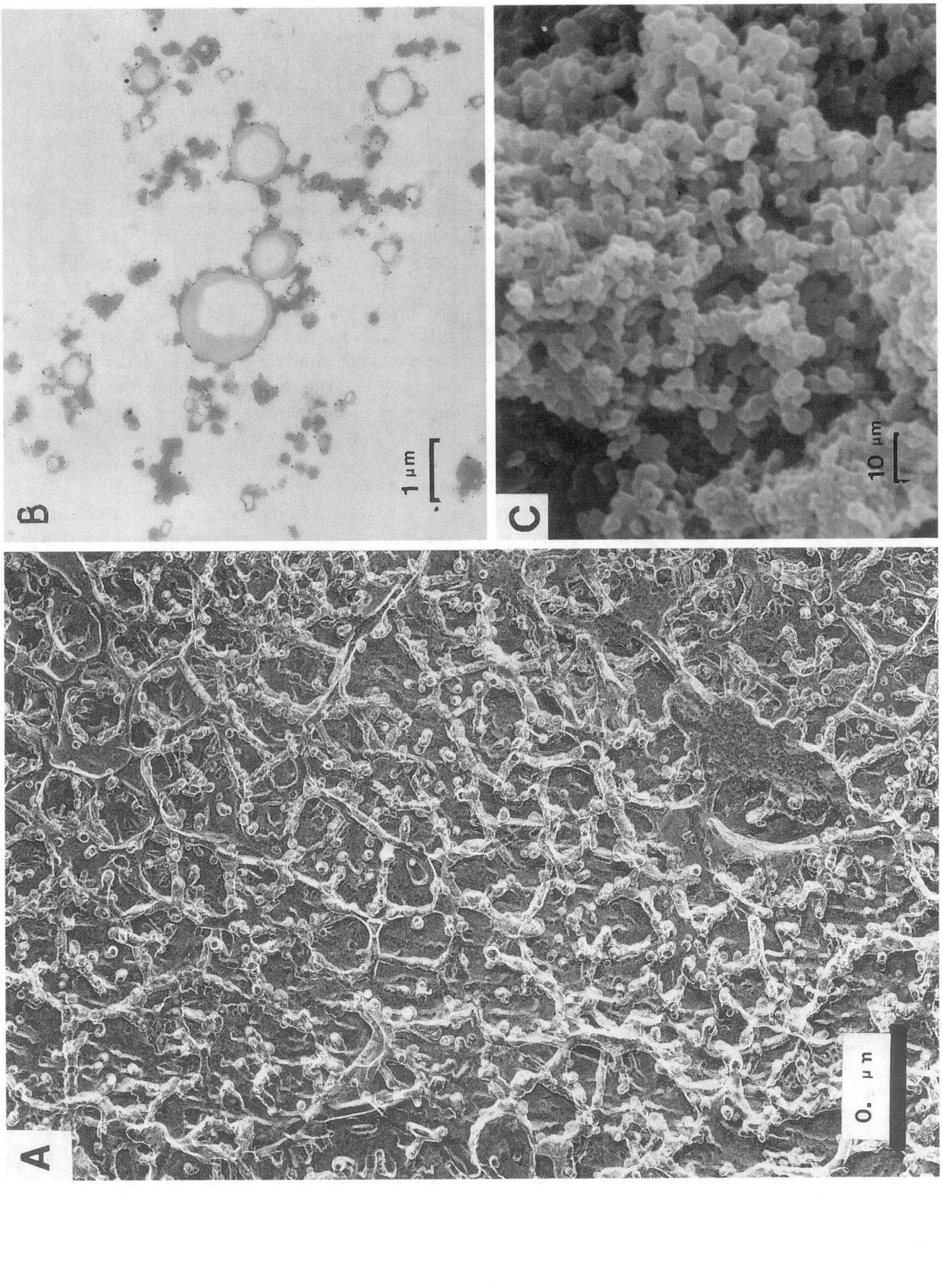
A
B
1 µm
C
10 µm

forced gels has been examined by scanning electron and transmission electron microscopy (Aguilera, 1992) and of starch-filled kamaboko gels by transmission electron and light microscopy (Verrez-Bagnis, Bouchet, & Gallant, 1993). Micrographs showing the microstructure of pure and mixed gels are shown in Figure 5–33.

Other techniques used to resolve the short-range molecular structure of gels include X-ray diffraction, NMR, and FTIR. A particularly useful technique for studying polymer-solvent interactions is neutron diffraction; its small-angle version is able to supply information on the chain structure within the gel. The application of some of these techniques to the study of gels is discussed in Guenet (1992).

## 5.12 STRUCTURE-PROPERTY RELATIONS IN GELS

This section intentionally follows the section devoted to gel microstructure. Ideally, acquaintance with the structural morphology of gels facilitates understanding of their transport and rheological properties and suggests modifications that might engender desired attributes.

### 5.12.1 Transport Properties

Gels are unique among "soft solids" in that their transport properties are similar to those of liquids (i.e., aqueous media). Diffusion of small solutes in gels occurs largely in the aqueous phase, and it is affected only by steric and electrostatic hindrance of the strands forming the gel matrix. Thus, the diffusivity is lower for large solutes and molecules carrying charges opposite to those of the polymer chains and when small pores are present (i.e., high solid concentration). Effective diffusivity values of solutes in gels ($D_{gel}$) as well as in pure water ($D_{water}$) are presented in Table 5–3. As observed, the ratio between both diffusivities is close to 1, making gels unique materials for immobilizing enzymes and cells (Doran, 1995). The sieve effect provided by the network is the basis for several molecular separation methods used in biochemistry and for sizing molecules (e.g., gel permeation and size-exclusion chromatography).

As expected, the thermal conductivity ($k$) of gels having low total solids (e.g., <1%) is similar to that of water (around 0.6 W/m·K at 20°C). However, the effective thermal conductivity of filled gels ($k_{gel}$) decays as the volume fraction of a dispersed phase increases ($k_{filler} < k_{water}$). The physical models presented in Table 9–1 can be used to find the exact dependence of transport properties on volume phase and architecture.

The thermal conductivity of high-moisture porous foods may vary with temperature during heat processing (e.g., during baking or drying) as water evaporates into air-filled pores. Since gels are good models for such foods, Sakiyama and Yano (1994) determined that the effective thermal conductivity of air-filled alginate and albumin gels decreased with the volume fraction of the air phase, as discussed before. However, the $k_{gel}$ for volume fractions between 0.40 and 0.67 showed interesting behavior as a function of temperature: $k_{gel}$ was lower than $k_{water}$ up to around 60°C and increased faster for lower volume fractions (i.e., when more moisture was present). Thus, in these structured porous systems there is a dynamic variation in thermal properties during heating as water vapor diffuses into the pores (bubbles), and

**Table 5–3** Effective Diffusivity of Selected Solutes in Gels

| *Solute* | *Gel* | *Temperature (°C)* | *Diffusivity × $10^9$ ($m^2s^{-1}$)* | $D_{gel}/D_{water}$ |
|---|---|---|---|---|
| Oxygen | 2% agar | 30 | 1.94 | 0.70 |
| Glucose | 3% Ca alginate | 30 | 0.62 | 0.87 |
| Sucrose | 2% Ca alginate | 25 | 0.48 | 0.86 |
| L-tryptophan | 4% κ-carrageenan | 30 | 0.58 | 0.88 |

Source: Doran (1995).

heat conduction through the solid phase may not be the only heat transfer mechanism present.

### 5.12.2 Viscoelastic Behavior

Because of their viscoelastic behavior, gels are best characterized in terms of their mechanical response to oscillatory shear under small deformations. In particular, the storage modulus G′, a measure of the solidlike behavior, is related to the connectivity of the polymeric network, and it is frequently used to characterize gels and to follow the kinetics of formation (Clark, 1992). A typical "curing" experiment showing the kinetics of G′ as the gel structure develops is shown in Figure 3–18. The loss modulus G″ is at least an order of magnitude smaller than G′, illustrating the predominantly solidlike response. Another feature of gels is that G′ is relatively independent of frequency (Ross-Murphy & McEvoy, 1986). This fact led Almdal, Dyre, Hvidt, and Kramer (1993) to propose that "solid-like gels are characterized by the absence of an equilibrium modulus, by a storage modulus G′ ($\omega$) that exhibits a pronounced plateau to times at least of the order of seconds, and by a loss modulus G″ ($\omega$) which is considerably smaller than G′ in the plateau region".

Typical of gels is that the G′ at the plateau region varies as a power function of the concentration of the gelling agent (if gelation and measurement conditions remain the same). In particular, for strong concentrated gels of many biopolymers, the power exponent is found to be close to 2 (Clark, 1992). This same square dependence has been observed for the elastic modulus $E$ (e.g., $E = kc^2$) for gelatin gels over a wide range of temperatures (Guenet, 1992). Although the value of the exponent has been related to the theory of rubber elasticity, it appears that is more directly related to the "cellular" or foamlike morphology present in gels (see Section 3.9.6).

### 5.12.3 Mechanical Properties of Gels

Gels are mechanically assayed by small and large deformation (fracture). The former mode aims at simulating small forces acting during handling of products, and it is appropriate when structural properties of the intact system are to be determined. The latter may represent the action of mastication or cutting. It has been estimated that the rate of jaw movement during mastication is 16–33 mm/s (Bourne, 1982). Care has to be taken in stating whether moduli (slope of the force-deformation curve in the elastic or linear range) or just a stress is calculated from experimental data, because in some gels proportionality between stress and strain is observed only at very small strains.

Not much work has been reported on gel failure and its relationship to microstructure, although it is known that fracture involves structural phenomena: presence of stress concentrators (inhomogeneities) and crack propagation. Some gels fail in brittle fracture under small deformation (e.g., 1% gellan gels) but most of them show a yield stress (flow and deformation), which helps in dissipating energy that otherwise would contribute to fracture. Observed yield stresses of gels range from $10^{-3}$ to $10^4$ Pa and depend on strain rate (Walstra, 1996). Fracture depends on the presence of inhomogeneities. The fracture properties of Gouda cheese were found to be time dependent, with inhomogeneities on the order of 0.1–0.3 mm, and the fracture energy varied between 1 and 19 $Jm^{-2}$ (Luyten, van Vliet, & Walstra, 1991). The failure characteristics of gellan gels measured in compressive, tension, or torsional modes were determined by Lelievre, Mirza, and Tung (1992).

In filled gels, gel strength often increases above a critical phase volume of the filler, particularly if there is good interaction or adhesion between the filler and the matrix. A relative stress at failure ($\sigma_r$), when fracture is produced through the continuous phase, can be determined from the following expression (Ross-Murphy & Todd, 1983):

$$\sigma_r = \frac{\sigma_c}{\sigma_o} = \frac{(1 - \phi^{1/3})}{(1 - \phi)^{5/2}} \qquad \textbf{Equation 5–1}$$

where the indices $c$ and $o$ refer to the filled gel and the gel without filler, respectively.

## 5.13 AN ILLUSTRATIVE EXAMPLE OF STRUCTURE-PROPERTY RELATIONSHIPS IN MIXED GELS

### 5.13.1 An Experimental Oddity

Several food polymers are known to form stronger gels in mixed than in pure form, and a variety of mechanisms have been proposed to explain such synergic behavior (see above and Morris, 1990). It is known that thermotropic protein-starch gels are stronger if the starch is gelatinized *in situ* prior to protein gelation. For example, different sources of starch are added to surimi-based products to improve their gel strength. It was found that gels of a whey protein isolate produced by heating to temperatures above 75°C presented a maximum in elastic modulus ($E$) when approximately 20% of the weight of the isolate was replaced by an equal amount of cassava starch (Aguilera & Rojas, 1996). Interestingly, cassava starch forms very weak gels upon heating and is much cheaper than a whey protein isolate. The obvious question is, why does replacing a portion of a strong gel-forming material by a poor-gelling material result in a stronger mixed gel? It was decided to use a microstructural approach to find the structure-property relationships (Aguilera & Baffico, 1997; Stanley, Aguilera, Baker, & Jackman, 1998). The microstructural approach aims at generating simultaneously the structural and physical data and linking both in order to generate structure-property relationships (Figure 5–34).

### 5.13.2 Structural Model and Mechanism: Use of Hot-Stage Video Microscopy

Hot-stage microscopy was instrumental in visualizing with minimum intrusion structural changes during gelation at a heating rate of 40°C/min. Phase-contrast microscopy provided images with excellent contrast and brightness owing to differences in the refractive index of the elements in the field of view (Figure 5–35).

Native cassava starch granules had almost circular area projections (5–15 $\mu$m in diameter) (Figure 5–35, parts A and B). During heating, larger granules started to swell at 60°C, imbibing water (C). At 65°C, there were granules at all stages of swelling (D), and the gelatinization process was completed at 70°C (E). The irregular appearance of the extragranule solution at 75°C (F) is due to the initiation of the aggregation-gelation of whey proteins, which was completed at 80°C (G). Real-time video microscopy was instrumental in demonstrating that the hydration and swelling of cassava starch granules removes water from the system before the whey protein isolate starts to gel and that gelatinization of the cassava starch and denaturation-gelation of the isolate are independent events. More important, hot-stage microscopy provided a structural view of the final mixed gel, which consisted of a continuous network of whey protein gel formed around swollen cassava starch granules (H). The structural changes observed coincided with peak temperatures of major thermal transitions detected by DSC: gelatinization of starch occurred at 65°C and denaturation of whey proteins occurred at 74°C (Aguilera & Rojas, 1996). Small deformation oscillatory rheometry permitted the researchers to follow the kinetics of gelation (curing experiment) at constant temperature (70–90°C) without distorting the developing structure. It was confirmed that the whey protein isolate gelled very fast and formed a strong gel (large $G'$ plateau); the cassava starch did not develop a significant $G'$ after heating at those temperatures and cooling (Aguilera & Rojas, 1997).

### 5.13.3 Interpretation

The microstructural interpretation of this synergism at low weight fractions of cassava starch ($<$0.3) is as follows. Water is removed from the system during swelling of the starch granules, increasing the "effective" concentration of the whey protein solution, which gels at a higher temperature. The modulus of the formed whey protein gel is then higher than predicted for the nominal concentration. For example, the modulus of the whey protein isolate increases by 50% if a nominal concentration of 10% is replaced by an "effective" concentration of 12% isolate (40.3 versus 62.0 kPa, respectively). Swollen starch granules, which have a low modulus themselves, immobilize large amounts of water and restrict the deformation of the whey protein matrix on compression. As the weight fraction of the starch increases ($>$0.3), formation and continuity of the strong whey protein network is progressively interrupted by swollen starch granules, which become the

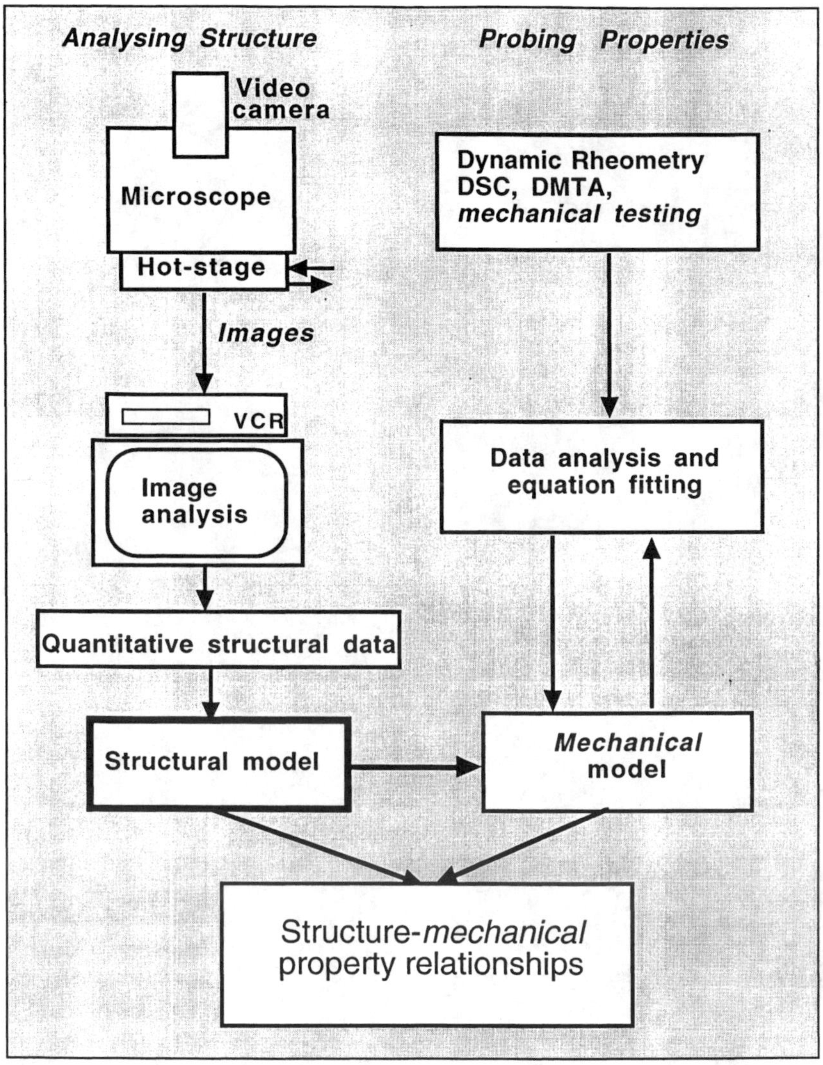


**Figure 5–34** Microstructural approach to the study of structure-property relationships in a mixed gel system. Analysis of structure formation in real time by video microscopy leads to a structural model. Complementary data are gathered by oscillatory rheometry and DSC. Both models contribute to unveil the structure-property relationships. *Source:* Reprinted from D.W. Stanley, J.M. Aguilera, K.W. Baker, and R.L. Jackman, Structure/Property Relationships of Foods as Affected by Processing and Storage, in *Chemical, Structual, and Rheological Changes During Phase/State Transitions in Foods*, M.A. Rao and R. Hartel, eds., pp. 1–56, by courtesy of Marcel Dekker, Inc.

continuous phase. The structural result of phase inversion is a weak matrix of contiguous starch granules having strong whey protein gel inclusions as filler (observed also by Svegmark & Hermansson, 1993).

### 5.13.4 The Mechanical Model: The Structure-Property Relationships

We now need equations to predict the mechanical properties of the gel (e.g., elastic modulus, $E$

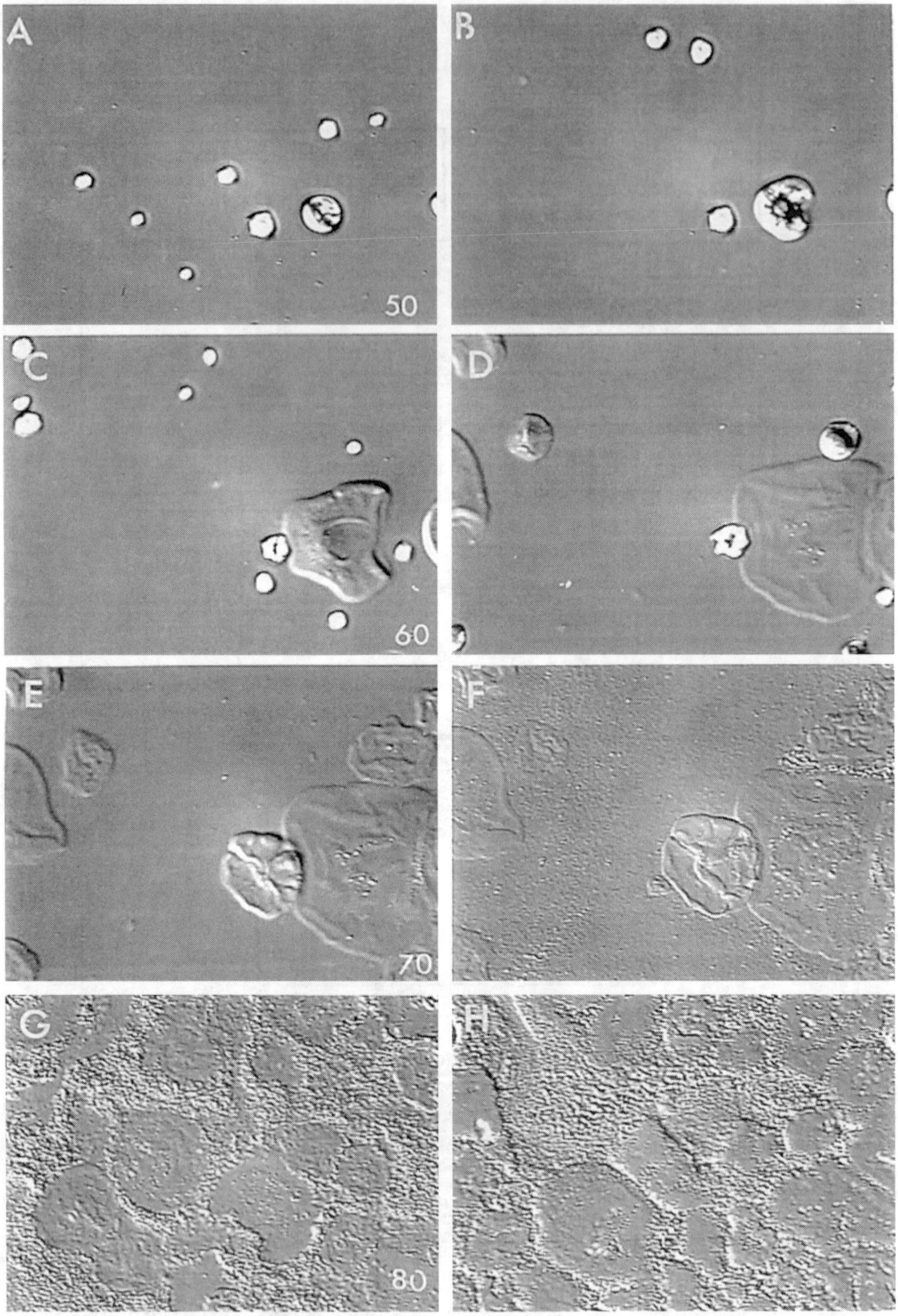


**Figure 5–35** Time-lapse photomicrographs taken by hot-stage videomicroscopy (every 5°C) during heating of a whey protein–cassava starch suspension (heating rate 40°C/min). Number in bottom right corner indicates the temperature. *Source:* Reprinted with permission from J.M. Aguilera and P. Baffico, Structure-Mechanical Properties of Heat-Induced Whey Protein/Starch Gels, *Journal of Food Science*, Vol. 62, pp. 1048–1053, 1066, © 1997, Institute of Food Technologists.

[storage modulus], and G′) for any formulation. Regression models are adequate to establish the dependence of $E$ (or $\sigma_{0.05}$) with concentration for whey protein and cassava starch gels:

$$E_i = k_i c_i^n \qquad \textbf{Equation 5–2}$$

where $E_i$ and $c_i$ are the moduli and concentration (g/g solution) of gels made of the pure materials and $k_i$ is a constant. The exponent $n$ is larger for the whey protein than for the starch (2.35 and 1.22, respectively), denoting a strong effect of concentration.

Since the mixed gel is a binary composite, models such as those proposed by Takayanagi can be used to predict the mechanical properties (see Section 3.9.3). We used the isostrain model because the composite C consists of a weak cassava starch filler X dispersed in a stronger continuous matrix Y:

$$E_C = \phi E_X + (1 - \phi)E_Y \qquad \textbf{Equation 5–3}$$

where $\phi$ is the volume fraction of the cassava starch. A severe limitation of the Takayanagi isostrain model when applied to food gels is that phase-separated mixed gels are really ternary systems in which water, present in a high proportion, is partitioned between both phases during gelation. Hence, the relationship has to be modified following the line of thought of Clark (1987), yielding this expression:

$$E_C = w_{WPI}E^*_{WPI} + w_{CS}E^*_{CS} \qquad \textbf{Equation 5–4}$$

where the indices *CS* and *WPI* refer to the cas-

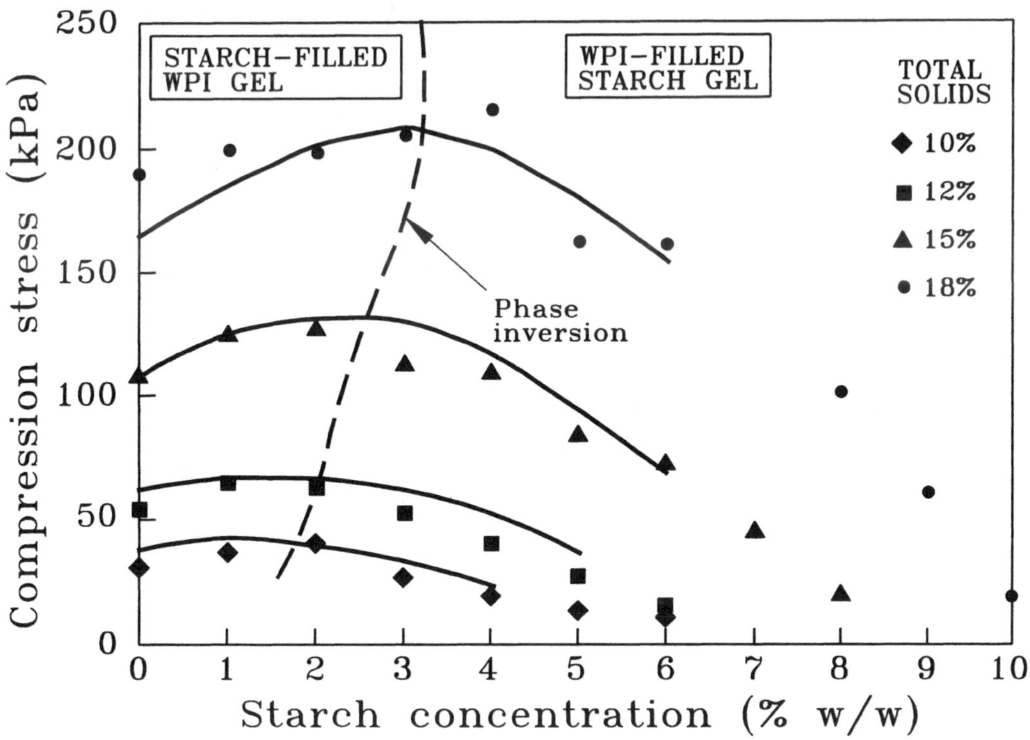


**Figure 5–36** Compressive stress ($\varepsilon < 0.05$) of mixed cassava starch–whey protein isolate gels of different total solids as a function of starch content. Dots show experimental data; lines represent prediction by the modified Takayanagi model. *Source:* Reprinted with permission from J.M. Aguilera and P. Baffico, Structure-Mechanical Properties of Heat-Induced Whey Protein/Starch Gels, *Journal of Food Science*, Vol. 62, pp. 1048–1053, 1066, © 1997, Institute of Food Technologists.

sava starch and whey protein isolate, respectively, and $E^*$ and $w$ represent the modulus at the effective concentration and the corrected weight fractions (after water partition), respectively (for details see Aguilera & Baffico, 1997). Figure 5–36 shows experimental values of a mechanical property of the mixed gel (compression stress under low deformation, $\varepsilon < 0.05$) as a function of starch concentration for mixed gels of different total solids. Solid lines represent fitting of the Takayanagi model modified for water partition.

This example suggests a method for researching the structure of food products that studies might follow in the future. First, the main mechanisms leading to structure formation are studied using the most appropriate microscopy techniques, with attention paid to other experimental data. Then, based on available structural data and previous findings, mathematical models are developed for the prediction of properties for any change in structure (e.g., those induced by a change in formulation). Nonlinearity in structure-property relationships of gels arise largely because of the architecture at the microstructural level (e.g., composites or cellular gels). In the case of homogeneous mixed gels (or alloys), the linear dependence of a property on composition is to be expected.

## BIBLIOGRAPHY


Abeysekera, R.M., & Robards, A.W. (1995). Microscopy as an analytical tool in the study of phase separation of starch-gelatin binary mixtures. In S.E. Harding, S.E. Hill, & J.R. Mitchell (Eds.), *Biopolymer mixtures* (pp. 143–160). Nottingham, England: Nottingham University Press.

Aboagye, Y., & Stanley, D.W. (1985). Texturization of peanut proteins by surface film formation. 1. Influence of process parameters on film forming properties. *Canadian Institute of Food Science Technology Journal, 18,* 12–20.

Aguilera, J.M. (1992). Generation of engineered structures in gels. In H.G. Schwartzberg & R.W. Hartel (Eds.), *Physical chemistry of foods* (pp. 387–421). New York: Marcel Dekker.

Aguilera, J.M. (1995). Gelation of whey proteins. *Food Technology, 49*(10), 83–86, 88–89.

Aguilera, J.M., & Baffico, P. (1997). Structure-mechanical properties of heat-induced whey protein/starch gels. *Journal of Food Science, 62,* 1048–1053, 1066.

Aguilera, J.M., & Karel, M. (1997). Preservation of biological materials under desiccation. *Critical Reviews in Food Science and Nutrition, 37,* 287–309.

Aguilera, J.M., & Kessler, H.G. (1989). Properties of mixed and filled-type dairy gels. *Journal of Food Science, 54,* 1213–1217, 1221.

Aguilera, J.M., & Kinsella, J.E. (1991). Compression strength of dairy gels and microstructural interpretation. *Journal of Food Science, 56,* 1224–1228.

Aguilera, J.M., & Rojas, E. (1996). Rheological, thermal and microstructural properties of whey protein-cassava starch gels. *Journal of Food Science, 61,* 962–966.

Aguilera, J.M., & Rojas, V.G. (1997). Determination of kinetics of gelation of whey protein and cassava starch by oscillatory rheometry. *Food Research International, 30,* 349–357.

Aguilera, J.M., & Stanley, D.W. (1985). A review of textural defects in cooked reconstituted legumes: The influence of storage and processing. *Journal of Food Processing and Preservation, 9,* 145–169.

Aguilera, J.M., & Stanley, D.W. (1986). Extrusion processing as applied to proteinaceous materials. In M. Le Maguer & P. Jelen (Eds.), *Food engineering and process applications.* Vol. 2. *Unit operations* (pp. 131–141). Barking, England: Elsevier Applied Science.

Almdal, K., Dyre, J., Hvidt, S., & Kramer, O. (1993). Towards a phenomenological definition of the term gel. *Polymer Gels and Networks, 1,* 5–17.

Andrews, C.J. (1996). How do plants survive ice? *Annals of Botany, 78,* 529–563.

Arbuckle, W.S. (1986). *Ice cream* (4th ed.). Westport, CT: AVI Publishing Co.

Autio, K., & Laurikainen, T. (1997). Relationship between flour/dough microstructure and dough handling and baking properties. *Trends in Food Science and Technology, 8,* 181–185.

Baer, E., Cassidy, J.J., & Hiltner, A. (1991). Hierarchical structures of collagen composite systems: Lessons from biology. In W. Glasser & H. Hatakeyama (Eds.), *Viscoelasticity of biomaterials* (pp. 2–23). ACS Symposium Series 489. Washington, DC: American Chemical Society.

Barbut, S. (1988). Microstructure of reduced salt meat batters as affected by polyphosphates and chopping time. *Journal of Food Science, 53,* 1300–1306.

Beckett, S.T. (1995). Chocolate confectionery. In S.T. Beckett (Ed.), *Physicochemical aspects of food processing* (pp. 347–367). London: Blackie Academic and Professional.

Bell, G., Halling, P.J., Moore, B.D., Partridge, J., & Rees, D.G. (1995). Biocatalyst behavior in low water system. *Tibtech, 13,* 468–473.

Bello, R.A., Luft, J.H., & Pigott, G.M. (1982). Ultrastructural study of skeletal fish muscle after freezing at different rates. *Journal of Food Science, 47,* 1389–1394.

Berger, K.G. (1990). Ice cream. In K. Larsson & S. Friberg (Eds.), *Food emulsions* (2nd ed.) (pp. 367–444). New York: Marcel Dekker.

Berger, K.G., Bullimore, B.K., White, G.W., & Wright, W.B. (1972a). The structure of ice cream. Part 1. *Dairy Industries, 37,* 419–425.

Berger, K.G., Bullimore, B.K., White, G.W., & Wright, W.B. (1972b). The structure of ice cream. Part 2. *Dairy Industries, 37,* 493–497.

Bernal, V.M., Bernal, W.V., & Stanley, D.W. (1987). Modifying the textural characteristics of formed beef products. In *Proceedings of 33rd International Congress of Meat Science and Technology* (Vol. 1) (p. 175).

Bernal, V.M., Smajda, C.H., Smith, J.L., & Stanley, D.W. (1987). Interactions in protein/polysaccharide/calcium gels. *Journal of Food Science, 52,* 1121–1125, 1136.

Bernal, V.M., & Stanley, D.W. (1989). Methylcellulose as a binder for reformed beef. *International Journal of Food Science and Technology, 24,* 461–464.

Blanshard, J.M.V. (1988). Elements of cereal product structure. In J.M.V. Blanshard & J.R. Mitchell (Eds.), *Food structure: Its creation and evaluation* (pp. 313–330). London: Butterworths.

Blanshard, J.M.V., & Franks, F. (1987). Ice crystallization and its control in frozen food systems. In J.M.V. Blanshard & P. Lillford (Eds.), *Food structure and behavior* (pp. 51–65). London: Academic Press.

Bomben, J.L., & King, C.J. (1982). Heat and mass transport in the freezing of apple tissue. *Journal of Food Technology, 17,* 615–632.

Borchet, L.L., Greaser, M.L., Bard, J.C., Cassen, R.G., & Briskey, E.J. (1967). Electron microscopy of a meat emulsion. *Journal of Food Science, 32,* 419.

Borgstrom, G. (1968). *Principles of food science* (Vol. 1). New York: Macmillan.

Bourne, M.C. (1982). *Food texture and viscosity: Concept and measurement.* London: Academic Press.

Brooker, B.E. (1995). Imaging food systems by confocal laser scanning microscopy. In E. Dickinson (Ed.), *New physicochemical techniques for the characterization of complex food systems* (pp. 53–68). London: Blackie Academic and Professional.

Brooker, B.E., Anderson, M., & Andrews, A.T. (1986). The development of structure in whipped cream. *Food Microstructure, 5,* 277–285.

Brownsey, G.J., & Morris, V.J. (1988). Mixed and filled gels: Models for foods. In J.M.V. Blanshard & J.R. Mitchell (Eds.), *Food structure: Its creation and evaluation* (pp. 7–23). London: Butterworths.

Burghardt, A., Rogut, J., & Gotkowska, J. (1988). Diffusion coefficients in bidisperse porous structures. *Chemical Engineering Science 43,* 2463–2476.

Caldwell, K.B., Goff, H.D., & Stanley, D.W. (1992). A low-temperature scanning electron microscopy study of ice cream. 1. Techniques and general microstructure. *Food Structure, 11,* 1–10.

Carrington, A.K., Goff, H.D., & Stanley, D.W. (1996). Structure and stability of the glassy state in rapid and slowly cooled carbohydrate solutions. *Food Research International, 29,* 207–213.

Clark, A.H. (1987). The application of network theory to food systems. In J.M.V. Blanshard & P. Lillford (Eds.), *Food structure and behavior* (pp. 13–34). London: Academic Press.

Clark, A.H. (1992). Gels and gelling. In H.G. Schwartzberg & R.W. Hartel (Eds.), *Physical chemistry of foods* (pp. 263–305). New York: Marcel Dekker.

Clark, A.H., Judge, F.J., Richards, J.B., Stubbs, J.M., & Suggett, A. (1981). Electron microscopy of network structures in thermally-induced globular protein gels. *International Journal of Peptide Protein Research, 17,* 380–392.

Clark, A.H., & Ross-Murphy, S.B. (1987). Structural and mechanical properties of biopolymer gels. *Advances in Polymer Science 83,* 57–192.

Colombo, V.E., & Späth, P.J. (1981). Structures of various types of gels as revealed by scanning electron microscopy (SEM). *Scanning Electron Microscopy, 3,* 515–527.

Colonna, P., Tayeb, J., & Mercier, C. (1989). Extrusion cooking of starch and starchy products. In C. Mercier, P. Linko, & J.M. Harper (Eds.), *Extrusion cooking* (pp. 247–319). St. Paul, MN: American Association of Cereal Chemists.

Coupland, J.N., & McClements, D.J. (1996). Lipid oxidation in food emulsions. *Trends in Food Science and Technology, 7,* 83–91.

Crowe, J.H., Carpenter, J.F., Crowe, L.M., & Anchordoguy, T.J. (1990). Are freezing and dehydration similar stress vectors? A comparison of modes of interaction of stabilizing solutes with biomolecules. *Cryobiology, 27,* 219–231.

Diller, K.R. (1981). Quantitative low temperature optical microscopy of biological systems. *Journal of Microscopy, 126,* 9–28.

Dimick, P.S. (1991). Principles of cocoa butter crystallization. *Manufacturer and Confectioner, 71*(5), 109–114.

Djabourov, M. (1991). Gelation: A review. *Polymer International, 25,* 135–143.

Doi, E. (1993). Gels and gelling of globular proteins. *Trends in Food Science and Technology, 4,* 1–5.

Doran, P.M. (1995). *Bioprocess engineering principles.* London: Academic Press.

Farnum, C., Stanley, D.W., & Gray, J.I. (1976). Protein-lipid interactions in soy films. *Canadian Institute of Food Science Technology Journal, 9,* 201–206.

Fennema, O.R. (1996). *Food chemistry* (3rd ed.). New York: Marcel Dekker.

Franks, F. (1985). *Biophysics and biochemistry at low temperature.* Cambridge: Cambridge University Press.

Gibson, W. (1992). Gellan gum. In A. Imeson (Ed.), *Thickening and gelling agents for food* (pp. 227–249). London: Blackie Academic and Professional.

Goff, H.D., & Jordan, W.K. (1989). Action of emulsifiers in promoting fat destabilization during the manufacture of ice cream. *Journal of Dairy Science, 72,* 18–29.

Goff, H.D., Liboff, M., Jordan, W.K., & Kinsella, J.E. (1987). The effects of polysorbate 80 on the fat emulsion in ice cream mix: Evidence from transmission electron microscopy studies. *Food Microstructure, 6,* 193–198.

Goff, H.D., & Sahagian, M.E. (1996). Glass transitions in aqueous carbohydrate solutions and their relevance to frozen food stability. *Thermochimica Acta, 280,* 449–464.

Gómez, M., & Aguilera, J.M. (1984). A physicochemical model for extrusion of corn starch. *Journal of Food Science, 49,* 40–43, 63.

Guenet, J.-M. (1992). *Thermoreversible gelation of polymers and biopolymers.* New York: Academic Press.

Guy, R.C.E., & Horne, A.W. (1988). Extrusion and co-extrusion of cereals. In J.M.V. Blanshard & J.R. Mitchell (Eds.), *Food structure: Its creation and evaluation* (pp. 331–349). London: Butterworths.

Harris, P. (1990). *Food gels.* London: Elsevier Applied Science.

Hartel, R.W. (1998a). Phase transitions in ice cream. In M.A. Rao & R.W. Hartel (Eds.), *Phase/state transitions in foods* (pp. 327–368). New York: Marcel Dekker.

Hartel, R.W. (1998b). Phase transitions in chocolate and coatings. In M.A. Rao & R.W. Hartel (Eds.), *Phase/state transitions in foods* (pp. 217–251). New York: Marcel Dekker.

Heertje, I. (1993). Structure and function of food products: A review. *Food Structure, 12,* 343–364.

Hermansson, A.-M. (1988). Gel structure of food biopolymers. In J.M.V. Blanshard & J.R. Mitchell (Eds.), *Food structure: Its creation and evaluation* (pp. 25–40). London: Butterworths.

Hermansson, A.-M., & Kidman, S. (1995). Starch: A phase-separated biopolymer system. In S.E. Harding, S.E. Hill, & J.R. Mitchell (Eds.), *Biopolymer mixtures* (pp. 225–245). Nottingham, England: Nottingham University Press.

Hiltner, A., Cassidy, J.J., & Baer, E. (1985). Mechanical properties of biological polymers. *Annual Review of Materials Science 15,* 455–482.

Howlett, M., & Hall, M. (1985). The cold stage in food research. *Laboratory Practice, 34*(10), 87–88.

Imeson, A. (1992). *Thickening and gelling agents for food.* London: Blackie Academic and Professional.

Järenbeck, L., & Liljemark, A. (1975). Ultrastructural changes during frozen storage of cod *(Gadus morhua L.).* 2. Structure of extracted myofibrillar proteins and myofibril residues. *Journal of Food Technology, 10,* 309–325.

Jewell, G.G., & Heathcock, J.F. (1988). Structured fat systems. In J.M.V. Blanshard & J.R. Mitchell (Eds.), *Food structure: Its creation and evaluation* (pp. 279–295). London: Butterworths.

Jolley, P.D., & Purslow, P.P. (1988). Reformed meat products: Fundamental concepts and new developments. In J.M.V. Blanshard & J.R. Mitchell (Eds.), *Food structure: Its creation and evaluation* (pp. 231–264). London: Butterworths.

Jones, K.W., & Mandigo, R.W. (1982). Effects of chopping temperature on microstructure of meat emulsions. *Journal of Food Science, 47,* 1930–1935.

Juriaanse, A.C., & Heertje, I. (1988). Microstructure of shortenings, margarine and butter: A review. *Food Microstructure, 7,* 181–188.

Kalab, M., Allan-Wojtasand, P., & Shea Miller, S. (1995). Microscopy and other imaging techniques in food structure analysis. *Trends in Food Science and Technology, 6,* 177–186.

Kanzawa, Y., Koreeda, A., Harada, A., Okuyama, K., & Harada, T. (1990). Significance of molecular association for polysaccharide gels in foods. *Journal of Home Economics (Japan) 41,* 951–956.

Karel, M., Buera, M.P., & Levy, G. (1993). In J.M.V. Blanshard & P.J. Lillford (Eds.), *The glassy state in foods* (pp. 133–156). Nottingham, England: Nottingham University Press.

Karel, M., Fennema, O.R., & Lund, D.B. (1975). *Principles of food science.* Vol. 2. *Physical principles of food preservation.* New York: Marcel Dekker.

Kent, N.L. (1975). *Technology of cereals* (2nd ed.). Oxford: Pergamon International Library.

Klibanov, A.M. (1995). What is remembered and why? *Nature, 374,* 596.

Kokini, J.L., Cocero, A.M., Madeka, H., & de Graaf, E. (1994). The development of state diagrams. *Trends in Food Science and Technology, 5,* 281–288.

Lai, L.S., & Kokini, J.L. (1991). Physicochemical changes and rheological properties of starch during extrusion. [Review]. *Biotechnology Progress 7,* 251–266.

Ledward, D.A., & Mitchell, J.R. (1988). Protein extrusion: More questions than answers? In J.M.V. Blanshard & J.R. Mitchell (Eds.), *Food structure: Its creation and evaluation* (pp. 219–229). London: Butterworths.

Ledward, D.A., & Tester, R.F. (1994). Molecular transformations of proteinaceous foods during extrusion processing. *Trends in Food Science and Technology, 5,* 117–120.

Lee, C.M. (1984). Surimi process technology. *Food Technology, 38*(11), 69–80.

Lee, C.M., Carrol, R.J., & Abdollahi, A. (1981). A microscopical study of the structure of meat emulsions and its relationship to thermal stability. *Journal of Food Science, 46,* 1789–1793, 1804.

Lelievre, J., Mirza, I.A., & Tung, M.A. (1992). Failure testing of gellan gels. *Journal of Food Engineering, 16,* 25–37.

Levine, H., & Slade, L. (1988a). Principles of cryostabilization technology from structure/property relationships of carbohydrate/water systems. [Review]. *Cryo Letters 9,* 21–63.

Levine, H., & Slade, L. (1988b). "Collapse" phenomena: A unifying concept for interpreting the behavior of low moisture foods. In J.M.V. Blanshard & J.R. Mitchell (Eds.), *Food structure: Its creation and evaluation* (pp. 149–180). London: Butterworths.

Lewis, D.F. (1988). An electron microscopist's view of foods. In J.M.V. Blanshard & J.R. Mitchell (Eds.), *Food structure: Its creation and evaluation* (pp. 367–385). London: Butterworths.

Little, A.C., Aguilera, J.M., Morales, A., & Kokini, J.L. (1997). Changes in soy protein during heating analyzed by pressure rheometry. *Food Science and Technology International, (Japan) 3,* 130–133.

Love, R.M. (1968). Ice formation in frozen muscle. In J. Howthorne & E.J. Rolfe (Eds.), *Low temperature biology of foodstuffs* (pp. 105–133). New York: Pergamon Press.

Lucca, P.A., & Tepper, B.J. (1994). Fat replacers and the functionality of fat in foods. *Trends in Food Science and Technology, 5,* 12–19.

Lugay, J.C., & Kim, M.K. (1981). Freeze alignment: A novel method for protein texturization. In D.W. Stanley, E.D. Murray, & D.H. Lees (Eds.), *Utilization of protein resources* (pp. 177–187). Westport, CT: Food and Nutrition Press.

Luyet, B. (1968). Basic physical phenomena in the freezing and thawing of animal and plant tissue. In D.K. Tressler, W.B. van Arsdel, & M.J. Copley (Eds.), *The freezing preservation of foods* (pp. 1–25). Westport, CT: AVI Publishing Co.

Luyten, H., van Vliet, T., & Walstra, P. (1991). Characterization of the consistency of Gouda cheese: Fracture properties. *Netherlands Milk Dairy Journal, 45,* 55–80.

Martino, M.N., & Zaritzky, N.E. (1988). Ice crystal size modifications during frozen beef storage. *Journal of Food Science, 53,* 1631–1637.

McGee, H. (1984). *On food and cooking.* New York: Scribner's.

Means, W.J., & Schmidt, G.R. (1986). Algin/calcium gel as a raw and cooked binder in structured beef steaks. *Journal of Food Science, 51,* 60–65.

Morris, E.R. (1990). Mixed polymer gels. In P. Harris (Ed.), *Food gels* (pp. 291–359). London: Elsevier Applied Science.

Moss, R., Stenvert, N.L., Kingswood, K., & Pointing, G. (1980). The relationship between wheat microstructure and flourmilling. *Scanning Electron Microscopy, III,* 613–620.

Murray, E.D., Myers, C.D., Barker, L.D., & Maurice, T.J. (1981). Functional attributes of proteins: A noncovalent approach to processing and utilizing plant proteins. In D.W. Stanley, E.D. Murray, & D.H. Lees (Eds.), *Utilization of protein resources* (pp. 158–176). Westport, CT: Food and Nutrition Press.

Nakai, S. (1983). Structure-function relationships of food proteins with an emphasis on the importance of protein hydrophobicity. *Journal of Agricultural and Food Chemistry, 31,* 676–683.

National Research Council. (1994). *Hierarchical structures in biology as a guide for new technology materials.* Washington, DC: National Academy Press.

Nielsen, B.J. (1973). Building and formation of ice cream microstructure during processing. *Milk Dairy, 52,* 10.

Oakenfull, D.G. (1991). The chemistry of high-methoxyl pectins. In R.H. Walter (Ed.), *The chemistry and technology of pectins* (pp. 87–108). New York: Academic Press.

Oelker, P. (1987). Elektronenoptische Beobachtungen uber ultrastrukturelle Bedingungen der Fettemulgierung und Wasserbindung in Bruhwurstbrat. *Fleischwirtschaft, 67,* 1405–1409.

Ohshima, T., Suzuki, T., & Koizuma, C. (1993). New developments in surimi technology. *Trends in Food Science and Technology, 4,* 157–163.

Onsoyen, E. (1992). Alginates. In A. Imeson (Ed.), *Thickening and gelling agents for food* (pp. 2–24). London: Blackie Academic and Professional.

Osada, Y., & Ross-Murphy, S.B. (1993). Intelligent gels. *Scientific American, 268,* 42–47.

Pomeranz, Y. (1970). From wheat to bread: A biochemical study. *American Scientist, 61,* 683–691.

Poppe, J. (1992). Gelatin. In A. Imeson (Ed.), *Thickening and gelling agents for food* (pp. 98–123). London: Blackie Academic and Professional.

Potts, M. (1994). Desiccation tolerance of prokaryotes. *Microbiology Reviews, 58,* 755–805.

Reid, D.S. (1978). A programmed controlled temperature microscopy stage. *Journal of Microscopy, 114,* 241–244.

Reid, D.S. (1980). Cryomicroscope studies of the freezing process in tissue and in model systems. *Revue Internationale du Froid, 3*(4), 226–228.

Ring, R.A., & Danks, H.V. (1994). Desiccation and cryoprotection: Overlapping adaptations. *Cryobiology, 15,* 181–190.

Ross-Murphy, S.B., & McEvoy, H. (1986). Fundamentals of hydrogels and gelation. *British Polymers Journal J., 18,* 2–7.

Ross-Murphy, S.B., & Todd, S. (1983). Ultimate tensile measurement of filed gelatin gels. *Polymer, 24,* 481–486.

Sakayima, T., & Yano, T. (1994). Temperature dependence of the effective thermal conductivity of food gels impregnated with air bubbles. *Journal of Chemical Engineering (Japan), 27,* 291–295.

Sime, W.J. (1990). Alginates. In P. Harris (Ed.), *Food gels* (pp. 53–78). London: Elsevier Applied Science.

Slade, L., & Levine, H. (1991). Beyond water activity: Recent advances based on an alternative approach to the assessment of food quality and safety. *Critical Reviews Food Science and Nutrition, 30*(2, 3), 1–360.

Smith, A.K., & Circle, S.J. (1972). *Soybeans: Chemistry and technology* (Vol. 1). Westport, CT: AVI Publishing Co.

Stading, M., Langton, M., & Hermansson, A.-M. (1993). Microstructure and rheological behavior of particulate β-lactoglobulin gels. *Food Hydrocolloids, 7,* 195–212.

Stanley, D.W. (1986). Chemical and structural determinants of texture of fabricated foods. *Food Technology, 40*(3), 65–68, 76.

Stanley, D.W. (1987). Food texture and microstructure. In H.R. Moskowitz (Ed.), *Food texture: Instrumental and sensory measurement* (pp. 35–64). New York: Marcel Dekker.

Stanley, D.W., Aguilera, J.M., Baker, K.W., & Jackman, R.L. (1998). Structure/property relationships of foods as affected by processing and storage. In M.A. Rao & R.W. Hartel (Eds.), *Phase/state transitions in foods* (pp. 1–56). New York: Marcel Dekker.

Stanley, D.W., Cumming, D.B., & deMan, J.M. (1972). Texture-structure relationships in texturized soy protein. 1. Textural properties and ultrastructure of rehydrated spun soy fibers. *Canadian Institute of Food Science Technology Journal, 5,* 118–123.

Stanley, D.W., Goff, H.D., & Smith, A.K. (1996). Texture-structure relationships in foamed dairy emulsions. *Food Research International, 29,* 1–13.

Stone, A.P., & Stanley, D.W. (1992). Mechanisms of fish muscle gelation. *Food Research International, 25,* 381–388.

Stone, A.P., & Stanley, D.W. (1994). Gelation of myofibrillar proteins at low ionic strengths measured using dynamic oscillatory techniques. *Food Research International, 27,* 155–163.

Svegmark, K., & Hermansson, A.-M. (1993). Microstructural and rheological properties of composites of potato starch granules and amylose: A comparison of observed and predicted structures. *Food Structures, 12,* 181–193.

Tanaka, T. (1981). Gels. *Scientific American, 110,* 124–138.

Tannahill, R. (1988). *Food in history.* New York: Crown Publishers.

Tanner, J.E. (1975). Observation of rapid freezing of salt solutions. *Cryobiology, 12,* 353–363.

Tolstoguzov, V.B. (1988). Creation of fibrous structures by spinneretless spinning. In J.M.V. Blanshard & J.R. Mitchell (Eds.), *Food structure: Its creation and evaluation* (pp. 181–196). London: Butterworths.

Tolstoguzov, V.B. (1995). Some physico-chemical aspects of protein processing in foods: Multicomponent gels. *Food Hydrocolloids, 9,* 317–332.

Tolstoguzov, V.B. (1997a). Thermodynamic aspects of dough formation and functionality. *Food Hydrocolloids, 11,* 181–193.

Tolstoguzov, V.B. (1997b). Multicomponent biopolymer gels. *Food Hydrocolloids, 11,* 159–170.

Tolstoguzov, V.B., Grinberg, V., & Gurov, A. (1985). Some physicochemical approaches to the problem of protein texturization. *Journal of Agricultural and Food Chemistry 33,* 151–159.

Urlacher, B., & Dalbe, B. (1992). Xanthan gum. In A. Imeson (Ed.), *Thickening and gelling agents for food* (pp. 202–226). London: Blackie Academic and Professional.

Vaeck, S.V. (1960, June). Cacao butter and fat bloom. *Manufacturer and Confectioner,* 35–46, 71–74.

van de Voort, F.R., Stanley, D.W., & Edamura, R. (1984). Improved utilization of dairy proteins: Coextrusion of casein and wheat flour. *Journal of Dairy Science, 67,* 749–758.

Verrez-Bagnis, V., Bouchet, B., & Gallant, D.J. (1993). Relationship between the starch granule structure and the textural properties of heat-induced surimi gels. *Food Structure, 12,* 309–320.

Vincent, J.F.V. (1993). Mechanical and fracture properties of fruits and vegetables. In E. Dickinson & P. Walstra (Eds.), *Food colloids and polymers: Stability and mechanical properties* (pp. 191–203). London: Society of Chemistry and Industry.

Visser, J. (1988). Dry spinning of milk proteins. In J.M.V. Blanshard & J.R. Mitchell (Eds.), *Food structure: Its creation and evaluation* (pp. 197–218). London: Butterworths.

Walstra, P. (1996). Dispersed systems: Basic considerations. In O.R. Fennema (Ed.), *Food chemistry* (pp. 95–135). New York: Marcel Dekker.

Watts, B., Elias, L.G., & Rios, B. (1987). Consumer utilization and acceptability of raw and cooked black beans in Guatemala. Ottawa, Canada: International Development Research Center.

Wielinga, W.C. (1983). Application of gum-based stabilization systems in ice cream and fruit ices. In G.O. Phillips, D.J. Wedlock, & P.A. Williams (Eds.), *Gums and stabilizers for the food industry.* Vol. 2. *Applications of hydrocolloids* (pp. 251–276). New York: Pergamon Press.

Windhab, E.J. (1995). Rheology in food processing. In S.T. Beckett (Ed.), *Physicochemical aspects of food processing* (pp. 80–116). London: Blackie Academic and Professional.

Wolf, W.J., & Cowan, J.C. (1975). *Soybeans as a food source.* Cleveland, OH: CRC Press.

Xiong, Y.L., Aguilera, J.M., & Kinsella, J.E. (1991). Emulsified milkfat effects on rheology of acid-induced milk gels. *Journal of Food Science, 56,* 920–925.

Zeelenberg-Miltenburg, M.J. (1995). *When fat and water change places: The development of a "fat free" margarine.* Vlaardingen, The Netherlands: Unilever Research Laboratorium.

Ziegler, G.R., & Foegeding, E.A. (1990). The gelation of proteins. *Advances in Food Science and Nutrition, 34,* 203–298.

# CHAPTER 6

# Food Microstructure and Quality

## 6.1 INTRODUCTION

Quality factors are those that determine the worth or value of a food product to the consumer. This is a broad definition that encompasses many different factors, including safety, nutritional attributes, and cost. The food scientist, however, has more influence over those factors that are perceived through the human sensory apparatus. These are commonly acknowledged to include appearance (e.g., size, shape, defects, and general condition), color, flavor, and texture. Structural organization plays an important role in determining all of these sensory properties, but nowhere is its influence more directly apparent than in regard to the attribute of texture. The closeness of structure-texture relationship is in fact indicated in the following well-known definition of texture:

> Texture is the sensory and functional manifestation of the structural and mechanical properties of foods, detected through the senses of vision, hearing, touch, and kinesthetics. (Szczesniak, 1998, p. 55)

According to another definition, texture is a manifestation of the organization of structural elements (deMan, 1976). While the textural properties of food are perceived by the consumer through the senses, this does not mean that texture cannot be studied, perhaps even accurately measured, through the use of instruments. While in the previous chapters we were interested in the building of structures in foods, here we focus on structure assessment and breakdown after the products have been made. Structure breakdown occurs during storage, sometimes during preparation, and obviously during mastication.

## 6.2 MEASUREMENT OF TEXTURE

A more general definition of texture than those above is that it is the composite of all physical characteristics sensed by the feeling of touch that are related to deformation under an applied force and are measured objectively in terms of force, distance, and time (Bourne, 1982). This implies that food texture is ultimately assessed by the consumer through sensory understanding. Although all textural perceptions do not arise in the mouth (some signals come from tactile and auditory organs), the main stimuli originate in the oral cavity as a result of nerve impulses from biting and crushing food with the teeth and the action of the tongue and palate. These stimuli are then processed in the brain and yield an integrated idea of texture.

It is not hard to understand why the most common approach to measuring food texture is to use human subjects. Customarily, a sensory panel is recruited and trained to respond to the particular textural attributes of interest. A guiding principle in this type of work is that one should try to control all extraneous signals or distractions that would confuse or hinder a human judge, such as sample properties (e.g., color, flavor, size, and temperature) and environmental factors (e.g., light, noise, odor, and surroundings).

By following this precept and adhering to a diligent training and selection regimen, it is possible, if the correct questions are asked, to obtain meaningful responses about the sensory properties of food texture. Then if appropriate statistical analyses are applied to the data, correct inferences can be drawn about, for example, the influence of various processing treatments on the textural properties of a product. Those who think this process has been overstated have likely never engaged in sensory analysis. It is, at first glance, one of the easiest techniques available to the food scientist, but in fact it is the most difficult to use properly. Particular care must be taken when applying statistical techniques, for, while the inquiring mind is rarely blessed with certainty, it is also true that data, if sufficiently tortured, can be made to confess to anything.

It is for this reason that instrumental methods of texture analysis have become so popular. Instruments offer the advantages of reproducibility and they are relatively impervious to their surroundings. On the other hand, they are capable of measuring, however accurately, only one facet of the complex set of mechanical properties that define food texture, and they are incapable of performing the kind of mental integration that allows a multitude of stimuli to yield a texture Gestalt. In texture measurement, as with many other things, it is often better to be approximately right than precisely wrong.

Because of the inherent difference in what is being measured, it is to be expected that sensory and instrument-based estimates of food texture often do not closely correlate. The lack of a high degree of correlation presents operational problems, since the usual goal is to attempt to replace the cumbersome, time-consuming sensory panel with a more convenient instrumental device. Rarely, however, will a single test provide the wide gamut of information delivered by sensory analysis. If replacement of human subjects by instruments is attempted, it is important to retain the panel to ensure that the devices chosen continue to measure the textural parameters identified as important by the judges.

Inevitably, the question arises as to which approach, sensory or instrumental, best gauges the "true" textural quality of the food. Often the debate centers around the "subjective" nature of sensory tests versus the "objective" nature of instrumental tests. This type of thinking is now outdated, and, as will be discussed, sensory methodologies are available that can yield objective data (free from human bias and reproducible) when judges are properly trained in measuring intensities of stimuli. On the other hand, it is sometimes desirable to ask a sensory panel to record hedonic or acceptability data in which the response is a measure of their liking or disliking of a food (product development and ingredient replacement decisions require this). As well, "objective" instrument-based tests are often less objective than researchers care to believe. The fact is that the human-instrument interface can result in much subjectivity. The overall conclusion should be that, when faced with a difficult task such as measuring texture in materials as complex as food, a correlative approach, using both sensory and instrumental methodologies, is required. The reader will find that successful food scientists employ this approach.

In the end, the issue of sensory versus instrumental food texture testing is unresolvable. Attempts to resolve it, however, have led to some interesting and useful perceptions. Since both sensory and instrumental data are usually collected, there must be some way in which the relationship between them can be expressed. Historically, simple linear correlation analysis has been the statistical procedure of choice, but it has been criticized for a number of reasons. First, it only indicates how one set of numbers varies with another. Second, it assumes, since this is a parametric statistic, that the data are normally distributed, but since sensory data are often gathered using ordinal (ranking) or nominal (category) scales, the assumption may be incorrect. Third, because several different parameters are usually measured by both sensory and instrumental techniques, a statistical test is required that will handle data in a matrix form.

Multiple regression and correlation analyses have also been used, but they too are open to criticism, since results are affected by the correlation of variables within each set. At the present time, the use of multivariate analyses would seem to be the best approach, since it allows for the efficient simplification and interpretation of a large number of variables simultaneously. Multivariate analysis techniques are concerned with the analysis of many variables, and because they can handle concurrently multiple dependent and independent variables derived from parametric and nonparametric data, they are well suited for sensory applications. Common procedures included in this group of statistical tests include cluster analysis, principal component analysis, and stepwise discriminant analysis.

As an example, let us consider the textural evaluation of a restructured beef product. Restructuring technology encompasses a set of techniques by which raw meat is particulated and then recombined into a uniform edible product (see Section 5.4.2). Restructuring can be profitable for meat processors, for they can utilize less desirable cuts of meat, particularly those with high amounts of connective tissue, to create steaklike products of higher market value and at the same time meet consumer demand for convenient individual entrees. Their main quality problem is that restructured beef products can have an unacceptable texture due to the presence of excess collagen and the overbinding of meat particles.

A study (Bernal, Bernal, Gullett, & Stanley, 1988) was undertaken to develop sensory and instrumental methods to evaluate the texture of laboratory-produced restructured beef steakettes and to determine if the instrument measurements could predict sensory responses. In this type of work, it is important to have a wide range of products available for evaluation, and thus processing variables included connective tissue, flake size, and salt levels. A trained panel sensory profile (discussed subsequently) was developed following an extensive training period. Ten attributes were selected that represented the major textural and flavor attributes of the samples. Five instrumental methods were used, including tensile tests both parallel and perpendicular to the surface plane, a shearing test, and a compression test. Evaporative drip and total cooking losses in cooking the product from the frozen state were also determined.

In generating the sensory profile, the panel carefully defined the attributes deemed to be of importance. Scoring was done by evaluating perceived intensity on a 10 cm unstructured scale anchored by opposite definitions (not very springy/very springy, not very hard/very hard, etc.). This type of sensory data collection has been reported to be suitable for parametric statistical analysis, and the data collection was validated by checking the adherence of the data set to the assumption of normality by examining skewness and kurtosis and the assumption of homoscedasticity; it was found that no transformations were required.

Correlation and regression procedures were used initially to determine if linear and predictive relationships existed among sensory attributes and sensory judgments. The objective of such analyses is to predict the outcomes of the expensive and time-consuming sensory tests by the rapid and convenient instrument-based methods. Accurate prediction can only occur, however, if the instrument-generated values provide reliable information regarding sensory attributes.

Since information was required about textural quality, the data sets were reduced by omitting all sensory flavor dimensions. Further reduction was achieved by removing the instrumental methods that were not, as indicated by the prior analyses, useful in detecting differences among the processing variables. The relationships among the remaining sensory attributes and instrumental methods, all directed toward textural quality, were first tested by visual examination of their scatterplots. None of these revealed relationships in which a variable required transformation to achieve linearity. Self–cross correlations of both data sets showed several highly significant correlations among the attributes, indicating that the tests were likely attempting to measure the same underlying property.

Principal components analysis was performed on the trained panel sensory data set in order to determine if the attributes of texture could be described by fewer dimensions. Two factors were extracted by this multivariate technique, together accounting for about 75% of the variance in sen-

**Table 6–1** Factors Extracted by Principal Components Analysis of Sensory Data for Restructured Beef Steakettes

| *Sensory Attribute* | *Factor 1* | *Factor 2* |
|---|---|---|
| Springiness | 0.22[a] | 0.92 |
| Hardness | 0.92 | 0.31 |
| Cohesiveness | 0.49 | 0.85 |
| Moisture/oil release | −0.11 | 0.94 |
| Chewiness of the mass | 0.96 | 0.16 |
| Size of chewed pieces | 0.09 | 0.30 |
| Chewiness | 0.98 | 0.12 |
| Amount of connective tissue | 0.98 | −0.09 |
| Cumulative variance | 48.2 | 75.1 |

[a] Rotated factor loadings.
*Source:* Adapted from Bernal et al., 1988.

sory response among the steakette-processing treatments (Table 6–1). Based on this and previous analyses performed on the data, it was concluded that the attributes in factor 1 were most influenced by the presence of connective tissue, and factor 1 was termed the connective tissue factor. Similarly, the second factor was thought to be representative of the strength of the particle bind of the product.

The relationship between these two principal factors and the instrumental methods was determined by stepwise regression analysis. Results of this regression (Table 6–2) showed that in each case one instrumental method could predict over 80% of the variance. For factor 1, a shear test contributed the majority to $r^2$, further implicating connective tissue; for factor 2, tensile tests, thought to reflect the strength of particle binding, were predominant. This shows how useful multivariate techniques can be used for reducing a confusing mass of data to workable proportions and for providing clues to the structural basis of quality factors.

**Table 6–2** Stepwise Regression of Instrumental Methods on Sensory Factors of Restructured Beef Steakettes

| *Sensory Factor* | *Step* | *Instrumental Method* | *Partial $R^2$* | *Model $R^2$* |
|---|---|---|---|---|
| 1 | 1 | Shear | 0.84 | 0.84 |
| | 2 | Tensile (1) | 0.04 | 0.88 |
| | 3 | Drip | 0.05 | 0.93 |
| 2 | 1 | Tensile (1) | 0.81 | 0.81 |
| | 2 | Tensile (2) | 0.06 | 0.87 |
| | 3 | Compression | 0.03 | 0.90 |

*Source:* Adapted from Bernal et al., 1988

### 6.2.1 Sensory Procedures

Although it is not the purpose of this section to detail procedures for analyzing the textural properties of food, it might be useful to outline the more commonly used methodologies available to the food scientist for textural property measurements. More complete information may be obtained from the works listed at the end of the chapter. Also, the discussion will be limited mainly to solid foods because of the obvious structural connections.

The approach taken to sensory analysis will be determined by the questions that need to be answered. For example, if a food manufacturer wishes to know if a new or "improved" product is acceptable to the public or, more importantly, if the degree of acceptability is high enough to ensure sufficient sales volume, a *consumer panel* will usually be employed. In this case, samples are provided to, hopefully, typical consumers of the product being evaluated. It is important that only simple questions be asked and that a wide range of responses be obtained. Success in this type of endeavor is dependent upon selection of a representative demographic sample and development of an appropriate questionnaire.

Consumer acceptance testing is often conducted at a commercial site, but a second type of sensory testing is done in a controlled laboratory setting. In this type, an *acceptance panel* is asked questions of a hedonic nature: How much is this sample liked/disliked and why? The panel is trained, and the data generated are often used for purposes of product improvement or ingredient replacement.

The third type of panel is more analytical and objective. An *analytical panel* attempts to quantitate intensity differences in the textural attributes of samples. The questions might include "how tough/tender is this meat sample?" or "how viscous/thin is this soup sample?" The objective is to obtain numerical data that can later be analyzed statistically. Obviously, such a formidable task requires a well-trained and dedicated panel of judges.

Although it is usually desirable to ensure independent judgments from panelists, members work together closely in the first stages of an analytical evaluation in order to develop a questionnaire that satisfactorily reflects the complex textural sensations encountered. One approach is to separate the sensations on a temporal scale so that the first response recorded reflects the initial impression made by the sample when bitten by the front teeth, the second indicates the sensation when the material is transferred to the back teeth, and the last describes the feel of swallowing the bolus. A panelist may be asked to record the number of chews required to prepare a given mass or volume for swallowing or to indicate the saliva flow produced by the material or to comment on the residual mouthfeel. Tactile, visual, and auditory responses may be elicited as well. Because of the difficulty in mentally cataloguing so many different impression, it is useful, if possible, to employ standard materials that members can refer to and utilize to gauge their responses.

The most comprehensive single system for measuring sensory texture is the sensory texture profile procedure based on the pioneering work of Dr. Alina Szczesniak at General Foods (Szczesniak, 1983, 1998). This procedure involves identifying the textural characteristics perceived in a food product, the intensity of each, and the order in which they are perceived. It is based on a series of classifications. The first grouping of textural characteristics is into three main classes: mechanical characteristics, geometric characteristics, and others (mainly related to moisture and fat content). The characteristics are later distributed into subclasses, with each characteristic defined in both physical and sensory terms.

Implementation of the sensory texture profile procedure demands a highly motivated, specially trained panel having access to reference standards for each textural property. For example, the hardness scale ranges from cream cheese (1) to rock candy (9) and the fracturability scale ranges from corn muffin (1) to peanut brittle (7). In order to achieve proper results, care must be taken at each step—selection of the panel, panel training, establishment of standard rating scales, and deciding upon a ballot. Texture profiling gives the most objective results of the sensory procedures now available. It allows the entire texture of a food product to be described analytically from initial impression through complete mastication and swallowing. One application of this expensive and exhaustive process is in product development, where it is employed to design foods and quantitate differences between prototypes and targets. Also, the data can often be used to discern components and structures responsible for texture and to identify the clusters of attributes that usually occur during sensory analysis.

### 6.2.2 Instrumental Procedures

Given our discussion of sensory analysis, one might ask, if we want to measure, say, "spreadability," why not just use a sensory test? Because a sensory test, though potentially valuable, is only an empirical test. It might tell us which sample is more spreadable, but it does not give numerical data in fundamental units, nor does it tell us why one sample is more spreadable than another. Instrumental tests can bring us one step closer to answering our questions.

As with sensory approaches to texture measurement, instrumental procedures fall into several major categories. They can attempt to imitate the human masticatory process, they can be used to measure some fundamental mechanical property of the food, or they can used to measure a nonmechanical physical property highly correlated with texture. All three of these approaches have their applications. Which one is chosen should depend on the problem being faced and the questions being asked.

The use of imitative devices is based on the notion that the closer texture-measuring equipment duplicates the action occurring in the mouth, the closer the results will be to those achieved by human subjects. It will be appreciated that mimicking the mouth is a formidable task. Mastication through chewing with the teeth is accompanied by many related actions, including saliva flow, pressure by the tongue and palate, and swallowing. Since mechanical devices are essentially limited to linear and circular motions, it is clearly beyond current engineering capabilities to duplicate such a biological process.

On the other hand, some significant attempts have been made. Again, the General Foods group has been the forerunner in this area. These researchers developed an instrument, the General Foods texturometer, designed to simulate mastication by means of a mechanical chewing device. An arm, moving in an arc, forces various designs of plungers into a food sample placed on a stationary platform attached to a strain gauge. The amplified results of strain gauge deformation are recorded for subsequent measurement. The instrument operates by partially compressing the sample twice, imitating the first two bites taken of a food. Several parameters are obtained from the resulting force-time plots (hardness, fracturability, cohesiveness, adhesiveness, springiness, gumminess, and chewiness), and these have been found to correlate highly with sensory ratings. The values for the various parameters make up the instrumental texture profile.

There are only a few of these instruments available, but modified texture profile analysis can be done using a so-called universal testing machine (an instrument characterized by a variable drive system, interchangeable test cells, and a force measuring and recording system). The sample is placed on a stationary flat horizontal plate and partially compressed twice by a linearly moving flat horizontal plate. The data collected are in the form of a force-distance curve. Again, numerous parameters are obtained. Both sensory and instrumental texture profiling are based on the premise that texture is the combination of a number of different properties, and multiple kinds of measurements are more apt to reflect the elements of texture (and thus texture itself) than simple one-point tests.

Imitation is one approach. Another is to subject a food to instrumental analysis in a fundamental device governed by strict engineering principles and obtain data in basic SI physical units. For example, it is relatively simple to compress a sample between two flat plates and note the Newtons of force required to reach the yield point or to measure in Pa·s the resistance to flow of a fluid food. Unfortunately, such one-point measurements of mechanical properties are rarely adequate predictors of sensory response. Food samples can, for example, exhibit identical yield points but differ widely in other textural attributes of great importance to overall quality perception (e.g., ability to induce saliva flow, mouthfeel, residue following chewing, etc.).

New and improved instruments for measuring the mechanical properties of food have proliferated over the past two decades. The basics are pretty much the same, however. Testing solid food instrumentally requires equipment consisting of a drive system that imparts controlled linear or circular motion to a probe that then makes contact with a sample held in a test cell. The force required to achieve a certain deformation is recorded, as is, in some cases, the actual deformation over time. The component that varies from machine to machine is the test cell. In the past, instruments were often designed specifically for a single product, but universal testing machines, with variable drive mechanisms and interchangeable test cells, have become available and popular.

Perhaps the most important realization that influenced how texture is measured is that food materials exhibit viscoelastic behavior; that is, their textural response is dictated by both elastic and viscous characteristics and is time dependent. Thus, textural parameters measured objectively under one set of test conditions will generally differ from those using another set of conditions. Also, any application of force or stress will induce a certain amount of permanent deformation or strain, the extent of which depends on its duration. The classification of objective methods that involve any amount of compression, tension, or shear as "nondestructive" is misleading. Owing to the viscoelastic nature of foods, no amount of deformation or force applied to a food over a measurable time period can be regarded as totally nondestructive. With this realization came the application of basic rheological techniques to food testing situations.

From a structural point of view, dynamic testing can be quite useful since it may enable the testing of various structures within the sample nondestructively (see Section 3.8.6). By varying the frequency of the oscillatory input, it is theoretically possible to probe the numerous elements contributing to structure.

Modern dynamic testing procedures are of growing importance to food analysts for the following reasons:

- They provide information not obtainable using large deformation tests, since dynamic methods are particularly sensitive to such events as glass transitions, cross-linking, phase separation, and molecular aggregation.
- They are nondestructive, they give results in basic units, and theory-based values exist to which these results can be compared. This is in contrast to the vast majority of textural evaluations of food, which are of an empirical nature.
- They are well suited to viscoelastic food materials, since the instrumental output can be separated into a viscous and an elastic component and the sample can be tested over a wide frequency range.
- They have predictive capabilities and are therefore of interest in research and development. Some studies have shown significant correlations between dynamic measurements and sensory analysis.
- The physical quantities measured support structural interpretations, and processes such as gel formation can be monitored in real time.

The basic strategy in dynamic testing is to apply a small amount of constant stress over a relatively long period of time (unlike large deformation testing, where a large amount of stress is applied over a relatively short period of time). Thus, instead of all the bonds holding the network together collapsing almost instantaneously, they collapse as a function of time, and hopefully they can be resolved into separate events and associated with distinct structures. But when we chew, we do so at a reasonably high rate of speed. So what's the point? It is not to correlate small deformation data with sensory analysis but to find out what forces and structures are involved in the food network so that they can be duplicated or improved.

Dynamic testing is being used more frequently in food laboratories. It is, however, a complex methodology, and a full account lies beyond the scope of the present discussion. Readers are directed to other works for a comprehensive treatment, including Stanley, Stone, and Tung (1996) and others cited in the "Suggested Reading" section. The remainder of this chapter will be restricted to the discussion of large deformation procedures, which are more commonly used in the routine examination of food texture.

In the instrumental testing of food texture using large deformations, fundamental properties are infrequently measured; rather, forces and deformations are observed during an arbitrary test period. Although some instruments are still being used that give only one reading, usually maximum force, these are either being replaced or connected to recorders to give a graph of the complete deformation process as a function of distance or time. Such a force-deformation plot (Figure 6–1B) differs markedly from a theoretical stress-strain curve (Figure 3–21). It obviously contains a great deal of information, and much care must be taken in interpreting a force-deformation or force-time curve, since it is quite easy to misjudge the significance of certain aspects of the data or to ignore useful parts of the curve. A wealth of information about texture can be hidden in such characteristics as yield point, initial angle of force uptake, work of rupture, compression or elongation required to produce yield, number of peaks observed, the jaggedness of the curve, and so on. Barret, Normand, Peleg, and Ross (1992) indicated that the ir-

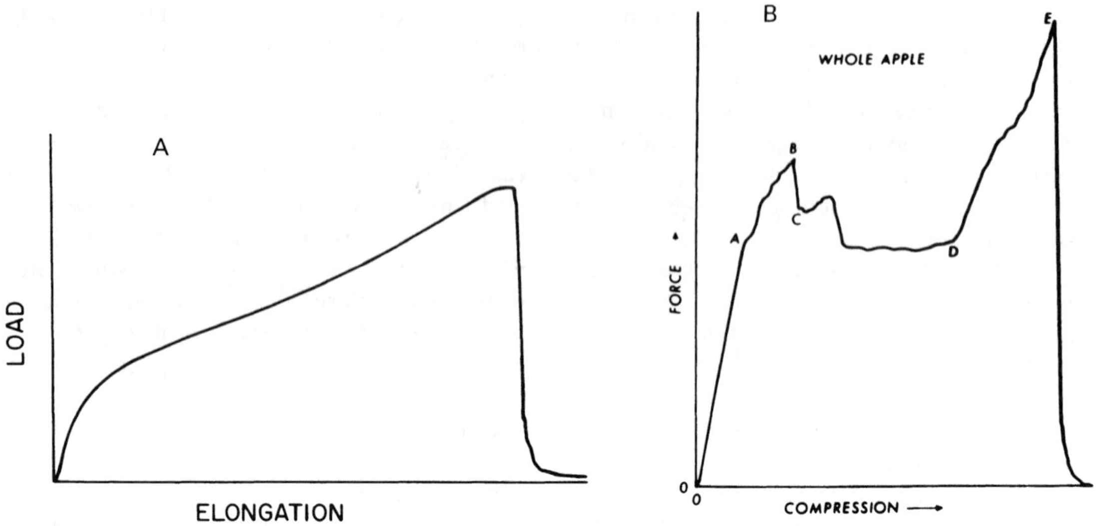


**Figure 6–1** (A) Theoretical stress-strain curve for a perfectly elastic body. (B) Force-deformation plot for compression of whole apple between two flat plates. *Source:* deMan and Stanley, 1984.

regularity of the pattern of the stress-strain curve of puffed extrudates was an intrinsic manifestation of the deformation and breakdown mechanism. These authors demonstrated that the apparent fractal dimension of the strain-stress curve was a convenient parameter for ruggedness assessment and that it increased as moisture content of the extrudate decreased. A particularly useful strategy is to examine the structure of the sample during deformation, either macro- or microscopically, as a means of revealing the nature of the material studied.

From a practical standpoint, perhaps the most important decision to be made by a researcher is the selection of an instrumental method. Several key factors play a role in this decision. The first consideration is the purpose of the test. For example, the aim may be to develop a quality control procedure. If this is the case, then a test should be sought that accurately measures the quality characteristic chosen, satisfactorily duplicates sensory evaluation, can be used routinely during all steps of the process, and can be performed within the time limits necessary for process control. Desired features will include ruggedness and versatility (within the boundaries set by cost). If, at the other end of the scale, the purpose is to undertake basic research on food texture, then precision, accuracy, and sophistication become primary.

Many other factors also influence method selection, factors too numerous and diverse to discuss fully. A valid approach is to judiciously apply all pertinent structural and compositional information, as well as information about the nature of the food, the objective of the testing, and the instrumentation available, to the selection of a group of tests that can be performed using the instrumentation available. This done, the next step is to evaluate preliminary tests performed on a broad spectrum of textures. In quality control situations, there are usually sensory evaluations involved, and, even though it is unlikely that a single instrumental test can replace integrated sensory responses, comparisons can be made during the selection process. As in the case of sensory analysis, instrumental data can often be reduced by multivariate statistical analysis to a few key independent measures, allowing subsequent testing to concentrate on these. It is recommended not to disband sensory panels upon the adoption of one or more instrumental tests if for no other reason than to assess intermittently the validity of the instrumental data.

Since what is usually required for the instrumental analysis of food texture is a test (or tests) that can accurately and quickly estimate sensory evaluations, it follows that, as long as the attribute selected correlates highly with texture, it does not have to be a mechanical attribute that is measured. In fact, quite a few physical properties measured with a nonmechanical device have been used to predict food texture. Chemical properties discovered through compositional analysis are also known to be associated with texture, but the time, expertise, and cost required to obtain the data make this option unpopular. Table 6–3 provides some examples of nonmechanical physical properties related to texture. Excluded from this category are mechanical properties revealed by measuring the behavior of a material under force. This restriction is more or less confining, depending upon whether gravity is included under the heading of "force" (if so, then consistometers, devices that measure flow, must be taken into account) and whether the effect of force is calculated in units other than length or time (if so, then water binding capacity techniques must be included). Also excluded are the properties of specific chemical constituents and sensory qualities perceived by the human senses. Obviously, nonmechanical devices can be useful in the study of food texture, and there are likely many more than might be employed if their relationship to texture could be established.

This brief examination of texture measurement has shown that a wide range of approaches is available, both direct and indirect. Nothing replaces experience with the testing procedures, however, as a guide to the selection of the test or tests most suitable for a particular problem. We now will explore the structural nature of texture in order to understand the basis of the texture measurement response.

## 6.3 STRUCTURAL ASPECTS OF FOOD TEXTURE

The term *texture,* it seems clear, refers to an elusive but important quality attribute of food that is difficult to measure or analyze. Nonetheless, whether the perturbing device is a human tooth or an engineered probe, what is being measured in textural testing is a response to the structural organization of the material. It follows that to fully understand texture, a knowledge of structure and its relation to force is required.

**Table 6–3** Examples of Nonmechanical Physical Properties Related to Texture

| *Property Type* | *Parameter(s) Measured* | *Example(s)* |
|---|---|---|
| Geometrical | Particle size and shape | Ground coffee, salt, milk powder |
| | Particle shape and orientation | Textured vegetable protein, tomato juice |
| | Volume | Loaf volume |
| Optical | Light diffraction | Sarcomere length of muscle tissue |
| | Reflected visible light | Ripeness of fruit |
| | X-ray, transmission | Lettuce quality, hollow heart in potatoes |
| | X-ray, diffraction | Crystal form of fat crystals |
| Thermal | Enthalpy change, transition temperature | Thermomechanical properties of meat |
| Electrical | Dielectric constant | Fruit texture and maturity |
| | Conductivity | Bean hardness |
| Auditory | Sound amplitude | Crispness, fruit ripeness |
| | Resonance frequency | Fruit and vegetable quality |

*Source:* Adapted from Szczesniak, 1983; Dull, 1986; Vickers, 1987.

Mastication is the process by which solid food is torn, ripped, crushed, and ground so that it can be swallowed and digested. Not only teeth but the softer tissues of the mouth (lips, cheeks, tongue, and palate) are involved. Also, chewing often releases entrapped liquids into the mouth, where they, along with saliva, are mixed with the bolus to facilitate swallowing. Considering that food texture is sensed through the response of nerve endings throughout the mouth (augmented by information gained through the eyes and ears) and that these signals are then integrated in the brain, where they are compared with other experiences, it is easy to understand why it is often difficult to correlate sensory data with that obtained through instrumental analysis. Real food in the mouth and selected samples undergoing controlled deformation in a test cell would seem to be widely dissimilar. Nevertheless, progress has been made in analyzing both these events through the realization that in each case forces are being brought to bear on a biological structure.

The term *structure* must be fairly broadly construed in its application to foods. It spans a wide range of dimensions, from the macrostructural, visible appearance of spongy air cells in bakery products down to the intricate cross-linking of structural proteins in muscle tissue that requires a transmission electron microscope (TEM) to be viewed. What links these seemingly disparate types of structures is their ability to elicit textural responses from humans and instruments. It is difficult to generalize, but one factor that seems important from a textural standpoint is structural anisotropy—the tendency to exhibit different physical properties depending upon the orientation or position of the material. For one thing, anisotropic structures tend to form anisotropic organizational networks. The filamentous proteins of muscle cells, for example, form into a structure with widely different properties depending upon the orientation of the axis, whereas the isodiametric, polyhedral parenchyma cells of fruits and vegetables are relatively uninfluenced by the direction from which force is applied.

It is worthwhile to consider what occurs when force is applied to these two kinds of structures. To a first approximation, the edible portion of fruits and vegetables may be considered liquid-filled spheres surrounded by rigid or semi-rigid cell walls, depending upon state of maturity (Figure 4–13). The cells are held together by the bonding action of the middle lamella. This material is generally isotropic so that orientation is not often considered in texture testing. Instrumentally, the most common methods for fruit and vegetable testing include puncturing, in which the maximum force required to push a probe tip into the commodity to a given depth is measured; extrusion-shear, in which the force required to make a sample flow through a hole or slit is measured; and compression, in which the force required to rupture a sample between flat plates is measured.

Now, when force is applied to this structure (or, more accurately, this network of interconnected cells), either in the mouth or instrumentally, several things can happen. It is to be expected that failure or rupture would occur at the point of least resistance. In this system, either the cells can separate through the middle lamella or the cells can rupture. Which of these events is observed depends upon the strength of the middle lamella. In general, raw fruit and vegetable cells tend to burst, while cooking leads to cell separation as a result of the thermal destabilization of pectic materials (Figure 6–17). This is one reason the sensory properties of cooked and uncooked foods differ. When raw cells are broken, the liquid contained within them escapes to the oral cavity, imparting the sensation of moistness. If the cells separate, however, cell sap stays within the walls. As an example, compare the mouthfeel of raw and cooked carrots or potatoes. Figure 6–2 and Table 6–4 illustrate the effects of heating on the microstructure and texture of plant tissue. Another factor influencing the effect of stress on these cells is turgor pressure; it seems that increased turgor reduces the compression force required for failure in raw produce, possibly because of a lowered tendency for crack propagation to occur in the intercellular spaces if the cell is dilated.

For the much more prevalent anisotropic food materials, a different and even more complicated situation exists. In both plant and animal tissue, fibrous polymers occur that primarily supply strength, rigidity, or elasticity to the microstructure. As we have seen, the preponderance of these

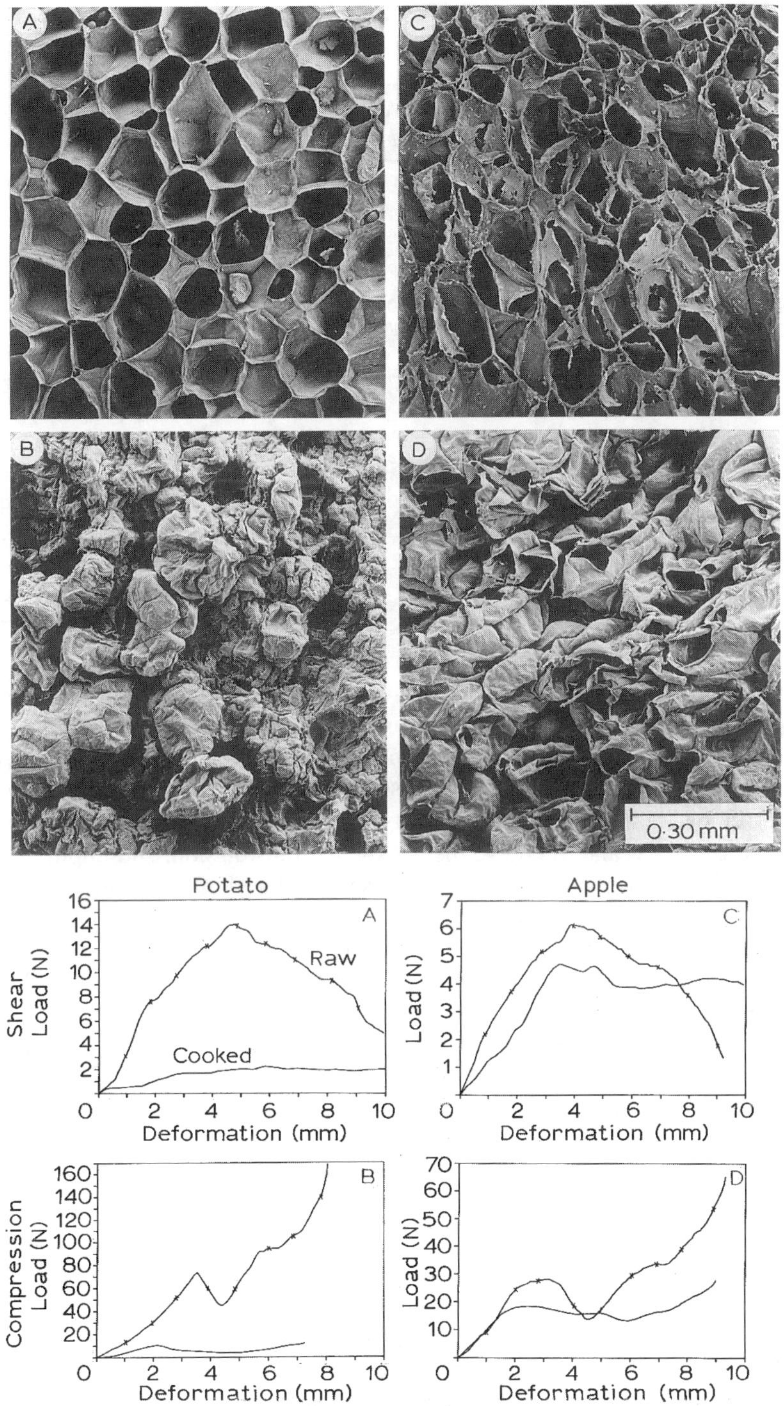


**Figure 6–2** Top: Scanning electron micrographs of (A) raw potato, (B) cooked potato, (C) raw apple, (D) cooked apple. Bottom: (A) Behavior of raw and cooked potato during shear, (B) behavior of raw and cooked potato during compression, (C) behavior of raw and cooked apple during shear, (D) behavior of raw and cooked apple during compression. *Source:* Smith and Stanley, unpublished data.

**Table 6–4** Influence of Cooking on Instrumental Texture of Potato and Apple Tissues

| | *Warner-Bratzler Shear* | | | *Flat Plate Compression* | | |
|---|---|---|---|---|---|---|
| | *Max. Force (N)* | *Compression (%)* | *Work (N-mm)* | *Max. Force (N)* | *Compression (%)* | *Work (N-mm)* |
| Raw potato | 13.8 | 49 | 90.6 | 143 | 91 | 453 |
| Cooked potato | 2.4 | 67 | 15.2 | 18 | 89 | 47 |
| Raw apple | 5.7 | 42 | 39.7 | 57 | 89 | 210 |
| Cooked apple | 4.2 | 52 | 24.9 | 32 | 81 | 101 |

*Note:* Apparent density of raw potato = 0.993 g/cm$^3$; raw apple = 0.815 g/cm$^3$. Note possible influence of intercellular air (reported as 20–25% in apple, 1% in potato. Bourne, 1982) on textural properties and apparent density.

*Source:* Smith and Stanley, unpublished data.

molecules are cellulose in plants and collagen in animals. Importantly, both cellulose and collagen molecules form extensive interacting networks that alter their physical properties significantly. In muscle tissue, the formation of networks is accomplished by intermolecular cross-linking, while in plants lignin serves to strengthen the cellulose. A third mechanism involves the interlocking of biopolymers such as pectins through the bridging action of polyvalent cations like calcium. These mesh- or weblike extracellular structures are most often found embedded in but not strongly bound to an amorphous matrix, and of course the surrounding environment is rich in water, although the structural elements themselves are hydrophobic and insoluble.

The role of these structures is to resist the actions of stress (compression, tensile, and shear) through strength and accommodate strain through flexibility and elasticity. They are remarkably successful at carrying out their responsibilities. It has been reported, for instance, that the load-carrying capacity of cellulose fibers approaches that of steel. The dual requirement of allowing more or less continuous, low-level stressing while guarding against catastrophic buckling, crushing, or tearing has led to some basic principles that seem to be in operation across all biological structures. Critical to the mechanical characteristics of polymerous biological structures is the degree of interaction between the fibers and the matrix. Consider a model in which strong parallel fibers are surrounded by an interconnecting, weaker homogenous network. Clearly less force is required to separate the microstructure along its fiber axis than across the fibers. It is easier to split wood (or meat), for instance, than to cut it across the grain. Now, the resistance of the model to compressive (and tensile) forces is improved as the relative strength of the lateral interactions in the matrix increases. However, at the same time the model loses its ability to withstand transverse shearing force. Ease of fracture across the fiber axis increases because the matrix material can transmit shear effectively, whereas if the interfaces between fibers are weak, it will absorb the force, slow its transmission, and improve the resistance to shear. Materials that exhibit poor communication between parallel elements are termed *notch-insensitive,* since their strength in compression or tensile testing is directly proportional to the load-bearing surface after it has been notched or cut. Materials that exhibit notch-sensitive behavior lose strength disproportionately more quickly with notching, presumably as a result of the effective transmission of shear force by the matrix.

To return to foods, the assumption is that their structural behavior under force would be influenced by the matrix-element interaction. Raw plant tissue of the type mentioned previously (apple, potato, carrot) is reported to be notch-sensitive, probably as a result of its microstructure, which can be represented as relatively weak cells embedded in a strong continuous matrix, the middle lamella. Cooking this tissue, however, would likely weaken the matrix and lead to notch-insensitivity (Figure 6–3), since the pectins of the middle lamella degrade, increasing the resistance to fracture.

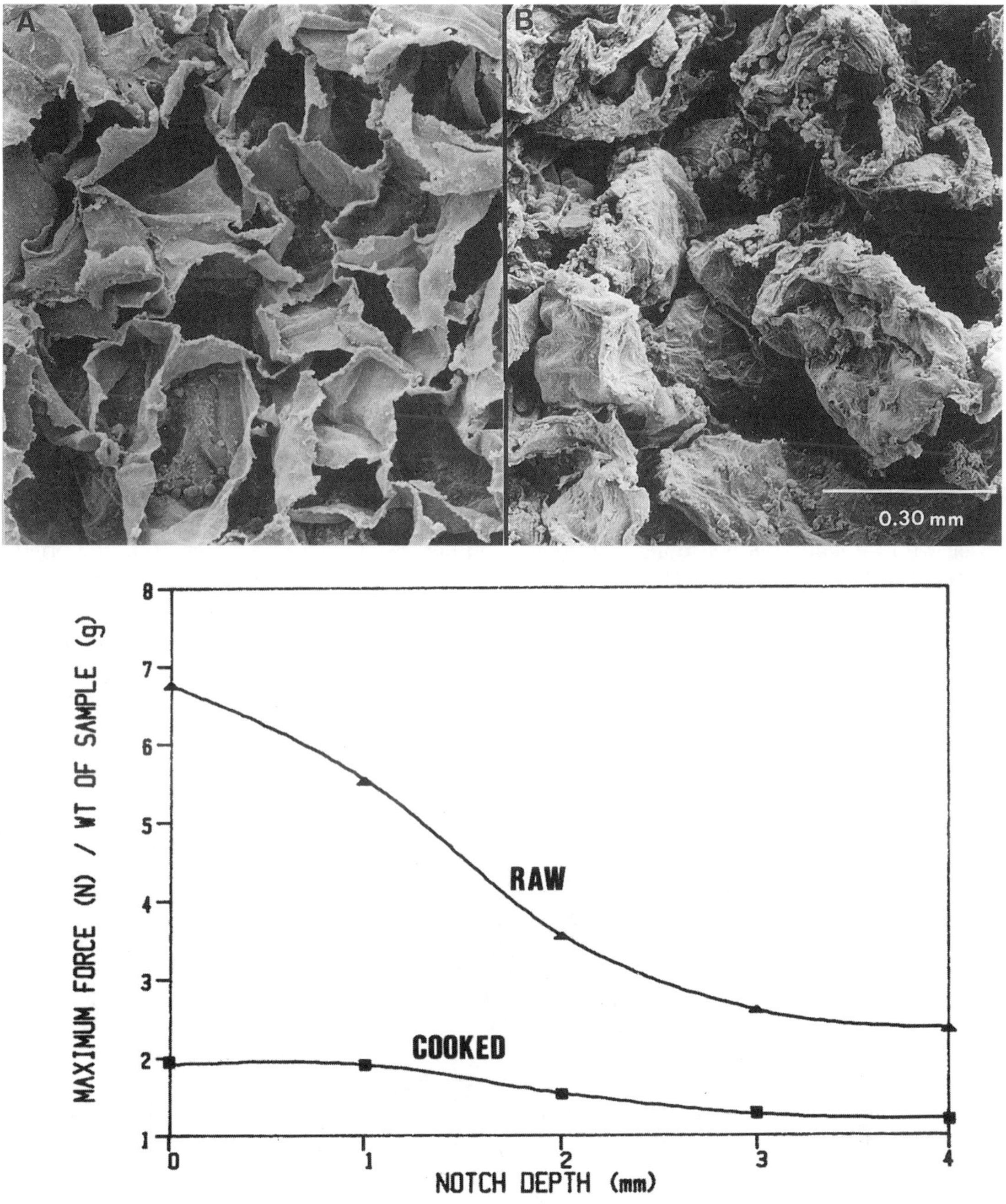


**Figure 6–3** Bottom: Notch behavior of raw and cooked turnip tissue. Top: Scanning electron micrographs of raw (A) and cooked (B) turnip tissue. *Source:* Smith and Stanley, unpublished data.

Meat, on the other hand, would be expected to exhibit a quite different response to force. It is highly anisotropic, and the force (corrected for area) required to break cooked muscle parallel to the muscle fibers is about 10 times higher than the force needed to pull apart the fibers perpendicularly. This indicates the matrix is inherently weaker than the fibrous elements. Consistent with this, cooked beef strips are reported to be notch-insensitive when cut perpendicular to the fiber axis. In fact, when tensile stress is placed on a notched specimen, cavitation occurs ahead of the crack tip, and when the advancing fissure reaches the cavity, it blunts itself and resists further splitting. The particular structural element of importance appears to be the perimysium connective tissue, and it is the low strength of this material that allows cooked meat to be easily split into individual fiber bundles but not to be easily fractured across the fibers (Figure 6–4).

A further consideration in applying tensile stress to muscle fibers is the progressive orientation of endomysial connective tissue fibers as stress is applied. Energy is required for this process, and this amount is subtracted from the total applied so that no irreversible structural damage occurs until the orientation is complete—low levels of stress are absorbed by the connective tissue but not the more fragile contractile apparatus. It has been reported that if uncooked muscle is stretched prerigor, the endomysial collagen fibers are pulled parallel to the longitudinal axis of the muscle fiber, but when unrestrained muscle is allowed to contract, the connective tissue arranges itself so that most of the fibers are perpendicular to the long axis. Unrestrained muscle tissue requires almost twice the elongation prior to rupture as does restrained tissue (Figure 4–5).

*Crispness* and *brittleness* are widely used textural descriptors frequently employed for characterizing foods. These terms generally have positive connotations. Consumers like crisp raw fruits and vegetables and brittle crackers and snacks. Thus, it is desirable to quantitate crispness and brittleness and understand their structural origin. As a first guess, *crispness* seems to describe foods composed of liquid-filled, turgid cells, while dry products containing thin-walled air cells appear to be brittle.

It would seem difficult to study crispness and brittleness together because they do not occur in the same foods. But what if a food was found that

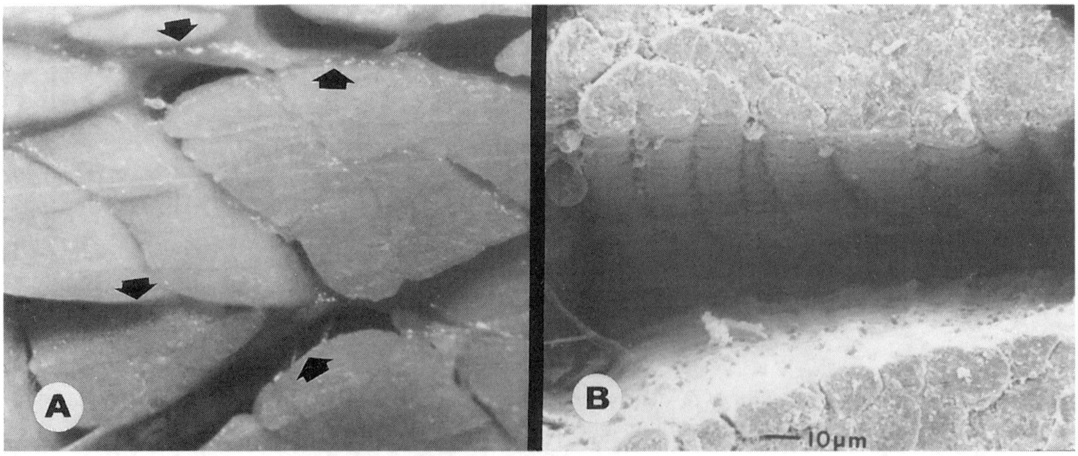


**Figure 6–4** Fracture behavior of cooked beef muscle. Meat cooked at 90°C and stressed perpendicular to the fiber direction. (A) Dissection microscope micrograph; arrows indicate fracture points. Note remnants of connective tissue between fibers. (B) Scanning electron micrograph. *Source:* Bernal and Stanley, 1987.

could be both crisp and brittle, one in which liquid-filled cells become air filled? Bacon adipose tissue appears to be just such a food. During cooking, a large amount of the water and fat is lost, so it might be possible to use this material as a model for both crispness and brittleness.

In order to test this idea, sensory, instrumental, chemical, and structural data were collected on bacon adipose cooked for various times (Table 6–5). It was found that, unlike raw fruits and vegetables, uncooked bacon had no characteristics of crispness: these develop during cooking, probably as a result of heating the connective tissue in the adipose cell walls. Cooked adipose responded to stress by a shearing of rigid cell walls, whereas less heated material exhibited cell separation by way of elastic and flow behavior, exactly the opposite of plant tissue (the difference is explained by the chemical composition of the cell wall material, collagen versus pectin).

Cooking produces important compositional changes in bacon adipose (but not in plant tissue). Taking the cooking losses into account, well-cooked adipose loses about 80% of its fat and almost all of its water, thus effectively converting liquid-filled cells with elastic walls into air-filled cells with rigid walls. As the transition happens, the adipose tissue shrinks, both on a micro- and macroscopic scale, as evidenced by a reduction in rasher thickness, and a decrease in cell diameter, while apparently cell wall width increases. Enormous textural changes accompany bacon cooking.

**Table 6–5** Influence of Cooking Time on Textural Chemical and Structural Properties of Bacon Adipose

| | *Cooking Time (min[a])* | | |
|---|---|---|---|
| | *5.0* | *6.5* | *8.0* |
| Sensory | | | |
| Physical/visual[b] | 2.4 | 5.2 | 6.3 |
| Brittleness[c] | 3.8 | 5.4 | 6.5 |
| Crispness[d] | 3.5 | 6.1 | 7.7 |
| Instrumental | | | |
| Warner-Bratzler | | | |
| Initial rupture force (kg) | — | 2.71 | 0.88 |
| Slope to maximum (kg/cm) | 30.2 | 40.1 | 93.2 |
| Peak angle (radians) | 0.34 | 0.22 | 0.06 |
| Maximum force (kg) | 3.63 | 4.19 | 2.93 |
| Number of peaks | 1.0 | 1.8 | 4.3 |
| Stress relaxation (%) | 70 | 38 | 18 |
| Compression | | | |
| Chemical | | | |
| Water removed (%) | 87.3 | 97.3 | 98.6 |
| Fat removed (%) | 39.1 | 67.5 | 79.2 |
| Structural | | | |
| Rasher thickness (mm) | 1.69 | 1.49 | 1.39 |
| Adipose cell diameter ($\mu$m) | 76 | 66 | 42 |
| Apparent cell wall width ($\mu$m) | 1.6 | 2.0 | 3.0 |

[a] Corresponding to sensory characterization of noncrisp, moderately crisp, and crisp, respectively.
[b] Defined by response to tactile force (bending, cracking, crumbling, shattering).
[c] Defined as force needed to produce first detectable deformation by incisor teeth or initial bite.
[d] Defined as force needed to produce cracking, crumbling, shattering by molar teeth on mastication.
*Source:* Stanley and Voisey, 1979.

Sensory judgments indicated increases in both crispness and brittleness, but a tactile test showed the transition occurred discontinuously. The material passed through progressive stages described by the judges as bending, cracking, crumbling, and then shattering, as would be expected in material passing from crisp to brittle.

Instrumental data reflect this trend as well. Warner-Bratzler shear force–deformation curves (Figure 6–5) can convey a great deal of information by their shape alone. In this study, the researchers observed the appearance of brittle fracture behavior, similar to that seen in other brittle foods such as extruded starch-based snacks, in the most well-done sample. Some force and slope measurements from the curves, recorded in Table 6–5, show considerable nonlinearity. Briefly, the data indicate that a longer cooking time is associated with the disappearance of an initial rupture force (although this can be seen in crisp vegetables; see Figure 6–2), increases in slope to maximum force and number of peaks, and decreases in peak angle and maximum force. Also, a stress relaxation test, in which the decay of stress under constant strain is measured, showed a decrease associated with longer cooking time, indicating a reduction in elasticity that probably results from the stiffening of cell walls. Again, these results are consistent with a crisp-brittle transition.

As has been repeatedly stressed, explanations for organoleptic and instrumental results lie in structural studies. Micrographs (light microscope and scanning electron microscope [SEM]) of uncooked bacon adipose (Figure 6–6) show this tissue to be composed of polyhedral cells (usually 4–6 sides in cross section) bound together by connective tissue and held in clusters by well-defined collagenous septa. Heat produces cracks in cell walls that allow the heat-expanded lipid to escape and accumulate in large pockets formed from ruptured cells. Shrinkage also results, a consequence of disassembly and collapse in the triple-helical collagen molecule, and adipose cells decreased in diameter by about 40%. This correlates with a macroscopic thinning of the whole rasher by about 20%. The apparent cell wall thickening may be a result of wrinkling, but the shrunken cell walls would be expected to be stronger and more rigid as a consequence of heating. This is reflected in the observed decrease in stress relaxation. Another feature of cooked bacon is the appearance of surface crust, characterized by a nondeformable surface coating sandwiching a rigid cellular matrix. The mechanical properties of this microstructure can be detected instrumentally and support the idea of a nonlinear transformation from a soft to a crisp to a brittle food as cooking progresses.

Not much has been said to this point about fluid foods, mainly because texture makes a less

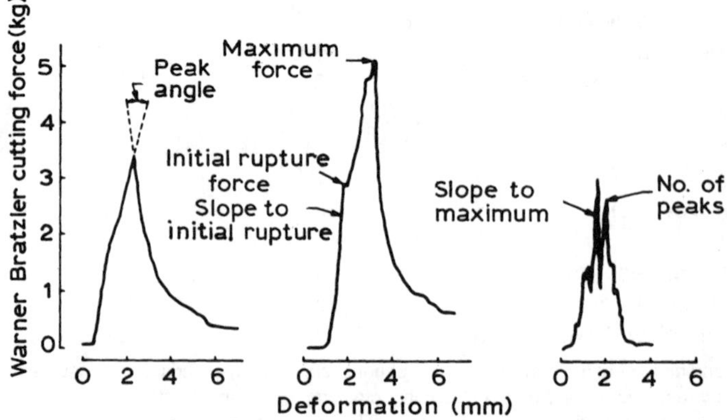


**Figure 6–5** Typical Warner-Bratzler force-deformation curves of bacon as a function of cook time. Left: 6.0 min. Center: 6.5 min. Right: 8.0 min. *Source:* Stanley and Voisey, 1979.

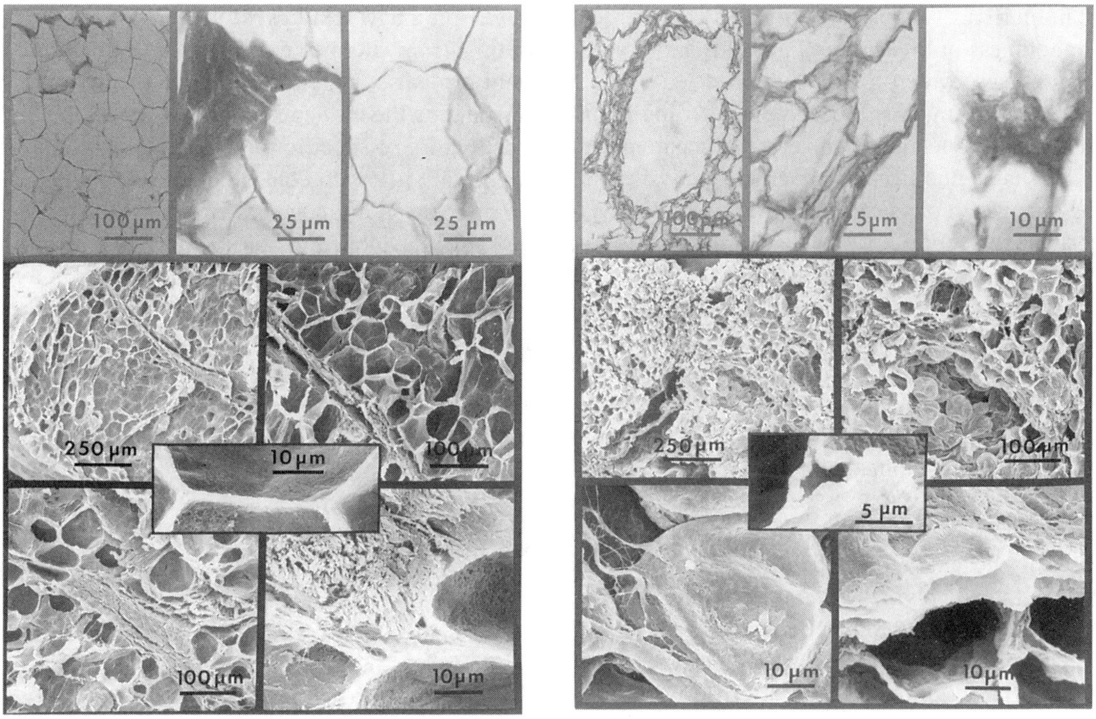


**Figure 6–6** Light and scanning electron micrograph of uncooked (left) and 8.0 min cooked (right) bacon adipose. *Source:* Stanley and Voisey, 1979.

significant contribution to the quality of these foods than is the case with most solid foods. Nevertheless, microstructure influences the texture of some fluids noticeably, as will be shown. Beverages include a wide variety of products. A partial listing includes soups; juices; milk and its derivatives; alcoholic, caffeinic, and carbonated drinks; emulsions such as gravies and salad dressings; syrups; oils; certain condiments; and, of course, water. More foods could be included in this group, since the distinction between fluid and solid foods is arbitrary. For example, hydrocolloid solutions, ketchup, fruit purees, condensed milk, mustard, and chocolate sauces all exhibit some liquid-like behavior (see Section 3.8.6). While beverages are frequently considered to be structureless, this is never the case if structure is defined broadly. Particulate inclusions and the interaction of these elements can have a significant influence on viscoelasticity and hence texture.

With liquid foods, the most common instrumental measure of texture is viscosity, a function of resistance to flow. To measure the viscosity of a liquid, shearing stress is applied to the liquid and causes flow or deformation, which can be expressed as shear rate (see Section 3.8). Most liquid foods, however, are non-Newtonian, and the shear stress–shear rate curves are either nonlinear or do not pass through the origin. For this reason, experimental values obtained at just one shearing stress are properly termed the *apparent viscosity* (quite different values will be obtained under different conditions). Thus, to the previously defined methods of applying force to a food (compression, tensile, and shear) must be added flow. The sensory evaluation of fluid food texture centers on eliciting judgments about viscosity-related characteristics (thickness-thinness), mouthfeel (smoothness-pulpyness-creaminess), and consistency (heavy-watery).

As can be imagined, particle-containing liquid foods respond to alterations in particle parameters (size, shape, number) by undergoing corresponding changes in physical properties, including viscosity (see Section 3.8.4). This relationship is not, however, as straightforward as it may appear. It has been reported that, at low particle concentrations in products such as fruit and vegetable juices, viscosity increases as a function of particle size. On the other hand, in emulsion-type foods, viscosity and particle size vary inversely. Tomato juice containing elongated particles demonstrated a higher viscosity than when spherical particles were present. At higher concentrations (>50%) viscosity became independent of particle size, as the biological polymers interact and deviate more and more from Newtonian behavior.

The structural interaction of particles suspended in a fluid produces interesting effects. Fluid milk contains casein proteins that, as a result of their chemical characteristics, form 30–300 nm spherical, porous, hydrated complexes called micelles. High temperature and acidic processing conditions lead to the interaction of destabilized micelles and to gel formation through the chemical binding of protein chains or submicelles. Gel formation is a nonlinear function of protein concentration (see Section 5.13.4), as shown in Figure 6–7. At lower solids concentrations, TEM examination reveals partially linked casein micelles, but, at the highest level, the spherical structures disappear and are replaced by highly fused particles. This structural change is accompanied by a nonlinear increase in gel firmness, reflecting the close correlation between microstructure and texture.

The discussion has thus far centered on traditional food materials—plant and animal tissue. Presently, there is economic pressure on the food industry to utilize alternative raw materials and new processing techniques to achieve novel fabricated products that meet both stringent legal regulations and consumer demands for high quality, improved nutrition, convenience, and low cost. Developing new textures—for example, those resulting from thermal extrusion of concentrated plant proteins—depends to a large extent on understanding the influence of processing on the microstructure of specific food materials and, more importantly, how to control and predict processing effects and ingredient alterations. Food scientists have developed the basic knowledge needed to produce a wide range of fabricated structures and textures, while food engineers continue to make the advances in processing equipment needed to implement new food-processing ideas. It remains for food companies to join the effort and provide the product development skills needed to bring these ideas to practical fruition.

## 6.4 TEXTURE-STRUCTURE: CONCLUSIONS

The conclusions to be drawn from the foregoing discussion seem indisputable. Textural responses are governed by structural organization; alterations in structures are reflected by corresponding shifts in texture.

A wide range of textures is available in nature, and many others can be fabricated. It is the organization of structural elements that is the primary determinant of texture. This fact underscores the need to accompany textural studies with structural examination, both of the native food and samples placed under stress.

One goal of textural studies is to determine to which microstructure(s) the instrumental probe or human tooth responds. The diverse spectrum of dimensions involved often make gathering the appropriate data difficult, but an effort should be made. Although it is not yet possible to fully predict textural responses from structural information alone, predicative ability should be sought. Moreover, the knowledge gained can go a long way in explaining processing effects and in selecting processing strategies.

## 6.5 QUALITY AND STRUCTURE

To see where we stand at this point, the reader may want to look at Figure 10–1. During the food

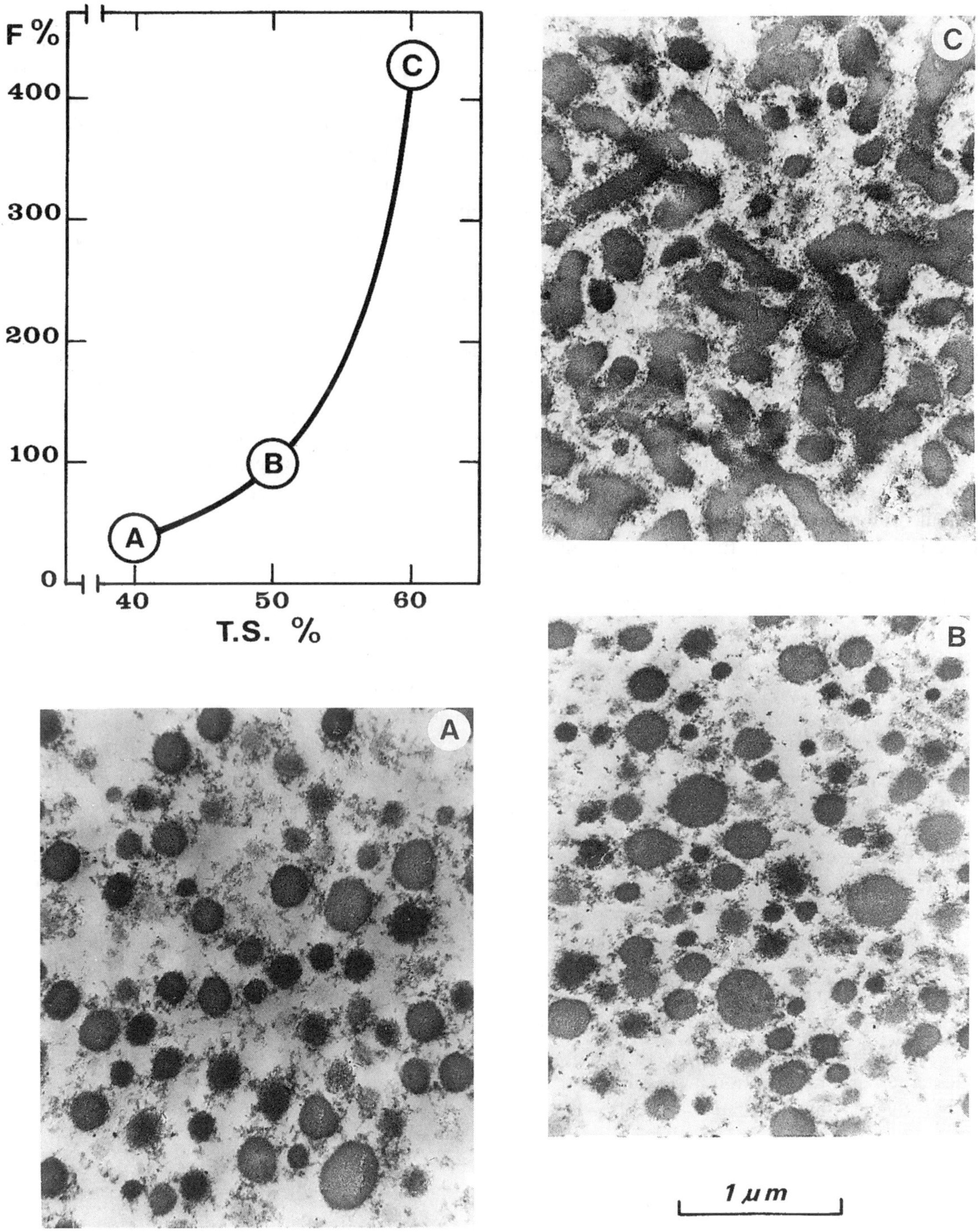


**Figure 6–7** Relationship between relative firmness (F%) and structure of milk gels at various concentrations of total skim milk solids (T.S.%). Points A, B, and C represent gels containing 40%, 50%, and 60% T.S. Transmission micrographs labeled correspondingly. *Source:* Kalab and Harwalkar, 1974.

fabrication process, every step contributes to imparting desirable attributes to the final product. As demonstrated in the last sections, the product's attributes can be measured and its quality assessed by sensory or objective methods. However, food structure and product attributes must also be preserved.

The next sections describe how susceptible plant and animal tissue are to natural postgathering and postprocessing deterioration and its effects on the quality of the final food product. Although food tissue is heterogeneous, being composed of mixtures of cell types, there is remarkable similarity in the structure and basic metabolism of plant and animal cells. Thus, understanding postharvest and post-mortem physiology is easier, for there are common biochemical pathways and structures, such as protein synthesis and membrane structure. The important point to remember is that successful food processing requires an understanding of the physiology of dead and dying tissue from both plants and animals.

### 6.5.1 Membrane Effects in Postharvest Physiology

The deterioration of postharvest plant tissue begins almost immediately. If there is one universal trigger for quality loss in both animal and plant material, it is the alteration of the physical and chemical characteristics of membranes, which in turn leads to changes in cellular metabolism and accelerated damage. According to present knowledge, plant membranes are fluid bilayers of phospholipids containing embedded proteins and sterols (Figure 4–10). In plant membranes, most phospholipids contain a saturated fatty acid (normally C16:0) in the sn-1 position of the glycerol backbone. The sn-2 position is usually filled with a long chain unsaturated fatty acid such as C18:1, 18:2, or 18:3, while sn-3 is occupied by a phosphate-containing polar group. Polar groups of sterols associate with the polar group of phospholipids, whereas proteins may be either immersed within the hydrophobic core of membranes or attached peripherally. Functional membranes are fluid, since phospholipids can move rather freely in lateral dimensions, but the presence of proteins and sterols influences membrane fluidity. The composition of constituent phospholipids can affect the fluidity of membranes, with unsaturated fatty acid–containing lipids being more fluid than saturated lipids (Stanley, 1991).

Fruits and vegetables go through three major physiological stages: growth, maturation, and senescence. Ripening and senescence may occur either on or off the plant, and they involve many structural changes that affect final quality. Of these, the most important appears to be membrane deterioration, since loss of membrane functionality initiates the secondary reactions that lead to quality loss. Senescence is a natural process in living organisms. Whether a plant is young or old, cells are continuously growing and dying. Plants exhibit diverse patterns of senescence, and within each plant some organs will senesce at different rates than others.

A picture is now beginning to emerge of the pattern of membrane deterioration in senescing plant tissue systems. Peroxidation of fatty acids and the resulting free radical formation constitute the major deteriorative process in membranes. Unsaturated fatty acids are prone to attack by lipoxygenase, and increased free radical production has been observed in a variety of senescent tissues. This reaction is essentially autocatalytic, since free radicals can continue their attack as long as $O_2$ and more substrate are available. Chemically, oxidation of membrane polyunsaturated fatty acids (PUFAs) can occur in both plants and animals (Figure 6–8). Since these reactions are autocatalyzing, they are referred to as PUFA cascades. A generalized scheme for membrane deterioration in plant tissue due to senescence is shown in Figure 6–9. Many of the lipid-degrading enzymes featured in these two figures are activated either by $Ca^{2+}$ or the $Ca^{2+}$-calmodulin complex. However, to date the "trigger" through which $Ca^{2+}$ is released into the cytosol and senescence initiated is still unclear.

One cause of membrane deterioration in plant tissue that affects food materials is chilling injury. Consumers, producers, and processors of fruits and vegetables routinely utilize refrigerated stor-

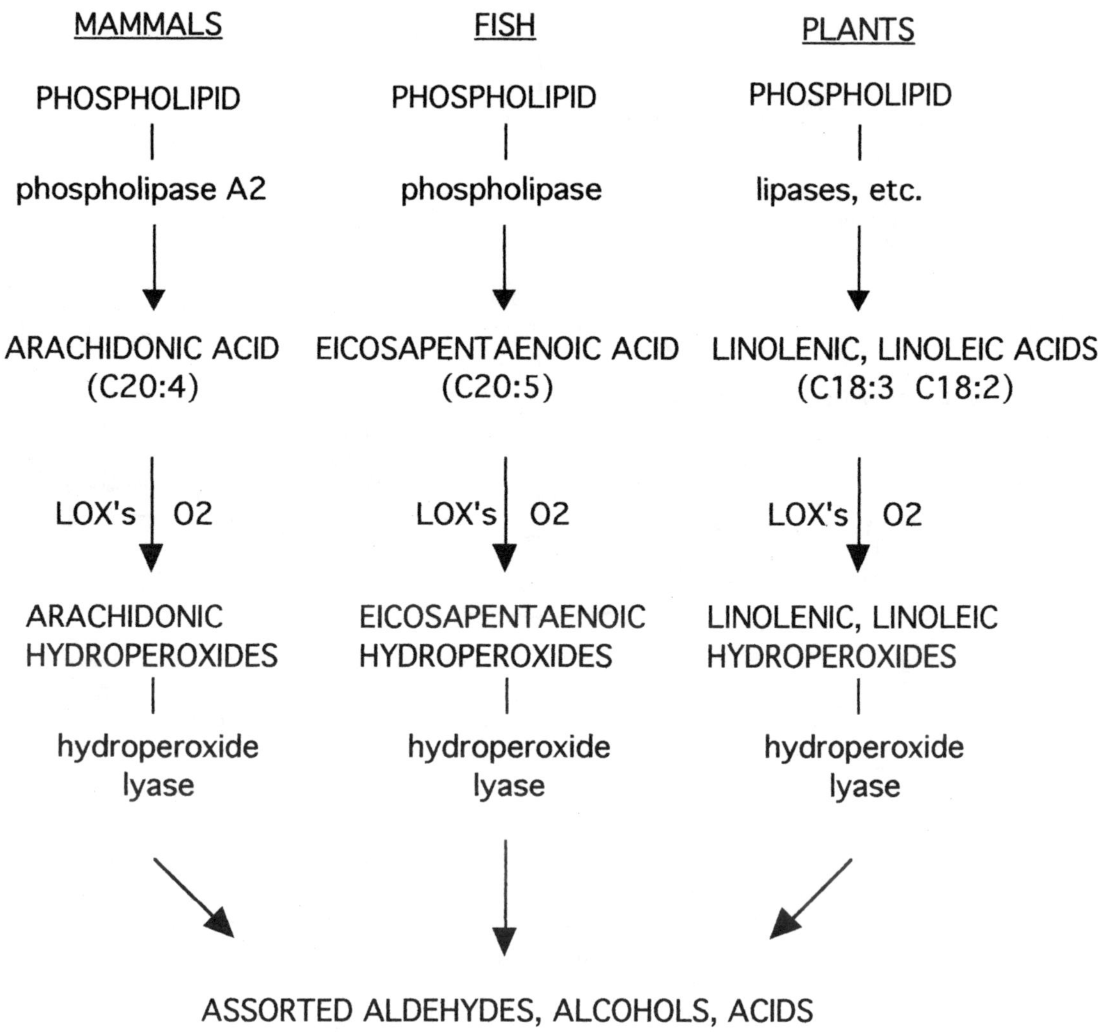


**Figure 6–8** Polyunsaturated fatty acid (PUFA) cascades resulting from cell injury. *Source:* Haard, 1995.

age to extend the shelf-life of produce. For most products, this yields the desired result, but some, especially those of tropical or subtropical origin, respond to chilling temperatures by displaying symptoms of physiological damage that can dramatically limit their acceptability as food. Chilling injury, which encompasses the physiological damage that many plants and plant commodities suffer as a result of exposure to low but nonfreezing temperatures, can occur almost anywhere along the food chain—harvesting, commercial storage, distribution, or storage in the home. Field chilling, transit chilling, and home refrigeration can all contribute to the problem. Chilling injury causes economic losses that, although difficult to quantify, are significant (Jackman, Yada, Marangoni, Parkin, & Stanley, 1988).

As an example, consumers in northern climates have come to demand a steady supply of fresh produce throughout the year, necessitating the importation of commodities from warmer areas during the winter months. Chief among these com-

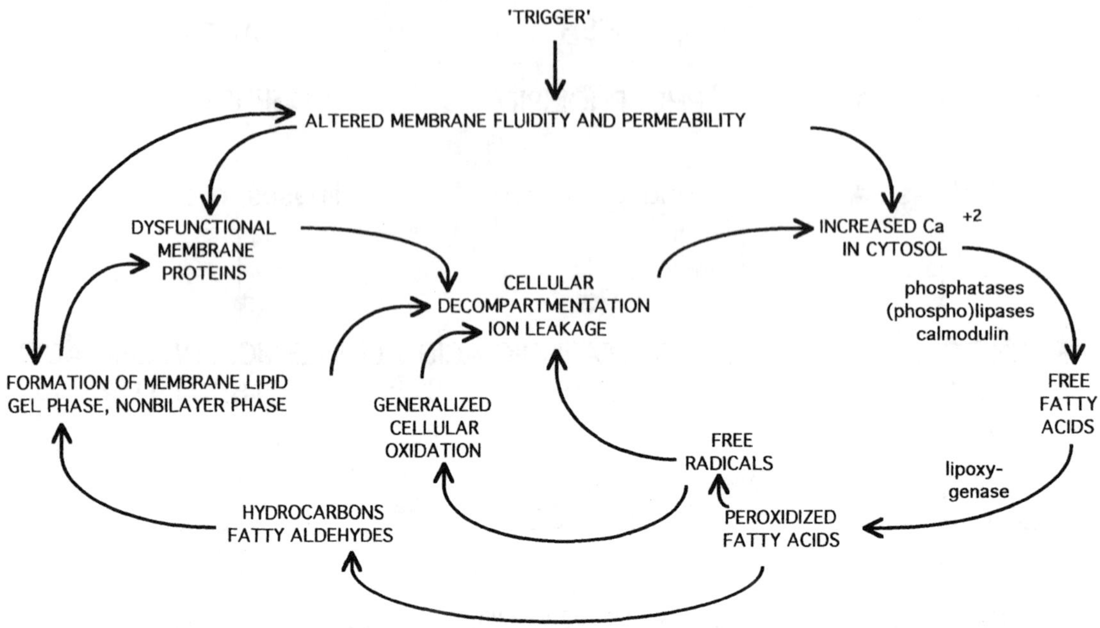


**Figure 6–9** Generalized scheme to account for membrane deterioration in plant tissue due to senescence or postharvest stress. *Source:* Reprinted from *Postharvest Biology and Technology*, Vol. 7, A.G. Marangoni, T. Palma, and D.W. Stanley, Membrane Effects in Postharvest Physiology, pp. 193–217, Copyright 1996, with permission from Elsevier Science.

modities are tomatoes, where chilling injury damage can occur during transportation and storage. The critical temperature (i.e., that temperature at which injury will occur) and the length of time necessary to induce symptoms is a function of the cultivar. The severity of symptoms increases and the time necessary for induction decreases as the temperature is reduced. As a broad generality, tomatoes will develop signs of this defect if exposed to temperatures between 8°C and 12°C for 10–14 days or more, depending on the variety and the conditions. Initial responses to chilling at the cellular level lead to quality defects only after a period sufficient to induce physiological breakdown. Quality-related chilling injury symptoms exhibited by tomato fruit include surface lesions in the form of pitting, water soaking of tissues, failure to ripen normally (e.g., development of a "mealy" texture, enhanced softening, and poor color development), and susceptibility to microbiological spoilage, typified by Alternaria rot. Chilling injury is thus a major problem in the postharvest handling of tomatoes, since it precludes both the cold storage of mature green fruit (which could be ripened later and used for consumption or processing) and the extension of the growing season into colder weather. It also contributes to imported produce.

Because of the economic losses caused by chilling injury, much research has gone into discovering the cellular mechanism by which it acts. Currently, most theories focus on the initial deterioration of biological membranes, an event chilling injury shares with senescence. Because of this similarity, some researchers view chilling injury as an induced form of rapid senescent breakdown in plants vulnerable to chilling. In any case, membrane dysfunction is involved in both senescence and chilling injury. A generalized scheme for chilling injury–associated membrane deterioration begins with hydrolysis of membrane phospholipids to free fatty acids through the action of phospholipases. Resulting free fatty acids are then attacked by lipoxygenase to generate free radicals as well as oxidized lipids. Free radicals act as destabilizing agents, altering membrane fluidity and leading to dysfunctional membranes. This in turn results in loss of permeability, increased ion leakage, and eventually deterioration of tissue quality.

Research on the membrane fraction of tomatoes has shown the validity of this scheme as regards the initial stages of chilling injury in this fruit. Electron micrographs of fruit tissue taken from chilled and unchilled tomatoes exhibited noticeable differences: the organelles of fruit tissue samples from chilled plants showed extensive damage compared with those from unchilled fruit (Figure 6–10).

Another economically important crop prone to chilling injury is potatoes. Chilling injury promotes the conversion of starch to sugar ("sweetening"), a reaction that is undesirable in tubers used for chipping or frying. Sugars accumulated during cold storage can react with free amino acids in the tissue during frying by way of the Maillard or nonenzymatic browning route to produce dark chips. This problem can be at least partially overcome by reconditioning potatoes at nonrefrigeration temperatures, which accelerates conversion of sugar back to starch. Chilling injury in potatoes is thought to involve membrane deterioration as well, since alteration of the amyloplast membrane structure is believed to be a prerequisite for carbohydrate metabolism in tubers. Also, the effects of chilling on the activity of enzymes involved in this process may be a function of their relationship to membranes, since the activity and conformation of many membrane-bound enzymes are influenced by chilling temperatures.

One way of determining membrane functionality is by measuring the phase transition temperature, a parameter inversely related to membrane fluidity. Decreased membrane fluidity (and the concomitantly increased phase transition temperature) in senescing tissues correlates with loss of biological integrity. Research has shown that phase transitions occur at higher temperatures in the microsomal membranes of chill-sensitive cultivars of tomatoes and amyloplasts of potatoes than in their chill-resistant counterparts. This suggests that chill-resistant membranes maintain their liquid state at lower temperatures than those from chill-sensitive varieties. Further support for the membrane damage theory comes from the fact that sensitive fruits can be acclimated, which is to say that, if they are exposed to nonchilling cold temperatures prior to refrigerated storage, their resistance is increased. Studies show the existence of a retailoring process in which saturated fatty acids, more abundant in injured membranes and associated with decreased fluidity, are replaced by unsaturated fatty acids with lower melting points. Higher unsaturated:saturated fatty acid ratios have also been observed in a variety of plant species during cold acclimation, along with an increase in adenosine triphosphate (ATP) activity. All of this evidence suggests that chilling injury could be viewed as accelerated senescence, since the processes observed in chill-injured fruits (ion leakage, lipid peroxidation, loss of phospholipids, and increased saturation) are the same as those observed during senescence. A great deal of work on preventing chilling injury has been reported. Most of it involves genetic modification, the manipulation of storage conditions, or the application of exogenous chemical treatments. While these strategies have proven useful in certain cases, no universally applicable remedy is yet available. This topic was reviewed by Marangoni, Palma, and Stanley (1996).

### 6.5.2 Respiration Effects in Postharvest Physiology

The cellular structure of plant tissue determines its physical properties, but the issue is complicated by physiological events and their consequences, such as the loss of turgor pressure mentioned previously. Just as post-mortem muscle tissue continues to undergo limited metabolism, so does postharvest plant tissue (and in both cases the metabolic activity is mostly deleterious to quality). Plant tissue, in fact, continues to metabolize longer than animal tissue. This is a consequence of the more highly organized microstructure of the latter and its reliance on a highly developed circulatory system. In postharvest plant tissue, gases such as oxygen and carbon dioxide can still exchange through diffusion, and waste products can be accommodated in vacuoles of parenchyma cells.

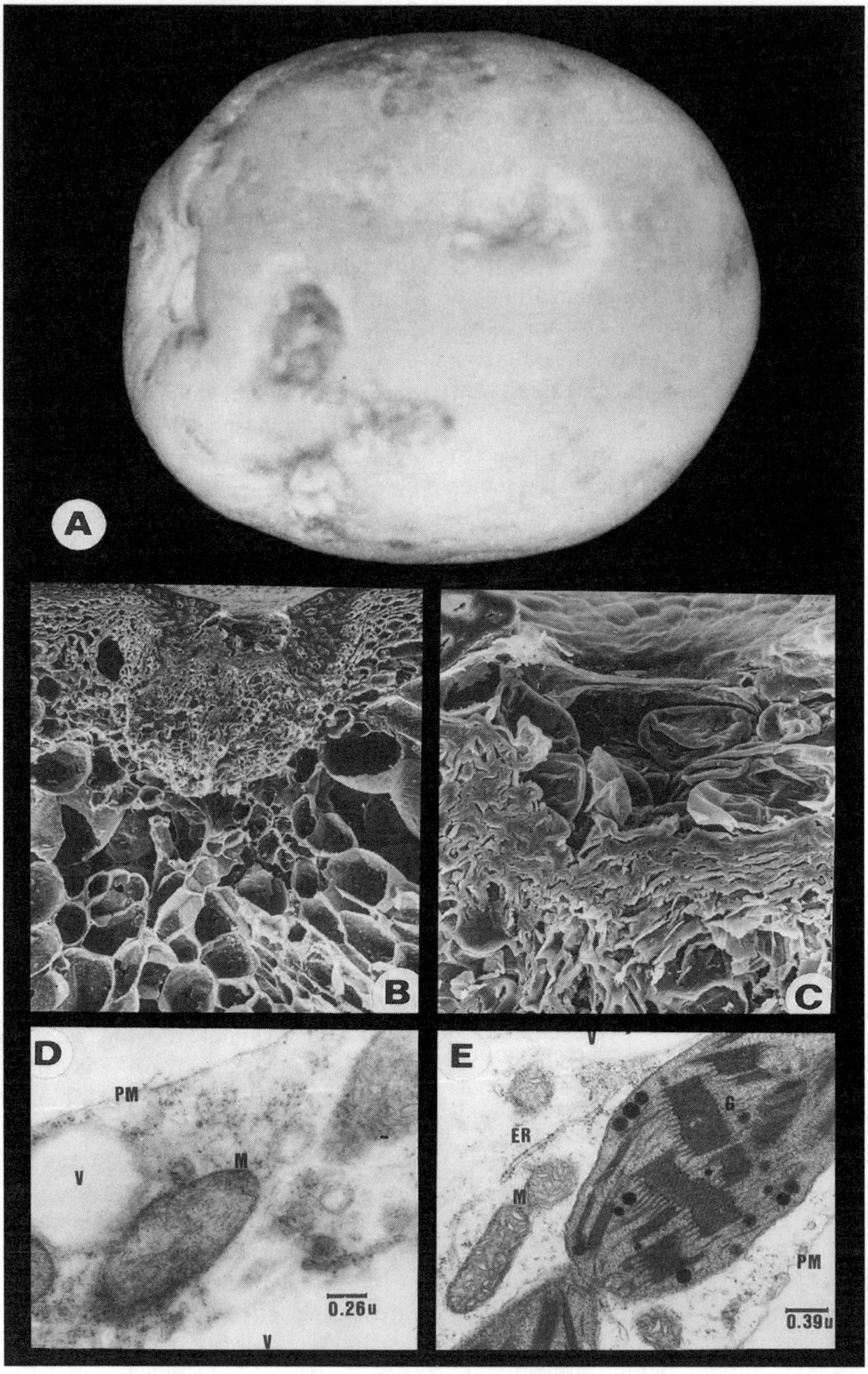
A
B
C
D
PM
V
M
0.26u
V
E
V
ER
M
G
PM
0.39u

Plant materials respire by taking up oxygen and giving off carbon dioxide and heat. Also, they transpire (or lose water). These processes continue after harvest and lead to quality deterioration. Respiration or biological oxidation is the oxidative breakdown of complex organic molecules to carbon dioxide and water, and it proceeds on the same basic principles in both plant and animal tissue. A further product of respiration is heat, and it has been estimated that in postharvest plant material over 50% of the total energy supplied by respiration is dissipated as heat, highlighting the importance of rapid cooling after harvest. Respiration can be aerobic or anaerobic depending on the oxygen concentration. Anaerobic conditions occur more rapidly in post-mortem animal tissue than in postharvest plant tissue owing to the failure of the more developed circulatory system in the former and the ability to exchange gases in the latter.

Postharvest respiration plays a major role in the quality of fruits and vegetables. For example, respiration can utilize as substrates certain food reserves such as sugars, whose disappearance can lead to a decline in flavors based on sweetness. Nutritional quality can also suffer. If an adequate concentration of oxygen is not maintained, a shift to anaerobic respiration can occur, leading to the conversion of sugars to alcohol, which, although sometimes desired in certain products, can also produce unwanted side reactions. On the other hand, the controlled reduction of oxygen concentration aids in regulating the respiration rate and slowing senescence, or anabolic degradation. As oxygen is utilized, carbon dioxide is produced, of benefit in certain crops because this production moderates respiration. Also, as mentioned previously, the heat produced by respiration is a major factor to be considered, since it affects refrigeration requirements and packaging demands.

Respiration rates in postharvest fruits and vegetables influence storage life, and there is a close inverse relationship between oxygen consumption and maximum storability for commercial commodities. Because of this relationship, the control of respiration is of paramount importance. Plant tissue respiration rates vary widely, with root, tuber, and bulb vegetables having the lowest rate. Foods that are harvested in the mature state, such as tomatoes and melons, respire at a lower rate than those picked immature, such as green beans, sweet corn, and peas. Edible tissues composed of growing vegetative or floral structures have high respiration rates. In general, the respiration rate is highest during the early stages of development and decreases as plant structures mature. Exceptions to this rule include certain fruits that exhibit a climacteric rise in respiration during ripening. Also, respiration rates generally decline with decreases in the water content of the tissue.

Ripening and senescence, which may occur either on or off the plant, involve many structural changes that affect final quality. The course of ripening depends upon respiratory patterns: fruits characterized as climacteric exhibit a typical rise in respiratory rate during maturation and tend to ripen more rapidly than those fruits that are nonclimacteric (and all vegetables). Ripening produces significant changes in cellular organization, especially intercellular binding, but the mechanisms involved are not totally known.

During the ripening of a typical climacteric fruit, parenchymous cells have expanded, result-

**Figure 6–10** Chill injury in tomatoes. (A) Chill-injured tomato exhibiting symptoms of severe pitting developed during cold storage. (B, C) Transverse section of chill-induced tomato pit depicting general morphology. (B) Collapsed large parenchyma cells near epidermis (×160). (C) Higher magnification of pit region (×830). (D, E) Transmission electron micrographs of chill-injured tomato fruit. (D) 14-day chilled mature green fruit showing loss of cytoplasmic integrity. Mitochondrial (M) cristae not evident. Mitochondrial (M) and vacuolar (V) membranes have degraded. Plasma membrane (PM) retains contrast (×38,600). (E) Nonchilled mature green fruit showing well-defined cellular ultrastructure. Endoplasmic reticulum (ER) and mitochondrial (M) and vacuolar (V) membranes exhibit high contrast. Mitochondrial cristae evident, chloroplast grana (G) visible, and plastoglobuli (dark inclusions in chloroplast) densely stained and intact. Plasma membrane (PM) in sharp relief (×25,900). *Source*: Jackman et al., 1988.

ing in narrowed cell walls, intercellular spaces, and cell organelles propelled to a thin layer between the cell membrane (plasmalemma) and the membrane of the swollen vacuole (tonoplast), held there by turgor pressure. Maturation initiates a progressive dissolution of the cell wall and middle lamella, probably as a result of softening enzymes triggered by the rise in respiration. As cells begin to separate, it is thought that turgor pressure acts to force them apart. On the molecular level, fruit softening is a consequence of the breakdown of pectic substances in the middle lamella via enzyme action, perhaps aided by the action of cellulases attacking cell walls, although whether this latter event occurs remains controversial. Ripening, during which a large number of other biochemical changes happen, such as starch degradation, sugar production, pigment synthesis, and the appearance of volatile compounds related to flavor and odor, may be in some species under the physiological control of ethylene, which is produced in large amounts by climacteric fruit and can also act exogenously.

As a way of exploring the nature of a respiring and ripening system, we will consider the cherimoya (Palma, Aguilera, & Stanley, 1993; Palma, Stanley, Aguilera, & Zoffoli, 1993). The cherimoya (*Annona cherimola*), an aggregate fruit belonging to the Annonaceae family, is native to the subtropical region of South America. Sometimes given the name "custard apple," it is a highly perishable climacteric fruit that has high respiration rates, ethylene production, and exhibits chilling injury, rapid enzymatic browning, and low tolerance to mechanical injury. Few North Americans know of this product, however, because it is not widely cultivated in that continent and cannot tolerate transportation. Exterior views of two varieties of cherimoya are shown in Figure 6–11.

Ripening is characterized in cherimoyas by the development of flavor and aroma, softening of the pulp, and slight changes in the color of the pulp and flesh. All these changes are paralleled by increases in the respiration rate and in ethylene production. Compared with other fruits, the cherimoya ripens rapidly. At room temperature, a cherimoya will ripen in 6–7 days. In this short period of time, the fruit softens, starch is reduced to sugars, and the acidity increases to give the customary cherimoya flavor.

Because it is a climacteric fruit with a high respiration rate, especially after harvest, the cherimoya is exceptionally perishable. Investigators have often found a second respiration peak after the initial climacteric (Figure 6–12, part A). This diffuse type of climacteric increase with more than one maximum has been observed in other Annonaceae fruits, and it has been suggested that the irregular increase in respiration may be due to the aggregate nature of these fruits, with different segments ripening at different times. Changes in ripening are associated with increased respiration. The development of aroma and flavor occurs after the onset of the climacteric increase, and optimal edibility is attained early in the second increase. Temperature has a marked effect on the respiration rate. Refrigerated storage ($\leq$12°C) both slows down the respiration increase and decreases $CO_2$ production. Although low temperatures delay the ripening process, the cherimoya is susceptible to chilling injury at or below 10°C. Fruit with chilling injury symptoms exhibit brachischlereid lignification and cell walls breakage, which results in high levels of starch and tannins in intercellular spaces as well as skin spotting.

Climacteric fruits are characterized by an increase in ethylene production at the onset of ripening. However, the ethylene rise in the cherimoya occurs after the climacteric rise of $CO_2$ (Figure 6–12, part B). External ethylene production rises after the climacteric $CO_2$ peak, and internal ethylene concentration is less than 0.1 ppm until respiration has increased substantially. No evidence exists for the induction of the climacteric rise by endogenously produced ethylene. Thus, ethylene does not appear to be a major controlling factor in respiration, even though external applications of this hormone can provoke an increase in respiration. Ripening appears to be more related to climacteric respiratory rise than to ethylene.

Cherimoya fruit accumulate starch during growth. Ripening induces starch hydrolysis, leading to an increase in glucose, fructose, and sucrose. Soluble solids rise concomitantly with the

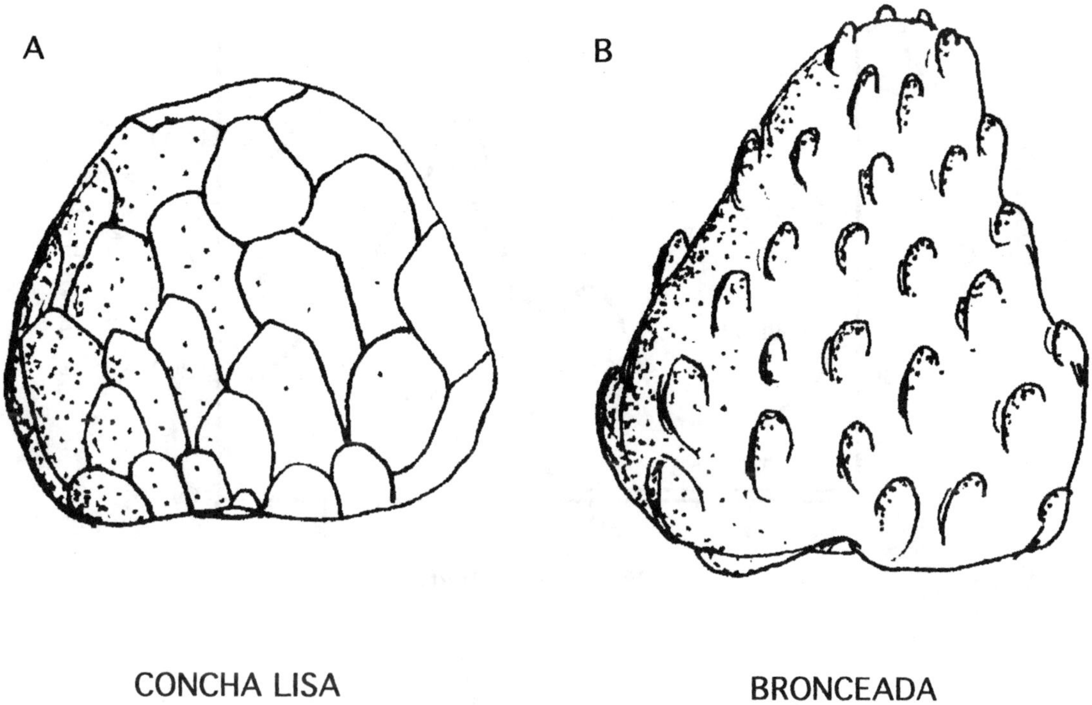


**Figure 6–11** Exterior views of two cherimoya varieties. Fruits weigh from 200 g to 2 kg. *Source:* Reprinted with permission from T. Palma, D.W. Stanley, J.M. Aguilera, and J.P. Zoffoli, Respiratory Behavior of Cherimoya (*Annona Cherimola* Mill.) under Controlled Atmospheres, *HortScience*, Vol. 28, pp. 647–649, © 1993, American Society for Horticultural Science.

respiratory increase and are at a maximum after the climacteric peak, at the onset of the second respiratory peak. This increase coincides with the maximum flavor and best sensory conditions; a ripened cherimoya should reach 18–24 °Brix values for total soluble solids. Acidity in the cherimoya increases during ripening, contrary to the behavior of most fruits. Titratable acidity increases, probably as a result of malic acid, until the onset of the second climacteric rise, coinciding with optimal edible condition, and then decreases. Cherimoya flesh loses its firmness quickly as the fruit ripens. This softening may correspond to an increase in the activity of pectinolytic enzymes, which attack cell walls and cause their separation as well as changes in structure. Softening begins with the onset of the respiratory rise, and firmness decreases dramatically until the first peak of respiration. At this point the fruit has attained its optimal eating firmness, and softening continues but at a slower rate.

It has been shown that controlling metabolic activity and respiration in postharvest tissue is critical for the retention of quality in plant tissue. Since metabolically more active tissues lose quality faster then less active tissues, storage life is a direct function of metabolic activity. Many storage and processing operations are aimed at controlling metabolism and so lengthening shelf-life.

### 6.5.3 Storage and Processing

The harvesting of plant materials invariably results in shock and stress and widespread changes in microstructure and physiology, and it can cause tissue to become so deteriorated that it is not suit-

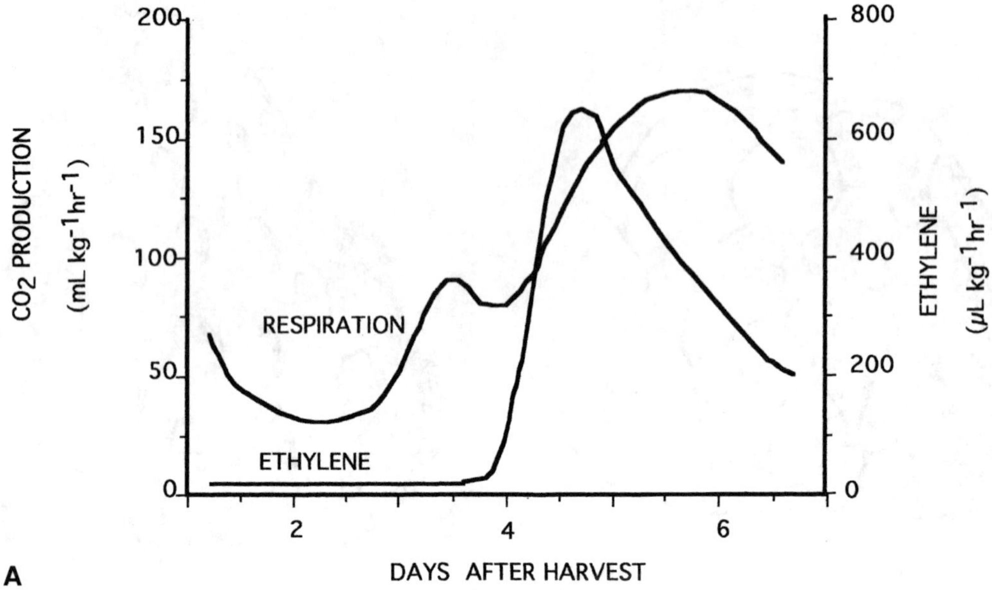


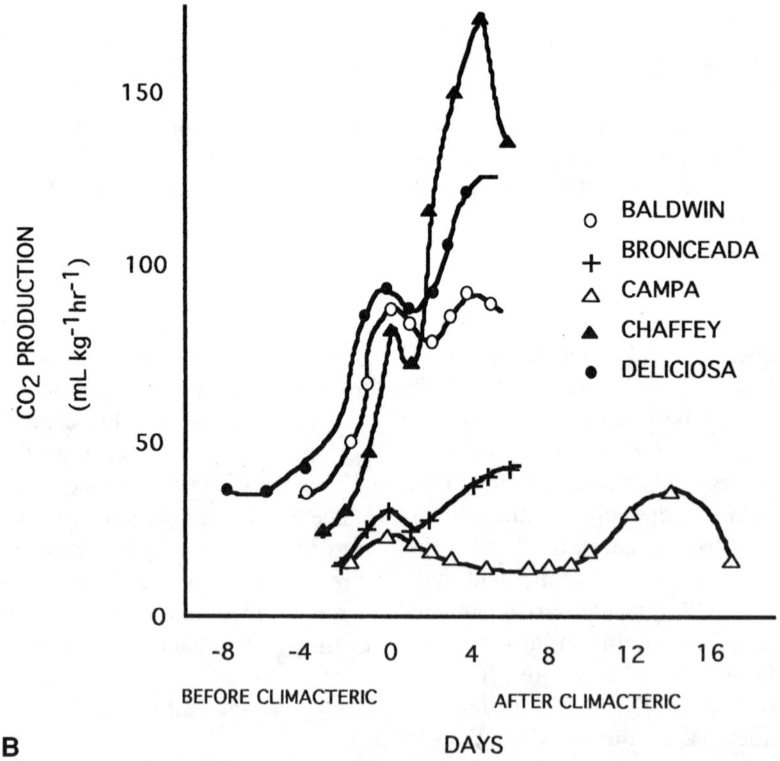


**Figure 6–12** (A) Ethylene rise and respiration in the Chaffey variety of cherimoya. (B) Respiration rate of several varieties of cherimoya fruit. *Source:* Reprinted with permission from T. Palma, D.W. Stanley, J.M. Aguilera, and J.P. Zoffoli, Respiratory Behavior of Cherimoya (*Annona Cherimola* Mill.) under Controlled Atmospheres, *HortScience*, Vol. 28, pp. 647–649, © 1993, American Society for Horticultural Science.

able for use as food. Processing aims at preserving the quality characteristics of the crop for as long as possible. Because there is always some delay in getting commercial produce to the consumer, storage of horticultural materials is inevitable. Storage times can vary from days to years, depending on the product and its end use, and maintenance of proper storage conditions is mandatory. The storage of fresh fruits and vegetables usually involves some combination of low but nonfreezing temperatures, minimization of water loss, and optimization of atmospheric composition. Product quality can be promoted by the application of suitable packaging technologies designed to protect against the damage caused by marketing and storage operations.

Nevertheless, significant deterioration can occur during the storage of fruits and vegetables regardless of the precautions taken. Wilting and pathological invasion are two of the most common disorders. From a structural and quality point of view, chilling injury and lignification are of particular interest, although microstructure is related to the former defects as well. For example, water transpiration rates depend on the microstructure of the tissue surface. Vegetables grown above ground often have surfaces designed to minimize evaporation; leafy parts transpire more than roots or stems. On the other hand, root crops, especially those grown in wet soils, have little protection against moisture loss. Also, the invasion of microorganisms into postharvest tissue is accelerated by structural damage but retarded by the formation of periderm, layers of lignified-type cells occurring at the site of injury.

### 6.5.4 Lignification

Lignification can occur during growth, during storage, or from wounding. The process results in toughened cell walls due to the deposition of lignin, a high molecular weight, three-dimensional polymer of phenolic subunits. Although present in small amounts, lignin can produce objectionable toughness and stringiness in crops with high respiration rates such as asparagus, broccoli, peas, corn, and beans. Lignin is thought to occur in close connection with cellulose fibrils, perhaps cross-linking them and adding to the ability of the tissue to resist compression forces. The mechanisms of lignin formation are far from fully understood, but initiation of lignification seems to follow wounding and/or ethylene treatment. The biosynthetic route contains both enzymatic and nonenzymatic steps and utilizes starting materials such as the aromatic amino acids phenylalanine and tyrosine, which arise from the shikimic acid pathway. Phenylalanine-ammonia lyase and peroxidase are two enzymes associated with lignin synthesis. Lignin polymerization usually begins in the middle lamella and spreads first to the primary wall and then to the secondary wall. Two important food crops for which lignification is a quality defect are asparagus and dried beans.

Asparagus is a rapidly respiring crop that is harvested as the growing spears emerge through the soil from underground crowns. Optimum storage temperature is 0°C to 2°C, and high relative humidity is essential to prevent desiccation. Storage temperatures higher than the optimum cause a large increase in the respiration rate and shorten shelf-life. The major quality defect encountered during storage is toughening or fibrousness. This occurs rapidly, usually within 2 days during room temperature storage. Lignification, the cause of this reaction, seems to be triggered by a wound response, for the fiber content increases from the base (cut end) of the spear toward the tip. Structural changes (Figure 1–7) observed include thickened cell walls with deposits of fluorescing material in the mechanical tissue layer and xylem vessel walls of the vascular bundles. The deterioration can be slowed, but not stopped, by rapid cooling after harvest.

### 6.5.5 Bean Hardening

Bean seeds stored at high temperatures and humidities are susceptible to the hard-to-cook defect. Beans with this defect do not soften sufficiently during cooking because they cannot imbibe sufficient quantities of water in the soaking step. Adverse storage conditions produce this defect, in particular, high temperature and relative humidity. These two factors interact to produce hardened beans if sufficient storage time has elapsed (Stanley & Aguilera, 1985). Bean hardening generally becomes a hindrance to consumption when instrumentally determined hardness values become twice the initial values. When beans are stored experimentally at a constant 30°C and 75% relative humidity, conditions not unusual in tropical climates, this threshold may be reached in only a few months, while noncontrolled, fluctuating field storage conditions will delay attainment of the threshold noticeably.

For growers, bean hardening means that some seeds will not germinate; for consumers, it means a prolonged cooking time is needed for beans to achieve suitable softness. While this latter problem may not sound serious, it can constitute a real hardship for poor people in developing countries, where bean consumption is high and most cooking energy is provided by wood fires. It has been estimated that, in some developing countries, per capita dried bean consumption among the poor averages around 5 kg/person-year and the cost of the energy used in food cooking represents as much as 11% of total income. Of additional concern is the documented reduction of protein efficiency as cooking time is lengthened. Following 10 months' storage in conditions where the mean maximum temperature ranged from 14°C to 31°C and the relative humidity ranged from 63% to 93%, beans approximately doubled in hardness and cooking time while the protein efficiency ratio decreased by about 20% to considerably less than 1.0.

Textural defects in stored beans originate in the structure of the tissue. The two major parts of a bean seed are the seed coat (or testa) and the cotyledons. Since water absorption is at the heart of the hard-to-cook defect, the seed coat is of obvious concern. Structural factors that have been shown to influence water absorption significantly include seed coat thickness, hilum size, seed volume, and color. Even though the seed coat makes up only a small percentage of seed volume (<1%), it contains components such as cellulose, hemicellulose, and lignin that would be expected to influence texture as well as water imbibition. Also, while seed coats soften during cooking, those from hard beans do so to a lesser degree, suggesting that seed coats harden under tropical storage conditions. The other major constituent of beans is the paired cotyledons. Legume cotyledons are composed primarily of parenchymatous cells containing starch granules and protein bodies and bounded by a cell wall and middle lamella.

Microscopic observation of cooked soft and hard beans demonstrates a major structural difference between these two tissues. Legume hardening is accompanied by a failure of the cotyledon cells to separate during cooking (Figure 6–13), the proximate cause of the hard-to-cook defect. Since the middle lamella material is known to be thermally labile in the presence of sufficient moisture, the question arises whether failure to soften during soaking and cooking is due to a lack of water imbibition or to an alteration of the middle lamella components. The answer is probably a combination of the two. It has been reported that hardened beans contain pectin with a lowered solubility, perhaps resulting from elevated levels of divalent cations (calcium and magnesium) and reduced pectin esterification, which would allow more attachment sites to be available for divalent cations to cross-link and insolubilize the pectin. Of at least equal importance, however, is the failure of moisture to penetrate the cotyledon structure. Water absorption has been shown to be significantly and negatively correlated with cooked bean hardness (del Valle, Stanley, & Bourne, 1992). Water is a vital chemical component of several reactions necessary for bean softening. Pectin solubilization, protein denaturation, and starch gelatinization all require the presence of water, and it follows that if this reactant is limited, softening will not occur. It is estimated that hard-to-cook beans bind about 25% less water in the cotyledon than soft beans.

In order to create effective control measures for bean hardening, it is necessary to have some idea of the biochemical mechanisms driving the hardening reaction. An early theory proposed that, under adverse storage conditions, hydrolysis of phytate, a storage form of phosphorus found in bean cotyledons, by the enzyme phytase, which is activated by these conditions, releases divalent

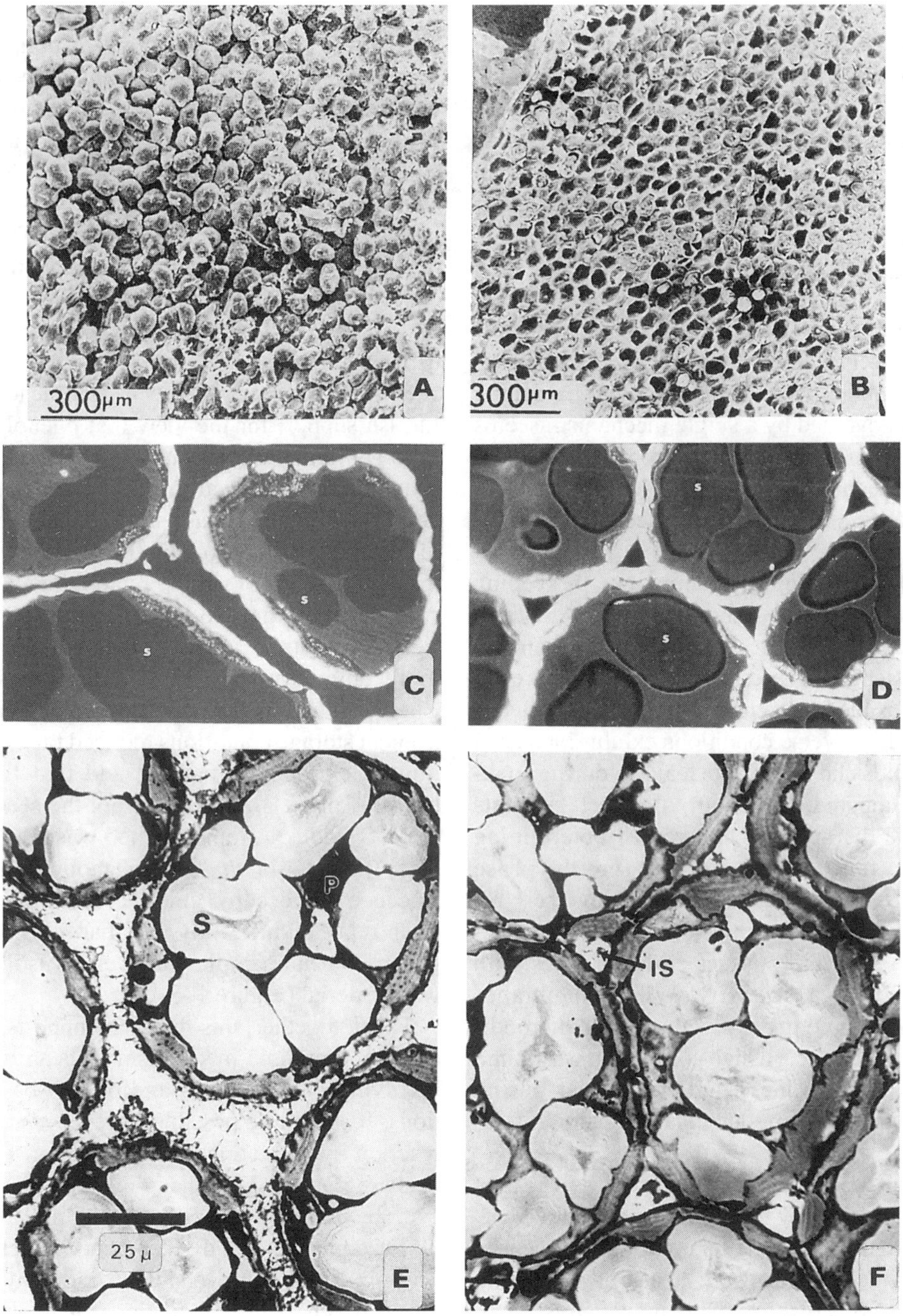


**Figure 6–13** Microstructure of cowpea cotyledon as influenced by heat and viewed in the SEM. (A) Heated 90 min at 100°C. Note cell separation. (B) Heated to 100°C. *Source:* Stanley and Aguilera, 1985. Influence of storage conditions on microstructure of cooked beans held 4 months as seen by fluorescence microscopy. (C) Sample stored at 15°C/35% RH. Note cell separation. (D) Sample stored at 30°C/85% RH. *Source:* Hincks and Stanley, 1986. Plastic embedded 1 $\mu$m toluidine blue–stained sections from cooked beans viewed in LM. (E) Sample stored at 15°C/35% RH. Note cell separation. (F) Sample stored at 30°C/85% RH. *Source:* Courtesy of L.C. Plhak.

cations previously chelated to the phytate molecule. These are the calcium and magnesium cations that then migrate to the middle lamella, where they function as cross-linking agents for the pectin molecules. This mechanism is employed usefully by food scientists when calcium salts are added to products such as canned tomatoes and potatoes to prevent cell sloughing during processing. The phytate hypothesis has been used widely to explain bean hardening, but more recent data suggest several other mechanisms that may assert themselves during this process. Indeed, that such a complex phenomenon as bean hardening would be governed by a single mechanism seems unlikely. Two alternative but probably not independent mechanisms for bean hardening have been advanced, and indeed both seem to be implicated.

The role of biological membranes in keeping solutes inside the cell and separating enzymes from potential substrates has previously been discussed at length. These functions are impaired if membranes lose their critical permeability. Beans stored under adverse conditions exhibit increased loss of solids and electrolyte leakage during soaking (as compared with controls), which suggests membrane disruption. One way of determining membrane functionality is to measure the phase transition temperature, a parameter inversely related to membrane fluidity. Fluidity, in turn, decreases in senescing tissues and correlates with loss of biological integrity. When membranes from beans stored under a wide variety of conditions were examined, a high correlation was found between both hardness and solids lost during soaking and the membrane phase transition temperature.

Another mechanism is suggested by the fact that when red kidney beans were stored under adverse conditions and the total soluble condensed tannins of the seed coat were measured along with bean hardness, a highly significant negative correlation was found between these two variables. That is, as the beans hardened, smaller amounts of soluble tannins were extracted. It is possible that the decline in soluble tannins represents an increase in condensed tannins in the seed coat and perhaps a migration of soluble tannins to the cotyledons. Tannins are found in both the seed coat and the cotyledons of common beans. It has been demonstrated that the deposition of condensed tannins occurs in conjunction with lignification to incrust and toughen cell walls, and it should be remembered that lignin itself is a phenolic polymer. Thus, it seems possible that the deposition of condensed tannins or lignin in the seed coat restricts its swelling, while a similar reaction in cotyledon cells strengthens this tissue and contributes to hardness.

Several lines of evidence have come together to furnish support for the view that phenol condensation or lignin deposition is a significant factor in bean hardening. Electron micrographs of cotyledon cells from hard beans showed that depositions of material reacting to a lignin stain were heavier than depositions of similar material isolated from soft beans (Figure 6–14). Isolated cell walls treated to extract tannins revealed significantly more tannins in hard than soft bean samples. Chemical analysis of bean extracts indicated that tropical storage conditions reduced the concentration of extractable tannins and that the unextractable material migrated from the seed coat to the cotyledons. A rapid method was developed to assay for total *in situ* seed phenolics using fluorescence microscopy, and the results were found to correlate significantly with hardness in a large group of bean samples, as well as isolated cell wall material (Figure 1–8).

Taken together, this research supports the concept of a multiple mechanism of bean hardening involving both phytate- and phenol-based reactions. It should be possible to reverse the phytate reaction by extracting the divalent cations with a chelating agent. The phenolic step should, however, be irreversible. With this is mind, a protocol was established that divided bean hardening into reversible and nonreversible portions by soaking beans in a chelating solution prior to cooking. This step removed divalent cations from hard-to-cook beans and replaced them with sodium so as to counteract the hardening effect of the pectin-phytate mechanism, leaving only the hardness due to phenols and condensed tannins. Both components

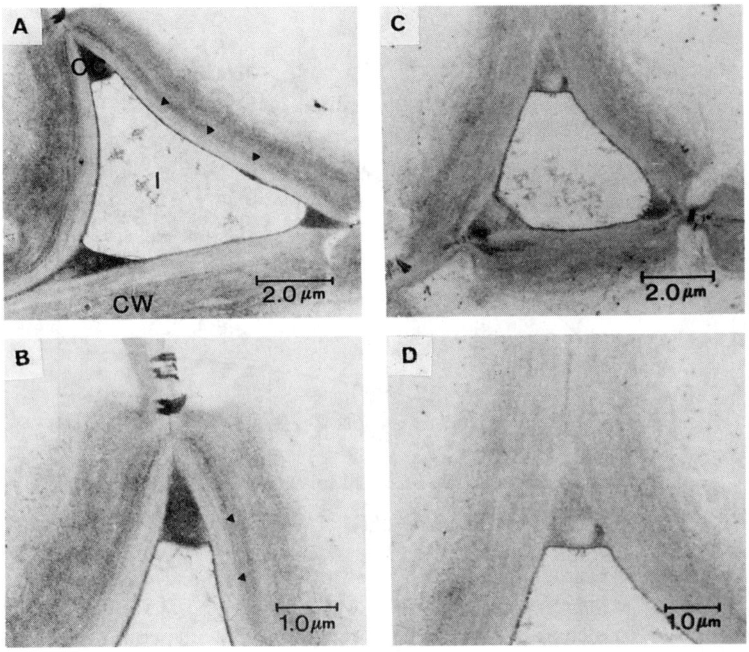


**Figure 6–14** Deposition of lignin stain–positive material in hard-to-cook (HTC) and control beans. (A, B) Intercellular space formed by joining of cotyledon cells of HTC sample produced by high-temperature/high-humidity storage. Note intense potassium permanganate staining in cell corners and secondary wall as compared with control (C, D). C = cell corner, CW = cell wall, I = intercellular space. Arrowheads denote boundary of primary and secondary walls. *Source:* Hincks and Stanley, 1987.

of hardening increased during storage, as indicated in Figure 6–15. Chemical data indicated that reversible hardening is associated with phytate metabolism whereas irreversible hardening is more related to the metabolism of phenolic compounds. Although the exact contribution of each mechanism to total bean hardness is not yet known, some understanding of the cause of the hard-to-cook defect in legumes has been achieved.

It would be ideal if the knowledge gained from studying the mechanism of bean hardening could be directed toward alleviating this problem. Although research in this area leaves unanswered questions, some methods have been found that might prove fruitful in ameliorating bean seed hardening. While genetic intervention does not seem promising at this time, not enough work has been done in screening the vast array of *Phaseolus* accessions to conclude that this approach may not be useful in the future. Producers may be able to manipulate agronomic factors to their benefit. Processors can overcome hardening completely by canning, a procedure that also can be mimicked in the home using a pressure cooker. Some softening can be attained during cooking by the addition of various salts during soaking or cooking, although more information is needed on the mechanism involved and the nutritional implications.

### 6.5.6 Controlled Atmosphere Storage

Many physiologically induced quality defects of fruits and vegetables can be restrained, to a greater or lesser extent, through controlled atmosphere storage (i.e., using an atmospheric composition different from that of normal air). Usually oxygen levels are decreased while carbon dioxide levels are increased to a precise balance in order to slow

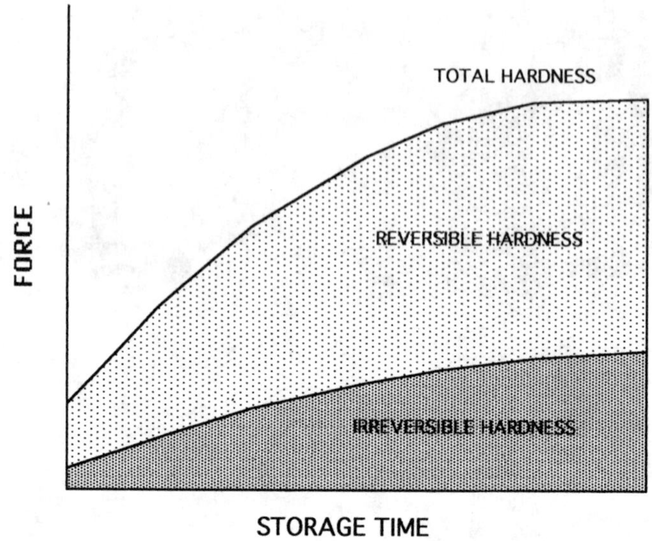


**Figure 6–15** Representation of bean hardening as a result of reversible and irreversible components. *Source:* Reprinted from *Food Research International*, Vol. 28, J.M. del Valle and D.W. Stanley, Reversible and Irreversible Components of Bean Hardening, pp. 455–463, Copyright 1995, with permission from Elsevier Science.

respiration and lengthen storage life by reducing the metabolic and structural damage mentioned previously. Other benefits accrue as well. These include a reduction in the rate of natural ethylene production, a decrease in sensitivity to ethylene, a lowering of the rate of chlorophyll degradation, the inhibition of pectic substance breakdown, and flavor improvement. Different commodities exhibit different responses to controlled atmosphere storage, however, and thorough experimentation is required to obtain the optimal storage atmosphere. Also, the increased cost of this type of storage demands careful evaluation of its benefits to ensure this technique is worth using. Further advantages may be obtained from the use of plastic films; hypobaric or reduced pressure storage is proving to be an effective but expensive control device for some crops.

To illustrate controlled atmosphere storage, let us return to the example of the cherimoya. Although the fruit is grown in many subtropical areas of the world, commercial trading of the cherimoya has been limited because it does not travel well. Traditional refrigerated storage below 10°C cannot be used with this fruit, since it is susceptible to chilling injury. Therefore, an alternative treatment that can be combined with moderate temperature (≥10°C) is necessary to delay ripening before this fruit can be successfully transported to world markets. Controlled atmosphere storage at 10°C would seem suitable, but the proper atmospheric conditions must be known. In order to determine this, fruit may be stored individually in chambers equipped with temperature control and gas flow and from which headspace gas can be sampled. Respiration patterns from an experiment of this type (Figure 6–16), where $O_2$ levels between 5% and 20% were maintained, show that fruit kept at 10°C with 20% $O_2$ exhibit a typical climacteric rise in $CO_2$ production and $O_2$ consumption. Cherimoya demonstrated normal behavior for climacteric fruit, in that lower levels of $O_2$ in the storage atmosphere reduced respiration. The respiration rate was clearly reduced in 5% $O_2$ and the climacteric rise was absent. These data, taken together with a lack of ethylene production and the retention of firmness after 43 days' storage, suggest that these samples may have started their climacteric period only after they were removed from the controlled atmosphere. Subsequently, these fruits de-

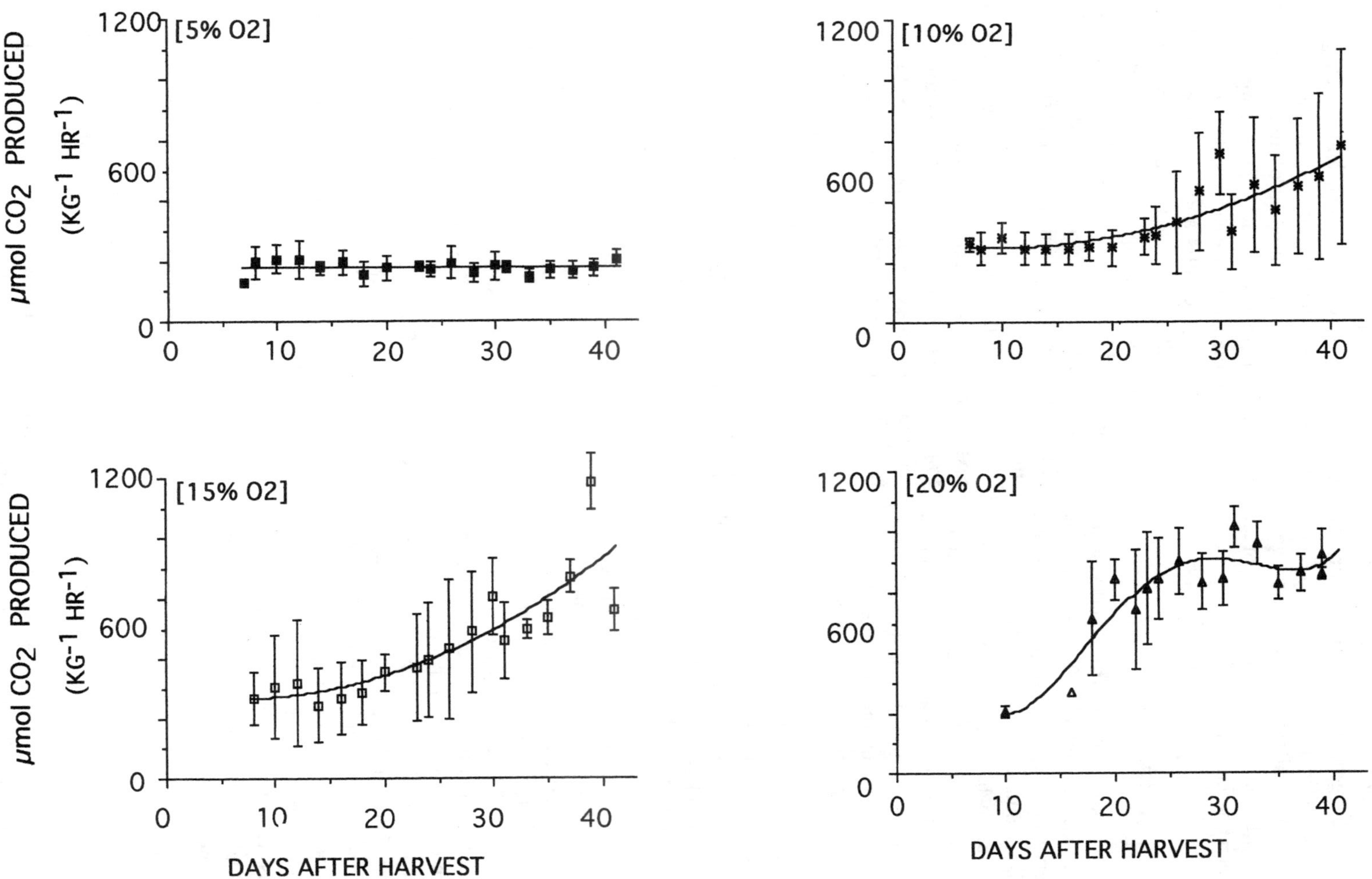


**Figure 6–16** Respiration rates of cherimoya fruit var. Concha Lisa held at 10°C and $O_2$ concentrations of (A) 20%, (B) 15%, (C) 10%, and (D) 5%. Curves represent best-fit regression of 6 replicates. Vertical bars represent standard deviations. *Source:* Reprinted with permission from T. Palma, D.W. Stanley, J.M. Aguilera, and J.P. Zoffoli, Respiratory Behavior of Cherimoya (*Annona Cherimola* Mill.) under Controlled Atmospheres, *HortScience*, Vol. 28, pp. 647–649, © 1993, American Society for Horticultural Science.

veloped characteristic aroma and flavor and softened normally during ripening at 20°C in air. Delay of the climacteric by the 5% $O_2$ treatment, coupled with the continued ability to ripen, suggests the suitability of controlled atmospheric storage for transporting cherimoya.

### 6.5.7 Canning and Freezing

Instead of being stored, many fruits and vegetables are processed immediately after harvest in order to preserve their nutritional quality, control microorganisms, and retain quality. The methods used are structurally harsh, as compared with storage, but the stabilized products have long shelf-lives, even if often original materials have undergone significant alteration. For instance, while stored potatoes can be mashed, canned potatoes roasted, and frozen potato products fried, all are designed for specific culinary applications and differ markedly from the original material.

Heating can produce profound structural alterations in the cell-cell adhesion of nonlignified tissue that depend upon the extent of the treatment. Interestingly, not all thermal effects are deleterious to physical properties. For example, the enzyme pectin methylesterase bound to the cell wall is activated during low temperature blanching (i.e., 50–60°C) and hydrolyzes methyl groups from pectin to yield pectic acid. This compound, with its free carboxyl groups, is relatively insoluble, remains in the cell wall, and leads to firmer texture through cross-linking with endogenous cations. If a more stringent heating regime is required, it is often possible to counteract unwanted softening through the use of firming agents—divalent cations—that cross-link and stabilize pectic substances with free carboxyl groups. However, some parenchyma-rich plant tissues whose cells are nonlignified do not soften, or soften to a limited extent, during heating (e.g., Chinese water chestnut), and a mechanism of phenolic cross-linking between cell wall polymers has been proposed to explain this phenomenon (Waldron, Smith, Parr, Ng, & Parker, 1997).

In general, however, heating softens fruits and vegetables. The softening is caused by loss of turgor pressure and occluded air, degradation of cell wall polysaccharides, and starch gelatinization. The final outcome is that the cells are able to separate, which has profound effects on texture. Whereas previously shearing action would produce fracture across the cell walls, breaking open the cells and releasing the cell contents, cooked material fractures through the middle lamellae, leaving cell walls intact (Figures 6–2, 6–13, and 6–17). Extended heating especially influences the integrity of cell walls (Figure 6–18), while lignified tissues show minimal changes.

Commercial freezing operations often result in textural damage so severe that many structurally fragile commodities cannot be processed by this method. To be successful, freezing must proceed rapidly to a low temperature, and storage should last a short time, the temperature should be kept low, and freeze-thaw cycles should be avoided. Failure to meet these requirements can lead to large needlelike ice crystal formation and concomitant tissue disruption and cell wall damage (Figure 4–6). In sum, the degree of freezing injury depends upon the susceptibility of the commodity, freezing rate, ultimate temperature, and storage conditions.

### 6.5.8 Post-Mortem Physiology

In living muscle, the biological oxidation of simple carbohydrates to carbon dioxide and water requires oxygen and generates ATP. Muscle contraction uses ATP as a chemically based source of energy. Glycogen is the primary form of storage carbohydrate in muscle fibers, and it appears in electron micrographs as single granules or clumps located in the sarcoplasm between myofibrils (Figure 1–12).

The first step in the conversion of muscle to meat is slaughter, and the cessation of circulation that occurs at that point directly influences the microstructure and physical properties of muscle tissue through a series of complex chemical reactions. These reactions are necessary for the development of flavor, color, and textural properties associated with this food product. Unwanted defects in post-mortem physiology can

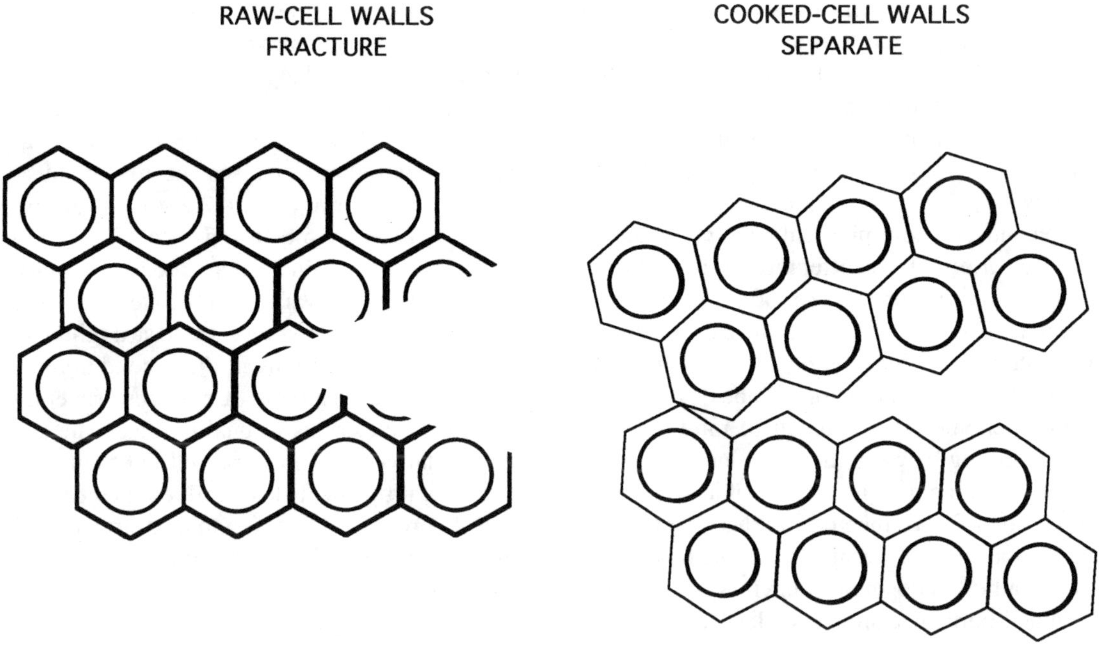


**Figure 6–17** Structural basis for fracture in raw and cooked plant tissue. *Source:* Reprinted from *Food Research International*, Vol. 27, D.W. Stanley, Understanding the Materials Used in Foods—Food Materials Science, pp. 135–144. Copyright 1994, with permission from Elsevier Science.

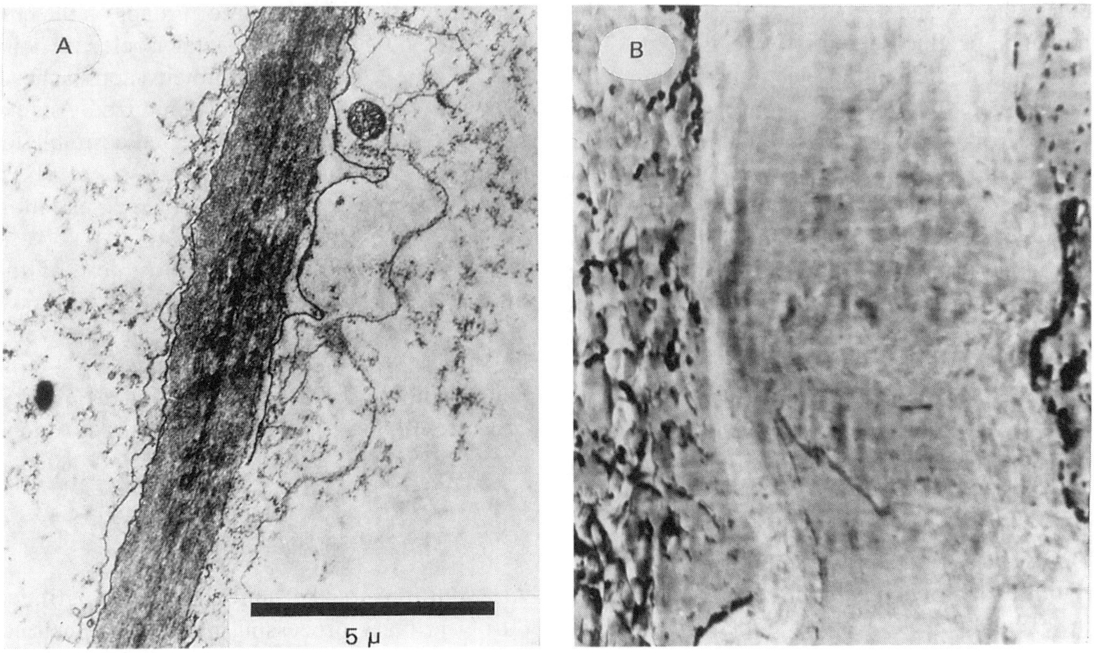


**Figure 6–18** Transmission electron micrographs of potato cell wall. (A) Raw. (B) Cooked. Note loss of membrane structure and separation of cells after heating. *Source:* Stanley and Tung, 1976.

lead to poor quality meat and associated economic loss. A common cause of such defects is an excessive stress level prior to slaughter. Key physiological consequences for post-mortem muscle cells include the inability to maintain temperature, pH level, $O_2$ concentration, and the supply of energy.

The most obvious physical post-mortem consequence for muscle is stiffening and loss of extensibility as a result of rigor mortis. As mentioned previously, if a muscle is restrained during this period, shortening does not occur, but if it is unrestrained, severe contraction will be observed. The rigor phenomenon is a result of multiple cross-links forming between the actin and myosin filaments, creating a rigid, inextensible structure. A resolution of rigor then occurs that produces tenderization but not because of cross-link separation. Rather, endogenous enzymes attack the Z-disc and perhaps also cytoskeletal elements and connective tissue to produce a loosening of muscle structure and more tender meat.

As the circulatory system ceases after death, so does the supply of glucose from the blood. Glucose is then supplied by glycogen degradation. When the $O_2$ supply becomes depleted, aerobic oxidation of glucose and fatty acids by the TCA cycle and mitochondrial electron transport halts, but the glycolytic pathway continues to convert glucose to pyruvate anaerobically. Pyruvate is then reduced to lactate, but this is much less efficient than producing ATP aerobically. The lactic acid that is formed lowers muscle pH. Stressful conditions prior to slaughter such as overexercise, extremes in temperatures, fasting, or fear lead to elevated rates of glycolysis, depleted glucose, and impaired pH drop. Also, increased glycolysis produces more heat and lowers the rate of the chilling operation. At low temperatures (0–5°C), membranes of the sarcoplasmic reticulum, a network of membranous channels surrounding each myofibril and capable of holding calcium ions, lose their sequestering ability, and if the ATP concentration is high enough, contraction will occur in the excised muscle, a defect termed *cold shortening* that can lead to tough, dry meat. In order to avoid cold shortening, lamb carcasses should, as a rule of thumb, not reach 10°C within 10 hours after death, and beef sides should not be exposed to air below 5°C or to a flow rate faster than 1 m/sec within 24 hours of slaughter.

Another consequence of post-mortem metabolism is a decrease in pH. ATP hydrolysis and lactate accumulation causes the pH of muscle to drop from around 7.0–7.2 to an ultimate pH of around 5.5–5.8 in red meat. The rate of pH decline and the final pH attained are governed by factors such as available substrate levels, prior nutritional status, stress levels, and struggling during slaughter. Variations in these conditions can lead to different patterns of pH drop, several of which are deleterious to quality (e.g., pale, soft, and exudative meat in pork and dark-cutting in beef). The ultimate pH level influences meat quality in many ways. For example, it affects resistance to growth of microorganisms, color, and water-holding capacity. Muscle contains about 75% $H_2O$, and it is important that $H_2O$ remain bound to the protein matrix to ensure high quality. Water binding occurs via ionic interactions and physical entrapment in muscle fibers; poor water-holding capacity is associated with the defects of dryness and drip. It is pH that determines the number and distribution of charged groups on the protein, so at low pH values, near the isoelectric point (4.5), the net charge on the proteins approaches 0 and water-holding capacity is minimized. An ultimate pH of around 5.8 is considered desirable for optimal eating quality. This is also the pH at which the proteases inducing tenderness are most active.

These examples indicate but a few of the many reactions occurring in post-mortem muscle. Imposed on these naturally occurring events are the consequences of processing steps, which, although designed to safeguard quality, can often have the opposite effect if not done properly.

### 6.5.9 Processing

An addition to the unit operations utilized during primary processing of meat is electrical stimulation. In this process, electrical voltage is applied directly to freshly slaughtered animals, mainly in order to improve the tenderness of the meat. This treatment accelerates the normal pH decline in muscle tissue and helps to prevent defects such as cold shortening. One way that electrical stimulation may improve tenderness is by activation of lysosomal cathepsins, a group of enzymes that make meat more tender during the aging period. It can also cause physical disruption of muscle structure. Stimulating carcasses after slaughter accelerates all the biochemical changes that normally occur. Thus, rigor develops earlier and the risk of cold shortening is reduced. Once rigor has developed fully, filament sliding is impossible, and it does not matter how cold the meat gets.

During the slaughtering process, the carcass is usually hung to use the weight of the animal to stretch the most commercially important muscles and prevent contraction during the rigor period that would lead to toughness. In species such as beef, where toughness is a consumer quality factor, the carcass may remain in a cooler for 10–14 days to increase tenderness. This process is known as conditioning or aging. During this period, certain proteolytic enzymes, such as cathepsins, attack structural proteins in muscle fibers, causing tenderization. Another enzyme involved in this process is a calcium-activated protease that slowly disrupts Z-discs by releasing a-actinin, a cytoskeletal protein that holds the thin filaments to the Z-disc. Longer conditioning times increase tenderness, but the benefit is counteracted by weight loss, surface spoilage, and the cost of refrigerated storage.

The most common secondary processing procedures for meat products involve the application of high temperatures. Not only does heating destroy potentially hazardous microorganisms, it causes textural change. Several mechanisms operate during cooking that influence physical properties: activation and subsequent deactivation of proteolytic enzymes, collagen gelatinization, and thermal denaturation of structural proteins, which is accompanied by loss of water and lipids and results in shrinkage. Impacting these reactions are simultaneous changes in muscle structure such as sarcomere shortening, myofibrillar fragmentation, and coagulation of proteins. The texture of a cooked piece of meat will depend upon the opposing actions of collagen softening and protein hardening; older, tougher muscles require longer heating to soften than younger tissue with less connective tissue.

## 6.6 CONCLUSIONS

This brief examination of food structures was intended to convey that foods and their components exhibit a wide range of incredibly complex structures. How rugged and persistent those structures are and how susceptible they are to natural postgathering and processing deterioration determine the best method of processing and the quality of the resulting food product. It is often not the structural elements that are of the greatest concern but how they interact to form an organization, frequently built up over a range of hierarchies. Structural information is necessary to characterize and control the properties of food materials and improve them through better approaches to processing. In the case of tissue foods, where structure is formed by nature, understanding the chemical and biochemical routes during synthesis and degradation and their relation to structure and texture is essential. Also, enchanced understanding may lead to the improvement of products through biotechnology.

## BIBLIOGRAPHY

Barret, A.M., Normand, M.D., Peleg, M., & Ross, E. (1992). Characterization of the jagged stress-strain relationships of puffed extrudates using the fast Fourier transform and fractal analysis. *Journal of Food Science, 57,* 227–232, 235.

Bernal, V.M., & Stanley, D.W. (1987). Effect of cooking temperature on the fracture behavior of pre-rigor bovine sternomandibularis muscle. *Canadian Institute of Food Science and Technology, 20,* 56–59.

Bernal, W.V.W., Bernal, V.M., Gullett, E.A., & Stanley, D.W. (1988). Sensory and objective evaluation of a restructured beef product. *Journal of Texture Studies, 19,* 231–246.

Bourne, M.C. (1982). *Food texture and viscosity.* New York: Academic Press.

del Valle, J.M., & Stanley, D.W. (1995). Reversible and irreversible components of bean hardening. *Food Research International, 28,* 455–463.

del Valle, J.M., Stanley, D.W., & Bourne, M.C. (1992). Water absorption and swelling in dry beans. *Journal of Food Processing and Preservation, 16,* 75–98.

deMan, J.M. (1976). Mechanical properties of food. In J.M. deMan, P.W. Voisey, V.F. Rasper, & D.W. Stanley (Eds.), *Rheology and texture in food quality* (pp. 8–27). Westport, CT: AVI Publishing Co.

deMan, J.M., & Stanley, D.W. (1984). Mechanical properties of food. In D.W. Gruenwedel & J.R. Whitaker (Eds.), *Food analysis, principles and techniques.* Vol. 1. *Physical Characterization* (pp. 221–245). New York: Marcel Dekker.

Dull, G.G. (1986). Nondestructive evaluation of quality of stored fruits and vegetables. *Food Technology, 40*(5), 106–110.

Haard, N.F. (1995). Food as cellular systems: Impact on quality and preservation. *Journal of Food Biochemistry, 19,* 191–238.

Hincks, M.J., & Stanley, D.W. (1986). Multiple mechanisms of bean hardening. *Journal of Food Technology, 21,* 731–750.

Hincks, M.J., & Stanley, D.W. (1987). Lignification: Evidence for a role in hard-to-cook beans. *Journal of Food Biochemistry, 11,* 1–58.

Jackman, R.L., Yada, R.Y., Marangoni, A., Parkin, K.L., & Stanley, D.W. (1988). Chilling injury. A review of quality aspects. *Journal of Food Quality, 11,* 253–278.

Kalab, M., & Harwalkar, V.R. (1974). Milk gel structure. 2. Relation between firmness and ultrastructure of heat-induced skim milk gels containing 40–60% total solids. *Journal of Dairy Research, 41,* 131–135.

Marangoni, A.G., Palma, T., & Stanley, D.W. (1996). Membrane effects in postharvest physiology. *Postharvest Biology and Technology, 7,* 193–217.

Palma, T., Aguilera, J.M., & Stanley, D.W. (1993). A review of postharvest events in cherimoya. *Postharvest Biology and Technology, 2,* 187–208.

Palma, T., Stanley, D.W., Aguilera, J.M., & Zoffoli, J.P. (1993). Respiratory behavior of cherimoya (*Annona cherimola* Mill.) under controlled atmospheres. *Horticultural Science, 28,* 647–649.

Stanley, D.W. (1991). Biological membrane deterioration and associated quality losses in food tissues. *CRC Critical Reviews in Food Science, 30*(5), 487–553.

Stanley, D.W. (1994). Understanding the materials used in foods: Food materials science. *Food Research International, 27,* 135–144.

Stanley, D.W., & Aguilera, J.M. (1985). Textural defects in cooked reconstituted legumes: The influence of structure and composition. *Journal of Food Biochemistry, 9,* 277–323.

Stanley, D.W., Stone, A.P., & Tung, M.A. (1996). Mechanical properties of food. In L. Nollet (Ed.), *Handbook of food analysis* (Vol. 1) (pp. 93–137). New York: Marcel Dekker.

Stanley, D.W., & Tung, M.A. (1976). Microstructure of food and its relation to texture. In J.M. deMan, P.W. Voisey, V.F. Rasper, & D.W. Stanley (Eds.), *Rheology and texture in food quality* (pp. 28–78). Westport, CT: AVI Publishing Co.

Stanley, D.W., & Voisey, P.W. (1979). Texture-structure relationships in bacon adipose tissue. In P. Sherman (Ed.), *Food texture and rheology* (pp. 395–424). London: Academic Press.

Szczesniak, A.S. (1983). Physical properties of foods: What they are and their relation to other food properties. In M. Peleg & E.A. Bagley (Eds.), *Physical properties of foods* (pp. 1–41). Westport, CT: AVI Publishing Co.

Szczesniak, A.S. (1998). Sensory texture profiling: Historical and scientific perspectives. *Food Technology, 52*(8), 54–57.

Vickers, Z.M. (1987). Crispness and crunchiness: Textural attributes with auditory components. In H.R. Moskowitz (Ed.), *Food texture* (pp. 145–166). New York: Marcel Dekker.

Waldron, K.W., Smith, A.C., Parr, A.J., Ng, A., & Parker, M. (1997). New approaches to understanding and controlling cell separation in relation to fruit and vegetable texture. *Trends in Food Science and Technology, 8,* 213–221.

## SUGGESTED READING

### Texture Determination

Dziezak, J.D. (1991). Viscosity measurement. *Food Technology, 45*(7), 82–96.

Kokini, J.L. (1987). The physical basis of liquid food texture and texture-taste interactions. *Journal of Food Engineering, 6,* 51–81.

Martens, H., & Russwurim, H., Jr. (1983). *Food research and data analysis.* London: Applied Science Publishers.

Moskowitz, H.R. (1987). *Food texture: Instrumental and sensory measurement.* New York: Marcel Dekker.

O'Mahony, M. (1986). *Sensory evaluation of food: Statistical methods and procedures.* New York: Marcel Dekker.

Pangborn, R.M.V. (1984). Sensory techniques of food analysis. In D.W. Gruenwedel & J.R. Whitaker (Eds.), *Food analysis: Principles and techniques.* Vol. 1. *Physical characterization* (pp. 37–93). New York: Marcel Dekker.

Szczesniak, A.S. (1987). Correlating sensory with instrumental texture measurements: An overview of recent developments. *Journal of Texture Studies, 18,* 1–15.

Voisey, P.W. (1976). Instrumental measurement of food texture. In J.M. deMan, P.W. Voisey, V.F. Rasper, & D.W. Stanley (Eds.), *Rheology and texture in food quality* (pp. 79–141). Westport, CT: AVI Publishing Co.

**Structural Aspects of Food Texture**

Atkins, A.G. (1987). The basic principles of mechanical failure in biological systems. In J.M.V. Blanshard & P. Lillford (Eds.), *Food structure and behaviour* (pp. 149–168). New York: Academic Press.

Brett, C., & Waldron, K. (1996). *Physiology and biochemistry of plant cell walls.* London: Chapman and Hall.

Gordon, J.E. (1978). *Structures or why things don't fall down.* London: Pitman Publishing.

Hamann, D.D. (1983). Structural failure in solid foods. In M. Peleg & E.B. Bagley (Eds.), *Physical properties of foods* (pp. 351–383). Westport, CT: AVI Publishing Co.

Jackman, R.L., & Stanley, D.W. (1995). Perspectives in the textural evaluation of plant foods. *Trends in Food Science and Technology, 6,* 187–194.

Purslow, P.P. (1987). The fracture behavior of meat: A case study. In J.M.V. Blanshard & P. Lillford (Eds.), *Food structure and Behaviour* (pp. 177–197). New York: Academic Press.

Stanley, D.W. (1987). Food texture and microstructure. In H.R. Moskowitz (Ed.), *Food texture: Instrumental and sensory measurement* (pp. 35–64). New York: Marcel Dekker.

Stanley, D.W., Aguilera, J.M., Baker, K.W., & Jackman, R.L. (1998). Structure/property relationships of foods as affected by processing and storage. In M.A. Rao & R.W. Hartel (Eds.), *Phase/state transitions in foods* (pp. 1–56). New York: Marcel Dekker.

Stanley, D.W., & Tung, M.A. (1976). Microstructure of food and its relation to texture. In J.M. deMan, P.W. Voisey, V.F. Rasper, & D.W. Stanley (Eds.), *Rheology and texture in food quality* (pp. 28–78). Westport, CT: AVI Publishing Co.

Vincent, J.F.V., & Lillford, P.J. (Eds.). (1991). *Feeding and the texture of food.* Cambridge: Cambridge University Press.

# CHAPTER 7

# Microstructural Aspects of a Fluid Food: Milk

## 7.1 INTRODUCTION

Milk and dairy products have been consumed since antiquity and are still highly regarded foods in most cultures. At the macrostructural level, they exhibit a wide assortment of physical states ranging from Newtonian suspensions to solid powders, including viscous fluids, gels, foams, emulsions, and plastic and thermoplastic materials. Milk products are rich in microstructural elements such as plastic globules, membranes, colloidal aggregates, and even crystals. All these elements interact to form the variety of textures characteristic of dairy products. Dairy processing is an excellent example of how restructuring of foods has operated for centuries.

This chapter reviews the microstructural elements in milk, their structure-building properties, and their interrelation during processing to conform dairy products.

### 7.1.2 Studying Dairy Food Microstructure

Electron microscopy has been extensively used to study the ultrastructure of milk components, the changes they undergo, and the interactions that occur among themselves and between them and other ingredients during processing. Good general reviews include those of Brooker (1979) and Kalab (1979a). A more basic presentation is found in Sara (1984). Holcomb (1991) has compiled a list of references on the structure and rheology of dairy products.

Dairy products are difficult to prepare for electron microscopy because of the simultaneous presence of liquid fat and water. Important technical developments have included the introduction of the cold-stage in SEM and the use of freeze-fractured replicas in TEM to study the ultrastructure of products based on fat. Kalab (1981) reviewed the electron microscopical techniques most widely used in the study of milk products. Table 7–1 is taken from this article.

## 7.2 BUILDING BLOCKS AND PHYSICOCHEMICAL PHENOMENA IN MILK

### 7.2.1 The Building Blocks

The main technological properties of milk derive not so much from its chemical composition as from the structure emerging as a consequence of the interaction between components. The physicochemical characteristics of the main components and microstructural elements found in fluid milk are shown in Table 7–2. It is amazing indeed that products spanning almost the entire processed food spectrum can be fabricated from a Newtonian liquid that basically has three building blocks: casein micelles, fat globules, and whey proteins. Figure 7–1 shows some of the structures present in dairy products: liquids, hard and plastic solids, foams, gels, powders, and so on. It is essential for understanding how structure is achieved that we realize these building blocks must achieve an

**Table 7–1** Electron Microscopical Techniques Used in the Study of Milk and Milk Products

| *Technique* | *Nature of Specimen* | *Example of Milk Product for Which Suitable* |
|---|---|---|
| | *Scanning Electron Microscopy* | |
| Conventional | Dry | Powdered milk, whey, buttermilk, etc. |
| | Dried | Low-fat milk products (yogurt, cottage cheese, some cheeses); cheese (fat extracted); cheese (trypsin-etched) |
| | Freeze-fractured and replicated with gold | Viscous and high-fat milk products (cream, butter, cheese, etc.) |
| Cold stage | Freeze-fractured and coated with carbon and/or gold | Viscous, whipped, high-fat products (ice cream, whipped cream, butter, etc.) |
| | *Transmission Electron Microscopy* | |
| Negative staining | Suspension | Fluid milk, cream |
| Metal shadowing | Suspension | Fluid milk, cream |
| Thin-sectioning | Suspension | Liquid products (microencapsulation) |
| | Solid | Products solid by nature (cheese) or solidified by mixing with agar (cream) |
| Freeze-fracturing (freeze-etching) | All products | All products |
| Replication of dried specimen | Solid | Milk products based on protein |

*Source:* Kalab, 1981.

"active state" to engage in nonspecific interactions. Attainment of this state is discussed in the following sections.

No food system has been studied in more detail, physicochemically and microscopically, than fluid milk. Milk components exist in a state of dynamic equilibrium in which colloidal phenomena predominate. These phenomena are highly complex, and this presentation only aims at facilitating the interpretation at ultrastructural and microstructural levels. Some of the subjects discussed here are still open to speculation. A short,

**Table 7–2** Structural Components of Milk and Milk Products

| | | *Protein* | *Lipids* | *Carbohydrates* | *Ash* |
|---|---|---|---|---|---|
| Structural elements | Casein micelles | Globular proteins | Fat globules | Lactose molecules | Minerals (Ca, P, Na, etc.) |
| Content (%) | 2.6 | 0.6 | 3.8 | 4.6 | 0.65 |
| Physicochemical state | Fine dispersion | Colloidal solution | Emulsion | True solution Crystals? | Colloidal and true solution |
| Particle dimension | 10–300 nm | 3–6 nm | 0.1–10 $\mu$m | 8 Å? | 4 Å |
| Number/ml | $10^{14}$ | $10^{17}$ | $10^{10}$ | — | — |
| Density at 20°C (g/ml) | 1.11 | 1.34 | 0.92 | 1.55 | — |

*Note:* Numerical values are approximate.

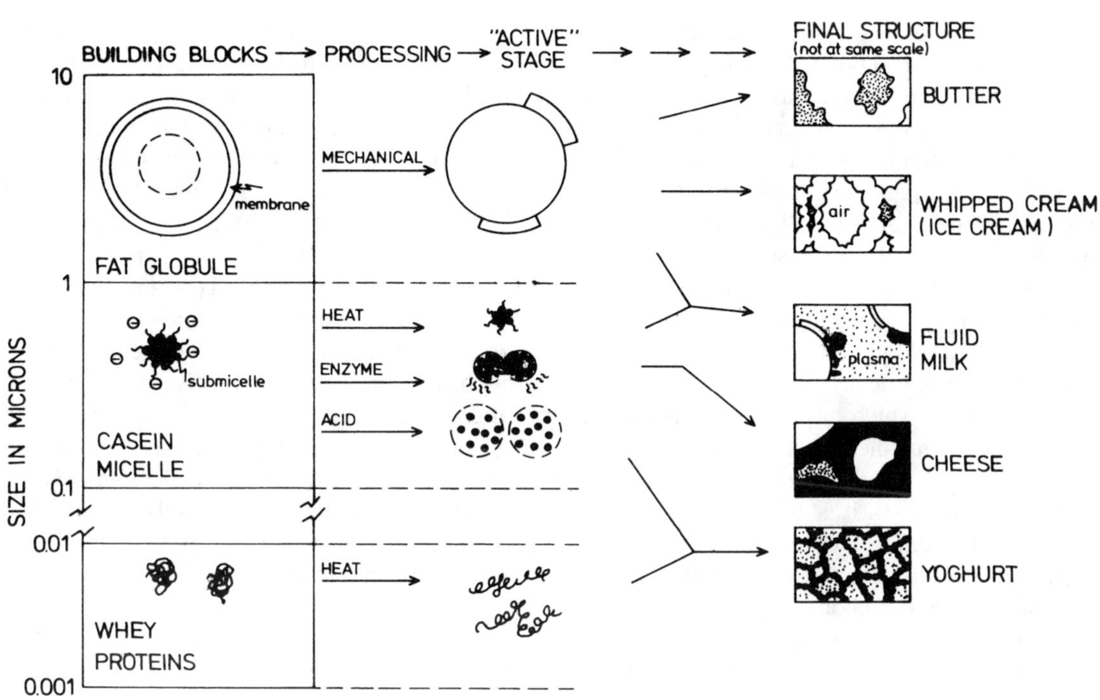


**Figure 7–1** Building blocks, active stages, and final structure of some dairy products. Several dairy products are formed when the three basic components of milk—casein micelles, fat globules, and whey proteins—achieve an "active state" that permits structure-building interactions to occur in the aqueous milieu.

formal approach can be found in the review by Walstra and van Vliet (1986). Excellent references include the books by Mulder and Walstra (1974) and by Walstra and Jenness (1984).

### 7.2.2 Milk Fat and the Fat Globule

The lipid fraction, which is mostly composed of triglycerides, represents about one-third of the total dry matter in milk. Fat is present in the form of globules (0.1–10 $\mu$m in diameter) surrounded by a complex lipoprotein layer or "membrane" about 10 nm thick. Fat globules in milk make up the dispersed phase of an emulsion, whose continuous phase is the milk plasma.

It seems ironic that in the study of the structural properties of the milk fat fraction, so much reference is made to the milk fat membrane, which is almost half protein, but represents only 2% of the weight of the fat globule. However, the total area of the fat globule membrane is some 80 $m^2$ per liter, which explains its important role in surface phenomena stabilizing the fat emulsion. The surface tension of the milk fat globule–plasma interface (1–2.5 mN/m or dynes/cm) is 6–15 times smaller than that of the liquid milk fat–plasma interface.

The fat globule membrane contains about 40–50% protein (about 1% of the total protein in milk) and 30% phospholipids. It is also the site of trace elements, about 10 different enzymes, and carotene. Electron microscopy and biochemical methods have been combined to study the ultrastructure of the fat globule membrane. (See Mather & Keenan, 1975; Oortwijn, Walstra, & Mulder, 1977; Patton & Keenan, 1975; Wooding, 1971. Buchheim, 1986, provides a review of the research.) The main ultrastructural element is a bilayer membrane 10–15 nm thick that surrounds the globule and has inner and outer layers (Keenan, Moon, & Dylewski, 1983). An inner protein layer, demonstrated using freeze-etching techniques, possesses a paracrystalline array and separates the membrane from the triglyceride core. This paracrystalline array is modified by the heating or cooling of raw milk and by changes in ionic strength (Buchheim, 1986).

No structural model yet exists that could account for the physicochemical behavior of the milk fat globule. Many decades ago, King (1955) proposed a concentrically layered structure based on composition, which has been extensively reproduced. Here another model, based on suggestions by Mulder and Walstra (1974), has been adopted (Figure 7–2). It is composed of an inner unit membrane that serves as a surface on which lipoprotein particles become associated by chelate bonding (Copius Peerebom, 1969).

Due to molecular diversity, milk fat inside the membrane exists as a mixture of oil and crystals in the temperature range between −40°C and 40°C. The extent of crystallization affects many technological processes, such as churning, homogenization, and creaming, and also some mouthfeel properties. Milk fat crystals can be present in needle and platelet shapes or as larger spherulites and in different polymorphic forms, with $\alpha$, $\beta$, and $\beta'$ being the most common. Crystal type and polymorphism affect structural and textural properties. The type, shape, number, and size of crystals depend on the number of nuclei formed, the rate of growth, and the temperature history. If crystallization is slow and very few nuclei exist, large crystals are formed. Large crystals can also be formed by solubilization of small ones and recrystallization on larger ones (Ostwald ripening), a process enhanced by temperature fluctuations. Later, fat crystals tend to flocculate into a network of considerable mechanical strength that behaves as a solid if the crystal content exceeds 80%. Further details on emulsion stability can be found in

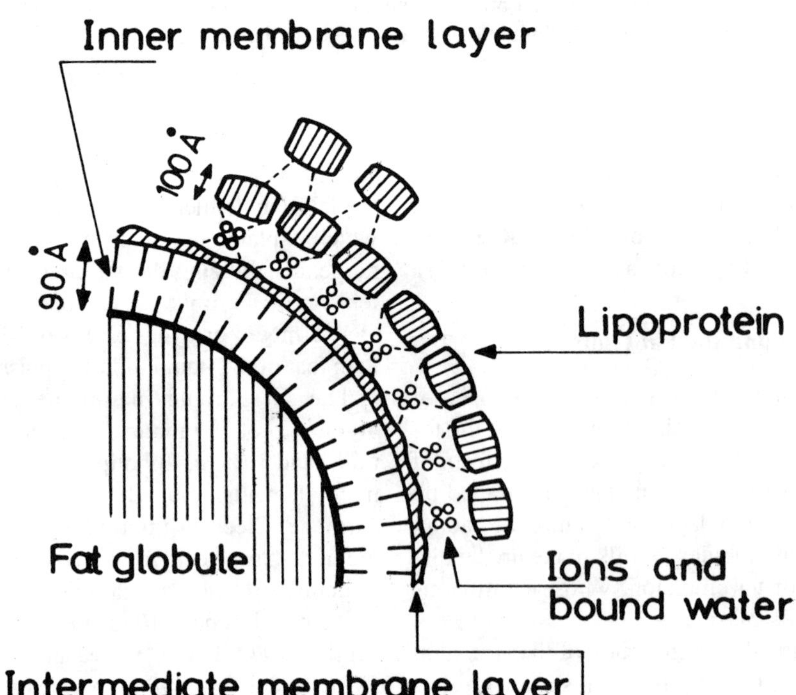


**Figure 7–2** One of several proposed models for the fat globule membrane. *Source:* Copius Peerebom, 1969.

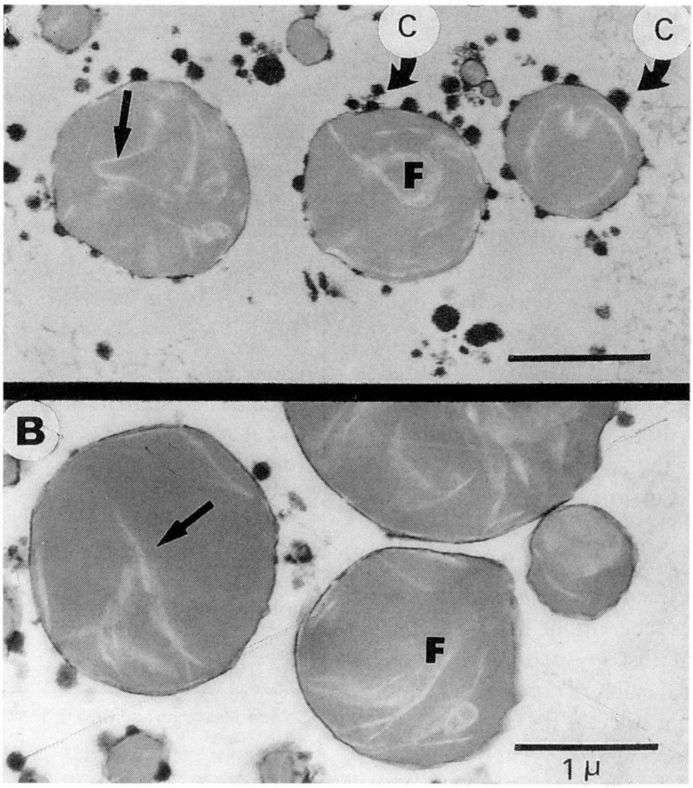


**Figure 7–3** Transmission electron microscopy of ice cream milk emulsion. Fat globule (F), casein micelle (C), and fat crystal within the globule (arrow). Marker = 1 $\mu$m. Source: Goff et al., 1987.

Carroll (1976) and on milk fat crystallization in Mulder and Walstra (1974). Figure 7–3 shows fat globules in an ice cream mix and the presence of solid fat within the globule.

### 7.2.3 Proteins and the Casein Micelle

In milk, the most versatile structural element involved in technological transformations is the protein conglomerate called *casein.* Casein represents almost 80% of the milk protein. Biochemically, it is composed of four fractions, $a_{S1}$-, $a_{S2}$-, $\beta$-, and $\kappa$-casein, in a ratio of approximately 3:1:3:1. The molecular weight of the individual casein fractions range from 19 to 25 kDa. Structurally, casein in milk is present as highly hydrated, nearly spherical clusters 0.02–0.30 $\mu$m in diameter called *micelles,* by association with colloidal chemistry. Casein micelles in fluid milk exist in the form of a suspension (i.e., a dispersion of "solid" particles in a liquid continuous phase).

Casein molecules are organized at a primary level as aggregates 10–20 nm in diameter and 250–2,000 kDa called *submicelles,* which contain between 15 and 25 casein molecules. Submicelles are visible with the aid of an electron microscope (Schmidt, 1980). Casein molecules are strongly hydrophobic and are held together inside the submicelles by salt linkages and hydrophobic bonds buried in the core, while a hydrophilic region exists on the surface. Hydrophobic chains are thicker (1–2 nm) than those in the hydrophilic surface; the latter are stretched by contact with water, according to a model derived from ion beam sputtering (Kimura, Taneya, & Kanaya, 1979).

The content of κ-casein appears to vary between different submicelles. Those containing a higher proportion of κ-casein occupy the outer portion of the micelle, as confirmed by a combination of antibody technique and electron microscopy (Carroll & Farrell, 1983). Once the surface of a micelle is uniformly covered with κ-casein, further addition of submicelles ceases. The N-terminal two-thirds of κ-casein is hydrophobic and interacts with $a_{S1}$-casein inside the submicelle. The C-terminal third, the caseinomacropeptide, is hydrophilic and protrudes as a flexible "hair" into the serum, producing steric and electrostatic repulsion between casein micelles.

Casein micelles also contain an inorganic moiety called *colloidal calcium phosphate* ($Ca_3(PO_4)_2$), which represents about 8% of the total weight. Electron microscopy of casein micelles suggests that colloidal calcium phosphate is finely divided throughout the micelle, and there is agreement that it acts as a cementing agent between casein submicelles (Schmidt & Buchheim, 1970). At the pH of milk, colloidal calcium phosphate is insoluble and protected from precipitation by casein. Upon acidification, it dissociates from the micelles until it is fully lost at pH 5 (Walstra & van Vliet, 1986). It also behaves as an ion exchanger and actively participates in the equilibrium with the serum phase. The stability of casein micelles is reviewed by Walstra (1990).

Casein micelles are also highly hydrated and contain about 3.7 g water/g protein, of which only 0.5 g/g protein is bound water (McMahon & Brown, 1984). Their diameter is between 40 and 300 nm (Figure 7–3). Other important characteristics of casein micelles are presented in Table 7–3. Figure 7–4 contains structural models of a casein submicelle and a casein micelle based on TEM evidence, colloidal behavior, and outstanding elements of other models (Schmidt, 1980; Schmidt & Payens, 1976; Walstra & Jenness, 1984).

Casein micelles exist in a state of dynamic equilibrium with the serum. If mechanically disrupted, they reassociate within minutes. Aggregation is prevented by steric and electrostatic repulsion of κ-casein fractions, which act as a "protective colloid." Casein micelles constitute a very stable system, particularly against heat treatments. Upon cooling, they become more resistant to flocculation, probably owing to steric repulsion by protrusion of polypeptide chains.

**Table 7–3** Average Characteristics of Casein Miscelles

| *Characteristic* | *Value* |
|---|---|
| Diameter | 130–160 nm |
| Surface | $8 \times 10^{-10}$ cm$^2$ |
| Volume | $2.1 \times 10^{-15}$ cm$^3$ |
| Density (hydrated) | 1.063 2 g/cm$^3$ |
| Mass | $2.2 \times 10^{-15}$ g |
| Water content | 63% |
| Hydration | 3.7 g $H_2O$/g protein |
| Voluminosity | 4.4 cm$^3$/g |
| Molecular weight (hydrated) | $1.3 \times 10^9$ daltons |
| Molecular weight (dehydrated) | $5 \times 10^8$ daltons |
| Number of peptide chains (mol. wt: 30,000) | $10^4$ |
| Number of particles per ml milk | $10^{14}$–$10^{16}$ |
| Whole surface of particle | $5 \times 10^4$ cm$^2$/ml milk |
| Mean free distance | 240 nm |

*Source:* McMahon and Brown, 1984.

Proteins remaining in solution after precipitation of casein at pH 4.6 (isoelectric point) are called *whey* or *milk serum proteins* and constitute about 20% of the milk proteins. A rather diverse group, they include α-lactalbumin, β-lactoglobulin, bovine serum albumin, and peptides derived from proteolysis of the caseins. Whey proteins are compact globular proteins molecularly dissolved and susceptible to heat denaturation.

### 7.2.4 Lactose and Salts

Chemically, lactose is a disaccharide of D-glucose and D-galactose joined in a β-1.4-glycosidic linkage that is less sweet than sucrose. Because at any time a certain proportion of free aldehyde is

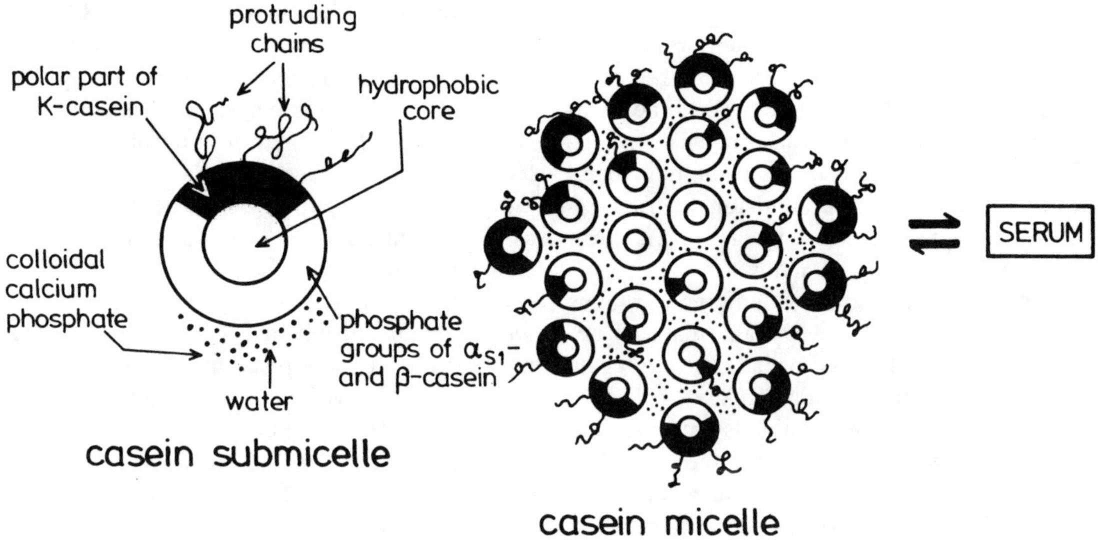


**Figure 7–4** Proposed models of casein submicelles and casein micelles.

present in the ring, it can undergo Maillard reactions with free amino groups, leading to desirable brown colors in toffees and mild flavors in evaporated milk. Nonenzymatic browning can also be deleterious, particularly in stored milk and whey powders.

Lactose can occur in two forms: crystalline and amorphous. Crystalline lactose presents an ordered three-dimensional structure that repeats itself. It exists in two crystal forms, α-hydrate and β-anhydrous, which differ in solubility. Crystal size should be controlled, since those larger than 10 μm are felt in the mouth and the product is said to be "sandy." Amorphous lactose (glass) is highly hygroscopic, and when the moisture content exceeds about 8% ($a_w$ around 0.4), it starts to crystallize. Dairy powders should be protected against moisture pickup during storage, or caking may occur. (See Section 9.5.4).

Salts are, indirectly, important structural elements in milk, since they define the physicochemical milieu that determines the conformation and stability of proteins. Salts in solution contribute the ionic strength that affects the thickness of the diffuse double layer of colloids (Walstra & Jenness, 1984). Almost two-thirds of the total calcium is complexed with phosphate and citrate in the micelles, while only 10% exists as free ionic $Ca^{++}$.

### 7.2.5 Milk as a Colloidal System

As we have seen, milk has many structural elements of colloidal dimensions. The concepts of colloidal science introduced in Section 3.6 will aid in the comprehension of this section. Those wanting to gain a deeper understanding of colloids may consult the classical text on surface chemistry by Adamson (1990) and the books on colloidal chemistry applied to foods by Mulder and Walstra (1974) and Dickinson and Stainsby (1982).

Technologically, we are interested in keeping colloidal particles apart (emulsions) or bringing them together (coagulation and gelling). Several interactions leading to structural changes in the protein and fat fractions will now be explained. Oil-in-water emulsions are lyophobic colloids. They are thermodynamically unstable and always tend to diminish their interfacial area. *Aggregation* is a general term for various types of interactions between colloidal particles. Milk fat globules can aggregate in different ways depending on the physical condition of the membrane. Normally, intact membranes produce steric repulsion between globules, but when they become damaged, bridging can occur via adsorbed plasma proteins, as is typically observed in homogenized cream. Shear thinning of emulsions can be explained as the progressive aggregate breakdown that occurs when weak attraction forces are overcome by shear forces. Closer interaction takes place when part of the membrane material between adjacent droplets is lost and the aggregate becomes fat continuous. Coalescence, or the formation of a single larger droplet, is prevented because solid fat provides rigidity to the individual globules; instead, "granules" are formed. Whipping of cream, butter making, and the formation of fat clumps in milk and cream are macroscopic results of this phenomenum (Darling, 1982). Stabilization of emulsions (i.e., avoidance of coalescence) is enhanced by a viscous continuous phase, the use of surface active agents, and steric repulsion of absorbed polymers or finely divided solids located at the interface. Different types of instability leading to creaming and coalescence as well as possible interactions between fat globules are shown in Figure 7–5.

A totally different process known as *cold agglutination* also results in milk fat globules coming together. It takes place by bridging through immunoglobulin M (mol. wt ≈ 900,000; diameter ≈ 30 nm) under low temperature conditions (Euber & Brunner, 1984). The process is reversed by warming. Protein solutions are lyophilic colloids and undergo different types of interactions. Flocculation is aggregation caused by weak forces

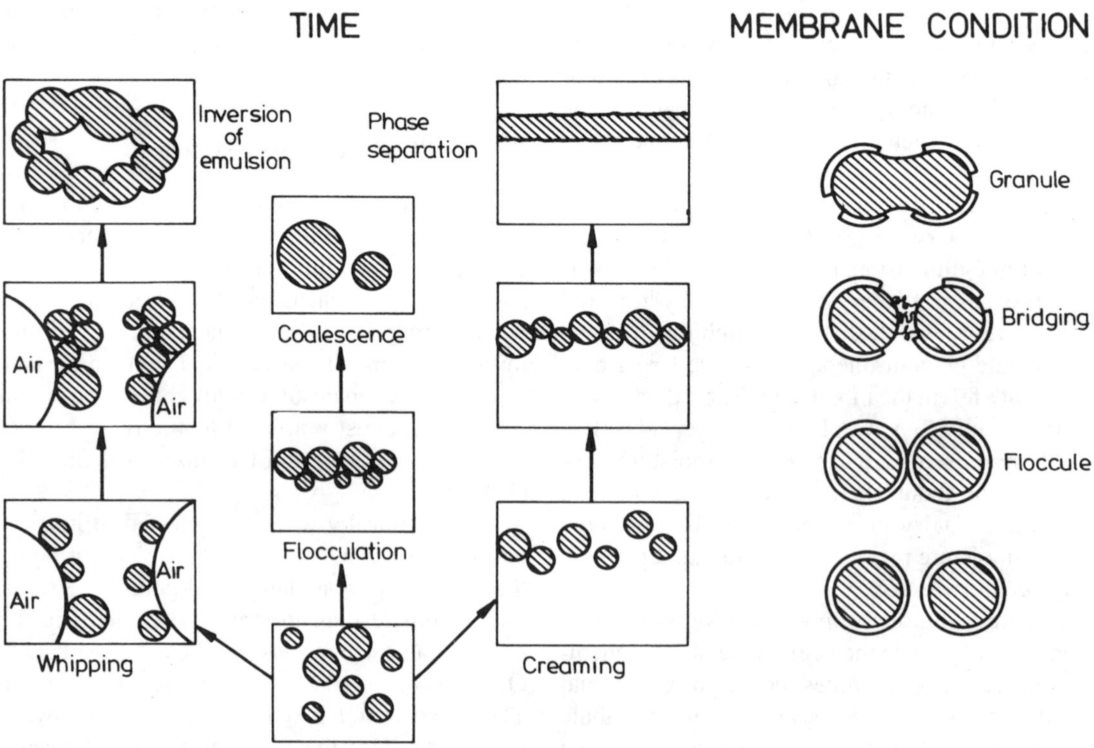


**Figure 7–5** Mechanisms leading to emulsion instability and forms of interaction between two membrane-bound fat globules.

(e.g., a few kT), as for example around the secondary minimum predicted by DLVO theory (see Section 3.6.7). Coagulation occurs when particles interact more strongly (e.g., at the primary minimum): it is usually prevented by an energy barrier (around 10 kT), as shown in Figure 3–14 (Payens, 1979). Several other interesting microstructural phenomena have their roots in colloidal and surface chemistry (e.g., droplet formation in spray drying of milk, spreading of butter, wetting of powder, and eye development in cheeses).

### 7.2.6 Milk Gels

As stated in Section 5.8, a gel is a three-dimensional network formed by the association or cross-linking of long polymeric molecules, which entraps and immobilizes the liquid solvent forming a rigid structure. Gelling is usually induced in milk in two ways by acidification (yogurt) or by enzymes (curd). The two types of gels possess different characteristics, as summarized in Table 7–4, but they share the properties of irreversibility and opaqueness (Green, 1980).

Upon acidification, colloidal calcium phosphate dissociates from the micelles, and aggregation occurs as the isoelectric point of casein is approached. A continuous network of casein strands is formed, as seen by electron microscopy (Kalab, Allan-Wojtas, & Phipps-Todd, 1983; Kalab & Harwalkar, 1973).

Several processing conditions can alter the properties and ultrastructure of acid gels. The state of the fat globule membrane greatly influences the rheological properties of milk gels. Homogenized globules probably interact with the gel matrix through the adsorbed casein molecules (van Vliet & Dentener-Kikkert, 1982) or by hydrostatically restraining movement without disrupting the gel architecture (Aguilera & Kessler, 1988). Direct cold acidulation (0°C) of skim milk to pH 5.5 with hydrochloric or citric acids followed by heating develops a core-and-lining ultrastructure in casein micelles, possibly by interaction of $\beta$-lactoglobulin and $\kappa$-casein (Harwalkar & Kalab, 1981). We will come back to acid-induced gelling when yogurt is discussed (Section 7.4.2).

Chymosin (rennet), on the other hand, removes the caseino-macropeptide from the surface of the casein micelles by precise hydrolysis at the bond between phenylalanine (105) and methionine (106), forming a hydrophobic moiety, paracasein. The negative surface potential of casein micelles is halved, and aggregation then proceeds irreversibly, favored by the diminished steric and electrostatic repulsions. The exact mechanism and kinetics of enzymatic gel formation are debated, but condensation and polymerization (Payens, 1979) is favored over cross-linking of polymeric units (Johnston, 1984a, 1984b).

The important microstructural outcome is that flocculation of paracasein particles leads to formation of a gel rather than a precipitate (e.g., separate flocs that sediment). This sequence of microscopic events was described by Walstra and van Vliet (1986). Gelling commences by the formation of irregular threadlike aggregates 1 to 4 micelles thick and some 10 micelles long. Aggregates grow until they touch each other, forming a continuous network alternated by thicker nodes and leaving openings up to 10 $\mu$m in diameter where the fat globules become en-

**Table 7–4** Properties of Dairy Gels Produced by Acidification or Enzymes

| *Property* | *Rennet* | *Acid* |
|---|---|---|
| Yield stress (Pa) | $\ll 1$ | $\approx 1$ |
| Loss tangent (at $10^2$ s) | 0.6 | 0.25 |
| Breaking stress (at 10 min) (Pa) | 10 | 100 |
| Deformation at breaking | 1.6 | 0.5 |
| Permeability coefficient B ($\mu m^2$) | 0.2 | 0.15 |
| $\frac{dB}{dt}$ ($nm^2\ s^{-1}$) | 20 | <1 |
| Rate of syneresis (arbitrary units) | 15 | <1 |
| Endogenous syneresis pressure (Pa) | 1 | $\ll 1$ |

*Source:* Walstra and van Vliet, 1986.

trapped. Apparently the structure of the gel network (regularity and pore size) is determined by the relative rates of the enzymatic reaction and the flocculation process. After the gel is formed, two phenomena occur: (1) more junctions between the reactive portions of the micelles are formed, and the network tends to shrink, expelling whey, a process known as syneresis, and (2) the touching points between micelles become fused. The net result is one giant paracasein matrix with holes where the fat and serum are entrapped.

Other types of dairy gels are formed by heating concentrated milk and the addition of ions. The ultrastructure of these gels varies from individual micelles linked by denatured $\beta$-lactoglobulin to thicker chains of fused micelles (Figure 6–7). Fusion of casein micelles and gel firmness are increased by the addition of calcium (Green, 1980; Kalab & Harwalkar, 1974).

Whey proteins form gels upon heating by unfolding and aggregation. The network structure is formed predominantly by disulfide bonding and the mediation of calcium (deWit & Klarenbeek, 1984). The presence of salt promotes aggregation and the formation of coarser gel textures (Hermansson, 1979). Aguilera and Kessler (1989) demonstrated the compatibility of whey and casein proteins in the formation of mixed dairy gels. Also, small fat globules can be accommodated in pores left in the protein matrix, reinforcing the structure of filled dairy gels. This is an area of great potential for the development of new fabricated dairy products.

Gelation can also be detrimental, as in the case of stored evaporated and ultra high temperature (UHT) concentrated milks. It has been postulated that gelation is produced by the cross-linking of casein micelles by denatured whey protein (Carrol, Thompson, & Melnychyn, 1971). Microstructurally, micelles in storage-gelled products show surface deformation and threadlike bridging between them; gelation is delayed by the addition of hexametaphosphate (Harwalkar, 1982). The exact nature of the phenomenon is still debated, but the involvement of enzymatic and polymerization reactions has not been ruled out.

## 7.3 THE EFFECT OF PROCESSING ON STRUCTURE

This section is not intended as a review of the processing aspects of fluid milk. Rather, those important microstructural modifications induced by the major unit operations in fluid milk processing are only briefly commented on. Changes in individual products are presented in the next section. Extensive treatment of processing and technological aspects and their physicochemical significance is found in Kessler (1981).

### 7.3.1 Centrifugation of Milk

Centrifugation enhances separation of the fat globules ($\rho_f = 0.92$ g/ml) from the milk plasma ($\rho = 1.03$ g/ml), producing cream and skim milk. The effect of different variables on rate of "creaming" ($v$) can be deduced from this formula:

$$v = \frac{(\rho - \rho_f)D_p^2}{18\eta} a_E \qquad \textbf{Equation 7–1}$$

where $D_p$ is the diameter of the fat globule, $a_E$ the acceleration of the field, and $\eta$ the viscosity of the plasma. If $a_E$ is changed from gravity (9.81 m/s$^2$) to values of 4,000 $\times$ g, similar to those in a commercial centrifuge, separation of cream can be attained in a few seconds. Milk components exit the centrifuge without major chemical or structural changes, although during vigorous flow of milk and cream, some disruption and partial coalescence of fat globules may occur. The coalescence rate is expected to be higher in cream than in whole milk, since it increases with fat concentration (Walstra & Jenness, 1984).

### 7.3.2 Homogenization of Fat Globules

Homogenization reduces the size of the fat globules and keeps whole milk or cream as a stable emulsion (see the effect of globule size on creaming rate in equation 7–1). It is accomplished by forcing the product at high speed under pressure through a slit slightly larger than the globule itself (see Section 3.6.6). This causes shearing, cavitation, and microturbulence, which leads to deformation and eventually breakage of globules

(Kessler, 1981). Associated with this reduction in size is a corresponding increase in surface area of 4 to 10 times. The total surface area formed when the average globule size produced by homogenization is at a minimum appears to be independent of pressure and has a value of 1.1 $m^2/mL$ (Phipps, 1983). Figure 1–20 presents TEM micrographs of commercially and experimentally homogenized milk.

After homogenization, a large proportion of the original membrane material remains on the globules, although redistributed over a larger surface area (Keenan et al., 1983). A new membrane is rapidly established in denuded areas, as has been documented by electron microscopy studies (Buchheim, 1986; Henstra & Schmidt, 1970; Oortwijn et al., 1977). The newly formed fat-serum interface preferentially adsorbs casein micelles and also submicelles and undenatured whey proteins. The makeup of the new membrane has been elucidated using sodium dodecyl sulfate polyacrylamide-gel electrophoresis in combination with freeze-etching and thin-sectioning TEM (Darling & Butcher, 1978). Interestingly, fat globules are more stable when they have a layer of pure plasma proteins (e.g., in recombined milk) than when the surface is composed of original membrane and adsorbed plasma proteins (Walstra & Jenness, 1984).

The technological consequences of the replacement of the original fat membrane by an adsorption layer of plasma proteins are extensive and derive mostly from the caseinlike behavior of fat globules. Buchheim (1986) has summarized some of them: (1) fat globules now participate in acid or enzymic coagulation processes; (2) foaming of cream is adversely affected; (3) clustering of fat globules occurs at low pH; and (4) reaggregation of disintegrated casein micelles in evaporated, heat sterilized milk begins at the interface of fat globules and results in protein-fat complexes.

### 7.3.3 Heating of Milk: Pasteurization and Sterilization

Thermal processing is extensively used in the dairy industry to provide extended shelf-life. Heating over 60°C generally induces undesirable chemical changes in milk, the extent of which depends on the temperature-time combination of the process. The two most prevalent heat processes are pasteurization (e.g., 62.8°C/30 min; 71.7°C/15 s or 90°C/1 s), which is intended to kill pathogenic bacteria, and sterilization, which is meant to kill all viable microorganism and hence is a more severe process (e.g., 118°C/20 min or a few seconds at 145°C or higher [UHT]).

Most changes induced by heat that affect the structure of milk are related to the protein fraction and can be summarized as follows (Walstra & Jenness, 1984): (1) denaturation of whey proteins and association with casein micelles (particularly $\beta$-lactoglobulin and $\kappa$-casein); (2) transfer of Ca and phosphate to casein micelles; (3) aggregation of casein micelles above 110°C; and (4) changes in the proteins of the fat globule membrane.

### 7.3.4 Other Processing Effects

As noted, flow in pipes under large shear gradients or high turbulence condition produces distortion of fat globules. Pumping between equipment pieces may cause mechanical damage, such as curd breakup. Vigorous agitation in the presence of air or vapor may lead to foaming and damage to the milk fat globules, as in the interaction between vapor bubbles and fat globules in a falling film evaporator. Prolonged exposure at interfaces may cause unfolding and denaturation of globular protein chains. Fouling or the formation of unwanted solid deposits over surfaces of processing equipment is a severe problem in the dairy industry. Milk deposits over steel surfaces, particularly in heat exchangers and evaporators, are mostly composed of protein and minerals. It appears that $\beta$-lactoglobulin partially unfolds under the effect of heat, exposing reactive sulfydryl groups and then polymerizes. Fouling impairs heat transfer, increases pressure drop in pipes, promotes corrosion and, by providing areas where pathogenic bacteria and spores can attach (biofouling), it threatens the sterility of foods (Fryer & Belmar-Beiny, 1991).

## 7.4 MICROSTRUCTURAL ASPECTS OF MILK PRODUCTS

### 7.4.1 Scope of the Presentation

Processed products derived from cow's milk are wide-ranging in nature and are created by various fabrication methods. There exist, for example, more than 2,000 names for cheeses throughout the world, and the International Dairy Federation acknowledges 510 different types (Tamime, 1986). Although the microstructure of milk products is already well reported in the scientific literature, sample preparation techniques and result interpretation are constantly being improved and constitute a very active area of research. Since it is impossible to cover in detail all microstructural aspects of milk and dairy products, only outstanding features of groups of products are described below. The reference list at the end of the chapter may compensate for this shortcoming.

### 7.4.2 Yogurt

Yogurt is a particle gel produced by acid from fermentation of heated milk (85–95°C) with a mixed culture of *Lactobacillus bulgaricus* and *Streptococcus thermophilus.* The *Lactobacillus* breaks down protein into small peptides, which are later utilized by the *Streptococcus* to produce acetaldehyde (flavor) and acid, lowering the pH to 4.3 and solubilizing colloidal calcium phosphate. A three-dimensional gel matrix is formed by casein micelles linked in long chains immobilizing the liquid phase, as clearly shown by classic studies using TEM and SEM (Kalab, Emmons, & Sargant, 1976; Modler & Kalab, 1983). TEM coupled with image analysis may assist in geometrical characterization of the protein network and its relation to changes in processing (Skriver, Hansen, & Qvist, 1997). Confocal laser scanning microscopy is being used to monitor structure formation of fully hydrated yogurt in real time and produce three-dimensional images (Hassan, Frank, Farmer, Schmidt, & Shalabi, 1995).

Heating of skim milk to 75–90°C before inoculation is essential for the proper development of yogurt structure. It results in firmer yogurts than those made from unheated milk and showing no syneresis. TEM has been instrumental in demonstrating that heating develops numerous appendages in casein micelles that facilitate linking into fine chains having about a 230 nm average cross section (Kalab et al., 1976). Appendages appear to be denatured $\beta$-lactoglobulin or $\beta$-lactoglobulin–$\kappa$-casein complexes (Davies, Shankar, Brooker, & Hobbs, 1978), and they provide stability against syneresis. Unheated micelles, on the contrary, show smooth contours and upon acidification become fused to form large aggregates with about a 460 nm average cross section (Davies et al., 1978; Kalab et al., 1976; Kalab et al., 1983). The $\beta$-lactoglobulin–$\kappa$-casein complex also coats the homogenized fat globules. A microstructural view of the particulate structure of the network of a yogurt gel is presented in Figure 7–6.

The degree of interaction between casein micelles can be controlled by incorporation of other proteins. The resulting structures range from fused micelles and short thin links when caseinate and skim milk powders are added to fine aggregates bridged by whey protein (Kalab et al., 1983; Modler & Kalab, 1983). The main microstructural changes during yogurt manufacture are schematically depicted in Figure 7–7. Bacterial colonies form pockets in the gel and attach to the protein matrix through filaments, probably of a polysaccharide nature, as documented by SEM (Kalab, 1979a; Kalab et al., 1983).

Rheological properties of the main types of yogurt (set-type, stirred, and drinking) are reviewed by Benezech and Maingonnat (1994). Oscillatory rheometry has shown that the gel is elastic at low strains (e.g., $G' > G''$), but at high strains the gel structure breaks down and the material becomes viscous (e.g., $G'' > G'$). In stirred yogurt there is some breakdown of the network structure, and the product may be induced to flow. Stirred yogurt shows thixotropic behavior, as its viscosity is shear-sensitive. Several stabilizers or thickening agents are used to modify viscosity and mouthfeel and to stabilize the gel against syneresis. Addition of milk-solids-non-fat (MSNF) to 12–14% of total solids is practiced to bring up texture to ac-

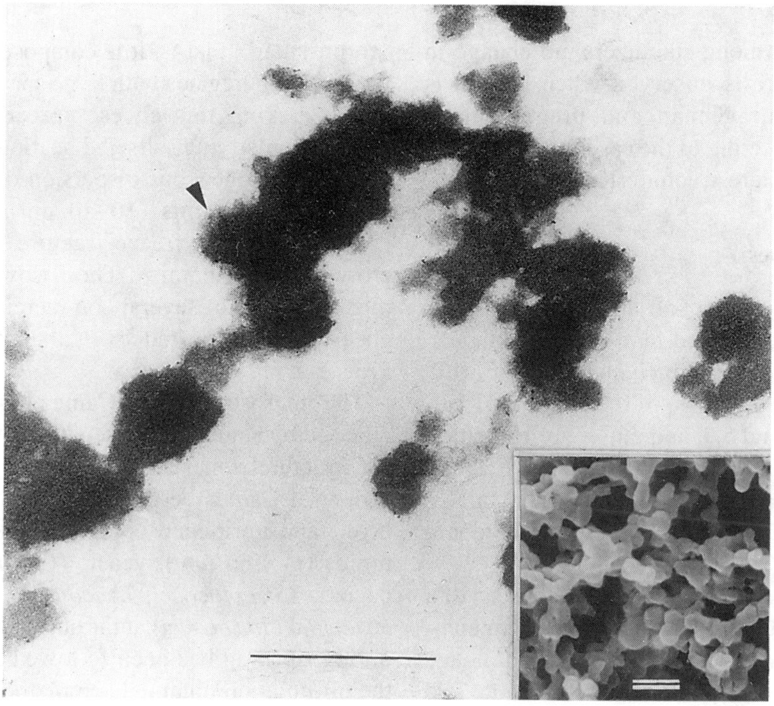

**Figure 7–6** Transmission electron micrograph of yogurt. Casein micelles linked together show a "fuzzy" outer surface (arrow). Marker = 0.5 $\mu$m. Inset: Scanning electron micrograph of the string-of-beads arrangement of casein micelles in yogurt. Marker = 1 $\mu$m.

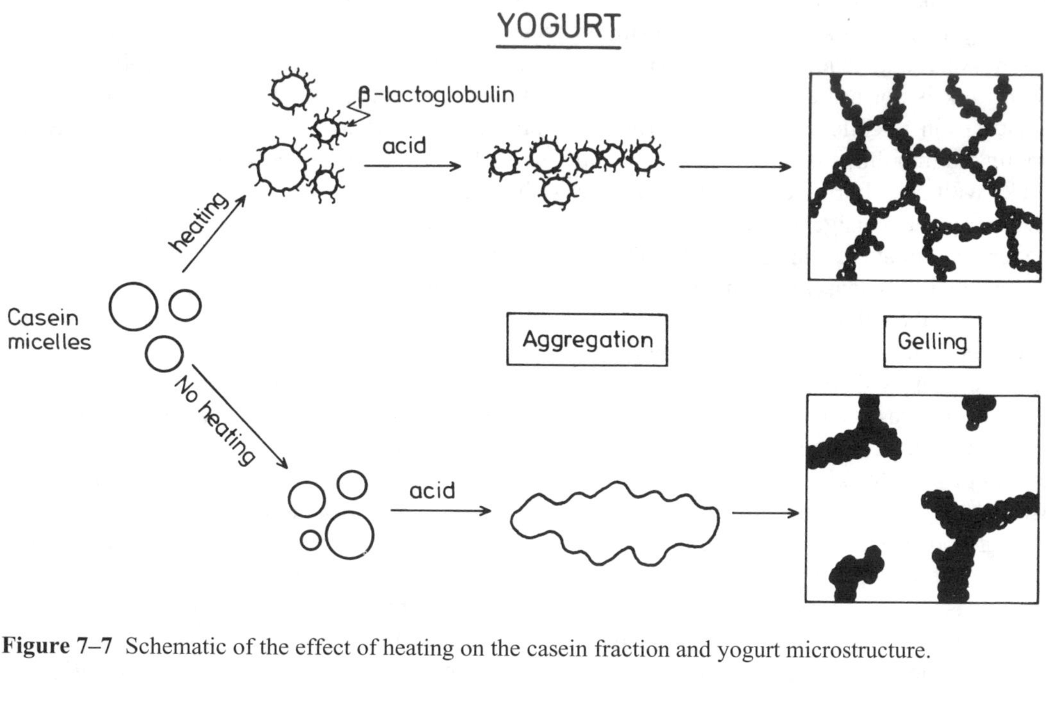


**Figure 7–7** Schematic of the effect of heating on the casein fraction and yogurt microstructure.

ceptable levels. Among stabilizers, no change in the microstructure is observed when gelatin is added, while carrageenan and pregelatinized starch lead to clustering of the casein micelles and a more fibrillar microstructure (Kalab, 1979a).

### 7.4.3 Soft Cheeses

The distinction between soft and hard cheeses is arbitrary, but it is adopted here for convenience. In general, soft cheeses (which include "semisoft" types) have $a_w$ in the range of 0.96–0.99, pH values between 4.3 and 5.0, and require refrigeration for shelf-life extension. Hard cheeses have $a_w$ below 0.98 and a pH range from 5.0 to 5.8 (Marcos, Alcala, Leon, Fernandez-Salguero, & Esteban, 1981).

Soft cheese manufacture basically consists of a reduction in pH brought about by lactic fermentation or the addition of acid followed by drainage of the whey and salting of the curd. The rate and quantity of lactic acid produced determines the solubilization of colloidal calcium phosphate in milk and the final structure of the cheese. Detailed processing conditions for individual cheeses may be found in Kosikowski and Mistry (1997).

Unlike yogurt, syneresis of the curd is desirable in cheeses like cottage cheese, and thus pasteurized skim milk is not heated. SEM micrographs show that casein micelles in a recently formed curd are linked into thin chains that later develop cross-links with neighboring chains. The average diameter of micelles change in milk from about 90 nm to over 200 nm at the end of cooking. The final protein network is denser than in yogurt owing to shrinkage and the tighter fusing of micelles. Casein micelles in the exterior of the granule are packed more densely than in the interior, but the presence of a skin has been undisputedly ruled out (Glaser, Carroad, & Dunckley, 1979; Kalab, 1979b). No differences are detected between the microstructure of acid- and culture-set curds (Glaser, Carroad, & Dunckley, 1980).

In cream cheese (fat content $> 30\%$), the protein network is absent since the curd is disrupted by stirring and homogenization. The microstructure of the finished product has been defined as "corpuscular"; that is, it is composed of compact fat-casein aggregates with large spaces filled with whey, making this cheese spreadable (Kalab, 1985). Emulsifying salts and heating tend to form a stable homogeneous dispersion of fat globules and casein subunits (10–20 nm) in whey, as demonstrated by freeze-fracture electron microscopy (Buchheim & Thomasow, 1984). The microstructure of several commercial cream-type cheeses was reported by Kalab, Sargant, and Froehlich (1981).

The microstructure of Camembert cheese was reported by Knoop and Peters (1971). Four different structural zones are simultaneously present in a ripened Camembert cheese: surface, rind, ripe layer, and unripened core. The rise and fall of molds (*Penicillium*), yeasts (*Geotrichum*), and bacteria (*Streptococci, Leuconostoc, Corneibacteria,* and *Micrococci*) at the surface of the cheese during ripening has been followed by SEM, and the information obtained complements microbiological data. The rind develops as superimposed layers of microorganisms and a yeast layer at the undersurface (Rousseau, 1984). Ripening causes disintegration of casein micelles into subunits the size of single submicelles (around 16 nm in diameter); these form thicker agglomerates within the surface and extending into the rind, following the direction of the ripening process. Ripened areas present a fine protein matrix with interspersed fat globules, while unripened areas resemble a young cheese.

The ripening of cheese usually results in the formation of a coarser structure of fused micelles. In Meshanger cheese, a ripened soft cheese with high moisture content, the microstructure changes from a thin network formed by individual paracasein micelles adhering together into a homogeneous network of thick strands with no internal organization. TEM examination reveals that the protein strands are composed of casein submicelles, probably formed by proteolysis of paracasein brought about by rennet. Photomicrographs also show that fat remains as a discontinuous phase during ripening. The disappearance of the internal structure caused by proteolysis is also related to changes in consistency (deJong, 1978).

The relative role of bacterial enzymes and rennet in casein micelle breakdown and microstructure formation during ripening was studied electron-microscopically by thin-sectioning and freeze-fracture techniques. Although in ripened Camembert cheese the breakdown units are the size of individual submicelles, the penetration of whey into the casein micelles in acid cheeses disintegrate them into still smaller units, 6 × 12 nm (Knoop & Buchheim, 1980).

### 7.4.4 Hard Cheeses

In general, hard cheese processing involves variations of the following events: (1) fermentation of lactose to lactic acid in whole milk by starter microorganisms; (2) coagulation of casein by rennet and acid; (3) formation of a continuous proteinaceous network (curd) in which fat is entrapped; (4) manipulation of the curd (cutting, cooking, squeezing out of whey, stretching, salting, pressing, etc.); and (5) ripening, during which proteolysis and lipolysis predominate.

The final structure of most cheeses may be analyzed at two levels (Kalab, 1979a): (1) by the form of the protein matrix within the curd granules, which determines the microstructure, and (2) by the way in which the curd granules are fused, which fixes the macrostructure. Differences between various mature cheeses are to a great extent due to their microstructure, which is largely determined by the composition and structure of the coagulum (Knoop, 1972).

The single most important stage in cheesemaking is the production of acid in the vat, because it largely dictates the basic microstructure of the cheese (Green & Manning, 1982; Lawrence, Heap, & Gilles, 1984). The composition and thermal history of the milk also affect the properties of the coagulum as well as microstructure development. Two variables that can be manipulated for fine-tuning of the firmness of the coagulum are the calcium content and the level of heat. The amount of calcium controls the extent of disruption of the casein submicelles. A better understanding of the relationship between these processing variables and structure development in cheeses would allow for better process control and the use of a wider variety of raw materials. Figure 4–2, part F shows the microstructure of processed Colby cheese.

Another important step in cheesemaking is to ensure that the chemical composition of the cheese curd encourages the microbial and enzyme activity required for textural overtones and typical flavor development during ripening. Critical stages in cheesemaking from the structural viewpoint are summarized in Figure 7–8.

#### *Microstructural Aspects*

The microstructure of curd obtained by rennet and acid coagulation develops by the chaining and clustering of casein micelles until a three-dimensional network is formed (readily apparent in TEM studies). This fine protein structure tends to disappear during ripening, and the final microstructure emerges (Kalab, 1979b).

Examples abound where alterations in raw materials and/or manufacturing conditions affect the final microstructure of cheese. Homogenization of milk generally results in a coagulum with a finer than normal protein network. Increasing the concentration of milk solids in Cheddar cheese produces a coarser protein network that is not fundamentally altered during the ripening process (Green, Turvey, & Hobbs, 1981).

Ripening alters the basic structural features of the curd to various extents. The three-dimensional structure in Gruyere cheese remains almost intact after 180 days, except for the appearance of fine granular zones (Rüegg, Moor, & Blanc, 1980). The fibrillar arrangement from the cheddaring process (Kalab & Emmons, 1978) undergoes further changes during ripening but never loses its oriented nature (Green et al., 1981; Kalab, 1977). Pickling disintegrates casein aggregates into small particles, loosening the structure of Domiati cheese (Abd El-Salam & El-Shibiny, 1973). The ripening of an acid-coagulated skim milk cheese (Kamiesh cheese) in NaCl solution for 6 weeks changes the single-branched chains of casein micelles into a matrix of thicker particles. The particle fusion is probably due to protein aggregation induced by ionic exchange of $Ca^{2+}$ for $Na^{+}$ in the

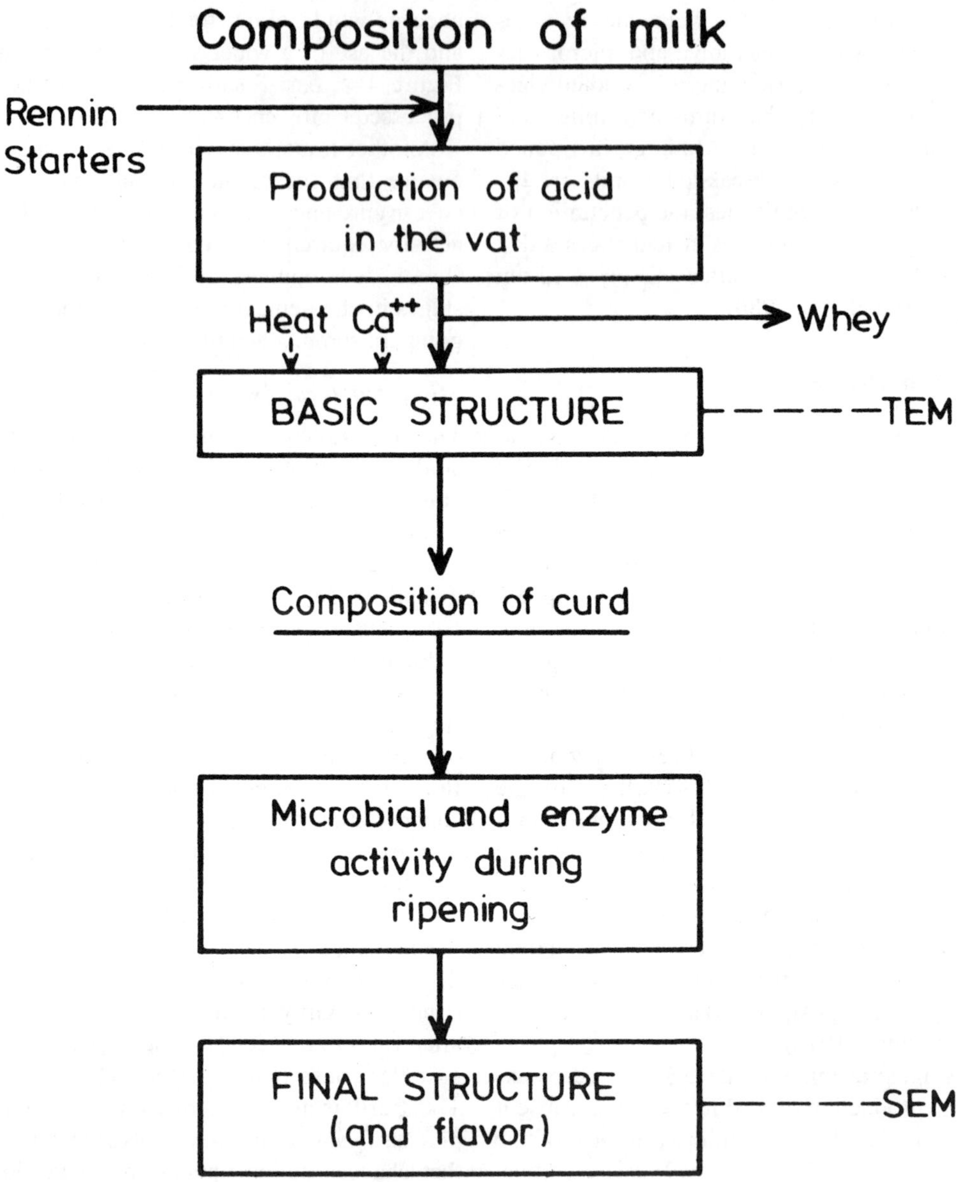


**Figure 7–8** Critical stages in cheesemaking and structure development.

colloidal calcium phosphate of caseinate (Abd El-Salam & Omar, 1985).

Processed cheese is prepared by heating a mixture of comminuted cheeses in the presence of melting or emulsifying salts. Desirable physical properties are meltability and firmness; the latter increases with temperature and the concentration of melting salts. Emulsifying salts split off calcium from the insoluble paracasein, aiding in swelling and dispersion of the protein fraction and emulsification with fat and water. Microstructurally, the protein matrix of processed cheese, seemingly unlike the protein matrix of normal cheese, is formed by submicelles or degradation products from paracasein breakdown (Buchheim & Thomasow, 1984; Heertje, Boskamp, van Kleef, & Gortemaker, 1981; Lee, Kilbertus, & Alais, 1981). Meltability is related to the presence of more or less isolated protein particles and short strands (10–15 nm in diameter), while harder cheeses show a more fibrous protein matrix (Taneya, Izutsu, & Buchheim, 1980). However, prediction of functional performance by observation of the product microstructure may not always be possible. Some mozzarella cheeses and analogs do not show a fibrous microstructure and yet have good melting properties (Taranto & Tom Yang, 1981; Taranto, Wan, Chen, & Rhee, 1979).

The appearance of microscopic and sometimes macroscopic crystalline inclusions during ripening of cheese is well documented for Roquefort (Swiatek & Jaworski, 1959), Cheddar (Brooker, Hobbs, & Turvey, 1975; Washam, Kerr, Hurst, & Rigsby, 1985), Grana (Bottazzi, Battistotti, & Bianchi, 1982), Emmental (Rüegg & Blanc, 1972), smoked processed Swiss (Washam et al., 1985), and processed cheese (Rayan, Kalab, & Ernstrom, 1980; Uhlmann, Klostermeyer, & Merkenich, 1983), among others. The crystals, as analyzed by SEM and X-ray technique, consist mainly of calcium lactate in Cheddar; potassium tartrate in processed Swiss cheese; calcium phosphate and calcium lactate in Grana; and lactose, tyrosine, melting salts, calcium phosphate, calcium citrate, or more complex compounds in processed cheese (Uhlmann et al., 1983). The exact role of crystals in determining the quality and intrinsic organoleptic properties is not well understood, although usually their presence is considered a defect.

***Macrostructural Aspects***

Curd granules are simply fused in brick, Edam, and Gouda cheese but they become stretched by various means in Cheddar, mozzarella, and provolone (Kalab, 1979b). Curd granules and milled curd junctions are permanent macrostructural features of Cheddar cheese (Lowrie, Kalab, & Nichols, 1982). Junctions seem to be interesting microstructural elements rich in casein and the site of bacteria microcolonies (Kalab, 1977).

An important macrostructural characteristic is the porosity of the cheese. In many cheese varieties, the presence of regular round holes or eyes (e.g., in Emmental and Gouda) or a uniform distribution of irregularly shaped openings (e.g., in Havarti and blue-veined cheese) is typical and desirable. The contribution of gas production by microorganisms to the open structure of cheeses was reviewed by Martley and Crow (1996). Eye development in Swiss cheese depends on the rate of $CO_2$ production by *Propionibacteria,* which is optimum in the pH range 5.2–5.4. Below this range the chance of getting a "blind" cheese increases, while excessive gas production at pH higher than 5.4 induces fracturing. The replacement of rennin by bovine pepsin gives a more open structure in both curd and ripened cheese (Eino, Biggs, Irvine, & Stanley, 1976, 1979).

***Textural Aspects of Hard Cheeses***

Texture of hard cheeses depends critically on the water and fat contents. Increases in the contents of these components weaken the protein structure while decreases result in hardening of the cheese. High-fat cheeses are smooth and those containing a high proportion of casein exhibit higher firmness. Water content is the principal factor affecting fracture during biting and mastication. A lower acidity during cheese preparation results in harder cheeses by weakening the bonding between molecules in the protein network (Jack & Paterson, 1992).

### 7.4.5 Butter

Several methods used in butter manufacture are described in the literature. The sequence of main steps during traditional butter making are clarification, separation of cream, ripening, pasteurization, cooling, churning, draining, washing, and working. Cream (about 30–33% fat) is an O/W emulsion (water is the continuous and fat the dispersed phase), while butter ($\approx$80% fat) is a W/O emulsion.

The condition of fat inside the globules is of utmost importance in butter making. The proportion of liquid and solid fat and the morphology of the lipid globule are partially set after ripening. Some globules show a peripheral crystal shell of up to 0.5 μm in thickness, while others are still completely liquid, with intermediate forms also present. Clustering of globules prevails in cultured cream but in sweet cream it is minimal (Buchheim & Precht, 1979).

Churning is a process of simultaneous agitation and beating air into the cream; it causes fat globules to come into contact at points where the fat globule membrane has been damaged by mechanical or surface tension effects. Temperature is a critical variable, since liquid fat must be squeezed out to cement the fat globules together but enough solid fat should be present to maintain their integrity. The churning temperature is usually kept between 9°C and 15°C (the higher the fat content of the cream, the lower the temperature). As churning proceeds, fat globules at the air-plasma interface are brought to partial coalescence when the bubbles break or shrink, and eventually a continuous fat phase develops and inversion of the emulsion takes place (Mulder & Walstra, 1974). Butter is technically a W/O emulsion, although part of the water appears to behave as a continuous phase.

A microscopic view of the structure of butter is presented in Figure 7–9 and a diagram in Figure 7–10. The diagram shows that some fat globules remain largely intact after processing. These survivors have been observed using electron microscopy to have roughly a 0.1–0.5 μm thick outer shell of fat crystals and a core of liquid fat and small crystalline aggregates (Precht & Buchheim, 1979). Some properties of the structural elements in butter are listed in Table 7–5.

The texture of butter is usually measured as "consistency." Under normal use, butter behaves as a pseudoplastic body. At relatively high stresses, the internal structure begins to break

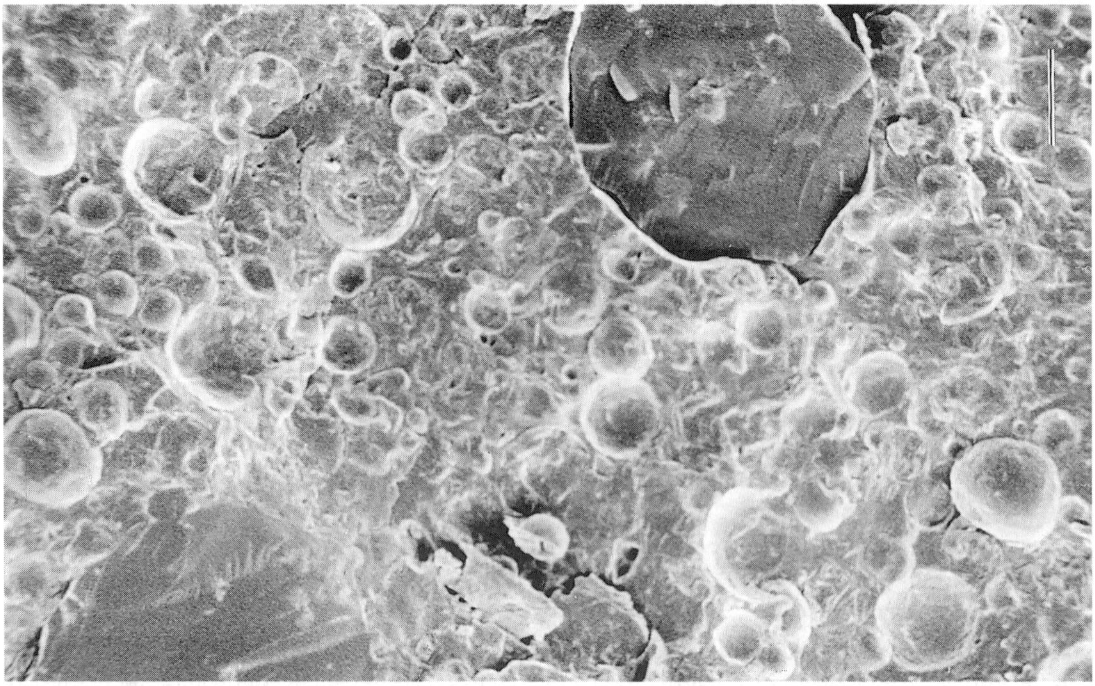

**Figure 7–9** Scanning electron micrograph of butter. Fat globules are embedded in a matrix of free fat. Marker = 5 μm.

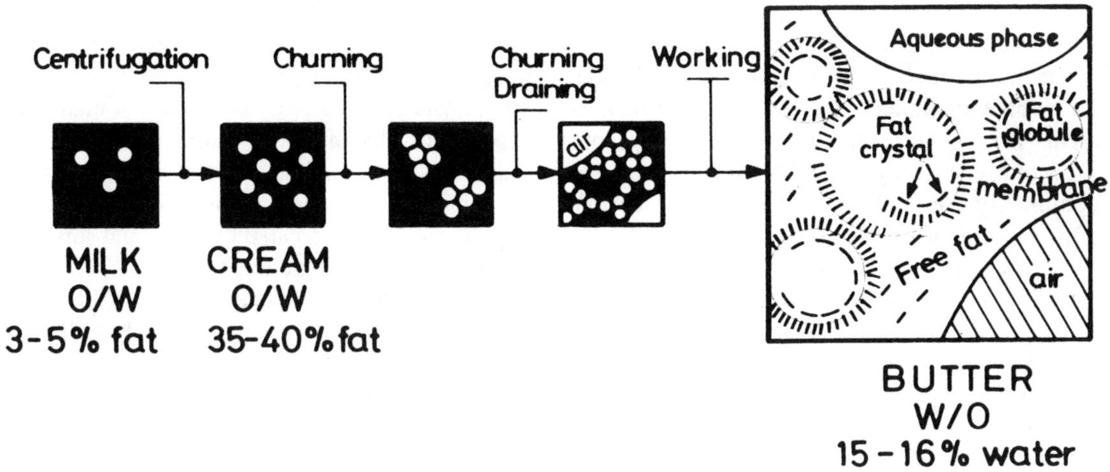


**Figure 7–10** Changes during butter manufacture and schematic of the microstructure of butter.

down and it shows shear thinning. Consistency (spreadability) is dependant on the proportion and physical distribution of solid and liquid fat inside the fat globule, which can be varied by the temperature of ripening, as demonstrated by electron microscopy (Precht & Peters, 1981).

### 7.4.6 Whipped Cream

Cream is an O/W (serum) emulsion containing 30–40% fat. Unlike whipping milk, whipping cream produces a foam that can be both lasting and rigid. Whipping incorporates air into the cream and forms a two-phase system or foam consisting of 50–60% air bubbles dispersed in a liquid phase. The interfacial processes relevant to the formation of a stable aerated structure in whipped cream are (1) a lowering of the surface tension at the air-serum interface fostered by proteins and phospholipids; (2) adsorption of globules to the air bubbles, which is facilitated by mechanically damaged globule membranes; and (3) floccula-

**Table 7–5** Structural Elements of Butter

| *Element* | *Approximate Number Concentration ($ml^{-1}$)* | *Proportion of Butter (%, v/v)* | *Dimension (μm)* | *Remarks* |
|---|---|---|---|---|
| Fat globules | $10^{10}$ | 10–50 | 2–8 | Differ in composition; with complete or partial membrane |
| Fat crystals | $10^{13}$ | 10–40 | 0.01–2 | Amount depends on temperature; at higher temperature, mainly in globules; at low temperature, solid networks |
| Moisture droplets | $10^{10}$ | 15 | 1–25 | Differ in composition |
| Air cells | $10^{7}$ | 5 | >20 | |

*Source:* Mulder and Walstra, 1974.

tion-coalescence of the fat phase (Birkett, 1983). Owing to the high spreading pressures at the air-serum interface, liquid fat drips through broken fat globule membranes and forms a layer that cements the globules together. SEM micrographs show that, eventually, the bubble surface becomes completely covered with packed globules that protrude into the interior (Buchheim, 1978).

Structure in whipped cream is made possible by semisolid fat particles in the cream exerting a stabilizing influence on the foam through a mechanism known as partial coalescence. When the milk fat globule membrane surrounding each fat particle is disrupted by shear and some of the fat is in the solid form (so that crystals are present), the shearing action of whipping causes the colliding globules to partially coalesce as a result of the fat crystals piercing the membranes. A coalesced network results that still retains some globular shape and has the ability to build structure during the whipping operation. Thus, this material can trap the air bubbles created by shear and surround them, leading to a stabilized structure. The general appearance of whipped cream is shown in Figure 7–11. Smooth, rounded air bubbles are seen embedded in a matrix formed from partially coalesced fat globules. Higher magnification of the lamellae area discloses the tendency of fat

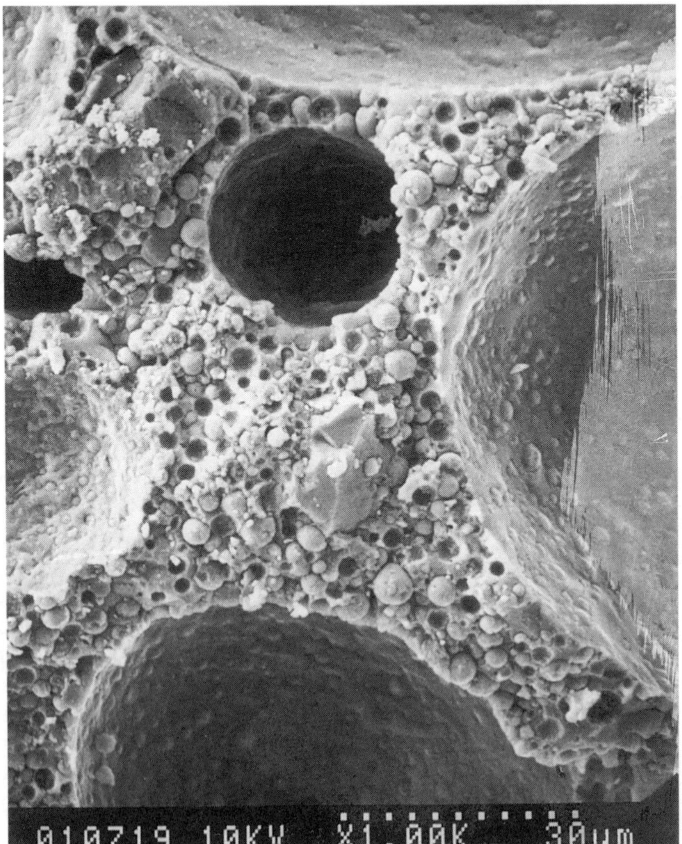


**Figure 7–11** Scanning electron micrograph of whipped cream. Void spaces represent the air cells of the foam structure. Marker = 30 $\mu$m.

globules to extend into the air bubbles, partially coalesce, and form bridges between adjacent bubbles. The phenomena have been studied with the aid of SEM by Schmidt and van Hooydonk (1980). Another important mechanism for adsorption to air bubbles appears to be the positioning of fat crystals on the surface of oil droplets at the air-liquid interface (Darling, 1982). In fact, this mechanism occurs in whipped topping foams, where large lipid crystals are present at the surface of air bubbles, as shown by electron microscopy (Buchheim, Banford, & Krog, 1985).

The stiffness of whipped cream depends on the structure of the foam lamellae. Fat globules must form clusters of 15–20 $\mu$m promoted by the presence of some "free" fat (Graf & Müller, 1965). Phenomena affecting the stability of aerated milk products, such as creaming, coalescence of gas bubbles, and drainage of liquid, are discussed by Prins (1986). Structural aspects of the air-serum interface of skim milk foams are discussed by Brooker (1985).

Considering the importance of rheological rigidity and stability to the commercial value of whipped cream foams, it is not surprising that emulsifiers and stabilizers are added to formulations. Emulsifiers migrate to the air-water interface so as to reduce excess surface tension and promote interaction of fat globules, while stabilizers increase the viscosity of the aqueous phase and slow diffusion of serum from the foam and also perhaps interact with the protein components of the cream.

Whipped cream and ice cream (to be studied next) are two examples of foamed dairy emulsions that require stabilization to prevent air cells from growing in size and adjacent cells from coalescing. As more becomes known about the action of polysaccharide stabilizers and surfactant emulsifiers it should be possible to improve their effectiveness. Foamed dairy emulsions have been reviewed by Stanley, Goff, and Smith (1996).

### 7.4.7 Ice Cream

Ice cream is revisited in this chapter on dairy products not only for the sake of completion but principally because it is a multicomponent, multiphase, and metastable system in which many structural phenomena take place. In Section 5.3.3 we looked at, in relation to ice cream, ways of structuring water at subfreezing temperatures through the formation of ice and a viscous glassy phase.

Ice cream may be regarded as a solid foam containing air as the dispersed phase. The solid fraction within the partially frozen continuous phase consists mostly of ice crystals, milk solids, fat, and lactose and sometimes sucrose crystals. It also contains an emulsion of butter fat globules, a micellar protein solution, and a dispersion of ice crystals in a continuous matrix or serum (Figure 7–12). The percentage increase in volume due to the incorporation of air or overrun varies from 100% for ice cream (10–12% fat) to 50% for sherbet (2% fat) and 0% for water ice.

The structure of ice cream has been studied extensively in the last 30 years (Berger, Bullimore, White, & Wright, 1972; Nielsen, 1973). Cryogenic SEM is now an established technique for examining ice cream (Caldwell, Goff, & Stanley, 1992), and details of the structure can be seen in Figures 7–12 and 5–11. Viewing ice cream nonintrusively can now be done using environmental SEM (ESEM) in the cryogenic mode (i.e., at −60°C), but imaging at around −20°C, where most microstructural changes take place over time, is not yet possible, since the structure becomes freeze-dried (Fletcher, 1997).

The processing sequence in the manufacture of ice cream consists of mixing, pasteurization and homogenization, cooling, aging, aeration-freezing, packaging, and hardening. Various ingredients perform different functions within the mixture. Stabilizers bind unfrozen water and maintain a stable distribution of air, water, and fat. Emulsifiers improve the degree of dispersion of the fat in the aqueous medium and also contribute to emulsion stability, particularly in the freezer. Fat promotes air dispersion, increases viscosity, and favors the formation of small fat crystals. Lactose and sugar in solution and salts lower the freezing point to the extent that the first ice crystals appear at −2 to −3°C. The mix is pasteurized, homogenized, and cooled to 4°C to supercool the fat and

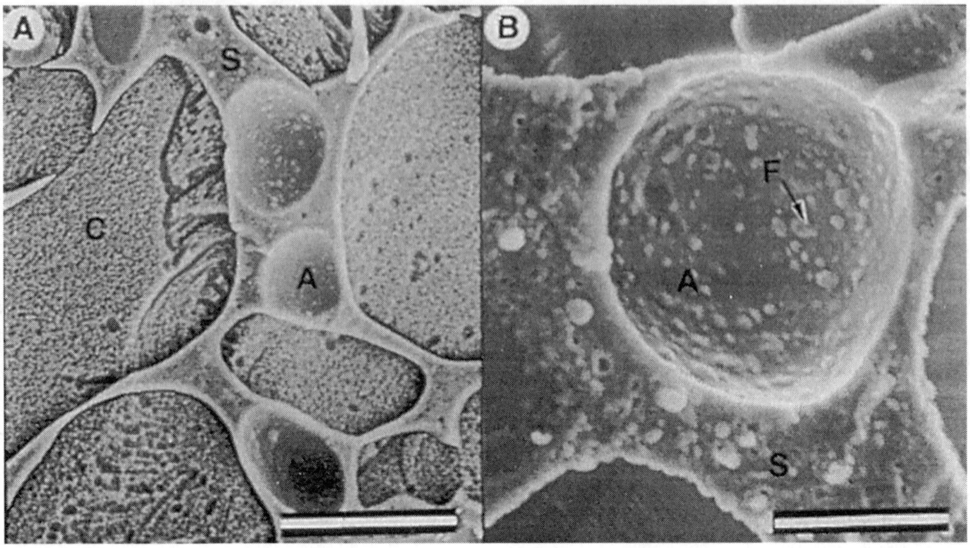


**Figure 7–12** (A) Cryo-SEM micrograph showing the four phases of ice cream. C = partially etched ice crystal socket, A = air bubble lined with fat globules, S = serum phase. Bar = 30 μm. (B) Cryo-SEM micrograph showing air bubble (A) lined with fat globules (F) and surrounded by serum phase. Bar = 7.5 μm. *Source:* Reprinted from *Food Research International*, Vol. 29, D.W. Stanley, H.D. Smith, and A.K. Smith, Texture-Structure Relationships in Foamed Dairy Emulsions, pp. 1–13, Copyright 1996, with permission from Elsevier Science.

induce nucleation. Globules in the mix range from 0.05 to 3 μm, with a mean diameter of 0.5 μm. The emulsion is "aged" for 2 to 3 hours, during which time the hydrocolloids become fully hydrated, fat becomes partially crystallized, and proteins become adsorbed on fat globules. Freezing to −6 to −7°C is accomplished in a scraped-surface heat exchanger in about 20 seconds of intensive mechanical agitation to assist in the incorporation of air. The ice cream mix, from which the frozen product is prepared, is a complex mixture of an oil-in-water emulsion containing emulsifiers, a colloidal solution formed by proteins and stabilizers, and a true solution of salts and sugars. As the cream mix is exposed to the cold barrel of a scraped-surface heat exchanger (−22 to −32°C), a high degree of supercooling occurs, which is the driving force for heterogeneous nucleation. The blade of the scraper shears off the nuclei from the barrel surface and crystal growth proceeds inside the cylinder (Hartel, 1998).

Aeration and freezing lead to the rupture of the membranes in most fat globules and to the release of liquid fat. The spreading of fat at the air bubble–globule interfaces cements the agglomerates of the remaining globules, similar to what occurs in whipped cream. The mean diameter of air cells ranges from 60 to 100 μm, and the lamellae between air cells are 10–20 μm and sometimes even thicker. On leaving the freezer, the mix contains about 50% air by volume, and almost half of the water is in the form of ice crystals, which have a mean size of about 30 μm. Lactose crystals are approximately 20 μm, but oversized crystals cause a textural defect called "sandiness." Sucrose microcrystals are occasionally present. Further temperature reduction to −30°C in a freezer produces hardening, and the ice cream is then ready for handling and distribution. The microstructure of ice cream is depicted in Figure 7–13.

The formation and preservation of small ice crystals in ice cream, highly desirable, is favored by rapid freezing. Temperature fluctuations during storage lead to the formation of larger crystals. A common cause of this temperature-associated textural defect found in stored ice cream is migratory recrystallization or Ostwald ripening, in which large agglomerated ice crystals grow big enough (>40–50 μm) to cause a noticeable coarse or rough mouthfeel. An example of this phenomenon is shown in Figure 7–14. The stabilizers used in ice cream manufacture, most often blends of hydrocolloids or gums, are known to arrest or delay ice crystal formation (Dickinson & Stainsby, 1982; Muhr & Blanshard, 1986.).

Stabilizers function as follows. Because of freeze concentration, the carbohydrate chains become entangled and increase the viscosity of the unfrozen continuous phase surrounding the ice crystals. In dilute polymer solutions, viscosity increases linearly with increasing concentration but eventually a point is reached at which the viscosity-concentration relationship deviates from linearity and increases sharply. This transition corresponds to the onset of coil overlap between the polymer chains, and the critical concentration ($c^*$) at which it occurs depends on the volume occupied by each molecule in isolation. The transition is attributed to specific associations (hyperentanglements) caused by the physical contact of individual molecules over a

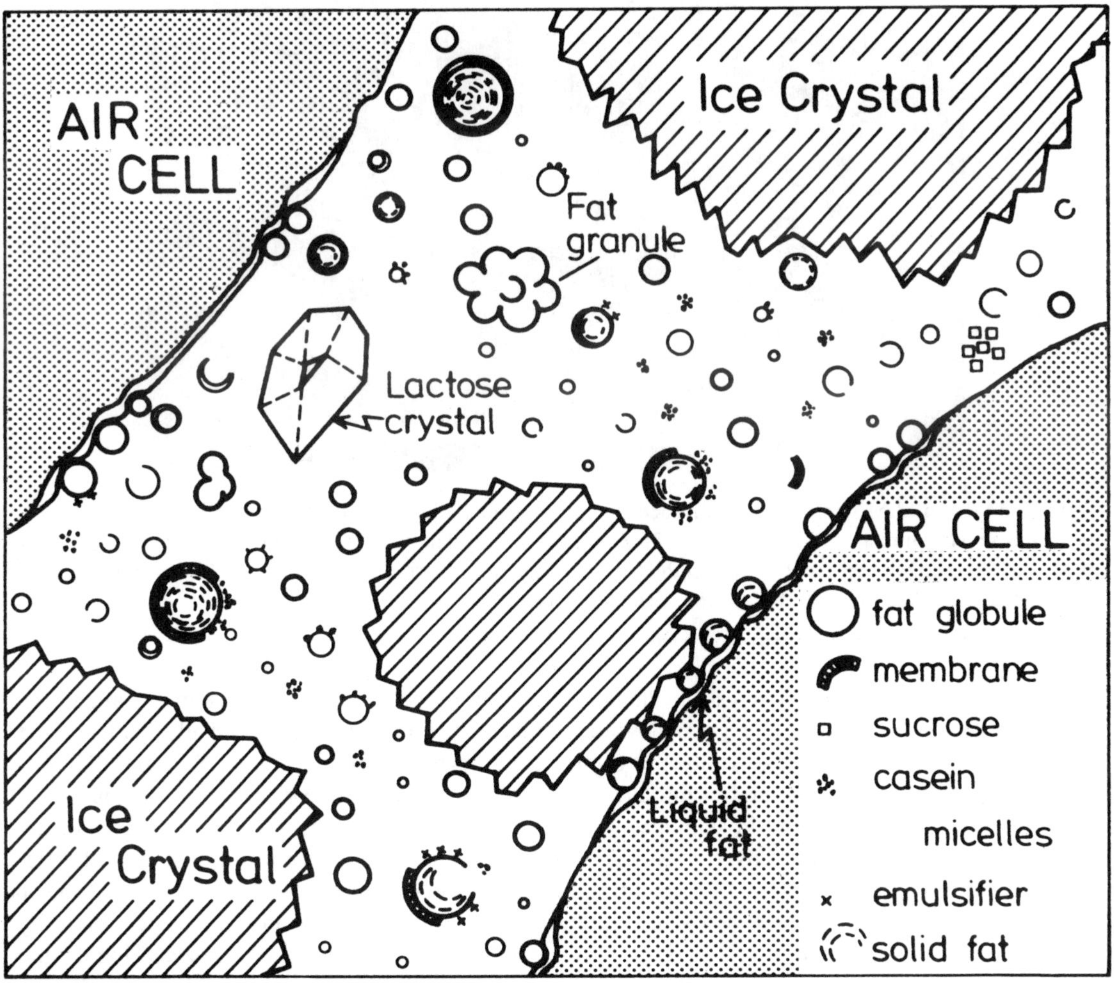


**Figure 7–13** Schematic of the microstructure of ice cream.

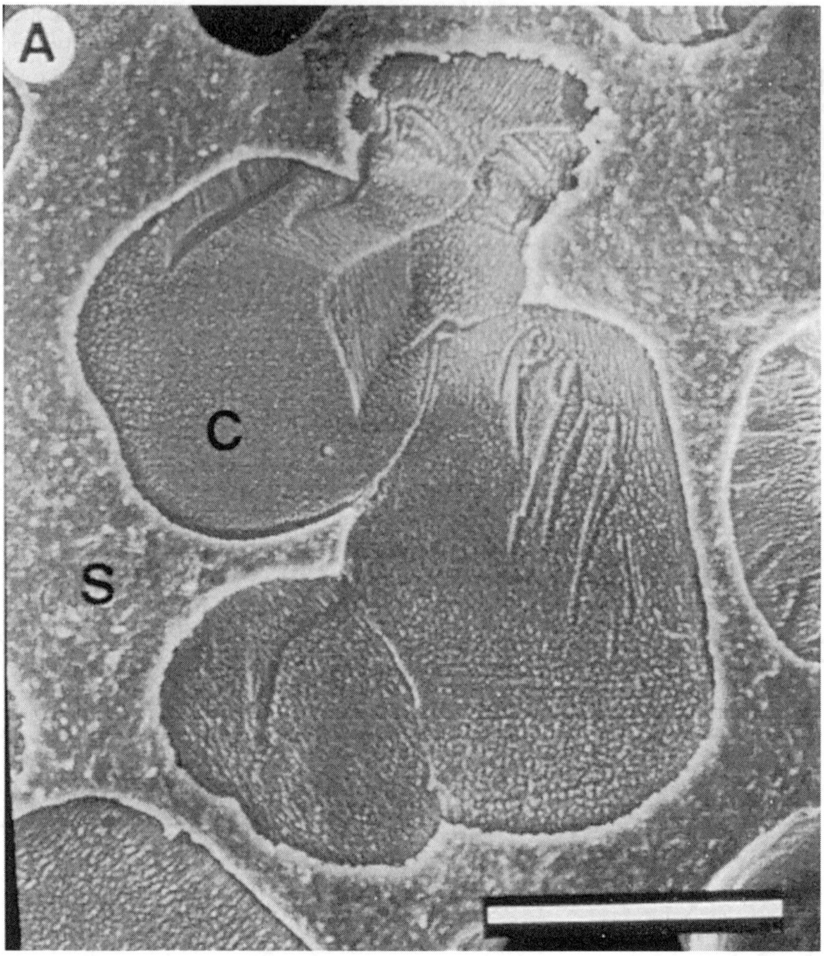


**Figure 7–14** Cryo-SEM micrograph of four ice crystals, surrounded by serum phase (S), that have merged into one (C) as a result of recrystallization. Bar = 25 $\mu$m. *Source:* Reprinted from *Food Research International*, Vol. 29, D.W. Stanley, H.D. Goff, and A.K. Smith, Texture-Structure Relationships in Foamed Dairy Emulsions, pp. 1–13, Copyright 1996, with permission from Elsevier Science.

longer time than the nonspecific physical entanglements, that lead to a dramatic change in flow behavior. Thus, the "micro" viscosity of the unfrozen phase of ice cream increases during freezing as a result of freeze concentration, but in stabilized systems an even higher viscosity results. As freeze concentration progresses during the manufacturing process, sugars, proteins, and stabilizers are forced into smaller areas surrounding the ice phase, increasing their concentration greatly. Recent research has demonstrated that the freeze concentration of ice cream mixes may cause the effective concentration to exceed $c^*$, leading to augmented viscosity effects thought to result from polymer chain overlap (Goff, Freslon, Sahagian, Hauber, Stone, & Stanley, 1995). These results strongly suggest that the textural benefits accruing from the use of stabilizers can be attributed to structural differences. Based on the previously mentioned concept of cryostabilization (Section 5.3.3), we might hypothesize that stabilizers act by decreasing the mobility of water in the serum phase surrounding ice crystals so as to prevent its migration to and recrystallization with existing ice crystals.

Structures of the type shown in Figure 7–14 are more prevalent in unstabilized than stabilized samples. A comparison of stabilized and unstabilized ice creams following a thermally abusive storage period shows that, while ice crystal size increased in all samples, the control samples contained more merged and contiguous crystals. Through image analysis of cryomicrographs, it was possible to quantitatively estimate the size of ice crystals in stabilized (mean diameter = 95 $\mu$m) and unstabilized ice creams (mean diameter = 114 $\mu$m). The estimates were then compared with the perceptions of sensory judges. Larger ice crystals in unstabilized ice cream were associated with significantly higher sensory icyness scores ($p < 0.05$) than coarser samples (Goff, Caldwell, Stanley, & Maurice, 1993).

In summary, these data demonstrate that the stabilizing ability of polysaccharides in ice cream results from the removal of water by freezing and from the concentration of the polysaccharide components of the mix in an unfrozen phase. When the critical concentration is reached, the polysaccharides begin to entangle and can thus greatly reduce the diffusion kinetics of water molecules in the unfrozen phase. The structural and textural studies of hydrocolloid stabilizers in ice cream lead to the conclusion that the growth of ice crystals in this product is controlled by the kinetic properties of the freeze-concentrated viscoelastic liquid surrounding them.

The structure of ice cream influences texture and rheological behavior. A creep compliance curve relates changes in the ratio of strain to shear stress over time. Mechanical models of the viscoelastic behavior of ice cream have been developed based on combinations of simple elements such as a spring, which obeys Hooke's law, and a dashpot containing a Newtonian liquid, as suggested in Section 3.8.6. Figure 7–15 shows a creep compliance curve for ice cream and the relevant mechanical model (Shama & Sherman, 1966).

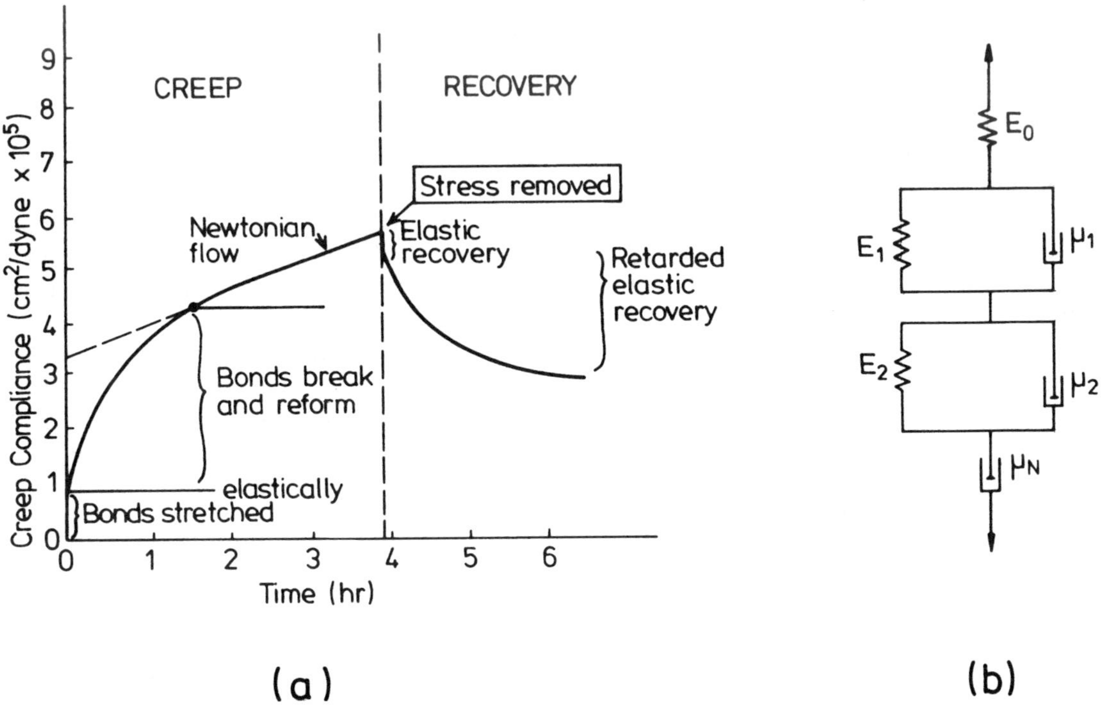


**Figure 7–15** Creeping compliance curve and mechanical model of ice cream. *Source:* Shama and Sherman, 1966.

Texture is largely determined by ice crystal size, the amount and dispersion of fat, and the proportion and size of air cells (Berger et al., 1972). The amount of water present as ice varies with temperature, while the size of crystals depends on the freezing rate and temperature history.

### 7.4.8 Concentrated Products

The main common manufacturing steps in the production of evaporated and condensed milk are standardization, preheating, concentration, and homogenization. Sweetened condensed milk has about 44% added sucrose and an $a_w$ of about 0.82.

Crystallization and gelation are two important microstructural processes in concentrated milk products. Less than half of the lactose is soluble in the available water, so crystallization must be controlled by seeding minute lactose crystals into the cooled, supersaturated lactose solution. Smooth-body, sweetened condensed milk may contain 300,000 lactose crystals smaller than 10 $\mu$m per cubic mm. High viscosity precludes deposition of crystals as well as separation of fat. A large quantity of sweetened condensed milk is used in the confectionery industry, where the interaction of casein with sugars during cooking produces the viscoelastic texture of toffees and fudges (deWit, 1984).

The addition of salts (phosphates or citrates) stabilize evaporated milks against protein precipitation during storage. Evaporated and condensed milks also tend to gel upon extended storage and this tendency increases with the time spent by the raw milk in cold storage.

### 7.4.9 Dried Dairy Powders

Removal of water increases the keeping properties of foods and extends shelf-life (see Chapter 9). Whole and skim milk powders are commercially produced currently by spray-drying. Whole or skim milk is dispersed as fine droplets into a stream of hot air, and the bulk of the water evaporates in a matter of seconds. Spray-drying skim milk preserves most of the properties of the casein micelles (Kalab, 1979a).

Spray-dried particles are predominantly hollow spheres, less than 100 $\mu$m in diameter, with small surface folds and a large central vacuole, sometimes with smaller particles trapped inside (see Figure 9–12). Other shapes and outer textures also exist, such as collapsed spheres and applelike structures that suggest an implosion in the last stage of drying (Buma & Henstra, 1971a, 1971b). Cross sections of dried particles observed with an SEM show multiple pores caused by thermal or mechanical stresses during drying or cooling (Buma, 1972). These studies also show that particles with high porosity have surface cracks and pores, whereas in less porous particles only surface folds are observed. As a consequence of fast drying, fat becomes distributed within the particle, with different accessibility to solvents (Buchheim, 1982; Buma, 1971).

Newer microscopy techniques are being used to study food powders. Figure 7–16 shows particles of a dairy creamer observed by a low-vacuum SEM (a type of ESEM), which does not require coating of the sample. It is not surprising that confocal laser scanning microscopy is now applied to milk powders to produce images of fat, lactose, and wetting agents with minimal intrusion (McKenna, 1997).

Scanning electron microscopy has been instrumental in elucidating the crystallization of lactose and its relationship to quality losses during the storage of milk powders. When milk is dried, amorphous lactose forms which is highly hygroscopic. Moisture uptake during storage induces the crystallization and cementing of powder particles into large lumps, a phenomenon known as caking (see Section 9.5.4). Lactose crystals present at the beginning of storage show the typical tomahawk shape and good keeping properties, while those formed during storage are needlelike and possess poor attributes (Roetman, 1979). Shifting from amorphous to crystalline lactose releases water and causes microstructural collapse, favoring nonenzymatic browning of whey powders (Saltmarch & Labuza, 1980).

The rehydration of powders is a complex surface and hydrodynamic process. Contrary

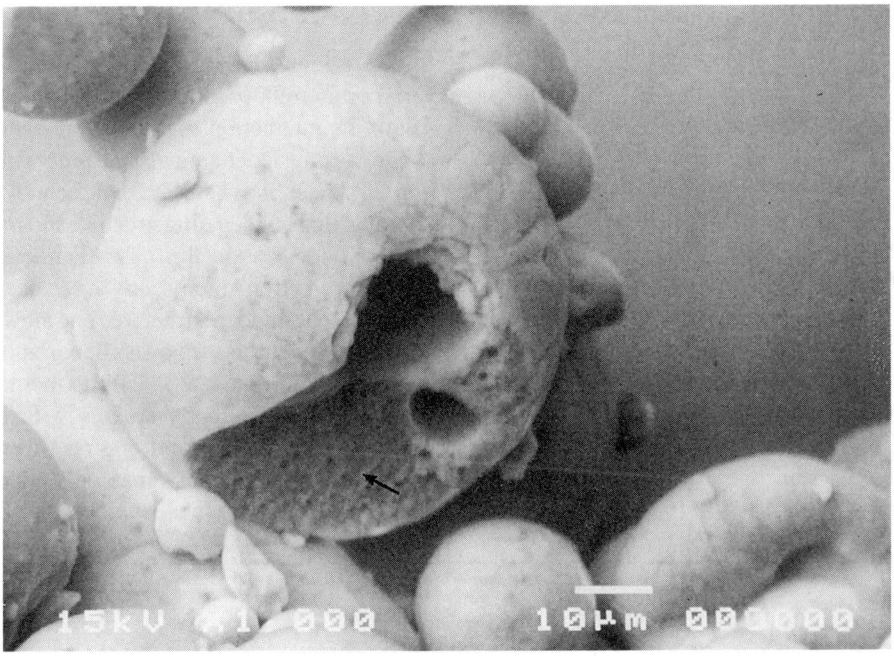


**Figure 7–16** Low-vacuum (LV) scanning electron micrograph of a dairy creamer. Arrow points at porous wall.

to intuition, the microstructure of the particles does not play a key role in the wetting and sinking of dairy powders in water, except for its effect on the solubilization of components (van Kreveld, 1974). Initial solubilization forms a viscous and sticky magma that holds unwetted particles together in lumps. Agglomeration exploits the same procedure to cement particles and form porous clusters or agglomerates (100–250 $\mu$m) with improved "instant" properties. The final product should have sufficient agglomerate porosity for fast liquid suction, a particle size between 0.2 and 2 mm and sufficient agglomerate strength to withstand handling and transportation (Schubert, 1993). Clustering is promoted by partial rewetting of dried particles or incomplete drying and is followed by drying with hot air. Coating with a surface active agent like lecithin also improves the wettability of powders.

### 7.4.10 Spun and Extruded Products

Meatlike analogs prepared by the spinning or thermoplastic extrusion of plant proteins have found interesting applications as meat substitutes or extenders. The basic structuring mechanism is the unfolding and stretching of proteins by heat and/or shear and the development of cross-links between the extended protein molecules, leading to a stable fibrillar matrix.

The properties of casein as a thermoplastic material have been known for many years, and important industrial applications were developed in Germany at the beginning of the century. Industrial uses of casein include the manufacture of fibers, papercoating, plastics, and adhesives. Casein glues were extensively used as water-resistant glues during World War I for the construction of wooden military aircraft (Browne & Brouse, 1939). Annual production of casein plastics or Galalith (from the Greek word for "milk stone") in 1925 amounted to 3,000 tons, mostly used in the manufacture of combs (Sutermeister, 1927).

Extrusion texturized milk proteins have been produced commercially in Poland (Szpendowski, Smietana, & Zuraw, 1983). Their production does not differ from that of soy products. An enzymatically produced protein coagulum adjusted to 15–30% moisture and extruded at 100–140°C results in a fibrous structure, as revealed in an SEM (Tuohy, Burgess, & Lambert, 1979; Smietana, Poznanski, Hosaja, & Kozlowska, 1978). The main disadvantage of casein fibers is their high solubility in water, and thus hydrocolloids have been added to improve the textural stability (Downey & Burgess, 1979a, 1979b).

## 7.5 CONCLUSIONS

Milk is a highly versatile food material. The wide variation in milk product microstructures is the result of the interaction of milk components brought about by physicochemical and processing conditions. Perhaps nowhere in food technology is the relationship between ultrastructure and microstructure so evident as in this case. Transmission and scanning electron microscopy have been used to peek into the hidden structure, and now confocal laser scanning microscopy is being adopted as a preferred noninvasive tool. Efforts should be made to correlate microscopy studies with chemical, physical, rheological, and engineering data. Results can be applied in raw materials selection, process control, and product development.

## BIBLIOGRAPHY


Abd El-Salam, M.H., & El-Shibiny, S. (1973). An electron microscope study of the structure of Domiati cheese. *Journal of Dairy Research, 40,* 113–115.

Abd El-Salam, M.H., & Omar, M.M. (1985). Microstructure of Kariesh cheese. *Journal of Dairy Research, 52,* 299–301.

Adamson, A.W. (1990). *Physical chemistry of surfaces* (5th ed.). New York: Wiley.

Aguilera, J.M., & Kessler, H.G. (1988). Physico-chemical and rheological properties of milk fat globules with modified membranes. *Milchwissenschaft, 43,* 411–415.

Aguilera, J.M., & Kessler, H.G. (1989). Studies on filled and mixed dairy gels. *Journal of Food Science, 54,* 1213–1217, 1221.

Benezech, T., & Maingonnat, J.F. (1994). Characterization of the rheological properties of yoghurt: A review. *Journal of Food Engineering, 21,* 447–472.

Berger, K.G., Bullimore, B.K., White, G.W., & Wright, W.B. (1972). The structure of ice cream [Part 1]. *Dairy Industries, 37,* 419–425.

Birkett, R.J. (1983). Role of interfacial processes in the whipping of dairy cream. In J.V. McLoughlin & B.M. McKenna (Eds.), *Proceedings of the Sixth International Congress of Food Science and Technology* (Vol. 2; pp. 149–150). Dublin: Boole Press.

Bottazzi, V., Battistotti, B., & Bianchi, F. (1982). The microscopic crystalline inclusions in grana cheese and their X-ray microanalysis. *Milchwissenschaft, 37,* 577–580.

Brooker, B.E. (1979). Milk and its products. In J.G. Vaughan (Ed.), *Food microscopy* (pp. 273–331). New York: Academic Press.

Brooker, B.E. (1985). Observations of the air-serum interface of milk foams. *Food Microstructure, 4,* 289–296.

Brooker, B.E., Hobbs, D.G., & Turvey, A. (1975). Observations on the microscopic crystalline inclusions in Cheddar cheese. *Journal of Dairy Research, 42,* 341–348.

Browne, F.L., & Brouse, D. (1939). *Casein and its industrial applications* (2nd ed.). New York: Reinhold.

Buchheim, W. (1978). Mikrostruktur von geschlagenem Rahm. *Gordian, 6,* 184–188.

Buchheim, W. (1982). Electron microscopic localization of solvent-extractable fat in agglomerated spray-dried whole milk powder particles. *Food Microstructure, 1,* 233–238.

Buchheim, W. (1986). Membranes of milk fat globules: Ultrastructural, biochemical and technological aspects. *Kieler Milchwirtschaftliche Forschungsberichte, 38,* 227–246.

Buchheim, W., Banford, N.M., & Krog, N. (1985). Relation between microstructure, destabilization phenomena and rheological properties of whippable emulsions. *Food Microstructure, 4,* 221–232.

Buchheim, W., & Precht, D. (1979). Elektronenmikroskopische Untersuchung der kristallisationsvorgange in den Fettkugelchen Wahrend der Rahmreifung. *Milchwissenschaft, 34,* 657–662.

Buchheim, W., & Thomasow, J. (1984). Structural changes in cream cheese induced by thermal processing and emulsifying salts. *North European Dairy Journal, 50,* 38–44.

Buma, T.J. (1971). Free fat in sprayed-dried whole milk, 8. The relation between free-fat content and particle porosity of spray-dried milk. *Netherlands Milk Dairy Journal, 25,* 123–140.

Buma, T.J. (1972). The cause of particle porosity of spray-dried milk. *Netherlands Milk Dairy Journal, 26,* 60–67.

Buma, T.J., & Henstra, S. (1971a). Particle structure of spray-dried milk products as observed by a scanning electron microscope. *Netherlands Milk Dairy Journal, 25,* 75–80.

Buma, T.J., & Henstra, S. (1971b). Particle structure of spray-dried caseinate and spray-dried lactose as observed by a scanning electron microscope. *Netherlands Milk Dairy Journal, 25,* 278–281.

Caldwell, K.B., Goff, H.D., & Stanley, D.W. (1992). A low temperature SEM study of ice cream. 2. Influence of ingredients and processes. *Food Structure, 11,* 11–23.

Carroll, B.J. (1976). The stability of emulsions and mechanism of emulsion breakdown. *Surface and Colloid Science, 9,* 1–67.

Carroll, B.J., & Farrel, H.M., Jr. (1983). Immunological approach to location of κ-casein in the casein micelle by electron microscopy. *Journal of Dairy Science, 66,* 679–683.

Carroll, B.J., Thompson, M.P., & Melnychyn, P. (1971). Gelation of concentrated skim milk: Electron microscopy study. *Journal of Dairy Science, 54,* 1245–1252.

Copius Peerebom, J.W. (1969). Modern views on the physical structure of the membrane of the fat globules in milk and cream and a possible relation with the migration of copper during butter manufacture. *Fette, Seifen, Anstrichmitted, 71,* 314–322.

Darling, D.F. (1982). Recent advances in the destabilization of dairy emulsions. *Journal of Dairy Research, 49,* 695–712.

Darling, D.F., & Butcher, D.W. (1978). Milk-fat globule membrane in homogenized milk. *Journal of Dairy Research, 45,* 197–208.

Davies, F.L., Shankar, P.A., Brooker, B.E., & Hobbs, D.G. (1978). A heat-induced change in the ultrastructure of milk and its effects on gel formation in yoghurt. *Journal of Dairy Research, 45,* 53–59.

deJong, L. (1978). Protein breakdown in soft cheese and its relation to consistency. 3. The micellar structure of Meshanger cheese. *Netherlands Milk Dairy Journal, 32,* 15–25.

deWit, J.N. (1984). Functional properties of whey proteins in food systems. *Netherlands Milk Dairy Journal, 38,* 71–89.

deWit, J.N., & Klarenbeek, G. (1984). Effect of various heat treatments on structure and solubility of whey proteins. *Journal of Dairy Science, 67,* 2701–2710.

Dickinson, E., & Stainsby, G. (1982). *Colloids in foods.* London: Applied Science Publishers.

Downey, G., & Burgess, K.J. (1979a). Texture studies on edible protein fibers produced by a wet spinning technique. 1. Fibers produced from casein and carageenan. *Journal of Food Technology, 14,* 21–31.

Downey, G., & Burgess, K.J. (1979b). Texture studies on edible protein fibers produced by a wet spinning technique. 2. Fibers produced from casein and alginate. *Journal of Food Technology, 14,* 33–40.

Eino, M.F., Biggs, D.A., Irvine, D.M., & Stanley, D.W. (1976). A comparison of microstructures of Cheddar cheese curd manufactured with calf rennet, bovine pepsin and porcine pepsin. *Journal of Dairy Research, 43,* 113–115.

Eino, M.F., Biggs, D.A., Irvine, D.M., & Stanley, D.W. (1979). Microstructural changes during ripening of Cheddar cheese produced with calf rennet, bovine pepsin and porcine pepsin. *Canadian Institute of Food Science and Technology Journal, 12,* 149–153.

Euber, J.R., & Brunner, J.R. (1984). Reexamination of fat globule clustering and creaming in cow milk. *Journal of Dairy Science, 67,* 2821–2832.

Fletcher, A.L. (1997). Cryogenic environmental SEM and its application to ice cream. *Microscopy & Analysis, 45,* 17–19.

Fryer, P.J., & Belmar-Beiny, M.T. (1991). Fouling of heat exchangers in the food industry: A chemical engineering perspective. *Trends in Food Science and Technology 2,* 33–37.

Glaser, J., Carroad, P.A., & Dunckley, W.L. (1979). Surface structure of cottage cheese curd by electron microscopy. *Journal of Dairy Science, 62,* 1058–1068.

Glaser, J., Carroad, P.A., & Dunckley, W.L. (1980). Electron microscopic studies of casein micelles and curd microstructure in cottage cheese. *Journal of Dairy Science, 63,* 37–48.

Goff, H.D., Caldwell, K.B., Stanley, D.W., & Maurice, T.J. (1993). The influence of polysaccharides on the glass transition in frozen sucrose solutions and ice cream. *Journal of Dairy Science, 76,* 1268–1277.

Goff, H.D., Fresion, B., Sahagian, M.E., Hauber, T.D., Stone, A.P., & Stanley, D.W. (1995). Structure development in ice cream: Dynamic rheological measurements. *Journal of Texture Studies, 26,* 517–536.

Graf, E., & Müller, H.R. (1965). Fine structure and whippability of sterilized cream. *Milchwissenschaft, 20,* 302–308.

Green, M.L. (1980). The formation and structure of milk protein gels. *Food Chemistry, 6,* 41–49.

Green, M.L., & Manning, D.J. (1982). Development of texture and flavor in cheese and other fermented products. *Journal of Dairy Research, 49,* 737–748.

Green, M.L., Turvey, A., & Hobbs, D.G. (1981). Development of structure and texture in Cheddar cheese. *Journal of Dairy Research, 48,* 343–355.

Hartel, R.W. (1998). Phase transitions in ice cream. In M.A. Rao & R.W. Hartel (Eds.), *Phase/state transitions in foods* (pp. 327–368). New York: Marcel Dekker.

Harwalkar, V.R. (1982). Age gelation of sterilized milks. In P.F. Fox (Ed.), *Developments in dairy chemistry.* Vol. 1. *Proteins* (pp. 229–269). London: Applied Science Publishers.

Harwalkar, V.R., & Kalab, M. (1981). Effect of acidulants and temperature on microstructure, firmness and susceptibility to syneresis of skim milk gels. *Scanning Electron Microscopy, III,* 503–513.

Hassan, A.N., Frank, J.F., Farmer, M.A., Schmidt, K.A., & Shalabi, S.I. (1995). Formation of yogurt microstructure and three-dimensional visualization as determined by confocal scanning laser microscopy. *Journal of Dairy Science, 78,* 2629–2636.

Heertje, I., Boskamp, M.J., van Kleef, F., & Gortemaker, F.H. (1981). The microstructure of processed cheese. *Netherlands Milk Dairy Journal, 35,* 177–179.

Henstra, S., & Schmidt, D.G. (1970). On the structure of the fat-protein complex in homogenized cow's milk. *Netherlands Milk Dairy Journal, 24,* 45–51.

Hermansson, A.M. (1979). Aggregation and denaturation involved in gel formation. In A. Pour-al (Ed.), *Functionality and protein structure* (pp. 81–103). ACS Symposium Series 92. Washington, DC: American Chemical Society.

Holcomb, D.N. (1991). Structure and rheology of dairy products. *Food Structure, 10,* 45–108.

Jack, F.R., & Paterson, A. (1992) Texture of hard cheese. *Trends in Food Science and Technology, 3,* 160–164.

Johnston, D.E. (1984a). Application of polymer cross-linking theory to rennet-induced milk gels. *Journal of Dairy Research, 51,* 91–101.

Johnston, D.E. (1984b). Application of polymer cross-linking theory to rennet-induced milk gels: a further comment. *Netherlands Milk Dairy Journal, 38,* 265–266.

Kalab, M. (1977). Milk gel structure. 4. Cheese texture and microstructure. *Milchwissenschaft, 32,* 449–458.

Kalab, M. (1979a). Scanning electron microscopy of dairy products: An overview. *Scanning Electron Microscopy, III,* 261–272.

Kalab, M. (1979b). Microstructure of dairy foods. 1. Milk products based on protein. *Journal of Dairy Science, 62,* 1352–1364.

Kalab, M. (1981). Scanning electron microscopy of dairy products: A review of techniques. *Scanning Electron Microscopy, III,* 453–472.

Kalab, M. (1985). Microstructure of dairy foods. 2. Milk products based on fat. *Journal of Dairy Science, 68,* 3234–3248.

Kalab, M., Allan-Wojtas, P., & Phipps-Todd, B. (1983). Development of microstructure in set-style nonfat yoghurt [A review]. *Food Microstructure, 2,* 51–66.

Kalab, M., & Emmons, D.B. (1978). Milk gel structure. 10. Cheddared curd. *Milchwissenchaft, 33,* 670–673.

Kalab, M., Emmons, D.B., & Sargant, A.G. (1976). Milk gel structure. 5. Microstructure of yoghurt as related to the heating of milk. *Milchwissenschaft, 31,* 402–408.

Kalab, M., & Harwalkar, V.R. (1973). Milk gel structure. 1. The application of SEM to milk and other gels. *Dairy Science, 56,* 835–842.

Kalab, M., & Harwalkar, V.R. (1974). Milk gel structure. 2. Relation between firmness and ultrastructure of heat induced skim milk gels containing 40–60% total solids. *Journal of Dairy Research, 41,* 131–135.

Kalab, M., Sargant, A.G., & Froehlich, D.A. (1981). Electron microscopy and sensory evaluation of commercial cream cheese. *Scanning Electron Microscopy, III,* 473–482.

Keenan, T.W., Moon, T.W., & Dylewski, D.P. (1983). Lipid globules retain globule membrane material after homogenization. *Journal of Dairy Science, 66,* 196–203.

Kessler, H.G. (1981). *Food engineering and dairy technology.* Freising, Germany: Verlag A. Kessler.

Kimura, T., Taneya, S., & Kanaya, K. (1979). Observation of internal structure of casein submicelles by means of ion beam sputtering. *Milchwissenschaft, 34,* 521–524.

King, N. (1955). *Milk fat globule membrane and some associated phenomena.* Farnham Royal, England: Commonwealth Agricultural Bureau.

Knoop, A.M. (1972). Strukturaufklrung durch elektronenmikroskopische Untersuchungen an Eiwei und Milchfett. *Milchwissenschaft, 27,* 364–373.

Knoop, A.M., & Buchheim, W. (1980). Die unterschiedliche Entwicklung der Substruktur bei der Reifung von Harzer-, Tilster- und Camembert-Kase. *Milchwissenschaft, 35,* 482–488.

Knoop, A.M., & Peters, K.H. (1971). Submikroskopische Strukturvernderungen im Camembert-Kase whrend der Reifung. *Milchwissenschaft, 26,* 193–198.

Kosikowski, F.V., & Mistry, V.V. (1997). *Cheese and fermented milk foods* (Vol. 2, 3rd ed.). Westport, CT: F.V. Kosikowski.

Lawrence. R.C., Heap, H.A., & Gilles, J. (1984). A controlled approach to cheese technology. *Journal of Dairy Science, 67,* 1632–1645.

Lee, B.O., Kilbertus, G., & Alais, C. (1981). Ultrastructural study of processed cheese: Effects of different parameters. *Milchwissenschaft, 36,* 343–348.

Lowrie, R.J., Kalab, M., & Nichols, D. (1982). Curd granule and milled curd junction patterns in Cheddar cheese made by traditional and mechanized process. *Journal of Dairy Science, 65,* 1122–1129.

Marcos, A., Alcala, M., Leon, F., Fernandez-Salguero, J., & Esteban, M.A. (1981). Water activity and chemical composition of cheeses. *Journal of Dairy Science, 64,* 622–626.

Martley, F.G., & Crow, V.L. (1996). Open texture in cheese: The contribution of gas production by microorganism and cheese manufacturing practices. *Journal of Dairy Research, 63,* 489–507.

Mather, I.H., & Keenan, T.W. (1975). Studies on the structure of milk fat globule membranes. *Journal of Membrane Biology, 21,* 65–78.

McKenna, A.B. (1997). Examination of whole milk powder by confocal laser scanning microscopy. *Journal of Dairy Research, 64,* 423–432.

McMahon, D.J., & Brown, R.J. (1984). Composition, structure and integrity of casein micelles [A review]. *Journal of Dairy Science, 67,* 499–512.

Modler, H.W., & Kalab, M. (1983). Microstructure of yoghurt stabilized with milk proteins. *Journal of Dairy Science, 66,* 430–437.

Muhr, A.H., & Blanshard, J.M.V. (1986) Effect of polysaccharide stabilizers on the rate of growth of ice. *Journal of Food Technology, 21,* 683–710.

Mulder, H., & Walstra, P. (1974). *The milk fat globule: Emulsion science as applied to milk products and comparable foods.* Farnham Royal, England: PUDOC, Wageningen, and C.A.B.

Nielsen, B.J. (1973) Building and formation of ice cream microstructure during processing. *Modern Dairy, 52,* 10–12.

Oortwijn, H., Walstra, P., & Mulder, H. (1977). The membranes of recombined fat globules. 1. Electron microscopy. *Netherlands Milk Dairy Journal, 31,* 134–147.

Patton, S., & Keenan, T.W. (1975). The milk fat globule membrane. *Biochemistry Biophysical Acta, 415,* 273–309.

Payens, T.A.J. (1979). Casein micelles: The colloid-chemical approach. *Journal of Dairy Research, 46,* 291–306.

Phipps, L.W. (1983). Effects of fat concentration on the homogenization of cream. *Journal of Dairy Research, 50,* 91–96.

Precht, D., & Buchheim, W. (1979). Elektronenmikroscopische Untersuchungen uber der physicalische Struktur von Streichfetten. 1. Die Mikrostrukur der Fettkugelchen in Butter. *Milchwissenschaft, 34,* 745–749.

Precht, D., & Peters, K.H. (1981). Die Konsistenz der Butter. *Milchwissenschaft, 31,* 616–620.

Prins, A. (1986). Some physical aspects of aerated milk products. *Netherlands Milk Dairy Journal, 40,* 203–215.

Rayan, A.A., Kalab, M., & Ernstrom, C.A. (1980). Microstructure and rheology of process cheese. *Scanning Electron Microscopy, III,* 635–643.

Roetman, K. (1979). Crystalline lactose and the structure of spray-dried milk products as observed by scanning electron microscopy. *Netherlands Milk Dairy Journal, 33,* 1–11.

Rousseau, M. (1984). Study of the surface flora of traditional Camembert cheese by scanning electron microscopy. *Milchwissenschaft, 39,* 129–135.

Rüegg, M., & Blanc, B. (1972). Beitrge zur elektronenmikroskopishen Struktur der Labgallerte un des Kseteiges. *Schweizer Milchwirtschaft Forschung, 1,* 1–8.

Rüegg, M., Moor, U., & Blanc, B. (1980). Vernderungen der feinstruktur von Greyerzerkse im verlauf der Reifung: Eine Studie mit dem Raster-Elektronenmikroskop. *Milchwissenschaft, 35,* 329–335.

Saltmarch, M., & Labuza, T.P. (1980). SEM investigation on the storage properties of spray dried whey. *Scanning Electron Microscopy, III,* 659–665.

Sara, M. (1984). Elektronenmikroskopische Untersuchung von Lebensmitteln. *Ernahrung, 8,* 14–18.

Schmidt, D.G. (1980). Colloidal aspects of casein. *Netherlands Milk Dairy Journal, 34,* 42–63.

Schmidt, D.G., & Buchheim, W. (1970). Elektronenmikroskopische Untersuchung der Feinstruktur von Caseinmicellen in Kuhmilch. *Milchwissenschaft, 25,* 596–600.

Schmidt, D.G., & van Hooydonk, A.C.M. (1980). A scanning electron microscopical investigation of the whipping of cream. *Scanning Electron Microscopy, III,* 653–658.

Schmidt, D.G., & Payens, T.A.J. (1976). Micellar aspects of caseins. *Surface and Colloid Science, 9,* 165–229.

Schubert, H. (1993) Instantization of powdered foods. *International Chemical Engineering, 33,* 28–45.

Shama, F., & Sherman, P. (1966). The texture of ice cream. 2. Rheological properties of frozen ice cream. *Journal of Food Science, 31,* 699–706.

Skriver, A., Hansen, M.B., & Qvist, K.B. (1997). Image analysis applied to electron micrographs of stirred yogurt. *Journal of Dairy Research, 64,* 135–143.

Smietana, Z., Poznanski, S., Hosaja, M., & Kozlowska, H. (1978). The structure analysis of texturized milk and plant proteins by scanning electron microscopy. *Milchwissenschaft, 33,* 601–603.

Stanley, D.W., Smith, H.D., & Smith, A.K. (1996). Texture-structure relationships in foamed dairy emulsions. *Food Research International, 29,* 1–13.

Sutermeister, E. (1927). *Casein and its industrial applications.* New York: Chemical Catalog Co.

Swiatek, A., & Jaworski, J. (1959). Histochemische untersuchungen uber die verteihung von mineralsalzen im Kase. In *Fifteenth International Dairy Congress* (pp. 1509–1518).

Szpendowski, J., Smietana, Z., & Zuraw, J. (1983). Production of extrusion texturized protein preparations. *Milchwissenschaft, 38,* 577–581.

Tamime, A.Y. (1986). Modern cheesemaking: Hard cheeses. In R.K. Robinson (Ed.), *Modern dairy technology* (Vol. 2, pp. 35–158). London: Elsevier Applied Science Publishers.

Taneya, S., Izutsu, T., & Buchheim, W. (1980). The submicroscopic structure of processed cheese with different melting properties. *Milchwissenschaft, 35,* 479–481.

Taranto, M.V., & Tom Yang, C.S. (1981). Morphological and textural characterization of soybean mozarella cheese analogs. *Scanning Electron Microscopy, III,* 483–492.

Taranto, M.V., Wan, P.J., Chen, S.L., & Rhee, K.C. (1979). Morphological, ultrastructural and rheological characterization of Cheddar and mozzarella cheese. *Scanning Electron Microscopy, III,* 273–278.

Tuohy, J.J., Burgess, K.J., & Lambert, I. (1979). A comparison of the texture of expanded milk protein meat extenders and TVP. *Journal of Food Technology, 14,* 473–481.

Uhlmann, G., Klostermeyer, H., & Merkenich, K. (1983). Kristallisationserscheinungen in Schmelzkseprodukten. 1. Phänomen und Ursachen. *Milchwissenschaft, 38,* 582–585.

van Kreveld, A. (1974). Studies on the wetting of milk powder. *Netherlands Milk Dairy Journal, 28,* 23–45.

van Vliet, T., & Dentener-Kikkert, A. (1982). Influence of the composition of the milk fat globule membrane on the rheological properties of acid milk gels. *Netherlands Milk Dairy Journal, 36,* 261–265.

Walstra, P. (1990). On the stability of casein micelles. *Journal of Dairy Science, 73,* 1965–1979.

Walstra, P., & Jenness, R. (1984). *Dairy chemistry and physics.* New York: Wiley.

Walstra, P., & van Vliet, T. (1986). The physical chemistry of curd making. *Netherlands Milk Dairy Journal, 40,* 241–259.

Washam, C.J., Kerr, T.J., Hurst, V.J., & Rigsby, W.E. (1985). A scanning electron microscopy study of crystalline structures on commercial cheese. *Developments in Industrial Microbiology, 26,* 749–761.

Washam, C.J., Kerr, T.J., & Todd, R.L. (1979). Scanning electron microscopy of blue cheese: Mold growth during maturation. *Journal of Dairy Science, 62,* 1384–1389.

Wooding, F.B.P. (1971). The structure of the milk fat globule membrane. *Journal of Ultrastructural Research, 37,* 388–393.

Wooding, F.B.P. (1975). Ultrastructure of the milk fat globule membrane with and without triglyceride. *Cell Tissue Research, 165,* 113–117.

# CHAPTER 8

# Microstructure and Mass Transfer: Solid-Liquid Extraction

## 8.1 INTRODUCTION

### 8.1.1 Food Engineering and Microstructure

The foundations of food technology and their relation to microstructure have been presented in previous chapters. We now turn to the relationship between food engineering and microstructure. The engineering concepts underlying food processing are merely introduced in this chapter, for they are discussed coherently and comprehensively in many excellent texts: Loncin (1969), Kessler (1981), Heldman and Lund (1992), and Fryer, Pyle, and Reilly (1997), among others.

Food engineering is based on the principles of chemical engineering and has adopted concepts such as unit operations and transport phenomena from that discipline. However, while chemical engineering deals mainly with liquids and gases (one exception being catalysis), the raw materials in food engineering are largely structured and heterogeneous solids. Moreover, these solid materials have a biological origin and sometimes are biochemically active. Liquids present in food processing are often non-Newtonian and therefore structured (see Section 3.8). It is surprising that food engineers have not made a more fundamental contribution to the engineering sciences, particularly to chemical engineering, by introducing the structure of materials as a critical parameter in transport phenomena and unit operations. Rather, they have applied chemical engineering principles and equations, avoiding the structural dimension and treating solids as "black boxes."

This chapter is a brief overview of mass transfer using the extraction of a solute from the interior of a solid by a solvent as a case study. It calls attention to the microstructural aspects involved while introducing some concepts of mass transfer. An excellent book for learning about mass transfer in biological systems is *Diffusion: Mass Transfer in Fluid Systems* (Cussler, 1997).

### 8.1.2 Solid-Liquid Extraction or Leaching

Solid-liquid extraction or leaching is a separation process effected by a solvent, and it involves the transfer of solutes from a solid to a fluid. It is extensively used as a unit operation to recover many important food components: sucrose from cane or beets, lipids from oilseeds and animals tissues, vegetable proteins from defatted meals, extracts from tea leaves and coffee grounds, hydrocolloids from algae and plant tissues, gelatin from hides and bones, and oleoresins and pigments from plants, among others (Table 8–1). Solid-liquid extraction (or simply extraction) is also used to remove undesirable contaminants and toxins present in foods and feeds, such as aflatoxins from oilseed meals and alkaloids from bitter lupins.

"Solid-liquid extraction" may be a misnomer, for it gives the impression that a mass transfer occurs at a sharp interface between a dry solid and the liquid phase. As explained later in the chapter,

**Table 8–1** Some Food Materials Involved in Commercial Leaching Processes

| *Product* | *Solids* | *Solute* | *Solvent* |
|---|---|---|---|
| Anthocyanins | Chokeberries, grapeskins | Anthocyanins | Ethanol, water |
| Apple juice solutes | Pressed apple pomace | Apple juice solutes | Water |
| Betalaines | Red beets | | Ethanol, water |
| Brewing worts | Malted barley | Sugars, grain solutes | Water |
| Carrageenan | Kelp | Carrageenan | Water |
| Carotenoid pigments | Leaves | Water first, then pigments | Ethanol, isopropanol |
| Cassava | Manioc | Cyanogenetic glycosides | Water |
| Citrus molasses | Juice pressing residues | Citrus sugars | Water |
| Gelatin | Collagen | Gelatin | Water or dilute acid |
| Cytoplasmic alfalfa protein | Coagulated alfalfa protein | Chlorophyll, chlorogenic acid | Acetone, ethanol, butanol |
| Decaffeinated coffee | Green coffee beans | Caffeine | Methylene chloride |
| Decaffeinated coffee | Green coffee beans | Caffeine | Supercritical $CO_2$ |
| Hop extracts | Hop flowers | Hop solutes | $CH_2Cl_2$ |
| Hop extracts | Hop flowers | Hop solutes | Supercritical $CO_2$ |
| Insulin | Beef or pork pancreas | Insulin | Acidic alcohol |
| Papain | Papaya latex | Papain | Water |
| Pectin | Desugared apple pomace | Pectin | Dilute acid |
| Pectin | Treated citrus peel | Pectin | Dilute acid |
| Pepsin | Hog stomachs | Pepsin | Aqueous HCl |
| Rennin | Calf stomach lining | Rennin | Aqueous NaCl |
| Soluble coffee | Ground roasted coffee | Coffee solutes | Water |
| Soluble tea | Dry tea leaves | Tea solutes | Water |
| Soybean oil | Soybeans | Soybean oil | Hexane |
| Soy protein concentrate | Defatted soy flour | Sugars, nonprotein solids | 70% ethanol at isoelectric point |
| Soy protein isolate | Defatted soy flour | Protein | pH 9 Aqueous NaOH |
| Spices extract | Paprika, cloves, pepper, thyme, marjoram, etc. | Spice solutes | 80% ethanol |
| Spice oleoresins | Paprika, etc. | Spice solutes | Methyl ethyl ketone |
| Sucrose | Sugar beets | Sucrose | Water |
| Treated citrus peel | Citrus peel | Flavonoids, hesperidin, sugars | Water |
| Vanilla | Vanilla beans | Vanilla | 65% ethanol |

*Source:* Reprinted with permission from H.G. Schwartzberg, Leaching—Organic Materials, in *Handbook of Separation Process Technology*, R.W. Rousseau, ed., pp. 540–577, © 1987, John Wiley & Sons.

in most food extractions, either the solid naturally contains a liquid phase (e.g., water in fruits) or it is impregnated by the extraction liquid; therefore, liquid-phase diffusion inside the solid is a major mass transfer mechanism during leaching.

### 8.1.3 Characteristics of Food Extraction

From an engineering viewpoint solid-liquid extraction of foods is a multicomponent, multiphase, and unsteady-state mass transfer operation. Commonly, it involves transfer of more than one chemical species (referred here to generically as the *solute*) from a solid to a solvent. During extraction, the concentration of the solute inside the solid varies continuously—hence, the nonstationary or unsteady state. Engineering aspects and equipment related to food extraction are described in Loncin (1969), Leniger and Beverloo (1975), Dousse (1978), and Schwartzberg (1980, 1987).

A series of phenomenological steps have to occur during the period of interaction between the solute-containing particle and the solvent effecting the separation (Loncin, 1969; Lysjanski, Popow, Redko, & Stabnikow, 1983; Zanderighi, 1983), as shown in Figure 8–1. These include

- entrance of the solvent into the solid matrix
- solubilization and/or breakdown of components
- transport of the solute to the exterior of the solid matrix
- migration of the extracted solute from the external surface of the solid into the bulk solution

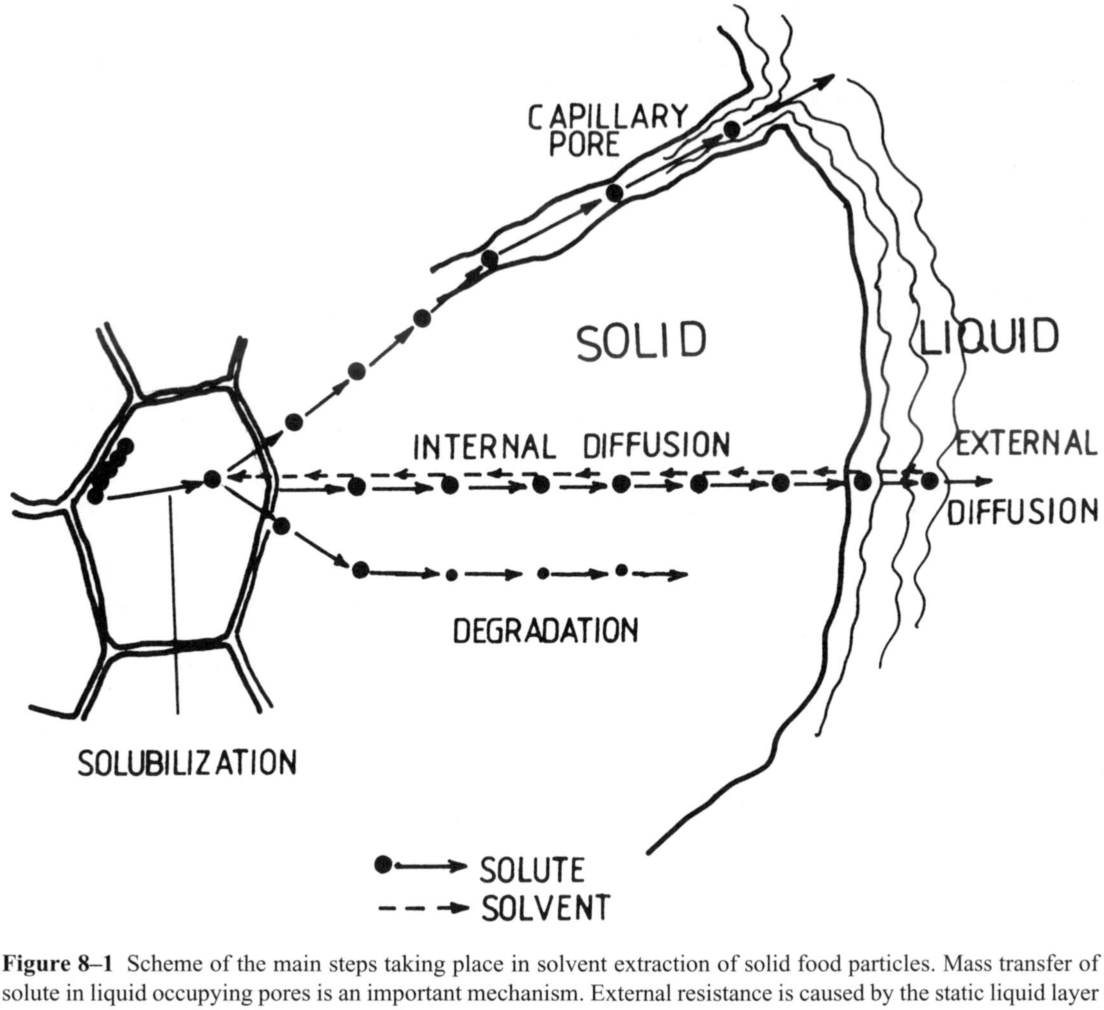


**Figure 8–1** Scheme of the main steps taking place in solvent extraction of solid food particles. Mass transfer of solute in liquid occupying pores is an important mechanism. External resistance is caused by the static liquid layer surrounding the particle.

- movement of the extract with respect to the solid (i.e., extract displacement)
- separation and discharge of the extract and solid

As a result of these phenomena, extraction takes place at a rate expressed in terms of mass of solute leached/unit time or, more commonly, as change in solute concentration in the solid/unit time *(dc/dt* or *dx/dt)*. In some food extractions, the relative rate at which different chemical species migrate through the solid may also be of concern (multicomponent extraction). Since each of the aforementioned elementary steps occurs at its own rate and in some cases sequentially, the overall rate of the extraction process is determined by the step having the slowest rate, termed the *rate-controlling step*. As discussed later, transport through the solid matrix is usually the rate-controlling step in food extractions.

In foods, the microstructure of the solid plays a major role in the rate and quality of the extraction process. To obtain a fundamental understanding of the relationship between rate and food microstructure, we must first consider some theoretical notions of mass transfer. A rigorous treatment on the subject is found in the classical chemical engineering texts by Sherwood, Pigford, and Wilke (1975) and King (1980). Gekkas (1992) presents concepts of transport phenomena and discusses their application to foods and biological materials. A good book on mass transfer suitable for beginners and experts is Cussler (1997).

## 8.2 FUNDAMENTAL ASPECTS OF EXTRACTION

### 8.2.1 Extraction as a Diffusional Process

Mass transfer can occur by convection, eddy diffusion, or molecular diffusion. In molecular diffusion, molecules are transported from one part of the system to another by random movement as a result of a concentration gradient. In leaching of foods, the interior of the solid cannot be agitated, and turbulence is unlikely to occur in small capillaries and pores, leaving molecular diffusion as the main transport mechanism. Solvent extraction may be considered a diffusion process in the liquid (fluid) phase, since solute transferred, even inside a solid, exists as a dilute solution.

In mass transfer, as well as in heat and momentum transfer, we are interested in the *rate* at which the process takes place. The common expression defining the rate of transfer is

$$Rate = \frac{1}{(Resistance)}(Driving\ force)$$

**Equation 8–1**

Fick's laws provide the semi-empirical basis for the analysis of molecular diffusion. Fick's first law, a version of equation 8–1, is useful for defining a diffusion coefficient or diffusivity *(D)* that is the inverse of the resistance to mass transfer. It establishes that under *steady-state* conditions the unidirectional flux of solute 1 ($J_1$ moles/s) in the $r$ direction is directly proportional to the diffusivity of the solute, the area traversed by the flux, and the gradient of solute concentration between two points (expressed in terms of absolute concentration [$dc/dr$] or molar fraction [$dx/dr$]).

Fick's first law describes diffusion in a fixed coordinate system under steady-state conditions (the concentration does not change with time). In the unidimensional case, it takes the form

$$j_1 = \frac{J_1}{A} = -cD\frac{dx_1}{dr} = -D\frac{dc_1}{dr}$$

**Equation 8–2**

where $j_1$ is the flux in moles per unit time and unit area, $r$ is the direction of flow, $A$ is the area across which diffusion occurs, and $c$ is the total molar concentration (moles/volume). The minus sign gives a positive flux term because the gradient is negative (flow occurs down a concentration gradient, from high to low concentration). This equation can be regarded as a limit for long times, when transient conditions have disappeared and a linear gradient is established in the system. It can be applied without problems whenever the solute is in high dilution in the solvent; diffusion in concentrated systems involves convection and requires a more complex mathematical treatment (see Cussler, 1997).

In practical situations and for short times, *unsteady* or nonstationary conditions exist, and the concentration of solute varies with time *(t)* and position *(r)* inside the solid. In such cases, Fick's second law (or the diffusion equation) applies. It has the general form

$$\frac{\partial c_1}{\partial t} = \frac{1}{r^{\nu-1}} \frac{\partial\left[r^{\nu-1} D \frac{\partial c_1}{\partial r}\right]}{\partial r} \quad \textbf{Equation 8–3}$$

where the index $\nu$ equals 1 for an infinite slab, 2 for an infinite cylinder, and 3 for a sphere. It is easier to understand Fick's second law when equation 8–3 is written as:

$$\frac{\partial c_1}{\partial t} = D\frac{\partial^2 c_1}{\partial r^2} = D\frac{\partial}{\partial r}\left(\frac{\partial c_1}{\partial r}\right) \quad \textbf{Equation 8–4}$$

This equation says that the flow of solute is directly proportional to the change of the concentration gradient with position, so when the gradient is constant (linear concentration profile), then $\partial c_1/\partial t = 0$, which means that steady-state conditions exist and we are back to equation 8–2.

Analytical solutions for equation 8–3 under several simple initial and boundary conditions are found in textbooks such as Bird, Stewart, and Lightfoot (1960) and King (1980). Solutions for more complex situations and for different geometries are analyzed in the classical book by Crank (1975). Schwartzberg and Chao (1982) present a detailed analysis of the assumptions and conditions under which solutions to equation 8–3 can be applied to the extraction of foods. In general, solutions relate the dimensionless extent of extraction of solute 1 as $X = (c - c_f)/(c_o - c_f)$ to time in a series expression represented by equation 8–5, where $c$ is the average solute concentration inside the solid at any time and $c_o$ and $c_f$ are the initial and final equilibrium concentrations, respectively.

$$X = \sum B_n \exp\left(-\frac{q_n^2 Dt}{L^2}\right) \quad \textbf{Equation 8–5}$$

$X$ depends on $\alpha = m\,E/R$ (where $m$ is the equilibrium distribution ratio between the solute concentration in the bulk solution and in the solid and $E/R$ is the extract-to-solid volume ratio) and on Fick's dimensionless number $Dt/L^2$ (where $L$ is a characteristic length, e.g., the particle size). Parameters $B_n$ and $q_n$ are functions of $\alpha$. When only the first term of the series is considered (e.g., for $Dt/L^2 > 0.06$), plots of log $X$ versus $t$ are straight lines with a slope equal to $-Dq_1^2/2.303L^2$, from which an *overall diffusion coefficient* can be obtained (effective or apparent diffusion coefficient).

The dimensionless number $Dt/L^2$ can be used as a criterion of closeness to the steady state. If it is much larger than unity (e.g., $Dt \gg L^2$), an equilibrium or steady-state condition may be assumed. Also, two parameters may be established for unsteady state conditions: a "velocity of diffusion," $\sqrt{D/\pi t}$, and a "penetration distance," $\sqrt{4\,Dt}$ (Cussler, 1997). Note that both parameters involve the square root of time.

### 8.2.2 Diffusion Coefficients

Data for diffusion coefficients are necessary to make calculations using the previous approach. Diffusivities may be determined experimentally or calculated. Experimental tools and methods for liquids include the diaphragm cell, the rotating disk (used for drug dissolution), NMR spin-echo techniques, and interferometer methods, including the Gouy interferometer (see Cussler, 1997). The orders of magnitude of these coefficients are important to remember (all units are in $cm^2/s$): for gases, $10^{-1}$; for liquids, around $10^{-5}$; and for solids, between $10^{-8}$ and $10^{-30}$ (if $m^2/s$ is used as unit, the values must be divided by $10^4$). Consequently, the aroma of tea in air diffuses much times faster than the tea extract in a cup with hot water! The diffusion coefficients of polymers in solvents under dilute conditions may be as low as $10^{-6}$ to $10^{-8}$ $cm^2/s$, and those of gases through synthetic membranes vary widely between $10^{-8}$ to $10^{-11}$ $cm^2/s$.

Numerous attempts have been made to predict diffusivities in liquids. According to the Stokes-Einstein equation, the diffusion coefficient of a large spherical solute species in a pure solvent of viscosity $\eta$ is given by

$$D = \frac{kT}{6\pi\eta r_s} \quad \textbf{Equation 8–6}$$

where $k$ is the Boltzmann constant (the gas constant divided by Avogadro's number, $R/N_o$), $T$ is the absolute temperature, and $r_s$ is the effective radius of the diffusing molecule. This expression is slightly misleading, because in practice viscosity and solute radius effects are more important than those of temperature. Also, adaptations of equation 8–6 for dilute solutions of macromolecules need to consider their size and shape. Nevertheless, the Stokes-Einstein equation has provided the foundation for several useful semi-empirical equations. For smaller molecules, a modified expression known as the Wilke-Chang equation relates $D$ to the molecular weight, the association parameter of the solvent, and the molar volume of the solute (King, 1980). Another equation based on the kinetic theory of diffusion has been proposed by Eyring (Loncin, 1980).

Most data for diffusion coefficients of liquids or gases (vapors) in solid foods come from solutions to the diffusion equation for standard geometries combined with experimental results, so they are "effective diffusion coefficients." The main problem is that all steps described in Section 8.1.3 for diffusion through a solid matrix are ignored, and the process is characterized by a single coefficient that is highly dependent on experimental conditions, solid geometry, and microstructural arrangement. It is not surprising, then, that "diffusion coefficients" for similar solute extractions vary by several orders of magnitude.

As explained before, effective or apparent diffusivities appear as calculated values of $D$ obtained from transient diffusion experiments. Defining $M$ as the total amount of solute that has diffused in (impregnation) or out (extraction) of a solid of regular geometry (slab, cylinder, or sphere) at any time and $M_\infty$ as the amount transferred after equilibrium is reached, the ratio $M/M_\infty$ may be calculated by integrating Fick's second law under appropriate boundary conditions (e.g., equation 8–5). For a slab of thickness $L$, when $1 - M/M_\infty$ is plotted versus $t$, the slope is $-\pi^2 D_{app}/4L^2$ (the expression changes for other geometries), from which a value of $D_{app}$ can be determined. The word "apparent" reafirms that we do not know exactly the mechanism of transport, which in most cases may be quite complex.

### 8.2.3 Relating Diffusion Coefficients to Structure

On examination of equation 8–4 and data previously presented, it can be concluded that food microstructure impacts the extraction rate predominantly through its effect on the diffusion coefficient. Chemical engineers have defined an *effective* (or apparent) diffusion coefficient $D_{eff}$ for use when dealing with impermeable porous solids with fluid-filled pores:

$$D_{eff} = D\frac{\varepsilon}{\tau} \qquad \textbf{Equation 8–7}$$

where $D$ is the diffusion coefficient within the pores, $\varepsilon$ is the void fraction or porosity of the solid, and $\tau$ is the tortuosity of the pores (which attempts to account for the longer distance traversed by the solute along a sinuous path). Porosity may be very low, as in potato tissue ($\approx$2%), or high, as in apples ($\approx$20%). See Table 6–4 for intercellular air contents. For solid materials used in chemical engineering (adsorbents, porous catalysts), tortuosity varies between 2 and 6, and porosity between 0.3 and 0.8; thus, $D_{eff}$ may be 6 to 15 times lower than $D$.

When the size of the pore and the solute are of comparable magnitude (as occurs in some membranes), equation 8–7 is corrected by a factor $\lambda$ that depends on the ratio of solute radius to pore radius. The restricted diffusion coefficient of spherical molecules within cylindrical pores ($D_p$) is usually modeled by the so-called Renkin equation (Lebrun & Junter, 1993):

$$D_p = D(1 - \lambda)^2 f(\lambda) \qquad \textbf{Equation 8–8}$$

The first term of the equation is a partition coefficient and accounts for steric hindrance at the pore entrance. The second factor, $f(\lambda)$, is a polynomial function of $\lambda$ and corrects for the friction between the diffusing molecule and the walls of the pore.

The effect of the structure and architecture of a solid composite on the overall diffusivity may be studied for simple geometries (Cussler, 1997). For example, if we assume that a soybean flake is a composite having parallel solid walls leaving straight pores ($\tau = 1$) perpendicular to the faces of the flake (case I in Figure 8–2). $D_{eff}$ is given by

$$D_{eff} = D(1 - \phi) \qquad \textbf{Equation 8–9}$$

where $\phi$ is the volume fraction occupied by the solid walls (which are impermeable to the solute) and $D$ is the diffusion coefficient in the pores. However, if the solid walls are arranged perpendicular to the diffusion path and staggered (case II in Figure 8–2), an aspect ratio $\alpha$ has to be defined

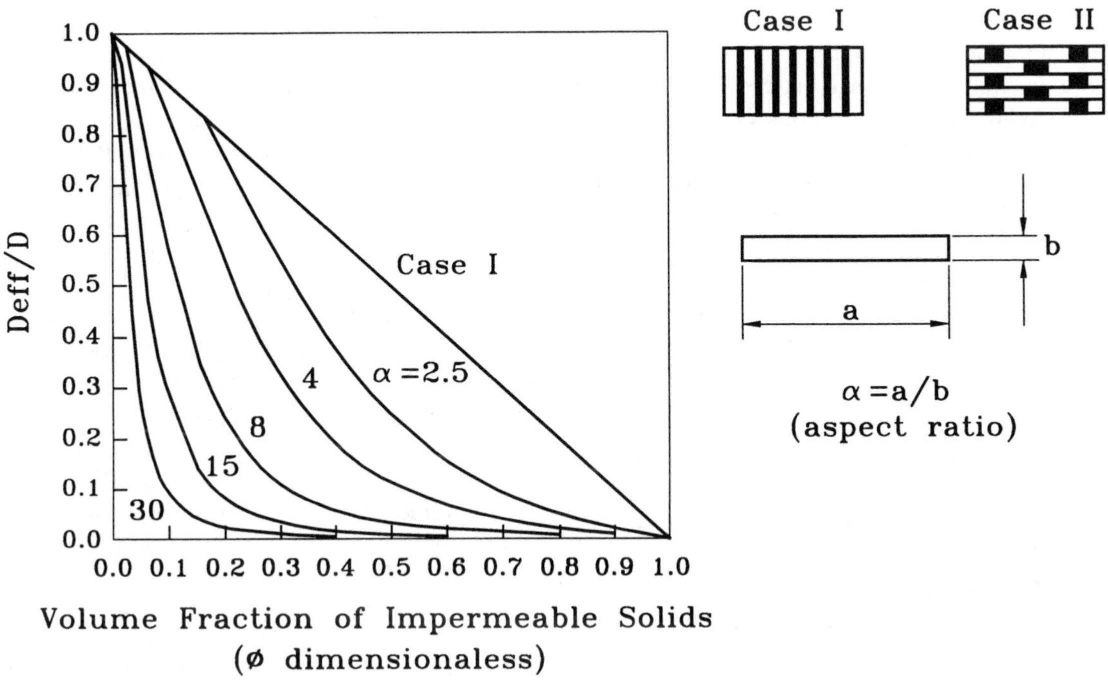


**Figure 8–2** Effect of architecture of a solid particle on the reduction of the effective diffusion coefficient ($D_{eff}$). Case I represents a parallel arrangement of pores and walls, and case II represents a staggered arrangement (higher tortuosity). In case I, $D_{eff}$ decreases linearly with the volume fraction of impermeable solids. In case II, the aspect ratio strongly affects $D_{eff}$ by increasing the distance that the solute has to diffuse in the interior of the solid.

(longest dimension divided by shortest dimension). In this case, the solute has to wiggle in between the narrow spaces left between the walls, and the previous relation changes to

$$D_{eff} = D\left(\frac{1}{1 + \alpha^2\phi^2/(1 - \phi)}\right)$$

**Equation 8–10**

Whereas in the first case there is a linear dependence of $D_{eff}$ on $\phi$, the latter equation shows a sharp decay in $D_{eff}$ with $\phi$ as $\alpha$ gets larger. In other words, for a constant $\phi$, the effective diffusion coefficient varies depending on the architecture of the solid composite (Figure 8–2). If the geometric arrangement of the solid was changed from walls to impermeable spheres occupying a volume fraction $\phi$, the corresponding expression for the diffusivity would be

$$D_{eff} = D\left[\frac{2(1 - \phi)}{2 + \phi}\right]$$

**Equation 8–11**

which shows a different dependence of the effective diffusivity on the volume fraction. In practice, this is a questionable result, because it implies that diffusion would not depend on the size of the spheres, only on their volume fraction. The point, however, is clear: the same material arranged into different architectures would extract at different rates.

The previous results may be expressed in a more general form. For a two-phase composite in which spherical particles of a material 1 are dispersed in a continuous phase 2, an effective diffusion coefficient may be obtained from this expression:

$$\frac{D_{eff} - D_1}{D_{eff} + 2D_1} = \phi\left(\frac{D_2 - D_1}{D_2 + 2D_1}\right)$$

**Equation 8–12**

in which $D_1$ is the diffusion coefficient through

the interstitial pores and $D_2$ is the diffusion coefficient through the particles.

So far we have assumed that the pores are quite large. When the size of the pores is on the order of magnitude of the distance between molecular collisions, Fick's laws do not apply anymore and so-called Knudsen diffusion takes place, in which molecules collide with the pore walls. Other diffusion mechanisms may also be present in porous materials, as discussed in Cussler (1997).

At this point, you may have gained the impression that Fick's laws could be better used to predict extraction rates if the microstructural architecture of a solid food particle was better understood. Microscopy can assist in developing the required structural models and finding parameters such as tortuosity, porosity, and pore sizes.

### 8.2.4 Use of Mass Transfer Coefficients

The problem with equation 8–2 is that the distance over which concentration changes occur *($\delta$)* needs to be known to determine the gradient. Discovering this distance is not an easy task for processes occurring inside process equipment. Moreover, interfacial mass transfer or the transfer between two different phases (solid and liquid) needs to be taken into account. Chemical engineers prefer to study interfacial mass transfer using *mass transfer coefficients* (individual or overall), which multiplied by a measurable driving force give the rate of mass transfer. Thus, the distance problem becomes hidden in the coefficient, which also implicitly contains the diffusivity. A practical expression for the rate of solute extraction takes this form:

$$\textit{Rate} = \textit{Mass transfer coefficient} \times \textit{driving force}$$

**Equation 8–13**

The driving force in extraction should be the difference between the chemical activity of the solute inside the solid and that in the bulk of the solution. For practical reasons, the rate is expressed as a mass transfer coefficient based on concentration multiplied by the difference between an external solute concentration ($c_{out}$) and a solute concentration in the interior of the solid ($c_{in}$) multiplied by a mass transfer coefficient based on the concentration. For this difference to be meaningful, it must be expressed in the same base; thus $c_{in}$ is usually taken as that concentration of the solute in the liquid phase that would be in equilibrium with the concentration inside the solid $c^*$. The driving force for extraction then becomes $(c^* - c_s)$, where $c_s$ could be measured in the bulk of the solution, and equation 8–13 takes this form:

$$N_1 = K_c(c^* - c_s)$$

**Equation 8–14**

This simple equation states that the rate of extraction of solute $N_1$ depends on the difference in a thermodynamic variable (expressed as a concentration) and a global mass transfer coefficient $K_c$ that includes all physical and microstructural parameters of the process. When the interfacial area for the transfer ($a$) is unknown, the coefficient takes the form $K_ca$. If each phase is taken separately, individual mass transfer coefficients $k_i$ can be defined for transport between the interface and the bulk of the respective phase (see King, 1980, or Cussler, 1997, for details).

If a solid is modeled as series of structures (e.g., a plant cell with protoplasm, plasmalemma, cell walls, etc.), the observed global mass transfer coefficient $K_c$ may be related to individual mass transfer coefficients inside each of the structures ($k_{ci}$) as a sum of resistances in series:

$$\frac{1}{K_c} = \sum \frac{1}{k_{ci}}$$

**Equation 8–15**

In this case, a rate-limiting step having the highest resistance (or the slowest flow) may be identified. The relevant structure then controls the mass transfer process, and efforts should be made to increase its rate within this phase. This approach can also be used when studying extraction as a series of sequential steps, as discussed in Section 8.1.3.

Theoretically, there is a relationship between $D$ and $K_c$ (e.g., $K_c$ is proportional to $D/\delta$, where $\delta$ is a thin film over which diffusion occurs), and, as previously concluded, study of the influence of microstructure on extraction rate is reduced to the analysis of its effects on the mass transfer coefficient.

We have not made much progress by introducing an ambiguous coefficient instead of the diffusivity, but at least the simpler expression for interfacial mass transfer velocity ($k_cL$) reminds us of chemical kinetics. Another advantage is that mass transfer coefficients are often correlated with dimensionless numbers, thus allowing predictions to be made for other experimental conditions. Dimensionless numbers are ratios of transport parameters and have physical meaning (e.g., Fick's dimensionless number, discussed in Section 8.2.1). For example, the most relevant of these numbers for solid-liquid extraction is the Sherwood number ($N_{Sh}$), which is the ratio of mass transfer velocity ($k_cL$) to diffusion velocity ($D$). A large $N_{Sh}$ in solvent extraction means that the controlling step is diffusion inside the solid, and reduction of particle size or restructuring into a porous material may prove effective in speeding extraction. $N_{Sh}$ is in turn related to several other dimensionless numbers, such as the Schmidt number, which represents flow effects (momentum transfer):

$$N_{Sh} = \frac{k_c L}{D} = f\left(\frac{\nu}{D}\right)^{1/3} \qquad \textbf{Equation 8–16}$$

where $\nu$ is the kinematic viscosity. See Cussler (1997) for a detailed discussion of dimensionless numbers in mass transfer.

### 8.2.5 Plant Cells and Extraction

Considerable confusion exists in the food-processing literature when reference is made to cellular components or organelles. This section describes the basic structural components of plant cells relevant to food extraction and also introduces a model of a parenchyma cell (Figure 8–3).

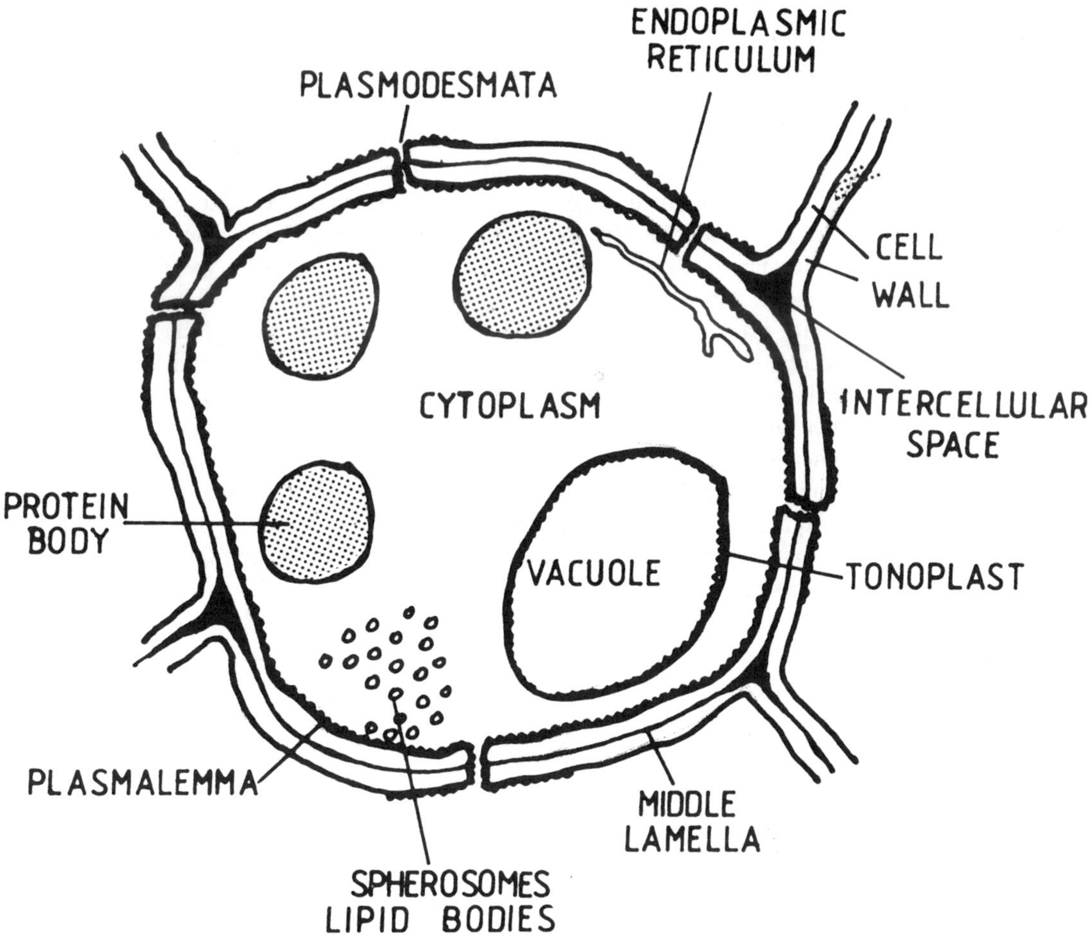


**Figure 8–3** Model of a parenchyma plant cell and its main structural components.

(See Section 4.6 for an account of the characteristics of the major substructures of cells.)

Membranes are composite structures formed by a phospholipid double layer infiltrated by protein; their thickness is on the order of 10 nm. They control the transport of water and solutes between the organelles that they surround. In plant cells, permeability is rapidly lost when membranes are denatured or become detached from the cytoplasm (Kramer, 1983). Membranes are discussed in Sections 4.6.2 and 4.8.

Characteristic of plant cells is the existence of a strong cell wall that limits expansion of the cytoplasm, resulting in turgor pressure. Cell walls are composite microstructures of cellulose microfibrils embedded in a matrix of polysaccharides and some protein (Aspinall, 1981). The wall thickness of a carrot cell is only 0.1 $\mu$m and can withstand a bursting pressure of 30 atm or a tensile stress of 66,000 psi (Carpita, 1985). Cell walls in lupin seeds are about 1 $\mu$m thick. The middle lamella, the outer layer of the cell wall, is composed mainly of pectic material and cements cells together. The primary wall is a more organized layer consisting of a skeleton of cellulose microfibrills (100–300Å) embedded in a matrix of pectic substances, hemicelluloses, and possibly proteins. A secondary wall lies inside the primary wall in contact with the lumen. It contains lignin associated with the matrix, which limits its plasticity. Cell walls contain numerous enzymes.

Plant cell walls are penetrated by strands of cytoplasm (plasmodesmata) or connections between adjacent plant cells, which create an intercellular continuum—the symplasm. Transport of materials between plant cells occurs across the matrix of the cell walls (*apoplastic transport*, which is slow) and via the plasmodesmata (*symplastic transport*, which is fast) (Lauchli, 1976; Spanswick, 1976). Apparently, the symplasm is the evolutionary response of the plant to overcome the limitations of diffusion through the apoplasm. Apoplastic transport depends on the molecular structure of the cell wall, but aqueous channels through the matrix may allow passage of macromolecules larger than 10 kDa. The importance of both types of transport in food processing will be discussed in connection with osmotic dehydration in Section 9.4.

Plasmodesmata have two main structures: (1) an inner tube (the desmotubule) running from cell to cell and sealed at both ends by the endoplasmic reticulum and (2) an annular tube (the cytoplasmic sleeve) surrounding the desmotubule. Plasmodesmata have been a favorite organelle of study for plant electron microscopists, and several minor subunits that help control the passage of molecules through them have been resolved (Robards & Lucas, 1990). Thus, there are two pathways for flow through plasmodesmata, one via the desmotubule and another through the cytoplasmic sleeve. Reported diameters of plasmodesmata range from 20 to 80 nm, and each plant cell may contain 1,000 to 100,000 of these conduits. Pits occur in the secondary wall where plasmodesmata are present in the underlying primary wall. Carpita, Sabularse, Montezinos, and Delmer (1979) showed with the aid of a phase contrast microscope that plasmodesmata in vegetable tissue easily allow the passage of salts, sugars, and amino acids. There is evidence of passage of large macromolecules, and molecular exclusion limits appear to be 700–900 Da, but viruses have learned how to penetrate plasmodesmata (Robards & Lucas, 1990). Plasmodesmata appear to be dynamic structures that are capable of undergoing considerable morphological modification. Several enzymes are associated with plasmodesmata, and they appear to play a role in forming large pores in cell walls after plasmolysis has set in (see Figure 8–11). This phenomenon should be studied in more detail, as it may provide a means for fast intercellular exchange.

Evidently movement of solutes from the interior of intact cells during extraction occurs through a liquid phase. Plant cells with their many compartments (vacuole, cytoplasm, etc.), surrounding membranes (tonoplast and plasmalemma), and natural transport pathways (intercellular as well as cell-intracellular spaces) provide many possible routes for solute transport, with instantaneous extraction being the result of the various contributions. Eventually, pores may

be reached that convey the solvent-solute mixture out into the extracting medium.

Plant tissue has not been constructed by nature to readily yield its valuable components. Often it has to be destroyed in controlled fashion to release solutes (e.g., by milling or flaking), but undesirable components are also extracted, unless the solvent is highly selective (as is hexane in the case of oil). This places an extra burden on downstream separation.

### 8.2.6 Structural Features of the Solid Matrix: Plant Tissue

As may be clear by now, food materials of plant origin have an intricate microstructure formed by cells, intercellular spaces, capillaries, and pores. There are four major types of mature plant tissues: (1) storage or parenchyma tissue; (2) conducting or vascular tissue, composed of phloem (transport of organic materials) and xylem (transport of water); (3) supporting tissue; and (4) protecting tissue. Only two of these contribute to the microstructure of edible parts of plant foods: parenchyma cells and the conducting tissue that form an intricate network throughout the material.

The solute may be present inter- or intracellularly, as shown in Figure 8–4. In the case of oilseeds, intact cell walls and adhering membranes constitute a major resistance to diffusion. They affect the permeability of solutes so that small molecules pass at a faster rate than larger ones, resulting in a selective transfer. The permeability of cell walls and membranes is increased and selectivity is reduced by heat-induced denaturation. Moreover, cell walls forming most of the supporting tissues and certain types of conducting cells (tracheids, vessel elements) are partially lignified, further reducing the passage of molecules.

During preparation for extraction, the solid is reduced to small particles, since, theoretically, extraction time varies inversely with the square of the characteristic dimension of the solid. As particle size decreases, the ruptured outer cells constitute a larger proportion of the particle volume, and extraction characteristics per unit mass of material change. Size reduction is limited because of pressure drops in the extractor, slow drainage rates, the presence of fines in the extract, flow instability, and entrainment. Keep in mind that extraction is just one of the unit operations in a separation process, and downstream purification depends strongly on how "clean" the extract is obtained.

### 8.2.7 A Little More Engineering

Food engineers design separation processes by contacting materials in a continuous fashion (as in packed columns) or by mounting a series of equilibrium stages. In either case, equilibrium relationships are required for adequate design. While in the first case, equilibrium is never attained within the column, in countercurrent stage separation, incoming flows are mixed until equilibrium is attained (at least theoretically) at each stage (the so-called "ideal stage"). Then, the resulting streams leave the stage (in equilibrium) and are moved forward or backward depending on whether they have been enriched or depleted. As usual, engineers utilize a stage efficiency factor to compensate for nonidealities during extraction (e.g., partial equilibration between the solution and the solute).

In leaching, three components are present: (1) the inert (I) or insoluble solids forming the matrix; (2) the extractable species, which for the sake of simplicity will be referred to as "the solute" (A); and (3) the solvent (S). To understand this system, think of a tea bag from which tea extract (A) is leached into hot water (S), leaving the wet spent solids that contain the insoluble components of the tea leaf (I).

Equilibrium conditions are normally represented in a right-triangular diagram that permits a choice of scale for the abscissa and the ordinate. In Figure 8–5, the mass fraction of solvent (separating agent) is represented in the $y$-axis while the mass fraction of solute is shown in the $x$-axis. This graph is a modification of the Ponchon-Savarit diagram used in distillation calculations. The hypotenuse of the triangle is the locus of all solutions (solvent + solute), which may show a saturation point (R). Any point inside the hatched area (e.g., point O) separates into a saturated solution (R)

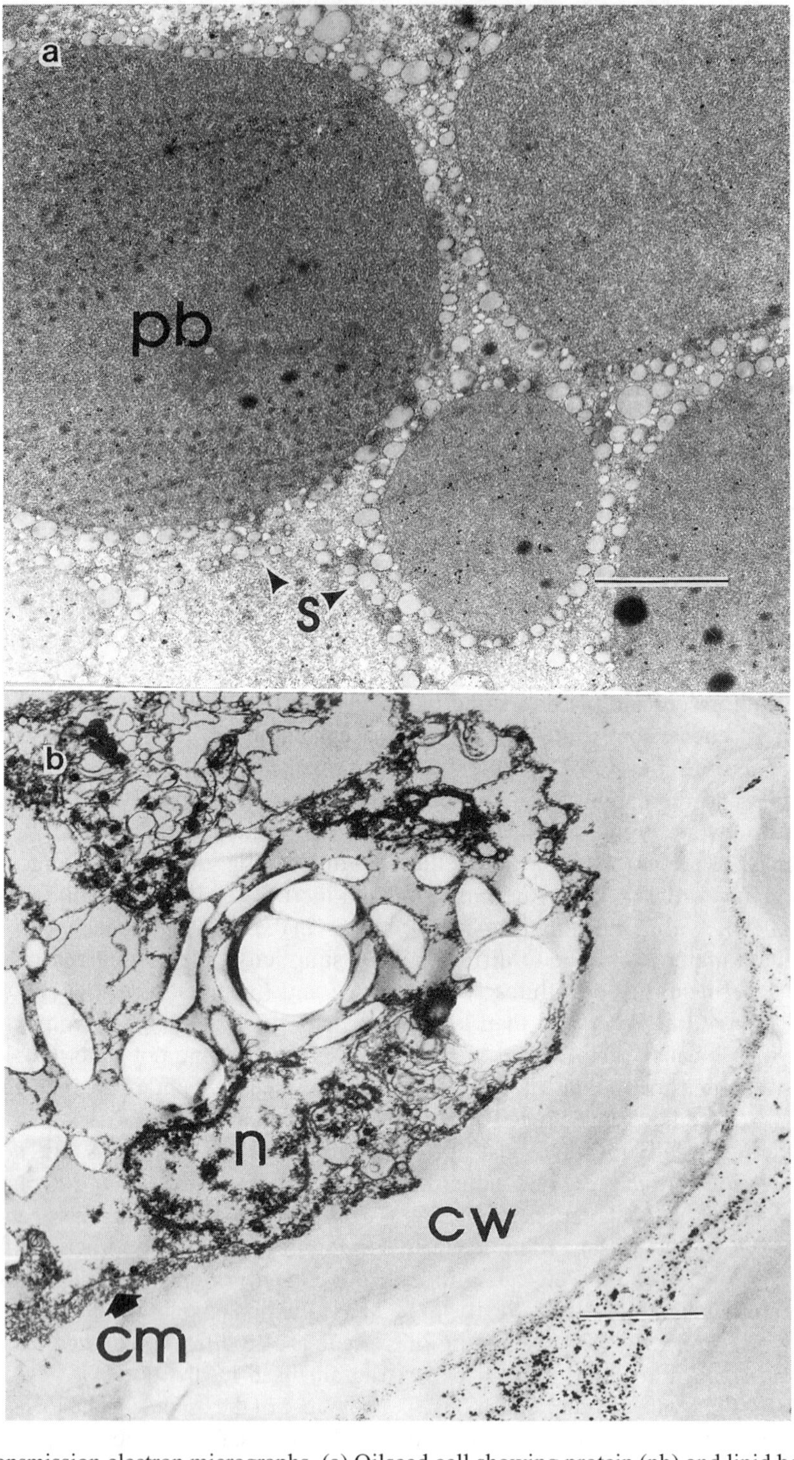


**Figure 8–4** Transmission electron micrographs. (a) Oilseed cell showing protein (pb) and lipid bodies or spherosomes (s). Marker = 1 $\mu$m. (b) Thick cell wall (cw) in an algae that contains the polysaccharide agar; cm = cell membrane, n = nucleus. Marker = 10 $\mu$m.

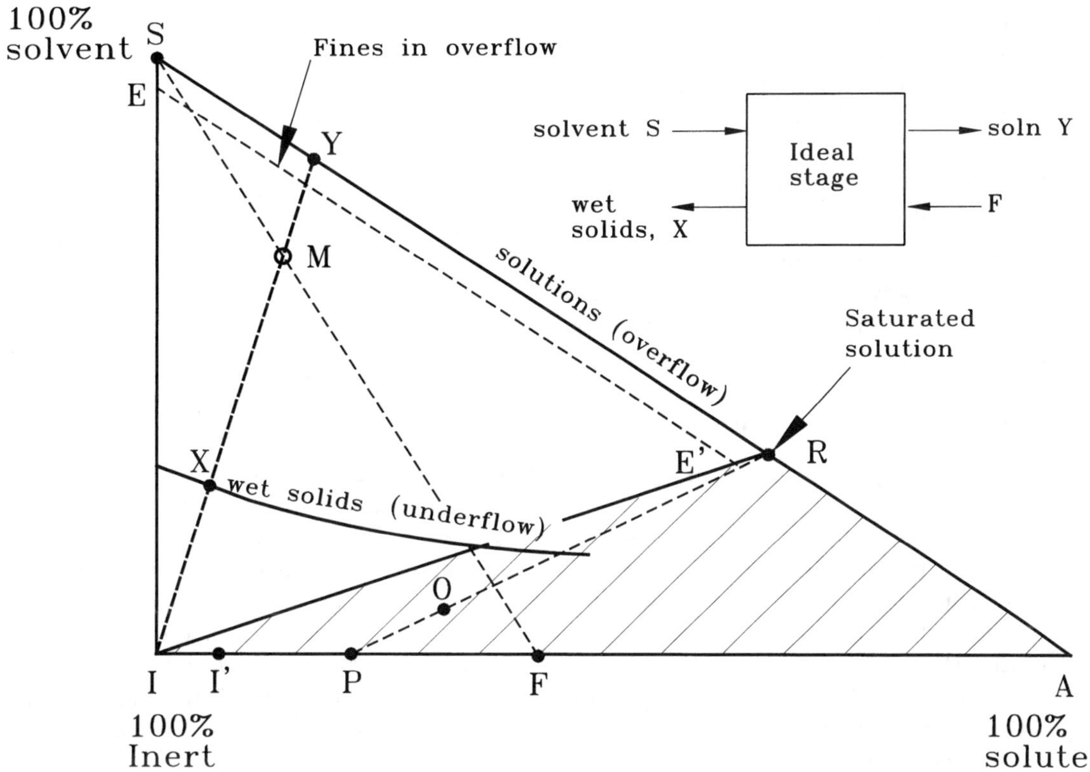


**Figure 8–5** Ternary diagram illustrating an ideal stage for solvent extraction. A solid (F) containing solute A is mixed with pure solvent (S), forming a fictitious mixture (M). The discharge streams are a solution (soln) containing the solute (Y) and wet spent solids (X).

and solid-containing solute (P). The origin represents the pure inert (I), such as the matrix holding the solute in the solid.

If we wanted to extract a solid F (mixture of I and A) with pure solvent S in a single contact stage (as depicted schematically in the upper part of Figure 8–5), the line FS would be the locus of all possible combinations of F and S, and their relative proportions would be given by the lever rule. In Figure 8–5, one part of F is mixed with 2 parts of S (verify using the lever rule) to give point M. After solute A has been extracted, two streams leave this contact stage: one that is the solution containing the extract (or overflow, with composition given by point Y) and another one (X) that represents the wet solids (inert + solution) or underflow. An ideal stage assumes that (1) all the solute comes into solution and (2) the solution wetting the spent solids has the same composition as the overflow (Y). Thus, the locus of wet solids is the underflow line.

How does the solid and its microstructure intervene in this otherwise "ideal" situation?

- First, all the solute may not be released from the solid (F) during the allowed contact time with the solvent. The causes may include low diffusivity, a slow solubilization rate, and poor solvent properties, among others. So some solute will remain occluded, and the disengaging solution will be in contact with a dry solid having a composition I′ rather than I.
- Disengagement of the solution from the extracted solids may be slowed by a large inter-

nal porosity and holdup or high solution viscosity. This is why you press a tea bag against the spoon before removing it from the cup.

- During extraction, fines may be produced by attrition if the wet solids are softened. As these fines leave with the solution, the locus for solutions is moved inside the triangle (line EE′).

There are many other nonidealities we may examine (e.g., the solvent becomes saturated with solute), but this is as much engineering as we need in order to understand solid-liquid extraction.

## 8.3 THE EXTRACTION PROCESS

### 8.3.1 Diffusion through the Solid Matrix

It should be evident from the previous analysis that microstructure influences mass transfer by molecular diffusion through its effect on the diffusion coefficient. Solute diffusivities can be defined in the liquid phase ($D_L$) or in the wet solid ($D_S$), and expressions for Fick's laws can be written accordingly. Differences between $D_L$ and $D_S$ can be attributed to factors such as membrane resistance, complexity of the diffusion path, sorption of the solute by the inert solid, and others discussed in Section 8.2.6. If all these factors were accounted for, $D_S$ could be related to $D_L$ by an expression of the form

$$D_S = F_m \times D_L \qquad \textbf{Equation 8–17}$$

where $F_m$ is a correction factor that includes all phenomena related to microstructure. Schwartzberg and Chao (1982) reviewed extensively the subject of solute diffusivities in foods. $D_L$ values for various food solutes at infinite dilution in water at 25°C vary from $5.4 \times 10^{-6}$ cm$^2$/s for sucrose to $1–7 \times 10^{-7}$ cm$^2$/s for some proteins. Selected $D_S$ values for cases in which microstructural effects appear important are presented in Table 8–2. The extraction experiments from which these data were obtained do not lead to easy determination of a single, constant diffusion coefficient. Typical curves for hexane extraction of oil from soybean grits and flakes of different particle sizes, as well as a theoretical curve, are shown in Figure 8–6 (Karnofsky, 1949). The slopes of the curves become less steep as extraction proceeds,

**Table 8–2** Diffusion Coefficients in Foods and Reference Values

| *Food Material* | *Solute* | *Solvent* | *Temperature (°C)* | *$D_{AB} \times 10^6$ cm$^2$/s* |
|---|---|---|---|---|
| Dilute solution[a] | Sucrose | Water | 25 | 5.4 |
| Sugar cane (across grain)[a] | Sucrose | Water | 75 | 5.1 |
| Sugar cane (with grain)[a] | Sucrose | Water | 75 | 3.0 |
| Sugar beets[a] | Sucrose | Water | 24 | 1.6–2.5 |
| Gelatin gel[b] | Sucrose | Water | 5 | 0.1–0.2 |
| Dilute solution[a] | NaCl | Water | 25 | 16.1 |
| Pickled cucumbers[c] | NaCl | Water | 25 | 5.3–11.0 |
| Dilute solution[a] | Lactose | Water | 25 | 4.9 |
| Small curds[d] | Lactose | Water | 25 | 3.0 |
| Peanut slices[e] | Oil | Hexane | 25 | 0.007 |
| Tungseed slices[f] | Oil | Hexane | 30 | 0.006 |
| Dry solid matrix[g] | Glyceride | — | 50 | $5 \times 10^{-4}$ |

[a]Schwartzberg and Chao, 1983.
[b]Geankoplis, 1978.
[c]Bomben, Durkee, Lowe, and Secor, 1974.
[d]Bressan, Carroad, Merson, and Dunckley, 1981.
[e]Fan, Morris, and Wakeham, 1948.
[f]Krasuk, Lombardi, and Ostrovsky, 1967.
[g]Naessens, Bresseleers, and Tobback, 1982.

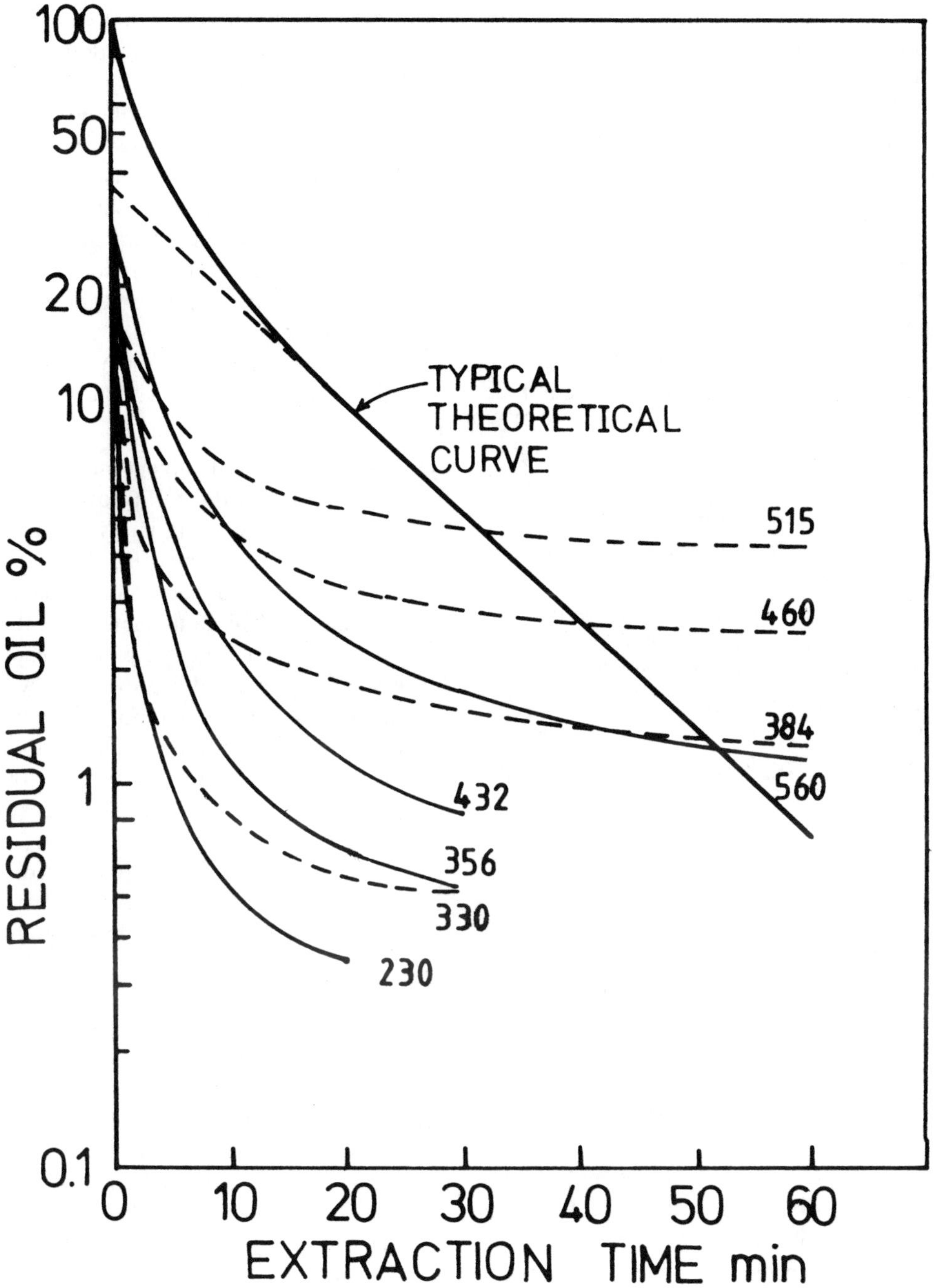


**Figure 8–6** Oil extraction curves of soybean flakes (– – –) and grits (—) of different particle size by percolation with hexane and theoretical behavior. Numbers on curves represent particle size or thickness in $\mu$m. *Source*: adapted from Karnofsky, 1949.

entailing a decrease in the value of $D_S$, so that usually an initial as well as a final diffusion coefficient have to be determined. The initial $D_S$ may be influenced by the previously mentioned release of surface solute (washing), and thus it is not truly representative of the diffusion of solute inside the solid. Consequently, the data on solid diffusivities listed in Table 8–2 should be analyzed with caution, always referring to the original work if further conclusions are to be drawn.

$D_L$ values from the literature and diffusivities calculated for extraction of the same component may be used to determine the order of magnitude of the correction factor $F_m$ in equation 8–17. The $D_L$ of caffeine in water is $6.9 \times 10^{-6}$ cm²/s, whereas the diffusivity of coffee solubles during extraction from grinds is on the order of $1.1 \times 10^{-6}$ cm²/s (Voilley & Simatos, 1980). In sucrose extraction from beets, the calculated diffusivities are one-half to one-third the $D_L$ at the same temperature. Diffusivities of pure linoleic and oleic acid in hexane are $3.6 \times 10^{-7}$ and $2.6 \times 10^{-7}$ cm²/s, respectively, while inside the seed they are reduced to $2.6–6.2 \times 10^{-8}$ cm²/s (Anderson, Osredkar, & Zagar, 1977). Hence, the correction factor in equation 8–17 appears to be 0.9–0.1 for small solutes, with higher values occurring when membranes have been denatured, as in sugar beet processing.

For macromolecules, $F_m$ is obviously smaller. In fact, Schwartzberg and Chao (1982) report that solutes exceeding a molecular weight of 2,000 cannot diffuse into intact cells of coffee grounds. The reduction of the diffusion coefficient for zein (corn protein) in dilute solution and inside the endosperm is about 1,000 times (Russel & Tsao, 1982).

Sorption of solvent and/or solutes by the solid also retards diffusion. Microstructural entrapment seems to play a role at least as important as the physicochemical sorption. Lignocellulosic materials are known to sorb water no longer available as solvent, but larger quantities of moisture are absorbed when the microstructure is intact than when it is ground into a fine powder (Bock & Ohm, 1983). Sorption is particularly critical when organic solvents must be removed from spent solids after extraction, in an operation known as desolventizing. Residual hexane left in rapeseed meals concentrates preferentially in the hulls, either dissolved in the residual oil entrapped inside thick-walled cells (Schneider & Rutte, 1984) or simply adsorbed. In the latter case, internal diffusion coefficients for hexane during adsorption into the solid matrix are very small ($\approx 10^{-10}$ cm²/s) and increase with hexane content, probably due to swelling of the structure (Roques, Naiha, & Briffaud, 1984).

Other problems related to the microstructure of the solid matrix that must be addressed in extraction of foods include these:

- The penetration of solvent into the solid matrix depends on the ability of the solvent to wet the solid. Wetting is favored by a large adhesion tension between solid and solvent and a low hydrodynamic resistance in internal passages. This latter factor depends on the mean width and tortuosity of the capillaries; sometimes the solid is previously swelled to facilitate solvent penetration and extraction.
- Changes occur in the effective particle size upon contact with the solvent, as, for example, swelling of tea leaves in water (Spiro & Sidiqque, 1981).
- Solvent has easier access to cells adjacent to pores and capillaries than others situated at the same relative position but surrounded by other cells, as shown in Figure 8–1 (del Valle & Aguilera, 1989).
- Certain cells or parts of them are impermeable by the solute. Milling of grits in beer manufacture breaks the outer cellulose protective layers and exposes more starch to the conversion process. Similarly, the epidermal layer of cells in a peanut kernel retards the extraction of oil by petroleum solvents and makes milling a necessary step (Fan, Morris, & Wakeham, 1948). The cuticula of the algae *Gracilaria* impedes extraction of agar, making enzymatic treatment or particle reduction a must (San Martin, Aguilera, & Hohlberg, 1988).
- Structural breakdown and disintegration of the solid matrix causes a decrease in the rate

of percolation or flow of solvent in deep beds, as when thin, friable flakes are used in oil extraction.

### 8.3.2 Solubilization and/or Breakdown

Occasionally, the solute is present in the form of an insoluble precursor that must be hydrolyzed before extraction can be carried out. The solubilization process is normally very rapid, but there are cases where this step is important, such as in the conversion of hemicellulose into oligosaccharides in coffee beans (Sivetz & Foote, 1963), the hydrolysis of collagen to gelatin (Idson & Braswell, 1957), the release of tannins from complexes during extraction of tea (Sanderson & Coggon, 1977), the conversion of prochymosin to chymosin (rennin) (Zayas, 1986), the hydrolysis of linamarin in cassava followed by extraction of the hydrocyanic acid released (Liener, 1977), and the release of agar bound to cellulose fibers (Rutman & Aguilera, 1985).

The solubles produced are often subject to secondary transformations or degradation. Agar is extracted commercially from a number of algae species, mainly of the genera *Gelidium* and *Gracilaria* and is a mixture of at least two polysaccharides: agarose, a neutral linear polymer, and agaropectin, a branched sulfated fraction. The presence of 3,6 anhydro derivatives induce gelling; where they are replaced by galactose 6-sulfate, the gelling power is considerably less, and gelation is completely suppressed by the presence of D-galactose 2,6-disulfate (Percival & McDowell, 1981). Treatment of the algae with NaOH simultaneously induces formation of 3,6-anhydro-L-galactose, which increases the structural stability of gels, and polymer chain breakdown, which reduces the gelling power (Nishinari & Watase, 1983). Similarly, pectins are solubilized by boiling with dilute acid, which converts protopectin to pectin but at the same time degrades the pectin molecule to some extent (Francis & Bell, 1975; Potter, 1966). In practice, this stage of the process is often a compromise to achieve maximum yield and minimum degradation. Last but not least, separation of a component from the cellular arrangement may hasten reactions that otherwise would be physiologically controlled (e.g., oxidative and enzymatic reactions).

Solid-liquid extraction of various food materials is controlled by internal diffusion except in the case of very small particles, poor agitation, or the presence of a skin (Cogan, Yaron, Berk, & Mizrahi, 1967; Leniger & Beverloo, 1975; Moo-Young & Andre, 1980; Spiro & Page, 1984). The extent of control between external and internal diffusion is indicated by the Sherwood number $N_{Sh} = k_c L/D$, where $k_c$ is the liquid-phase mass transfer coefficient, $L$ is a characteristic dimension of the solid (e.g., particle size), and $D$ is the internal diffusion coefficient (see Section 8.2.3). If $N_{Sh} > 200$, internal control can be safely assumed (Schwartzberg & Chao, 1982).

## 8.4 EXTRACTION OF FOOD MATERIALS

### 8.4.1 Solvent Extraction of Oilseeds

Oilseeds contain oil, protein, and sometimes starch in neatly packed microstructures inside cells 20–50 $\mu$m in diameter. Protein bodies (or aleurone grains) are relatively large, nearly spherical particles 2–10 $\mu$m in diameter that accumulate storage protein. Starch granules are larger than protein bodies, and although common in cereals and legumes, they are, among major oilseeds, significant only in peanuts. Oil is located in smaller units called spherosomes or oil bodies, 0.2–0.5 $\mu$m in diameter. Apparently, spherosomes are the precursors of oil bodies, which exist in oil-bearing plants simply as sites of oil storage (Smith, 1979). Oil bodies in situ are bound by half-unit membranes of about 3 nm thickness and composed in part of protein (15%) and phospholipids (Bair & Synder, 1980a; Yatsu & Jacks, 1972). Intact soybean cotyledons have an ordered system of central and lateral channels (Schneider, 1978). Microstructural aspects of different oilseeds are covered by Vaughan (1970), Smith (1979), and Arnott and Webb (1983). Detailed microscopy studies of individual oilseeds are available for soybeans (Wolf & Baker, 1972), peanuts (Vix, Gardner, Lembou, & Rollins, 1972), rapeseed (Stanley, Gill, deMan, & Tung, 1976),

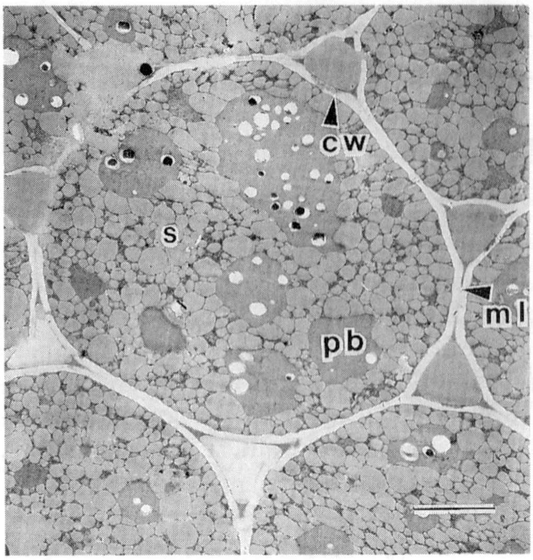


**Figure 8–7** Transmission electron micrograph of a cottonseed cell showing protein bodies (pb), spherosomes (s), cell wall (cw), and middle lamella (ml). Marker = 10 $\mu$m.

and cottonseed (Yatsu, 1965). Figure 8–7 shows important microstructural characteristics of a cottonseed cell as revealed by transmission electron microscopy.

Solvent extraction of oilseeds was developed in Europe in the 1920s. Since then it has been considered important to achieve residual oil contents in the meal of 1% or less, as dictated by economic constraints. It is not surprising, then, that the first researchers in studying the effect of microstructure on solvent extraction were those interested in oil removal from oil-bearing materials.

Technological aspects of the process are comprehensibly treated in books (Bernardini, 1984; Swern, 1976), updated in special issues of the *Journal of the American Oil Chemists' Society* (e.g., Vol. 53, No. 6, 1976; Vol. 54, No. 6, 1977; and Vol. 60, No. 2, 1983), and reviewed in short articles (e.g., Becker, 1978; Pohl & Mieth, 1984).

Oil-bearing materials for extraction may be divided into low-oil materials (18–22%, dry basis) or high-oil materials (>22%). As a general rule, those belonging to the first group, such as soybeans (18–22%), grapeseed, rice bran, and corn germ, are subject only to solvent extraction. Prepared high-oil materials like cottonseed (30–35%), rapeseed (40–45%), peanut (45–50%), sunflower (50–55%), palm kernel (50–55%), and copra (65–68%) are prepressed and later solvent extracted. Main processing steps in preparation and extraction of oil from soybeans are shown in Figure 8–8. Processing of oilseeds is initiated by preparation of the seeds where anatomical and microstructural aspects of seeds play an important role (Stein & Glaser, 1976). Cleaning, cracking, dehulling or decorticating, and flaking are common operations for most oilseeds (Galloway, 1976). Dehulling is practiced to lower the fiber content of the residual meal and to avoid possible losses of oil entrapped in the fiber. Soybeans are easily dehulled, since the internal compressed cells of the seed coat are detached from the outer cotyledon cells, as evident in the scanning electron micrographs of Wolf and Baker (1972). Scanning electron micrographs of the seed coats of sunflower and canola seeds and soybeans are presented in Figure 9–8. Soybeans are normally cracked in corrugated rolls into 6 to 8 cotyledon pieces that are then separated by aspiration of the hulls through exploiting differences in size and density. Cottonseed is decorticated with a bar huller, consisting of a rotating cylinder with protruding knives that cut the seeds into pieces. Rapeseed (canola) and sunflower are more difficult to decorticate and require special equipment. The anatomical characteristics of rapeseed demand high pressures to crack the hulls while avoiding disintegration of the meat particles and impregnation of the hulls with oil (Schneider, 1979a, 1979b).

Oil in whole seeds or large pieces is not extracted to a major extent by nonpolar solvents in which the lipid phase is otherwise completely soluble, proving the impermeability of at least some intact cells (Othmer & Agarwal, 1955). As long known, cells must be ruptured and membranes denatured to achieve high permeation rates. Conditioning of cotyledon pieces with steam raises their temperature to 70–75°C, denatures proteins, lowers the viscosity of the oil, inactivates some en-

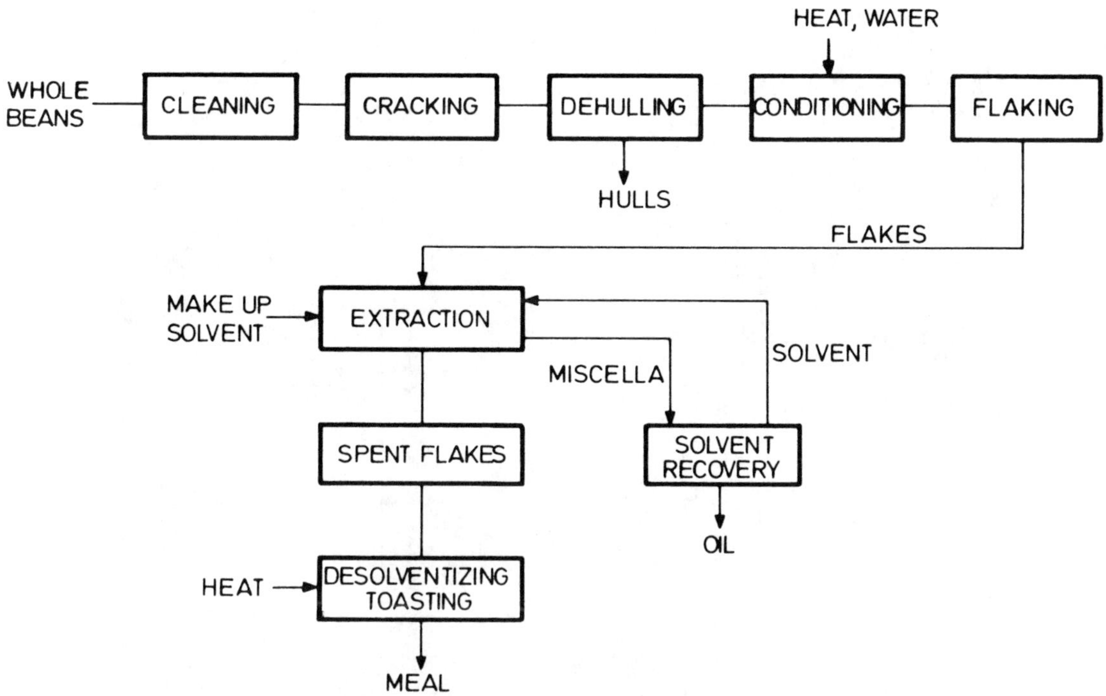


**Figure 8–8** Flow diagram of the oil extraction process for soybean.

zymes, provides the plasticity needed for good flake formation, and probably binds some phosphatides to the meal, thereby decreasing refining losses. Thin flakes (250–500 $\mu$m in thickness) are formed by flattening cotyledon pieces through compression and shear between smooth rolls rotating at differential speeds. Thus, the small characteristic particle dimension necessary for rapid extraction is obtained without the production of troublesome fines, as would be the case if milling was employed. Severe microstructural distortion occurs during flaking to conform to the new geometry and increased surface area. Conformation requires extensive rupture and deformation of the cell contents and separation of the cell wall from the cytoplasm, as shown in Figure 8–9. Deterioration takes place more rapidly after flaking because a larger surface is available for oxidation and the cell compartments have been ruptured, bringing enzymes and substrates in contact. Increase in free fatty acid content by hydrolytic action of lipase should be expected while flakes are left unextracted. High free fatty acid content in intact seeds is a sign of poor storage and mishandling.

Solvents for extraction of foods must have high solvent power, low toxicity, good selectivity, and should be safe to use. Several solvents, pure or in mixtures, have been studied to ascertain their oil extraction capacity, including normal and chlorinated hydrocarbons, alcohols, ketones, and water, among others, but a petroleum naphtha rich in hexane is used almost exclusively (Johnson & Lusas, 1983). Extraction is customarily performed by percolation of hexane through a bed of flakes and rarely by immersion. The equipment used for solid-liquid extraction is reviewed in Schwartzberg (1987) and Dousse (1978).

A typical extraction curve for soybean flakes and grits extracted with hot hexane is presented in Figure 8–6. The bulk of the oil, perhaps up to 80%

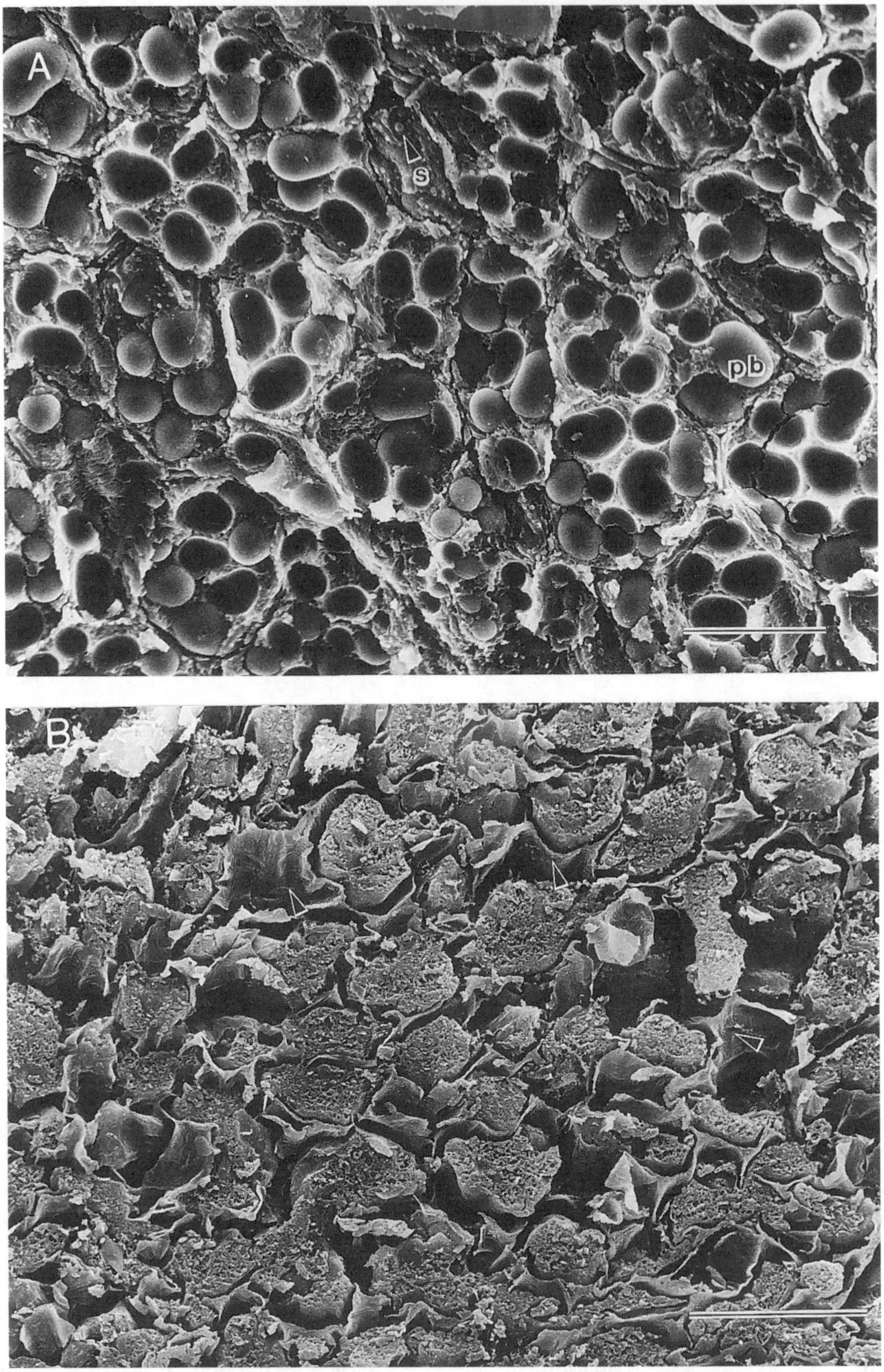


**Figure 8–9** Scanning electron micrographs. (A) Cross section of an intact soybean cotyledon. Protein (pb) and oil bodies (s) are visible. (B) Cross section of a soybean flake. Arrows point at zones where cell walls and cytoplasm have come apart. Complete obliteration of the cellular contents can be appreciated. Markers = 50 $\mu$m.

(depending on the particle size), is extracted rapidly, but the last portions are increasingly difficult to remove. The shape of extraction curves has been explained by the different locations occupied by the oil in the microstructure of the flake. Oil at the surface of the flake is washed rapidly while that next to capillaries, entrapped in the fused cytoplasm or in the interior of the cells, is extracted predominantly by diffusion (Karnofsky, 1949). The washing-diffusion sequence occurs also in thin peanut slices having a great proportion of intact internal cells (Fan et al., 1948). Almost pure washing occurs when seeds are ground with solvent to average particle sizes of 130 μm (equivalent to a few cells). Up to 85% of the oil is removed in a single grinding stage of 1 minute duration and a seed-solvent ratio of 1.66 (Diosady, Rubin, Ting, & Tasso, 1983). Figure 8–10 shows an idealized microstructural model of a flake based on the photomicrograph presented in Figure 8–9 and the possible extraction mechanisms: (1) washing from broken outer cells, (2) diffusion through fused protoplasm, (3) diffusion through cell walls in intact cells, and (4) diffusion into pores or capillaries.

The mixture of oil and hexane is called the miscella. The rate of extraction is independent of the miscella concentration up to 20% oil in hexane (Coats & Karnofsky, 1950). Extraction time for soybeans and peanuts to a specific residual oil level varies directly with flake thickness raised to the 2.2–2.5 power, while for cottonseed and flaxseed the index is 1.5 and 7, respectively (Becker, 1978; Coats & Karnofsky, 1950). All these facts prove that actual extraction of oilseed flakes deviates from the theoretical diffusion model presented previously.

As extraction proceeds, phospholipids and free fatty acids also become solubilized in the miscella (Arnold, Choudhury, & Chang, 1961). A trade-off

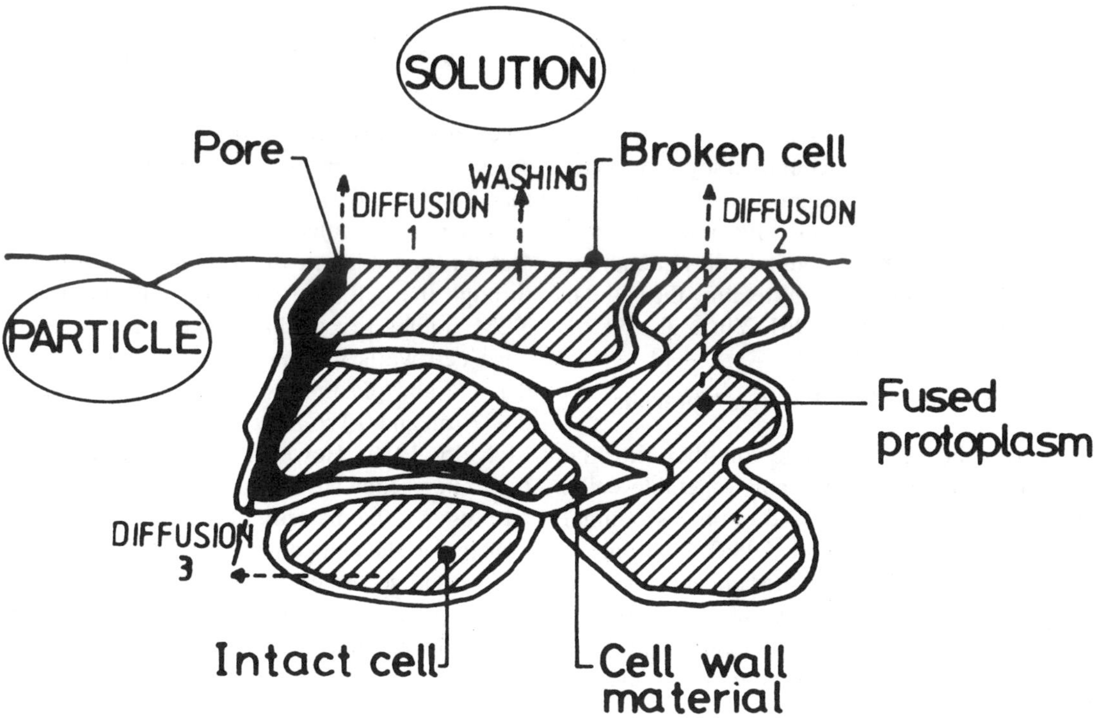


**Figure 8–10** Extraction mechanisms contributing to oil release from an oilseed flake or particle.

exists between the cost of the residual oil left in the meal and the extra cost of refining the crude oil. The marc (spent flakes plus occluded miscella) is separated from the final miscella by gravity or mechanical means (Christiansen, 1983), desolventized with live steam, and sometimes toasted to destroy antinutritional factors. Crude oil is recovered from the miscella by distillation of the solvent. Due to the low selectivity of the hexane extraction process, impurities (waxes, free fatty acids, phospholipids, and coloring matter) must be later separated from triglycerides by several refining operations.

Changes in cellular and subcellular structure during conditioning, cracking, flaking, hexane extraction, and desolventizing and toasting of soybeans have been studied using light, transmission electron, and scanning electron microscopy. Cellular structure is slightly modified by crushing but thoroughly disrupted by flaking. Lipid bodies are removed during hexane extraction and are not observed in spent flakes (Bair & Snyder, 1980b; Yiu, Altosaar, & Fulcher, 1983). Cytoplasmic disruption seems to occur during extraction of thin (300 $\mu$m) cottonseed slices with water-containing solvents but not with apolar ones; however, TEM studies showed no effects on the cell walls (Hensarling, Yatsu, & Jacks, 1970).

Full-fat and low-heated defatted soybean flours present many of the structural features of cotyledons. Toasting of the defatted meal destroys cell walls and agglomerates protein bodies into amorphous masses, completely obliterating the cellular structure (Sipos & Witte, 1961; Yiu et al., 1983).

### 8.4.2 Expression of Oil-Bearing Materials

Other methods are available for liberating the fatty material from animal and vegetable cells. Wet and dry rendering is used to process animal residues and wastes (Field, 1984). Cooking to rupture the cells is accomplished under pressure at 160–180°C for 1.5 to 3 hours. The free fat is separated by skimming or centrifugation, and the residue is pressed to obtain more grease and tallow.

Expression is the process of mechanically separating liquid out of liquid-containing solids using pressure. It is commonly used in many food processes either before or after solvent extraction (Schwartzberg, 1983). The advantages of expression over solvent extraction are that materials pressed out generally have native properties better preserved, end products are free of chemicals, and it is a safer process; the main disadvantage is that yields are seldom higher than 80–90%. Screw pressing has almost completely replaced hydraulic pressing for extraction of oil from tree nuts, olives, jojoba, cocoa, and coffee grounds. Prepressing is used on high-oil vegetable materials to extract most oil (80–86%) prior to solvent extraction of the rest (Khan & Hanna, 1983). Three processing steps are involved in the expression of oilseeds (Bredeson, 1977):

1. rolling of decorticated oilseeds to rupture the greatest number of cells and provide an homogeneous flake
2. cooking the flakes without overcooking to coagulate the protein and rupture the remaining intact oilseed cells
3. pressing effectively so that the capillaries through which the oil is expelled are not sealed promptly by the increased pressure

There is some controversy about the transport mechanism of oil within the material inside the press. A major route appears to be the plasmodesmata (Smith, 1979). It has been postulated, based on laboratory results of mechanical expression, that it is possible to extract up to 85% of the oil through the cell wall pores or plasmodesmata. TEM studies reveal that cashew and rapeseed have cell walls with porosities of 0.093% and 0.171% and plasmodesmata with diameters of 0.087 $\mu$m and 0.126 $\mu$m, respectively (Mrema & McNulty, 1980). However, scanning electron and light microscopy studies of peanuts show that after pressing, the majority of the cell walls appear broken and the cell contents are depleted of lipid bodies and separated from the wall. Light photomicrographs of intact peanuts show "pits" in the cell wall that may be sites of plasmodesmata (Schadel, Walter, & Young, 1983). Similar observations are made after cooking and screw pressing of rapeseed, namely, that protein bodies fuse into large masses while storage lipids coalesce into larger droplets (Yiu et al., 1983). Material microstructure and changes occurring during operation do play a definite role in expression, as emphasized by Schneider and Khoo (1986). Further work is needed to elucidate the relative roles of pore transport and cell disruption in the mechanical pressing of oilseeds.

### 8.4.3 Extraction of Vegetable Proteins

The water extractable proteins of defatted meals are derived from protein bodies that, in the case of soybeans, contain more than 70% of the protein (Saio & Watanabe, 1966; Tombs, 1967). Protein bodies contain 82–83% protein (N × 5.8) and are 2–10 μm in diameter (Wolf, 1970; Wolf & Baker, 1972). Interestingly, the 2S fraction of water extractable proteins that contains the trypsin inhibitors is located in the cytoplasm rather than in the protein bodies (Wolf, 1972).

Refined vegetable protein products can be produced by aqueous solubilization of protein from low heat–treated, defatted meals. The process is similar for the various residues left after oil extraction. Production of soybean protein isolates exploits variations in solubility with pH and/or salts to extract and subsequently precipitate the protein fraction (Smith & Circle, 1972). At about pH 4.5, the globular protein fraction has its point of minimum solubility and aggregates, forming a curd (Lee & Rha, 1978), which is separated from the whey by centrifugation and later spray-dried. When protein and lipids are extracted jointly by aqueous processing, the resulting isolate entraps fat in its microstructure, as shown in TEM photomicrographs (Dieckert & Dieckert, 1982). The microstructures of soybean concentrates (at least 70% protein) and isolates (more than 90% protein) powders present structural features different from those in intact cotyledons, predominantly as a result of the drying method (Smith, 1979; Wolf & Baker, 1972).

Storage proteins in soybeans are globulins with molecular weights of 100,000 to 180,000 that form quaternary structures like the 11S and 15S ultracentrifuge fractions of 350,000 and 600,000 daltons. A puzzling question is, how are protein molecules extracted from the interior of intact cells? Some clues for answering this question may be provided by the way in which protein is transported and accumulated in the mature seed. According to Dieckert and Dieckert (1976), protein in oilseeds is produced in the endoplasmic reticulum and transported by way of the cavities to vacuoles. The membranous structures forming the endoplasmic reticulum have been shown by transmission electron microscopy to be connected to the plasmodesmata. The most plausible answer, then, is that proteins are extracted by way of the symplasm, in which pores become enlarged during processing. Microstructural evidence supporting this hypothesis has been presented by Aguilera (1989) for the aqueous extraction of protein from lupins. Large pores in the interior wall of a lupine cell after protein extraction can be observed in Figure 8–11.

Under the assumption that internal transport is limited during protein extraction, Aguilera and García (1989) studied the effect of microstructural modifications induced by flaking and explosion on the rate and extent of protein extraction from lupins. As the size of dry particles increased from 128 to 1,425 μm, the total protein extracted from untreated (control) particles after 2 hours decreased from close to 100% to slightly over 50% of the theoretical extractable protein, while the largest particles with modified microstructure already released 82–86% of the total protein. Another important finding was that over 90% of the protein solubilized after 2 hours was extracted in the first 40 minutes regardless of the internal structure of the particle. More work is needed to understand the basic mechanism of extraction of large molecules from cellular tissue, and microscopy techniques can be of valuable assistance.

### 8.4.4 Malting and Extraction of Wort

Brewing is a good example of the fruitful application of microscopy to the study of an industrial food process (McCaig, 1984). The brewing process begins with controlled germination of the barley grain at about 40% moisture. Malting induces *de novo* synthesis and transport of α-amylase through the whole endosperm. Examination

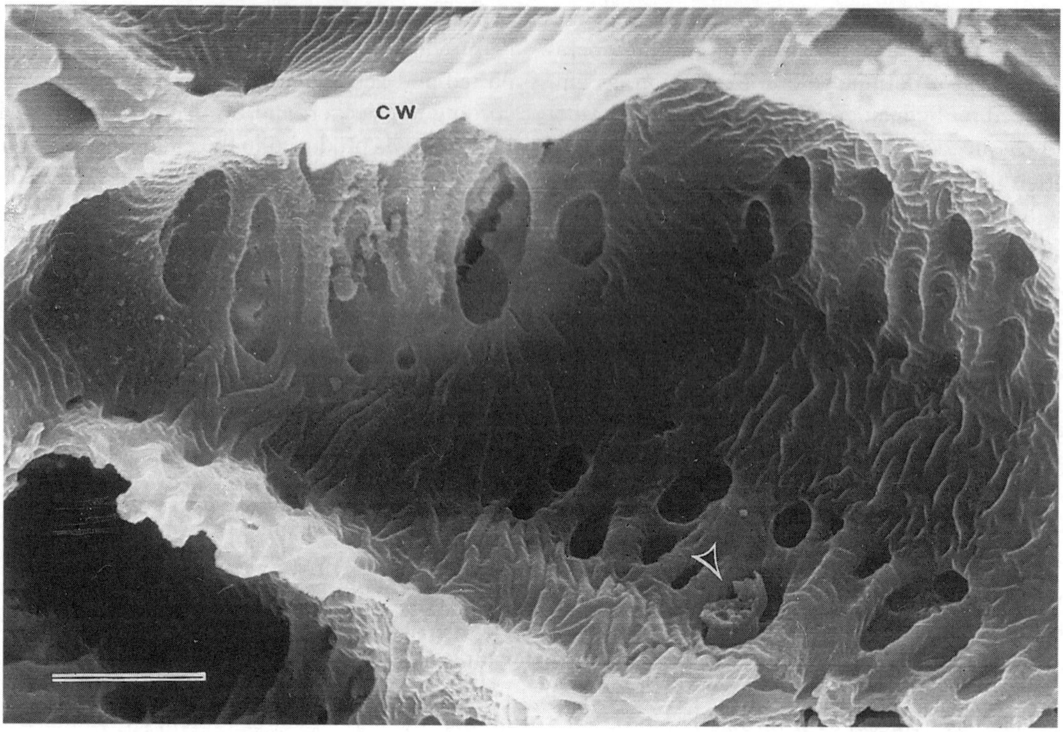


**Figure 8–11** Scanning electron micrograph of the interior of a cell of sweet white lupine after protein extraction. The remnants of the cytoplasm have been removed. The cell wall (cw) presents several pores, possibly the sites of plasmodesmata. Arrow points at a broken piece of cytoplasm protruding from a pore. Marker = 10 $\mu$m.

by scanning electron microscopy shows that the endosperm of malt loses much of the cell wall material present in barley, the protein matrix is dissolved, and starch granules are loosened from the matrix but remain intact. The controversy about the site of enzyme synthesis and the pattern of transport within the grain has been resolved to a great extent with the aid of microscopy. A macrofluorescence microscopy technique has been used to observe degradation of major endosperm cell wall components and follow the enzyme transport from the scutellum of the embryo throughout the starchy endosperm (Gibbons, 1981). Analytical techniques and scanning electron microscopy have been combined to show that breakdown of starch granules, $\beta$-D-glucans, pentosans, and proteins of the endosperm of germinated barley is brought about by enzymes released from the aleurone layer (Palmer, Gernah, McKernan, Nimmo, & Laycock, 1985). Fredtzdorff, Pomeranz, and Betchtel (1981) present evidence gained by scanning electron and fluorescence microscopy that hydrolytic enzymes may be formed initially in the scutellum but later diffuse into the entire endosperm, also via the aleurone layer.

Mashing is the process of extracting grinds (ground malt and adjuncts) with water at a temperature of 63–68°C for 1 to 2 hours, providing the brewer's extract or wort. The first application of the scanning electron microscope in brewing was in the study of the solubilization of starch granules during mashing. Degradation of malt starch apparently results from enzymic attack both outside and inside the granules, particularly in small granules, which are slower to gelatinize during mashing (Palmer, 1972).

The sweet wort is run off in a process known as *lautering*, a combination of filtration and extraction. A trade-off exists, since the rate of filtration is proportional to the square of the particle size and decreases with bed depth, while the leaching efficiency increases with bed depth but is inversely proportional to the square of the particle size. Wilkin (1983) discusses various processes for recovery of the fermentable solute from the grain particles and their efficiency (plant yield versus laboratory or theoretical yield), which in all cases is higher than 97%.

### 8.4.5 Extraction of Sugar from Beets

In the case of sugar beets, the aim of the extraction process is to extract the maximum amount of sucrose and the minimum amount of impurities. The dominant feature of the beet root architecture is the alternating rings of vascular or conducting tissue and storage parenchyma (60 to 80% of the beet tissue). Thin-walled parenchyma cells store sucrose solution almost exclusively, leaving intercellular spaces filled with liquid (Merva, 1975). Several microscopy techniques have been used to characterize the cell wall structure of sugar beets at the point of industrial maturity (Steinert, Galling, & Buttersack, 1990).

Beets are cut into long, thin slices or cossettes with triangular or v-shaped cross sections, leaving a great proportion of internal undamaged cell walls permeable to sucrose but not to macromolecules. The particle size is limited by destruction of thin cossettes during movement in the extractor and the release of fine pulp that plugs screens. To permit the extraction of the sugar, the cytoplasm and the plasmalemma are made permeable to the cell juice by a process known as plasmolysis or denaturation, induced by heating at temperatures above 50–60°C (McGinnis, 1971). Denaturation may occur during rise of the temperature in the diffuser or in a separate operation. The permeability of the cell wall increases and proteins coagulate, favoring the selectivity of the extraction process (Genie, 1982). A microstructural view of sugar beets cell walls is presented in Figure 8–12.

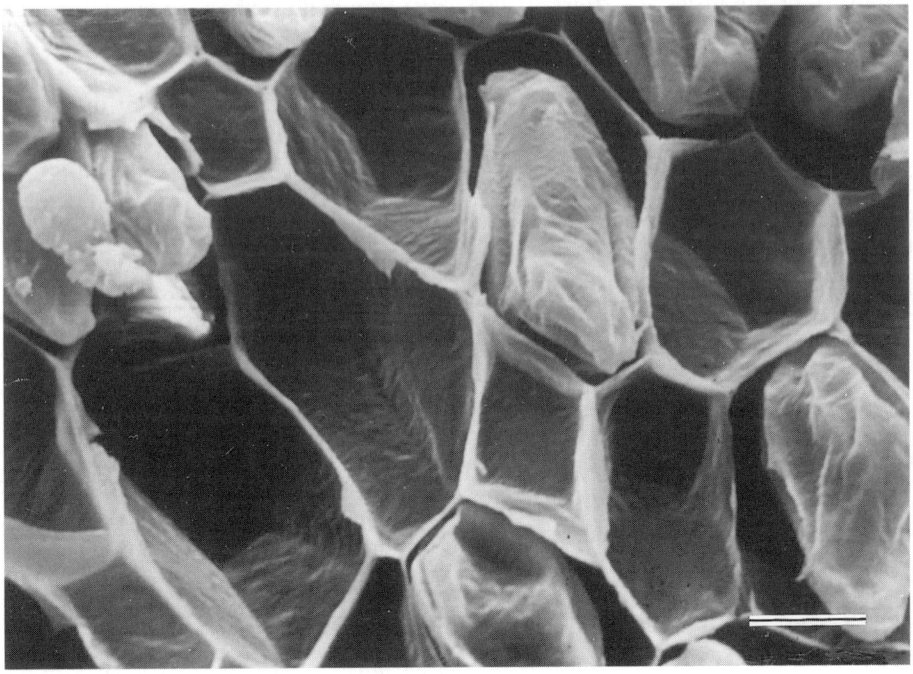

**Figure 8–12** Scanning electron micrograph of cells and cell walls of raw sugar beet. Cytoplasms and the internal smooth surface of intact cell walls can be observed. Marker = 20 $\mu$m.

Several transport mechanisms have been suggested during hot water extraction of sucrose. An obvious one is dialysis from cells to the intercellular liquid through the cell walls, followed by free diffusion through the tubular vascular system (Soddu & Gioia, 1979). The microstructure of intact sugar beet cell walls present round pores 0.3–1.0 $\mu$m in diameter, possibly sites of plasmodesmata. Upon thermal denaturation (75°C), they become oval and expand to a size much larger than sucrose and protein molecules. Their role during sugar extraction seemingly is to form a microscopic porous system in the cell wall for convective transport of liquid (Shokrani & Delavier, 1978). This view is in accord with the mechanism postulated for protein extraction by Aguilera (1989) but contradicts the proposed "osmotic pump" theory of extraction, which holds that sucrose-laden liquid moves owing to hydrostatic pressure caused by osmosis of water into the cell rather than to diffusion (Rathje, 1970). In view of the good quantitative agreement between extraction data for sucrose and the diffusion equation (Schwartzberg & Chao, 1982), it may be possible that the osmotic pressure inside unextracted cells promotes enlargement of the pores but that the controlling mechanism is bulk diffusion.

Microscopy is viewed as an irreplaceable complement to conventional analytical methods used in the factory laboratories of sugar beet refineries (Cleriot, 1994). Light microscopy provides information on the state of cellular tissues of beet and cossets and their quality for processing. It gives a way of rapidly assessing the microbial load of juices in the diffuser, after purification, and in stored sugars. Also, it permits examination of particle size and shape—those of calcium carbonate and of sucrose crystals as well as deposits of extraneous matter during manufacture. The sucrose crystal has more than 15 simple forms, and its morphology, which is directly related to its growth kinetics, depends on many factors, including the presence of impurities (such as potassium chloride) and the blocking of growth in some faces by raffinose (Mantovani, 1991).

### 8.4.6 Extraction of Fruit Juices

Fruit extraction aims at separating the fruit components soluble or dispersible in water, commonly termed "juice," from the fibrous matrix or pomace. Some fruits contain 95–97% juice and only 3–5% insoluble material. The juice components are mainly located in the vacuoles but some may be associated with the cytoplasm. For extraction through intact tissue, they must diffuse through the plasmalemma, a step that controls the mass transfer rate (Emch, 1980). Normally, fruits are pressed or squeezed to release the juice from the cells. Mechanical expression of fruits and vegetables is reviewed by Cantarelli and Riva (1983).

There are two ways in which extraction processes may be used in the production of fruit juices: (1) adequately prepared fruit slices may be extracted directly, and (2) residues from conventional pressing operations, containing up to 20% of the original juice, may be subjected to secondary extraction.

In the first case, the fruit must be cut into neat slices, avoiding grinding or pulping. In practice, slicers that cut fruit into slices of 3 mm thickness with undulated surfaces are preferred because they expose more surface area to the water (Possmann, 1981). The fruit slices are then heated to overcome the semipermeability of the cell membrane. Optimal temperature-time combinations for plasmolysis in apples vary between 55–70°C and 7–10 minutes (Binkley & Wiley, 1978; Emch, 1980). Slices are then fed into an extractor in which the solid and liquid phase travel countercurrently in approximately a 1:1 ratio (Osterberg & Smith Sorensen, 1981). In the DDS extractor, two interlocking screw conveyors within a steam-heated jacket transport the solids through an incline of 6° in 60–90 minutes. The wet pomace is pressed and the press water returned to the extractor. Yields of juice by this extraction procedure are 92–93%, compared with about 80% for pressing operations. Tests performed using apple as raw material showed that extracted juices have a lower content of total acid, substantially higher polyphenol content, and about 10% more mineral components than pressed juices (Possmann, 1981).

### 8.4.7 Percolation of Coffee

Roasted coffee beans have cells of 20 $\mu$m average diameter and a porous structure that is easily apparent in the SEM (Figure 8–13). Length-to-di-

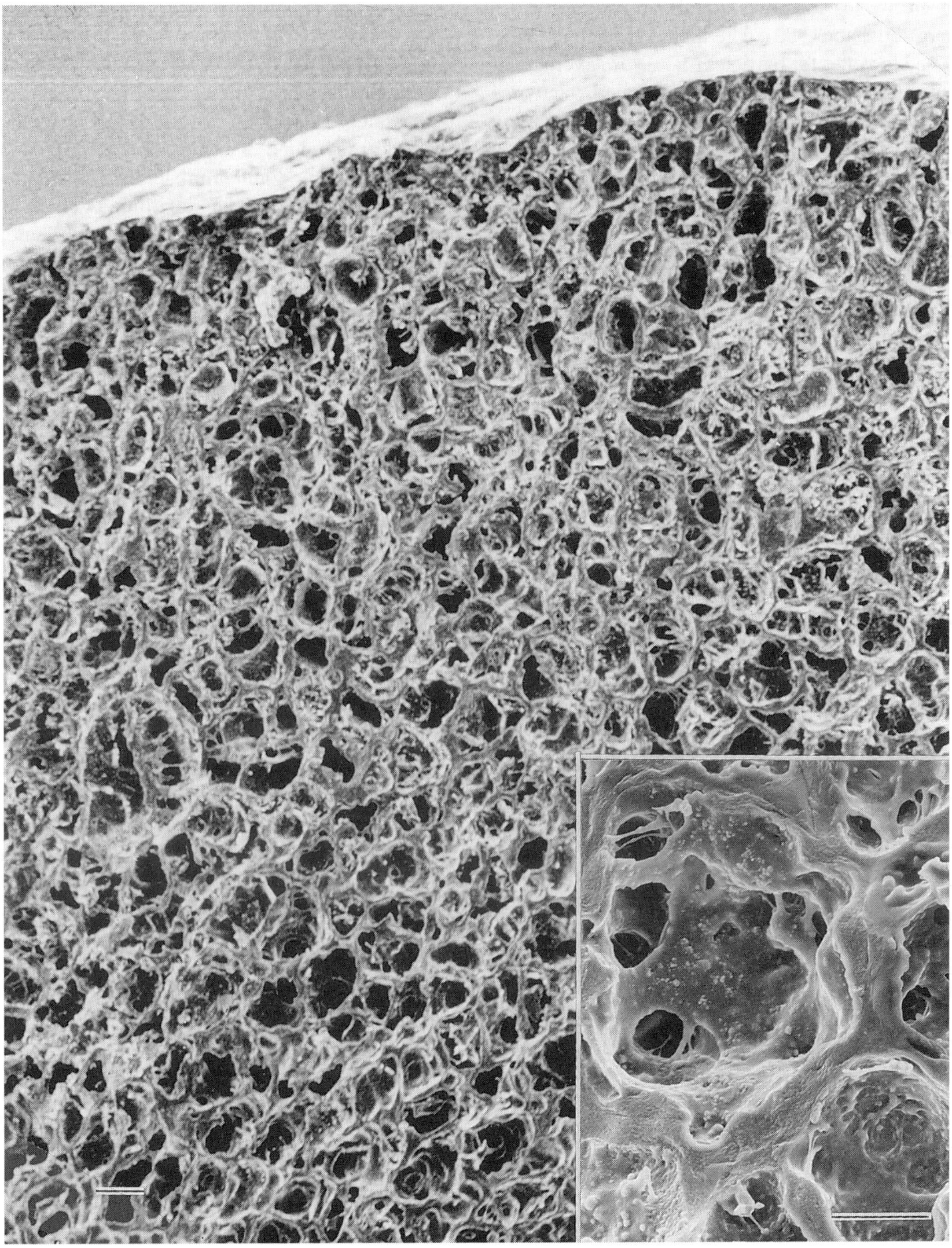

**Figure 8–13** Scanning electron micrograph of a cross section of a roasted coffee grain showing the porous and tortuous microstructure. Marker = 20 $\mu$m. Inset: Enlarged view of pores. Marker = 1 $\mu$m.

ameter ratios for pores are about 1:4 (Schwartzberg & Chao, 1982). The actual surface area exposed to water for particles of 330 $\mu$m average diameter may be 6 times the external surface area (Clo & Voilley, 1983). Roasting of coffee beans at temperatures over 200°C accomplishes chemical and structural changes. Among the latter, the most important are the expansion of the grain due to gas and vapor production, increased porosity, and migration of coffee oil to the surface.

Percolation of roasted coffee grains encompasses three distinct stages. First occurs wetting of the coffee particles, filling of pores with hot extract, and displacement of gases. Simultaneously, water is absorbed by the fibrous structure raising the solubles concentration. The second stage involves hydrolysis of water insoluble carbohydrates into soluble molecules. Finally, solubles are diffused through the extract, filling the pores (Sivetz & Foote, 1963; Voilley & Simatos, 1980).

Extraction of coffee solubles occurs rapidly, in less than 20 minutes, with more than half of the solutes extracted in 5 minutes. The high extraction rate is due to the porous microstructure and the small number of cells in a particle: a 20-mesh grind (mean diameter 800 $\mu$m) is 30 to 40 cells across. Individual chemical species extract at different rates. Simple sugars soluble in water and small molecules that contribute to bitterness extract first, together with caffeine, trigonelline, chlorogenic acid, and free salts. Larger molecules produced by hydrolysis, polymerized sugars, caramelized carbohydrates, and proteins are last to diffuse out. The rate of solubles extraction is controlled by the concentration of free extract in the particle, and temperature has a greater effect on coarse than on fine grinds (Sivetz & Foote, 1963). Separation of the extract from the grind is usually accomplished by filtration.

### 8.4.8 Extraction of Spices and Pigments

Oleoresins are solvent-prepared extracts that contain the aroma and important coloring elements of spices. They should not be confused with essential oils, which contain only substances that can be volatilized; they are generally obtained by steam distillation. Technological information on extraction processes for oleoresins is found in a paper by Sabel and Warren (1973) and in books of Pruthi (1980) and Lewis (1984). Several solvents, including alcohol, acetone, hexane, ethylene dichloride, and methylene chloride, are used for oleoresin extraction. Water immiscible solvents are preferred, since they do not get diluted with moisture, and the extraction of sugars, resins, and gums is prevented. However, stringent regulations exist on the use of solvents when spices are to be used in foods, and some solvents are banned in some countries.

A study by Aguilera, Escobar, del Valle, and San Martin (1987) shows the effect of microstructural changes induced by blanching and flaking on ethanol extraction of paprika. Blanching induces extensive destruction of cell walls, but flaking completely obliterates the microstructure of dried red peppers, as shown in light photomicrographs. Consequently, the rate of extraction and the total amount of extract was higher in flaked than in blanched and intact materials. Controlled release of oleoresin fractions can be effected by adequate microstructural modifications and the use of different solvents.

Interest in natural coloring agents has increased as questions regarding the safety of artificial colors have been raised. Solid-liquid extraction has been used to extract coloring matter such as betadines from red beets (Wiley, Lee, Saladini, Wyss, & Topalian, 1979) and anthocyanins (Bronnum-Hansen, Jacobsen, & Flink, 1985; Markakis, 1982).

### 8.4.9 Extraction of Toxic and Antinutritional Factors

Toxic and antinutritional components either occur naturally in plants or become associated with them during agricultural production or postharvest practices. They are usually present in low concentrations relative to the main food components and must be reduced to still lower and safer levels. If heat, chemical, or other processing methods fail to produce the desired results, they may be extracted with appropriate solvents. An important fact to keep in mind is that leaching will always leave a finite, although small, amount of undesirable component associated with the solution occluded in the inert matrix, even if the contaminant becomes completely solubilized.

An interesting example is the extraction of gossypol from defatted glanded cottonseed meal using various solvents (Cherry & Gray, 1981; Gardner, Hron, & Vix, 1976). Gossypol, found in the pigment glands of cottonseed, is toxic to most monogastric animals and imparts undesirable color to oil and protein products. The gossypol gland is ruptured almost instantaneously by water, complicating its extraction. However, to increase the release of the gland and the effectiveness of the solvent, the moisture of the meal is adjusted to weaken the membrane surrounding the gland. Microscopic studies have shown that gossypol is actually "enmeshed" in a water soluble matrix (possibly arabinogalactan) within the lumens of the glands and that these are broken during comminution (Yatsu, Jacks, Kircher, & Godshall, 1986).

Other cases with important nutritional implications include the removal of alkaloids from lupins and the detoxification of cassava. Bitter lupin grains containing 2–3% poisonous alkaloids have been leached with water for centuries by inhabitants of the Andean altiplano of Peru and Bolivia and subsequently consumed for their high protein (40%) and oil (20%) content. Soaked beans are cooked for extraction to reduce protein solubilization and facilitate alkaloid release. Aqueous extraction and simultaneous fractionation of protein, oil, and alkaloids from lupins has been proposed (Aguilera, Gerngross, & Lusas, 1983), as has been the use of organic solvents (Lucisano, Pompei, & Rossi, 1984).

Cassava contains linamarin ($\beta$-glucoside of acetone cyanohydrin) that releases hydrocyanic acid (HCN) upon hydrolysis by the endogenous enzyme linamarase. Significant reduction is achieved by peeling and thorough washing, but if the pulp is allowed to undergo fermentation, additional release of HCN is favored. Subsequent cooking or sun drying readily volatilizes the HCN (Liener, 1977). Retting is another traditional form of detoxification. It consists of immersing fresh cassava roots in water so that the tissue breaks down, enzymes cleave the HCN, and the soluble materials are leached out. This process removes up to 98% of the initial cyanide and is much more efficient than sun drying (Ayernor, 1985).

Aflatoxins are mycotoxins produced by fungi invading grain under hot and moist weather conditions. Scanning electron microscopy has been instrumental in locating the mycellia and spores of *Aspergillus flavus* in cottonseeds. The site of invasion is just beneath the seed coat, so presumably most of the toxins are concentrated close to the hull (Lee, Koltun, & Buco, 1983). Proper dehulling and removal of fines reduces contamination, but the defatted meal may still contain an appreciable amount of toxins. Three factors influence solvent extraction of aflatoxins from oilseed meals: (1) the use of an appropriately polar solvent, (2) adequate moisture to release the aflatoxins, and (3) high temperatures to effectively solubilize the toxins. Azeotropes like propanol-water and acetone-hexane-water have been used to reduce the aflatoxin content of prepressed–solvent-extracted cottonseed meal from 300 to 2 parts per billion (Rayner, Koltun, & Dollear, 1977). Multicomponent extraction of oil, gossypol, and aflatoxins with isopropanol-water azeotrope is presented in Figure 8–14 (Aguilera, 1982). Another area of potential important application is the removal of flatulence-inducing oligosaccharides and phytates from soybeans.

### 8.4.10 Extraction with Supercritical Fluids

In recent years there has been considerable interest in the extraction of natural products with supercritical fluids (Brunner, 1994; King & Bott, 1993; King & List, 1996). A supercritical fluid is a substance that is above its critical temperature and pressure and possesses characteristics intermediate between a liquid and a gas (Table 8–3). Thus, a supercritical fluid has a density higher than a gas and consequently more solvent power. It also has higher diffusivity, lower viscosity, and lower surface tension than liquids, allowing it to penetrate faster through solid matrices (Brunner,

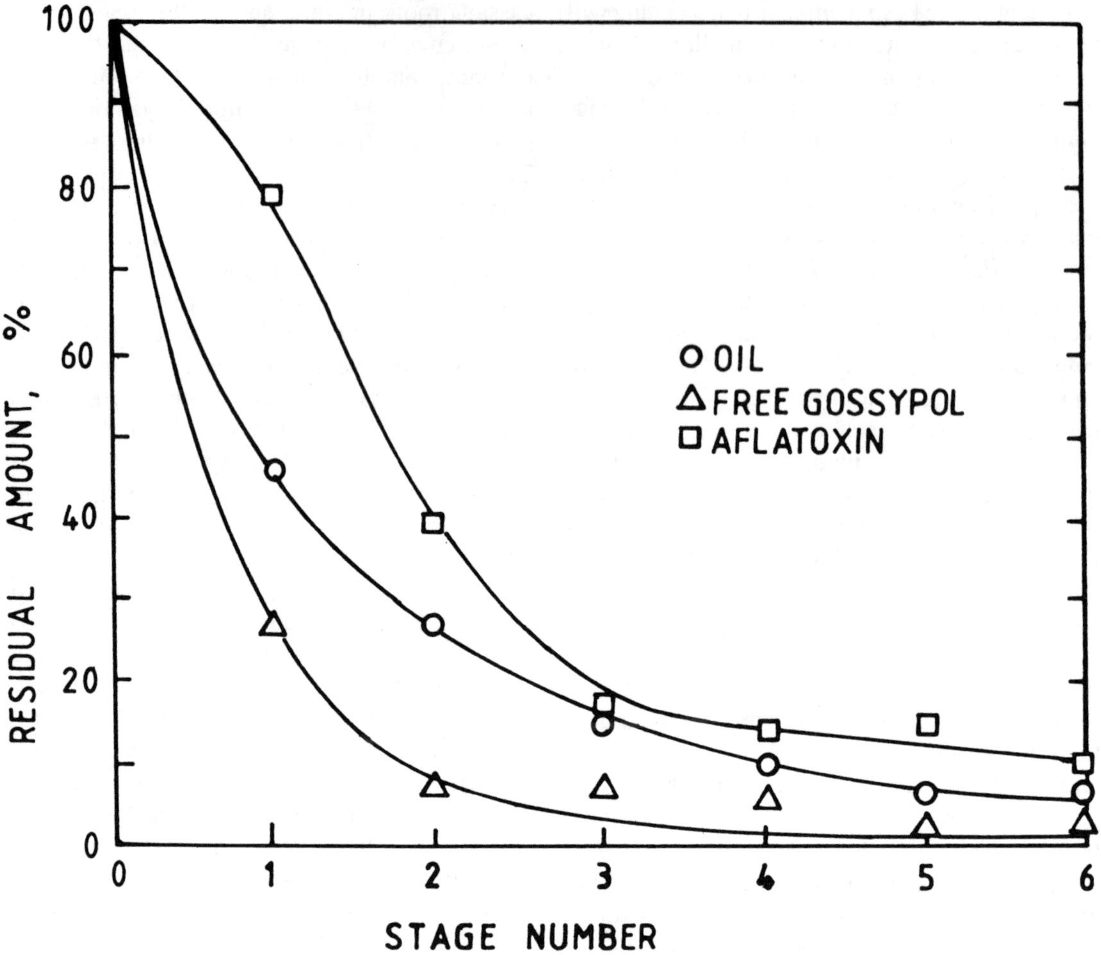


**Figure 8–14** Multicomponent extraction of oil, gossypol, and aflatoxins from cottonseed using a water-isopropanol azeotrope as solvent. The abscissa represents equilibrium contact stages of "marc" and new solvent.

1994). The extraction rate (which depends directly on the diffusivity) of a component within a plant material is said to be at least 2.5 times higher with a supercritical fluid than with liquid carbon dioxide (Hubert & Vitzthum, 1980). The density of a supercritical fluid changes dramatically near (but above) the critical point. Since solubility depends on density and increases with pressure, the solute can be recovered simply by reducing the density of the supercritical through phase lower-

**Table 8–3** Diffusivity, Density, and Viscosity of Gases, Liquids, and Supercritical Fluids

| | *Diffusivity ($cm^2s^{-1}$)* | *Density ($g\ cm^{-3}$)* | *Viscosity Pass· (Pa .s)* |
|---|---|---|---|
| Gas | $1–4 \times 10^{-1}$ | $10^{-3}$ | $1–3 \times 10^{-5}$ |
| Liquid | $0.2–2 \times 10^{-5}$ | 0.6–1.6 | $0.2–3 \times 10^{-3}$ |
| Supercritical fluid | $0.2–0.7 \times 10^{-3}$ | 0.2–0.8 | $1–9 \times 10^{-5}$ |

ing the pressure at constant temperature. Fractionation of solutes can be similarly effected (Brogle, 1982; Mangold, 1983). The main drawbacks of supercritical fluid extraction are the relatively high initial investment and energy costs of a high pressure plant, although presently batch and continuous extraction are possible. Actual applications in the food industry include decaffeination of green coffee and extraction of hops and spices (King & Bott, 1993). A third area of significant commercial development has been in the flavor and fragrances industry (Palmer & Ting; 1996).

Supercritical carbon dioxide is ideally suited for the food industry, as it is nontoxic, nonflammable, and can be removed easily from the miscella and the marc. The critical temperature ($T_c = 31.3°C$) is just above the ambient temperature, while $P_c = 72.9$ atm. Carbon dioxide does not dissolve polar compounds, and to achieve solubility it is necessary to add a co-solvent or modifier, which must be completely miscible in $CO_2$.

Extraction of vegetable oils with supercritical $CO_2$ has been actively studied (King & List, 1996; Stahl, Schutz, & Mangold, 1980). As in the case of liquid solvents, only the surface oil released by fracturing is removed from cracked soybeans. Grinding (94% particles $< 100$ mesh) or flaking (to 250 $\mu$m), increasing surface area and cell wall breakage, respectively, is required for extraction of more than 98% of the oil (Snyder, Friederich, & Christianson, 1984). Similarly, the extent of oil extraction from oilseeds depends not only on the solvent:solid ratio but also on the degree of cell damage induced by pretreatment of the seed (Brunner, 1994; Eggers, 1996). Decortication and depressurization do not sufficiently damage the cell walls, but mechanical flaking breaks open the cells and more oil is released. Press cake is most readily extracted after extensive shearing and disruption of the cellular arrangement (Eggers, Sievers, & Stein, 1985).

The characteristics of the sample matrix can have a profound effect on supercritical fluid extraction (King & France, 1992). The rate of extraction is a function of the solute solubility in the supercritical fluid and the mass transport out of the matrix. As was the case in solid-liquid extraction, smaller particle size and higher porosity favor higher extraction rates and completeness. Many times the solid matrix swells in contact with the supercritical fluid, facilitating internal mass transport. A major parameter is the moisture content of the substrate, which affects the type of solute being preferentially extracted and the rate (partial dehydration increases the rate of extraction). The explanation is that hydrophilic matrices inhibit contact between the supercritical fluid and the target solutes. However, in some cases water may act as an "internal co-solvent" and assist in extraction. Coupling of solubilization and diffusion has been noticed in supercritical extraction. Part of the nicotine in raw tobacco is bound by the matrix, which limits the rate of extraction. Desorption of the solute from the surface of the matrix is often the rate-limiting step, and use of a co-solvent such as water or methanol before extraction may accelerate desorption. In other cases, formation of a condensed surface layer of the dense fluid may retard the extraction rate at the solid-fluid interface. Lastly, free convection currents due to the variable density of the supercritical fluid in a vertical reactor promote mass transfer in the fluid phase.

## 8.5 MODIFYING MICROSTRUCTURE

### 8.5.1 Rationale

Since the rate-controlling step in food extraction is diffusion within the solid matrix, efforts have been devoted to decreasing internal resistances related to the microstructure. Some promising methods for modifying the microstructure are described below.

### 8.5.2 Structural Degradation by Thermal Energy and Radiation Waves

The histological effects of heat processing on fruits and vegetables are well documented (Reeve, 1970; Weier & Stocking, 1952) and exploited to effect biochemical and microstructural changes prior to extraction. Thermal processing of plant material may cause starch gelatinization, protein insolubilization, plasmolysis or separation of the plasmalemma, breakdown of pectins in the middle lamella, and cell separation (Jewell, 1979). Heating causes degradation of membranes and increased permeability but also loss of selectivity. For example, heating of rapeseed markedly increases extraction of chlorophyll and related undesirable pigments (Johansson & Appelqvist, 1984). The use of microwaves, known to partially degrade cellulosic materials (Ooshima, Aso, Harano, & Yamamoto, 1984), has not been reported in connection with preparation of materials for extraction.

Ultrasonic energy causes two important phenomena in the liquid phase of solvent-soaked particles: cavitation and microstreaming. Cavitation is the formation of tiny gas-vapor bubbles (e.g., in the interstices between cells), which oscillate in the ultrasonic field and eventually collapse. Microstreaming is turbulence at a microscopic level in the area surrounding a solid object. Enhanced diffusion by ultrasonics has been reported for extraction of bitter principles in hops, flavorings, cocoa butter, and enzymes (Sokolov, 1966).

The effect of ultrasonic energy has been studied in solvent extraction of oilseeds. Disruption of tissue and release of hexane-soluble lipids are hypothesized to result from pseudocavitation due to trapped bubbles in the intercellular interstices of cells and acoustic streaming. Marked increases in fines accompany longer periods of sonication, as do different microstructural effects in hulls and cotyledons (Schneider, Rutte, & Khoo, 1985). Greater power input magnifies the phenomena and increases the diffusion of oil and solvent within the seed. The diffusivity increases linearly with power, from $0.8 \times 10^{-7}$ to $2.0 \times 10^{-7}$ cm/s$^2$ at 0.18 W/cm$^2$ g of inert solids (Schurig & Sole, 1967). Disintegration of oilseeds at the microstructural level can also be induced by dielectrically generated heat. Apparently, heat causes evaporation of water inside microstructures, and the resulting increase in pressure leads to disintegration of the material (Gondar, 1968).

Application of ultrasonic waves to heated soy flakes increases the efficiency of protein extraction from 16% to 58%. The ultracentrifuge pattern of proteins extracted after sonication is the same as that of proteins isolated from low-heated flakes by conventional processes (Wang, 1975). The process has been implemented continuously at a pilot scale, with yields similar to those of commercial plants (Moulton & Wang, 1982). Improvement of rennin extraction by application of ultrasonic energy increased efficiency through tissue dispersion, destruction of cells, intensive blending, separation of particles, and an increase in surface area (Zayas, 1986).

The use of pulsating hydrodynamic action generating steep local velocity gradients has been proposed as a way to facilitate pectin extraction and diffusion. Microstructural effects include flexing and disruption of cell walls, accelerating pectin release and improved diffusion through the gelatinous layer surrounding fruit particles. Claimed benefits include increased yields (30–60% higher), 2 to 5 times faster extraction rates, and increased extract concentration (Kratchanov, Marev, Kirchev, & Bratanoff, 1986).

The use of electric fields (e.g., <40 kV) to cause electroplasmolysis, denaturate cell membranes, and aid diffusion of sucrose in beet cossettes was studied in the Soviet Union beginning in 1948. Developments were frequently reported in abstracts of Soviet scientific literature, but no commercial applications are known to be in operation. Low-dose radiation is known to induce cell wall breakage in fruits and vegetables (Maxie & Sommer, 1968) and should be studied as a pretreatment before extraction, including its possible side effects.

### 8.5.3 Structural Degradation by Enzymes

Enzymes have been utilized to overcome the resistance to extraction of cell walls and membranes in plant tissues. Pectic substances are largely responsible for coherence and integrity of plant tissue, although they represent less than 1% (w/w) of the weight on a wet basis. Polysaccharide-degrading enzymes play an important role in the production of foods. A variety of pectolytic enzymes occur in nature, and commercial preparations ("pectinase") contain different proportions of them.

Pectic enzymes are used to control the cloud stability of fruit and vegetable juices and nectars. They have also been used to facilitate pressing, extract colored material, and increase yield in juice extraction from soft fruits and to increase efficiency in apple and citrus juice processing. Enzymes have been applied in wine production to augment the release of juice from white grape skins and to degrade pectin, which extracts jointly with coloring matter in red wine production (Neubeck, 1981; Rombouts & Pilnik, 1978). A novel development is the preparation of suspensions of intact cells from fruits and vegetables (maceration) using special enzyme products that degrade predominantly the middle lamella without attacking the cell wall (Schmitt, 1983). Complete liquefaction of the fruit tissue is rapidly accomplished using a mixture of pectic enzymes and cellulolytic enzymes (Voragen, Krist, Heutnink, & Pilnik, 1980).

Cellulases and proteinases have been used to hydrolyze portions of the plant matrix and allow more efficient dispersion and extraction of proteins, particularly when they have been heat denatured. A combination of ultrasonic and enzymatic processing allows extraction of 90% of the available protein in 1 minute (Childs, Forte, & Ku, 1977). Cellulases are added in instant tea manufacture to break down cell wall material in leaves and increase yield (Sanderson & Coggon, 1977). They have also been instrumental in removing the outer "skin" and releasing agar from the cell walls of *Gracilaria* algae (San Martin et al., 1988; see Figure 8–15).

The use of microbial enzymes has also been proposed as a way to facilitate extraction of oilseeds and reduce refining operations (Fullbrook, 1983; Voragen et al., 1980). An enzyme "cocktail" that includes a cellulase and a pectinase has been used to enhance the oil extractability of sunflower seeds at the pressing stage as well as during solvent extraction. Improvement in the protein digestibility of the meal was also claimed (Dominguez, Sineiro, Nuñez, & Lema, 1996).

Cell wall–degrading enzymes are already present in fruits and vegetables, and thus fermentation is sometimes exploited as a means of modifying the internal microstructure prior to extraction. Withered tea leaves are fermented to cause endogenous enzymes (catechol oxidases) and tea flavonols to come in contact and form tea pigments and aroma. Cocoa beans are also fermented for aroma development, but fermentation is not capable of causing the extensive cell wall and membrane destruction needed for fat liberation and for facilitating expression (Kuznetzova & Yuskova, 1985).

### 8.5.4 Restructuring

Restructuring involves increasing the porosity of the solid matrix by breaking down the original microstructure and forming a new one by extrusion or pelleting. During the process, the cellular architecture is obliterated and the solutes (and contaminants too) are made more available for extraction. The extraction of contaminants can be minimized if the solvent is highly selective for the solute.

Extruders have been used to enhance oil extraction from cottonseed and soybeans (Hendrick, 1983; Watkins, Johnson, & Doty, 1989). The process consists of conditioning, flaking, steam injection extruding, and drying the extrudate before solvent extraction (Marchand, 1984). Major advantages of extrusion restructuring are as follows: (1) the density and strength of the pellet are augmented, so extraction capacity and draining are enhanced; (2) less solvent is used per unit weight of seed (because of reduced solvent holdup); and (3) more complete cooking of the components is achieved (Bredeson, 1983; Williams, 1997).

The presence of starch may preclude the use of steam injection, since it becomes gelatinized, encapsulating much of the oil (an efficient encapsulating operation). Studies of high-oil corn (19% oil) demonstrate, however, that conventional high-shear or dry extrusion is adequate to free the oil from the spherosomes and produce a porous pellet after expansion at the extruder die, thus increasing the amount of "surface" oil and the extraction rate (Aguilera & Lusas, 1986). The results of laboratory stage extraction experiments

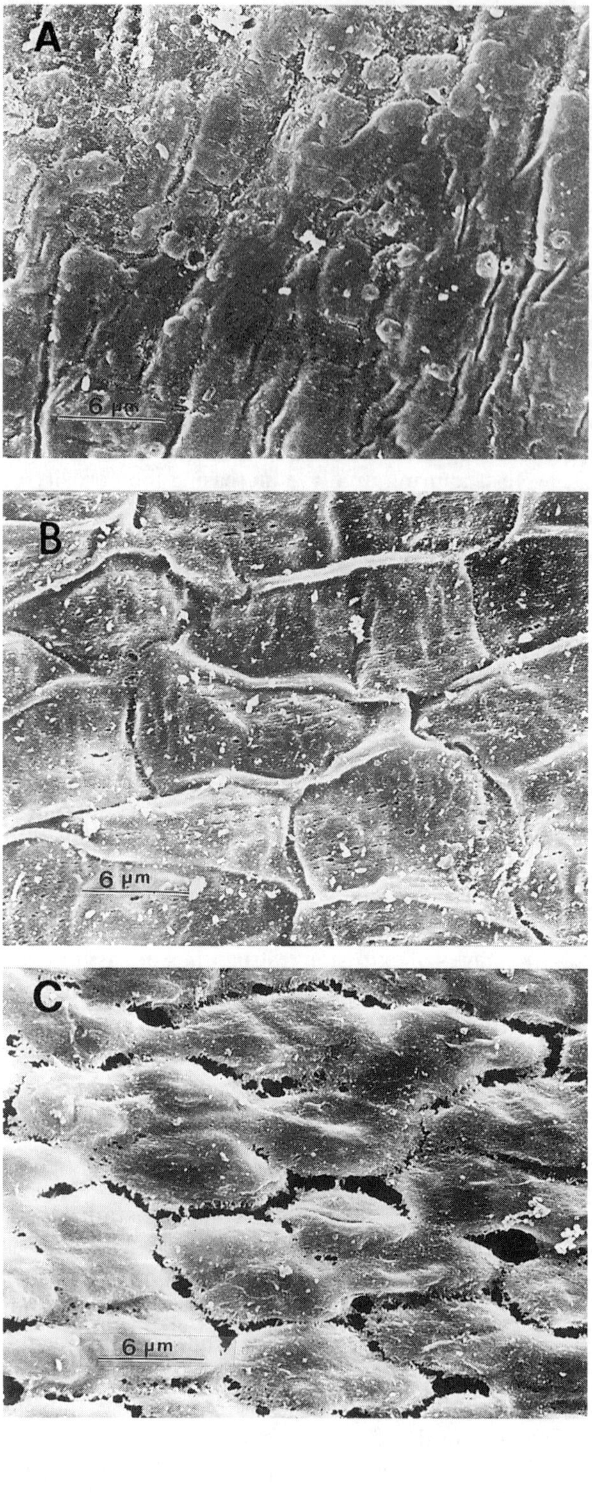


**Figure 8–15** Action of cellulases on the outer skin of *Gracilaria* algae, facilitating diffusion of NaOH into the inner cells and extraction of agar from the cell walls. (A) Untreated skin. (B) After washing with nonpolar solvent. (C) After treatment with cellulase.

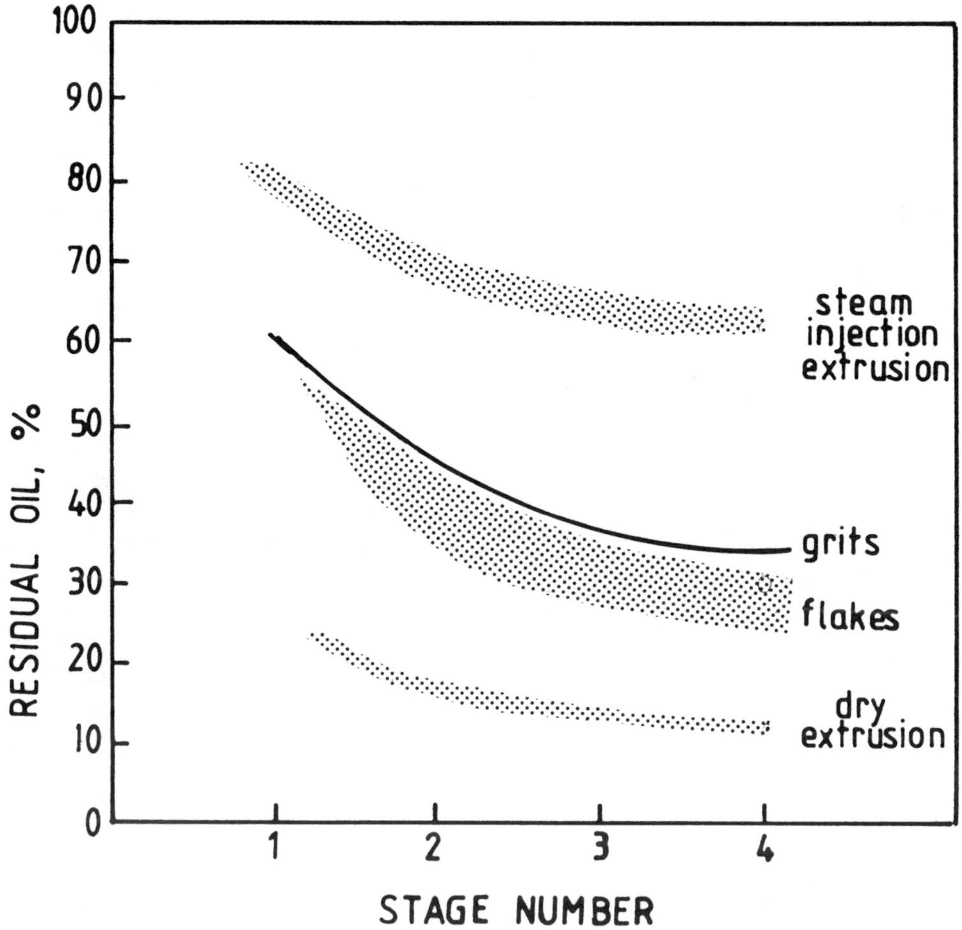


**Figure 8–16** Effect of pretreatment on the extractability of oil from high-oil corn using hexane. Dry extrusion yields an expanded porous structure that facilitates oil removal. Steam injection extrusion induces starch gelatinization and "encapsulation" of oil. Curves for grits and flakes are shown for reference.

on extruded samples, flakes, and grits are depicted in Figure 8–16. More than 70% of the total oil in dry extruded pellets is extracted after one contact stage with hexane, compared with 40–50% for grits and flakes. Thus the main mechanism for oil removal in the case of extruded pellets appears to be washing of the oil, which has been relocated at the surface of the porous pellets.

Mechanical shear in the form of milling and pelleting can also be used to liberate solutes and create a new structure for extraction. Dried hop cones are milled, pelleted, and remilled into a powder to increase the bulk density and ensure maximum rupture of glands containing bitter principles that provide unique flavors to beer. Prepared hops extracted with liquid carbon dioxide show yields 3–8% above those of intact hops (Daoud & Kusinski, 1986).

## 8.6 MODELING THE EXTRACTION PROCESS

### 8.6.1 Background and Scope

Thus far we have assumed that a better understanding of microstructural effects would lead to improved models for predicting and optimizing solid-liquid extraction of foods. As Schwartzberg (1983) stated, "The formulation of mass-transfer models which adequately account for the structural features of the solid being extracted, cell wall selectivity and structural breakdown (and others), are needed for extraction to be a well understood unit operation."

Food materials prepared for extraction depart from the ideal homogeneous solids and simple geometrical shapes for which solutions to the differential equations of Fick's second law exist. The extraction process itself is phenomenologically more complex than the unidirectional diffusion of a single solute. Thus, the analytical solutions for homogeneous solids and constant diffusivity extensively used in the past are too simplistic to model the extraction process of food particles satisfactorily. Some of the models and predictive equations that are currently used in food extraction are presented below.

### 8.6.2 Levels of Analysis

The level of analysis of the extraction problem must be established when defining the system under study and/or when selecting an approach for interpreting the phenomena.

When defining a system under examination, the first level of analysis and modeling is the cell. At this level, the cell architecture—the ways in which the building blocks (cell wall, cytoplasm, protein bodies, spherosomes) are arranged and interact—defines the system. Histological studies and other techniques used in botany to measure physical properties of cells may be used to advantage. A second level of complexity encompasses the microstructural arrangement of cells, intercellular spaces, and pores within a particle, including changes induced during preparation. Microscopy is an invaluable tool for describing structural features at these first two levels. At the third level of analysis, macrostructural aspects predominate (e.g., the interaction of particles in a bed and the hydrodynamic and other engineering aspects of the process).

As for the approach used to represent the phenomena, several alternatives exist. The choice between mass transfer and diffusion models is one of preference rather than precision. A diffusion model makes sense when concentrations need to be described as a function of time and position. Mass transfer models are simpler to use and perfectly adequate when average concentrations are required. A kinetic approach may also be employed to describe the system, but it is more empirical (Cussler, 1997).

### 8.6.3 Diffusion Models

Since internal diffusion from undamaged cells is the slowest step in extraction, the release of solute from simple solids with intact cells has been analyzed. Thin slices of peanuts, carefully prepared to simulate conditions imposed by Fick's second law, had extraction curves that became straight lines after a transient period of less than 30 minutes. During that period, oil was extracted by hexane more rapidly than predicted by the theoretical relation, and the difference was accounted for by the release of oil from ruptured surface cells. The apparent diffusion coefficient of the straight-line portion was independent of the thickness of the slices, as predicted by diffusion theory, with an average value of $0.67 \times 10^{-8}$ cm$^2$/s (Fan et al., 1948). Thicker slices of tung seeds (0.7 to 2 mm thick, with cells of 22 $\mu$m average diameter), sealed on all sides except for the two major opposites, were used to study diffusion from whole cells (Krasuk, Lombardi, & Ostrovsky, 1967). A model with constant diffusivity did not represent the data adequately, but a diffusion coefficient varying exponentially with concentration fitted the experimental results with satisfactory precision. In both studies, the oil diffusivity increased as moisture content decreased. Note that "satisfactory" fitting of experimental data to a straight line in a plot of log $X$ versus $t$ (Section 8.2.1) is necessary but not sufficient to prove the existence of a single, constant diffusion coefficient (Aguerre, Gabitto, & Chirife, 1985).

Few researchers support the idea that intact cells of oilseeds are permeable to a triglyceride-hexane miscella. Rac (1967) characterized the histological structure (cell dimension, surface, and wall thickness) of 10 oil-bearing materials and studied their extraction behavior. Washing of broken outer cells was a very important extraction mechanism in particles of 800–1250 $\mu$m extracted with petroleum ether, in particular, for raw materials having large cells. Between 40% and 90% of the oil was removed in the first 2 minutes of extraction, and after 48 hours less than 3% of the original oil remained in the seeds (Figure 8–17). Based on this study, Schneider (1980), in an enlightening paper, presented a model for solvent extraction from intact cells in which the solvent-permeable cell membranes allow first the formation of an internal miscella that develops intracellular pressure, deforming the cell wall and enlarging the membrane pores. Solubilized triglycerides diffuse out through the expanded pores into the intercellular miscella, from which they are transported to the bulk solution (Figure 8–18). No microstructural evidence for the model has been presented except for the already mentioned paper by Aguilera (1989). The whole area of transport of chemical components through cells and cell walls as related to food applications is still poorly understood, and more basic research is needed.

When the solid microstructure is considered as a mixed system of ruptured and unruptured cells, solute extraction may be modeled as the weighted

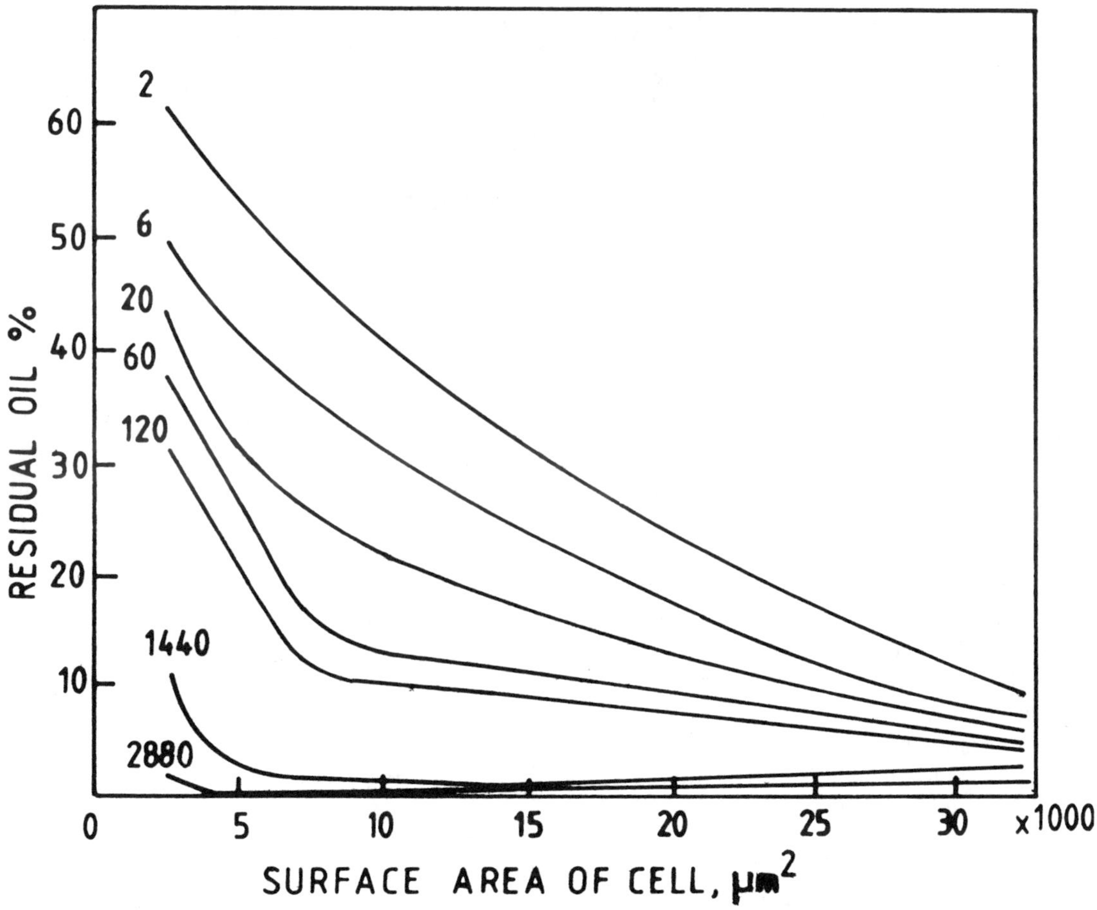


**Figure 8–17** Effect of time and cell size on extraction characteristics of particles from oil-bearing materials. The numbers next to the lines represent extraction times in minutes. *Source*: Adapted from Rac, 1967.

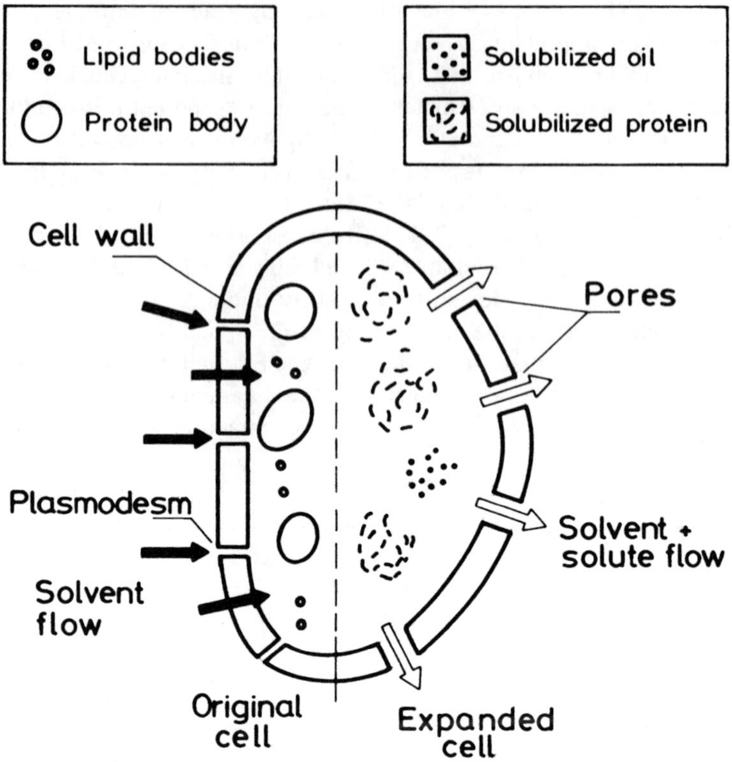


**Figure 8–18** Model for solvent and solute transport from intact cells. In the case of macromolecules, the main transport route during extraction would be diffusion through expanded pores.

summation of diffusion through two parallel structures, one of which represents fast extraction from ruptured or surface cells and the other slow transfer from intact cells. Agreement of such model with actual oilseed extraction was better than that of single-diffusion models, and the $D$ values for the two types of structures differed by a factor of 10 (Osborn & Katz, 1944).

In oil extraction from oilseeds, crushing and flaking largely disrupt the cell architecture, creating a random system of pores and capillaries. An oil-bearing flake is assumed to be composed of four parts: (1) non-oil-bearing solids (or inert material), (2) water, (3) oil, and (4) air space. During extraction, the air spaces and the spaces occupied by the extracted oil are filled by the surrounding solution. This void-filling solution, termed *static holdup*, would not drain. Undissolved oil was bound in the structure in a slowly soluble resistance that was not reformed by replacing the extracted oil. The rate of extraction was almost independent of the miscella concentration (Coats & Karnofsky, 1950).

Models have also been developed that take into account the parenchymatous-vascular nature of vegetable tissue. Based on microstructural observations, it was assumed that the sugar contained in sugar beets was either immobilized inside the cells or mobile in the vascular bundles. Free sugar was extracted by diffusion through the intercellular liquid, and an equilibrium relationship between both solute states existed, given by a sorption isotherm of the Langmuir type (dual-sorption model). Both theory and experimental data showed that sugar extraction was strongly dependent on the cellular integrity of the sugar beet (Soddu & Gioia, 1979).

Several approaches have been used to derive extraction parameters that avoid the problem of cellular microstructure heterogeneity. One used frequently is the application of an "effective diffusion coefficient" ("apparent diffusivity") determined experimentally from unidimensional solutions to Fick's equation. The macrostructural effect of a peel or thin skin covering the solid has been accounted for by using a boundary layer model (Loncin, 1980). Likewise, lumped mass transfer coefficients obtained from equation 8–14 are often used for analysis and design purposes (Zanderighi, 1983). It is obvious that, because internal transport controls most of the practical extraction processes, better models could be derived if phenomena at the microstructural level were better understood.

### 8.6.4 Kinetic Models

Under this heading have been grouped models that use expressions similar to those derived from the study of chemical kinetics (i.e., based on the principle that the instantaneous rate is dependent on the concentration of the reacting species [solute]).

If extraction is assumed to be mainly a process of viscous flow of miscella through capillaries and not a diffusional process, the rate of extraction can be expressed as

$$\frac{dc_1}{dt} = k\left(\frac{\rho\gamma}{\eta}\right) f(c_1) \qquad \textbf{Equation 8–18}$$

where $f(c_1)$ is a function of the instantaneous concentration of the solute in the solid ($c_1$), and the rate constant depends on the surface tension $\gamma$, the density $\rho$, and the viscosity $\eta$ of the solvent (Othmer & Agarwal, 1955).

The fact that equation 8–18 has the form of a first-order reaction and that during extraction it is relatively simple to monitor the variation of solute concentration with time has prompted the development of kinetic models. The following first-order equation has been proposed to represent extraction of solubles from tea and coffee (Long, 1979; Spiro & Jago, 1982; Spiro & Selwood, 1984; Spiro & Siddique, 1981):

$$\ln\left(\frac{c_\infty}{c_\infty - c}\right) = kt \qquad \textbf{Equation 8–19}$$

where the constant $k$ represents the contribution of various resistances and has to be determined experimentally. A similar expression was derived from a materials balance for the extraction of oil (Kminek, Vackova, & Zajic, 1981).

Any particle reduction process (e.g., grinding) will directly expose solute material at the surface of particles. In this case, washing of the solute from the surface of the solid is an important extraction mechanism, since it occurs almost instantaneously upon immersion and may transfer considerable amounts of solute. A semi-empirical model based on simple reaction kinetics of order $n$ has been used to analyze the effect of different preparation conditions (flaking and blanching) on the extraction rate of oleoresins from paprika (Aguilera et al., 1987). A value of $n = 1.5$ was determined by fitting the experimental data, and values for $k$, $c_\infty$ (an equilibrium concentration), and $c_o$ (a time 0 concentration due to washing) were calculated. The model predicts closely the actual extraction data, and all three parameters were greater for flaking than for blanching and also greater for blanching than for the controls.

### 8.6.5 Combined Diffusional-Kinetic Models

Particularly in the case of prepressed oilseeds, the oil in the surface layers is easily washed when immersed in the solvent. The overall extraction process can be kinetically modeled as the addition of two mechanisms treated separately, washing and diffusion (Patricelli, Assogna, Casalaina, Emmi, & Sodini, 1979). The expression derived for this model is

$$c = c_\infty^w(1 - e^{-k_w t}) + c_\infty^d(1 - e^{-k_d t})$$

**Equation 8–20**

where $c_\infty^w$ is the final hypothetical concentration of oil in the miscella due to washing, $c_\infty^d$ is the hypothetical concentration for diffusion, and $k_w$ and $k_d$ are the kinetic coefficients for washing and diffusion, respectively. Experimental data showed that as flake thickness decreased, the contribution of washing became larger and the total oil extracted increased. Typical extraction data are shown in Figure 8–19.

The structural model of Patricelli has been extended to include two diffusional processes instead of one (So & MacDonald, 1986). The two mechanisms supposedly originate from two types of cells produced during preprocessing: (1) ruptured cells unobstructed by membranes or other barriers, resulting in unhindered diffusion, and (2) unruptured cells involving transfer across membranes and a slow, hindered diffusion process. The fitting of this modified model to the data is better than for the previous model only in the case of thick flakes, where the proportion of unruptured cells is higher, but its solution requires the determination of more parameters.

The washing effect has frequently been overlooked in modeling. Voilley and Simatos (1980) used the equation for unicomponent diffusion through a sphere to model the entire extraction process of solubles from coffee grounds. The diffusion model could not account for the high soluble concentration at times less than 1 minute, when almost 90% of the final concentration was reached, and it behaved satisfactorily only for longer times. A kinetic model that incorporates instantaneous washing and time-dependent extraction of soluble protein from lupin has been presented by Aguilera and García (1989). The protein washed from the surface of particles and the total protein extracted increased as the particle radius decreased, while the first rate constant increased slightly or remained constant. The washed protein accounted for about 80% of the total extracted

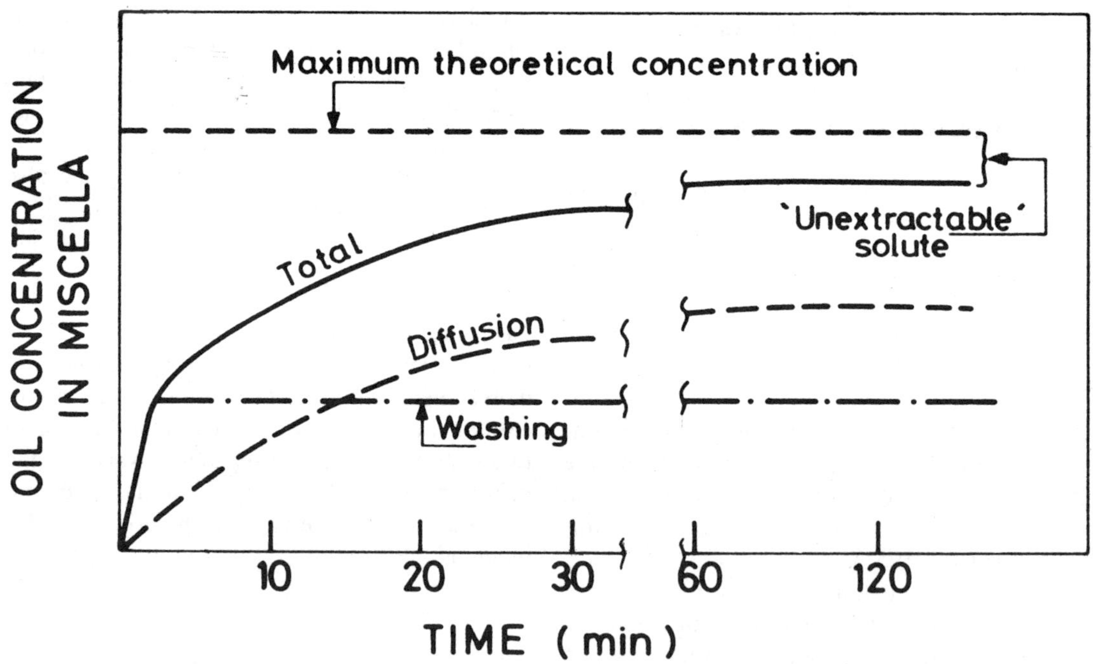


**Figure 8–19** Extraction kinetics for oil extraction from rapeseed flakes. Initial washing from broken outer cells represents a considerable amount of the extractable oil. Diffusion proceeds more slowly and eventually stops, leaving "unextractable" oil in the flake.

protein when the hydrated radius of the particle (83 $\mu$m) was on the order of 3 cell diameters, and microstructural modifications such as flaking played no role at such dimensions. For larger particle sizes, the relative effect of washing decreased, as expected, but nevertheless, for particles of nearly 2 mm, it represented 6% and 14% of the total extracted protein for untreated and flaked material, respectively.

The problem of simultaneous solubilization and diffusion of solutes has been analyzed mathematically by coupling the solutions for a first-order reaction (similar to equation 8–19) with that for diffusion (equation 8–5) (Schwartzberg, 1975). When $Dq_1^2/kL^2 > 3$ the following expression is obtained:

$$\frac{c_\infty - c}{c_\infty} = \exp(-kt)\left(1 + \frac{B_1}{(Dq_1^2/kL^2) - 1}\right)$$

**Equation 8–21**

These combined-mechanism models reveal the complex nature of some extraction processes and point toward a more phenomenological approach in which lumped coefficient solutions are substituted for coupled multiple-mechanism solutions.

### 8.6.6 Equations for Extractor Design and Performance

In solvent extraction practice, equations to predict the time of extraction as a function of various operational parameters abound. Silin's equation for cossettes extraction is one of them, and despite its shortcomings it has been used by industry for more than 20 years. A computer model has been developed to simulate the performance of rotary and tower diffusers (Genie, 1986). A calculation method for designing percolation and immersion extractors based on laboratory data is presented by Karnofsky (1986).

### 8.6.7 Effect of Temperature

The influence of temperature on the diffusion coefficient or the rate constant is usually modeled after an Arrhenius equation. The relationship between the rate constant and the absolute temperature is

$$k = k_o \exp\left(\frac{-E_a}{RT}\right)$$

**Equation 8–22**

where $E_a$ is an "apparent" activation energy of the process, $R$ is the gas constant, and $k_o$ is a pre-exponential or frequency factor. Plotting ln ($k$) versus $1/T$ should give a straight line with slope $-E_a/R$, from which a value for $E_a$ can be obtained.

Loncin (1980) reported that the activation energy calculated for the diffusion of cyclohexanol in potatoes was 35.7 kJ/mol, about twice the value of cyclohexanol diffusion in water but similar to values measured in lipid membranes. Similarly, Spiro and Selwood (1984) reported the $E_a$ for caffeine diffusion through swollen coffee bean particles between 25°C and 85°C to be 32 kJ/mol, again twice as large as for free diffusion in water. These values are in the range of activation energies for diffusional processes (8–40 kJ/mol) commonly listed in the literature (Thijssen & Kerkhof, 1977).

A change in the mechanism of extraction with temperature may lead to a broken curve in the plot. This is the case in extraction of protein from corn, where at temperatures lower than 40°C solubilization controls, and there is an exponential relationship between rate and temperature, with an $E_a$ of 16 kJ/mol. At higher temperatures, the dependency is reduced to the typical linear function of diffusivity of macromolecules, with the temperature represented by equation 8–6 (Russell & Tsao, 1982).

### 8.6.8 A Comment on Numerical Techniques

New numerical methods, such as finite elements, are being increasingly implemented in digital computers to solve diffusion differential equations in nonhomogeneous solids of odd shapes and with complex boundary conditions. It must be remembered that the solution obtained will be no better than the assumptions under which the equations were developed. Microstructural analysis combined with reliable experimental data of the process may greatly assist in developing correct phenomenological models for computer implementation.

## BIBLIOGRAPHY


Aguerre, R.J., Gabitto, J.F., & Chirife, J. (1985). Utilization of Fick's second law for the evaluation of diffusion coefficients in food processes controlled by internal diffusion. *Journal of Food Technology*, *20*, 623–629.

Aguilera, J.M. (1982). Alcohol-based processes for oilseeds extraction. Paper presented at the 42nd Annual Institute of Food Technology Meeting. June, 1982, Las Vegas, NV.

Aguilera, J.M. (1989). Protein extraction from lupin seeds: Microstructural aspects and hypothesis of a mechanism. *International Journal of Food Science and Technology*, *24*, 29–37.

Aguilera, J.M., Escobar, G.A., del Valle, J.M., & San Martin, R. (1987). Ethanol extraction of red peppers: Kinetic studies and microstructure. *International Journal of Food Science and Technology*, *22*, 225–232.

Aguilera, J.M., & García, H.D. (1989). Protein extraction from lupin seeds: A mathematical model. *International Journal of Food Science and Technology*, *24*, 17–27.

Aguilera, J.M., Gerngross, M.F., & Lusas, E.W. (1983). Aqueous processing of lupin seed. *Journal of Food Technology*, *18*, 327–333.

Aguilera, J.M., & Lusas, E.W. (1986). Laboratory and pilot solvent extraction of extruded high oil corn. *Journal of the American Oil Chemists' Society*, *63*, 239–243.

Anderson, R., Osredkar, R., & Zagar, C. (1977). Diffusion properties of lipids: Pure and in seeds. *Journal of the American Oil Chemists' Society*, *54*, 487–489.

Arnold, L.K., Choudhury, R.B.R., and Chang, H.Y. (1961). Relation of the quality of solvent-extracted soybean oil and residual oil content in the meal. *Journal of the American Oil Chemists' Society*, *38*, 336–338.

Arnott, H.J., & Webb, M.A. (1983). Scanning electron microscopy and histochemistry of oilseeds. In D.B. Bechtel (Ed.), *New frontiers in food microstructure* (pp. 149–198). St. Paul, MN: American Association of Cereal Chemists.

Aspinall, G.O. (1981). Constitution of plant cell wall polysaccharides. In W. Tanner & F.A. Loewus (Eds.), *Encyclopedia of plant physiology* (New Series, Vol. 13, Part B; pp. 3–8). Berlin: Springer-Verlag.

Ayernor, G.S. (1985). Effects of the retting of cassava on product yield and cyanide detoxication. *Journal of Food Technology*, *20*, 89–96.

Bair, C.W., & Snyder, H.E. (1980a). Electron microscopy of soybean protein bodies. *Journal of the American Oil Chemists' Society*, *57*, 279–282.

Bair, C.W., & Snyder, H.E. (1980b). Ultrastructural changes in soybeans during processing to yield desolventized toasted meat. *Journal of Food Science*, *45*, 529–533.

Becker, K.W. (1978). The solvent extraction of soybeans. *Journal of the American Oil Chemists' Society*, *55*, 754–761.

Bernardini, M. (1984). *Technical encyclopedia on oilseeds, oils and fats*. Vol. 1. *Raw materials and extraction techniques*. Rome: B.E. Oil Publishing House.

Binkley, C.R., & Wiley, R.C. (1978). Continuous diffusion-extraction method to produce apple juice. *Journal of Food Science*, *43*, 1019–1023.

Bird, R.B., Stewart, W.E., & Lightfoot, E.N. (1960). *Transport phenomena*. New York: Wiley.

Bock, W., & Ohm, G. (1983). Einfluβ der gewachsenen biologischen Struktur auf die Wasserbindungkapazitat ausgewahlter Obst- und Gemusepraparate. *Die Nahrung*, *27*, 205–210.

Bomben, J.L., Durkee, E.L., Lowe, E., & Secor, G.E.J. (1974). A laboratory study on countercurrent desalting of pickles. *Journal of Food Science*, *39*, 260–265.

Bott, F.R. (1980). Supercritical gas extraction. *Chemistry and Industry (London)*, *15*, 228–232.

Bredeson, D.K. (1977). Mechanical pressing. *Journal of the American Oil Chemists' Society*, *54*, 489A–490A.

Bredeson, D.K. (1983). Mechanical oil extraction. *Journal of the American Oil Chemists' Society*, *60*, 211–213.

Bressan, J.A., Carroad, P.A., Merson, R.L., & Dunkley, W.L. (1981). Modeling of isothermal diffusion of whey components from small curd cottage cheese during washing. *Journal of Food Science*, *47*, 84–88.

Brogle, H. (1982). $CO_2$ as a solvent: Its properties and applications. *Chemistry & Industry (London)*, 385–390.

Bronnum-Hansen, K., Jacobsen, F., & Flink, J.M. (1985). Anthocyanin colorants from elderberry (*Sambucus nigra L.*) 1. Process considerations for production of liquid extract. *Journal of Food Technology*, *20*, 703–711.

Brunner, G. (1994). *Gas extraction: An introduction to the fundamentals of supercritical fluids and the application to separation processes.* New York: Springer.

Cantarelli, C., & Riva, M. (1983). A survey of mechanical extraction of vegetables. In C. Cantarelli & C. Peri (Eds.), *Progress in food engineering* (pp. 75–94). Switzerland: Forster-Verlag AG.

Carpita, N.C. (1985). Tensile strength of cell walls of living cells. *Plant Physiology*, *79*, 485–488.

Carpita, N.C., Sabularse, D., Montezinos, D., & Delmer, D.P. (1979). Determination of pore size of cell walls of living plant cells. *Science*, *205*, 1144–1147.

Cherry, J.P., & Gray, M.E. (1981). Methylene chloride extraction of gossypol from cottonseed products. *Journal of Food Science*, *46*, 1726–1733.

Childs, E.A., Forte, J.L., & Ku, Y. (1977). Utility of enzymes in solubilization of seed and leaf proteins. In R.L. Ory & A.J. St. Angelo (Eds.), *Food and beverage processing* (pp. 304–316). ACS Symposium Series 47. Washington, DC: American Chemical Society.

Christensen, P.L. (1983). Solvent extraction: Recent developments. *Journal of the American Oil Chemists' Society, 60*, 214–215.

Cleriot, J. (1994). Sugar manufacturing seen through the microscope. *International Sugar Journal, 96*, 129–142.

Clo, G., & Voilley, A. (1983). Evaluation of ground coffee particles for optimum extraction. *Lebensmittel-Wissenschaft und Technologie, 16*, 39–42.

Coats, H.B., & Karnofsky, G. (1950). Solvent extraction. 2. The soaking theory of extraction. *Journal of the American Oil Chemists' Society, 27*, 51–53.

Coats, H.B., & Wingard, M.R. (1950). Solvent extraction. 3. The effect of particle size on extraction rate. *Journal of the American Oil Chemists' Society, 27*, 93–96.

Cogan, U., Yaron, A., Berk, Z., & Mizrahi, S. (1967). Isolation of soybean protein: Processing conditions on yields and purity. *Journal of the American Oil Chemists' Society, 44*, 321–324.

Crank, J. (1975). *The mathematics of diffusion*. London: Oxford University Press.

Cussler, E.L. (1997). *Diffusion: Mass transfer in fluid systems* (2nd ed.). Cambridge: Cambridge University Press.

Daoud, I.S., & S. Kusinski (1986). Process aspects of the extraction of hops with liquid carbon dioxide. *Journal of the Institute of Brewing, 92*, 559–567.

del Valle, J.M., & Aguilera, J.M. (1989). Effect of densification and solvent conditions on supercritical extraction of mushrooms. *Journal of Food Science, 54*, 135–141.

Dieckert, J.W., & Dieckert, M.C. (1976). The chemistry and biology of the vacuolar proteins of seeds. *Journal of Food Science, 41*, 475–482.

Dieckert, J.W., & Dieckert, M.C. (1982). Ultrastructural analysis of a soybean protein isolate. *Journal of Food Science, 47*, 1291–1293.

Diosady, L.L., Rubin, L.J., Ting, N., & Tasso, O. (1983). Rapid removal of canola oil. *Journal of the American Oil Chemists' Society, 60*, 1658–1661.

Dominguez, H., Sineiro, J., Nuñez, M.J., & Lema, J.M. (1996). Enzymatic treatment of sunflower kernels before oil extraction. *Food Research International, 28*, 537–545.

Dousse, R. (1978). L'extraction solide-liquide dans l'industrie alimentaire: Apercu theorique et pratique. *Brauerei-Rundschau, 89*, 60–65.

Eggers, R. (1996). Supercritical fluid extraction of oilseeds/lipids in natural products. In J.W. King & G.R. List (Eds.), *Supercritical fluid technology in oil and lipid chemistry* (pp. 35–64). Champaign, IL: AOAC Press.

Eggers, R., Sievers, U., & Stein, W. (1985). High pressure extraction of oil seed. *Journal of the American Oil Chemists' Society, 62*, 1222–1228.

Emch, F. (1980). Extraction of fruits and vegetables. In P. Linko, Y. Malkki, J. Olkku, & J. Larinkari (Eds.), *Food process engineering* (Vol. 1; pp. 424–433). London: Applied Science Publishers.

Fan, H.P., Morris, J.C., & Wakeham, H. (1948). Diffusion phenomena in solvent extraction of peanut oil: Effect of cellular structure. *Industrial and Engineering Chemistry, 40*(2), 195–199.

Field, J. (1984). Rendering: The beginnings of the story. *Journal of the Society of Leather Technologists and Chemists, 68*, 71–74.

Francis, B.J., & Bell, J.-M.K. (1975). Commercial pectin: A review. *Tropical Science, 17*, 25–44.

Fredtzdorff, B., Pomeranz, Y., & Betchtel, D.B. (1981). Malt modification assessed by histochemistry, light microscopy and transmission and scanning electron microscopy. *Journal of Food Science, 47*, 786–791.

Fryer, P.J., Pyle, D.L., & Rielly, C.D. (1997). *Chemical engineering for the food industry*. London: Blackie Academic and Professional.

Fullbrook, P.D. (1983). The use of enzymes in the processing of oilseeds. *Journal of the American Oil Chemists' Society, 60*, 476–478.

Galloway, G.P. (1976). Cleaning, cracking, dehulling, decorticating and flaking of oil bearing materials. *Journal of the American Oil Chemists' Society, 53*, 271–274.

Gardner, H.K., Jr., Hron, R.J., Sr., & Vix, H.L.E. (1976). Removal of pigment glands (gossypol) from cottonseed. *Cereal Chemistry, 53*, 549–560.

Geankoplis, C.J. (1978). *Transport processes and unit operations*. Boston: Allyn and Bacon.

Gekkas, V. (1992). *Transport phenomena of foods and biological materials*. Boca Raton, FL: CRC Press.

Genie, G.V. (1982). Juice extraction in the beet sugar factory. *Sugar Technology Reviews, 9*, 119–270.

Genie, G.V. (1986). Computer simulation and mathematical modelling of diffusion. *Zuckerindustrie, 111*, 149–154.

Gibbons, G.C. (1981). Visualization of $\alpha$-amylase movement and cell wall breakdown during barley malting: Practical application of current research. *Journal of the American Society of Brewing Chemists, 39*, 55–59.

Gondar, J. (1968). Uber die mechanische und dielektrishe Zerleinerung von Getreide- und Olsaaten. *Fette, Seifen, Anstrichmittel, 70*, 433–438.

Heldman, D.R., & Lund, D.B. (1992). *Handbook of food engineering*. New York: Marcel Dekker.

Hendrick, W.B. (1983, May). Expander preparation for solvent extraction. *Oil Mill Gazetteer*, pp. 43–44.

Hensarling, T.P., Yatsu, L.Y., & Jacks, T.J. (1970). Extraction of lipids from cottonseed tissue. 2. Ultrastructural effects of lipid extraction. *Journal of the American Oil Chemists' Society*, *47*, 224–225.

Hubert, P., & Vitzthum, O.G. (1980). Fluid extraction of hops, spices and tobacco with supercritical gases. In G.M. Schneider, E. Stahl, & G. Wilke (Eds.), *Extraction with supercritical gases* (pp. 25–43). Weinheim, Germany: Verlag Chemie.

Idson, B., & Braswell, E. (1957). Gelatin. *Advances in Food Research*, *7*, 235–338.

Jewell, G.C. (1979). Fruits and vegetables. In J.G. Vaughan (Ed.), *Food microscopy* (pp. 1–34). London: Academic Press.

Johansson, S.A., & Appelqvist, L.A. (1984). Effect of seed pretreatment and extraction conditions on the amount of minor components of seed oils. 3. Chlorophyll and related pigments. *Fette, Seifen, Anstrichmittel*, *86*, 304–309.

Johnson, L.A., & Lusas, E.W. (1983). Comparison of alternative solvents for oil extraction. *Journal of the American Oil Chemists' Society*, *60*, 229–242.

Karnofsky, G. (1949). The theory of solvent extraction. *Journal of the American Oil Chemists' Society*, *26*, 564–569.

Karnofsky, G. (1986). Design of oilseed extractors. 1. Oil extraction. *Journal of the American Oil Chemists' Society*, *63*, 1011–1014.

Kessler, H.G. (1981). *Food engineering and dairy technology*. Freising, Germany: Verlag A. Kessler.

Khan, L.M., & Hanna, M.A. (1983). Expression of oil from oilseeds: A review. *Journal of Agricultural Engineering Research*, *28*, 495–503.

King, C.J. (1980). *Separation processes* (2nd ed.; pp. 550–556). New York: McGraw-Hill.

King, J.W., & France, J.E. (1992). Basic principles of analytical supercritical fluid extraction. In B. Wenclawiak (Ed.), *Analysis with supercritical fluids: Extraction and chromatography* (pp. 32–60). Berlin: Springer-Verlag.

King, J.W., & List, G.R. (Eds.). (1996). *Supercritical fluid technology in oil and lipid chemistry*. Champaign, IL: AOAC Press.

King, M.B., & Bott, T.R. (Eds.). (1993). *Extraction of natural products using near-critical solvents*. London: Blackie Academic and Professional.

Kminek, M., Vackova, E., & Zajic, J. (1981). Untersuchungen zur Extraktion ölhaltiger Rohstoffe. Theoretisches Modell eines Gegenstrom-Extrakteurs. *Fette, Seifen, Anstrichmittel*, *83*, 260–261.

Kramer, P.J. (1983). *Water relations in plants* (pp. 23–56). Orlando, FL: Academic Press.

Krasuk, J.H., Lombardi, J.L., & Ostrovsky, C.D. (1967). Diffusion extraction of oil containing materials. *Industrial and Engineering Chemistry, Process Design and Development*, *6*(2), 187–195.

Kratchanov, C., Marev, K., Kirchev, N., & Bratanoff, A. (1986). Improving pectin technology: Extraction using pulsating hydrodynamic action. *Journal of Food Technology*, *21*, 751–761.

Kuznetzova, L.S., & Yuskova, E.P. (1985). Microstructure of fresh and retailable cocoa beans. *Khlebopekarnaya i Konditerskaya Promyshlennost*, *4*, 33–34. (Abstract No. 10K1, *FSTA*, 18(10), 1986).

Lauchli, A. (1976). Apoplasmic transport in tissues. In V. Luttgge & M.G. Pitman (Eds.), *Encyclopedia of plant physiology* (New Series, Vol. 2, Part B; pp. 3–34). Berlin: Springer-Verlag.

Lebrun, L., & Junter, G.A. (1993). Diffusion of sucrose and dextran through agar membranes. *Enzyme and Microbial Technology*, *15*, 1057–1062.

Lee, C.H., & Rha, C. (1978). Microstructure of soybean protein aggregates and its relation to the physical and textural properties of the curd. *Journal of Food Science*, *43*, 79–84.

Lee, L.S., Koltun, S.P., & Buco, S. (1983). Aflatoxin distribution in fines and meats from decorticated cottonseed. *Journal of the American Oil Chemists' Society*, *60*, 1548–1549.

Leniger, H.A., & Beverloo, W.A. (1975). *Food process engineering* (p. 502). Dordrecht, Netherlands: D. Reidel.

Lewis, Y.S. (1984). *Species and herbs for the food industry* (pp. 27–31). Orpington, England: Food Trade Press.

Liener, I.E. (1977). Removal of naturally occurring toxicants through enzymatic processing. In R.E. Feeney & J.R. Whitaker (Eds.), *Food proteins: Improvements through chemical and enzymatic modifications* (pp. 7–57). Advances in Chemistry Series 160. Washington, DC: American Chemical Society.

Loncin, M. (1969). *Die Grundlagen der Verfahrenstechnik in der Lebensmittelindustrie* (pp. 245–256). Germany: Verlag Sauerlander.

Loncin, M. (1980). Diffusion phenomena in solids. In P. Linko, Y. Malkki, J. Olkku, & J. Larinkari (Eds.), *Food process engineering* (Vol. 1; pp. 354–363). London: Applied Science Publishers.

Long, V.D. (1979). Aqueous extraction of black leaf tea. 3. Experiments with an stirred column. *Journal of Food Technology*, *14*, 449–462.

Lucisano, M., Pompei, C., & Rossi, M. (1984). Oil and alkaloid removal from *Lupinus mutabilis* by extraction with hexane. *Lebensmittel-Wissenschaft und Technologie*, *17*, 324–327.

Lysjanski, V.M., Popow, W.D., Redko, F.A., & Stabnikow, W.N. (1983). Verfahrenstechnische Grundlagen der Lebensmitteltechnik (pp. 437–441). Darmstadt, Germany: Steinkopff Verlag.

Mangold, H.K. (1983). Liquified gases and supercritical fluids in oilseed extraction. *Journal of the American Oil Chemists' Society*, *60*, 226–268.

Mantovani, G. (1991). Growth and morphology of the sucrose crystal. *International Sugar Journal*, *93*, 23–27.

Marchand, D.E. (1984). Utilization of expander type machines in crushing: Importance and outlook. *Revue Francaise du Corps Gras*, *31*, 13–17.

Markakis, P. (1982). Anthocyanins as food additives. In P. Markakis (Ed.), *Anthocyanins as food colors* (pp. 245–253). New York: Academic Press.

Maxie, E.C., & Sommer, N.F. (1968). Changes in some constituents in irradiated fruits and vegetables. In *Preservation of fruit and vegetables by radiation* (p. 39). Vienna: International Atomic Energy Agency.

McCaig, R. (1984). The development of the electron microscope and its application to the brewing industry. *Journal of the American Society of Brewing Chemists*, *42*, 152–159.

McGinnis, R.A. (1971). *Beet sugar technology* (2nd ed.; pp. 125–135). Fort Collins, CO: Beet Sugar Development Foundation.

Merva, G.E. (1975). *Physioengineering principles*. Westport, CT: AVI Publishing Co.

Moo-Young, M., & Andre, G. (1980). An appraisal of mass transfer phenomena in food processing operations. In P. Linko, Y. Malkki, J. Olkku, & J. Larinkari (Eds.), *Food engineering progress* (Vol. 1; pp. 342–353). London: Applied Science Publishers.

Moulton, K.J., & Wang, L.C. (1982). A pilot plant study of continuous ultrasonic extraction of soybean protein. *Journal of Food Science*, *47*, 1127–1129, 1139.

Mrema, G.C., & McNulty, P.B. (1980). Mechanisms of mechanical oil expression from rapessed and cashew. 1. Solids Separation Processes. *Chemical Engineering Symposium Series* No. 59.

Naessens, W., Bresseleers, G., & Tobback, P. (1982). Diffusional behavior of dipalmitin in a freeze-dried model system of different water activities. *Journal of Food Science*, *47*, 1245–1249.

Neubeck, C.E. (1981). Recent development in juice extraction processes. In *Proceedings of the Symposium on Technological Problems of Fruit Juice Concentrates* (pp. 80–92). Oregon State University *Agricultural Experimental Station* Technical Pub. No. 6234. Corvallis, OR.

Nishinari, K., & Watase, M. (1983). Effect of alkali pretreatment on the rheological properties of concentrated agar agar gels. *Carbohydrate Polymers 3*, 39–52.

Ooshima, H., Aso, K., Harano, Y., & Yamamoto, T. (1984). Microwave treatment of cellulosic materials for their enzymatic hydrolysis. *Biotechnology Letters*, *6*, 289–294.

Osborn, J.O., & Katz, D.L. (1944). Structure as a variable in the application of diffusion theories to extraction. *Transactions of the American Institute of Chemical Engineers*, *5*, 511–531.

Osterberg, N.O., & Smith Sorensen, T. (1981). Apple juice extraction in a counter-current diffuser. *Journal of Food Technology*, *16*, 379–402.

Othmer, D.F., & Agarwal, J.C. (1955). Extraction of soybeans: Theory and mechanism. *Chemical Engineering Progress*, *51*, 372–378.

Palmer, G.H. (1972). Morphology of starch granules in cereal grains and malts. *Journal of the Institute of Brewing*, *78*, 326–332.

Palmer, G.H., Gernah, D.I., McKernan, G., Nimmo, D.H., & Laycock, G. (1985). Influence of enzyme distribution on endosperm breakdown (modification) during malting. *Journal of the American Society Brewing Chemists*, *43*, 17–22.

Palmer, M.V., & Ting, S.S.T. (1996). Applications of supercritical fluid technology in food processing. *Food Chemistry*, *52*, 345–352.

Patricelli, A., Assogna, A., Casalaina, A., Emmi, E., & Sodini, G. (1979). Fattori che influenzano l'estrazione dei lipidi da semidecorticati di girasole. *Rivista Italiana delle Sostanze Grasse*, *56*, 136–142.

Percival, E., & McDowell, R.H. (1981). Algal walls: Composition and biosynthesis. In W. Tanner & F.A. Loewus (Eds.), *Encyclopedia of plant physiology* (pp. 277–316). Berlin: Springer-Verlag.

Pohl, J., & Mieth, G. (1984). Ubersichartikel Aufbereitung und Extraktion von Olsaaten. *Die Nahrung*, *28*(5), 517–532.

Possmann, P. (1981). Recent development in juice extraction processes. In *Proceedings of the Symposium on Technological Problems of Fruit Juice Concentrates* (pp. 64–79). Oregon State University *Agricultural Experimental Station* Technical Pub. No. 6234. Corvallis, OR.

Potter, R.S. (1966). Extraction of pectin. *Process Biochemistry*, *1*, 378–384.

Pruthi, J.S. (1980). Spices and condiments: Chemistry, microbiology, technology. *Advances in Food Research* (Suppl. 4), pp. 228–242.

Rac, M. (1967). Influence de la structure cellulaire sur quelques propriétes technologiques de certain matières premières oleagineuses. *Revue Francaise Corps Gras*, *14*, 223–228.

Rathje, W. (1970). Extraktion der Sukrose aus Rubenschnitzeln durch osmotisches Pumpen, nicht durch Diffusion. *Zuckerind*, *20*, 410–413.

Rayner, E.T., Koltun, S.P., & Dollear, F.G. (1977). Solvent extraction of aflatoxins from contaminated agricultural products. *Journal of American Oil Chemists' Society*, *54*, 242A–244A.

Reeve, R.M. (1970). Relationships of histological structure to texture of fresh and processed fruits and vegetables. *Journal of Texture Studies*, *1*, 247–284.

Robards, A.W., & Lucas, W.J. (1990). Plasmodesmata. *Annual Review of Plant Physiology Molecular Biology*, *41*, 369–419.

Rombouts, F.M., & Pilnik, W. (1978). Enzymes in fruit and vegetable juice technology. *Process Biochemistry*, *13*, 9–13.

Roques, M., Naiha, M., & Briffaud, J. (1984). Hexane sorption and diffusion in rapeseed meals. In B.M. McKenna (Ed.), *Engineering and food* (Vol. 1; pp. 13–21). London: Elsevier Applied Science Publishers.

Russell, M.K., & Tsao, G.T. (1982). Protein removal from corn endosperm by solvent extraction. AIChE *Symposium* Series No. 218, *78*, 83–89.

Rutman, M., & Aguilera, J.M. (1985). Intercellular modifications for improved processing of marine products. Presented at the 45th Annual *Institute of Food Technologists* Meeting. Atlanta, GA, June 1985.

Sabel, W., & Warren, J.D.F. (1973). Theory and practice of oleoresin extraction. In *Species. Tropical Product Institute Conference Proceedings* (pp. 189–197). London: Tropical Products Institute.

Saio, K., & Watanabe, T. (1966). Preliminary investigation on protein bodies of soybean seeds. *Agricultural and Biological Chemistry* (*Tokyo*), *30*, 1133–1138.

San Martin, R., Aguilera, J.M., & Hohlberg, A. (1988). Effect of cellulase pretreatments on red algae agar extractability. *Carbohydrate Polymers*, *8*, 33–43.

Sanderson, G.W., & Coggon, P. (1977). The use of enzymes in the manufacture of black tea and instant tea. In R.L. Ory & A.J. St. Angelo (Eds.), *Enzymes in food and beverage processing* (pp. 12–26). ACS Symposium Series 47. Washington, DC: American Chemical Society.

Schadel, W.E., Walter, W.M., & Young, C.T. (1983). Cotyledon structure of resting peanut (*Arachis hypogaea* L. cv Florunner) seed before and after hydraulic pressing. *Journal of Food Science*, *48*, 1405–1407.

Schmitt, R. (1983). Neue wege zur enzymatischen Verflussigung von Obst and Gemuse. *Flussiges Obst.*, *50*, 23–28.

Schneider, F.H. (1978). Zur Mikrostruktur von Olsaaten unter dem Aspekt der extraktiven Trennung. *Fette, Seifen, Anstrichmittel*, *80*, 265–271.

Schneider, F.H. (1979a). Schalung von Rapssaat durch definierte Vermorung. 1. Untersuchungen zum Saatanatomie. *Fette, Seifen, Anstrichmittel*, *81*, 11–16.

Schneider, F.H. (1979b). Schalung von Rapssaat durch definierte Vermorung. 2. Untersuchungen zum Schalverhalten. *Fette, Seifen, Anstrichmittel*, *81*, 53–65.

Schneider, F.H. (1980). Zur extraktiven Lipid-Freisetzung aus pflanzlichen Zellen. *Fette, Seifen, Antrichmittel*, *82*, 16–23.

Schneider, F.H., & Khoo, D. (1986). Trennpressen: Versuch einer Bestandsaufnahme experimenteller Arbeiten. *Fette, Seifen, Anstrichmittel*, *88*, 329–340.

Schneider, F.H., & Rutte, U. (1984). Zur Bindung der Resthexans in Raps-Extraktionsschroten. *Fette, Seifen, Anstrichmittel*, *86*, 331–339.

Schneider, F.H., Rutte, U., & Khoo, D. (1985). Ultraschall-Extraktion vegetabiler Fetstoffe: Untersuchungen zur Lipidfreisetzung aus Rapssaat. *Fette, Seifen, Anstrichmittel*, *87*, 66–74.

Schurig, W.F., & Sole, P. (1967). Alcoholic extraction of oilseed with the aid of ultrasonics. *Journal of American Oil Chemists' Society*, *44*, 585–591.

Schwartzberg, H.G. (1975). Mathematical analysis of solubilization kinetics and diffusion in foods. *Journal of Food Science*, *40*, 211–212.

Schwartzberg, H.G. (1980). Continuous countercurrent extraction in the food industry. *Chemical Engineering Progress*, *76*, 67–85.

Schwartzberg, H.G. (1982, November 1). Progress and problems in solid-liquid extraction. Paper presented at the First Latin American Congress on Heat and Mass Transfer, La Plata, Argentina.

Schwartzberg, H.G. (1983). Expression-related properties. In M. Peleg & E.B. Bagley (Eds.), *Physical properties of foods* (pp. 423–471). Westport, CT: AVI Publishing Co.

Schwartzberg, H.G. (1987). Leaching. Organic materials. In R.W. Rousseau (Ed.), *Handbook of separation process technology* (pp. 540–577). New York: Wiley.

Schwartzberg, H.G., & Chao, R.Y. (1982). Solute diffusivities in leaching processes. *Food Technology*, *36*(2), 73–86.

Sherwood, T.K., Pigford, R.L., & Wilke, C.R. (1975). *Mass transfer*. New York: McGraw-Hill.

Shokrani, R., & Delavier, H.J. (1978). Elektronenmikroskopische Studien von Zuckerrüben- und Zuckerrohrgewebe. *Zuckerind*, *103*, 301–307.

Sipos, E., & Witte, N.H. (1961). The desolventizer toaster process for soybean oil meal. *Journal of American Oil Chemists' Society*, *38*, 11–12, 17–19.

Sivetz, M., & Foote, H.E. (1963). *Coffee processing technology* (pp. 320–349). Westport, CT: AVI Publishing Co.

Smith, A.K., & Circle, S.J. (1972). *Soybeans*: *Chemistry and technology*. Vol 1. *Proteins* (Chap. 9). Westport, CT: AVI Publishing Co.

Smith, C.J. (1979). Oilseeds. In J.G. Vaughan (Ed.), *Food microscopy* (pp. 35–74). London: Academic Press.

Snyder, J.M., Friederich, J.P., & Christianson, D.D. (1984). Effect of moisture and particle size on the extractability of oils from seeds with supercritical $CO_2$. *Journal of American Oil Chemists' Society*, *61*, 1362–1365.

So, G.C., & MacDonald, D.G. (1986). Kinetics of oil extraction from canola (rapeseed). *Canadian Journal of Chemical Engineering*, *64*, 80–86.

Soddu, A., & Gioia, F. (1979). Diffusion through living systems: Sugar loss from sugar beets. *Chemical and Engineering Science*, *34*, 763–773.

Sokolov, A.Y. (1966). Nomenclature, classification and standarization of technological equipment for the food industry. In *Proceedings of the 2nd International Congress of Food Science and Technology* (pp. 221–232). Warsaw, Poland.

Spanswick, R.M. (1976). Symplastic transport in tissues. In V. Luttgge & M.G. Pitman (Eds.), *Encyclopedia of plant physiology* (New Series, Vol. 2, Part B; pp. 35–53). Berlin: Springer-Verlag.

Spiro, M., & Jago, D.S. (1982). Kinetics and equilibria of tea infusion. *Journal of the Chemical Society Faraday Transactions 1*, *78*, 295–305.

Spiro, M., & Page, C.M. (1984). The kinetics and mechanism of caffeine infusion from coffee: Hydrodynamic aspects. *Journal of the Science of Food and Agriculture*, *35*, 925–930.

Spiro, M., & Selwood, R.M. (1984). The kinetics and mechanism of caffeine infusion from coffee: The effect of particle size. *Journal of the Science of Food and Agriculture*, *35*, 915–924.

Spiro, M., & Siddique, S. (1981). Kinetics and equilibria of tea infusion: Kinetics of extraction of theaflavins, thearubigins and caffeine from koonsong broken pekoe. *Journal of the Science of Food and Agriculture*, *32*, 1135–1139.

Stahl, E., Schutz, E., & Mangold, H.K. (1980). Extraction of oilseeds with liquid and supercritical carbon dioxide. *Journal of the Agriculture and Food Chemistry*, *28*, 1153–1157.

Stanley, D.W., Gill, T.A., deMan, J.M., & Tung, M.A. (1976). Microstructure of rapeseed. *Canadian Institute of Food Science and Technology Journal*, *9*, 54–60.

Stein, W., & Glaser, F.W. (1976). Continuous solvent extraction of sunflower seed, groundnuts, palm kernels, rapeseed and copra. *Journal of American Oil Chemists' Society*, *53*, 283–285.

Steinert, P., Galling, G., & Buttersack, C. (1990). The cell wall in the storage tissue of the sugarbeet root. *Zuckerind*, *115*, 95–102.

Swern, D. (1976). *Bailey's industrial oil and fats products* (3rd ed.). New York: Interscience.

Thijssen, H.A.C., & Kerkhof, P.J.A.M. (1977). Effect of temperature-moisture content history during processing on food quality. In T. Hoyem & O. Kvale (Eds.), *Physical, chemical and biological changes in food caused by thermal processing* (pp. 10–30). London: Applied Science Publishers.

Tombs, M.P. (1967). Protein bodies of the soybean. *Plant Physiology*, *42*, 797–813.

Vaughan, J.G. (1970). *The structure and utilization of oil seeds*. London: Chapman and Hall.

Vix, H.L.E., Gardner, H.K., Jr., Lembou, M.G., & Rollins, M.L. (1972). Ultrastructure related to cottonseed and peanut processing and products. In G.E. Inglett (Ed.), *Seed proteins* (pp. 212–230). Westport, CT: AVI Publishing Co.

Voilley, A., & Simatos, D. (1980). Modeling the solubilization process during coffee brewing. *Journal of Food Processing and Engineering*, *3*, 185–198.

Voragen, F.G.J., Krist, R., Heutnink, R., & Pilnik, W. (1980). Apple cell wall digestion by polysaccharide degrading enzymes. In P. Linko, Y. Malkki, J. Olkku, & J. Larinkari (Eds). *Food process engineering* (Vol. 2; pp. 264–275). London: Applied Science Publishers.

Wang, L.C. (1975). Ultrasonic extraction of protein from autoclaved soybean flakes. *Journal of Food Science*, *40*, 549–552.

Watkins, L.R., Johnson, W.H., & Doty, S.C. (1989). Extrusion-expansion of oilseeds for enhancement of extraction, energy reduction and improved oil quality. In T.H. Applewhite (Ed.), *Vegetable protein utilization in human foods and animal feedstuffs* (pp. 41–46). Champaign, IL: American Oil Chemists' Society.

Weier, T.E., & Stocking, C.R. (1952). Histological changes induced in fruits and vegetables by processing. *Advances in Food Research*, *2*, 297–342.

Wiley, R.C., Lee, Y.N., Saladini, J.J., Wyss, R.C., & Topalian, H.H. (1979). Efficiency studies of a continuous diffusion-extraction apparatus for recovering betaines from red table beets. *Journal of Food Science*, *44*, 208–212.

Wilkin, G.D. (1983). Raw materials: Milling, mashing and extract recovery. In E.G. Priest & I. Campbell (Eds.), *Current developments in malting, brewing and distilling* (pp. 35–44). London: Institute of Brewing.

Williams, M.A. (1997). Preparation of oil-bearing materials for extraction. In P.J. Wan & P.J. Wakelyn (Eds.), *Tecnology and solvents for extracting oilseeds* (pp. 121–136). Champaign, IL: AOCS Press.

Wolf, W.J. (1970). Scanning electron microscopy of soybean protein bodies. *Journal of American Oil Chemists' Society*, *47*, 107–108.

Wolf, W.J. (1972). Soybean ultrastructure and its relationship to processing. In G.E. Inglett (Ed.), *Symposium: Seed proteins* (pp. 231–241). Westport, CT: AVI Publishing Co.

Wolf, W.J., & Baker, F.L. (1972). Scanning electron microscopy of soybeans. *Cereal Science Today*, *17*, 124–126, 128–130, 147.

Wolf, W.J., & Baker, F.L. (1975). Scanning electron microscopy of soybeans, soy flours, protein concentrates and protein isolates. *Cereal Chemistry*, *52*, 387–396.

Yatsu, L.W. (1965). The ultrastructure of cotyledonary tissue from *Gossypium hirsutum* L. seeds. *Journal of Cell Biology*, *25* (Part I), 193–199.

Yatsu, L.W., & Jacks, T.J. (1972). Spherosomes and membranes: Half unit membranes. *Plant Physiology*, *249*, 937–943.

Yatsu, L., Jacks, T.J., Kircher, H.W., & Godshall, M.A.

(1986). Chemical and microscopic studies of the matrix substance in pigment glands of cotton (*Gossypium hirsutum* L.) seeds. *Journal of the American Oil Chemists' Society*, *63*, 534–536.

Yiu, S.H., Altosaar, I., & Fulcher, R.G. (1983). The effects of commercial processing on the structure and microchemical organization of rapeseed. *Food Microstructure*, *2*, 165–173.

Zanderighi, L. (1983). Mass transfer in solvent extraction. In C. Cantarelli & C. Peri (Eds.), *Progress in food engineering* (pp. 101–107). Switzerland: Forster-Verlag AG.

Zayas, J.F. (1986). Effect of ultrasonic treatment on the extraction of chymosin. *Journal of Dairy Science*, *69*, 1767–1775.

# CHAPTER 9

# Simultaneous Heat and Mass Transfer: Dehydration

## 9.1 INTRODUCTION

Preservation by water removal was probably humankind's first "technology" for extending the stability of foods. In fact, nature relies on the reduction of water content by evaporation to preserve the viability of seeds and continue the life cycle on Earth. The most common limitation on the shelf life of foods is microbial growth; hence, several preservation processes are aimed at achieving stability of foods by reducing the moisture content to levels below those required by microorganisms for survival and reproduction. The terms *dehydration* and *drying* will be used interchangeably hereafter to refer to the controlled removal of water from a food.

Water is the most important constituent of all forms of life. Every organism requires the presence of water, which in some cases amounts to as much as 95% of the organism's total weight. However, there are some organisms that survive under desiccation conditions at moisture contents less than 0.7% (Aguilera & Karel, 1997). From this it can be inferred that not all the water present in living matter plays an equally important role. In the life sciences, a distinction is made between biological active water, that needed for peak biological activity, and structural water, which protects biological life such as in seeds or bacterial spores (Karmas, 1973). Engineers discriminate between bound and unbound water; the former exerts less vapor pressure than pure liquid water at the same temperature. Materials containing bound water are called hygroscopic (McCabe, Smith, & Harriott, 1985).

Drying is a complex process because it involves removal of large quantities of water that plays very important chemical, physical, and structural roles. As early as the late 1950s, Kuprianoff (1958) referred to these possible adverse effects of removing bound water from foods: (1) denaturation of proteins by concentration of solutes, (2) irreversible structural changes leading to textural modifications after rehydration, and (3) storage stability problems. As a unit operation, drying requires simultaneous heat and mass transfer, which may be effected by multiple mechanisms often acting concurrently. Several other unit operations and processes in foods involve the exchange of water vapor, when the environment in which we live is not in equilibrium with the product moisture content. In other cases, dehydration is an integral part of processing, although it is not always fully recognized, as in the case of frying.

The objective of this chapter is to review physicochemical and microstructural aspects that influence the water-binding properties of a food, to relate heat and mass transfer to the microstructure of foods, and to present applications of dehydration that create new food microstructures with added functionality.

## 9.2 BASIC CONCEPTS

### 9.2.1 Water Activity and Sorption Isotherms

A consideration of water sorption phenomena and their relation to microstructure is essential to understanding their role in dehydration and in dried products. About four decades ago, it was realized that it was not the actual moisture content (e.g., g water/g dry matter) but another factor related to the "state" of water that determined the stability of foods. The concept of water activity ($a_w$) was introduced by the microbiologist Scott (1957), and it is a preferred term indicating the "quality" of the water content of a food. It describes the degree of "boundness" of water and hence its availability to participate in physical, chemical, and microbiological reactions. Since the introduction of $a_w$, the subjects of water activity and moisture sorption have been of prime concern to food scientists, and comprehensive treatments have been published by Duckworth (1975), Troller and Christian (1978), Labuza (1984), Simatos and Multon (1985), and Rockland and Beauchat (1987), among others. Probably the best and most updated information regarding water properties in foods comes from the International Symposium on the Properties of Water (ISOPOW) meetings and their published proceedings.

The definition of water activity and its relation to the equilibrium relative humidity (%ERH) of the environment are given by the expression

$$a_w = \frac{p}{p^o} = \frac{\%ERH}{100} \qquad \textbf{Equation 9–1}$$

where $p$ is the partial pressure of water in the food and $p^o$ the vapor pressure of pure water at the same temperature. Note that the measured parameter is $p/p^o$, or the "relative vapor pressure" (RVP), and that sometimes this is not equal to $a_w$. Since the term *water activity* is common in the food science literature, it is used in this book, but the reader needs to understand its true meaning.

At constant temperature and under equilibrium conditions a unique relationship exists between the moisture content and water activity (or equilibrium relative humidity) of a food, depending on whether equilibration is achieved by adsorption or desorption. This relation is called the *sorption isotherm*, and the shape of the curve depends strongly on structural, physicochemical, and chemical properties of the food components. Sorption isotherms of food products are generally sigmoidal in shape, a result of several basic interacting mechanisms of water binding, to be discussed later. Some typical moisture sorption isotherms are shown in Figure 9–1. Ample variability in sorption behavior can be observed among the principal food components. The sorption isotherms of many foods are given in Iglesias and Chirife (1982), and Wolf, Spiess, and Jung (1985) contains a comprehensive bibliography on the subject.

As noted, the term *water activity* as defined above is valid only at equilibrium. Since in many processes and during storage of foods, equilibrium may not prevail, water activity has been challenged as predictor of the stability of low- and intermediate-moisture foods. Where a glassy metastable state exists physicochemical attributes are time dependent, and $T_g$ may be considered an alternative tool for predicting food stability and processability (Slade & Levine, 1991). However, water activity has been used advantageously by food scientists as a key parameter for assessing the microbial stability of high-moisture foods (Chirife & Buera, 1996).

### 9.2.2 Interpretation of Sorption Isotherms

Knowledge of the thermodynamic equilibrium between the surrounding air and a solid is a prerequisite for analyzing drying, as it defines the driving force for mass transfer. A sorption isotherm depicts the relation between the equilibrium moisture content in the solid and the corresponding water activity (or relative humidity) of the surrounding atmosphere at a constant temperature.

To comprehend the different types of interactions between water (either molecules or in bulk) and the food, it is useful first to analyze the theoretical mechanisms that depress the vapor pressure of water. Such analysis distinguishes three regions in the sorption isotherm, defined by whether the prevailing effect on $a_w$ is due to pure sorption, capillary condensation, or solute effect (Figure 9–2). Although simplistic (e.g., all three effects may be present simultaneously in a food), this analysis is presented for the value in relating physicochemical concepts to the ob-

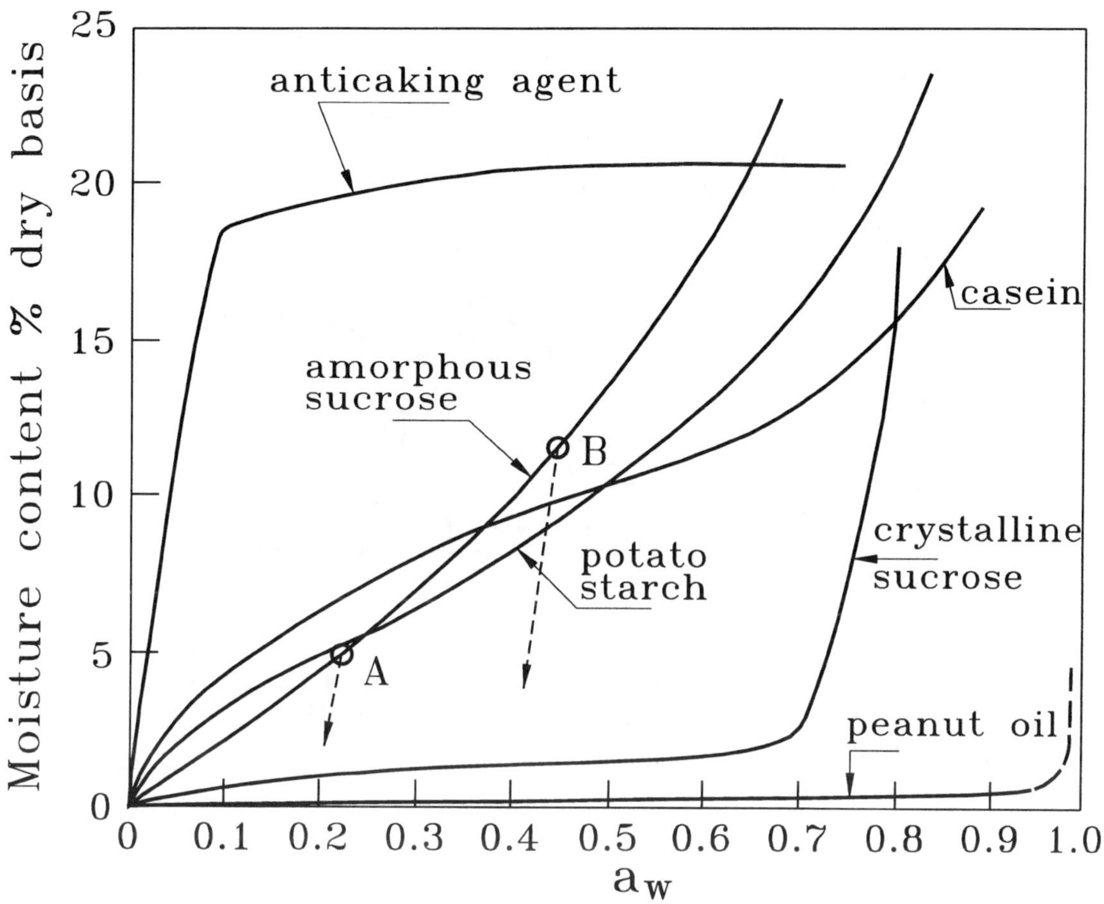


**Figure 9–1** Typical moisture sorption isotherms of important food components (temperature 20–25°C). In some sugars the transition from the amorphous to the crystalline form involves a drastic decrease in moisture (dotted lines).

served behavior of food materials and to their microstructure.

In region I, minimal water is contained in the product, and the water molecules present are tightly bound to active sites (e.g., polar groups in molecules) mainly by hydrogen bonding. A typical case is sorption of water onto highly hydrophilic biopolymers such as proteins and polysaccharides. The moisture content theoretically representing the adsorption of the first layer of water molecules (point A in the isotherm) is called the *monolayer value*, and it is generally found when $a_w$ is 0.2–0.4 and the moisture content is below 0.1 g/g solids.

Region II represents water more loosely bound, initially as multilayers above the monolayer; later, as moisture content increases, this water successively fills micropores and macropores in the system. In this region, chemical and biochemical (e.g., enzymatic) reactions requiring solvent water start to take place because of the increased mobility of solutes. The depression of water activity is mainly due to capillary condensation (see Section 3.6.5) and can be estimated by the Kelvin equation (equation 3–45).

In region III, excess water is present in macrocapillaries or as part of the fluid phase in high-moisture materials, exhibiting nearly all the prop-

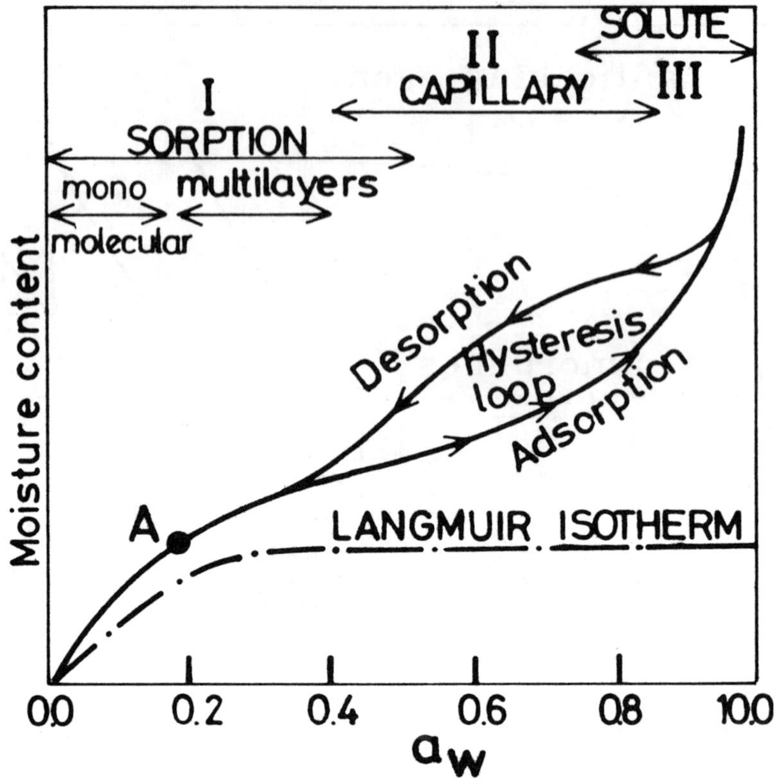


**Figure 9–2** Idealized moisture sorption isotherm of a food and water activity–depressing mechanisms. Point A represents the so-called monolayer value derived from models for sorption isotherms.

erties of bulk water. Microbial growth becomes a major deteriorative reaction in this region, and $a_w$ (or RVP) gives a rough indication of the lower limit for growth of salt-tolerant microorganisms (0.60), most molds (0.80), yeasts (≈0.87), and pathogenic bacteria (0.91). In some instances, the minimum $a_w$ for growth depends on the type of solute. At high water content, the soluble components of the system are in solution (or exist as saturated solutions), and the water-depressing action is described by Raoult's law for solute effects in ideal systems:

$$a_w = x_w = \frac{n_w}{n_w + n_s} \qquad \textbf{Equation 9–2}$$

where $x_w$ is the mole fraction of water and $n_w$ and $n_s$ are the number of moles of water and solute, respectively. This equation accurately predicts $a_w$ for dilute aqueous solutions of low molecular solutes such as sugars and salts, but care has to be taken since it does not consider the solubility limit of the solute in the system. Electrolytes deviate strongly from Raoult's law. A schematic representation of the three main water activity–depressing mechanisms that may act at the microstructural level in a food is presented in Figure 9–3.

### 9.2.3 Hysteresis

In practical sorption experiments, the desorption isotherm usually lies above the adsorption branch, a phenomenon called *hysteresis* (Figure 9–2). In hysteresis, a lower vapor pressure is needed to

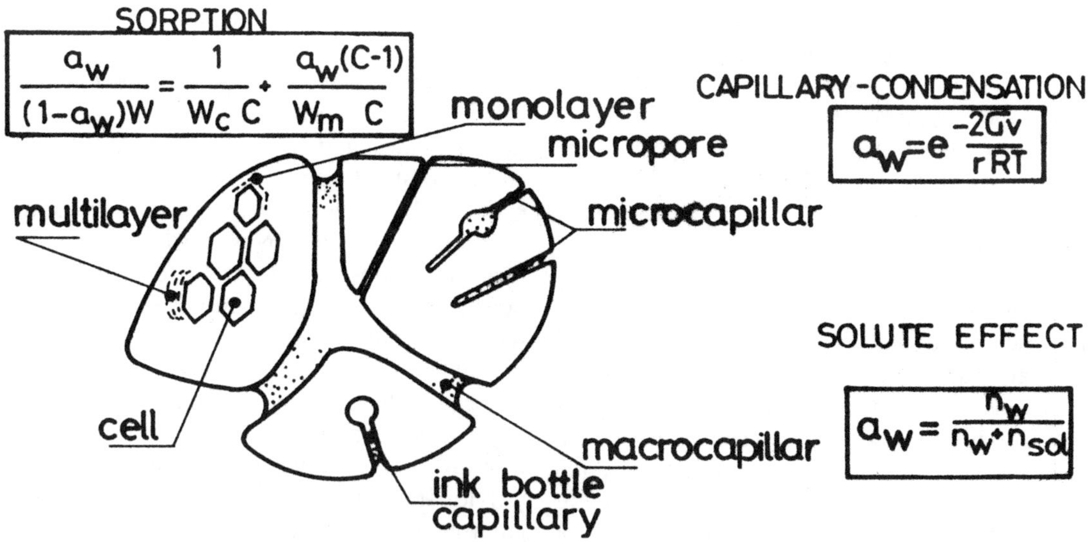


**Figure 9–3** Schematic diagram representing the basic water activity–depressing mechanisms acting at the microstructural level. *Source:* Adapted from Schneider (1981).

achieve a given moisture content by desorption than by adsorption. Possible explanations for this thermodynamical oddity are as follows: (1) solutions may become supersaturated during desorption (drying), holding more water than the crystals present in adsorption from the dry state; (2) due to the presence of impurities, the contact angle of the water film during desorption is smaller than that during adsorption, and therefore the adsorption branch will be at a higher vapor pressure; (3) the microstructure of the solid in the desorption branch may be different from that in the adsorption branch, as it is known, for example, that the shape of capillaries directly influences the form of the hysteresis loop (de Boer, 1958). All these explanations for hysteresis assume an equilibrium situation, which may not exist if an amorphous solid is above its $T_g$ (as may be the case when $a_w$ is high) and structural modifications take place in the rubbery state. Thus, the microstructure of the solid during desorption may be different from that in the adsorption branch.

Because of its historical value and to exemplify another possible effect of microstructure on hysteresis, the ink bottle neck theory is reviewed. This theory treats capillaries as closed bodies with larger radii than those of the connecting necks (an ink bottle capillary is shown in Figure 9–3). Since the Kelvin formula predicts that water activity in capillaries is proportional to $e^{-1/r}$, when water is being adsorbed and fills the pore body, it exerts a larger vapor pressure than in desorption, when the pore is full and the narrow neck is blocked by the water meniscus.

### 9.2.4 Sorption Phenomena in Foods

Most foods are complex, multicomponent systems consisting of many phases and complex microstructures. Consequently, deviations of sorption phenomena from the theoretical treatment previously presented are to be expected. Also, in actual food materials, the various mechanisms depressing $a_w$ overlap at certain intermediate points in the isotherm, complicating its interpretation.

At high-moisture levels, an aqueous environment prevails and many components, in particular salts and low molecular weight sugars, exist in solution. The undissolved components are high molecular weight materials (e.g., polysaccharides and proteins) forming the structural matrix of the food in which the aqueous phase is contained. Liquid water itself has a structure created by hydrogen bonding that is perturbed every time a solute is introduced. Depending on the type of solute added, deviations from Raoult's law arise, and other equations are available to calculate water activity for actual solutes in the high-moisture range (i.e., for $a_w > 0.75$).

High molecular weight polymers such as polysaccharides and some proteins may immobilize large quantities of water as gels. Sorption isotherms for gels extend over the whole range of $a_w$ and often show marked hysteresis interpreted in terms of capillary condensation theories and steric rearrangements (Texter, Kellerman, & Klier, 1975). Semi-rigid agar gels with $a_w = 0.8$ can hold 0.43 g water/g dry matter, equivalent to about half the amount of water of a glycerol solution of the same $a_w$ (Johnson, Busk, & Labuza, 1980). The contribution of gels to food microstructure and their depression of water activity deserve further research aimed at fabricating shelf-stable moist products, in particular, those that exhibit a crisp texture.

The Kelvin equation predicts significant lowering of the vapor pressure of water only when capillaries are the size of the absorbing molecules; hence its applicability is dubious. Most pores in foods are in the 10–300 $\mu$m range; assuming complete wetting, the predicted $a_w$ values would be in the 0.989–0.999 range. Only a small percentage of the pores are expected to be 0.01–0.001 $\mu$m, resulting in $a_w$ values of 0.340–0.889 (Labuza, 1984). Prediction of $a_w$ depression by capillary effects in foods is complicated by difficulties in determining pore sizes, swelling of the matrix during sorption, and lack of information about the actual radius of curvature of the meniscus (involving the wetting effect). To circumvent the last problem, a very interesting application of SEM was introduced by Gvirtzman, Magaritz, Klein, and Nadler (1987), who utilized a cold stage and cryoscopic chamber to rapidly freeze a wet sample of porous soil and investigate the morphology of the water menisci in situ. Additional work is necessary to demonstrate and quantify capillarity effects in foods, an area where microscopy can be a quite useful tool.

As moisture is reduced, the situation is complicated by supersaturation and delayed crystallization of solutes, particularly low molecular weight carbohydrates (sugars such as lactose, sucrose, etc.). At these intermediate moisture levels, mobility is diminished by viscosity effects, and amorphous molecules are probably in the nonequilibrium glassy state but slowly shifting into a more stable crystalline condition. Because crystals bind no internal water (chemically bound water in hydrates is not relevant to food processing), or substantially less than the amorphous state (see Figure 9–1), water is released. Accordingly, any low-moisture food containing sugars in an amorphous state (such as dry whey, nonfat dry milk, dried fruit powders, etc.) will crystallize at a rate that increases with $a_w$ and consequently with $(T - T_g)$. Crystallization leads to a complete change in physical structure and stability of the product.

### 9.2.5 Water Activity and Structural Stability

Low-moisture foods may be regarded at the microstructural level as nonequilibrium, heterogeneous systems. At this scale, a baked cookie may be viewed as an alloy of many phases or domains having different compositions and preferential concentrations of some of the major components (e.g., fat, protein, starch, and sugar). Segregation or concentration of components at the microstructural scale during processing may result from incomplete dispersion of an ingredient (e.g., starch granules), intrinsic phase immiscibility (fat and aqueous phases), crystallization or the formation of a high viscosity amorphous state during water removal, or thermodynamic incompatibility of polymers.

The moisture content in the microregions (domains) of a food need not be in equilibrium after processing, and water migration will occur until thermodynamic equilibrium is reached at the microstructural level. Since a difference in water activity between domains is the driving force for moisture migration, average or gross determination of the water activity of the product gives little information when changes are analyzed at the microstructural level. It is conceivable that the glass-rubber transition and its concomitant effects proceed at this level, although the product may be in thermodynamic equilibrium with the surrounding atmosphere [e.g., $a_w$ (product) = $a_w$ (head space)]. The fact that macroequilibrium does not necessarily imply equilibrium at the microstructural level has been disregarded in the analysis of stability of foods. Ways of measuring moisture content (or $a_w$) at the microstructural level are urgently needed, and electron energy loss spectroscopy in combination with scanning transmission electron microscopy may be a way of mapping water distribution (see Section 1.6.7).

An interesting phenomenon results from the difference in the water sorption behavior of amorphous and crystalline sugars. In the amorphous state, sugars are hygroscopic and tend to absorb large amounts of water at low relative humidity (Figure 9–1). In the crystalline state, sugars absorb moisture only at high relative humidity and as a result of solubilization. Above a critical water activity, the amorphous material releases water as it crystallizes, and the water content decreases to that of the crystallized form. This phenomenon is time dependent and may take years at low relative humidity (point A) or hours at higher relative humidity (point B) (Roos, 1995). The rate of crystallization also increases as a function of the difference between the actual temperature and the glass transition temperature ($T - T_g$). Crystallization of amorphous sugars has two important consequences. First, the water released during crystallization is absorbed by the rest of the material, which in a closed system may lead to increased storage instability (e.g., higher rates of nonenzymatic browning). Second, the amorphous-crystalline transition may involve important structural changes (e.g., collapse of the structure and changes in texture).

## 9.3 THE DRYING PROCESS

### 9.3.1 Heat and Mass Transfer Mechanisms

Drying of foods is a process involving simultaneous interface transfer of heat and mass (water vapor). In a typical air drying operation, a moist solid is placed in a closed environment in which hot air is circulated, causing evaporation of water from the body. Various transfer mechanisms present in drying are displayed in the physical model presented in Figure 9–4 (King, 1980). Heat and mass transfer effects are represented separately to facilitate comprehension, but they take place simultaneously.

Heat is transported from the surroundings to the surface of the material by radiation, convection, or conduction. In the common case of air drying, convection is the predominating mechanism. Heat reaching the surface is transported to the evaporation zone generally by conduction and radiation, but the prevailing mode depends on microstructural characteristics such as porosity; some of the heat is used to warm up the dried layer. The total heat flow also depends on the microstructural arrangement of chemical components with different thermal properties (e.g., fat and wet lean tissues in meats). Heat arriving at the interface is available for the vaporization of water, and the vapor produced must be transported through the interior of the solid to the surface. Liquid water and water vapor transport mechanisms important in drying and their corresponding equations are presented in Bruin and Luyben (1980). Once the water vapor reaches the surface of the solid it must be removed from the immediate surroundings; otherwise, it would stop the diffusion process from the inside.

This physical model clearly shows that heat and mass transfer between the surface of a solid and the evaporation zone is highly dependent on the product microstructure. The rate of transfer can be expressed mathematically as

$$Rate = transfer\ coefficient \times driving\ force$$

**Equation 9–3**

The driving force for heat transfer is a temperature difference whereas that for mass transfer can be expressed as a moisture or partial pressure difference. Since the instantaneous driving forces for heat and mass transfer are thermodynamic variables, all microstructural effects are lumped in the rate transfer coefficient. At any instant during drying, one of the transfer mechanisms has the small-

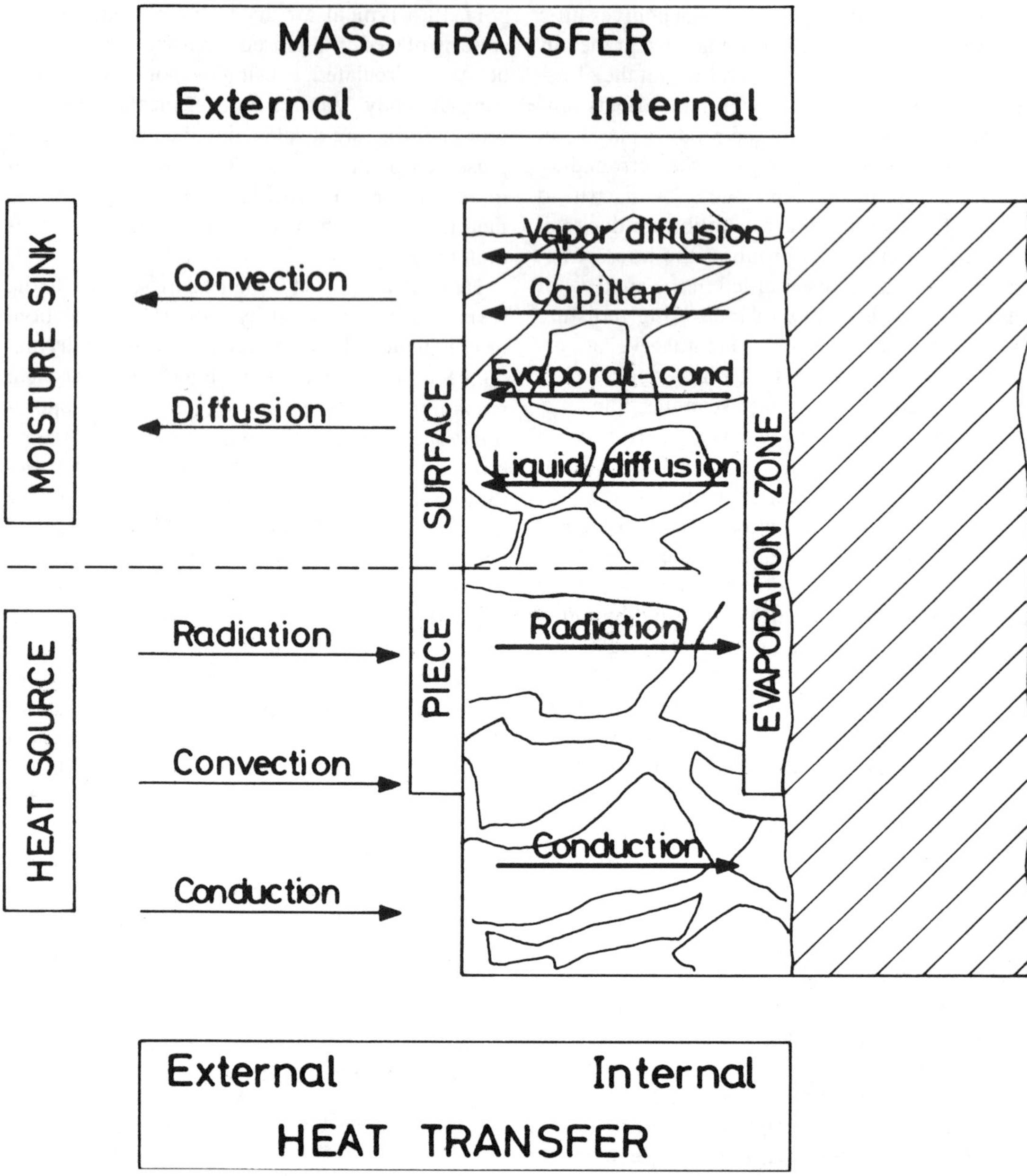


**Figure 9–4** Heat and mass transfer during drying of foods and major transport mechanisms. *Source:* Adapted from King (1980).

est heat or mass transfer coefficient and requires the largest temperature or concentration driving force. This is the rate-limiting or -controlling step, and efforts should be aimed at enlarging its transfer coefficient.

Initially, the drying rate is controlled by external resistances, but as drying proceeds, internal resistances usually increase and become rate limiting. In dense products, it is more likely that the transfer of water from the interior of the drying piece to the surface is more rate limiting, while in highly porous foods (foam or freeze-dried), internal heat transfer may be dominant (King, 1980). This means that no matter how fast the other steps are, the drying rate will be controlled by the capacity to transport water from the inside or heat to the center. Further details on basic aspects of transport phenomena can be found in King (1980) and Cussler (1997). Fundamental as well as practical aspects of drying are reviewed in *Handbook of Industrial Drying* (Mujumdar, 1995) and in the book by Barbosa-Canovas and Vega-Mercado (1996).

### 9.3.2 Drying Periods

Attempts have been made to gain a phenomenological understanding of the drying process. Since products being dried vary widely in physical and structural properties, and drying conditions are also quite variable, only general consequences can be derived. Nevertheless, such an attempt can be instructive, because it moves us away from the "black box" approach and switches the focus to possible mechanisms operating at different times during drying. The quantitative basis for analysis is the drying curve depicted in Figure 9–5.

The drying process is usually divided into an initial constant rate period and subsequent falling rate periods. If the product is initially wet, its surface can be assumed to be covered by a thin film of water, and evaporation takes place from the surface at a temperature close to the wet-bulb temperature. Moisture exerts nearly full vapor pressure and is held on the surface and in large capillaries. During this initial period, the rate of evaporation or the rate of drying remains constant until the average moisture content of the product reaches a value $W^*$, the *critical moisture content* shown in Figure 9–5. The critical moisture content is not a property of the food but depends on particle size and on the conditions of the drying air. It should be pointed out that only seldom is the constant rate period observed in industrial food drying.

In the constant rate period, the main mass transport mechanism is capillary flow of liquid water, although some liquid diffusion may exist. The internal mechanism of moisture flow does not affect the drying rate, which from the viewpoint of the product is a surface-wetting phenomenon. However, as the diameter of pores and capillaries decreases, shrinkage sets in. As expected, the flow of liquid water carries accompanying solutes, which become deposited on the surface because they are nonvolatile. The result is a condition known as *case hardening* that greatly impairs water removal in later stages.

When the outer surface of the product becomes "unsaturated" with moisture, one or more falling rate periods may set in, and the temperature rises continuously from the wet-bulb point. A first falling rate period begins when the continuous water layer is replaced by threads of moisture over the entire evaporating surface. This is sometimes called the *funicular state*, and since the surface occupied by liquid water decreases, the evaporation rate also diminishes. In the first falling rate period, vapor diffusion from the evaporating zone to the surface is the predominant mechanism.

In the second falling rate period the evaporation surface has receded into the solid, a situation termed the *pendular state*. The drying rate falls sharply and is controlled by the internal rate of moisture movement. Moisture may be held in fine capillaries, and it migrates to the vapor phase mostly by evaporation-condensation.

A third period starts when only bound water exists in the inner portions of the material. The process slowly approaches equilibration as evaporation of water equals condensation and the partial pressure difference between interior and air is small. In porous solids, heat transfer may be the controlling mechanism. Figure 9–5 shows the variation of the drying rate as drying proceeds through different periods.

### 9.3.3 Microstructure and Moisture Distribution During Drying

Drying curves are derived from variations of the *average* moisture content of the drying piece with time. More important, however, is how moisture is distributed inside the body and varies with time, since many chemical and structural reactions (e.g., those related to $T_g$) are a function of

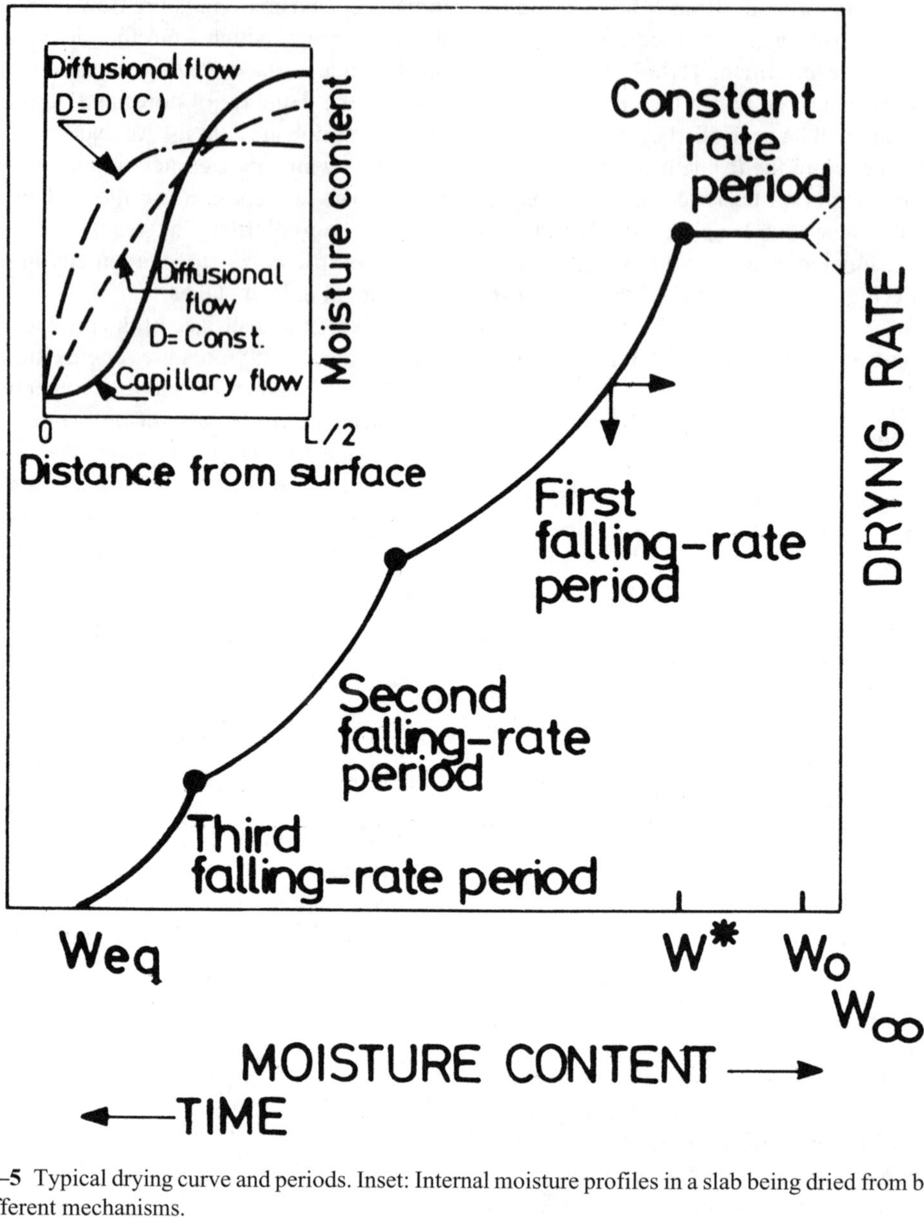


**Figure 9–5** Typical drying curve and periods. Inset: Internal moisture profiles in a slab being dried from both sides under different mechanisms.

the water activity. Figure 9–6 shows hypothetical moisture profiles in a hygroscopic body as drying proceeds.

Initially, a wet material of a certain size (e.g., a droplet of milk) may have a very high moisture content ($W_\infty$). As water evaporates freely from the surface, appreciable shrinkage occurs, until a solid structure is formed with an average moisture content of $W_o$. The moisture content is still uniformly distributed in the interior of the piece or droplet, but water must now be transported from the interior to the surface. No

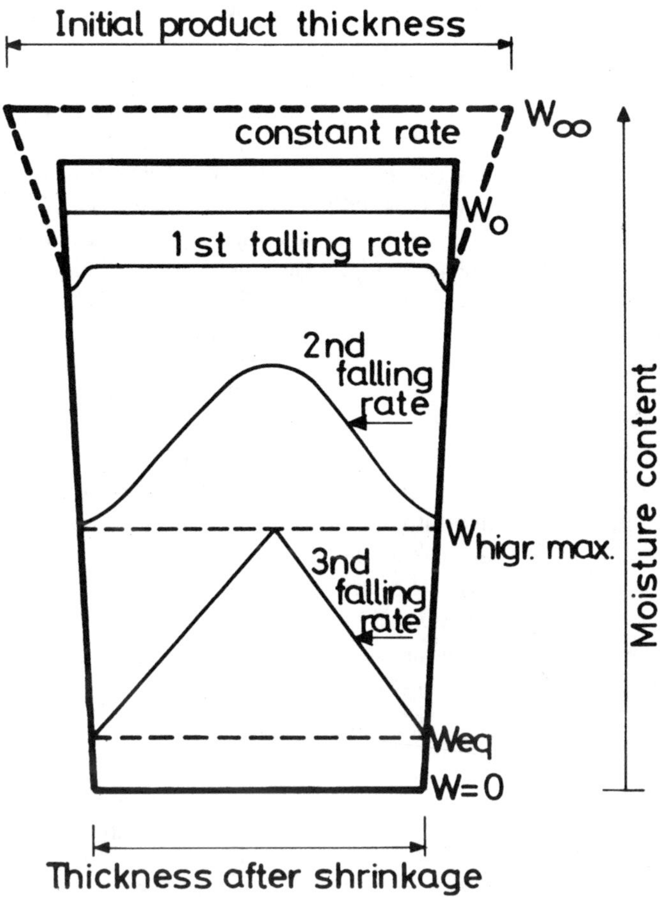


**Figure 9–6** Drying periods and moisture profiles during drying of a slab. *Source:* Adapted from Kessler (1981).

significant further shrinkage takes place from here on.

Depending on the prevailing moisture transport mechanism and the microstructure into which the solid has set, different moisture distribution profiles develop during the falling rate period. If diffusion is the predominant mechanism, as it would be in nonporous solids, parabolic profiles arise in accordance with Fick's law. However, if flow by capillarity predominates, as occurs in porous microstructures, a point of inflection divides the profile in two parts, one concave upwards and the other concave downwards. The actual moisture profile inside a food slab when mixed mechanisms are present is different from both theoretical profiles (see inset of Figure 9–5).

Drying rate and moisture profiles as well as the interactions between microstructure and drying conditions are critical in the dehydration of long pasta products. Microscopic examination of a cross section of freshly extruded spaghetti reveals a compact structure with intact starch granules deeply embedded in a protein matrix. The outer surface is also dense and is coated with a continuous protein film (Matsuo, Dexter, & Dronzek, 1978). The extreme compactness of the microstructure results in a low rate of drying controlled by internal diffusion, which imposes several restrictions on the process:

- absence of unwanted steep moisture gradients due to liquid movement that may lead to structural stresses and internal cracking (Ollivier, 1985)
- control of temperature to avoid starch gelatinization of the hydrated granules and to ensure a gluten structure suitable for satisfactory cooking consistency
- minimization of browning and enzymatic and microbial (acidification) reactions highly favored by intermediate- and high-moisture conditions (Cantarelli, 1985)

The parts closer to the surface dry more quickly, so they have a greater tendency to shrink (see Section 9.3.6) than the inner layers, resulting in the development of shear stresses. Shearing appears parallel to the surface while tensile stresses develop at right angles to the surface, but only the latter stresses can cause cracking or crazing (Gorling, 1958). Traditional drying of spaghetti consists of several stages of time, temperature, and relative humidity combinations aimed controlling the evaporation of water. In the first stage, lasting 1 to 1.5 hours, the moisture content is reduced from 32% to 21% using high-temperature air. Then follows a 6-hour second stage to further reduce moisture to about 15%. Lastly occurs an equilibration period of about 6 hours to adjust the moisture content to approximately 12.5%. It is not uncommon for the total drying time to be 15–20 hours. Microwave drying has circumvented many of the limitations of conventional drying of pasta. Since high-frequency heating evaporates water from the interior of the product and causes convecting cooling on the surface, the outside surface remains wet during the drying process, reducing the risk of cracking.

The main microstructural features of a dry spaghetti noodle are shown in Figure 9–7. The outer surface presents intact starch granules that did not undergo gelatinization during drying and that are embedded in a dense protein matrix (A). The compact inner structure and cracks produced during drying are depicted in part B.

### 9.3.4 Mathematical Modeling

Mathematical modeling is important for quantitating the effect of changes in variables and parameters on the drying rate and the moisture content of the material. The ultimate objective is to be able to predict the final conditions of the dried product and the course and extent of many reactions taking place during drying, such as browning (Aguilera, Chirife, Flink, & Karel, 1975).

Since the falling rate periods account for a major proportion of the drying time and are usually controlled by internal mass transfer, modeling has commonly applied simple Fickian diffusion to the last stages of drying. For a semi-infinite slab with initial uniform moisture distribution, the extent of water removal $W^*_{avg}$ is given by

$$W^*_{avg} = \frac{W_{avg} - W_{eq}}{W_o - W_{eq}} = \frac{8}{\pi^2} \sum_{i=0}^{\infty} \frac{1}{(2i+1)^2} \times exp\left[-\frac{(2i+1)^2 \pi^2 Dt}{L^2}\right] \quad \textbf{Equation 9–4}$$

where $W_{avg}$ is the average moisture content (dry basis) at any time; $W_o$ is the initial moisture content; $W_{eq}$ is the equilibrium moisture content (in equilibrium with the relative humidity of the air); $t$ is the drying time; $L$ is the thickness of the slab; and $D$ is the diffusion coefficient, assumed to be constant. Similar expressions are available for other simple geometries.

When $W^*_{avg} < 0.6$, the previous equation reduces to this simplified expression (Vacarreza & Chirife, 1978):

$$W^*_{avg} = \frac{8}{\pi^2} exp\left(-\frac{\pi^2 Dt}{L^2}\right) \quad \textbf{Equation 9–5}$$

Consequently, the extent of moisture removal, according to Fick's law, is directly proportional to the diffusion coefficient and drying time and inversely proportional to the square of the characteristic product dimension. A plot of $ln\,(W^*_{avg})$ versus $t$ is a straight line with a slope proportional to $D/L^2$, from which a constant value for $D$ can be obtained.

Simple equations like equation 9–5 are attractive to use because all phenomenological parameters are hidden in a single coefficient, the diffu-

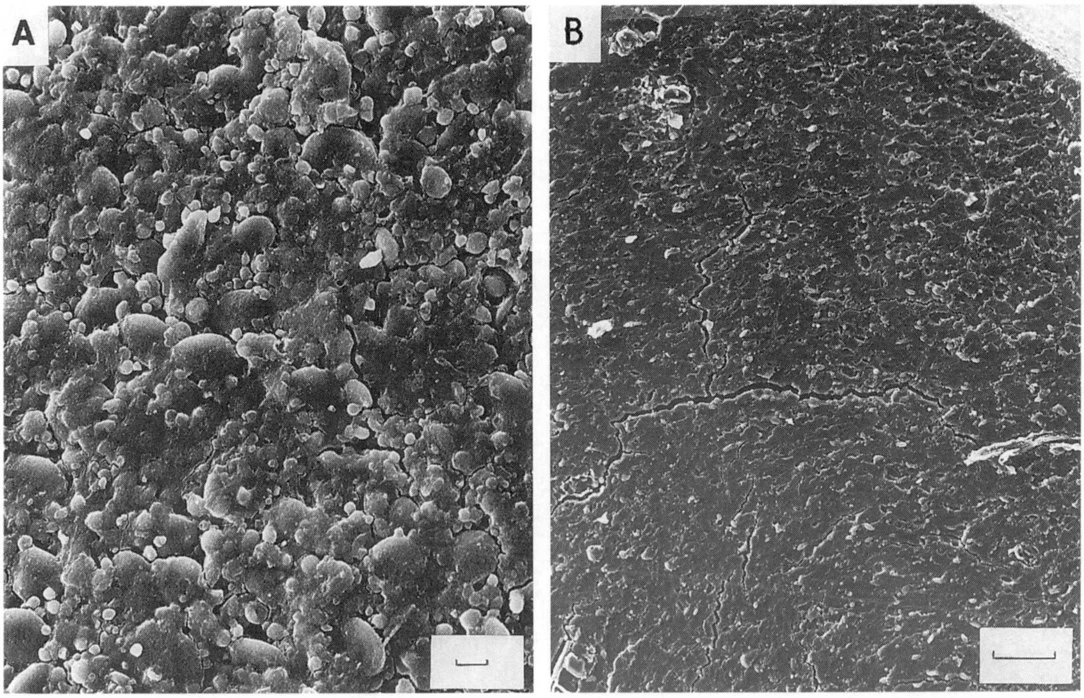


**Figure 9–7** Scanning electron micrographs of a dry spaghetti noodle. (A) Outer surface showing intact starch granules embedded in a dense protein matrix. (B) Cross section showing dense structure and presence of cracks produced during drying. Scale bars = 10 $\mu$m.

sivity. Unidirectional molecular diffusion of a single species is an ideal situation present only in the evaporation of water from solutions and gels. The *effective* or *apparent* diffusivity, $D_{eff}$, derived from experimental data plotted as described above, encompasses all mass transfer mechanisms. Yet, the basic assumptions involved in the derivation of equation 9–5 must be fulfilled (simple geometry, no shrinkage, product homogeneity, etc.). According to Karel (1975), they usually are not, and variations in reported values for $D$ (or $D_{eff}$) from $10^{-8}$ to $10^{-5}$ cm$^2$/s, often for similar materials, stem from violations of the assumptions underlying the application of Fick's law. Marinos-Kouris and Maroulis (1995) have compiled extensive data for the effective moisture diffusivities of several foods. They report that moisture diffusivity in foods ranges from $10^{-9}$ to $10^{-2}$ cm$^2$/s, with most values (82%) gathered in the interval $10^{-7}$ to $10^{-4}$ cm$^2$/s. The authors surmise that this large variation is due to the complex structure of foods and the strong binding of water to food polymers. Gekkas (1992) has tabulated values for the moisture diffusivity of several foods and gels during air drying. For apples, diffusivity values vary from $36 \times 10^{-6}$ in the initial stages to $0.065 \times 10^{-6}$ cm$^2$/s at the end, a five-hundred-fold change. Apparent moisture diffusivity in fish varies between $0.13 \times 10^{-6}$ and $2.6 \times 10^{-6}$ cm$^2$/s in the temperature range 30–40°C. The moisture diffusivity during drying of potato gels differs by a factor of 10,000 ($1.5 \times 10^{-6}$ to $1.0 \times 10^{-10}$ cm$^2$/s). In a study of the effect of structure, starch particles of the same origin and size (50 $\mu$m) but subject to different pretreatments exhibited moisture diffusivities varying from $5 \times 10^{-8}$ to $8 \times 10^{-5}$ cm$^2$/s. Thus, values of effective moisture diffusivities can at best characterize the drying of a material under specific conditions. In conclusion, many factors affect the moisture diffusivity in a food, ranging from those related to its composition and the distribution of components to processing parameters such as temperature and rate of drying. Most likely, all these factors ultimately alter the structure of the food matrix and thus physical parameters (e.g., porosity) and obstacles to water transport (e.g., membranes) are relevant to the process.

Mass transfer is accompanied by heat transfer. Important thermal properties of foods include thermal conductivity ($k$), specific heat ($c_p$), and thermal diffusivity ($\alpha = k/c_p\rho$), which substitutes for $D$ in the differential equations related to unsteady state heat transfer. For Lewis numbers ($\alpha/D$) greater than 60 or a characteristic dimension smaller than 3 cm, thermal gradients can be safely neglected and the temperature considered uniform throughout the sample but varying with time. When heat transfer controls or thermal gradients inside the particle become important, solutions to Fourier's law can be calculated; these are similar to the equations presented above for mass transfer. Expressions for calculating parameters relevant to heat transfer as a function of water content and composition are reviewed by Sweat (1986). Sweat's article is the basis of the first part of the following section, and the reader is referred to the original for a detailed treatment of the topic.

### 9.3.5 Heat and Mass Transfer Properties and Microstructure

We have discussed, in Section 8.2.3, how models can assist in defining an effective diffusion coefficient based on the architecture of a solid and on individual properties of the phases. Several parameters related to heat transfer are needed to model drying of foods. Specific heat (kJ/kg°C) and thermal conductivity (W/m°C) are commonly modeled as a linear function of water content ($W$ = kg water/kg total weight). For example:

Above freezing $c_p = 0.837 + 3.349W$

**Equation 9–6**

Below freezing $c_p = 0.837 + 1.256W$

**Equation 9–7**

For fruits and vegetables $k = 0.148 + 0.493W$

**Equation 9–8**

For meats and fish $k = 0.080 + 0.52W$

**Equation 9–9**

Additive models based on composition are also available, but care should be taken before using them, since they are derived from correlations involving a wide variety of foods and do not take into account the effect of microstructural arrangements discussed below:

For liquid foods $c_p = 4.180X_w + 1.711X_p + 1.928X_f + 1.547X_c + 0.908X_a$

**Equation 9–10**

For nonporous or liquid foods $k = 0.61X_w + 0.2X_p + 0.175X_f + 0.205X_c + 0.135X_a$

**Equation 9–11**

where $X$ is the weight fraction and the subscripts for various components are as follows: $w$ = water, $p$ = protein, $f$ = fat, $c$ = carbohydrate, and $a$ = ash.

As noted, knowledge of the microstructural arrangement of heterogeneous food should lead to improved physical models and better modeling and simulation. Figure 9–8 presents the complex microstructure of the hull of three oilseeds, a structural element that controls the flow rate of water during soaking and drying as well as the diffusion of gases. The same sort of distinctive physical outer resistance is found in the skins of fruits and vegetables (see Section 9.3.8) and in impervious films or layers formed during drying (e.g., case hardening).

In the case of heterogeneous biphasic materials, several structural models can be used to calculate an overall value for a transport property, such as thermal conductivity $k$ (Table 9–1). The series model assumes that heat conduction is perpendicular to alternating layers of the two phases whereas the parallel model assumes that it is parallel to both phases. In the mixed model, the process takes place by a combination of series and parallel phases. As its name implies, the random model supposes that the two phases are randomly mixed, while in the Maxwell-Eucken model, one phase is continuous and the other is dispersed in the form of spheres. The term $\phi$ represents the volume fraction of phase 2, and, in the mixed

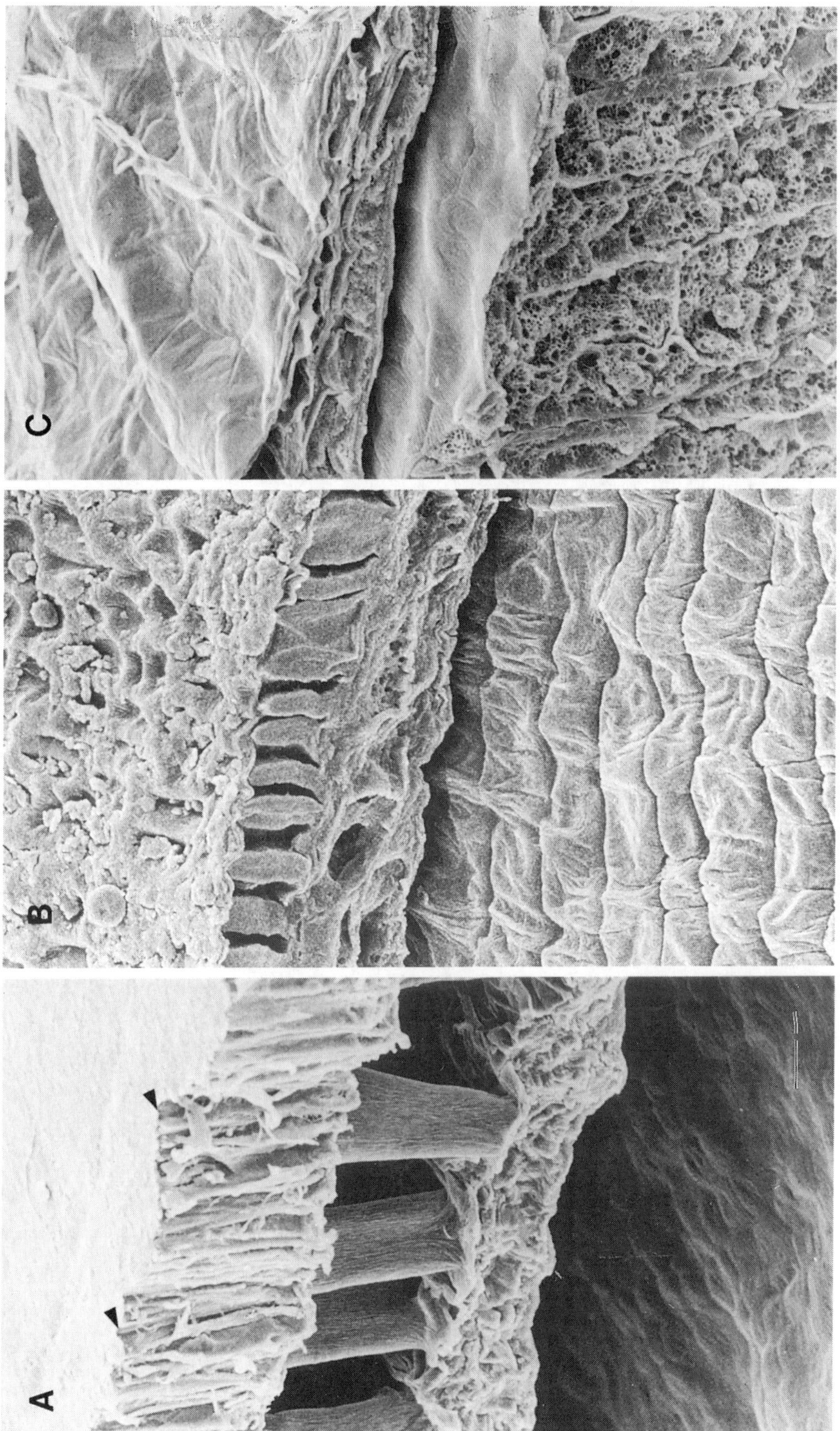


**Figure 9–8** Scanning electron micrographs of hulls from sunflower seed (A), canola seed (B), and soybean (C). The complex architecture of the seed coat contrasts with the apparently regular cellular arrangement of the cotyledon. Notice in (A) the thin outer layer resembling an anisotropic membrane (arrows). Marker = 10 $\mu$m.

model, F is the fractional contribution of the series term.

An important feature of these relationships is that they depend on the volume fraction of one phase and not on the size of dispersed elements (e.g., in the Maxwell-Eucken model there is no effect of the size of the spheres). The parallel model gives the upper value of the property and the perpendicular model gives the lower bound.

### 9.3.6 Improved Drying Models

The transport of water in structured food materials is difficult to describe mathematically. Several correction criteria have been introduced to compensate for the complexity in mass transfer and microstructural effects. Improved results are obtained when variable diffusion coefficients are used. Moisture distribution profiles similar to those present in real foods (and those shown in the inset of Figure 9–5) can be obtained by modeling Fick's second law with a diffusion coefficient that varies with the moisture content (Husain, Chen, & Clayton, 1973). Experimental data demonstrated that indeed the diffusivity decreases sharply as moisture content is reduced, as shown in Figure 9–9.

Better moisture profiles can also be obtained by postulating a parallel dual diffusion model with different diffusivities. A theory was devised for water vapor transport in porous bodies, where permeability of the porous structure is dependent on the moisture content (Harmathy, 1969). The simplest microstructural effects in diffusion of water vapor in porous solids are those due to tortuosity ($\tau$) and porosity ($\varepsilon$). $D_{eff}$ values are obtained as a function of $f(\varepsilon D/\tau)$, where $\tau$ varies between 1.5 and 10. Considering all previous factors affecting moisture diffusivity, a general equation may be proposed that takes into account structural, moisture ($W$), and temperature ($T$) effects:

$$D(W\,T) = a_o(\varepsilon, \tau)\,exp(a_1 W)\,exp(-a_2/T)$$

**Equation 9–12**

where $a_o$ is a parameter accounting for structural effects and $a_1$ and $a_2$ are constants.

For the first falling rate period of dense products having very small pores, a diffusion resistance factor has been defined that encompasses the reduction in cross-sectional area for flow and the increase in tortuosity (Kessler, 1981). The value of the resistance factor indicates how much less water vapor diffuses compared to a free layer of equal cross-sectional area. Values for some dried products vary from 6.8 (chocolate pudding powder) to 1.6 (roasted coffee). Correlations also exist between $D_{eff}$ and the diffusivity of a solute in bulk phase for liquids as a function of porosity and the ratio of the diffusing molecule to the pore radius (Ternan, 1987).

Mathematical models are also available for drying where particular physical or geometrical characteristics are relevant. Loncin (1980) developed a model for heterogeneous materials

**Table 9–1** Structural Models for Thermal Conductivity ($k$) in Heterogeneous Biphasic Materials

| *Model* | *Equation* |
|---|---|
| Parallel | $k = (1 - \phi)k_1 + \phi k_2$ |
| Series | $1/k = (1 - \phi)/k_1 + \phi/k_2$ |
| Mixed | $\frac{1}{k} = \frac{1 - F}{(1 - \phi)k_1 + \phi k_2} + F\left(\frac{1 - \phi}{k_1} + \frac{\phi}{k_2}\right)$ |
| Random | $k = k_1^{(1-\phi)} k_2^{\phi}$ |
| Maxwell-Eucken | $k = \frac{k_2[k_1 + 2k_2 - 2(1 - \phi)(k_2 - k_1)]}{k_1 + 2k_2 + (1 - \phi)(k_2 - k_1)}$ |

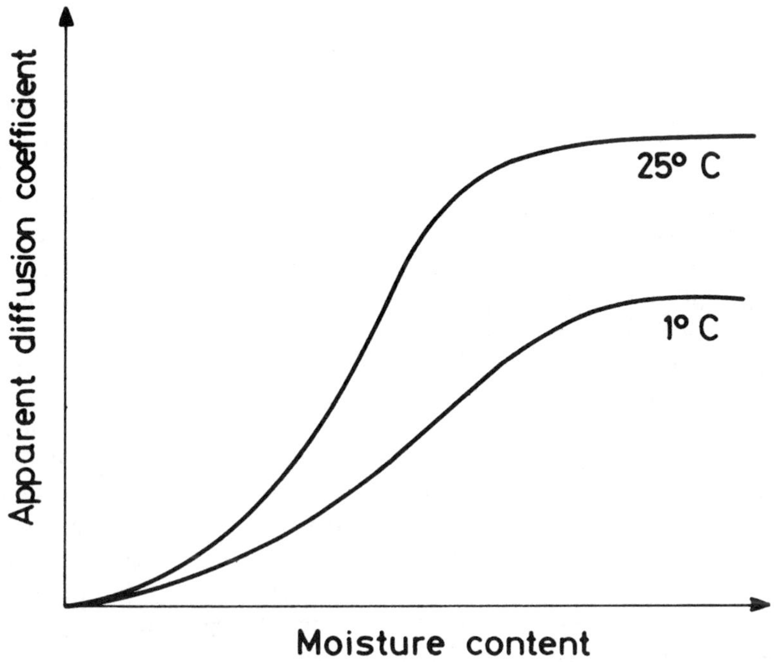


**Figure 9–9** Variation of the apparent diffusion coefficient of water with moisture content during drying of foods.

having a surface resistance. Drying of fruit while the cellular structure is maintained was modeled as a series-parallel arrangement, with water fluxes between cells and along the walls being a similar order of magnitude (Crapiste, Rotstein, & Urbicain, 1985). Numerical methods, such as finite element analysis, have become of great assistance in solving cases of irregularly shaped particles, variable geometry, heterogeneity, and moisture-dependent diffusion coefficients. A better knowledge of the phenomenology involved and of the physical and structural properties of food materials is necessary if these powerful computational procedures are to deliver their full potential benefit.

### 9.3.7 Microstructural Changes During Drying

Drying of long pasta introduced the subject of interactions between drying conditions and product microstructure. These interactions give rise to several phenomena: loss of cellular structure, shrinkage, changes in macromolecules, and changes in sugars.

#### *Loss of Cellular Structure*

In a high-moisture cellular material such as plant tissue, transport of water during drying occurs in a highly heterogeneous medium. Moisture must migrate from the protoplasm through the cell membrane and surrounding wall and across the porous structure of the tissue. Loss of cellular structure may be either induced (to facilitate water migration from the interior of a piece) or preserved (if textural properties similar to those of the intact tissue are required after rehydration).

Blanching or scalding prior to dehydration is sometimes done to inactivate enzymes and denature cell membranes. The drying rate is increased and the drying is faster and more complete (Lazar & Rasmussen, 1964). Fish (1958) demonstrated that blanched potato pieces had diffusion coefficients similar to those of a potato starch gel, with the role of membranes minimized after scalding. However, by mathematical modeling, Rotstein and Cornish (1978) showed that for apples and similar products the controlling resistance in air drying was not the cell membrane but either the tissue itself or the boundary layer surrounding the food piece.

### *Shrinkage*

Shrinkage of food pieces during air drying adversely affects the quality of dried products, and in spite of its technological and economic importance, it is not well understood. In some fruits and vegetables, shrinkage is extensive and affects the rate of drying as well as physical and functional properties of the product. The shrinking behaviors of different food materials (resulting in different particle shapes and microstructures) were observed and classified by Luyben, Olieman, and Bruin (1980) as follows: type I, no shrinkage (animal feed); type II, perfect homogeneous shrink (glucose solution); type III, perfect homogeneous shrink, first external and later internal (skim milk and coffee extract); type IV, homogeneous shrinkage, first perfect and later rumpling but maintaining a constant surface area (apple tissue); and type V, a combination of III and IV (potato tissue). Types III and V develop internal voids owing to early setting of the outer structure while internal shrinking proceeds.

It is believed that moisture gradients within a particle induce microstructural stresses that lead to shrinking. The rate of water removal influences the extent of shrinkage during air drying of tissue foods. If drying is slow, internal stresses are minimized and the product shrinks nearly uniformly into a solid core. However, if drying is fast so that the surface is much drier than the center, a permanent tension sets in that preserves the original dimensions of the piece but causes numerous cracks and voids in the interior. Water imbibition during rehydration as well as product density can be controlled by adjusting the drying rate. To summarize, air drying of vegetable tissues is characterized by shrinkage and a slow falling rate period, leading to several microstructural changes (see Figure 9–10).

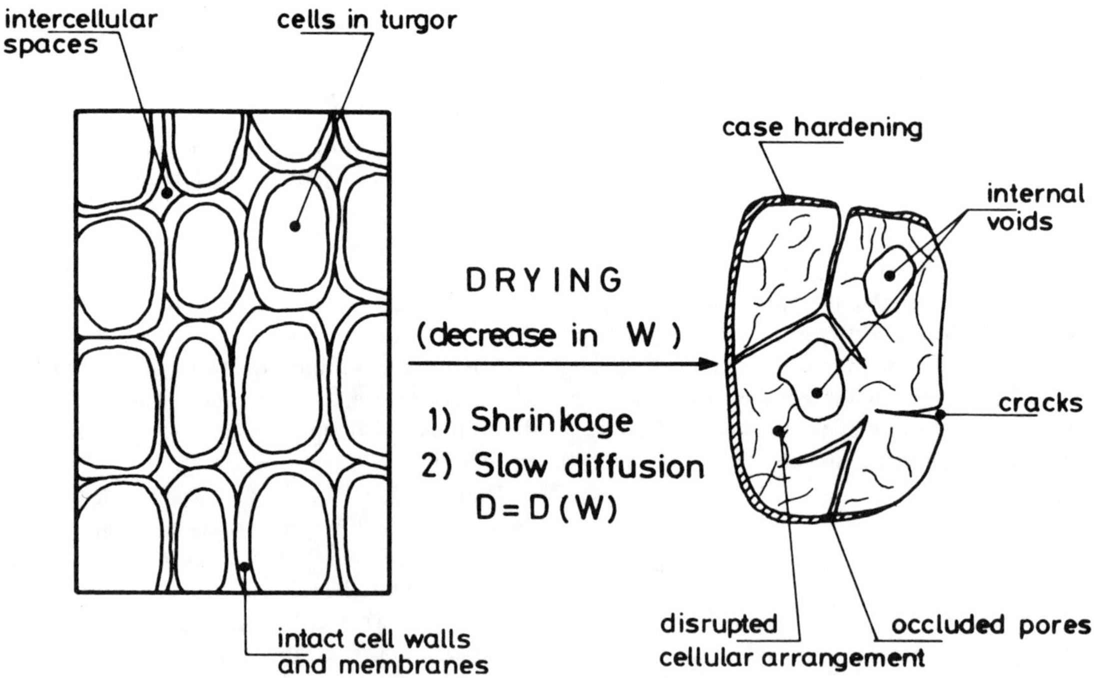


**Figure 9–10** Microstructural changes during air drying of vegetable tissues.

Shrinkage in fruits and vegetables, according to some authors, is related to the glass transition of the matrix of soluble components (mostly sugars), but other authors assign a major structural role to the cell walls. The view that the glass-rubber transition explains structural changes during air drying appears to be based on studies of freeze-drying of sugars and polysaccharide solutions. Freeze-drying induces minimal product shrinkage unless the temperature of the dry matrix exceeds a "collapse" temperature ($T_c$). Bellows and King (1973) used sugar solutions as models for fruit juices and proposed that collapse during freeze-drying occurred when the viscosity of the freeze-concentrated phase was reduced 4–7 log cycles below that in the glassy state ($10^{11}$ Pa·s). $T_c$ is quite low for citrus juices (−24°C to −36°C) but higher for potato tissue (−1.5°C). Levine and Slade (1986) postulated that collapse is a $T_g$-governed phenomena, and a close correlation was found between $T_g$ and $T_c$ for maltodextrins (e.g., $T_c$ occurs 40°C to 70°C above $T_g$, depending on the moisture content). Anglea, Karathanos, and Karel (1993) found that fresh potato and apple tissues had a single $T_g$ of −45°C when scanned from −100°C to 20°C, and they assigned this value to the $T_g$ of the concentrated amorphous sugar solution. However, Karathanos, Anglea, and Karel (1993) found that significant shrinkage occurred during drying of celery no matter how low the temperature of the drying air, although ($T - T_g$) was 75°C higher for air at 80°C than at 5°C.

An alternative hypothesis states that the edible fleshy structures of plants largely consist of cells filled with a sugar-acid solution and that the cell wall material, accounting for only 1–3% of the total weight, is responsible for the rigidity and solidlike behavior of the fresh tissue. Consequently, in cellular material the role of soluble sugars in shrinkage is likely to be minor compared with the deformation and flow of the cell walls. This fact has largely been disregarded in research on the shrinkage of plant material during drying. So far, there are no reliable data for the $T_g$ of cell wall material (Aguilera, Cuadros, & del Valle, 1998).

***Changes in Macromolecules***

As the moisture content decreases during the latter part of drying, several reactions occur at the macromolecular level. Among the most important are crystallization of polysaccharides in plant material and aggregation of proteins in muscle tissue, leading in the latter case to a loss in tenderness. Aggregation of macromolecules at low-moisture levels—particularly proteins—results in loss of solubility and biological activity (Aguilera & Karel, 1997).

***Changes in Sugars***

Rate of drying can affect the final state of a sugar. Typically, rapid drying is associated with an amorphous rather than a crystalline state. For example, hygroscopic whey powders contain mostly amorphous or glassy lactose whereas in nonhygroscopic powders most of the lactose is crystallized. As hygroscopic particles take up moisture from the air, mobility is increased and lactose molecules rearrange themselves into regular crystal lattices at a rate that depends on $a_w$ and temperature (Saltmarch & Labuza, 1980).

### 9.3.8 Use of Microscopy for the Interpretation of Microstructural Changes During Drying

It should be evident by now that the complexity of foods at the microstructural level affects dehydration. Models such as those discussed in Section 9.3.4 should be built after correct interpretation of the phenomenological steps and consideration of the microstructural changes. Microscopy can be of valuable assistance in model building, as claimed by Gejl-Hansen and Flink (1976). These authors described the predominant features of dehydrated food materials using different microscopy techniques (stereo, transmitted bright-field, scanning electron, and polarized optical microscopy). Spiess (1969) used a dark-field microscope to develop heat and mass transfer models for foods. Flink, Gejl-Hansen, and Karel (1973) designed a microscopy stage to follow changes during freeze-drying and to study the influence of microstructure on aroma retention. Lee and Rha (1979) utilized SEM to study the microstructure of spray-dried soy protein isolates and freeze-dried semisolid foods.

Light microscopy is a simple but useful technology for assessing changes at the cellular level after drying. Photomicrographs of fresh and air-dried carrot show how the original tissue is significantly disrupted after air drying (Figure 9–11). Light microscopy technique was used to confirm the absence of tissue damage and the disruption of cell walls after air drying in rehydrated celery treated with a glycerol solution (Shipman, Rahman, Segars, Kapsalis, & Westcott, 1972). Similarly, transmission electron microscopy was used to correlate changes in the cell membranes and textural changes in celery. Dehydration effects were virtually reversed by rehydration to $a_w$ 0.987 (Willis & Teixeira, 1988).

Jason (1958) performed a thorough study of the falling rate period during the drying of cod muscle. No capillary or porous structure was observed, but a "continuous gel structure" appeared that behaved like an isotropic medium. It was concluded that the main mechanism of water removal was molecular diffusion in a solid medium. Effective diffusion coefficients parallel to the three principal axes varied 10% at most. The falling rate period took place in two distinct phases characterized by constant effective diffusion coefficients independent of shrinkage, which were considerably higher in the first than in the second phase ($2.2–3.6 \times 10^{-6}$ and $0.3–1.0 \times 10^{-6}$ $cm^2/s$, respectively). The effect of fat was to decrease the values of the effective diffusion constants in both phases of the falling rate period. Other authors have reported that during drying of fish, changes in dimension along the main axis are less than half those occurring in width and thickness (Balaban & Pigott, 1986).

Scanning electron microscopy is an ideal tool to study the microstructure of spray-dried milk

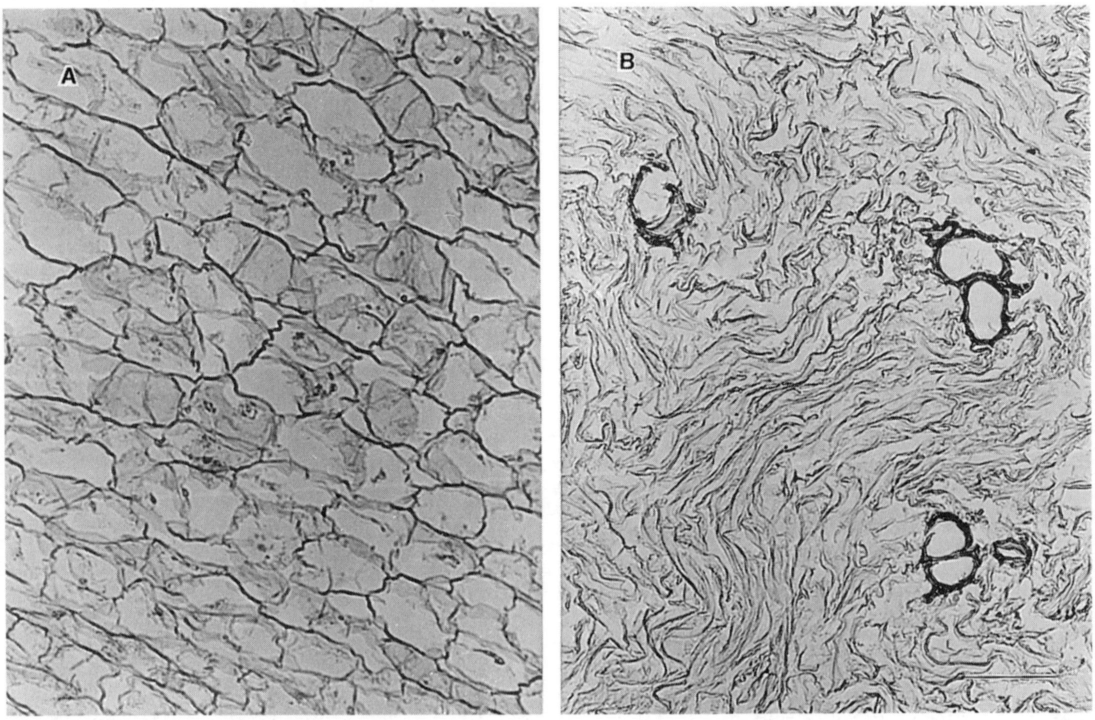


**Figure 9–11** Light microphotograph of sections of carrot tissue. (A) Fresh. (B) Air-dried in a convection oven at 70°C. Marker = 50 $\mu$m.

particles (Buma & Henstra, 1971a). Spray-dried materials are usually hollow spheres, corresponding to type III in the previous classification. Vacuole formation originates from a shrinking process that occurs after case hardening of the outer surface and subsequent expansion of air bubbles trapped inside the droplet (Verhey, 1972). Morphological features of spray-dried milk particles have been attributed to different components: a smooth or feathery surface to the presence of amorphous or crystallized lactose, respectively; deep surface dents and folds to the amount of casein; and porosity to the free fat content (Buma, 1971; Buma & Henstra, 1971b). Scanning electron microscopy was used to show that when lactose crystallizes after spray-drying (e.g., during storage), needlelike crystals are formed, while precrystallization leads to the usual tomahawk shape (Roetman, 1979). Figure 9–12 shows a particle of spray-dried skim milk powder with deep surface folds (A) and another particle containing several vacuoles and smooth surfaces typical of amorphous lactose (B). See Section 2.6 for a modern application of microscopy to particle characterization.

Drying of sultana (white) grapes provides another example of the importance of microstructural elements in drying. Each berry is covered by an external waxy layer consisting of a series of overlapping, hydrophobic platelets that protect against transpiration (Figure 9–13). Underneath this waxy layer is a cuticle that can be peeled off as a sheet and is semipermeable to water. Both structures (referred to here as skin) offer considerable external resistance to mass transfer during drying, and significant increases in the drying rate are obtained if they are removed, as shown in

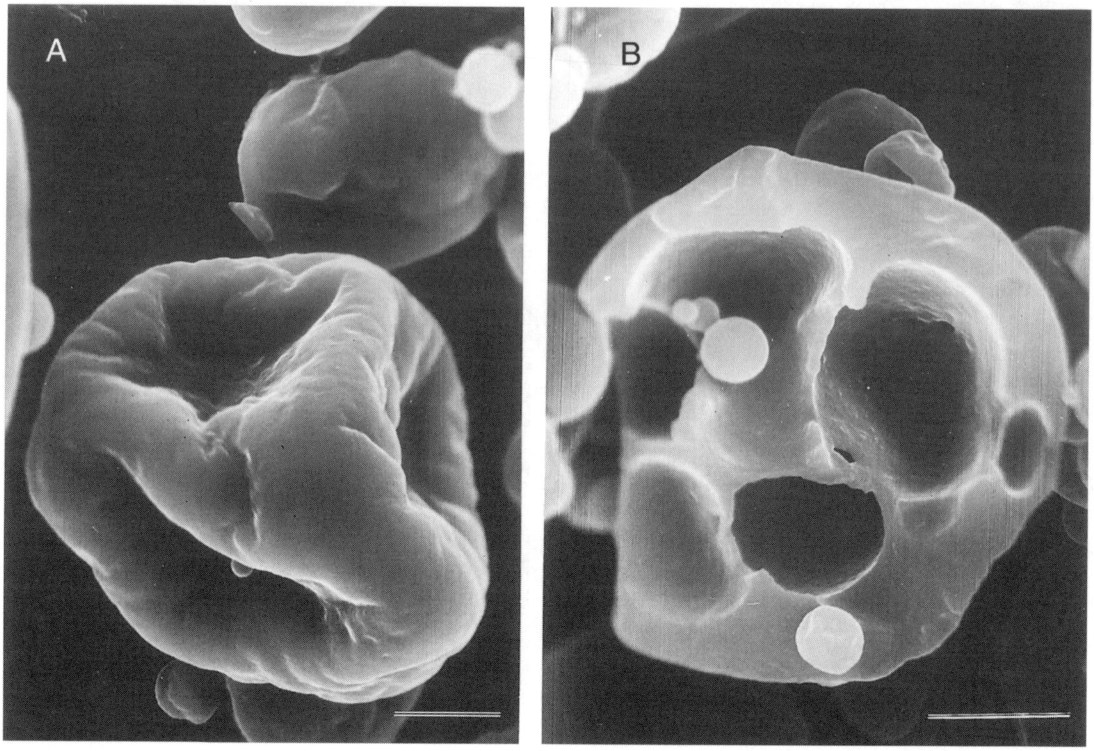


**Figure 9–12** Particles of spray-dried skim milk powder. (A) Deep folds in the outer surface. Marker = 10 $\mu$m. (B) Cross section of a particle with several vacuoles showing solid and smooth walls with amorphous lactose. Marker = 10 $\mu$m.

Figure 9–14. Treatment with petroleum ether dissolves the waxy layer; NaOH additionally induces microscopic cracks in the cuticule. Of more interest, dipping grapes in the surfactant ethyl oleate not only decreases the resistance of the skin tissues but also facilitates internal diffusion during the latter stages of drying. These effects have been demonstrated by a combination of drying experiments and microscopy by Riva and Peri (1983) and Aguilera, Oppermann, and Sanchez (1986). Data from Riva and Peri (1983) have been used to calculate apparent diffusion coefficients during drying of red grapes at 50°C at the beginning and end of the process (see Table 9–2). Diffusivities decreased significantly in the latter stages of drying when the moisture content was low. Presence of the skin reduced the *D* value in all cases, particularly at the beginning of drying. Dipping in surfactant resulted in almost the same diffusion coefficient in the final drying stage under low-moisture conditions as when the skins were removed.

**Table 9–2** Moisture Diffusion Coefficients ($cm^2/s \times 10^{-6}$) During Air Drying of Grapes

| *Pretreatment* | *High W* | *Low W* |
|---|---|---|
| Without treatment | 0.5 | 0.6 |
| Dipping in NaOH | 2.1 | 1.1 |
| Dipping in surfactant | 4.0 | 3.1 |
| Without skins | 10.0 | 3.3 |

*Source:* Adapted from Riva and Peri, 1983.

So far only postexperiment applications of microscopy have been discussed. Ideally, microstructural changes during dehydration should be studied dynamically under the microscope. Shrinkage of thin layers of fruit can be observed in real time under controlled experimental con-

**Figure 9–13** Scanning electron micrographs. (A) The outer waxy layer of sultana grape. (B) After treatment with surfactant. Marker = 40 $\mu$m.

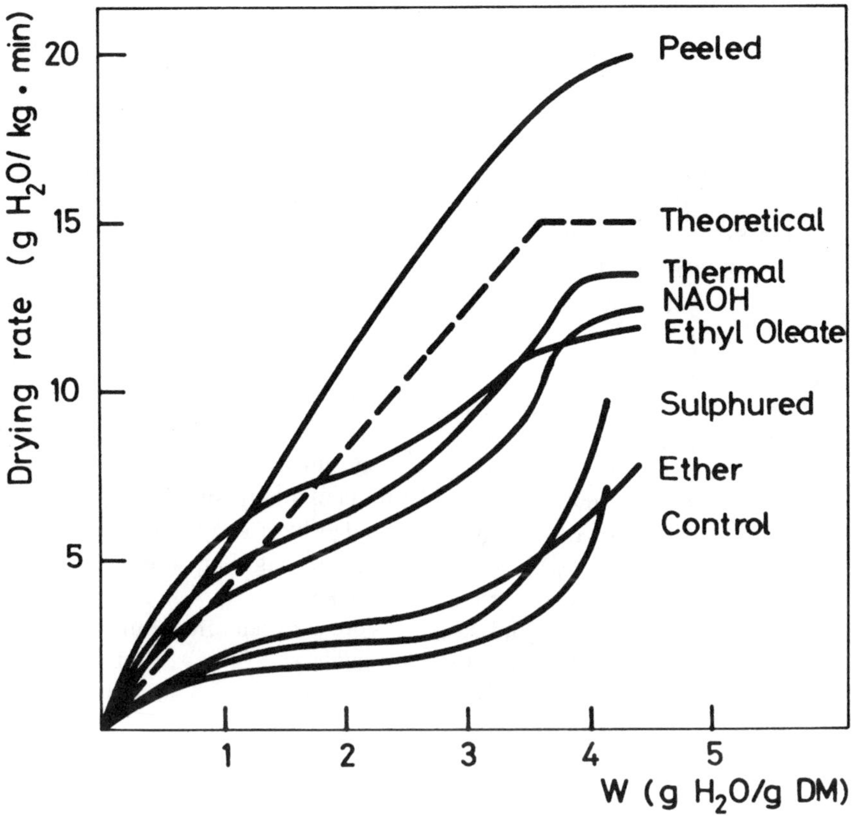


**Figure 9–14** Drying rates of sultana grapes after various pretreatments that affect moisture transfer through the skin.

ditions (air velocity, temperature, and relative humidity) by videomicroscopy if a stage is built where these variables can be monitored. The possibility of using an ESEM for larger pieces and its ability to control the surrounding atmospheric conditions and the heating rate should be assessed. ESEM images can be binarized and processed to obtain changes in geometrical parameters of cells as a function of time (Figure 9–15).

The influence of microstructure on drying and the properties of the dried products has been more fully analyzed in the case of freeze-drying than any other food dehydration method. Freeze-drying (removal of water as vapor from a frozen material) produces the highest quality product obtainable by any drying operation. Prominent characteristics include the structural rigidity afforded by the frozen material at the surface where sublimation occurs and the absence of liquid water, which results in a porous, nonshrunken food structure that is the primary quality factor in freeze-dried foods (King, 1971).

Another major benefit of freeze-drying over conventional drying methods is its superior ability to retain volatile flavors and aroma components. This is interesting, since many flavor compounds exhibit vapor pressures greater than that of water. The high degree of volatile retention in freeze-dried products has been explained by two theories:

1. The selective diffusion theory holds that the diffusion coefficients of organic aroma compounds decrease more rapidly with increasing solute contents than do the solute

water diffusion coefficients (Thijssen & Rulkens, 1968).

2. The microregion theory postulates that volatile components become trapped in microregions formed by a carbohydrate matrix (Flink & Karel, 1972). One interpretation of this theory is that an amorphous glassy phase is formed surrounding regions containing the volatiles.

When freeze-drying is conducted above a certain critical temperature (the collapse temperature), the solute matrix loses its shape (collapses), and the resulting product has reduced aroma retention and rehydration capacity and uneven dryness (Bellows & King, 1973). Loss of entrapped flavor volatiles does not occur either by grinding or by high vacuum but takes place when samples are heated or humidified (i.e., $T - T_g$ increases), indicating that encapsulation is dependent on the freeze-dried structure (Karel, 1985). Using microscopy, To and Flink (1978) were able to visualize entrapped volatile precursors in the matrix of a model freeze-dried system and to correlate volatile loss and collapse after heating or exposure to moisture.

Several other properties of freeze-dried products are affected by microstructure. It is not unusual for freeze-drying to be controlled by internal heat transfer rather than internal mass transfer, as in air drying. Slow freezing produces big crystals and leaves large pores (porosities of most freeze-dried products range from 0.65 to 0.90) that favor mass transport during drying and reconstitution but result in low thermal conductivity during processing. However, if freezing produces many small isolated ice crystals surrounded by a solid matrix, then the vapor must diffuse through the solid, and usually the matrix is cracked.

Pore size can affect other product characteristic such as color. A cold-stage optical microscope was used to determine freezing conditions

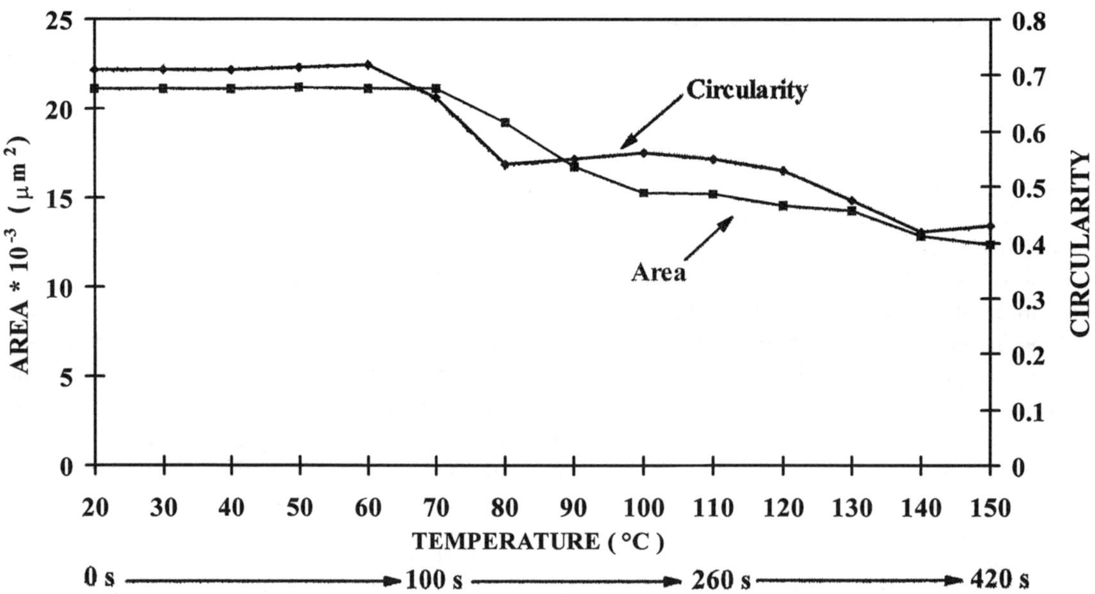


**Figure 9–15** Changes in geometrical parameters during shrinkage of a potato cell. Binary images obtained by video microscopy in real time and image processing are shown in Figure 1–10. *Source:* Courtesy of P. Bouchon.

and ice structure. A darker freeze-dried coffee, which is preferred by consumers, can be obtained by slow freezing, which produces large ice crystals with random structure. The darker particle surface is due to the light scattering properties of the large voids left by the crystals after sublimation (Barnett, 1973). The porous microstructures of three freeze-dried food materials (strawberries, coffee, and meat) are presented in Figure 9–16.

## 9.4 OSMOTIC DEHYDRATION

### 9.4.1 Principles of Osmotic Dehydration

In simple terms, osmosis is a process in which a solvent flows from a diluted to a concentrated solution through a semipermeable membrane to equalize the chemical potential of the solute. In theory, osmotic dehydration is a dewatering process in which cellular materials (mostly fruits and vegetables) are immersed in an hypertonic solution and water is removed from inside the cells by osmosis. In practice, in addition to dewatering, solution penetrates into the product, solutes (organic acids and sugars) are removed, and considerable shrinkage occurs. The process may occur at low temperatures and does not involve the phase change usual in conventional drying.

Understanding the role of microstructural elements such as cell membranes, pores, and tissue architecture is obviously necessary for modeling the osmotic dehydration process. Poor comprehension of the mechanism involved in the countercurrent water removal and solute penetration inside a complex structure has hindered the development of industrial applications of the process (Yao & Le Maguer, 1996). Several food technologists and engineers have made important contributions to unveiling the bases of osmotic dehydration, as this technique has not been analyzed in traditional chemical engineering.

First, some basic physical chemistry. In osmosis, the excess pressure needed to reach the state of equilibrium between a pure solvent and a solution is called osmotic pressure ($\pi$):

$$\pi = -\frac{RT}{V} \ln a \qquad \textbf{Equation 9–13}$$

where $V$ and $a$ are the molar volume and activity of the solvent, respectively. This formula is easy to remember because it resembles the law of ideal gases. Since in foods the solvent is water, equation 9–13 can be expressed as (Lewicki & Lenart, 1995)

$$\pi = -4.6063 \times 10^5 T \ln a_w \qquad \textbf{Equation 9–14}$$

where $a_w$ is the water activity ($T$ is in °K and $\pi$ in MPa). Hence, at equal concentrations the osmotic pressure increases as the molar mass of solute decreases. Each ion of an electrolyte behaves as one chemical species and contributes independently to the osmotic pressure.

Theoretically, undamaged cell membranes in tissues act as semipermeable membranes permitting the passage of water from the interior of the cell to the hypertonic solution. Practically, however, water loss is always accompanied by solute exchange (and net gain by the product); thus osmotic dehydration involves the simultaneous countercurrent flow of water and solutes. Although not a preservation process itself, osmotic dehydration needs further stabilization to reduce the water activity and hinder proliferation of microorganisms. The subject of osmotic dehydration is reviewed by Raoult-Wack (1994), Chaudhari, Kumbhar, Singh, and Narain (1993), and Torreggiani (1993).

### 9.4.2 Modeling Osmotic Dehydration

Modeling osmotic dehydration is not simple. Analysis at the microstructural level is required, since multiple mechanisms operate simultaneously. The main structures of a piece of food are the outer layer of broken cells produced by cutting, intercellular spaces and natural pores, internal cells, and intracellular compartments. Conceptually, the interaction between the osmotic solution and the food as well as transport phenomena should be analyzed at least at four hierarchical levels (Figure 9–17): whole tissue (including outer broken cells), internal porous structure,

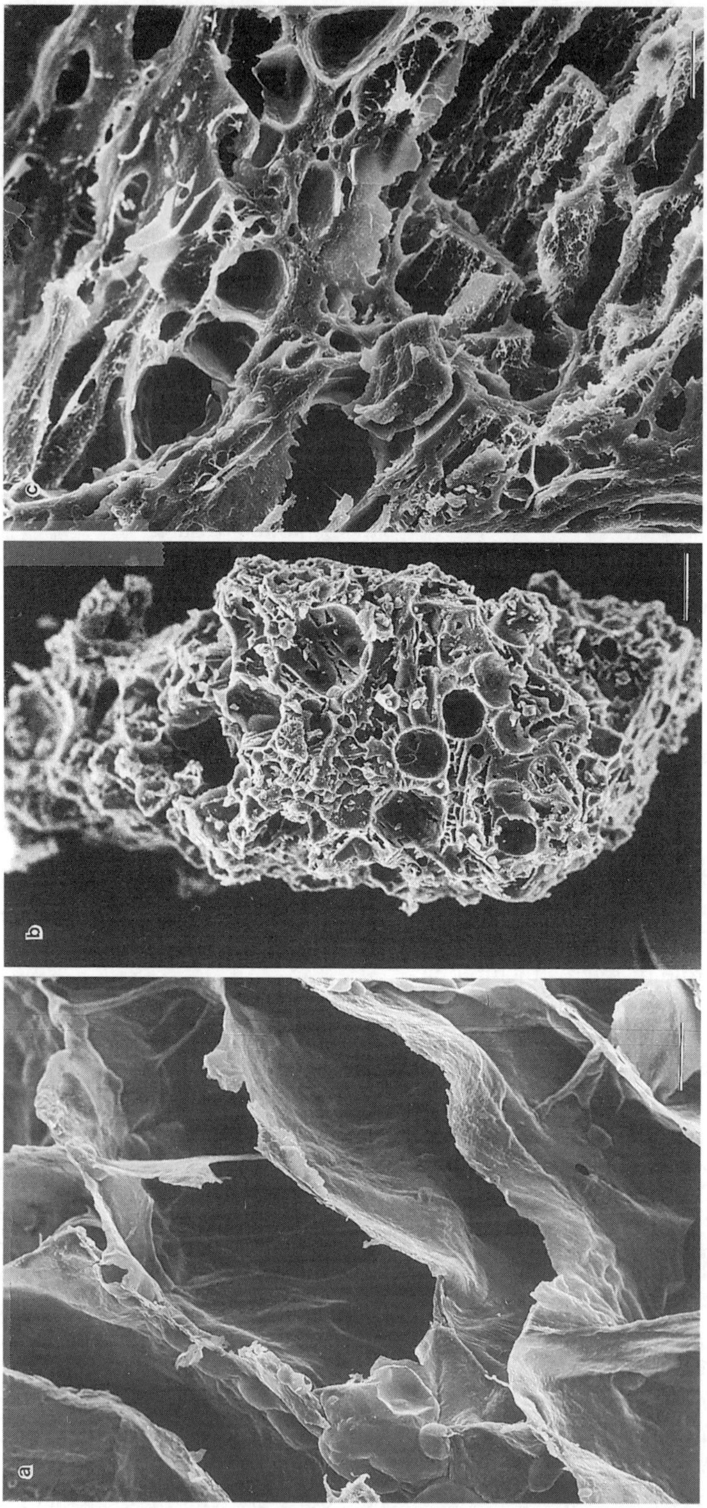


**Figure 9–16** Scanning electron micrographs of freeze-dried products. (A) Strawberry. (B) Coffee. (C) Meat. Porous structure is the result of water sublimation from ice crystals. Markers = 100 $\mu m$.

cell walls and intercellular spaces, and membranes.

A first level of analysis considers the food piece and external surface. During immersion in the hypertonic solution, the contents of the outer broken cells are washed away and solution penetrates into open pores. Upon removal of the piece after dehydration, the outer surface will have solution adhering to it, particularly if the solution is concentrated and viscous. Depending on the surface:volume ratio, this effect will be more or less pronounced. The second structural level of analysis concerns penetration of bulk solution into open pores initially filled with air (or possibly gases), which occurs by various mechanisms (predominantly by capillary suction) until an equilibrium sets in. Pores and intercellular spaces filled with air account for 2–3% of the volume in potato, around 20% in apple, and 25% in sugar beet. Impregnation of the pores by solution is facilitated if a vacuum is drawn on the pieces before immersion in the solution (Fito & Pastor, 1994). Eventually, spaces between cells will come into contact with the hypertonic solution, and osmotic flow will start. Up to this point, the process has involved impregnation of the tissue by the hypertonic solution. The third level of analysis concentrates on the cell itself (usually parenchymatous cells) and diffusion of water and solutes through the cell wall. The cell wall may be regarded as an aqueous gel filled with loose microfibrils leaving open spaces large enough to allow passage of water, ions, and small solutes. The presence of plasmodesmata should also be taken into account. The fourth level concerns transport through intact or denatured membranes. As the protoplasm loses water, its volume decreases, the plasmalemma becomes detached from the cell wall (plasmolysis), and the hypertonic solution fills the volume between the cell walls and plasmalemma. At this point, the plasmalemma becomes a barrier to mass transfer. If it is assumed that this membrane remains intact, actual osmosis may occur between the diluted protoplasm and the occluded hypertonic solution. However, most of the water to be lost from a cell is initially located in the vacuole, which is surrounded by another membrane, the tonoplast. Ultimately, it seems, osmosis has to occur between the solution of (mostly) minerals and sugars in the vacuole and the existing solution in the proto-

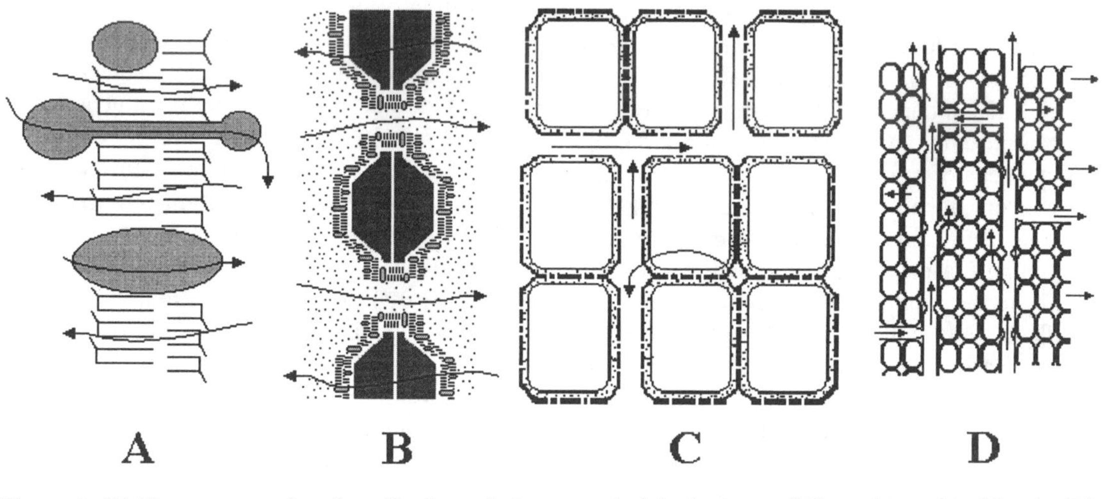


**Figure 9–17** The mass transfer of small solutes during osmotic dehydration at different hierarchical levels. (A) Membranes (diffusion, active, and passive transport through proteins). (B) Cell walls (symplasmic and apoplasmic transport). (C) Intercellular spaces and pores. (D) Whole tissue. Note at the right the broken cells produced by cutting. Courtesy of F. Valenzuela.

plasm. Needless to say, the relative importance of these transport mechanisms during osmotic dehydration has never been demonstrated, although they are key to understanding the kinetics of water outflow and solute(s) inflow during this process. According to this theoretical view, solutes need not transverse any membrane just to accumulate in the space between the plasmalemma and cell wall.

Most attempts to model water and solute transport between the osmotic solution and the food piece are based on Fick's second law for unsteady state diffusion (Raoult-Wack, 1994). However, the models are too simplistic and rely (as usual) on a constant and overall "apparent diffusivity." There are at least three factors that strongly affect the diffusivity values: (1) microstructural effects, (2) counter-flux associated with convective effects, and (3) changes in physical properties of the food as a result of osmotic dehydration. Partial dehydration and solute uptake would be expected to increase local viscosity and decrease local diffusivity in the food. In addition, fluxes of water and osmotic agent within the food matrix occur in a countercurrent fashion, and it has been suggested that interactions exist between the two. Indeed, conditions favoring fast initial dehydration, such as high solute concentration or temperature, do not necessarily hasten solute uptake (Raoult-Wack, 1994). Under these conditions, solute is carried by the aqueous phase leaving the solid to the food surface where it concentrates.

Lewicki and Lenart (1995) examined two paths for transportation of water and the osmotic agent through cells of a fruit or vegetable: the continuous network of cell walls and intercellular spaces (the apoplasm) and the system of protoplasm and connecting plasmodesmata (the symplasm). According to these authors, the hydraulic conductivity of cell walls is two orders of magnitude larger than for plasmalemma, while the diffusive permeability coefficient for ions is six orders of magnitude higher in cell walls. Interestingly, the authors disregard the symplasmic transport between cells in relation to the apoplasmic transport.

An elaborate transport model at the cellular level has been proposed by a group at the University of Guelph (Marcotte, Toupin, & Le Maguer, 1991; Toupin, Marcotte, & Le Maguer, 1989; Yao & Le Maguer, 1996). This physical model focuses on extracellular and intracellular volumes separated by a semipermeable membrane. The mathematical model incorporates diffusion, bulk flow, transmembrane flux, and shrinkage of the matrix. Symplasmic transport is ignored and the solution is considered to be diluted.

### 9.4.3 Applications

Since water losses do not require a phase change and take place at near ambient temperatures, osmotic dehydration has some advantages over freeze-drying or conventional drying from the standpoint of operating costs and final product quality. On the other hand, the maximal water removal capacity of osmotic dehydration is limited by the solubility of the osmotic agent in water and the solute uptake by the food. The latter effect may be advantageous from a stability standpoint but undesirable when fresh food attributes are required. Osmotic dehydration is being used for the processing of intermediate-moisture and minimally processed fruits (Alzamora, Tapia, Argaiz, & Welti, 1993).

There are several general factors that may affect osmotic dehydration kinetics and the final composition of osmodehydrated foods: the composition of the osmotic solution, process conditions, and food pretreatment. The selected osmotic agent and its concentration influences the $a_w$ of the solution and solute uptake by the food. Low molecular weight ionizable solutes are best from an $a_w$-depressing standpoint but are more easily picked up by the food owing to their increased diffusivity. This may limit the use of an osmotic agent for sensorial reasons—common salt (NaCl) for the osmotic dehydration of fish, meat, and vegetable tissue, and sucrose for the osmotic dehydration of fruits (sometimes mixtures of these solutes prove to be effective).

Process conditions that affect osmotic dehydration include the solution:food ratio, the temperature of the osmotic solution, and agitation. During osmotic dehydration, the solution becomes diluted because of water gains and solute diffusion from or to the food material, so that high solution: food ratios are favored from a mass transfer standpoint. However, the use of high ratios results in increased energy requirements for agitation and reconstitution of the osmotic solution. Mass transfer can also be improved by increasing processing temperatures or reducing external resistance by increased agitation, but care should be taken to avoid thermal and mechanical damage to the food. The economics of the process require that the used solution be cleaned and recycled.

As in other dehydration processes used for fruits and vegetables, blanching can increase the permeability of cell membranes and remove gases from pores. Size reduction (cutting or slicing) reduces the influence of internal mass transfer resistances at the expense of increased solute uptake by the food. As mentioned before, the osmotic solution adheres to the food surface and/or penetrates the porous network of intercellular spaces and vascular tissue under the influence of capillarity, so the composition of small pieces of osmodehydrated materials is highly influenced by posttreatment washing. Details of specific osmotic dehydration procedures, energy aspects, and product characteristics can be found in Lewicki and Lenart (1995).

## 9.5 INFLUENCE OF DRYING ON STRUCTURAL PROPERTIES: EXAMPLES

### 9.5.1 Texture

Attaining a desirable texture after reconstitution is often a target for the dehydration process. Such is the case with potato flakes, a form of dehydrated mashed potato that must be capable of reconstitution to a consistency comparable to that of the mash prepared from freshly cooked potatoes.

Cooking of potatoes results in two important microstructural alterations: initially, starch becomes gelatinized between 58°C and 70°C, and then, after the process is complete, the middle lamellae break down and cell separation occurs. If there is a pronounced swelling of the gelatinized starch, excessive cell wall rupturing occurs that releases an appreciable amount of starch, a common cause of stickiness or gumminess in mashed potato products. In fact, microscopic counts of ruptured cells provide an index for textural quality in dehydrated mashed potatoes (Reeve & Notter, 1959). A series of technological steps must be adhered to in order to achieve the proper final texture, all of which pertain to the formation of the correct microstructure (i.e., minimally ruptured potato cells). Precooking of potatoes for about 20 minutes at 71°C causes adequate starch gelatinization, cell separation and stabilization of the middle lamellae and cell walls (by the activation of the pectin methylesterase enzyme and the release of calcium from gelatinized starch), and the formation of calcium bridges between pectin molecules (Andersson, Gekas, Lind, Oliveira, & Oste, 1994). Cooling of the precooked slices retrogrades extracellular amylose and reduces the pastiness that results from cell breakage. Cell damage must also be prevented during the drum-drying operation (Feustel, Hendel, & Juilly, 1964). Figure 9–18 shows the outer surface of a potato flake in which many individual cells have been flattened by the action of the drying rolls.

### 9.5.2 Microencapsulation

Microencapsulation is a process designed to apply thin polymeric coatings around droplets of liquid. It consists of three stages: (1) establishment of a three-phase system (material to be coated, liquid vehicle, and coating material), (2) deposition of the liquid-polymeric wall material over the product (vitamins, aroma-containing materials) by emulsification, and (3) solidification of the wall material in a glassy or amorphous state. Spray-drying is a common method of removing the liquid vehicle (water) and forming the protective wall of microcapsules, which range dimensionally from 1 to 5,000 $\mu$m (Bakan, 1973). Microencapsulation of flavors by spray-drying is based on the same principles of volatile retention already discussed for freeze-drying. The effectiveness of encapsulation depends heavily on the formation of a metastable amorphous structure with low perme-

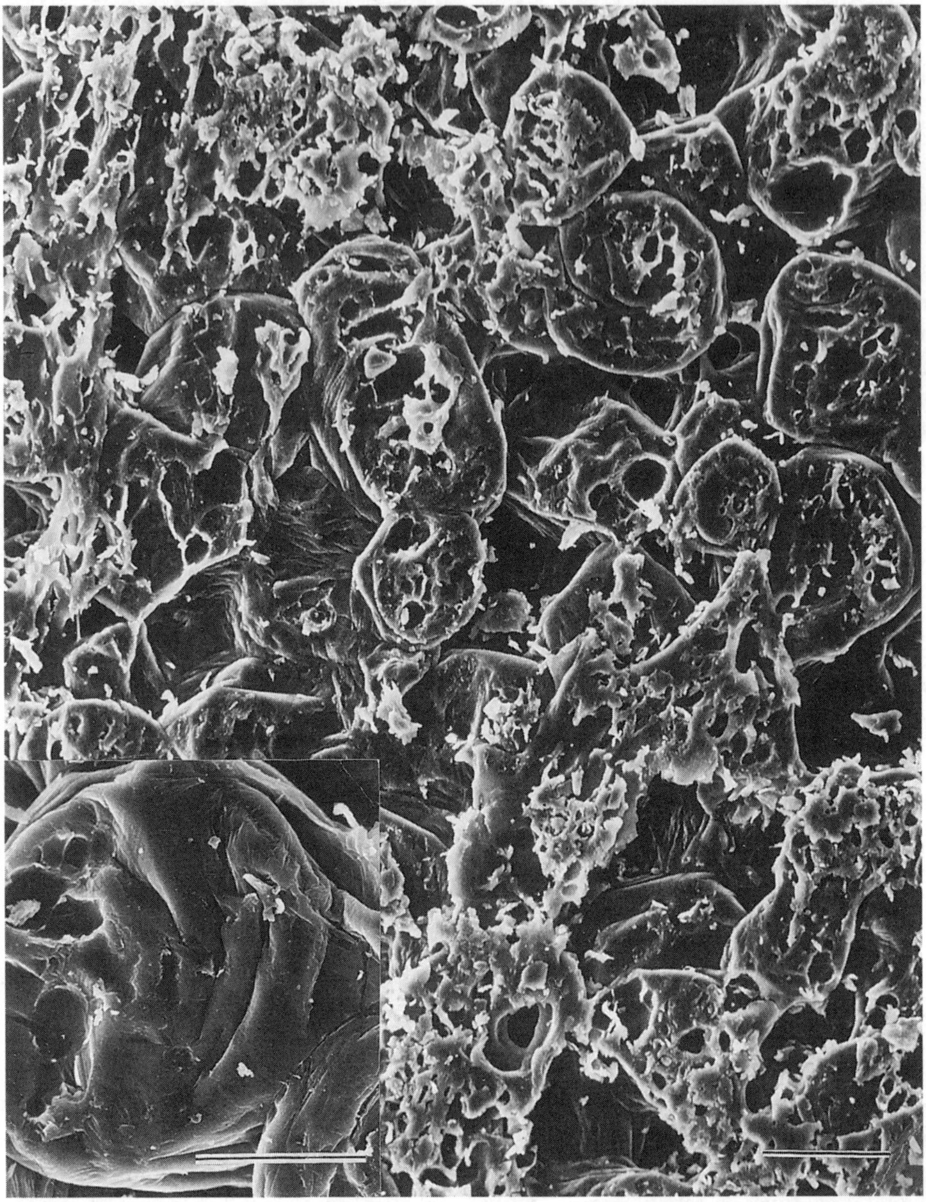

**Figure 9–18** Scanning electron micrograph of the surface of an instant potato flake. Gelatinized starch granules appear flattened by the action of the flaking roll. Marker = 10 $\mu$m. Inset: Detail of a starch granule. Marker = 5 $\mu$m.

ability to organic compounds encapsulated within it. Common polymeric walls used in foods include sugars, dextrins, vegetable gums, and proteins such as casein and gelatin.

Scanning electron microscopy has been proved particularly suitable for rapidly determining the encapsulating ability of various polymers (Beatus, Raziel, Rosenberg, & Kopelman, 1985; Rosenberg, Kopelman, & Talmon, 1985). If the outer surfaces of the particles are very porous or cracked, volatile losses or oxidation of entrapped lipids may occur. Collapse of the wall matrix may expose the encapsulated material to the environment and promote deteriorative reactions such as lipid oxidation. The presence of dents on the outer surface has an adverse effect on flow properties of the powders.

### 9.5.3 Rehydration

Many dehydrated products are consumed after rehydration. In the extreme case, they must not only pick up water but rapidly dissolve and form a uniform solution (e.g., instant powders like dried milk).

Instantization of powders is a dynamic process consisting of several stages: wetting, sinking, dispersion, and suspension. Product properties affecting the instantizing properties may be divided into surface and internal factors. Among the former, wettability (the ability of the liquid to spread over the surface of the solid particle) is the most important. Wettability requires that the contact angle between water and solid be as low as possible or that, in other words, the polarity of the surface be high. In cases where free fat covers the surface, addition of surfactants such as lecithin may counteract the effect. Water penetration into the interior of a porous particle is generally impaired by viscosity effects at the wet front (as sugars and salts dissolve) and by swelling of the matrix and attendant capillary contraction. This type of hydrodynamic resistance can be overcome by agglomeration, in which small particles are fused together into greater aggregates with a large external area and wider pores or channels. Sinkability is a concurrent property of an instant powder; it can only occur if the density of the particle or agglomerate is greater than unity. This implies that interstitial air in the pores must be removed and water penetrated into the interior. The final stage in the interaction between the powder and water is solubilization into a colloidal solution. Details on instantization of food powders can be found in Schubert (1993).

Dried fruits and vegetables should possess two properties for proper rehydration: (1) the capacity to imbibe water rapidly and uniformly and (2) the ability to retain the water in a form similar to that in the natural product. As discussed for milk powders, the first property depends on physicochemical interactions (e.g., wetting) and on diffusion-related events, such as case hardening and the setting of the product into a dense microstructure (e.g., as occurs in slow drying). An alternative dehydration process is puff-drying, where the sudden depressurization accompanying water removal expands the product and yields a porous microstructure. In extreme cases, a porous microstructure may be achieved by slow freezing prior to drying. Cavities formed by large ice crystals permit easy imbibition of water, but the water-binding properties may not necessarily be enhanced. Surprisingly enough, controlled compression of freeze-dried fruits and vegetables to a fraction of their original volume does not impair their normal appearance after rehydration. Reversibility is achieved by increasing the moisture content to the plastic state (e.g., 9% in freeze-dried spinach) or by spraying the dehydrated food with water, glycerine, or propylene glycol before compression (Rahman, Shafer, Taylor, & Westcott, 1969).

### 9.5.4 Caking

Caking is a deleterious phenomenon in which a low-moisture, free-flowing powder is first transformed into soft lumps, then into an agglomerated solid, and ultimately into a sticky material as it picks up moisture. Caking is ubiquitous in food, feed, fertilizer, pharmaceutical, and related industries. This section deals with the caking of amorphous powders (e.g., those prepared by spray-drying), although there are other forms of caking induced by fat melting and recrystallization, wetting of surfaces, and so on (Peleg, 1983). Starch, amylose, dextrins, low molecular weight sugars, proteins, and their hydrolysates (e.g., soy sauces) form amorphous phases and so their powders are susceptible to caking. The subject is reviewed by Aguilera, del Valle, and Karel (1995) and by Roos (1995).

Amorphous food powders exist below the glass transition temperature ($T_g$) as metastable solids because the high viscosity ensures that the material supports its own weight. As the product temperature exceeds $T_g$ they enter a rubbery condition and the decreasing viscosity induces flow and deformation often within observable times. The viscosity of amorphous powders may decrease from $10^{12}$ Pa·s at or below $T_g$ to $10^6$–$10^8$ Pa·s in the rubbery state. Collapse, stickiness, and caking appear to be related phenomena. The collapse temperature, sticky point, and $T_g$ are highly correlated and show similar dependence on moisture content (Aguilera et al., 1995).

An obvious goal for food technologists is to be able to predict the degree of caking during storage. After a dry product is packed, it picks up moisture from the moist air introduced by renewal of the headspace during use (e.g., in an impervious container), by diffusion through a permeable package material, or from a faulty seal. Another mechanism by which free moisture becomes available in a closed system is sugar crystallization. Crystallization releases water, which in closed containers becomes absorbed by the amorphous phase of the powder. A thorough review of structural transitions in dried carbohydrates is presented in Flink (1983). In all cases, the extent of moisture pickup depends on the sorption characteristics of the food.

The rate of caking will depend on the instantaneous moisture content (through its effect on $T_g$) and the ambient temperature. Applying these concepts and the WLF equation, Aguilera et al. (1995) demonstrated a strong effect of $\Delta T = (T - T_g)$ on the kinetics of the caking of fish hydrolysate powders. When $\Delta T \approx 20°C$, caking occurred within days; when $\Delta T$ was doubled to about 40°C, loss of flowability was observable within hours.

Scanning electron microscopy has been used to assess caking qualitatively although quantitation by stereo microscopy and image analysis is now the preferred practice. Since mobility after the glass-rubber transition is increased, it is not uncommon to observe extensive flow, the bridging of particles, and the presence of crystals in caked products (Lloyd, Dong Chen, & Hargreaves, 1996). Figure 9–19 shows the effect of $a_w$ on the caking of milk powder stored at room temperature for 20 days. Note the dramatic change in structure at $a_w = 0.79$ and the presence of bridges in the caked product.

As should be evident at this point, low moisture content and low temperature during storage are key factors in minimizing the caking of amorphous powders. However, in many cases anticaking agents are used. These act by (1) competing with the host powder for available moisture, (2) increasing the $T_g$ of the amorphous phase, (3) interfering in the liquid bridging process by weakening or disrupting links formed by dissolved solids (e.g., silicon dioxide, powdered cellulose), or (4) covering and lubricating the surface of particles (e.g., stearates). The subject of flow conditioners and anticaking agents is reviewed by Peleg and Hollenbach (1984).

### 9.5.5 Agglomeration

Agglomeration refers to a process of controlled particle size enlargement through the aggregation of small particles. The purpose is to improve some property of the system (Schubert, 1981): bulk density, dispersability, solubility, sinkability, dispensability, dust formation avoidance, and so on. Commonly used agglomeration processes can be divided into three types: (1) pressure agglomeration (e.g., tableting), (2) growth agglomeration (e.g., granulation, pelleting), and (3) agglomeration by drying (e.g., during spray-drying) (Schuchmann, 1995). Instantization of powders is aimed at facilitating dissolution in the liquid.

## 9.6 FRYING OF FOODS

### 9.6.1 The Frying Process

Frying is a unit operation unique to food processing and a singular example of simultaneous heat and mass transfer and surface dehydration (Singh, 1995). Through frying, foods are cooked and bestowed with desirable textures and flavors. In spite of the widespread use of frying around the

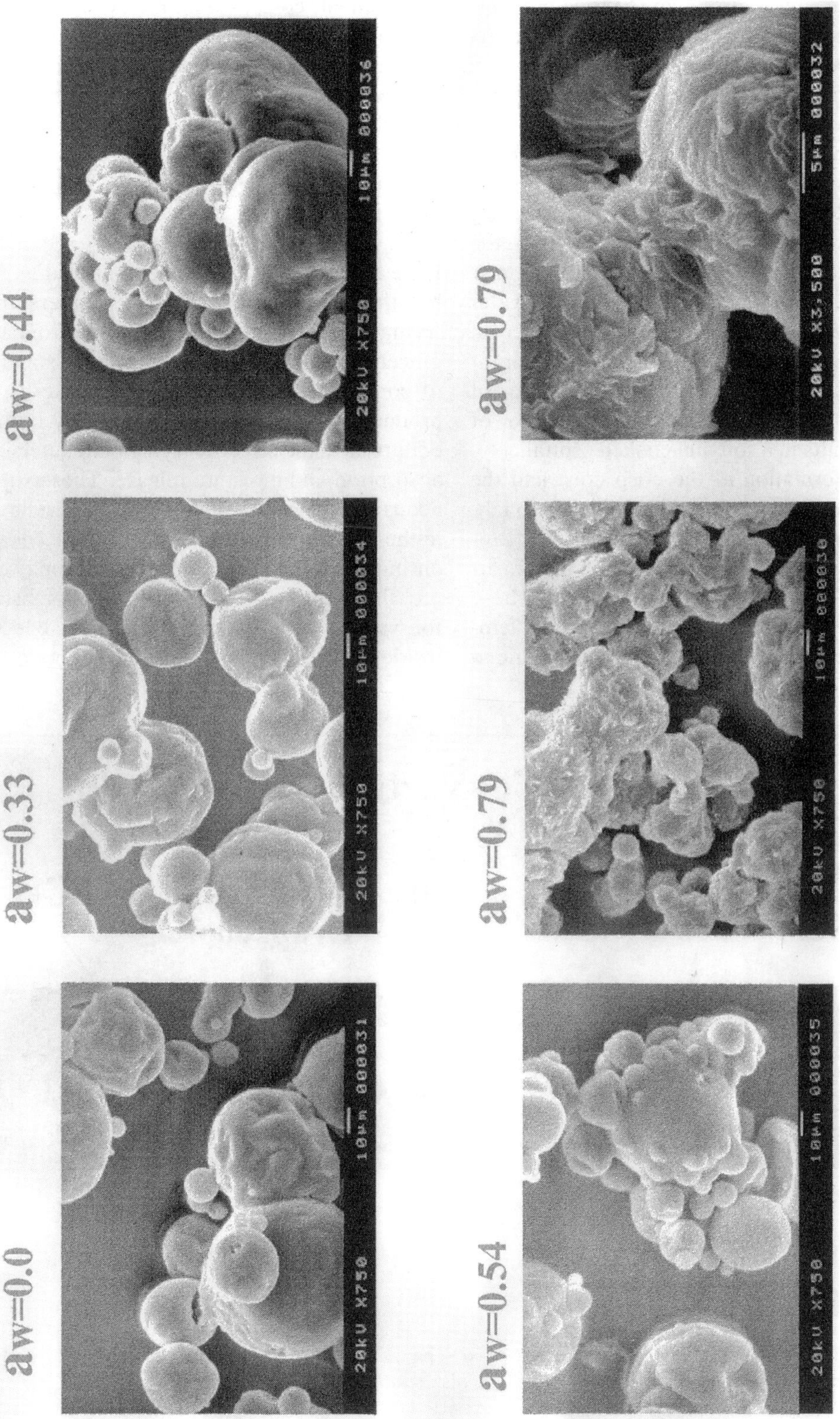


**Figure 9–19** Caking of food powders. Scanning electron micrographs of a dry milk powder (possibly agglomerated for instantization) after exposure to several $a_w$ levels for 20 days at room temperature.

world and the enormous volume of fried products sold as snacks and in fast-food operations, its study by food engineers began only recently. What follows is a description of the frying operation as applied to potato pieces.

Immersion or deep-fat frying is an unsteady state process. It involves convection heating from the surrounding oil (≈150–200°C) to the surface of the product and heat conduction into the interior of the piece (Figure 9–20). Since raw materials have a high moisture content, water evaporates from a sharp receding interface in the form of steam bubbles that rise to the top of the oil, inducing dehydration and the formation of a thin outer crust. Heat transfer to the interior of the piece results in a soft and cooked central core. In a sense, formation of the crisp crust and the mealy core in potato products is equivalent to fabricating a composite food. A review has been prepared by Saguy and Pinthus (1995) on factors affecting fat absorption during frying. These include oil quality and composition, frying temperature, duration of frying, product shape, moisture content, composition, prefrying and surface treatments, gel strength, initial interfacial tension, porosity, and crust formation. As may be concluded from the previous list, frying is an extremely complex process and is still not completely understood.

### 9.6.2 Frying Kinetics

Excessive fat consumption is regarded as major health risk in modern society. Understanding frying kinetics is important not only from an engineering standpoint but also from the standpoint of controlling and reducing oil uptake by fried products. There is a wide discrepancy in the literature as to the effect of frying parameters on oil absorption and moisture release. These simultaneous and apparently linked phenomena have led to the widespread notion that "oil replaces water during frying." The unaware reader may interpret this wrongly as meaning that oil substitutes for water quantitatively and in the same location inside the product. In the frying of potato chips,

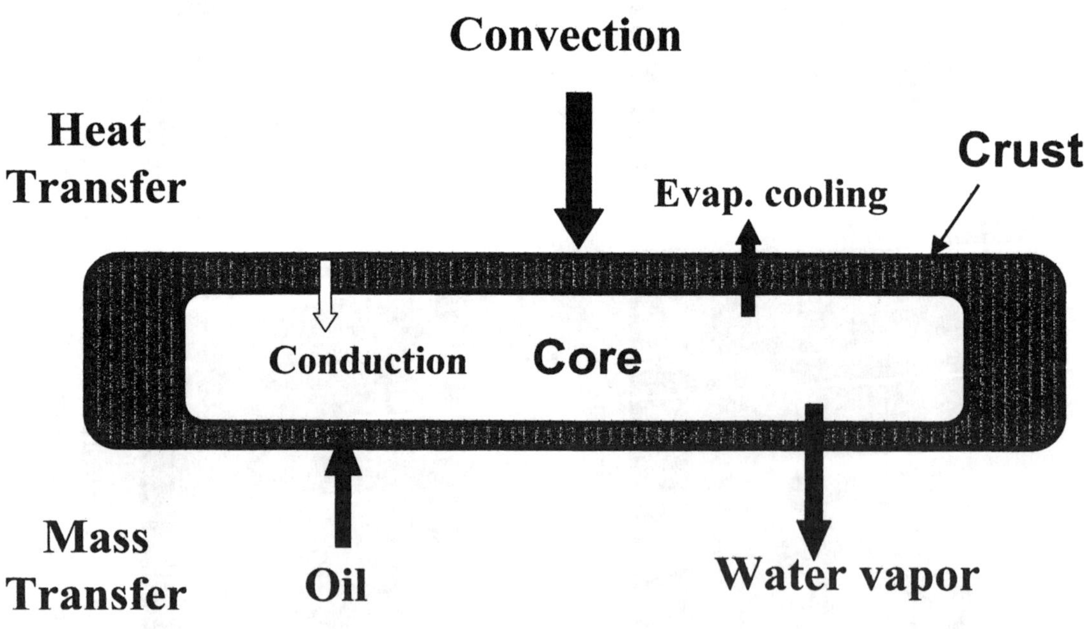


**Figure 9–20** Heat and mass transfer during frying of food pieces. Note the formation of a composite structure in a fried potato product: a dried, crisp outer crust and a moist, soft inner core.

more moisture is lost than oil absorbed, and several studies have found no clear relationship between these two variables (Saguy & Pinthus, 1995). Water is continuously removed during frying but at a decreasing rate; it is very intense at the beginning and then drops considerably as the dehydration front recedes into the interior. Macroscopically, this is perceived as an initial massive generation of steam bubbles from many sites on the surface and a subsequent slower release of large bubbles from only a few locations.

The kinetics of oil uptake are obscured by the fact that oil sticks or wets the surface of a piece from the moment it is introduced into the frying oil. Thus, oil "uptake" is high even over short frying times. The total oil in French fries can be divided into structural oil, located inside the crust, and surface oil. Structural oil is absorbed by two mechanisms: "migration" during the course of frying and absorption from the surface after the product is removed from the frying pan (Figure 9–21). The latter mechanism appears to be significant in the frying of potato and tortilla chips, and it is based on suction through pores in the crust by the vacuum produced by condensing steam in internal spaces (Moreira, Sun, & Chen, 1997; Ufheil & Escher, 1996).

These findings show frying to be an unusual mass transfer operation and lead to at least three important conclusions that provoke imaginative hypotheses regarding the mechanisms of oil uptake: (1) water is released as steam bubbles from specific locations (chimney formation), (2) oil uptake is strongly dependent on the wetting properties of the surface of the piece, and (3) most of the oil migrates (rather than diffuses) into the crust after removal from the fryer.

### 9.6.3 Microstructural Aspects of Frying

Since the first histological studies of deep-fat fried potatoes by Reeve and Neel (1960), it has been noticed that cells in the uncut portion of potato sections remain fairly intact after frying. Cells in the crust are dehydrated, shrunken, and contain in their interior gelatinized starch granules occupying the whole volume and pushing against the cell walls. In the core, the general aspect of the tissue is similar to that of cooked potatoes. Hot-stage and video microscopy have confirmed that these changes occur in thin potato sections exposed to hot oil as well. Blistering in potato chips takes place after separation of neighboring intact cells alongside cell walls (van Marle, Clerkx, & Boekstein, 1992).

A better visualization of the location of oil in the crust of fried potatoes is obtained by confocal laser scanning microscopy because of its optical sectioning capability and nonintrusive nature. Frying in hot oil with a thermoresistant fluorochrome (Nile red) permits detection of the oil inside the crust by this type of microscopy with no sample preparation. A three-dimensional reconstruction of the gallery of images generated at steps of 10 $\mu$m in depth using the software Imaris is presented in Color Plate 8. Oil in the crust is located as an egg-box structure around intact cells, and almost no oil is present inside the cells (Pedreschi & Aguilera, 1998). Kinetic data suggest that oil migrates from the surface to the interior of the crust after frying through fine capillary openings in the intercellular spaces, probably by two mechanisms: pressure flow promoted by a vacuum from condensing steam and capillary flow. Passages for oil uptake are dynamically formed between cells after the middle lamella has softened and steam bubbles are released. One hypothesis is that the steam pressure inside the crust prevents the massive entrance of oil during frying while the condensing steam provides a negative pressure "sucking" the oil into the crust during cooling. Indeed, frying may be regarded as a dynamic process in which the porosity of the crust is higher in the fryer than in the raw or the finished product. Interconnected intercellular spaces surrounding cell walls are natural passages for moisture and oil transfer and places for oil deposition in the fried product (egg-box structure). These mechanisms of oil impregnation and water release make Fickean diffusion a poor model for mass transfer during frying.

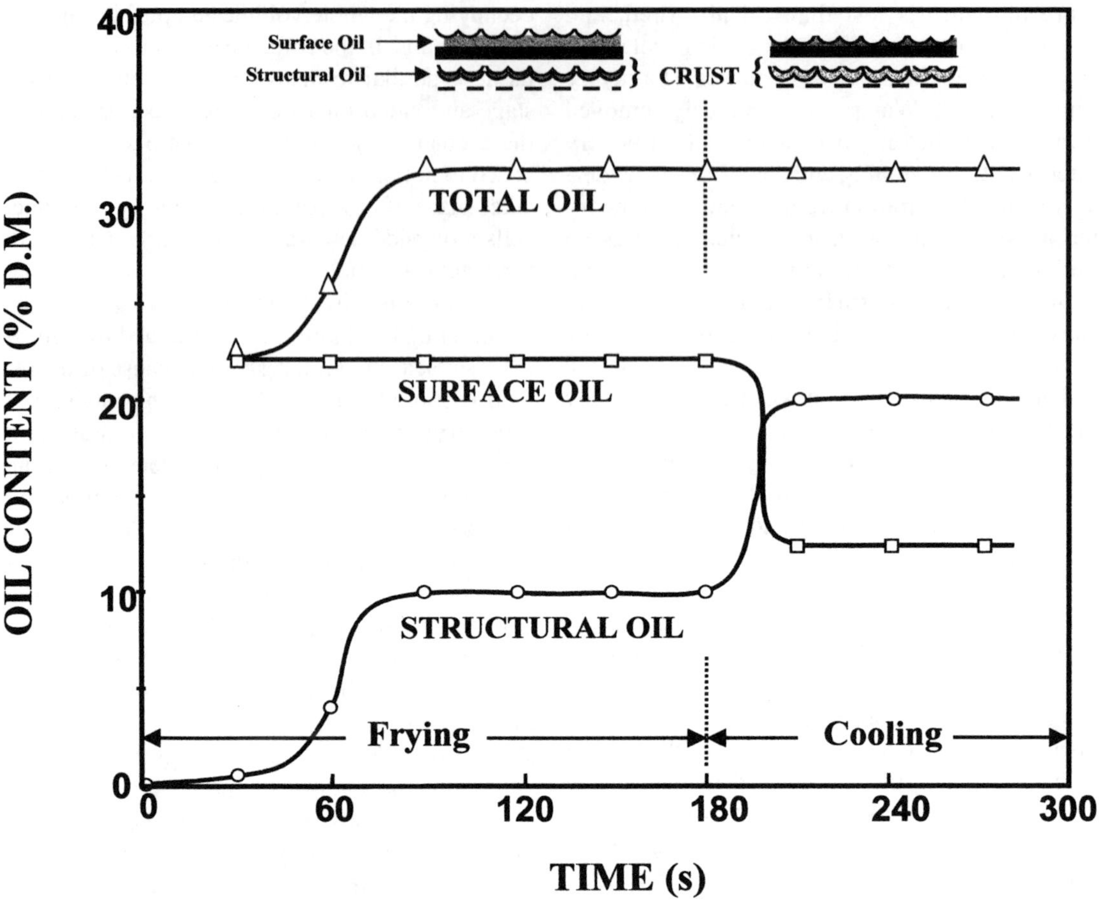


**Figure 9–21** Kinetics of oil uptake in the crust of a potato product during deep-fat frying. Semiquantitative data show that structural oil (inside the crust structure) increases dramatically after the product is removed from the fryer, as oil is pulled in from the surface by differential pressure (vacuum).

---

**BIBLIOGRAPHY**

Aguilera, J.M., Chirife, J., Flink, J.M., & Karel, M. (1975). Computer simulation of nonenzymatic browning during potato dehydration. *Lebensmittel-Wissenschaft und Technologie*, *8*, 128–133.

Aguilera, J.M., Cuadros, R.M., & del Valle, J.M. (1998). Differential scanning calorimetry of low-moisture apple products. *Carbohydrate Polymers*, *37*, 79–86.

Aguilera, J.M., del Valle, J.M., & Karel, M. (1995). Caking phenomena in amorphous food powders. *Trends in Food Science and Technology*, *6*, 149–155.

Aguilera, J.M., & Karel, M. (1997). Preservation of biological materials under desiccation. *Critical Reviews of Food Science and Nutrition*, *37*, 287–309.

Aguilera, J.M., Levy, G., & Karel, M. (1993). Effect of water content on the glass transition and caking of fish protein hydrolyzates. *Biotechnology Progress*, *9*, 651–654.

Aguilera, J.M., Oppermann, K., & Sanchez, F. (1986). Kinetics of browning of Sultana grapes. *Journal of Food Science*, *52*, 990–993, 1025.

Alzamora, S.M., Tapia, M.S., Argaiz, A., & Welti, J. (1993). Applications of combined methods technology in minimally processed fruits. *Food Research International*, *26*, 125–130.

Andersson, A., Gekas, V., Lind, I., Oliveira, F., & Oste, R. (1994). Effect of preheating on potato texture. *Critical Reviews of Food Science and Nutrition*, *34*, 229–251.

Anglea, S.A., Karathanos, V., & Karel, M. (1993). Low-temperature transitions in fresh and osmotically dehydrated plant materials. *Biotechnology Progress, 9*, 204–209.

Bakan, J.A. (1973). Microencapsulation of foods and related products. *Food Technology, 27*(11), 34, 35, 38–44.

Balaban, M., & Pigott, G.M. (1986). Shrinkage of fish muscle during drying. *Journal of Food Science, 51*, 510–511.

Barbosa-Canovas, G., & Vega-Mercado, H. (1996). *Dehydration of foods*. New York: Chapman and Hall.

Barnett, S. (1973). Freezing of coffee extract to produce a dark colored freeze-dried product. *AIChE Symposium Series No. 132, 69*, 26–32.

Beatus, Y., Raziel, A., Rosenberg, M., & Kopelman, I.J. (1985). Spray-drying microencapsulation of paprika oleoresin. *Lebensmittel-Wissenschaft und Technologie, 18*, 28–34.

Bellows, R.J., & King, C.J. (1973). Product collapse during freeze drying of liquid foods. *AIChE Symposium Series No. 132, 69*, 33–41.

Bruin, S., & Luyben, K.C.A.M. (1980). Drying of food materials: A review of recent developments. In A.S. Mujumdar (Ed.), *Advances in drying* (Vol. 1; pp. 155–215). Washington, DC: Hemisphere.

Buma, T. J. (1971). Free fat in spray dried whole milk. 8. The relation between free-fat content and particle porosity of spray-dried whole milk. *Netherlands Milk Dairy Journal, 25*, 123–140.

Buma, T.J., & Henstra, S. (1971a). Particle structure of spray-dried milk products as observed by a scanning electron microscope. *Netherlands Milk Dairy Journal, 25*, 75–80.

Buma, T.J., & Henstra, S. (1971b). Particle structure of spray-dried caseinate and lactose as observed by a scanning electron microscope. *Netherlands Milk Dairy Journal, 25*, 278–281.

Cantarelli, C. (1985). Technological developments in the pasta industry. In C. Mercier & C. Cantarelli (Eds.), *Pasta and extrusion cooked foods* (pp. 33–51). London: Elsevier Applied Science Publishers.

Chaudhari, A.P., Kumbhar, B.K., Singh, B.P.N., & Narain, M. (1993). Osmotic dehydration of fruits and vegetables. [A review]. *Indian Food Industry, 12*, 20–27.

Chirife, J., & Buera, M.P. (1996). Water activity, water glass dynamics, and the control of microbial growth in foods. *Critical Reviews of Food Science and Nutrition, 36*, 465–513.

Crapiste, G.H., Rotstein, E., & Urbicain, M.J. (1985). Drying of foods while the cellular tissue structure prevails. *Latin American Journal of Chemistry, Engineering, and Applied Chemistry, 15*, 77–91.

Cussler, E.L. (1997). *Diffusion, mass transfer in fluid systems* (2nd ed.). Cambridge: Cambridge University Press.

de Boer, J.H. (1958). The shapes of capillaries. In D.H. Everett & F.S. Stone (Eds.), *The structure and properties of porous materials* (pp. 68–94). London: Butterworths Scientific Publications.

Duckworth, R.B. (1975). *Water relations in foods*. London: Academic Press.

Feustel, I.C., Hendel, C.E., & Juilly, M.E. (1964). Potatoes. In W.B. Van Arsdel & M.J. Copley (Eds.), *Food dehydration*, Vol. 2, *Products and technology* (pp. 301–398). Westport, CT: AVI Publishing Co.

Fish, B.P. (1958). Diffusion and thermodynamics of water in a potato gel. In *Fundamental aspects of the dehydration of foodstuffs* (pp. 143–157). London: Society of Chemical Industry.

Fito, P., & Pastor, R. (1994). Non-diffusional mechanisms occurring during vacuum osmotic dehydration. *Journal of Food Engineering, 21*, 513–519.

Flink, J.M. (1983). Structure and structure transitions in dried carbohydrate materials. In M. Peleg & E.B. Bagley (Eds.), *Physical properties of foods* (pp. 473–521). Westport, CT: AVI Publishing Co.

Flink, J.M., Gejl-Hansen, F., & Karel, M. (1973). Microscopic investigations of the freeze-drying of volatile-containing model food solutions. *Journal of Food Science, 38*, 1174–1179.

Flink, J.M., & Karel, M. (1972). Mechanism of retention of organic volatiles in freeze-dried systems. *Journal of Food Technology, 7*, 199–204.

Gejl-Hansen, F., & Flink, J.M. (1976). Application of microscopic techniques to the description of structure of dehydrated food systems. *Journal of Food Science, 41*, 483–489.

Gekkas, V. (1992). *Transport phenomena of foods and biological materials*. Boca Raton, FL: CRC Press.

Gorling, P. (1958). Physical phenomena during the drying of foodstuffs. In *Fundamental aspects of the dehydration of foodstuffs* (pp. 42–52). London: Society of Chemical Industry.

Gvirtzman, H., Magaritz, M., Klein, E., & Nadler, A. (1987). A scanning electron microscopy study of water in soil. *Transport Porous Media, 2*, 83–89.

Harmathy, (1969). Simultaneous moisture and heat transfer in porous systems with particular references to drying. *Industrial and Engineering Chemistry Fundamentals, 8*(1), 92–103.

Husain, A., Chen, C.S., & Clayton, J.T. (1973). Simultaneous heat and mass diffusion in biological materials. *Journal of Agricultural Engineering Research, 18*, 343–354.

Iglesias, H.A., & Chirife, J. (1982). *Handbook of food isotherms: water sorption parameters for food and food components*. New York: Academic Press.

Jason, A.C. (1958). A study of evaporation and diffusion processes in the drying of fish muscle. In *Fundamental aspects of the dehydration of foodstuffs* (pp. 103–134). London: Society of Chemical Industry.

Johnson, J.L., Busk, G.C., & Labuza, T.P. (1980). Examination of the crystallinity of food gels by X-ray diffraction. *Journal of Food Science*, *45*, 77–83.

Karathanos, V., Anglea, S.A., & Karel, M. (1993). Collapse of structure during drying of celery. *Drying Technology*, *11*, 1005–1023.

Karel, M. (1975). Dehydration of foods. In M. Karel, O.R. Fennema, & D.B. Lund (Eds.), *Physical principles of food preservation* (pp. 309–357). New York: Marcel Dekker.

Karel, M. (1985). Effects of water activity and water content on mobility of food components, and their effects on phase transitions in food systems. In D. Simatos & J. L. Multon (Eds.), *Properties of water in foods* (pp. 153–169). Dordrecht, The Netherlands: Martinus Nijhoff.

Karmas, E. (1973). Water in biosystems. *Journal of Food Science*, *38*, 736–739.

Kessler, H.G. (1981). *Food engineering and dairy technology* (pp. 269–328). Freising, Germany: Verlag A. Kessler.

King, J.C. (1971). *Freeze-drying of foods*. Cleveland, OH: CRC Press.

King, J.C. (1980). *Separation processes*. New York: McGraw-Hill.

Kuprianoff, J. (1958). Bound water in foods. In *Fundamental aspects of the dehydration of foodstuffs* (pp. 14–23). London: Society of Chemical Industry.

Labuza, T.P. (1984). *Moisture sorption: Practical aspects of isotherm measurement and use*. St. Paul, MN: American Association of Cereal Chemists.

Lazar, M.E., & Rasmussen, C.L. (1964). Dehydration plant operations. In W.B. Van Arsdel & M.J. Copley (Eds.), *Food dehydration*, Vol. 2, *Products and technology* (pp. 132–221). Westport, CT: AVI Publishing Co.

Lee, C.H., & Rha, C.K. (1979). Application of scanning electron microscopy for the development of materials for food. *Scanning Electron Microscopy*, *III*, 465–470.

Levine, H., & Slade, L. (1986). A polymer physico-chemical approach to the study of commercial starch hydrolysis products (SHPs). *Carbohydrate Polymers*, *6*, 213–244.

Lewicki, P.P., & Lenart, A. (1995). Osmotic dehydration of fruits and vegetables. In A.S. Mujumdar (Ed.), *Handbook of industrial drying* (2nd ed., Vol. 1; pp. 691–713). New York: Marcel Dekker.

Lloyd, R.J., Dong Chen, X., & Hargreaves, J.B. (1996). Glass transition and caking of spray-dried lactose. *International Journal of Food Science and Technology*, *31*, 305–311.

Loncin, M. (1980). Diffusion phenomena in solids. In P. Linko, Y. Malkki, J. Olkku, & J. Larinkari (Eds.), *Food process engineering* (Vol. 1; pp. 354–363). London: Applied Science Publishers.

Luyben, K.C. A.M., Olieman, J.J., & Bruin, S. (1980). Concentration dependent diffusion coefficients derived from experimental drying curves. In A.S. Mujumdar (Ed.), *Drying '80* (Vol. 2; pp. 233–243). Washington, DC: Hemisphere.

Marcotte, M., Toupin, C.J., & Le Maguer, M. (1991). Mass transfer in cellular tissues. 1. The mathematical model. *Journal of Food Engineering*, *13*, 199–220.

Marinos-Kouris, D., & Maroulis, Z.B. (1995). Transport properties in the drying of solids. In A.S. Mujumdar (Ed.), *Handbook of industrial drying* (2nd ed., Vol. 1; pp. 113–159). New York: Marcel Dekker.

Matsuo, R. R., Dexter, J.E., & Dronzek, B.L. (1978). Scanning electron microscopy study of spaghetti processing. *Cereal Chemistry*, *55*, 744–753.

McCabe, W. L., Smith, J.C., & Harriott, P. (1985). *Unit operations of chemical engineering* (4th ed.). New York: McGraw-Hill.

Moreira, R., Sun, X., & Chen, Y. (1997). Factors affecting oil uptake in tortilla chips in deep-fat frying. *Journal of Food Engineering*, *31*, 485–498.

Mujumdar, A.S. (1995). *Handbook of industrial drying* (2nd ed., Vols. 1, 2). New York: Marcel Dekker.

Ollivier, J.-L. (1985). Pasta drying at very high temperature: A fact. In C. Mercier & C. Cantarelli (Eds.), *Pasta and extrusion cooked foods* (pp. 90–97). London: Elsevier Applied Science Publishers.

Pedreschi, F., & Aguilera, J.M. (1998). Use of CLSM in the study of frying of potato strips. *Analysis and Microscopy*, in press.

Peleg, M. (1983). Physical characteristics of food powders. In M. Peleg & E.B. Bagley (Eds.), *Physical properties of foods* (pp. 293–323). Westport, CT: AVI Publishing Co.

Peleg, M., & Hollenbach, A.M. (1984). Flow conditioners and anticaking agents. *Food Technology*, *38*(3), 93–102.

Rahman, A.R., Shafer, G., Taylor, G.R., & Westcott, D.E. (1969). *Studies on reversible compression of dehydrated vegetables* (Technical Report 70-36-FL). Natick, MA: United States Army Natick Laboratories.

Raoult-Wack, A.L. (1994). Recent advances in the osmotic dehydration of foods. *Trends in Food Science and Technology*, *5*, 255–260.

Reeve, R.M., & Neel, E.M. (1960). Microscopic structure of potato chips. *American Potato Journal*, *37*, 45–57.

Reeve, R.M., & Notter, S.K. (1959). An improved microscopic method for counting ruptured cells in dehydrated potato products. *Food Technology*, *13*, 574–578.

Riva, M., & Peri, C. (1983). *Etude du sechage des raisins*. I. Effet de traitements de modification de la surface sur la cinetique du sechage. *Sciences des Aliments*, *3*, 527–550.

Rockland, L.B., & Beauchat, L.R. (1987). *Water activity: Theory and applications to foods*. New York: Marcel Dekker.

Roetman, K. (1979). Crystalline lactose and the structure of spray-dried products as observed by scanning electron microscopy. *Netherlands Milk Dairy Journal*, *33*, 1–11.

Roos, Y.H. (1995). *Phase transitions in foods*. San Diego: Academic Press.

Rosenberg, M., Kopelman, I.J., & Talmon, Y. (1985). A scanning electron study of microencapsulation. *Journal of Food Science, 50*, 139–144.

Rotstein, E., & Cornish, A.R.H. (1978). Influence of cellular membrane permeability on drying behavior. *Journal of Food Science, 43*, 926–934, 939.

Saguy, I.S., & Pinthus, E.J. (1995). Oil uptake during deep-fat frying: Factors and mechanisms. *Food Technology, 49*(4), 142–145, 152.

Saltmarch, M., & Labuza, T.P. (1980). SEM investigation of the effect of lactose crystallization on the storage properties of spray dried whey. *Scanning Electron Microscopy, III*, 203–209.

Schneider, F.H. (1981). Zur Wasserbestimmung in vegetabilen Olrohstoffen. *Fette, Seifen, Anstrichmittel, 83*, 329–337.

Schubert, H. (1981). Principles of agglomeration. *International Chemical Engineering, 21*, 363–377.

Schubert, H. (1993). Instantization of powdered foods. *International Chemical Engineering, 33*, 28–45.

Schuchmann, H. (1995). Production of instant foods by jet agglomeration. *Food Control, 6*, 95–100.

Scott, W.J. (1957). Water relations of food spoilage microorganisms. *Advances in Food Research, 7*, 83–127.

Shipman, J.W., Rahman, A. R., Segars, R. A., Kapsalis, J.G., & Westcott, D.E. (1972). Improvement of the texture of dehydrated celery by glycerol treatment. *Journal of Food Science, 37*, 568–571.

Simatos, D., & Multon, J. L. (1985). *Properties of water in foods*. Dordrecht, The Netherlands: Martinus Nijhoff.

Singh, R.P. (1995). Heat and mass transfer in foods during deep-fat frying. *Food Technology, 49*(4), 134–137.

Slade, L., & Levine, H. (1991). Beyond water activity: Recent advances based on an alternative approach to the assessment of food quality and safety. *CRC Reviews of Food, Science and Nutrition, 30*, 115–359.

Spiess, W.E.L. (1969). *Uber den Transport des Wasserdampfes bei der Grefiertrocknung von Lebensmitteln*. Unpublished doctoral dissertation, Universitat Karlsruhe. Cited by Gejl-Hansen & Flink (1976).

Sweat, V.E. (1986). Thermal properties of foods. In M.A. Rao & S.S.H. Rizvi (Eds.), *Engineering properties of foods* (pp. 49–87). New York: Marcel Dekker.

Ternan, M. (1987). The diffusivity of liquids in pores. *Canadian Journal of Chemical Engineering, 65*, 244–249.

Texter, J.A., Kellerman, R., & Klier, K. (1975). Water sorption of a dextran gel. *Carbohydrate Research, 41*, 191–210.

Thijssen, H.A.C., & Rulkens, W.H. (1968). Retention of aromas in drying of food liquids. *Ingenieur, 80*, 45.

To, E.C. & Flink, J.M. (1978). Collapse: A structural transition in freeze-dried carbohydrates. I. Evaluation of analytical methods. *Journal of Food Technology, 13*, 551–565.

Torreggiani, D. (1993). Osmotic dehydration in fruit and vegetable processing. *Food Research International, 26*, 59–68.

Toupin, C.J., Marcotte, M., & Le Maguer, M. (1989). Osmotically-induced mass transfer in plant storage tissues: A mathematical model. Part 1. *Journal of Food Engineering, 10*, 13–38.

Troller, J.A., & Christian, J.H.B. (1978). *Water activity and food*. New York: Academic Press.

Ufheil, G., & Escher, F. (1996). Dynamics of oil uptake during deep-fat frying of potato slices. *Lebensmittel-Wissenschaft und Technologie, 29*, 640–644.

Vacarreza, L.M., & Chirife, J. (1978). On the application of Fick's law for the kinetic analysis of air drying of foods. *Journal of Food Science, 43*, 236–238.

van Marle, J.T., Clerkx, A.C.M., & Boekstein, A. (1992). Cryo-scanning electron microscopy investigation of the texture of cooked potatoes. *Food Structure, 11*, 209–216.

Verhey, J.G.P. (1972). Vacuole formation in spray powder particles. 1. Air incorporation and bubble expansion. *Netherlands Milk Dairy Journal, 26*, 186–202.

Willis, C.A., & Teixeira, A.A. (1988). Controlled reduction of water activity in celery: Effect on membrane integrity and biophysical properties. *Journal of Food Science, 53*, 111–116.

Wolf, W., Spiess, W.E.L., & Jung, G. (1985). *Sorption isotherms and water activity of food materials*. England: Science and Technology Publishers.

Yao, Z., & Le Maguer, M. (1996). Mathematical modeling and simulation of mass transfer in osmotic dehydration processes. 1. Conceptual and mathematical models. *Journal of Food Engineering, 29*, 349–360.

# CHAPTER 10

# The Microstructural Approach

## 10.1 FOODS AND STRUCTURE

### 10.1.1 Introduction

The dedicated reader who has traveled through the micropores of these chapters may benefit from an attempt to synthesize the important messages they contain. Figure 10–1 summarizes the authors' present views on foods and food microstructure. This diagram represents a new perspective on the way food scientists and engineers can integrate microstructure into their problem-solving processes.

Chemical components play significant roles in the structure, stability, and nutritional value of a food. They are so important that the discipline of food science evolved earlier this century from the study of chemicals in foods, called *bromatology*. Macromolecules (polysaccharides, proteins, and lipids) form assemblies that, embedded in or intertwined with other phases, give rise to the native or *basic structure* of food raw materials. Humankind has learned to break down natural structures and produce purified ingredients. Thus, there exists a *primary* food industry that extracts and refines agricultural raw materials and converts them from their native structure into functional molecules (sugars, oils, and fats) and macromolecules (polysaccharides and proteins), which are the basic building blocks used by a high-volume restructuring food industry that mixes and reassembles them into products.

Processing, or the controlled incorporation of materials and energy into a food, is really a restructuring or reassembling operation, whether desired or not. As a result, an "engineered" structure is obtained that has probably undergone modifications at four structural levels: molecular, ultrastructural, microstructural, and macrostructural. Lillford (1998) has suggested that transformations leading to the engineered structure contribute to the novelty of the product. The term *engineered structure* connotes a controlled transformation of a native structure into any desirable structure worthy of consumption as a recognizable food. During processing, attributes are imparted to the mix of molecules and assemblies until a product is created. The term also implies that we can determine and quantify a food's property as if it were an engineering material. Food technologists now have the tools to evaluate almost any property of a food by several means (mechanical, thermal, chemical, etc.) and at each of the four structural levels. Advances have also been made in the sensorial dimension of this evaluation and in uncovering its relation to structure.

The food industry aims at selling high-quality goods. When a finished product leaves the factory (a slaughterhouse or, for that matter, a tree), the metastable nature of food structures begins to be evident. Structural changes during distribution and storage are a function of time and environmental conditions. Optimally, the engineered structure will have built-in retardation mechanisms to minimize changes during this often uncontrolled stage. Some products require further processing or preparation before consumption

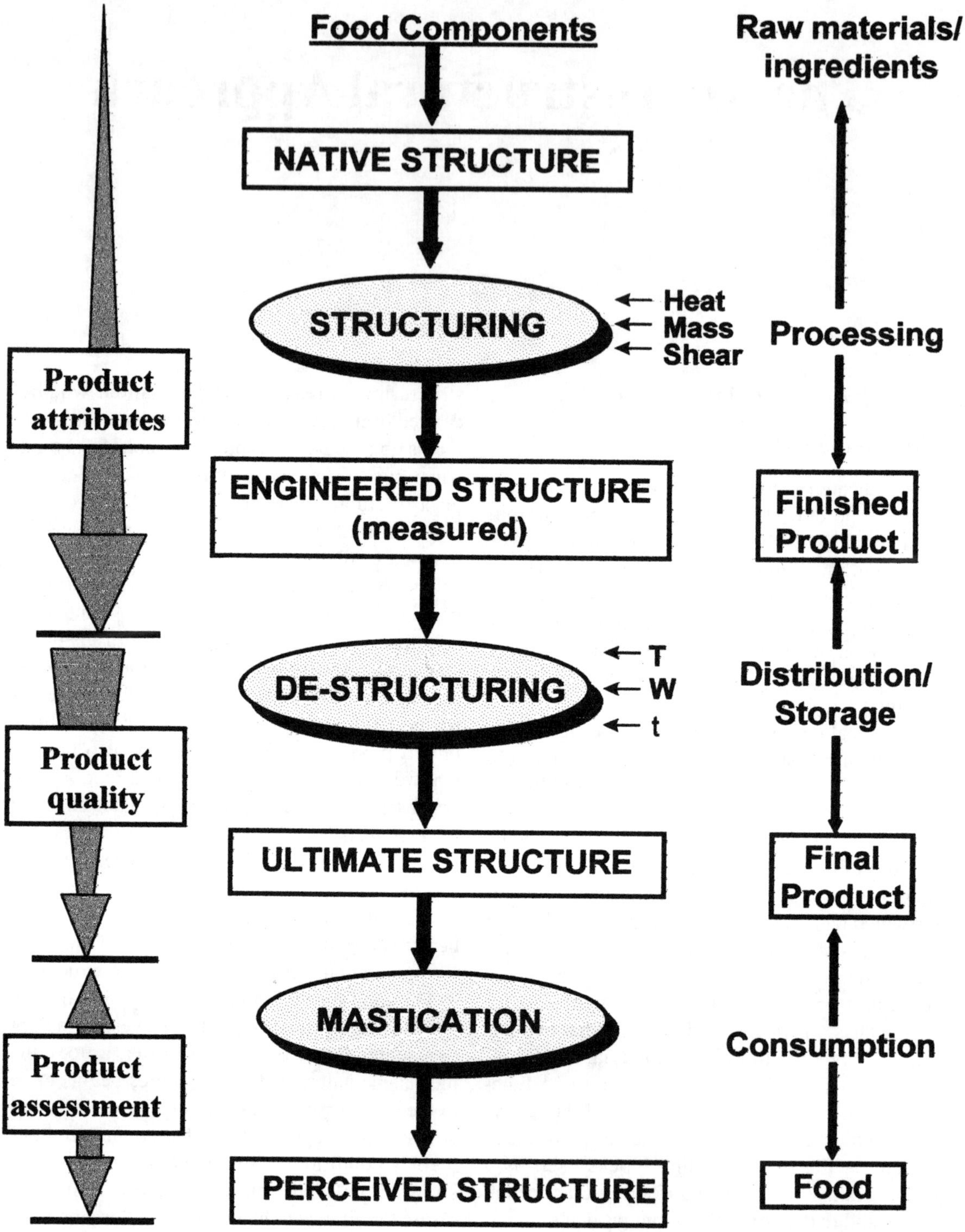


**Figure 10–1** Microstructural view of food processing, storage, and consumption. *Source:* Redrawn from Aguilera and Stanley (1990).

(thawing, heating, whipping, cooking, etc.). Therefore, "destructuring" (and in some cases even a secondary "restructuring") occurs to various extents before the *ultimate structure* (and quality) reaches the consumer. Lastly, structure is destroyed and perceived during mastication and swallowing. The product has become a food. Little is known of the breakdown of structure in the mouth or the physics and sensorial aspects involved. According to Lillford (1998), the eating stage leads to quality perception and allows the food to realize its potential value.

Food specialists see this sequence of events from their own perspective, and contributions to achieving the engineered, ultimate, and perceived structures differ. The food technologist gathers and distributes information on the behavior and interactions of food components under different processing conditions. Considerable advances have been made in the last 20 years in gathering data on the "functionality" of ingredients in model systems, but we are still far from being able to extrapolate from these results to the complex multicomponent systems found in a food. The food chemist focuses on basic molecular structure, physicochemical interactions, safety, and stability—in short, on the way molecules contribute to the creation of a stable microstructure. The food engineer views microstructure as an opportunity to return to the traditional engineering concept of assembling building blocks through the judicious use of heat, mass, and momentum transfer. To the food engineer, foods are not inert homogeneous substances but dynamic structured materials. The nutritional scientists sees in structure an opportunity to protect nutrients during processing, modify the mix of nutrients, and convey them in an appealing form. The thesis presented in this book is that food microstructure is a compelling converging point for food specialists, nutritionists, and marketing personnel involved in the production and distribution of foods.

### 10.1.2 Composition versus Structure

In general, composition gives only partial information regarding food structure. Imagine a pyramid having as its vertices the four main components present in foods: water (at the apex), protein, polysaccharides (sugars), and lipids. Inside the pyramid's volume any food may be placed according to its gross chemical composition. Most foods are crowded into the upper vertex of the pyramid (pure water), so it is better to unfold the sides and obtain a large triangle with four inner triangles in which most foods may be viewed as ternary diagrams (Figure 10–2). Resolution is vastly improved, yet overlapping is still inevitable, for foods that we know are physically different are located close together. (This reminds us of the two-dimensional world of *Flatland*: in this world the concept of a three-dimensional body does not exist, and any plane figure—a triangle, square, or circle—looks to a Flatlander like a straight line [Abbott, 1984].)

The added dimension that makes foods recognizable to us is structure. If new pyramids were now constructed above each ternary system having as a third dimension some structural parameter (e.g., viscosity or a modulus), individual food products would become discernible. The structure-composition pyramid for a few dairy products is shown in Figure 10–2. What this indicates is that food composition may have the potential for structure formation but it is not sufficient to characterize a food.

### 10.1.3 A Paradigm Shift in Food Technology

An obvious question at this point is, how might microstructure relate to future developments in the food industry? Advances in food science and technology during the last two decades have multiplied the diversity of products on the supermarket shelves and brought their quality to unprecedented levels. Never in the past have more people in the world been provided so well with wholesome, plentiful, and nutritious food. Some economists believe that freedom in the modern industrial society is epitomized by the fact that anyone can buy a nutritious, appetizing, and wholesome meal for a fraction of a daily wage. Yet, undernutrition still afflicts a large part of the developing world, and obesity is an expanding prob-

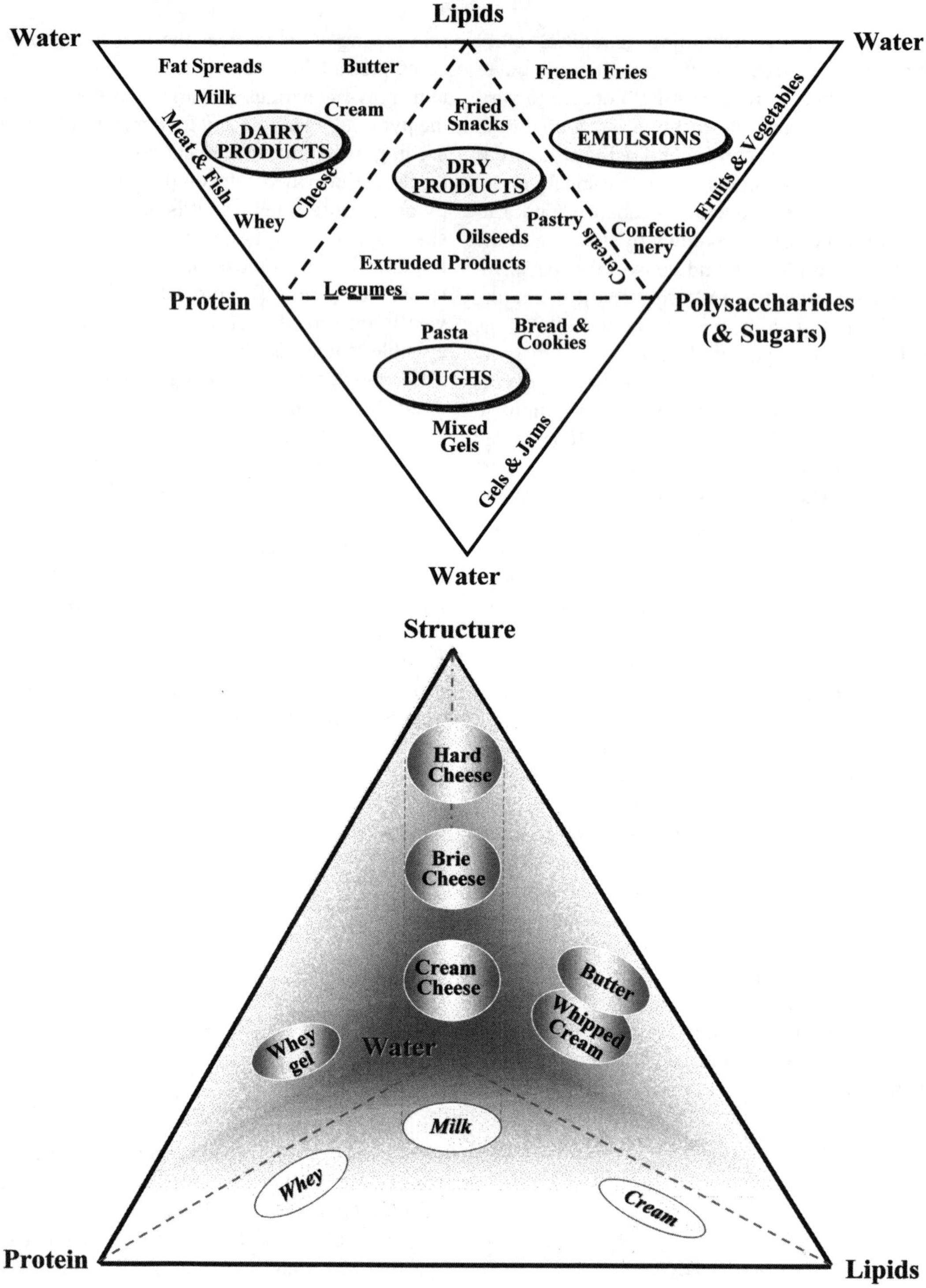


**Figure 10–2** Top: The food composition pyramid with unfolded sides. Each of the four triangles represents a ternary system in which foods are located according to their gross composition. Bottom: The structure-food composition pyramid for dairy products. Only for qualitative interpretation.

lem as cheaper food products coexist with sedentary habits. As migration from rural to urban areas continues and food is increasingly consumed away from home, people all around the world will rely more and more on processed food products. The mission of food technology is to provide improved nutrition worldwide through high-quality, wholesome, and affordable products that consumers demand. In this scenario, the goals of the modern food business include:

- minimizing postharvest losses, extending the shelf life of foods, and adding convenience
- reducing costs through increased efficiency and novel processing techniques
- stabilizing fresh products by minimal processing and reducing the use of additives or replacing them with natural products
- developing new or improved foods, including high-quality analogs (e.g., second-generation products) that fulfill specific nutritional or dietary roles
- introducing new ingredients that are safer, cheaper, and/or more functional than conventional ones

Improved quality, the use of new ingredients, and innovative product development in the 21st century will demand a paradigm shift in the food industry. Gone will be the days of trial-and-error food innovation. We now have the tools to identify picogram quantities of almost any chemical species, probe the mobility of molecules inside foods, and look into them with minimal intrusion and in real time. Advances in high-technology areas will eventually find applications in food science: genetic engineering and biotechnology will open the route to naturally synthesized components and raw materials with tailor-made properties, and "nanotechnologies" will allow the manufacture and positioning of specific molecules (e.g., anti-freeze polymers) in microstructures. As the advances occurring in the modern metal, plastics, and ceramics industries diffuse into the food industry, we will observe a paradigm shift in the food industry from the macrostructural to the molecular and microstructural fabrication of foods and the development of food materials science.

## 10.2 STRUCTURE-PROPERTY RELATIONSHIPS

Food materials science is a search for basic knowledge about the properties of foods and their relation to processing, stability, and structure. Over the last 10 years, ideas arising from research into polymer physics, composite materials, and product microstructure have been applied with increasing success to foods, augmenting the respectability of food science, as discussed by Stanley (1994).

Throughout this book, we have emphasized the importance of structure in foods. In food processing and engineering, structure matters because it is responsible for many of the desirable properties with which we wish to endow a food. In Section 3.7, we reviewed the relationships between the structure and the rheological-mechanical properties of foods and introduced some physical and mathematical models that describe the link between structure and behavior. In this closing chapter, we recapitulate and expand on some important structure-property concepts because they are essential to the development of a science of food materials.

### 10.2.1 What Are Structure-Property Relationships?

As discussed throughout this book, the physicochemical, functional, technological, and nutritional properties of foods are related to their structure (see Section 3.7.2). What is implicit in the concept of a "structure-property relationship"? That there is a *causal connection* between the structure and the way the product behaves.

Structure-property relationships reflect the interactions occurring at the molecular, ultrastructural, microstructural, and macrostructural levels during food processing. (See also Section 3.1). The first level includes molecules and polymers of different sizes and types as well as specific interactions between functional groups (e.g., enzymatic cleavage, hydrogen bonding, and chemical cross-linking). Ultimately, the "potential" for structure formation depends on the molecules available. The ultrastructural level includes changes in conformation, association and breakdown of macromolecules, formation of natural assemblies, and so on. At the microstructural level, colloidal phenomena predominate, and the first recognizable elements in processed foods become evident: droplets, crystals, segregated phases, air cells, aggregates and fibers, among others. In the ultra- and microstructural levels, nonspecific interactions predominate and yield the nonequilibrium structures of real food systems. Finally, at the macrostructural level, the assemblage and bonding of major structures or phases become manifest.

Throughout this book we have emphasized the kinetic nature of molecular and other types of interaction and the fact that foods are not in thermodynamic equilibrium. The old concept of a "functional property" of isolated proteins and polysaccharides is of little relevance, owing to the greater role assigned to nonspecific interactions and the kinetic component introduced by processing. Structure building through the "nonspecific" interactions predominating in foods appears not to be used by nature, as high levels of organization are achieved by highly specific interconnectivity (Baer, Cassidy, & Hiltner, 1992).

### 10.2.2 Which Is the Structure?

Heertje (1993) defined structure as the spatial arrangement of elements in a food and their interaction, stressing that visual observation is important in the analysis of the formed structure. Using different microscopy techniques, he identified several structural elements in foods: water and oil droplets, gas cells, fat crystals, strands, granules, micelles, and interfaces. Another definition of structure comes from a discussion paper by Raeuber and Nikolaus (1980), which states that structure "is the organization of a number of similar or dissimilar elements, their binding into a unit, and the interrelationship between the individual elements and their groupings." In both definitions, two aspects are emphasized as crucial: the *organization of elements* and their *interaction*. Thus, the microstructure of a food can only be understood when its elements (solid, liquid, and gaseous) and architecture are considered in a dynamic and interrelated condition.

Definitely, most foods possess structure and are not homogeneous or isotropic materials, although as a first approximation a few foods may be regarded as homogeneous alloys (e.g., soda crackers) or simple solutions. This bring us to the problem of the scale of observation, since food elements are viewed differently at different scales. For instance, a starch granule may be a filler in a structure at a scale of 200 $\mu$m but a stratified composite of amylose and amylopectin at a scale of 1 $\mu$m. A related problem, rarely addressed by food material scientists, is the microheterogeneity of foods. Most tissue foods maintain the hierachical structure bestowed by nature.

Finally, a crucial reason that the study of foods structure differs from the study of engineering materials is that foods are often in a metastable state and/or exhibit biological activity. Structure in foods changes with time, either perceptibly (e.g., during frying) or unnoticed (during storage).

### 10.2.3 Which Are the Properties?

The next question is, what kind of properties are we interested in? A property is a particular trait. Most physical properties of foods—including mechanical, rheological, electrical, optical, and transport properties (heat and mass diffusivity)—are common to all industrial materials. Additional desirable attributes of foods are somewhat more specific, such as emulsifying capacity, surface roughness, and sound. Other desirable characteristics are even more difficult to define or measure, for example, flavor, texture, nutritional value, and shelf-life stability. All of them are important, and in each case structure may play a role in the property's manifestation. If it does, the structure-property relationship needs to be found.

Consider texture (see Section 6.1). The term *texture* encompasses all physical and structural characteristics sensed by touch that are related to deformation under an applied force. In their discussion paper on microstructural bases of the texture of plant foods, Jackman and Stanley (1995) start by saying that "the literature is replete with examples of texture having been measured using inappropriate methods." They emphasize that the relationship between objective measurements and sensory properties is poorly understood. That texture is the external manifestation of the arrangement of microstructural elements and their breakdown in the mouth appears to be clear. A simple but illustrative example is the detection of oversized fat or ice crystals in the tongue as defects in poorly tempered chocolate or frozen ice cream, respectively. The ultimate challenge is to find the relationships for the triad structure-property-sensorial effect.

### 10.2.4 What Kind of Relationship?

Assuming that structural issues have been resolved and the property has been correctly measured, the next question is, what kind of relationship are we after? There are multiple answers to this simple question. It is easy, for example, to fit a mathematical function to conform with experimental data. Yet this approach not only has failed to win for food science respect from other scientific disciplines but also has been ineffective at illuminating the relationship under investigation. However, in a few instances, such as in the case of standardized product or process, it may be all that is required.

According to Bailey (1998), modeling does not make sense without defining beforehand what problem it is intended to resolve. Models should be able to bring order to experience, clarify which components and interactions are important in a complex system, generate new hypotheses, and lead to the proper interpretation of qualitative observations.

Kinetic models are frequently used because they provide a way of describing how a property varies with time and temperature. Note that in simple chemical kinetics the extent of a reaction is a function of these two variables, and since a change in a food often results from a chemical reaction (or an overall reaction at least), there seems to be an appeal to this fundamental principle. The justification for using this kind of approach is that "foods are much too complex to suggest that they can be mechanistically modeled. Again, data can be fitted to the appropriate kinetic model by a variety of statistical techniques. The main problems with this approach are that data for heterogeneous materials exhibit lack of precision and are often insufficient to determine the order of the reaction.

At the next level of functional dependence are semiempirical (or semitheoretical) models. Outstanding among these models are those used in food rheology to describe the behavior of non-Newtonian fluids (e.g., suspensions, pastes, etc.); these translate into equations linking stress, yield stress, and rate of deformation. Constitutive models that aim at relating rheological properties to molecular architecture are more fundamental (See Section 3.8.4). Such models are well developed for dilute solutions of random coils and polymers with rodlike conformations (Kokini, 1994).

Structural models, on the other hand, are preferred by engineers because they are based on the architecture and structural properties of the intervening elements (matrix, fillers, walls, etc.) and their interactions. Structural models of the mechanical properties of composites, foams, and cellular materials are being applied to foods with increasing success (Peleg, 1997). Similar models of the rheological behavior of suspensions have also been proposed (Windhab, 1995). Through structural models we strive to find the scaling laws that translate microstructural data into macrostructural behavior.

## 10.3 PROBING MICROSTRUCTURE IN FOOD SCIENCE, TECHNOLOGY, AND ENGINEERING

Food processing may be viewed as a controlled effort to preserve, transform, destroy, or create structure. Unlike the architects or structural engineers, food technologists can neither observe directly the mortar and bricks of their structure nor precisely control the way in which they are laid down during construction. Thanks to microscopy and related techniques, we can now visualize and identify food components in a structure with minimal intrusion (Aguilera & Lillford, 1996). We are also developing a qualitative understanding of how these engineered microstructures are formed and stabilized by nonspecific interactions induced by processing. The application of scientific principles from materials, polymer, and colloid science will continue to unveil these phenomena, but we are still far from attaining comprehensive quantitative knowledge. Nonetheless, food processing is slowly gaining respectability as a form of controlled manufacturing.

We foresee the following trends in the study of food microstructure. Spurred by requirements in medicine, biology, and materials science, a multitude of analytical and microscopy techniques have become available. Many of these could be adapted to the study of foods, as discussed by Kalab, Allan-Wojtas, and Shea-Miller (1995). Technologies already being applied include confocal laser scanning microscopy, which is capable of optically sectioning a sample with minimum intrusion; environmental scanning electron microscopy, which allows examination of surfaces of uncoated "wet" specimens at ambient conditions; and scanning probe microscopy and related high-resolution techniques, which scan specimen surfaces at the molecular level.

Simultaneous visualization of structures and identification of food components *in situ* is an added benefit of microscopy. Several current techniques allow direct identification and eventually quantification of ions or molecular species within specific structures. Coupling microscopy and spectroscopy may unveil the composition of a specimen as small as a few microns by measuring the absorption of energy at different wavelengths. Immunolabelling—the use of probes (antibodies or lectins) that bind to specific sites on individual molecules—is now a standard method in most histology laboratories.

Direct microscopic observation of controlled experiments in a stage mounted under the microscope can provide structural information in real time. There is now commercial equipment that allows "miniaturization" heating and cooling (freezing) of thin samples or solutions at controlled rates and visualization through charge-coupled device (CCD) cameras linked to a TV monitor and VCR.

Development of a food materials science requires understanding the physical properties or function of biopolymeric microstructures. Several techniques used in mechanical and polymer science may be coupled with microscopy to simultaneously observe a structure and measure rheological behavior, thermal transitions, and/or mechanical properties (Figure 10–3). Micromanipulation techniques borrowed from biology may be adapted to assess the properties of cells and tissues.

Computer-mediated image analysis can help quantify many of the features revealed by microscopic examination of foods: sizes and shapes of cellular components, thickness of cell walls or particle networks in gels, pore size and size distributions in gels, relative proportions of various phases, and others as well. Quantitative image analysis is the extraction of information from data that are in the form of images or pictures.

## 10.4 THE MICROSTRUCTURAL APPROACH

Let us review our progress through the chapters of this book. First we considered the main alternatives existing today for "seeing" and probing into the microstructure of foods. Machine-enhanced imaging is a rapidly expanding technological field, and beneficial contributions will come from applications in biology and materials science. Food technologists and engineers should become familiar with the principles and potential applications of microscopy and related technologies and select the most "appropriate" one or the best combinations. The next topic covered was image analysis. Then we were introduced to the fundamentals of polymer, colloid, and materials science. After that we looked at the main physicochemical transformations of food molecules and assemblies. Then we reviewed traditional and newer structuring techniques and their relation to texture and quality. In Chapters 7–9 we analyzed some products and processes from the microstructural viewpoint.

Although advances have been made in conceptualizing how food structure is formed and in relating structure to chemical and physical properties (in particular, mechanical and rheological properties), food scientists still lag behind their materials and polymer science counterparts. One

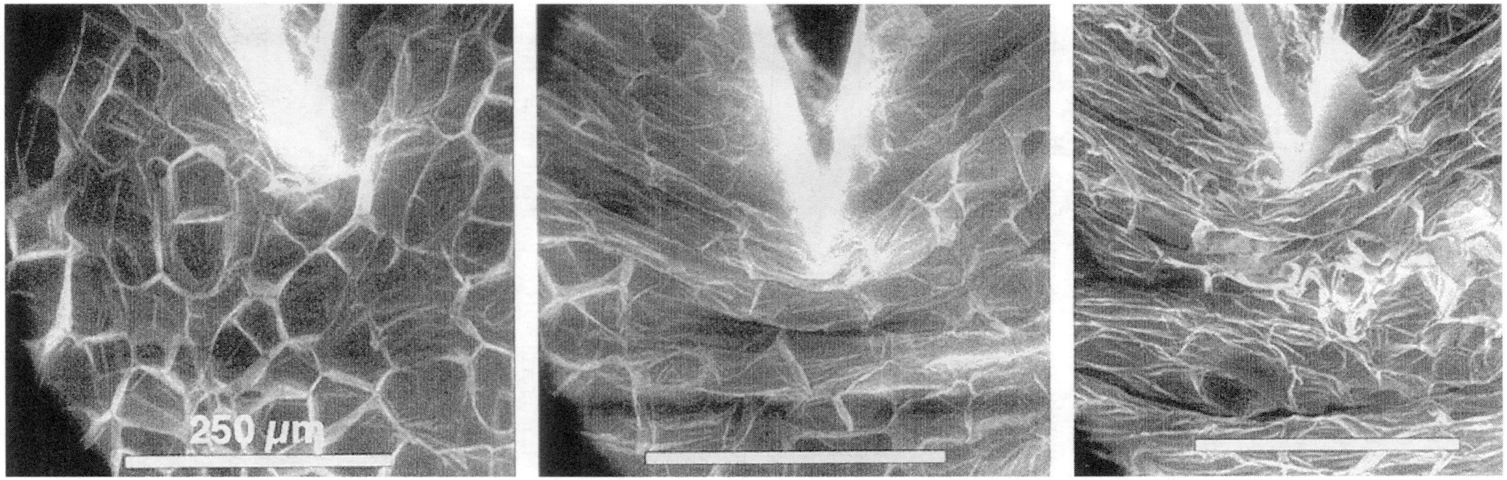


**Figure 10–3** ESEM micrographs of fresh (left), one-week-old (center), and three-week-old (right) carrot parenchymal tissues during a slice test. This is a prototype method for observing mechanical responses of cellular materials, in the absence of a conducting coating, during *in situ* dynamic testing. The method involves forcing a scalpel blade though the sample while monitoring the deformation both visually and instrumentally. It can be seen that as the carrot tissues age, the cell walls become more difficult to fracture and the stress fields around the scalpel tip increase. *Source:* Reprinted with permission from B.L. Thiel and A.M. Donald, In Situ Mechanical Testing of Fully Hydrated Carrots (Daucus carota) in the Environmental SEM, *Annals of Botany*, Vol. 82, pp. 727–733, © 1998, Academic Press, Ltd.

obstacle to progress is the complex nature of foods. As indicated in the previous paragraphs, some of the interaction forces participating in structure formation and stabilization are beginning to be understood qualitatively, but quantitative data and models are not yet available.

The microstructural approach recognizes that many problems in food research are rooted in events that are beyond the resolution of the naked eye and that structure plays an important role in food quality. Hence, food technologists ought to consider microstructure as a key element in process and product development. Similarly, food researchers should always examine the possible role of structure even when studying apparently simple systems such as macromolecules in solutions. Food science students should be exposed to this microstructural approach in their lectures, projects, and research work.

The microstructural approach is based on the realization that food technology is largely a controlled effort to preserve, create, transform, or destroy microstructure. Consequently, it incorporates structure as a main parameter and identifies key structural elements. It dissects foods into key structural elements from the molecular level to that of major elements (droplets, air cells, strands, interfaces, micelles, and crystals) and assemblies (fibers, cell walls, cells, and tissue). It uses the most appropriate microscopy and imaging techniques in a *systematic* effort to provide physical and morphological information of the system under study at the relevant scale (see Exhibit 10–1). It aims from the start at integrating microstructural information and data generated by other physicochemical methods (e.g., rheology) to derive structure-property relationships in foods. It provides value-added content to microscopy stud-

**Exhibit 10–1** The Role of Microscopy in the Microstructural Approach

**The Problem**. Define the problem to be studied. Is there enough evidence that structure plays a key role? Remember that microscopy may be expensive.

**Relevant Scale**. Define the relevant scale(s). Which are the structures most likely to be affected and at which level? How do they relate to the problem?

**Background**. Gather support microstructural information (review journal papers) and other data (physical, chemical). Which are the specific hypotheses? Prepare for discussion with the microscopist.

**Information**. What would you like to see? Images, compositional data, or numerical data? Make a sketch of what is expected based on available information. How would this information support or complement your other data?

**Techniques**. Select the most appropriate microscopy techniques for the task. Discuss them with microscopist. Don't engage in overkill!

**Experiments**. Are your current experiments appropriate for microscopy work and representative of sample preparation? Remember, you are responsible for the samples for microscopy work.

**Sampling**. Ask how samples have to be gathered, handled, and preserved.

**Specimen Preparation**. This is a specialized process and involves lots of time. Stay away from it and trust the technician. Here is where most artifacts arise.

**Microscopy Session**. Try to be at the microscopy session. Take your own notes and enough images from several portions of the specimen. Save images in a computer for later image analysis and mounting.

**Data**. Now you have it. It is time to apply the proper image analysis programs. Are the images and data satisfactory? Most times, new sessions have to be scheduled. Overcome the temptation to present nice pictures but poor evidence.

**Interpretation**. This is the toughest task but may be extremely rewarding if everything goes well. How does the microscopy information fit with other data? Which are the new hypotheses?

**Reporting**. Use images only to make specific points, remembering that a picture is worth a thousand words. Avoid the temptation to add an extra figure to your report just because it looks nice.

ies and aims at generating numerical data as well (e.g., geometrical features, color changes, and chemical information). It is a focused approach and leads naturally to teamwork and to the generation of new working hypotheses. In sum, food material science is a converging point for food scientists, technologists, engineers, nutritionists, and marketing people involved in food production.

## BIBLIOGRAPHY


Abbot, E.A. (1984). *Flatland: A Romance of many dimensions.* New York: New American Library.

Aguilera, J.M., & Lillford, P. (1996). Microstructural and imaging analyses as related to food engineering. In P. Fito, E. Ortega, & G. Barbosa-Canovas (Eds.), *Food engineering 2000* (pp. 23–38). London: Chapman and Hall.

Aguilera J.M., & Stanley, D.W. (1990). *Microstructural principles of food processing and engineering.* London: Elsevier Applied Science.

Baer, E., Cassidy, J.J., & Hiltner, A. (1992). Hierarchical structure of collagen composite systems. In W. Glasser & H. Hatakeyama (Eds.), *Viscoelasticity of biomaterials* (pp. 2–23). ACS Symposium Series 489. Washington, DC: American Chemical Society.

Bailey, J.E. (1998). Mathematical modeling and analysis in biochemical engineering: Past accomplishments and future opportunities. *Biotechnology Progress, 14*, 8–20.

Heertje, I. (1993). Structure and function of food products: A review. *Food Structure, 12*, 343–364.

Jackman, R.L., & Stanley, D.W. (1995). Perspectives in the textural evaluation of plant foods. *Trends in Food Science and Technology, 6*, 187–194.

Kalab, M., Allan-Wojtas, P., & Shea-Miller, S. (1995). Microscopy and other imaging techniques in food structure analysis. *Trends in Food Science and Technology, 6*, 177–186.

Kokini, J.L. (1994). Predicting the rheology of food biopolymers using constitutive models. *Carbohydrate Polymers, 25*, 319–329.

Lillford, P. (in press). *Food composites.* Second Ibero-American Congress on Food Engineering. Bahia Blanca, Argentina. New York: Chapman & Hall.

Peleg, M. (1997). Review: Mechanical properties of dry cellular solid foods. *Food Science and Technology International, 3*, 227–240.

Raeuber, H.J., & Nikolaus, H. (1980). Structure of foods. *Journal of Texture Studies, 11*, 187–198.

Stanley, D.W. (1994). Understanding the materials used in foods: Food materials science. *Food Research International, 27*, 135–144.

Vincent, J. (1990). *Structural biomaterials* (rev. ed.). Princeton, NJ: Princeton University Press.

Windhab, E.J. (1995). Rheology in food processing. In S.T. Beckett (Ed.), *Physicochemical aspects of food processing* (pp. 80–116). London: Blackie Academic and Professional.

# Index

**O**



**P**



**Q**



**R**